AF181060

Dahlem Workshop Reports
Life Sciences Research Report 33

The goal of this Dahlem Workshop is:
to evaluate the role of chemical "speciation" in environmental processes and to establish criteria for identifying key species of elements and for developing "speciation" methods

Life Sciences Research Reports
Editor: Silke Bernhard

Held and published on behalf of the
Stifterverband für die Deutsche Wissenschaft

Sponsored by:
Senat der Stadt Berlin
Stifterverband für die Deutsche Wissenschaft
Stiftungsfonds Schering AG

The Importance of Chemical "Speciation" in Environmental Processes

M. Bernhard, F. E. Brinckman,
P. J. Sadler, Editors

Report of the Dahlem Workshop on
The Importance of Chemical "Speciation"
in Environmental Processes
Berlin 1984, September 2–7

Rapporteurs:
E. A. Jenne · D. R. Kester · W. R. Wolf · J. M. Wood

Program Advisory Committee:
M. Bernhard · F. E. Brinckman · U. Förstner
K. J. Irgolic · E. A. Jenne · W. Mertz · P. J. Sadler
H. J. Tobschall

Springer-Verlag
Berlin Heidelberg New York
London Paris Tokyo 1986

Copy Editors: K. Geue, J. Lupp

Photographs: E. P. Thonke

With 4 photographs, 126 figures, and 68 tables

Library of Congress Cataloging-in-Publication Data
Dahlem Workshop on the Importance of Chemical "Speciation"
in Environmental Processes (1984: Berlin, Germany)
The importance of chemical "speciation" in environmental processes.
(Dahlem workshop reports) (Life sciences research report; 33)
Includes indexes.
1. Chemistry, Analytic–Congresses.
2. Environmental chemistry–Congresses.
I. Bernhard, Michael
II. Brinckman, Frederick E.
III. Sadler, Peter J.
IV. Title.
V. Series.
VI. Series: Life sciences research report; 33.
QD71.D35 1984 574.5'222 86-26009

ISBN-13: 978-3-642-70443-7 e-ISBN-13: 978-3-642-70441-3
DOI: 10.1007/978-3-642-70441-3

This work is subject to copyright. All rights are reserved, whether the whole or part of the material is concerned, specifically those of translation, reprinting, re-use of illustrations, broadcasting, reproduction by photocopying machine or similar means, and storage in data banks. Under § 54 of the German Copyright Law, where copies are made for other than private use, a fee is payable to „Verwertungsgesellschaft Wort", Munich.

© Dr. S. Bernhard, Dahlem Konferenzen, Berlin 1986.

Softcover reprint of the hardcover 1st edition 1986

The use of registered names, trademarks, etc. in this publication does not imply, even in the absence of a specific statement, that such names are exempt from the relevant protective laws and regulations and therefore free for general use.

2127/3020-543210

Table of Contents

The Dahlem Konferenzen

Founders

Recognizing the need for more effective communication between scientists, especially in the natural sciences, the Stifterverband für die Deutsche Wissenschaft*, in cooperation with the Deutsche Forschungsgemeinschaft**, founded Dahlem Konferenzen in 1974. The project is financed by the founders and the Senate of the City of Berlin.

Name

Dahlem Konferenzen was named after the district of Berlin called "Dahlem", which has a long-standing tradition and reputation in the arts and sciences.

Aim

The task of Dahlem Konferenzen is to promote international, interdisciplinary exchange of scientific information and ideas, to stimulate international cooperation in research, and to develop and test new models conducive to more effective communication between scientists.

Dahlem Workshop Model

Dahlem Konferenzen organizes four workshops per year, each with a limited number of participants. Since no type of scientific meeting proved effective enough, Dahlem Konferenzen had to create its own concept. This concept has been tested and varied over the years, and has evolved into its present form which is known as the *Dahlem Workshop Model*. This model

---

* The Donors Association for the Promotion of Sciences and Humanities
** German Science Foundation

provides the framework for the utmost possible interdisciplinary communication and cooperation between scientists in a given time period.

The main work of the Dahlem Workshops is done in four interdisciplinary discussion groups. Lectures are not given. Instead, selected participants write background papers providing a review of the field rather than a report on individual work. These are circulated to all participants before the meeting to provide a basis for discussion. During the workshop, the members of the four groups prepare reports reflecting their discussions and providing suggestions for future research needs.

Topics

The topics are chosen from the fields of the Life Sciences and the Physical, Chemical, and Earth Sciences. They are of contemporary international interest, interdisciplinary in nature, and problem-oriented. Once a year, topic suggestions are submitted to a scientific board for approval.

Participants

For each workshop participants are selected exclusively by special Program Advisory Committees. Selection is based on international scientific reputation alone, although a balance between European and American scientists is attempted. Exception is made for younger German scientists.

Publication

The results of the workshops are the Dahlem Workshop Reports, reviewed by selected participants and carefully edited by the editor of each volume. The reports are multidisciplinary surveys by the most internationally distinguished scientists and are based on discussions of new data, experiments, advanced new concepts, techniques, and models. Each report also reviews areas of priority interest and indicates directions for future research on a given topic.

The Dahlem Workshop Reports are published in two series:
1) Life Sciences Research Reports (LS), and
2) Physical, Chemical, and Earth Sciences Research Reports (PC).

Director

Silke Bernhard, M.D.

Address

Dahlem Konferenzen
Wallotstrasse 19
1000 Berlin (West) 33

The Importance of Chemical "Speciation" in Environmental Processes,
eds. M. Bernhard, F. E. Brinckman, and P. J. Sadler, pp. 1–5.
Dahlem Konferenzen 1986. Springer-Verlag Berlin, Heidelberg
© Dr. S. Bernhard, Dahlem Konferenzen

Introduction

M. Bernhard*, F. E. Brinckman**, and P. J. Sadler***

* Centro Studi Ambiente Marino
ENEA
19100 La Spezia, Italy
** Institute for Materials Science and Engineering
NBS
Gaithersburg, MD 20899, USA
*** Dept. of Chemistry
Birkbeck College
London WC1E 7HX, England

The purpose of the workshop covered by this volume was to increase the general awareness of the important role that chemical species play in environmental processes and to further advancement in species analysis, the understanding of species transformation and species reactivity, and to investigate whether different environmental disciplines are at a comparable state of the art and whether a transfer to knowledge and technology between them is feasible.

The idea of holding such a workshop had been contemplated by one of the editors (MB) for some years. His meeting the two other editors of this Report at the NATO Conference on Environmental Inorganic Chemistry held in San Minato (Italy) in June 1974, as well as many colleagues who also turned out to be interested in these problems, helped to bring this workshop into being. It was clear from the outset that fruitful discussions of such complex issues could not proceed within the framework of a conventional meeting. Rather, it seemed evident that only the *Dahlem Workshop Model* would provide an appropriate context for this project.

The workshop participants were selected by a Program Advisory Committee according to two categories: "users" who were likely to have a need for better information about chemical species, and "donors" who were likely to be able to provide knowledge to "users" about physical, physicochemical, chemical, or biological analytical techniques for obtaining that information, although for some participants no such clear-cut distinction could be made. On the whole, "users" were environmental scientists who have found that they cannot fathom the environmental phenomena of interest without study of the chemical species involved. Hence in this workshop we limited our interest to the study of chemical species (which, naturally, finds application in many disciplines), in environmental processes, according to the following four topics, of which each was discussed by one group:

1. Biota (or living systems in general)
   Moderators: W. Mertz and P.J. Sadler,
   Rapporteur: W.R. Wolf;
2. Freshwater and terrestrial systems
   Moderators: G.K. Müller and H.J. Tobschall,
   Rapporteur: E.A. Jenne;
3. Marine and estuarine systems
   Moderators: M.O. Andreae and S.G. George,
   Rapporteur: D.R. Kester; and
4. Systems under stress
   Moderators: B.A. Fowler and S. Silver,
   Rapporteur: J.M. Wood.

The fourth topic was chosen because of the special importance of chemical species in pollution management and abatement and because systems under stress may be considered "experiments in the environment." Atmospheric systems were excluded as a topic, since a recent Dahlem Workshop dealt with atmospheric chemistry in its various aspects (Goldberg, 1982).

The participants were selected so as to have representatives of both "users" and "donors" in each of the four groups. The authors of the background papers were chosen among the participants in each group and one referee was assigned to each paper. It goes without saying that the "users" were not the only ones who profited from the discussions and that the "donors" did not feel that they were exploited by the "users". On the contrary, the "donors" also greatly profited from these joint discussions, and it became obvious that only via a close collaboration between both categories will we be able to bring most of the important environmental problems into sufficient focus and find meaningful and effective ways for their solution.

From the very outset of the workshop it had become clear that in the context of chemistry the participants used the term "speciation" in at least four different meanings. Hence, in order to assure textual coherence of this

Report, the editors replaced the term "speciation" wherever it occurred by "identification and quantification," or "description of abundance," or "reactivity," or "transformation" of a chemical species, according to whichever one of the four meanings the author had evidently meant to convey.

In line with the *Dahlem Workshop Model,* this Report comprises the background papers written in advance of the meeting on the current status of problems in environmental research and on advanced analytical techniques for the identification and quantification of chemical species, as well as the group reports summarizing the results of the discussions held during the meeting. Each group report was prepared during the meeting by one "rapporteur" with the help of members of that group and finalized by the rapporteur (listed as the first author of the group report) after the meeting, taking into account both verbal comments made during the presentation of the reports in the plenary session at the end of the workshop and written comments received afterwards.

The advance in the understanding of the role played by chemical species is very different for various chemical elements. In some cases it is possible to describe the complete biogeochemical cycles of the transformations of element species, but in the majority of cases the most that can be done is to illustrate by means of model experiments the influence of different chemical forms (very often defined only operationally) on the fate of element species in the environment. At present it seems that the greatest bottleneck in understanding the environmental role of chemical species arises from the lack of suitable chemical methods and procedures for their identification and quantitative analysis at the concentrations at which they occur in nature. Furthermore, nondestructive *in situ* measurements of chemical species are very rarely performed and most species are studied under *in vitro* conditions or in samples. Since the coverage of relevant topics in the background papers and group reports could not be comprehensive, of course, only hopefully typical examples are presented in this Report.

The editors thank the authors of the background papers and group reports for their cooperation in responding to editorial enquiries about their manuscripts. Special thanks are due the rapporteurs, who took on the very demanding job of summarizing the discussions and integrating the different views expressed by the participants, often having to work late into the night during the workshop. It is a pleasure for us to acknowledge also the efficiency and charm of the members of the Dahlem Konferenzen staff, who managed to create a relaxed atmosphere that greatly fostered the free and easy exchange of ideas between participants of such different professional backgrounds (Fig. 1). All participants will long cherish the memory of this workshop. Last, but not least, we extend our thanks to the sponsors of the Dahlem Konferenzen, the Stifterverband für die Deutsche Wissenschaft and the Senat der Stadt Berlin, without whose generous financial support this project could not have been undertaken.

Fig. 1. Solemn oath taken by the moderators and editors. From left to right: F. Brinckman, H. Tobschall, P. Sadler, M. Bernhard, A. Andreae, B. Fowler

THE "SPECIATION" SOCIETY

$Fe^{II} \rightleftharpoons Fe^{III} + e^-$

SPECIARE NECESSITAS EST

This is to confirm that the participants of this workshop have solemnly sworn in the name of Lucifer, honorable great wizard, master and patron of all alchemists and those who would like to be alchemists, to speciate at all times, in all climes, and in all unlikely places; unless either he, his colleagues or his sponsors do not understand what he is doing. Further, the solemn promise has been given not to embark on any new speciation enterprises without first bowing to the four winds and repeating **"SPE, IN SPE, SPECIATION"**.

As signified by the seal below, on this day of
Friday, September 7, in the year of 1984 Anno Domini.

In testament thereof:

Peter Sadler

The Importance of Chemical "Speciation" in Environmental Processes, eds. M. Bernhard, F. E. Brinckman, and P. J. Sadler, pp. 7–14. Dahlem Konferenzen 1986. Springer-Verlag Berlin, Heidelberg
© Dr. S. Bernhard, Dahlem Konferenzen

Why "Speciation"?

M. Bernhard*, F. E. Brinckman**, and K. J. Irgolic***

* Centro Studi Ambiente Marino
ENEA
P.O. Box 316
19100 La Spezia, Italy
** National Bureau of Standards
Gaithersburg, MD 20899, USA
*** Dept. of Chemistry
Texas A&M University
College Station, TX 77843, USA

Abstract. The terms "species" and "speciation" have been borrowed by chemists from biology. In biology "species" refers to a population of organisms which possess hereditary traits and which interbreed, conserving these. In chemistry "species" refers to the molecular form (configuration) of atoms of an element or cluster of atoms of different elements. In biology the term "speciation" denotes the process(es) by which species arise in the course of evolution. However, in chemistry the term "speciation" has been used in many different contexts, of which chemists are often not even aware. Chemical speciation has been used to describe:

1. the analysis of species leading to their identification and quantification in a defined area or in a volume (e.g., a sample) (species analysis);
2. the description of the abundance (or numerical distribution) of species of an element in a volume (species distribution or abundance);
3. the reactivity of a given species (species reactivity);
4. the transformation of one species into another (species transformation).

In order to increase the awareness that different categories of chemical species exist, the terms "parent or conceptional species," "matrix species," and "analyte species" are introduced. Parent species refers to the idealized concept of the species. It refers to that part of the species which will not be changed when the species comes into contact with different matrices. Since species exposed to different matrices will change their reactivity, the resulting species becomes a matrix species. During chemical analysis the species present in a given volume will often be transformed into a single species for which the analytical instrument is sensitive. This is the analyte species.

The role which species play in environmental processes and chemical analysis is briefly discussed. An effort should be made to develop chemical methods suitable for species identification and quantification. This development is considered one of the greatest challenges to modern chemical analysis.

Introduction

The literature discussing the concentration of inorganic trace substances in the environment uses the term "element" very frequently for the description of their composition. This usage tends to obscure the well-known fact that chemical "elements" are part of molecules (such as covalently bound groups of atoms, complexes, and weak associations of chemical species). Every element is capable of forming at least a few molecular species. Depending on the conditions, an element may occur in the environment just as a single molecular species, or if more species of an element are present the concentration of one of them may be higher than the concentration of the others. The total concentration of the element can be used to assess its environmental impact only if it is present in the environment as a single species. But this condition will be met only very rarely. Since the chemical reactivity (behavior) of the element depends on the particular species in which it is present, and since often the most abundant species is not the most reactive one, it is not surprising that the behavior of an element in the environment (e.g., bioavailability, toxicity, distribution) on the basis of its total concentration cannot be reliably predicted. Moreover, many analytical procedures thought to assay total element concentrations are now known to measure only single species or groups of previously unrecognized species.

Unfortunately, the identification of individual species and their quantificative determination in the environment is still difficult at present, being possible only for certain species. Analytical procedures of sufficient sensitivity and specificity are not available to determine the species of many environmentally significant elements, which frequently occur at very low concentrations.

Hence, advances in the understanding of environmental processes are presently hindered by the lack of suitable methods for the identification and quantification of species and for the analysis of their behavior in the environment. The development of new methods for these purposes poses a great challenge to analytical chemists and environmental scientists.

The Term "Speciation"

In the chemical context, the term speciation is borrowed from biology. In biology "species" refers to a population of organisms possessing a set of common characteristics or hereditary traits. Before the publication of Darwin's "Origin of Species", it was generally believed that a species has fixed characteristics and does not change. Then it came to be recognized that a species comes into being during evolution through a process called "speciation." During speciation a population becomes reproductively isolated, i.e., interbreeding between the original population and the isolated population no longer occurs. The isolated population has an opportunity to develop new hereditary traits through spontaneous variation that distinguish it from the parent population. A novel biological species thus arises which maintains its integrity through reproductive isolation.

When chemists began to appreciate the importance and ubiquitous nature of different molecular forms of an element in environmental processes, they felt the need to distinguish between investigations concerned with the determination of the total amounts of elements of flux and those dealing with the different molecular forms of an element. Unfortunately, however, chemists began using the term speciation in a variety of senses, some of them ill-defined.

A biological species is a population of organisms with stable characteristics. The members of this population interbreed only with specimens belonging to the same species (reproductive isolation). Biological speciation is the process of species evolution and, therefore, has only a kinetic connotation. Chemists, however, use the term "speciation" both with "static" and "kinetic" notations. The following are examples quoted from recent literature: "speciation of iron," "speciation properties," "measurement of metal speciation," "original speciation pattern," "nature of the speciation type," "information about general speciation," "respeciation," "biological effects of metal speciation."

In its "static" implication the term "speciation" refers to either the operational process of "identification and quantification of the abundance (numerical distribution) of the species of an element" (species analysis) or the results obtained from this species analysis: "the distribution or abundance of the species in the environment or a sample" (species distribution). In its "kinetic" implication it refers to "transformation of chemical species" (species transformation). The "kinetic" implication is similar to that used in biological evolution theory. The reproductive isolation of the biological species has its analog in the "reaction specificity" of a chemical species. The specific reactivity of a species has also been called speciation. The term "speciation" has therefore been used to imply four different meanings.

Because of the confusion in the different uses, we make a plea that in a chemical context authors either avoid the use of the term "speciation" altogether or define how they want the term to be understood.

Chemical Species

The term "species" refers to the molecular form (configuration) of atoms of an element or a cluster of atoms of different elements in a given matrix. For example, phosphate, diphosphate, and adenosine triphosphate are some of the species of the element phosphorus. The term "compound" can be considered synonymous with "stable species." Often, however, the term "species" is used to indicate uncertainty or lack of knowledge about the nature of the species one expects to find in an environmental sample. Each species has its characteristic chemical reactivity and lifetime and it will react differently in different matrices or environments. Many species have in their molecules some groups that are strongly bound to the central atom and others groups that are only weakly bound. When the matrix changes the weakly bound groups can be replaced. The part of the species that remains unchanged through a matrix-generated transformation is called the "parent species," or "conceptional species." The species that actually exists in the environment is the "matrix species" or "functional species" (see Zeppezauer, this volume). This means that the term "chemical species" is actually an idealized concept. Analytical procedures generally convert the matrix species into a stable species (compound) that can be detected and quantified. The species which is detected is called the "analyte species." The different terms are illustrated for methylmercury in Table 1.

The term chemical species should refer to a precise molecular configuration of atoms of an element, but quite often the species analyzed is operationally defined. For instance, when an element is said claimed to occur in a particulate or colloidal form, it is not clear whether the particle consists entirely of one species (e.g., Fe(III)hydroxide colloid) or the species is adsorbed onto or included in the particle. As long as an operational definition

Table 1. Parent species of methylmercury, some of its matrix species, and the most common analyte species

Parent species	Matrix	Matrix species	Analyte species
CH_3Hg^+	Air	CH_3HgL (L = weakly bound ligand)	$\xrightarrow{Cl^-}$ CH_3HgCl
	Water	$[CH_3Hg(OH)]^+$	
	Biol. tissue	CH_3Hg-S-protein	
	Soil	CH_3Hg-humic acid	

of the species is considered only a first approach to species quantification, it is acceptable. If, however, it becomes only a convenient convention, every effort must be made to identify and quantify the species. Only then can an understanding of the process under study be obtained.

The Determination of Species

Sample contamination, but not the "species problem," is now widely recognized as a problem during analysis. This is probably due to the fact that many analyses still serve only descriptive purposes and that they are not (or can not yet be) part of a conceptional design of environmental studies. Therefore, the "validity" of the analytical results cannot always be checked against a hypothesis or a model of the environmental processes under study.

Although "total element" concentrations are rarely sufficient to describe environmental processes adequately, "total element" amounts are useful for the estimation of the quantitative importance of the species present. The precise and accurate determination of "total element" amounts is not an easy task. Very few analytical methods measure total element amounts intrinsically (e.g., instrumental neutron activation analysis). Most methods determine concentrations of only certain species or groups of species (e.g., anodic stripping voltammetry). Therefore, with these latter methods total amounts are only determinable if all species present are transformed quantitatively and irreversibly by a suitable pretreatment into that species for which the method is specific and sensitive. Unfortunately, enough attention has not always been given to this problem, and in many cases it was not realized that even the determination of the total amount can be a species problem.

The determination of species is a more complex task than the determination of total element concentrations. At present, most efforts in "species analysis" are directed towards the identification and quantification of chemical species in isolated samples. Since during sampling, storage, and analysis of the sample species transformation may occur, appropriate action must be taken to assure that the species to be determined does not change during the interval between sampling and analysis. Reliable, generally applicable methods for the preservation of the original distribution (abundance) of the species do not yet exist, and for each species and matrix the preservation of the original distribution of species during sample storage must be studied.

Often the species to be determined have to be preconcentrated to reach concentrations sufficiently high for detection. During these preconcentrations the original matrix species will generally be transformed, and the analysis will provide information only about parent species. The ideal analytical method would determine species directly *in situ* and *in vivo,* noninvasively and nondestructively, and would allow the transformations of the species to be monitored over time.

Role of Species in Environmental Processes

Early studies of environmental processes used almost exclusively "total element" concentrations or unspecified fractions of total concentrations which depended on the analytical procedure used. Often analysts were not concerned at all whether they did determine of total amounts or fractions of total amounts. Rarely were species considered. Stating the analytical method used seemed to be sufficient. For example, specifications such as "reactive phosphate" referred to the phosphate reacting with the reagents and not to the phosphate participating in the environmental processes. Although there might be a connection, none was sought.

As the necessity arose to predict the toxicity and fate of pollutants released into the environment, first for the disposal of radioisotopes and later for the evaluation of the impact of trace elements on ecosystems and humans, it soon became evident that neither total element concentrations nor "total element concentrations" in terms of generally undefined groups of species were adequate for predicting and assessing environmental processes. The determination of the abundance and distribution of trace elements in the environment, their accumulation by organisms, their bioavailability, and their toxicity to terrestrial and aquatic organisms (including humans) can be understood only in terms of trace element species.

"Ecodisasters," such as the Minamata methylmercury incident, serve as tragic reminder of the importance of trace element species for the mobilization and bioaccumulation of toxic elements. Since that incident our understanding of environmental processes has advanced gradually from crude approximations, to investigations, to experimentally better-founded hypotheses concerning the behavior of some elements in the environment. Environmental assessments proceeded from considerations of acute toxicity via evaluation of subchronic effects to estimations of mutagenicity. Investigations of the behavior of toxicants examined initially the bioaccumulation and degradation of these substances and then explored biogeochemical pathways and cycles.

As long as analytical methods were unable to determine concentrations much lower than one g L^{-1} or g kg^{-1}, only bulk properties (chemical oxygen demand, concentrations of major and minor elements) could be determined and correlated with acute and chronic toxicities and the essentiality of some elements could be established. With increasing sensitivity of the analytical methods (one µg L^{-1} and one µg kg^{-1}) "total trace element concentrations" could be determined and accumulation and release of trace elements by organisms be investigated. Also biogeochemical pathways and cycles of trace elements could be studied. To understand the interactions of trace element species with biologically important molecular species (such as enzymes, DNA, and RNA) even more sensitive and analytical techniques able to determine concentrations below one µg L^{-1} are necessary. Investigations of these interactions must consider the kinetic aspects of relevant reactions. The progress achieved in these areas is not the same for all elements. Environmental and health protection needs have markedly influenced the direction and research intensity of trace element research. Greatest advances seem to have been made when interdisciplinary teams were active in assessing the impact of toxicants. Much progress has recently been made correlating molecular structures or topologies determined or inferred for matrix species of a given element, with physicochemical or toxicological properties, such as aqueous solubility, octane(???)-water partition coefficients, bioaccumulation factors, and carcinogenicity.

Need for the Investigation of Chemical Species

Numerous examples are given in the following background papers, which clearly demonstrate the importance of species identification for an understanding of environmental processes. Elements may be present in different species that are formed through natural processes or are introduced by man.

Several species of elements such as As, Hg, and Se have been shown to be transformed in the environment during weathering, uptake by biota, elimination from biota, fixation in sediments, and remobilization. Essential elements such as Cu and Fe take part in biogeochemical cycles. Some elements (e.g., Mg, Zn, Fe) are cofactors in enzymes and vitamins. Even isomers of metal compounds can differ markedly in their pharmacological properties (e.g., Pt anticancer drugs). It is obvious that total element analyses will not give sufficient information on the specific reactivity and function of these element species. During the past decades the determination of total elemental concentrations has sometimes been followed by the determination of trace element species. When element species are determined, mass balances should be established to ascertain that all species have been detected. The change from total trace element determination to the determination of species is driven by the deeper insight which molecular-based information can provide about the interactions between trace element species and the environment.

Recommended Readings

Brinckman FE, Bellama JM (eds) (1978) Organometals and Organometalloids: Occurrence and Fate in the Environment. ACS Symposium Series, vol 82. Washington, DC: American Chemical Society

Brinckman FE, Fish RH (eds) (1981) Environmental Speciation and Monitoring Needs for Trace Metal-containing Substances from Energy-related Processes. NBS Special Publ 618. Gaithersburg: National Bureau of Standards

Harrison PM, Hoare RJ (1980) Metals in Biochemistry. London: Chapman and Hall

Hughes MN (1981) The Inorganic Chemistry of Biological Processes, 2nd ed. Chichester: J Wiley and Sons

Irgolic KJ, Martell AE (eds) (1985) Environmental Inorganic Chemistry. Deerfield Beach, FL: VCH Publishers, Inc

Leppard GC (ed) (1983) Trace Element Speciation in Surface Waters and Its Ecological Implications. New York: Plenum Press

Lippard SJ (ed) (1983) Platinum, Gold and Other Chemotherapeutic Agents. ACS Symposium Series, vol 209, Washington, DC: American Chemical Society

Nriagu JO (ed) (1984) Changing Metal Cycles and Human Health. Dahlem Konferenzen. Berlin, Heidelberg, New York, Tokyo: Springer-Verlag

Thayer JS (1984) Organometallic Compounds and Living Organisms. New York: Academic Press

Standing, left to right:
Michael Zeppezauer, Roger Thorneley, Magnus Piscator, Walter Mertz, Wayne Wolf

Seated (center), left to right:
Peter Sadler, Rolf Mehlhorn, Colin Mills, Ulrich Oehmichen

Seated (front), left to right:
Günther Weber, Jan Ludwicki, Kurt Irgolic

The Importance of Chemical "Speciation" in Environmental Processes, eds. M. Bernhard, F. E. Brinckman, and P. J. Sadler, pp. 17–38. Dahlem Konferenzen 1986. Springer-Verlag Berlin, Heidelberg
© Dr. S. Bernhard, Dahlem Konferenzen

Importance and Determination of Chemical Species in Biological Systems

Group Report

W. R. Wolf, Rapporteur
K. J. Irgolic
K. J. Ludwicki
R. J. Mehlhorn
W. Mertz
C. F. Mills
U. Oehmichen
M. Piscator
P. J. Sadler
R. N. F. Thorneley
G. Weber
M. Zeppezauer

Introduction

Biological systems constantly synthesize, change, and degrade organic and inorganic chemical species. These processes and the interactions among chemical compounds maintain and propagate life. The major "organic elements" (C, H, N, O, P, S) combine in many ways to give compounds such as amino acids, lipids, carbohydrates, and nucleotides. These simple compounds serve as building blocks for proteins, enzymes, membranes, structural tissues, and genetic materials. Knowledge of the "organic" chemistry of biological systems is quite advanced. However, the organic chemistry of life would not be possible without certain inorganic elements. The beneficial and detrimental influences of inorganic chemical species, in this report called chemical species or simply species, on biological systems was the main theme of this group report.

To understand the significance and role of such chemical species, their movements through ecosystems (Fig. 1) and the changes they experience on this journey must be known. This report concentrates on the function, utilization, and transformations of inorganic chemical species in higher animals and in man. Chemical species enter the animal/human systems with food, water, and air. The processes of digestion, absorption, transport, distribution, incorporation into active sites, storage, and excretion influence

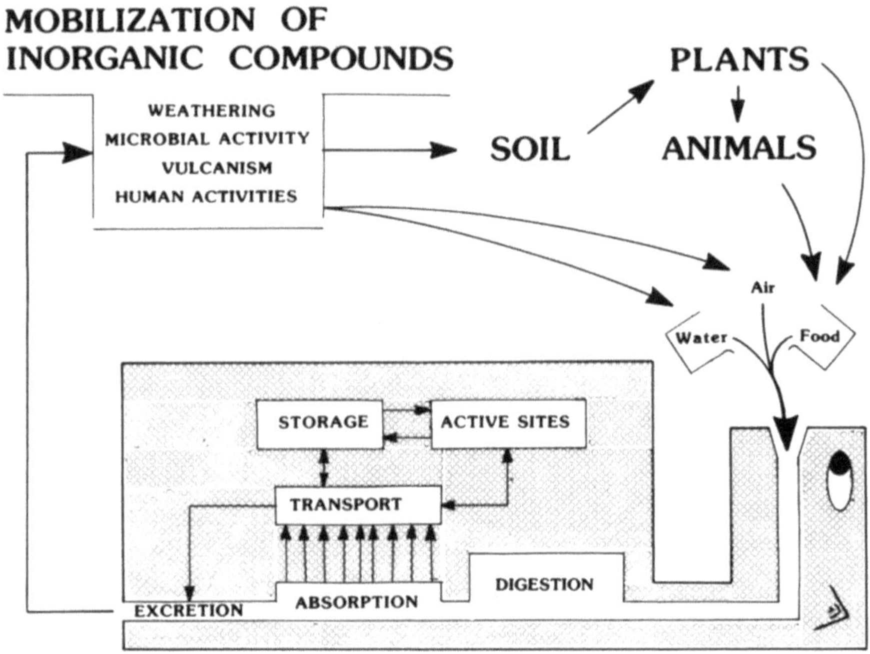

Fig. 1. A general scheme of the environmental cycles of elements

the biological utilization of the chemical species and their movement through the animal/human systems. Any experimental approach to answer questions regarding the biological action of chemical species depends on the accurate measurement of their concentrations in complex matrices. This report assesses the current incomplete knowledge of the importance of trace elements for animals and humans, the transformation of trace elements, their interactions with biological systems, and the analytical methodologies required to increase this knowledge.

Health Effects of Trace-Element Compounds

Strong evidence links the health effects of trace elements with their chemical forms and the concentration of each chemical species to which an organism is exposed (Merian, 1984). The species dependence of health effects is observable at all stages of the environmental cycle of an element, from the soil-plant interface through absorption and metabolism by an organism to excretion and recycling.

In spite of this evidence, national and international efforts to solve environmental health problems of man, plants, and animals have made little use of the idea of the species dependence of health effects. Environmental exposures from soil, water, air, and food and dietary recommendations are still expressed in most cases in terms of "total elements" and not in terms of "elemental species." It is encouraging that several recommendations for environmental exposures were given recently in terms of individual species or groups of species such as chromium(VI), arsenic(III), and methylmercury.

The following examples demonstrate that the nature and concentrations of species are important to assess the impacts of elements on health:

- Chromium as Cr(VI) in zinc chromate dust inhaled with air is a strong carcinogen. Chromium as a Cr(III) compound present in food is an essential element (National Academy of Sciences, 1974).
- Adequate dietary concentrations of "total zinc" cause severe zinc deficiences when sufficiently high concentrations of calcium and phytic acid are present in the diet (Prasad, 1976).
- The same arsenic intake that produces toxic effects when arsenic is in the form of arsenite is tolerated without deleterious effects when arsenic is provided as arsenobetaine (Vahter et al., 1983).
- Generally, trace-element compounds are changed chemically during digestion. High-molecular-weight species, for instance, are broken down to small molecules. The question as to whether the resulting species can be used by the organism is determined by the availability of suitable ligands in the food and the intestinal secretions. Absorption of a species from the intestinal tract via transport through the intestinal wall is associated with a change of ligands. The molecular events of this process are poorly understood. However, the process is very important because its impairment or cessation causes serious diseases such as acrodermatitis enteropathica, a serious zinc deficiency (Prasad, 1976).
- Once an element in its appropriate molecular form is absorbed, transport proteins form new species that are carried to sites of action or to storage sites. Several transport proteins have been described. When these proteins are abnormal, absent, or present in insufficient concentrations, diseases occur (Martell, 1980). How these proteins function is not known in detail. The determination of chemical species in blood appears to be feasible. Such determinations (e.g., Fe saturation of transferrin, concentration of serum ferritin, concentration of zinc metallothionein) would help to establish the trace-element status of men and animals.

- The chemical environment of elements such as zinc, copper, iron, and molybdenum in active sites of enzymes is fairly well known (Xavier, 1986). Additional work on metalloenzymes that have lost their function should link the nature of the metal species in the enzyme with clinically established signs of metal deficiencies.
- Elements are excreted by organisms in appropriate chemical forms in the urine, in feces, in sweat, and in exhaled air. Very little is known about the nature of the excreted species. The species excreted might change with the level of exposure. For instance, animals exposed to selenium at levels not drastically higher than the physiological amount excrete trimethylselenonium ion in the urine. At higher exposure, volatile dimethyl selenide is expelled with the exhaled air (National Academy of Sciences, 1976).

The literature provides many examples of the connection between trace elements and health and disease (Merian, 1984). However, the understanding of these connections on a molecular scale is quite limited. Only few biochemical pathways are known for trace elements.

Chemical Species in Foods

The macromineral elements chlorine, calcium, magnesium, potassium, and sodium form ionic compounds, most of which are easily soluble in water. Therefore, in most cases knowledge about species, in which these elements occur in foods, is considered not to be important. The determination of "total element" concentrations in foods is sufficient in these cases. Minor concerns exist about insoluble compounds of calcium and magnesium that might not be usable by organisms.

The trace elements have a much richer chemistry than the macromineral elements and form many compounds. Knowledge about trace-element compounds in foods is therefore of great importance. The elements As, Co, Cr, Cu, Fe, Hg, Mo, Se, and Zn have been studied in some detail. Information about the metabolic importance of species of these elements is available, and knowledge about the occurrence of these species in foods is accumulating. Much less is known about the species of Al, Cd, F, I, Mn, Ni, Si, and V in foods and the biochemical roles of these species. Most of the information available on species is qualitative. Whereas composition and structure of some species are known, few data exist about their concentrations in foods (Chesters, 1981; Pickford, 1981). The effects of food processing and cooking on species has received little attention.

The analytical methodologies for the determination of the macromineral elements Ca, Cl, K, Mg, and Na are satisfactory for most applications. Species of these elements are generally not determined. The methodologies for the determination of "total" minor and trace elements are adequate for As, Cd, Co, Cu, F, Fe, Hg, Mo, and Zn, fair for Cr, Ni, Pb, and V. Methodologies for the determination of species are barely adequate for compounds of As, Hg, and Se and poor for species of all other elements. Very little information is available about the stability of most species under the conditions required by the analytical procedures (extraction from foods, chromatography). Standard materials for chemical species in foods do not exist; only a few homogeneous food samples that could be used in interlaboratory studies are available.

Chemical Species in Digestive and Absorptive Processes

The absorption of trace elements in the intestines is dependent on many factors including the chemical nature of the trace-element compounds and the nature of the ligands available in the intestines. Bioavailability is the ability of a chemical species of a trace element to meet the physiological requirement of an organism. It is influenced by the efficiency of intestinal absorption, by transport of species to sites of action, and by the utilization of active sites. For example, the gastrointestinal absorption of iron in humans is influenced by the solubility and absorbability of various forms of iron such as heme iron, non-heme iron, divalent iron, and trivalent iron and by the status of the body's iron reserve. Iron(II) is easily absorbed, whereas iron(III) is not. Reducing agents such as ascorbic acid might enhance the absorption of iron (Bernat, 1983).

The efficiency with which trace-element compounds are absorbed can vary markedly. For instance, absorption of soluble forms of zinc or iron by a deficient animal is extremely efficient, whereas copper from diets rich in inhibitory factors and metallic mercury is hardly absorbed at all. Homeostatic processes are partly responsible for regulating absorption.

The nature of the ligands in the intestinal environment and the partitioning of these ligands among the elements will be influenced by the composition of the diet and by changes in the digestive environment (pH, redox potential, ionic strength). The digestive environment may be changed by bacterial activity within the digestive tract. Such changes and concomitant modifications of trace-element species are likely to occur in the rumen of herbivores and the caecum and colon of monogastric animals. Although only few such processes have been adequately characterized, their relevance is undeniably demonstrated by the following examples:

- The phosphorus storage compound phytic acid (myo-inositol with all six hydroxyl groups esterified with dihydrogen phosphate) present in many foods of plant origin can profoundly affect the absorption of zinc (Prasad, 1976). Phytic acid probably forms within the digestive tract Zn/Ca phytates that are insoluble at the intestinal pH and thus not absorbable. Zinc deficiency can be induced in this manner by phytate concentrations commonly present in diets. The precise nutritional significance of the zinc-phytic acid interactions is controversial because the kinetics of formation, the stability, and the solubility of mixed alkaline-earth/zinc phytates under physiological conditions prevalent in the intestines has not been investigated in sufficient detail.
- The absorbability of copper species is strongly linked with the susceptibility of ruminant animals to copper deficiency in many parts of the world (Venugopal and Luckey, 1979). Microbial action and ligand-exchange reactions influence the biological availability of copper. The inadequately characterized copper species in plants eaten by ruminants are immediately subjected to microbial "digestion" in the rumen. Sulfide generated in the rumen by bacterial reduction of sulfate and by degradation of thioamino acids may react with copper to form physiologically unavailable copper sulfides. The availability of ingested copper is largely determined by the efficiency with which reactive sulfide is removed from the digestive tract and by the presence of copper species that are sufficiently stable to react with sulfide or hydrogen sulfide but sufficiently labile to yield absorbable copper species. The bioavailability of copper is also influenced by other trace elements. For instance, molybdenum and iron prevent the removal of sulfide by forming metal sulfides that react at pH 2 to 4 with copper at later stages of the digestion making copper physiologically unavailable. These interactions cannot be understood until the Cu/Mo/Fe/S species have been characterized.
- Methylmercury is easily absorbed from the intestines, mercury(II) chloride only partially, and metallic mercury not at all. Metallic mercury, however, passes easily through the lung-blood barrier (Friberg and Vostal, 1972).

These examples illustrate the importance not only of the initial forms of elements in the diet or in the environment but also of the species formed during digestion and absorption. Many important aspects of the transformation of species, such as binding to membranes, cell walls, and macromolecules, transport across membranes, and diffusion through membranes, have not received sufficient attention. These interactions influence the uptake and efflux of chemical species. To be absorbed from the intestines a species must pass through a viscous mucous layer surrounding the intestinal microvilli, through membranes of the brush border, and through epithelial cells. Little is known about these processes for major nutrients and nothing for trace elements. Specific uptake mechanisms have been studied with isolated brush-border residues for a few substrates. However, experimental knowledge of the diffusion and transport of trace-element species is so limited that only theoretical ideas about transport phenomena (Mehlhorn, this volume) are available to make any predictions about the uptake of these species. Little is known about the role of endocytosis and vesicular absorption phenomena, some of which may be initiated by immunoreactive macromolecules in secretions from tissue or by molecules serving as relatively nonspecific carriers present in foods such as milk.

Chemical Species and Active Sites

Many different molecules are synthesized in biological systems by only a limited number of enzymatic reactions. This "synthetic" diversity is achieved by combining molecular building blocks (amino acids, carbohydrates, fatty acids, and a few other organic molecules) containing only few types of elements to form the mostly macromolecular biomolecules.

Chemical Species in Structure-Function Relationships

Most enzymes catalyze proton transfer (mediated by side chains of amino acids) or electron transfer (mediated by metal ions and a few organic cofactor molecules). A very limited number of organic cofactors, coenzymes, and side chains of amino acids participate in group transfer reactions. Zinc enzymes take part in all of these types of reactions. More than 100 zinc enzymes are known and approximately 30 are well characterized (Spiro, 1983). The zinc enzymes carbonic anhydrase, carboxypeptidase, and alkaline phosphatase catalyze hydrolytic reactions. DNA and RNA polymerase facilitate group transfers. Alcohol dehydrogenase is involved in redox reactions. Zinc is also involved in the storage of insulin as $Zn_2Insulin_6$.

The active sites of the enzymes and the binding sites for zinc in insulin determine the reactivity of zinc via the coordination geometry and the nature of the ligands coordinated to zinc. Zinc in the hydrolytic enzymes carbonic anhydrase and carboxypeptidase is coordinated to hard ligands (Bertini, 1981). Three histidine-nitrogen atoms serve as donor atoms in carbonic anhydrase. Carboxypeptidase provides two nitrogen atoms from histidine and one oxygen atom from a glutamic acid. Exchangeable ligands such as water and organic substrates with carbonyl groups complete the coordination sphere of zinc. Alcohol dehydrogenase has two zinc-binding sites in each unit of the dimeric liver enzyme or tetrameric yeast enzyme. The noncatalytic site offers predominantly soft ligands such as four sulfur atoms from cysteine. The catalytic site provides two hard histidine-nitrogen atoms, two soft cysteine-sulfur atoms, and one or perhaps two hard oxygen atoms from water or substrates as exchangeable ligands.

The coordination geometry provided to a metal ion by a site in an enzyme is very important. A distorted tetrahedral environment is provided for zinc by the active sites of carbonic anhydrase, carboxypeptidase, and alcohol dehydrogenase, whereas an almost regular tetrahedral geometry is characteristic for zinc in the noncatalytic site of liver alcohol dehydrogenase.

The ability to rapidly change the coordination environment of a metal ion in an enzyme is responsible for the catalytic effectiveness of that enzyme. These changes may occur in the inner sphere of the complex by bonding of a substrate molecule or release of a bound substrate molecule. Inner-sphere changes are associated with an increase or a decrease in coordination number. Outer-sphere changes, which do not disturb the ligands bonded to the metal ion, may be caused by disturbances in the environment of the protein via hydration or alteration of polarity.

The metal-binding sites in liver alcohol dehydrogenase differ with respect to the kinetics and thermodynamics of the metal binding. For example, zinc in the noncatalytic site is exchanged much faster for a cadmium or cobalt(II) ion than is zinc in the catalytic site. New active enzymes can be prepared by replacement of zinc in the catalytic site by Cu(I), Cd(II), Ni(II), or Co(II). Enzymes with Cu(II) or Pb(II) in the catalytic site are inactive. The activities of the enzymes with Fe(II) or Fe(III) are not known.

The steps in the biosynthesis of metalloproteins are not well understood. In addition to "catalytic" and "structural" metal-binding sites, other peripheral sites do exist. These other sites may be important for the mechanism of incorporating the metal into the protein and the folding of the protein. Very little is known about the regulatory influence of metals on the biosynthesis of metalloproteins. Almost nothing is known about biologically important molecules that are most easily affected by metal deficiencies and about the consequences of such effects that may be lethal to an organism within a short time.

Relevance of Active Site Modelling

Metal complexes with a well-defined coordination geometry and ligand environment are the simplest models for active sites that can yield information about structure-function relationships for metal ions in biological systems. The success of a model is judged by how closely structures, catalytic rates, substrate affinities, and transport properties approach the corresponding characteristics of the isolated, naturally occurring molecules. However, an isolated protein is also only a model for the *in vivo* compound.

The four most important aspects of model and *in vivo* metalloproteins are structure and stability of the active site(s), catalytic activity, substrate specificity, and regulation of activity. These properties will be influenced by the nature of the metal ion, its oxidation state, and its inner coordination sphere. Some successful models have used metal ions other than those found in the natural metalloprotein in order to allow certain experiments to be carried out.

The inner coordination sphere of a model complex can be "coarse-tuned" by the choice of ligands generally containing sulfur, oxygen, or nitrogen as donor atoms that impose a particular coordination geometry. Coordinatively unsaturated metal sites, which are often an essential feature in natural proteins, are difficult to model. The coordination sphere is "fine-tuned" by subtle changes of the structure and composition of the ligands. The groups bridging donor sites in multidentate ligands are important for the coordination geometry and coordination number of a complex. The success of Palermo et al. (1984) in modelling the molybdenum center in sulfite oxidase is a good example for this approach. The most successfully modelled sites are the iron-sulfur core clusters that are found in a number of proteins associated with the respiratory chain, photosynthesis, and nitrogen fixation (Holm, 1981; Thomson, 1985).

The investigation of model complexes by a variety of physicochemical techniques generate data bases that are very valuable for the interpretation of results obtained with natural systems. For example, the N-15 NMR data from molybdenum complexes containing dinitrogen and other nitrogen species ($=N-NH_2$, $-NH-H_2$, $=NH$, $-NH_2$) as ligands have been essential for the studies of nitrogen fixation. Models will never achieve the specificity and degree of regulation shown by the natural proteins, but complementary and synergistic studies with model complexes, purified proteins, and whole cells or organs will provide much information about the nature of metal species in metalloproteins.

Synergistic and Antagonistic Interactions among Elements

Inorganic species with similar properties may compete for binding sites on proteins and influence metabolic processes. For example, sulfate (SO_4^{2-}) and tungstate WO_4^{2-}) compete with molybdate (MoO_4^{2-}) for carriers responsible for transport through membranes. A similar competition exists between Mg^{2+} and K^+. Inorganic species may also compete for relatively nonspecific sequestration sites of proteins, whose synthesis is induced by changes in the concentrations of certain elements in tissues.

Binding of cadmium, copper, zinc, and perhaps mercury by metallothioneins is an example of nonspecific sequestration (Kägi and Nordberg, 1979). The cysteine-rich metallothionein initially thought to serve only as a scavenger of cadmium and thus protect the organism from cadmium intoxication is now known to be involved in the storage and mobilization of zinc. Metallothionein-related effects are reduced absorption and retention of copper when the zinc supply is excessive and decreased tolerance for copper when the zinc supply is limited. Although much about the metabolic role of metallothioneins is not known, it is clear that metallothioneins are involved in the prenatal storage of zinc and copper in tissues, in the transient sequestration of zinc liberated from tissues in a catabolic state, and in the sequestration of high concentrations of copper and cadmium in tissues. The present, limited understanding of the role of metallothioneins is sufficient to suggest that the tolerances to excesses of copper, cadmium, and zinc can be improved by optimizing the conditions for the synthesis of metallothionein. The concentration of metallothionein in blood plasma may be used as an indicator of the zinc status of tissues because zinc is required for the synthesis of metallothionein. The importance of metallothionein for the translocation of cadmium, copper, and zinc will not be clarified until the tissue distribution and the biological half-lives of metallothioneins have been determined.

Molybdenum-thiol complexes are involved in the metabolism of copper and iron and the etiology of ruminant copper deficiencies (National Academy of Sciences, 1977). The inadequate understanding of the properties and kinetics of the formation of insoluble, physiologically unavailable thio- and oxythiomolybdates of copper makes it difficult to predict the occurrence and evaluate the significance of copper deficiencies. Model compounds for such molybdates have identified likely candidates for reactive groups but do not explain the biochemical phenomena that are associated with high intakes of molybdenum. As the most potent inhibitors of copper metabolism yet discovered, the thiomolybdates have great potential as therapeutic agents for the control of chronic and acute copper intoxication in copper-intolerant animals such as sheep.

Whereas copper and molybdenum are linked in biological systems through their preference for the same types of ligands, iron and copper are connected via common metabolic processes. For instance, severe copper deficiency reduces the absorption of iron and the mobilization of iron from the liver because low copper concentrations reduce the formation of copper enzymes that are necessary for the transfer of iron across membranes. A synergistic relationship has recently been discovered between selenium and copper, probably associated with protection of tissues from reactive oxygen species by glutathione peroxidase (Se) and superoxide dismutase (Cu). These elements might also have roles in the synthesis of connective tissues.

Transport and Storage of Chemical Species

As an element moves through ecosystems (Fig. 1) it will become part of a variety of compounds. The properties of these compounds will, for instance, determine the mode and extent of their interactions with membranes (Mehlhorn, this volume), their biological availabilities, and their metabolic transformations. The ability to detect, isolate, and characterize a particular compound is determined by its concentration in the tissues. Species whose rates of formation or uptake are much faster than their rates of metabolic conversion or excretion will be present in high concentrations and will be easy to detect.

Lipophilic molecules tend to accumulate in fatty tissues where they are protected from metabolic degradation. These stored molecules may change membranes and disrupt membrane functions or catalyze destructive oxidation of lipids through interaction with metabolic electron donors (Mehlhorn, this volume). Because of the potential toxicity of lipid-soluble trace-element compounds, their characterization is of considerable interest. However, remarkably little work has been done in this area. No lipid-soluble trace-element compound that is extractable by an organic solvent from a mammalian tissue has yet been adequately characterized.

Polymeric materials that accumulate in lysosomes are very persistent in mammalian tissues. These polymers are formed by cross-linking reactions using malondialdehyde, a product of the peroxidation of lipids, or by free-radical reactions. The age pigment of the brain has been shown to contain copper and iron. Perhaps other trace elements are associated with such polymers. The products of these random cross-linking reactions are not recognized by enzymes and therefore are not catabolized. The chemical characterization of the random polymers will be difficult.

Another group of persistent compounds are known as metal-ion traps that function as ion exchangers. For example, radioactive manganese is trapped in rats for months. However, when nonradioactive manganese is administered the radioactive manganese will be excreted rapidly (Cotzias et al., 1960). These traps need to be characterized. Metalloporphyrins, fulvic and humic materials, solid tissues such as bone and teeth, and connective tissues also retain trace-element species. The stability of these materials might aid the characterization of trace-element compounds.

Chemical species with sufficient polarity for high solubility in water and low membrane permeability have residence periods similar to the time scale of blood circulation. If these species are not recycled by the kidney they may be detectable in the urine. Their detection in the plasma may be possible. Redox-active and pH-sensitive species with some water solubility yet high membrane permeability have short lifetimes. These species are probably altered whenever they cross membranes separating environments of different redox potential or pH.

The most transient chemical species are free radicals. Radicals are generated during enzyme-catalyzed reactions and during spontaneous oxidations such as lipid peroxidation and quinone autooxidation. Despite the great interest in free radicals as highly damaging species in biological environments, the prospects for the direct detection of radicals are not good.

Inorganic Compounds as Chemotherapeutic Agents

The biological activities of inorganic compounds with kinetically and thermodynamically stable bonds are remarkably dependent on the central atom (Lippard, 1983). For instance, tetrachloroplatinate(II) and hexachloroplatinate(IV) are potent skin allergens. Sensitized persons respond to femtograms of these compounds. The analogous palladium compounds are inactive. The chemical basis of these immunological reactions are not known. Tris(ethylenediamine)platinum(IV) is a neuromuscular blocking agent.

Platinum(II) and platinum(IV) complexes with two primary or secondary amine ligands in *cis*-position are anticancer agents. The *trans* isomers are inactive. Cisplatin, *cis*-[$(NH_3)_2PtCl_2$], is used for the treatment of ovarian and testicular tumors. Carboplatin, *cis*-[$(NH_3)_2$Pt(1,1-dicarboxycyclobutane)] is a promising second-generation platinum drug with fewer side effects than cisplatin (Hacker et al., 1984).

The pharmacological activities of the platinum complexes are associated with their ability to exchange ligands. Inside the cell, where the chloride concentration at 4 mM is much lower than in the plasma (104 mM), one of the chloride ions in cisplatin is replaced by a water molecule which ionizes in a neutral medium and forms the complex *cis*-[$(NH_3)_2$PtCl(OH)]. This complex is more reactive toward DNA than cisplatin. The $Pt(NH_3)_2$ group probably links two guanine bases in the same DNA strand via the N-7 nitrogen atoms. Antibodies have been identified that recognize such platinum species. Tumor cells apparently cannot remove these platinum bridges, but normal cells can. Results of recent N-15 NMR work suggest that sulfur-containing ligands in the cells replace one of the ammonia molecules in the platinum complex before it reaches and reacts with DNA.

Several other metal complexes have now been shown to possess anticancer activity. The U.S. National Cancer Institute has tested 385,000 organic compounds but only 13,000 metal and metalloid compounds for antitumor activity. Close collaboration between testing groups and inorganic chemists is needed because inorganic compounds are more likely to react with solvents and culture media than are organic compounds. For instance, cisplatin is solvolyzed in dimethyl sulfoxide in a few hours. Some organic drugs and their metabolites will be active only after complexation with a metal. For example, bleomycin cleaves DNA in the form of a bleomycin-iron(II) complex (Dabrowiak et al., 1980).

The ligands in a metal complex serving as a therapeutic agent may govern its absorbability. The gold-containing, water-soluble antiarthritic drug aurothiomalate is administered by injection because it is not absorbed when taken orally. However, the lipophilic, noncrytalline gold-phosphine drug (tetraacetyl-β-D-thioglucose)(triethylphosphine)gold(I) is absorbed when administered orally. Aurothiomalate, despite having been in use since the 1930s, has only recently been properly characterized. X-ray absorption, proton and carbon-13 nuclear magnetic resonance, and electronic absorption spectroscopy suggest a polymeric structure for aurothiomalate with thiolate-sulfur atoms bridging gold(I) ions. Aurothiomalate deposits gold in the lysosomes (aurosomes). The chemical nature of these deposits is not known. More than 80 percent of the circulating gold is carried by albumin perhaps bound to Cys-34 in this protein. The gold-phosphine drug is readily taken up by the cells. Triethylphosphine is displaced from the complex and eliminated as triethylphosphine oxide in the urine long before gold is excreted. It is important to relate ligand exchange and redox reactions of such complexes to their biological activities. Some of these needed investigations can be performed in the intact cells using NMR spectroscopy.

The oxidation state of a metal can be very important for its therapeutic value. Colloidal gold has no therapeutic value. Gold(I) complexes are good antiarthritic drugs. Gold(III) (for instance, in tetrachloroaurate(III)) is a strong oxidizing agent and will be quickly reduced in the cellular environment. Antimony(V) and the more toxic antimony(III) compounds are effective against the parasitic disease leishmaniasis. Antimony can be delivered to reticuloendothelial cells by liposome vesicles.

Old, established inorganic pharmaceutical preparations need to be reinvestigated. How do zinc-oxide ointments or antidandruff preparations containing selenium disulfide work? Radiopharmaceutical preparations also need to be better characterized. The oxidation state of technetium in many technitium formulations is still doubtful. Chemically bound tin – introduced via tin(II) chloride used for the reduction of pertechnate – was recently discovered in some Tc-99m products.

It will be a challenge for pharmaceutical research to design molecules that can control the uptake, distribution, and concentration of the bulk and trace elements currently thought to be essential. Such research – based on molecular-scale knowledge of inorganic biochemistry – promises to produce exciting results.

Determination of Chemical Species in Biological Systems

That inorganic and organometallic compounds play an important role in biological systems is evident. As part of their metabolic activities biological systems take up and produce a variety of these compounds that may be gases, may be dissolved in cellular or extracellular liquids, or may be present as insoluble solids. The concentrations at which such compounds are present in organisms range from millimolar to picomolar. The identification and quantative determination of these compounds, especially those present at very low concentrations, cannot yet be accomplished in a routine manner (Irgolic, 1985).

Determinations of "total" trace-element concentrations in biological matrices have been carried out for a number of decades. These analytical endeavors have produced scores of data, many of which are unreliable. Accurate and precise "total" trace-element data are not easy to obtain routinely. Interferences caused by biological matrices are difficult to predict. Analysts are often unaware of many of these interferences. Even though trace-element compounds are of no concern when "total" trace-element concentrations are to be determined, the interferences may be compound-specific. In such cases a matrix-matched calibration curve that could account for interferences cannot be constructed because the chemical species, in which an element is present in a sample, are not yet identified. An analytical method to be used for the determination of an element in a particular matrix must be carefully checked to assure precision. Once the precision of a method is established, the accuracy must be determined for each particular matrix, for instance, by analysis of a pool sample with various independent methods (Wolf, this volume).

The determination of trace-element species is plagued by the same difficulties as total element determinations and must – in addition – be carried out in a manner that prevents chemical changes in the species during initital concentration procedures and during the analytical process. If such undesirable chemical changes cannot be prevented, the analytical procedure must at least proceed in a well-defined way that will allow conclusions to be drawn about the chemical nature of the original species.

To evaluate the risk associated with exposures to certain elements the determination of oxidation states may be all that is required. Examples of such elements are chromium and arsenic. Chromium(VI) compounds are highly toxic, whereas Cr(III) is essential for body functions (Piscator, this volume). Arsenic(III) in the form of arsenite is more toxic than arsenic(V) compounds. Organisms are capable of oxidizing As(III) and reducing As(V) (Vahter and Enxall, 1983). The oxidation stattes of elements can often be determined without great difficulties. The preservation of species with differing oxidation states in environmental samples is a problem for which acceptable solutions are not available. On-the-spot analyses are possible only under rare circumstances.

To determine organometallic and organometalloidal compounds, methods must be used that uniquely identify entire molecular entities. The carbon-element bond(s) in these compounds must remain intact during the analytical procedures. Other atoms, groups, or ligands associated with the central atom should also remain bonded, although valuable information is obtained even if only the carbon-element bonds are unaffected. Methylmercury compounds, CH_3-Hg-X, are examples of species with a stable carbon element and a labile element-X bond. Arsenic forms a variety of stable organic compounds such as methylarsonic acid, dimethylarsinic acid, arsenocholine, arsenobetaine, and arsenic-riboses (Irgolic, 1986). Methods do exist for the identification and determination of stable organometallic and organometalloidal compounds.

Metal complexes that are thermodynamically or kinetically stable under the conditions required by the analytical procedures can – in principle – be determined without difficulties. However, many complexes are labile. Ligands may exchange rapidly or slowly depending upon the stability of complexes, the availability of compounds capable of serving as ligands, the composition of the medium (pH, ionic strength), and the physical conditions of the medium (temperature, pressure). The identification and determination of labile species is a complex task.

Two different analytical approaches are available for the determination of chemical species in biological systems. The first approach extracts and concentrates the species, separates the compounds, for example, by chromatographic methods, and determines the elements in the fractions. The second approach uses *in situ* spectroscopic techniques to identify and determine chemical species.

Extraction/Concentration/Separation/Detection

Trace-element species can be divided into three classes in increasing difficulty with respect to their determination.

- Volatile species: Volatile species are generally excreted by the organism, appear in the atmosphere, and can be collected in closed systems. The natural excretion process produces a clean separation from the matrix. Generally the gaseous species can easily be determined by gas chromatography, mass spectrometry, GC-MS, and other methods of "headspace" analysis. Examples of compounds belonging to this class are dimethyl selenide and methylarsines.
- Species that can be converted to volatile derivatives: Some species of tin, lead, mercury, arsenic, antimony, bismuth, selenium, and tellurium can be reduced in a controlled manner to well-defined volatile derivatives. The volatile derivatives can then be separated from the matrix and determined. Examples of such reducible compounds are selenite, arsenite, arsenate, and methylarsenic acids.
- Nonvolatile species: Most of the trace-element species in living organisms are neither volatile nor convertible to volatile derivatives. In most

cases these species must be separated from the matrix and concentrated. Presently, liquid chromatography with element-specific detection is a good method for the determination of these compounds. Ion chromatography and other variants of liquid chromatography have been used to separate trace-element-containing compounds (Brinckman and Irgolic, this volume). The element-specific detectors most frequently used are graphite furnace atomic absorption spectrometers, flame absorption, flame emission, and plasma emission spectrometers, mass spectrometers, and electrochemical devices. The detection limits of these element-specific chromatography systems vary considerably with the element but reach low $\mu g\ L^{-1}$ levels under favorable conditions. All of these systems are sufficiently developed to be used in environmental work although not usually in a routine manner. Each analytical endeavor directed toward the determination of chemical species is a research project rather than a routine operation.

All of the above methods have been applied primarily to organometallic and organometalloidal compounds that have at least one hydrolytically stable carbon-element bond in the molecule. Coordination compounds can also be determined by these methods if the ligands in the complexes are not easily exchanged. For instance, metal dithiocarbamates have been separated by liquid chromatography.

In Situ Methods

In situ methods for the identification and determination of trace-element compounds avoid extraction, concentration, and separation procedures that may change the species originally present in the sample. Spectroscopic and other techniques useful for *in situ* analyses are listed in Table 1. These methods are capable of identifying species or classes of species. Some methods such as electronic absorption spectrometry, circular dichroism, infrared spectroscopy, and NMR spectroscopy can be used in kinetic investigations. Most of these techniques are unfortunately not sufficiently sensitive to detect trace-element compounds at the concentrations found in organisms. Sometimes the properties of chemical species can be determined by biological methods (bioassays) based on the determination of growth rates, growth characteristics (induction, repression), or mutation rates. The study of interactions between antigens and antibodies, enzymes and substrates, and hormones and receptors will provide information about biological properties of chemical species.

Table 1. Spectroscopic and other techniques useful for *in situ* analyses

Technique	Application
Electronic Absorption Spectrometry (Ultraviolet-Visible)	Determination of organic and inorganic chromosphores
Circular and Magnetic Circular Dichroism (CD, MCD)	Characterization of chiral chromophores
Fourier Transform Infrared Spectroscopy	Determination of organic and inorganic species; identification of functional groups
Luminescence (Fluorescence) Spectroscopy	Determination of organic and inorganic species and functional groups
x-ray Absorption Spectroscopy (Near-Edge, XANES; Fine structure EXAFS)	Structure of metal complexes, electronic and chemical environment of central atom
Resonance Raman Spectroscopy	Organic compounds, metal complexes, certain functional groups in organic molecules
Nuclear Magnetic Resonance Spectroscopy	Determination of structure, geometry, and binding conditions in organic and inorganic species
Electron Spin Resonance Spectroscopy	Characterization of organic radicals and paramagnetic metal compounds
Light scattering, X-ray Scattering	Diffusion behavior of macromolecules
Mössbauer Spectroscopy	Determination of oxidation states, spin states, and coordination geometry of a few metals
Analytical Electron Microscopy, Proton-induced X-ray Emission (PIXE)	Mapping of elements and determination of their concentrations in very small volumes or areas
Photoelectron Spectroscopy	Characterization of solids
X-ray Diffraction	Determination of the molecular structure of crystalline compounds

Recommendations for Research

Knowledge about the biochemistry of inorganic and organometallic compounds is fragmentary. Much needs to be learned about the effects of trace elements on organisms, the cycling of micronutrients, the transformation of trace-element compounds, biochemical pathways to metalloproteins, and interactions among trace elements. Knowledge about these topics will be most valuable when it comes from investigations that explore the molecular basis of the biochemistry of trace elements. Techniques to carry out such investigations are now available and should be used for studies of trace elements whenever possible. The following recommendations for research do not cover all the areas that need more attention. The list contains suggestions that appeared to be important, timely, and reasonable with respect to available techniques.

- Much of what has been learned in studying sorption, precipitation, transformation, and exchange phenomena in the soil, freshwater, and marine environments is applicable to investigations of phase-boundary and surface phenomena that modulate the flux of elements in plants and animals. Steps should be taken to promote interdisciplinary exchange of pertinent information. The analytical expertise and the instrumentation required for the identification and determination of trace-element compounds is often not available to the biologists solving the essential questions about the role of chemical species in organisms. Progress in biological studies of medical and nutritional aspects of trace elements in animals and men and the productive use of complex and costly analytical instruments will not be optimal unless good communication and collaboration exists among the specialists that are needed for work with trace elements. Expertise to construct often complex mathematical models of trace elements in biological systems and computer time to use such models are urgently needed.
- Biology students should be given a sound base of knowledge about bioinorganic chemistry. Other physical-science students should also receive the necessary interdisciplinary education with a mathematical orientation sufficient to allow problems associated with trace elements to be solved quantitatively.
- Compartments within organisms need to be identified that are especially sensitive to the trace-element status of the organism. Similarly, metalloenzymes that are strongly dependent on trace-element concentrations need to be found. Reliable tests can then be developed for the trace-element status of an organism. Methods need to be developed for the determination of chemical species in urine as a liquid diagnostic for the metabolism of elements.
- The molecular forms of phosphorus in foods need to be identified. Of special importance are the concentrations of phytic acid, which affect the availability of zinc and copper. Physicochemical studies of the formation, stability, and solubility of alkaline-earth/zinc phytates as a function of pH and amino acid concentrations should be undertaken. The ex-

perimental conditions should simulate the environment in the digestive tract.

- Knowledge about the forms of sulfur in foods must be improved. The metabolic cycles of sulfur require in-depth studies.
- The transformation of trace-element species caused by cooking and other methods of preparing food must be explored.
- Methods need to be developed for the efficient extraction and preconcentration of trace-element compounds from foods. The stability of trace-element compounds under the extraction conditions must be established.
- The analytical methodology for the determination of minor and trace elements and their compounds in foods must be improved.
- Appropriate standards for trace-element compounds occurring in foods must be developed.
- Studies of metalloproteins in the cell and in isolated form and of model compouns are needed to advance knowledge about the structure, kinetics, and mechanism of action of metalloproteins. Techniques are needed to study their flux in cells.
- Methods allowing real-time observation of intermediates in catalytic and metabolic cycles involving metal centers need to be improved. Methods need to be found to trap such intermediates.
- The active sites of metalloproteins need to be better characterized and the inorganic systems in these compounds studied in detail. The pathways leading to iron-sulfur and other clusters must be elucidated. Their relation to the metabolic sulfur cycle must be established. Their occurrence in foods needs to be ascertained. The Cu/Mo/Fe/S species formed under physiological conditions of high iron intake need to be characterized.
- The nonproteinaceous species of iron and copper that promote free-radical reactions in biological systems and especially the formation of hydroxyl radicals from hydroperoxides (Fenton reaction) must be characterized.
- Lipophilic species of trace elements that can be extracted with organic solvents need to be identified in normal and diseased tissues.
- Studies of the molecular pharmacology of inorganic compounds using the most up-to-date analytical and biotechnological methods should be carried out to improve our understanding of their mechanisms of action.
- More inorganic and organometallic compounds need to be tested for their biological activity. Pharmacologists, biologists, and clinicians need to work closely with inorganic chemists on this task.

- Long-used inorganic pharmaceutical preparations need to be characterized and their mode of action established.
- Improved chemotherapeutic agents based on inorganic compounds containing essential and nonessential elements should be synthesized. The dependence of function on the structure of these compounds should be established.
- The analytical methodologies for the determination of trace-element species in animal and plant tissues, in body fluids such as blood, serum, and urine, in air, and on respirable particles need to be improved to achieve higher specificity and sensitivity. The systems using inductively coupled plasmas as ion sources for mass spectrometers promise to enhance the sensitivity of the liquid chromatography/optical plasma atomic emission spectrometer systems by three orders of magnitude.
- Special attention should be given to *in situ* methods for the characterization of trace-element compounds to improve the currently troublesome low sensitivity and lack of general applicability.
- The effects of extraction/concentration/isolation procedures on the chemical nature of trace-element compounds must be defined.
- The available methods for the identification, determination, and characterization of trace-element compounds must be used by more investigators to generate a sufficiently large data base on which models can be built with proper attention to intra- and extracellular conditions (pH, redox potential, ligands, ionic strength) and the stability of trace-element species.

References

Bernat I (1983) Iron Metabolism. Plenum Press, New York

Bertini I (1983) The coordination properties of the active site of zinc enzymes. In: Bertini I, Drago RS, Luchinat C (eds) The Coordination Chemistry of Metalloenzymes. D. Reidel Publishing Co., Dordrecht, Holland, pp 1–18

Chesters JK (1981) Nutritional chemistry of inorganic trace constituents in the diet. Chem Soc Rev 10:255–269

Cotzias GC, Borg DC, Bertinchamps AJ (1960) Clinical experiences with manganese. In: Feren MJ (ed) Metal Binding in Medicine. J. Lippincott, Philadelphia, pp 50–58

Dabrowiak JC, Greenaway FT, Santillo FS, Crooke ST, Essery JM (1980) The iron bleomycins. ACS Symp Ser 140:237–248

Friberg L, Vostal J (eds) (1972) Mercury in the Environment. CRC Press, Cleveland, Ohio

Hacker MP, Double EB, Krakhuff IH (eds) (1984) Platinum Coordination Complexes in Cancer Therapy. Martinus Nijhoff, Boston
Holm RH (1981) Metal clusters in biology: quest for a synthetic representation of the catalytic site of nitrogenase. Chem Soc Rev 10:455–490
Irgolic KJ (1985) Environmental inorganic analytical chemistry. In: Irgolic KJ, Martell AE (eds) Environmental Inorganic Chemistry. VCH Publishers, Deerfield Beach, FL, pp 547–564
Irgolic KJ (1986) Arsenic in the environment. In: Xavier AV (ed) Frontiers in Bioinorganic Chemistry. VCH Verlagsgemeinschaft, Weinheim, Germany, pp 399–408
Kagi JHR, Nordberg M (1979) Metallothionein. Birkhauser Verlag, Basel, Switzerland
Lippard SJ (ed) (1983) Platinum, Gold, and Other Metal Chemotherapeutic Agents. ACS Symp Ser 290. American Chemical Society, Washington, DC
Martell AE (ed) (1980) Inorganic Chemistry in Biology and Medicine. ACS Symp Ser 140. American Chemical Society, Washington, DC
Merian E (ed) (1984) Metalle in der Umwelt: Verteilung, Analytik und biologische Relevanz. Verlag Chemie, Weinheim, Germany
National Academy of Sciences/NRC (1974) Chromium. U.S. National Academy of Sciences, Washington, DC
National Academy of Sciences/NRC (1976) Selenium. U.S. National Academy of Sciences, Washington, DC, p 56
National Academy of Sciences/NRC (1977) Copper. U.S. National Academy of Sciences, Washington, DC
Palermo RE, Singh R, Bashkin JK, Holm RH (1984) Molybdenum atom ligand substitution reactions of $MoFe_3S_4$ cubane type clusters: synthesis and structures of clusters containing Mo-bound pseudosubstrates of nitrogenase. J Am Chem Soc 106:2600–2612
Pickford CJ (1981) Sources of, and analytical advances in, trace inorganic constituents in food. Chem Soc Rev 10:245–254
Prasad A (ed) (1976) Trace Elements in Human Health and Disease, vol I. Zinc and Copper. Academic Press, New York
Spiro TG (1983) Zinc Enzymes. Wiley Interscience, New York
Thomson AJ (1985) Iron-sulphur proteins. In: Harrison P (ed) Metalloproteins. Part 1: Metal Proteins with Redox Roles. Verlag Chemie, Weinheim, Germany, pp 79–120
Vahter M, Enxall J (1983) *In-vivo* reductions of arsenate in mice and rabbits. Envir Res 32:14–24
Vahter M, Marafante E, Dencker L (1983) Metabolism of arsenobetaine in mice, rats, and rabbits. Sci Total Envir 30:197–211
Venugopal B, Luckey TD (1979) Metal Toxicity in Mammals. Plenum Press, New York
Xavier AV (ed) (1986) Frontiers in Bioinorganic Chemistry. VCH Publishers, Weinheim, Germany

The Importance of Chemical "Speciation" in Environmental Processes,
eds. M. Bernhard, F. E. Brinckman, and P. J. Sadler, pp. 39–58.
Dahlem Konferenzen 1986. Springer-Verlag Berlin, Heidelberg
© Dr. S. Bernhard, Dahlem Konferenzen

Approaches to the Determination of Chemical Species in Biological Materials

W. R. Wolf

Nutrient Composition Laboratory
U.S. Dept. of Agriculture
Beltsville, MD 20705, USA

Abstract. Living systems exist in dynamic equilibrium, wherein levels and interconversions between different chemical species is constantly changing. To gain information regarding the biological function and utilization of these chemical species, a concerted collaborative effort using a wide variety of techniques and disciplines is required. At present most analyses of chemical species rely upon *in vitro* measurements, which must then be used to extrapolate to real-time measurements of the fluctuation of these species *in vivo*. Analytical approaches to the determination of chemical species generally include various separation techniques followed by specific spectroscopic methodologies to quantitate the trace element content of the separated fractions. A number of specific studies have covered a wide range of separation and atomic detection techniques. Applications of molecular spectroscopic techniques, which have been limited due to sensitivity, can give a great deal of information on the chemical species itself rather than just quantitation of the metal content. These molecular techniques may be rejuvenated in the future due to significant advances in sensitivity combined with continuous flow analysis techniques capable of very rapid, high-precision, low blank-level, and multisample analyses. Use of biocatalytic and ion-sensitive microelectrodes for *in vivo* determination of chemical species has considerable promise for relatively noninvasive measurements in single cells. Further development and use of computer simulation models of equilibria in human blood plasma between a number of trace elements with naturally occurring and administered chelating agents will be very beneficial to an understanding of these complex systems. The growing awareness of the important multiple roles of trace elements in human health has led to increased medical-clinical interest in diagnosis and therapy involving specific chemical species of these elements.

Introduction

The analysis of chemical species in biological systems is not an end goal in itself. The main goal is to gain information about the function or utilization of these species within the entire system from which specific samples are taken. A "living" system is one in constant change. Most biological systems exist in dynamic, equilibrium rather than static states. Thus, levels and interconversions between different chemical species or forms can be constantly changing, dependent upon a multitude of chemical and physical parameters subject to change in the system. The sample is (hopefully) representative of the state of the system at the instant in which it is taken.

At the present time most analyses of species studies rely upon *in vitro* measurements of the specific chemical forms of the elements. Samples are removed from the system of interest. Various extraction, digestion, and separation schemes are carried out to remove the species from the matrix. The species is then usually measured as the total amount of the element in separated fractions using an atomic property of the element as a probe. What is required is more sensitive, accurate, and species-specific methods. These must then be used to extrapolate from the *in vitro* samples to real-time measurements of the fluctuation of chemical species *in vivo*.

It is not enough to fractionate a sample of food or biological tissue and identify quantitatively all the chemical species therein, which, although not a trivial task, can potentially be done. Understanding the specific biological action of the species is the ultimate goal of this effort. Reaching this requires a concerted, collaborative effort employing acknowledged techniques and a variety of disciplines including analytical separations, trace level analyses, physical inorganic chemistry, biochemistry, physiology, nutrition, coordination chemistry, radiochemistry, enzymology, luck, and many others.

This paper will concentrate on the first step of these studies, analytical approaches for *in vitro* measurement of chemical species with some speculation on future potential for *in vivo* techniques.

Background Studies

Role in Human Health

Awareness of the important multiple roles of trace elements in human health has generated a need to define much more closely their levels of exposure and intake. This generally means better definition of content in foods as this is the main pathway of functional exposure for humans to these elements. It is no longer sufficient merely to identify the presence of these elements or to say that a "trace" exists. The exact levels must be defined and the variations in these levels quantitated. These multiple roles have led to a public health interest and duality of concern towards toxicological overexposure and/or nutritional underexposure or subadequate intake of essential elements. Table 1 lists those elements of either nutritional or toxicological concern. This duality of concern has previously been discussed in some detail (Wolf, 1981).

Past studies of intestinal absorption have generally looked at the total amount of metal in the foods supplied in the diets. The first step towards more complete understanding of absorption is a detailed knowledge of the chemical species of trace elements in foods or diets, although very few reports have appeared which discuss these individual chemical species. A recent paper has described preliminary studies on trace element extractability from foods following enzymolysis with gastric juice (pepsin) and intestinal juice (pancreatin and bile salts in saline solution) to release the metal species (Crews et al., 1983). Differences in behavior between endogenous and added inorganic forms of the elements were also investigated. It was concluded that often only a portion of the trace elements are released and solubilized after enzymolysis. In this approach the assumption is made that any element which remains in an insoluble form is unavailable and that soluble forms are at least potentially available for absorption. Other factors may lead to additional solubilization in the stomach. These factors vary from element to element and according to food type. These preliminary studies point the way for much further work on the bioavailability of metal-containing species in foods.

Table 1. Micro-inorganic elements of interest in human health (Wolf, 1981).

Nutritional	Toxic
As, Co, Cr, Cu, F, Fe, I, Mn, Mo, Ni, Se, Si, Sn, V, Zn	As, Be, Cd, Co, Cr, F, Hg, Mn, Mo, Ni, Pb, Pd, Se, Sn, Tl, V, Zn

More knowledge is also becoming available regarding a wide range of biological and metabolic interactions between trace elements. A complex network of interactions has been reported. Some specific examples of interactions are a) the exaggeration of lead toxicity as a consequence of deficiencies of calcium, phosphorus, iron, zinc, and, in some cases, copper; b) the interaction and protective effects of selenium with mercury; and c) interactions of zinc, selenium, and cadmium. Protective interactions of some essential elements with toxic elements might be attributed to competition for binding sites on ligands involved in homeostatic capabilities. These are only a few of the many reported interactions which have been detailed in the proceedings of a recent symposium on this topic (Micronutrient Interactions, 1980).

Many of the interactions between elements, and the movement of the elements through the environment, depend upon the chemical species or form of the element. For example, the element selenium can exist in soils as elemental selenium (Se^0), as selenites (SeO_3^{2-}), as selenates (SeO_4^{2-}), or be bound to heavy metals as selenide (Se^{2-}). Each of these species can react differently in movement from soil through plants and animals in the environment. Some of the species, such as insoluble elemental selenium, are not as "biologically active" or must be biochemically changed into more active species before utilization (NAS, 1976).

Only when the individual species can be identified and quantitated can its complex biochemical role be studied and understood. For many of the elements, the level of some particular species, which may be in small proportion to the whole amount, may be far more important than the total amount of the element. Since the total level of these elements may be very minute, parts per million down to subparts per billion, the level of individual species may be close to or below the present state-of-the-art detection limits. This points towards the need to develop even more sensitive and more selective analytical methods for metal-containing species in specimens from living systems and biological materials which influence those systems. Development and application of these analytical techniques will lead to an increase in knowledge greatly exceeding that of the roles of "trace" elements which occurred in the 1950s and 1960s in conjunction with the growth and development of sensitive analytical techniques.

Medical-Clinical Interests

The main goal in the identification and quantification of chemical species is to gain knowledge about the entire living system. For medical-clinical interests, the major purpose of obtaining this knowledge is to keep the system living by translating the knowledge into progress in therapy. This progress is related to improvements in diagnosis which in turn are partly the result of better analysis of specific components in tissues and body fluids (Van den Hamer and Houtman, 1980). The physiology and interactions must be studied at molecular, cellular, and macro levels including distribution over various binding entities.

Requirements of medicine for trace element analysis in the 21st century including extensive discussion of the requirements of individual species have also been reviewed by Bowen (1980). Analysis of the total concentrations of elements such as Ca, Cl, Fe, K, and Na are useful for clinical diagnosis. In a number of cases, analysis of other elements such as Cu, I, Mg, Mn, and Zn may be useful. At present only a few diagnostic clinical measurements involve specific chemical species; these include transferrin saturation or hemoglobin content of whole blood for iron, iodine as thyroglobulin for confirmation of iodine deficiency with goitrous symptoms, and selenium in platelets or glutathione peroxidase. The existence of so few diagnostic applications is probably due to lack of understanding of the specific biochemistry of the elements and the resultant absence of analytical techniques for specific species. Comprehensive information regarding the known roles of trace elements in a variety of human diseases is available in the literature. A good, concise overview is given by Faulkner (1981).

General Analytical Approaches

Most analyses of species have been carried out using various separation techniques for the species of interest followed by specific detection of the total element in separated fractions, mainly of three distinct types: 1) determination of different ions or oxidation states of the metals, e.g., Cr^{3+} and Cr^{6+} using HPLC and ion-exchange techniques, with detection by either atomic spectroscopy (AAS, ICP)* or potentiometric techniques; 2) use of natural volatility properties of low-molecular-weight species utilizing gas chromatography for separation, with detection by AAS; and 3) studies of higher-molecular-weight metal complexes using HPLC, gel permeation chromatography, selective extraction, or other organic or biochemical separation procedures, with analysis of fractions by atomic spectroscopic techniques.

---

* AAS: atomic absorption spectroscopy.
ICP: inductively coupled plasma (spectrometry).

The first of these types of study is of potential interest in environmental toxicity situations or in aqueous systems where the elements may occur in ionic form. In biological systems most of the elements are present as complexes rather than as free ions. Oxidation states of some elements are often readily transformed or changed in living systems and measurement of free metal ions often provides little information, except for the major electrolytes such as Na^+, K^+, Cl^-, and possibly Ca^{2+} and Mg^{2+}. Uses of ion-selective electrodes for these analyses will be addressed later.

The second type of study deals mainly with the elements which form low-molecular-weight organometallic compounds such as the methyl derivatives of Se, As, Sn, and Hg. These compounds have been shown to be biotransformed and their subsequent biopotency is strongly affected by the chemical form. These systems have been extensively studied by use of GC-AAS or sophisticated GC-MS techniques. These studies are specifically detailed elsewhere in this volume and will not be further discussed.

The third type of study involving large-molecular-weight, nonvolatile metal complexes is of more general interest in living systems. Separation techniques ranging from size exclusion column chromatography to highly sophisticated HPLC have been utilized. The separated fractions are usually analyzed for metal content by atomic spectroscopic techniques such as AAS, AES (ICP), or other atomic techniques such as NAA* or X-ray fluorescence. Analytical methods and strategies are being devised to look at the various physicochemical forms of elements (Schwedt, 1983). Several selected examples of these types of study will be detailed below.

Specific Studies

A number of studies have been carried out to determine different chemical forms (species) of specific elements in foods and biological tissue samples. Some examples of the wide range of techniques applied to these types of studies will be discussed.

Iron in Foods

One important question in nutrition is the assessment and treatment of iron deficiency anemia. This is a widespread health problem throughout the world in spite of the fact that average diets often contain adequate levels

---

* Neutron activation analysis.

of total iron. Biological availability of this iron is generally low and is greatly dependent upon the different chemical forms of the iron in foods and their interactions with other dietary components such as phytate.

Iron bioavailability has been extensively studied by use of radiolabels. Intrinsic tagging, food growth incorporating a radio-iron label, has become less common as extrinsic tagging procedures have increased in application. The extrinsic tagging techniques involve addition of an exogenous radiolabeled iron spike, usually as $^{59}FeCl_3$ or $^{59}FeSO_4$, to a test diet. The assumption is made that a complete exchange takes place between the spike and an "isotopic exchange pool" consisting of non-heme iron naturally occurring in foods. Heme iron is assumed to behave differently and to be more bioavailable than non-heme iron. This isotopic exchange has been verified under some conditions but has not been proven for all conditions (Consaul and Lee, 1983). These studies generally look at the influence of selected foods or other factors such as ascorbic acid (which increase Fe bioavailability) or chelating agents (which decrease bioavailability).

Added extrinsic iron may act differently toward various enhancers or inhibitors if it has a different oxidation state (ferrous versus ferric) than that of the endogenous iron. A factor which significantly enhances Fe absorption in diets containing meat has been reported (Monsen et al., 1978). These observations all depend upon the assumption of complete exchange of the added iron. In general these studies have not looked extensively at specific forms of iron in this "exchange pool." Detailed information of the specific forms involved in these studies must be obtained.

Heme iron is a more available source of iron than non-heme iron sources in foods. Historically heme iron content of foods has been determined by the difference between the total iron content and non-heme iron content based upon different solubilities of the two physicochemical forms. The distribution of iron between five main components in muscle meats has recently been reported (Hazell, 1982): an insoluble fraction, ferritin, hemoglobin, myoglobin, and a low-molecular-weight fraction. Absorbance spectra of various gel filtration fractions confirmed the presence of hemoproteins, which have a very distinct absorbance band at approximately 410 nm (Soret band). This study showed that approximately equal levels of non-heme iron were present in these various meats, with much increased levels of heme iron in the darker meats.

One of the major factors affecting iron bioavailability is the interaction of iron and phytate in the diet. The iron, combined with phytate in aqueous salt extracts of wheat bran, has been characterized as monoferric phytate by use of gel chromatography. The bioavailability of these complexes and synthetic monoferric phytate is similar. The synthetic product exchanges with the non-heme iron pool of meals consumed by human subjects, but wheat bran reduced the absorption of iron from meals consumed by human subjects. In order to clarify the chemical nature of these types of compounds, *in situ* studies of the iron in wheat were carried out by Mössbauer spectroscopy (May et al., 1980). Mössbauer spectroscopy is sensitive to changes in oxidation, spin state, and the conformation of the ligands around the iron. The spectra of the iron in seeds and bran were the same as the spectrum of solid synthetic monoferric phytate, indicating that most of the iron in wheat is present in this chemical form. The Mössbauer parameters indicated that the iron is in the high-spin form. This study (May et al., 1980) is a good example of the application of classical physical inorganic chemistry to more complex biological systems.

Zinc-Binding Ligands in Biological Fluids

Trace elements are rarely found in "free form" (aquated ions) in biological systems. They are predominately bound to proteins or low-molecular-weight ligands such as amino acids. In order to characterize these compounds, attempts are often made to separate them from their biological matrices. Gel filtration chromatography has been used in several fields to separate low-molecular-weight compounds. This technique may not be suitable for those elements which form labile metal-ligand species. In solution many metal complexes take part in ligand exchange, Eq. 1, or metal exchange reactions, Eq. 2. Since gel filtration chromatography separates substances based upon molecular size, the metal-ligand complex migrates ahead of the metal and the ligand(s). The complex may constantly dissociate and reestablish equilibria dependent upon the stability constants, kinetics, and concentrations of the metal(s), ligand(s), and metal-ligand complex(es) as it passes through the chromatographic medium. This could result in a very complicated elution profile for these metal species. It has also been shown that the dextran gels commonly used for these separations can bind metal ions, especially zinc. The problems associated with metal-ligand dissociation during gel filtration chromatography can be minimized by use of solvents that contain a significant concentration of the metal of interest. This ensures that equilibrium conditions are maintained as the ligand and metal-ligand complex migrate through the column. This method has been adapted for the detection of labile zinc complexes in biological fluids (Evans et al., 1979). These same authors also pre-equilibrated the dextran gels with a zinc-containing buffer in order to eliminate the problems with zinc binding to the gel.

$$ML + L^* \rightleftharpoons ML^* + L \quad (1)$$

$$ML + M^* \rightleftharpoons M^*L + M \quad (2)$$

Reported studies have indicated the presence of a specific zinc-binding ligand that increases bioavailability of zinc in human breast milk. Studies done to isolate this ligand have led to considerable controversy regarding its identity. The study cited herein (Evans et al., 1979) indicates that most of this controversy was probably due to the anomalous behavior of labile zinc complexes on dextran gel columns unbuffered with zinc ion.

Various techniques have been used to study the binding of zinc to specific proteins in blood plasma, including salt fractionation, electrophoresis, gel filtration, and sucrose density gradient centrifugation. But they can be time-consuming and liable to contamination. A study directed towards methods suitable for clinical laboratories, which may not have specialized apparatus suitable for some of the more sophisticated techniques, has been reported (Dawson et al., 1980). These workers examined several types of gel filtration and ion-exchange chromatography techniques combined with atomic absorption spectrometry to quantitate the metal content of the various fractions.

Biologically Active Chromium Complexes in Yeast Extracts

A role for chromium in mammalian nutrition was first suggested in 1957 and a chromium-containing fraction which influenced glucose metabolism in rats was subsequently isolated from brewer's yeast and from acid hydrolysates of pork kidney powder (Schwarz and Mertz, 1959). Since then many studies have been carried out on the effects of chromium in glucose metabolism and a number of efforts have been made to isolate and identify chromium-containing compounds from yeast that exhibit appropriate biological activity. A recent report describes a procedure, based upon ion-exchange chromatography, that allows the separation of eleven apparently homogeneous chromium-containing fractions from a brewer's yeast extract (Haylock et al., 1983). Four of the fractions are amphoteric and show no biological activity, three are classified as negative (two of which are biologically inactive, while the third one shows slight inhibition of biological activity), whereas the four cationic chromium-containing fractions all show varying degrees of biological activity. Biological activity was determined using a bioassay measuring CO_2 production from glucose metabolism by yeast cells incubated in suitable buffer containing the samples to be assayed. A rather detailed separation procedure was developed for commercial yeast extracts including cation and anion-exchange columns and gel filtration chromatography. Chromium analyses of each fraction in the separation scheme were carried out by atomic absorption spectrometry. This report is significant because it shows that the biologically active fractions represent only a small portion of the total initial chromium (1.9% in the most active fraction isolated from the yeast). This combination of biochemical separation techniques, care in trace element quantitation, and utilization of an appropriate biological test for activity of the fractions is an excellent example of the type of effort required to determine trace metal species in biological materials.

Atomic Detection Techniques

As described earlier, most investigations of different chemical species dealing with high-molecular-weight metalloorganic complexes utilize biochemical separation techniques followed by detection of the metal of interest in separated fractions. A wide range of potential techniques for determining the metals exists. General reviews describing these techniques in detail are readily available (IAEA, 1980).

Most of these techniques inherently have satisfactory accuracy and precision for this type of study when sensitivity is adequate. Not all have the required sensitivity to study fractionation of species when the total naturally occurring level is very low. The techniques available include atomic spectroscopy (atomic absorption, atomic emission, and atomic fluorescence), nuclear-radio (neutron activation, X-ray fluorescence, substoichiometric extraction), mass spectrometry (isotope dilution), and spectrophotometry of metal complexes. A short discussion of some of the advantages and disadvantages of several of these techniques will be given in the following sections.

Atomic Spectroscopy

The topic of atomic spectrometry for inorganic elements in foods has recently been reviewed (Harnly and Wolf, 1984). Choice of a suitable technique to carry out the analysis of species of interest to this workshop probably depends more upon the individual preference and expertise of the analyst involved in the project than upon inherent superiority of the individual technique. There are advantages and limitations to all techniques. The skill and expertise of the analyst is of utmost importance in carrying out these types of analysis.

The most sensitive atomic spectroscopic technique at present is *furnace atomization atomic absorption*. Due to its sensitivity, however, it is probably the most difficult to handle reliably.

In *furnace atomic absorption spectrometry* a major area of potential error is the problem of differentiating between true analyte absorption at very low levels of analyte and spurious background absorption from condensed particles (smoke) or gaseous molecules. These problems are particularly severe with biological samples, due to the large excess of organic material and/or major electrolytes such as Na, K, Ca, or PO_4 relative to the very low levels of trace element (ppb). In the past five years significantly improved background correction methods have become commercially available. The Zeeman method is based upon splitting atomic spectral lines into several components by a magnetic field (De Galan, 1982). The Smith-Hieftje method (1983) uses line reversal in a periodically over-driven hollow cathode lamp to measure intensity both at and to either side of the analytical absorption line. A third method, while not yet commercially available, utilizes a continuum source-wavelength modulation procedure to monitor background. This continuum source procedure has been adapted to develop a simultaneous multi-element atomic absorption spectrometer (Harnly et al., 1979).

An area which requires fuller discussion is that of the introduction of the sample into a suitable atomization source for atomic spectroscopy (Browner, 1983). This has often been neglected in the past. However, with the great advance in instrumentation has come a need to utilize more effectively methods for more rapid and more efficient introduction of the sample into the atomization source.

The "classic" species analyses have been carried out by *coupling a gas or liquid chromatograph to an AA or ICP system*. Chromatography is then used to provide species separation followed by specific in-line detection of the element. Gas chromatography-AA has been used to separate species of lead, organo-arsenic, mercury, selenium, and other metals (VanLoon, 1979). The utilization of volatile hydrides of elements such as Se, Te, As, and Sb, using either gas chromatographic separation in AAS or direct introduction (Wolnick et al., 1981) into an ICP, has been very successful in the analysis of species of these elements. Problems of volatility and stability of most organometallic species makes liquid chromatography a more attractive general approach.

Liquid chromatographic separation coupled in-line with specific atomic spectroscopy detection will become the major thrust of development of analytical approaches to chemical species in biological systems. One limitation of this approach is the efficiency of transport of liquid samples to flames or plasmas. This efficiency of aerosol formation and transport is on the order of 1%–5%, i.e., 95%–99% of the sample is deposited into waste through the spray chamber. A great deal of basic work needs to be devoted to this area (Browner, 1983).

Another feature of today's sophisticated instrumentation is the potential for extensive *automation and continuous analysis*. One approach to this is the wide-ranging technique of flow injection analysis (Van der Linden, 1982). This has been used for sample introduction for AAS and ICP. Samples can be as small as 5–20 μL per injection and detection limits are within a factor of two of continuous sampling, which requires up to 2 mL of sample. Other aspects of flow injection will be described later. Other microsampling techniques have also been described (Browner, 1983).

For cases where the naturally occurring species are too low for a particular atomic spectroscopic technique, *preconcentration* of the elements after chelation-extraction is necessary. Use of reversed-phase liquid chromatography of β-hydroxyquinoline complexes and β-diketone chelates can give concentration factors of 10–500 after extraction.

Another novel procedure for introduction of metals into atomic spectroscopy sources involves the use of *volatile metal chelates*. Materials of biological interest can be digested and reacted or reacted directly with chelates which form stable volatile complexes. The volatile metal chelate can then be either directly flushed into an atomization source such as ICP (Browner, 1983) or introduced via gas chromatography into a variety of detection systems. Volatile chelates have been used to determine selenium by gas chromatography and have recently been utilized in very sensitive, high-precision techniques for stable isotope dilution mass spectrometry analysis of Cr (Veillon et al., 1979) and Se (Reamer and Veillon, 1981) in biological materials. This will be discussed in more detail below. However, all these procedures have one great disadvantage for analysis of species in that the *in situ* species has been destroyed before final measurement by atomic spectroscopy. Information about these *in situ* species is inferred by knowledge of the separation procedures used to obtain the various fractions of the endogenous trace elements.

Neutron Activation Analysis and Radiotracers

Severe analytical difficulties in experimentation at trace element levels in biological media can arise from contamination during analysis. Use of highly sensitive neutron activation analysis (NAA) techniques and use of radioanalytical techniques employing highly specific radiotracers to trace and measure elements at the ppb level in various tissues can help to overcome some of these difficulties. Use of NAA presupposes that the experimentalist has access to resources such as a source of neutrons and possesses or has collaboration with suffficient analytical expertise in this area.

Measurement of trace elements in cellular components at nanogram or subnanogram levels can be achieved in these studies. Appropriate biochemical techniques for separation of components of interest are developed taking into account requirements imposed by the NAA detection procedures. An excellent review of an extensive program (Sabbioni et al., 1985) details several of the specific studies carried out. a) *Vanadium:* the influence of the valence state of vanadium on its metabolic behavior was studied using ^{48}V-labeled compounds injected into rats. b) *Arsenic:* inorganic trivalent (AsO_2^- and pentavalent (AsO_4^{3-}) as well as organic As-species such as arsenobetaine and arsenocholine were radiolabeled and administered to rabbits and rats to evaluate biotransformation. Species differences were observed in these biotransformations. c) *Chromium:* different metabolic patterns of Cr were observed after administration of cationic $^{51}Cr^{3+}$ and anionic $^{57}CrO_4^{2-}$ ions. d) *Thallium:* metabolism of thallium *in vivo* is not influenced by the oxidation state.

Another aspect of the use of radioisotopes is the possibility of using *substoichiometric analysis* techniques in trace level analysis of biological materials (Kudo and Suzuki, 1984). In this method a known amount of radiolabeled element is added as a spike to the sample. The endogenous "cold" element in the sample is then allowed to equilibrate with the spike, usually by digesting the sample. At this point *quantitative* separation of a known amount of metal, which is less than the total, is accomplished by using an amount of chelating agent less than that required for complete reaction (substoichiometry). The radioactivity is then measured in the extracted sample. The amount of dilution of this radioactivity by the "cold" endogenous metal allows a calculation of the endogenous metal in the sample by isotope dilution techniques. Highly selective chemical separations can be conducted. This method can be very sensitive and can be applied to over 50 elements (Kudo and Suzuki, 1984).

Another technique which might be included at this point is the *radioimmunoassay* (RIA). Many specific antigen-antibody assays are available or are being developed. Use of these techniques for highly specific studies on chemical species would be possible. It has been speculated that newer advances in even more specific monoclonal antibody assays will ultimately replace RIA techniques (Thompson, 1982).

Mass Spectrometry: Stable Isotope Dilution

This method can be used for those elements which contain at least two stable or long-lived radioactive isotopes. A known quantity of spike solution containing enriched amounts of one isotope is added to samples containing natural ratios of the stable isotopes. The spike must then be completely exchanged with the endogenous metal in the sample, usually by digestion. A portion of the element is then isolated from the matrix and the isotope ratio measured by mass spectrometry. This measurement can be made with extremely *high accuracy* and *precision* of $\pm 1\%$ or less. Since addition of a known amount of the enriched spike has changed the isotope ratio in the sample from its natural state, calculation using the new ratio due to isotope dilution will give the unknown amount of endogenous element in the sample. The element(s) to be determined need not be isolated quantitatively *after* isotope dilution has taken place since a ratio, not an absolute measurement, is being made.

One disadvantage of this technique has previously been the extremely high cost and sophistication of suitable mass spectrometry equipment for inorganic solids using thermal ionization sources. Recently other types of sample introduction-ionization methods have been used for this type of study. Separation of the element of interest by conversion into a volatile metal chelate complex and analyses of the complexes by electron bombardment ionization MS following a GC separation have been reported for Cr (Veillon et al., 1979) and selenium (Reamer and Veillon, 1981). Several studies of trace elements in animal and human subjects have been carried out by these workers using stable isotope tracers. These open up a potentially wide-ranging area of *in vivo* use of stable isotope tracers to study metabolism of very low levels of elements in humans where radiotracers would be contraindicated.

The use of isotope dilution mass spectrometry has recently been reviewed (Heumaun, 1982). Recent commercial introduction of an ICP source mass spectrometer capable of measuring isotope ratios using liquid solutions should open up many potential areas of study of chemical species following biochemical separations (Date, 1983).

Spectrophotometry

Previous sections have described techniques based upon atomic properties to determine total metal in various biochemical media or separated fractions. Spectroscopic techniques such as UV, visible, or infrared spectroscopy can be species-specific but often require relatively pure compounds in solutions of μM or higher concentrations. Most element-containing species of interest in biological systems occur in concentrations of nM or less.

Some biologically important molecules exhibit quite distinctive UV-Visible absorption spectra, for example, porphyrins and corrins in heme proteins and Vitamin B_{12}. These compounds exhibit a very distinctive absorption at approximately 410 nm, called the Soret band, which has been utilized to analyze or characterize these compounds.

The chemistry of formation and extraction of metal ions with specific complexing agents has been known for a long time. Determination of these types of complexes by spectrophotometric techniques was routinely performed prior to the widespread commercial introduction of atomic absorption spectroscopy techniques in the mid-1950s.

A rejuvenation of spectrophotometry may occur. There have been advances in detectors with adequate sensitivity and in continuous flow analysis methods capable of very rapid, high-precision, low-blank-level, and multisample analysis. A report of a study using flow injection analysis to handle sample transfer and reagent mixing while minimizing blank levels and optical perturbations has shown successful analysis of trace levels of iron, chromium, and copper (Leach et al., 1983). Reported detection limits were 0.7 ppb, 0.3 ppb, and 1.0 ppb, respectively, comparable with the most sensitive atomic spectroscopic techniques.

Other Spectroscopic Techniques

These include Raman and Resonance Raman (RR), electron paramagnetic or spin resonance (EPR, ESR), electron nuclear double resonance (ENDOR), Mössbauer, and nuclear magnetic resonance (NMR) spectroscopy. Description of theoretical bases and advances in these areas is beyond the scope of this paper. Selective applications of several of these techniques will be highlighted below. A description of NMR applications appears elsewhere (Sadler, this volume).

About 95% of the copper in blood serum is associated with the metalloprotein ceruloplasmin, and approximately half of this copper is in the paramagnetic state. Only copper in the paramagnetic state is detectable by *EPR*, whereas the total copper content is detected by atomic absorption spectroscopy. Therefore, the use of EPR in conjunction with AAS makes it possible to determine if the copper elevation generally observed in neoplastic diseases is due to an elevation of either Cu^{+}, Cu^{2+}, or both. A comparison of serum copper measurements by EPR and AAS has been made (Kleinhaus et al. 1983).

Studies of Mn^{2+} as a regulatory ion for phosphenol pyruvate carboxykinase, a regulated enzyme of hepatic gluconeogenesis, required the knowledge of free or exchangeable Mn^{2+} in hepatocytes (Schramm, 1982). The free Mn^{2+} was estimated by placing freshly prepared cells in an EPR instrument. This is nondestructive and the cells remained viable throughout. The experiment showed that the free Mn^{2+} concentration was approximately 1.0 μM in fed rats and 0.2 μM in fasted rats.

EPR has also been used to study levels of 20 μM VO^{2+} in tissues and subcellular fractions (Chasteen, 1983). The same report also reviews the use of NMR and ENDOR as probes in studies of the biochemistry of vanadium. Detailed use is made of the VO^{2+} species complexed to proteins or other biochelators as an *in vivo* EPR probe to the coordination environment of the metal.

EPR can be applied to a wide range of metals by binding of the metal to a complex compound with a radical-containing (spin-labeled) reagent. The intensity of the EPR signal produced by the peripheral organic part of the complex compound is then measured (Nagy, 1983).

The vibrational spectra of molecules, either infrared or Raman (absorbed or scattered light), provide fingerprints of and are very sensitive to molecular conformation. Raman spectra can be measured in aqueous solution with very small amounts of materials (e.g., 10 μL of liquid). Modern advances in intense laser light sources, improved spectrometers, and multichannel detectors to allow very fast acquisition of complete Raman spectra have led to a field of Raman bioanalysis (Carey, 1982). Water molecules have intense absorption bands in IR, whereas they have very weak absorption using Raman techniques. Thus aqueous systems, which are most important in biological applications, are more suitable to Raman studies. Another major difference is the existence of the Resonance Raman effect, to which there is no IR analog. This occurs when the excitation wavelength lies under an intense electronic absorption band of a chromophore. A resonance is set up and a considerable intensity enhancement of certain Raman bands may occur with absolute intensities increased by a factor of 10^3–10^5. Since these electronic absorption bands are a significant feature in many metal-containing chromophores, this technique of Resonance Raman spectroscopy is particularly suitable for studies of low levels of metals in the concentration range of 10^{-4}–10^{-6} M. This makes Resonance Raman suitable for many applications of speciation of elements in biological systems.

Resonance Raman spectra have been obtained from chromophores in live tissues and bacterial cells under physiological conditions. Information on purified materials has included oxidation and spin state of hemoproteins and the chemistry of the ligands around the chromophoric metal center of a variety of metalloproteins. A Resonance Raman labeling technique has been developed to explore biological sites not containing a suitable chromophore (Carey, 1982).

Enzyme and Ion-Selective Electrodes

Although not yet applied to extensive study of elemental species, potentiometric methods employing selective electrodes have significantly advanced in recent years much beyond the routine determinations of pH measurements. Progress in the area of enzyme and ion-selective electrodes (Lubbers et al., 1981) includes a wide variety of ion and enzyme systems and significant advances in sensitivity and microelectrodes suitable for intracellular use. A brief review of several of these areas is of interest for *in vivo* species analysis studies.

The field of ion-selective electrodes (ISEs) encompasses a wide range of various types of sensors. IUPAC* has published recommendations concerning the nomenclature of ISEs, including a classification scheme based upon the materials used to prepare the selective membrane. The review by Martin (1982) summarizes this scheme and gives common examples of each type of electrode. Commercially available primary electrodes cover a wide range of ions including electrodes for F^-, Cl^-, Br^-, I^-, Cu^{2+}, Pb^{2+}, and Cd^{2-}, in addition to H^+, Na^+, K^+, Ca^{2+}, ClO_4^-, NO_3^-, and others. Sensitized or double membrane electrodes include gas-sensing electrodes commercially available for NH_3 and CO_2. Bioselective electrodes are based upon enzyme-substrate reactions and are very selective, although at present none are commercially available. Immunochemically based electrodes which show promise as highly selective sensors for antibodies have only recently been described. A large amount of future effort will be devoted to these sensors. Detection limits for these ISEs are around 1 mM for the primary and gas-sensing electrodes and are presently about 100 μM for the biocatalytic sensors.

Biocatalytic electrodes are based upon immobilizing a layer of enzyme or biocatalyst at the surface of a cation-selective glass electrode using a dialysis or semipermeable membrane. Thus an ISE is made sensitive to a nonionic chemical species by coupling the ISE with a biocatalyst (Martin, 1982). Biocatalytic electrodes for dozens of biochemical species have been

* International Union of Pure and Applied Chemistry.

described (Kobos, 1980). Not only enzymes but more complex biological systems, including tissue slices and even intact living organisms, may be used as the biocatalyst. These types of systems open up many possibilities for analysis of biological and biochemical species.

One important development in selective electrodes has been the technology of ion-sensitive microelectrodes. These electrodes can be manufactured with tips less than 1 µm in diameter allowing relatively noninvasive measurements in single cells. Two types are commercially available: ion-sensitive glass electrodes for H^+ and Na^+ activity, and liquid ion-exchange resin electrodes to measure K^+, Ca^{2+}, and Cl^-. This technology will allow studies of the transport of chemical species into single cells (Thomas and Moody, 1980).

Computer Modeling

Computer simulation models of the low-molecular-weight equilibria in human blood plasma have identified the predominant complexes of Ca^{2+}, Mg^{2+}, Mn^{2+}, Fe^{3+}, Cu^{2+}, Zn^{2+}, and Pb^{2+} with naturally occurring ligands and a variety of administered chelating agents. These computer simulation models have been extended to examine the metal-binding effects of both natural and synthetic chelating agents administered by intravenous infusion. Use of these types of computer simulation procedures to develop models of complex systems of species could be beneficial in attempting to understand these very complex systems and can be useful in translating inorganic biochemistry research into successful therapies by producing tactical leads for animal screening (May and Williams, 1981). However, they apply to systems in thermodynamic equilibrium and do not take into account possible kinetic factors.

References

Bowen HJM (1980) The requirement for trace element analysis in the 21st century. In: Braetner P, Schramel P (eds) Trace element analytical chemistry in medicine and biology. Berlin, W. De Gruyter and Co, pp 783–800

Browner RF (1983) Sample introduction for inductively coupled plasmas and flames. Trends Anal Chem 2:121–124

Carey PR (1982) Biochemical applications of Raman and resonance Raman spectroscopes. Academic Press, New York

Chasteen ND (1983) The biochemistry of vanadium. In: Structure and bonding – 53, ch. 3. Springer, Berlin

Consaul JR, Lee K (1983) Extrinsic tagging in iron bioavailability research: a critical review. J Agric Food Chem 31:684–689
Crews HM, Burrell JA, McWeeny DJ (1983) Preliminary enzymolysis studies on trace element extractability from food. J Sci Food Agric 34:997–1004
Date AR (1983) An introduction to inductively coupled plasma source spectrometry. Trends Anal Chem 2:225–230
Dawson JB, Bahreyni-Tosi MH, Hodgkinson A, Troughton PR (1980) Determination of copper and zinc and plasma protein fractions by means of column chromatography and atomic absorption spectrophotometry. In: Braetner P, Schramel P (eds) Trace element analytical chemistry in medicine and biology. W. de Gruyter and Co, Berlin, pp 461–467
De Galan L (1982) Zeeman atomic absorption spectrometry. Trends Anal Chem 1:203–205
Evans GW, Johnson PE, Brushmiller JG, Ames RW (1979) Detection of labile zinc binding ligands in biological fluids by modified gel filtration chromatography. Anal Chem 51:839–843
Faulkner WR (1981) The trace elements in laboratory medicine. Lab Management 21–35
Harnly JM, O'Haver TC, Golden BM, Wolf WR (1979) Background corrected simultaneous multielement atomic absorption spectrometer. Anal Chem 51:2007–2014
Harnly JM, Wolf WR (1984) Atomic spectrometry for inorganic elements in foods. In: Charalambous (ed) Analysis of foods and beverages, ch 14. Academic Press, New York, pp 451–481
Haylock SJ, Buckley PD, Blackwell LF (1983) Separation of biologically active chromium-containing complexes from yeast extracts and other sources of glucose tolerance factor (GTF) activity. J Inorg Biochem 18:195–211
Hazell T (1982) Iron and zinc compounds in the muscle meats of beef, lamb, pork and chicken. J Sci Food Agric 33:1049–1056
Heumaun KG (1982) Isotope dilution mass spectrometry for micro- and trace-element determination. Trends Anal Chem 1:357–361
IAEA (1980) Elemental analysis of biological materials. Technical Report 197. International Atomic Energy Agency, Vienna
Kleinhaus FW, Kline SC, Dugan WM, Williams JG (1983) Comparison of electron paramagnetic resonance and atomic absorption serum copper measurements in human normal controls and cancer patients. Cancer Res 43:3447–3450
Kobos RK (1980) In: Freiser H (ed) Ion selective electrodes in analytical chemistry, vol II, ch 1. Plenum Press, New York
Kudo K, Suzuki N (1984) Substoichiometry in trace analysis. Trends Anal Chem 3:20–24
Leach RA, Ruzicka J, Harris JM (1983) Spectrophotometric determination of metals at trace levels by flow injection and series differential detection. Anal Chem 55:1669–1673
Lubbers DW, Acker H, Buck RP, Eisenman G, Kessler M, Simon W (eds) (1981) Progress in enzyme and ion selective electrodes. Springer, Berlin
Martin CR (1982) Current trends in ion-selective electrodes. Trends Anal Chem 1:175–179
May L, Morris ER, Ellis R (1980) Chemical identity of iron in wheat by Mössbauer spectroscopy. J Agric Food Chem 28:1004–1006

May PM, Williams DR (1981) Complex speciation for absorption and excretion. In: Howell MC et al. (eds) Trace element metabolism in man and animals (TEMA-4). Australian Academy of Sciences, Canberra, pp 149–151 (and references therein)
Micronutrient Interactions: Vitamins, Minerals, and Hazardous Elements (1980) Ann NY Acad Sci 355 (and references therein)
Monsen ER, Hallberg L, Lagrisse M, Hegsted DM, Cook JD, Mertz W, Finch CA (1978) Estimation of available dietary iron. Am J Chem Nutr 31:134–141
Nagy VY (1983) Application of stable radical containing reagents to the ESR determination of metals. Trends Anal Chem 2:136–142
NAS (1976) Selenium, Washington, DC: National Academy of Sciences, Committee on Medical and Biological Effects of Environmental Pollutants (ISBN-0-309-02503-6)
Reamer DC, Veillon C (1981) Determination of selenium in biological materials by stable isotope dilution gas chromatography – mass spectrometry. Anal Chem 53:2166–2169
Sabbioni E, Edel J, Goetz L (1985) Trace metal speciation in environmental toxicology research. In: Abdulla M (ed) Health effects and interactions of essential and toxic elements. Nutrition Research (Suppl 1), pp 32–43
Schramm VL (1982) Metabolic regulation: could Mn^{2+} be involved? Trends Biochem Sci 7:369–371
Schwarz K, Mertz W (1959) Chromium (III) and the glucose tolerance factor. Arch Biochem Biophys 292–295
Schwedt G (1983) Species analysis of elements in foods. Trends Anal Chem 2:39–42
Smith SB, Hieftje GM (1983) A new background correction method for atomic absorption spectrometry. Appl Spectr 37(5):419–424
Thomas RC, Moody WJ (1980) Ion-sensitive microelectrodes for intracellular use. Trends Biol Sci April: 86–87
Thompson RJ (1982) Are monoclonal antibodies the end of radioimmunoassay? Trends Biol Sci, December: 419–420
Van den Hamer CJA, Houtman JPW (1980) Special forms of bound trace elements; their analysis and interest in medicine. In: Braetner P, Schramel P (eds) Trace element analytical chemistry in medicine and biology. W. de Gruyter and Co, Berlin, pp 233–242
Van der Linden WE (1982) Flow injection analysis; the manipulation of dispersion. Trends Anal Chem 1:188–191
VanLoon JC (1979) Metal speciation by chromatography/atomic spectroscopy. Anal Chem 51:1139–1150A
Veillon C, Wolf WR, Guthrie BF (1979) Determination of chromiumin biological materials by stable isotope dilution. Anal Chem 51:1022–1024
Wolf WR (1981) Duality of concern with trace metals in the environment: toxic and/or essential. In: Brinkman FE (ed) Special publication 618. Bethesda, MD:NBS
Wolnick KA, Friche FL, Hahn MH, Caruso JA (1981) Sample introduction system for simultaneous determination of volatile elemental hydrides and other elements in foods by inductively coupled argon plasma emission spectrometry. Anal Chem 53:1030–1035

The Importance of Chemical "Speciation" in Environmental Processes,
eds. M. Bernhard, F. E. Brinckman, and P. J. Sadler, pp. 59–70.
Dahlem Konferenzen 1986. Springer-Verlag Berlin, Heidelberg
© Dr. S. Bernhard, Dahlem Konferenzen

The Dependence of Toxic Reactions on the Chemical Species of Elements

M. Piscator

Dept. of Environmental Hygiene
Karolinska Institute
10401 Stockholm, Sweden

Abstract. A distinction is made between the toxicology of metal and metalloid elements which form stable oxyanions and those which do not. As examples of the latter cadmium and nickel are discussed. Species analysis is important for detection of nickel subsulfide, which can cause lung cancer. Mercury can exist in different oxidation states and in a methylated form, which indicates great differences in metabolism and toxicity between different forms of mercury. As(III), As(V), and Cr(VI) can appear as oxyanions. Arsenic can cause systemic effects as well as lung and skin cancer, whereas Cr(VI) compounds are known to be able to cause lung cancer but only minor systemic effects due to the rapid reduction to Cr(III).

Introduction

The mad hatter in *Alice in Wonderland* illustrates well what happens when one metal species, i.e., mercury nitrate, is handled in such a way that a more poisonous species, mercury vapor, is formed. Such problems are now recognized in industrial environments, but it took some time before the mechanisms were understood. Today, due to advances in analytical methodology, our understanding of the conversion processes in man and the environment has improved considerably. However, there is no consistent pattern. Methylation of Hg(II) gives persistent and highly toxic compounds, whereas methylation of arsenic produces less toxic and even nontoxic compounds. Most sulfides are regarded as harmless since they have low solubilities and low reactivities, but nickel sulfide is known to cause lung cancer.

Among the elements of interest two main groups can be distinguished. The first group consists of compounds with metals in low oxidation states: +2 or +3. Examples include aluminum, beryllium, cadmium, chromium, cobalt, copper, iron, lead, manganese, mercury, nickel, tin, and zinc. Some of them are essential elements, others are nonessential, but all can cause toxic effects if exposure is excessive. These metal ions may form complexes with body constituents and compete with each other. For some, specific transport mechanisms exist, for others the mechanisms are nonspecific or have not been investigated in detail. In body fluids, albumin may be a carrier of both essential and nonessential metals. One of these metals, mercury, can exert its action in many different forms (species), from the vapor to the methylmercury. Synthetic organometallic compounds of lead, manganese, and tin have been widely used and metabolism and toxicity will differ from other complexes of these metals.

The second group consists of metals or metalloids which can form oxyanions. Arsenic, Cr(VI), molybdenum, selenium, and vanadium are in this group. The oxyanions can easily pass through cell membranes along the same pathways as sulfate or phosphate. Manganese can also form an oxyanion (MnO_4^-) but it is a very strong oxidizing agent and does not generally exist in the media to which human beings are exposed.

Platinum seems to be in a special class since the structure of certain platinum complexes decides how the cells handle them. The *cis* forms of platinum diamine complexes have an anticancer effect as well as mutagenic effects, whereas the *trans* forms are inactive.

In the following, a few elements will be discussed in more detail, with special emphasis on arsenic, since this element causes a wide spectrum of effects and during recent years much work has been done on identification and transformations of its species.

Arsenic

Occurrence, metabolism, and effects of arsenic have been extensively described in recent EPA and WHO documents (USEPA, 1983a; WHO, 1981). In occupational environments there is mainly exposure to As(III) as arsenic trioxide, but in certain processes, i.e., manufacturing and handling of arsenic pesticides, there is mainly exposure to As(V) compounds.

For members of the general population the main exposure to arsenic occurs via water and food. In drinking water, arsenic concentrations are generally low, but in certain areas of the world high concentrations,

$>100\,\mu g\,l^{-1}$, have been found (USEPA, 1983a; WHO, 1981). The analytical data on arsenic species in water are still few and there have been large problems in identifying the oxidation states due to the ease of oxidation of As(III) to As(V) in water. Both have been identified as well as small amounts of methylated forms.

The fate of inhaled arsenic will mainly depend on solubility and particle size. Thus arsenic trioxide will be rapidly cleared from the lungs, whereas the less soluble sulfide will be retained longer (Pershagen et al., 1982). Analyses of lung tissues from decreased workers have shown relatively high arsenic concentrations, even many years after last exposure, but the species have not been identified. Since arsenic trioxide is eliminated rapidly, it may be that a more insoluble compound is retained which may be responsible for the lung cancer seen in occupationally exposed workers. No studies have been made on how pulmonary cells handle arsenic, and no data are available on the species present in the arsenic particles in air. It could well be that a small part of the total arsenic in air is a compound with low solubility, e.g., the sulfide, which could be the carcinogenic agent.

In the gastrointestinal tract small amounts of arsenic are probably absorbed to almost 100 per cent. Absorbed compounds of arsenic, from either the lung or the gut, will be partly excreted unchanged via urine, partly methylated in the liver, and partly taken up in some soft tissues, e.g., the skin.

After a dose of an As(III) or As(V) compound, four species of arsenic have been identified in urine from laboratory animals (Vahter, 1981) and human beings (Braman and Foreback, 1973): As(III) and As(V) complexes, as well as monomethyl and dimethyl arsenic. It is well established that monomethyl and dimethyl arsenic compounds are formed, but it is less clear to what extent and where reduction or oxidation of As(V) and As(III) occurs. In the organo-compounds arsenic is As(V), but it has been claimed that arsenic must be As(III) to be methylated, as in bacterial systems (Challenger, 1978; McBride et al., 1978; Mushak, 1983).

In Table 1 data from a study on mice (Vahter and Envall, 1983) are shown. Two dose levels of As(III) and As(V) were used. It can be seen that already one hour after oral administration of arsenite or arsenate a considerable part of the dose has been excreted as dimethylarsinic acid, and that the transformation is dose-dependent: the smaller the dose, the higher the percentage of methylated arsenic. It can also be seen that after exposure to arsenate, As(III) was identified in the urine, in fact in larger amounts than after exposure to As(III) itself. In the same study plasma was also examined, and As(III) was also identified there one hour after oral or intravenous administration of As(V). Less than one percent of the arsenic in the

Table 1. Arsenic metabolites in bladder urine of mice 1 hr after oral administration of (^{74}As)-arsenite or (^{74}As)-arsenate. DMA is dimethylarsenic acid. Four to six animals in each group. (Modified from Vahter and Envall (1983))

	Dose mg/kg b.w.	Urinary metabolites % of dose			
		As(III)	As(V)	DMA	As(III)/As(V)
As(V)	0.04	0.04	3.8	9.3	0.11
	0.4	2.9	10.4	2.0	0.28
As(III)	0.04	0.3	1.3	7.7	0.23
	0.4	1.4	1.3	2.8	1.08

Table 2. Cumulative average excretion of As(III), As(V), and dimethylarsinic acid (DMA) in urine of seven rabbits given ^{74}As intravenously as arsenate (0.04 mg As/kg b.w.) (modified from Vahter and Envall (1983)).

Hours after dosing	As(III)	% of dose As(V)	DMA
0.5	1.2	5.5	0.02
1	2.7	16.6	0.09
2	4.5	21.9	0.64
3	6.7	29.1	2.5
4	10.4	32.4	6.5

original arsenate solution was in the trivalent form. Further evidence for the reduction of arsenate was seen in a study on rabbits (Vahter and Envall, 1983) as shown in Table 2. There was a delay in the excretion of dimethylarsonic acid (DMA) in relation to the excretion of As(III), indicating that As(III) must be formed before methylation occurs, since DMA is more rapidly excreted than As(III) (Buchet et al., 1981 a).

Studies on human beings have shown that with increasing doses the relative proportion of methylated species decreases (Buchet et al., 1981 b).

In human beings monomethylarsenic constitutes a relatively large part of the arsenic in urine, increasing with increases in dose, whereas in laboratory animals only a few percent of the total arsenic in urine is in the monomethyl form. Due to the sampling difficulties, present data on As(III) and As(V) in human urine (Braman and Foreback, 1973; Yamamura and Yamauchi, 1980) must be regarded as unreliable because the oxidation state may change before urine is voided.

It should also be mentioned that the most common laboratory animal, the rat, metabolizes arsenic quite differently from human beings and other mammals, making the rat unsuitable for evaluating arsenic toxicity. An intriguing finding is that the marmoset monkey was unable to methylate arsenic (Vahter et al., 1982).

The main effect seen after exposure by inhalation is lung cancer and, after exposure via drinking water, skin cancer. Arsenic can also cause effects on the cardiovascular and nervous systems. Arsenic(III) is generally regarded as more toxic than As(V). As mentioned above, knowledge about the pulmonary handling of arsenic is scarce and there is a great need for more data on the species of arsenic in industrial environments and in the lung. Epidemiological studies have indicated that at high exposure levels there is a dose-response relationship, but it is not clear if there is a threshold. More data on how arsenic is metabolized by lung cells would certainly help to support or reject the threshold concept. Studies in Taiwan, where a large population was exposed to high doses of arsenic via drinking water, have shown that there is a dose-response relationship between arsenic intake and skin cancer (USEPA, 1983a; WHO, 1981). The arsenic is probably in the As(V) form. There are no data on the species distribution of arsenic in urine of this population. The concentrations of arsenic in drinking water were from about $100\,\mu g\,l^{-1}$ to more than $1\,mg\,l^{-1}$. The Taiwan data have been used to predict the risk for skin cancer at arsenic concentrations below $1\,\mu g\,l^{-1}$ (USEPA, 1983a) but it is obvious that a linear relationship may not be applicable if the biotransformation is relatively more efficient at low intakes, i.e., a higher proportion of arsenic is methylated, so reducing the toxicity (Hood et al., 1982; USEPA, 1983a). Also, the rate of reduction of As(V) to As(III) must play a role.

In marine organisms arsenic occurs mainly in the form of "shrimp arsenic" or "fish arsenic," a complex mixture of organo forms which is regarded as nontoxic. It is easily absorbed from the gut but excreted rapidly without any transformation (Tam et al., 1982). Both lipid-soluble and water-soluble arsenic compounds have been found; among the latter are arsenobetaine and arsenocholine (Norin and Christakopoulos, 1982). Recent studies on the fate of arsenobetaine show that it is absorbed and excreted without biotransformation (Vahter et al., 1983).

Cadmium

Cadmium in the environment and in human beings exists only as Cd(II). Organo cadmium compounds are unstable and exist only under laboratory conditions.

After exposure to cadmium via air, generally as the oxide or sulfide, or after oral exposure, absorbed cadmium is retained in the body. The biological half-time is more than ten years. Most of the cadmium is bound to metallothionein (Kägi and Nordberg, 1979). The initial storage occurs in the liver, which has a high capacity to synthesize this low-molecular-weight protein which has a very high affinity for cadmium. From the liver, cadmium is then slowly released as a cadmium-metallothionein complex to the kidneys, where it is filtered and reabsorbed by the proximal tubule cells. The kidneys can synthesize metallothionein, but the capacity is limited and when it is exceeded renal damage will occur. Metallothionein thus has extremely important roles in mediating cadmium toxicity. Hitherto deficiencies in metallothionein synthesis have not been reported. If such were to occur, a situation similar to that of Wilson's disease might result, in which a defect in ceruloplasmin synthesis prevents normal transport and storage of the essential metal copper.

In studies on the effects of cadmium, it is thus important to identify metallothionein and establish how much cadmium is bound to this protein. Several quantitative methods now exist for determination of metallothionein (Onosaka and Cherian, 1982; Tohyama and Shaik, 1981). Data on cadmium toxicity obtained after acute exposures to large doses cannot be used for evaluating chronic risks since acute effects are related to cadmium not bound to metallothionein, whereas chronic effects are caused by exceeding the renal capacity to bind cadmium to metallothionein and synthesize it. Then small amounts of low-molecular-weight cadmium complexes are released into the renal tissue.

Chromium

This metal occurs in nature as Cr(III). Chromium, especially as the complex "glucose tolerance factor" which may be a Cr(III) complex of glutathione (Wolf, this volume), is regarded as an essential element (Mertz et al., 1974). There may be occupational exposures to Cr(VI) compounds as well as Cr(III) compounds. Chromium(VI) occurs as chromates, i.e., oxyanions which easily penetrate cell membranes. In the cells they are reduced to

Cr(III). Such reductions also occur in the gastrointestinal tract and ingestion of small amounts of Cr(VI) compounds will not necessarily lead to their absorption. Chromium(III) salts are often poorly absorbed, whereas the "glucose tolerance factor" is more efficiently absorbed.

The most important health effect after inhalation exposure is cancer of the lung. Cr(VI) is generally regarded as the causative agent, even if Cr(III) has also been implicated. The counter cation in chromates, e.g., sodium, lead, or zinc, and its solubility seem to be factors of importance (USEPA, 1983 b).

Chromium(VI) may be absorbed into the body if exposure via air is high. It has been claimed that tissues from an autopsied worker contained Cr(VI) a long time after last exposure (Hyodo et al., 1980), but it is not clear if the analytical procedures were accurate. During exposure it is conceivable that Cr(VI) may be circulated and excreted.

Systemic effects have generally not been reported. A slight influence on renal tubular function has been noticed in workers exposed to Cr(VI). It is known that in experimental animals chromates produce tubular damage (USEPA, 1983 b).

To understand the action of chromium, its reduction in cells is of interest. There is only one study where lung cells were investigated (Petrilli and de Flora, 1978). A microsomal preparation from such cells had a low capacity to reduce Cr(VI) to Cr(III). On the other hand, there are a large number of studies on how liver cells handle Cr(VI). Chromium(VI) is mutagenic in bacterial systems, but if microsomal preparations from liver are used a decrease in mutagenic activity is seen. The reducing capacity of microsomal systems from liver was much greater than that of microsomal systems from lung cells (Petrilli and de Flora, 1978).

Since the lung is the organ of greatest interest, it is obvious that lung cells should be studied more. The efficient reduction of Cr(VI) to Cr(III) by liver cells has led to theories about Cr(III) being the carcinogenic agent, but the data on lung cells indicate that the action on DNA may be exerted by Cr(VI). Since chromium is classified as a carcinogen, occupational data have been used to estimate risks at low-level exposure using linear models. Such an extrapolation assumes that nanogram amounts of Cr(VI) are handled in the same way as milligram amounts. Even if the reducing capacity in the lungs is limited, it is conceivable that small amounts of Cr(VI) will be rapidly reduced in the cells.

Mercury

As stated in the introduction, the hazards from exposure to mercury vapor have been recognized for some time. Mercury salts (e.g., the chloride) are known to be highly toxic in large oral doses and have been popular suicidal agents. The modification of mercury toxicity by synthesizing organo-mercury compounds was used for the production of both therapeutic agents (e.g., diuretics) and pesticides. The list of mercury species with different properties was thus quite long even before the discovery that organo-mercury compounds could be synthesized in nature by certain microorganisms.

The metabolism and effects of different mercury compounds are fairly well-known (WHO, 1976). Inhaled mercury vapor is absorbed to about 80 percent and taken up in the blood, where oxidation will occur. Some mercury vapor penetrates the blood-brain barrier and oxidation then occurs in brain tissue, where Hg(II) is trapped. Elimination is slow and eventually systemic effects may result. Inhaled or ingested Hg(II) compounds are taken up to a lesser degree, occur in water-soluble forms in the blood, and are excreted relatively rapidly. The kidney is the critical organ (sublimate nephrosis).

Organo-mercury compounds of interest are phenylmercury and methylmercury. Phenylmercury is absorbed to a higher degree than, for example, $HgCl_2$, but the mercury-carbon bond is easily cleaved and the metabolism and effects are similar to those involving other mercury compounds. Methylmercury compounds contain stable covalent Hg-C bonds, have a high lipid solubility, and pass through membranes easily. Absorption from the gut is close to 100 percent and the distribution in the body is relatively even. Thus the brain will also receive a part of the absorbed dose. Methylmercury affects the brain, but the symptoms are different from those caused by mercury vapor.

The available analytical methods permit determination of the different species of mercury, and both external and internal exposures can be determined accurately, especially with regard to methylmercury compounds, for which a good metabolic model exists. The determination of total mercury in hair is an indicator of the body burden of methylmercury, and total mercury in urine is an indicator of exposure to mercury vapor or mercury salts.

Nickel

In the general environment nickel occurs as Ni(II). In certain industrial operations there may also be exposure to nickel carbonyl [$Ni(CO)_4$]. Nickel is nowadays regarded as an essential metal, but the requirement is probably very low for human beings. Systemic effects have not been reported in nickel workers. The main hazards are nasal cancer and lung cancer.

The available data indicate that nickel compounds with low solubility, especially the subsulfide, can cause cancer (Abbracchio et al., 1982). In animal experiments nickel subsulfide has produced respiratory tract cancer, whereas soluble nickel compounds do not. Occupationally exposed persons have been exposed to high concentrations of dust, but analysis of the different nickel species in the dust has generally not been performed.

Nickel has been declared a carcinogen, and estimates of risk at low-level exposure have been made using linear extrapolation from occupational data. Nickel concentrations in ambient air are very low. There are no data on the percentage of nickel subsulfide in relation to total nickel. Furthermore, as was discussed in relation to Cr(VI), it is doubtful whether nanogram amounts of nickel subsulfide behave in a manner similar to milligram amounts. The carcinogenic action is thought to be caused by the uptake of nickel subsulfide particles by cells in the respiratory tract and a continuous release of nickel from the particles (Abbracchio et al., 1982). Very small amounts of nickel subsulfide will probably be dissolved by the fluids in the respiratory tract. The problem of species identification is further complicated by the fact that particles of crystalline nickel subsulfide are more easily phagocytized than the amorphous form (Abbracchio et al., 1982).

Conclusion

In this paper a distinction has been made between metals and metalloids appearing in living systems as oxyanions and those in other complexes. Nonessential cations, e.g., cadmium and mercury, may compete with essential elements for binding sites. For a metal such as cadmium, a distinction between different species is of interest with regard to exposure, via inhalation or ingestion, since that will give information about the likelihood of systemic absorption. In the body cadmium will mainly be bound to metallothionein. Nonspecific binding to albumin occurs for cadmium as for other nonessential or essential divalent cations. A distinction between different Hg species is of great importance for evaluation of both external and internal exposure. The present knowledge about the mercury cycle in the environment and the existence of good analytical methods for identification of the different species make it possible to make relatively good risk estimates. Some metal ions cause only local effects; as an example nickel has been discussed. Nickel subsulfide is thought to cause cancer in the upper and lower respiratory tract, and identification of this species is thus of great importance. In industrial environments, only total metal concentrations in the air are generally determined; identification and quantification of the metal species would be of great value.

The oxyanions behave quite differently. They can easily pass through membranes and compete with other anions such as sulfate and phosphate. As examples arsenic and chromium have been discussed. Arsenic appears as arsenite or arsenate, and identification of the oxidation state is thus of great interest for understanding and predicting toxicity. Furthermore, methylation occurs and this is thought to be a detoxification mechanism. Since arsenic exposure can cause lung cancer or skin cancer after inhalation or ingestion, respectively, it is of great interest to obtain more information as to how small amounts of arsenic are handled in the body. There are data on how metabolism differs at high and low doses. It is questionable whether cancer risks for very low exposures can be extrapolated from high-level exposures without taking metabolism into account. A comparison can be made with vinylchloride, for which risk estimates are based on the metabolized amounts, since the metabolites, not the orginal compound, are carcinogenic. If a non-organo form of arsenic is the carcinogenic agent, the possibility of methylation should be taken into account. It was also pointed out that the analysis of the species composition is needed to identify the individual compounds making up the total arsenic in air. Of special interest is the identification of arsenic species with low solubility and long retention times in the lung, e.g., the sulfide species.

Another oxyanion is chromate or dichromate. Sensitive methods are needed to study how lung cells handle those ions, in solution or as particles. Especially the low doses are of interest, since ambient air levels are very low and cancer risks have been estimated for these low levels. This problem is the same as for low-level nickel exposure. There is also a need to identify Cr(VI) in ambient air.

Other oxyanions of interest are formed by vanadium, molybdenum, and selenium, but from a toxicological point of view they have attracted less attention.

Finally, is should be kept in mind that not only relevant species but also dose and exposure route should be taken into account when using toxicity data to predict risks.

References

Abbracchio MP, Simmons-Hansen Y, Costa M (1982) Cytoplasmic dissolution of phagocytized crystalline nickel sulfide particles: a prerequisite for nuclear uptake of nickel. J Toxicol Envir Health 9:663–676

Braman RS, Foreback CC (1973) Methylated forms of arsenic in the environment. Science 182:1247–1249

Buchet JP, Lauwerys R, Roels H (1981 a) Comparison of the urinary excretion of arsenic metabolites after a single oral dose of sodium arsenite, monomethylarsenate and dimethylarsinate in man. Intl Arch Occup Envir Health 48:71–79

Buchet JP, Lauwerys R, Roels H (1981 b) Urinary excretion of inorganic arsenic and its metabolites after repeated ingestion of sodium metaarsenite by volunteers. Intl Arch Occup Envir Health 48:111–118

Challenger F (1978) Biosynthesis of organometallic and organometalloidal compounds. In: Brinckman FE, Bellama JM (eds) Organometal and organometalloids: occurrence and fate in the environment. American Chemical Society, Washington, DC, pp 1–22

Hood RD, Harrison WP, Vdel GC (1982) Evaluation of arsenic metabolites for prenatal effects in the hamster. Bull Envir Contam Toxicol 29:679–687

Hyodo K, Suzuki S, Furuya N, Meshizuka K (1980) An analysis of chromium, copper and zinc in organs of a chromate worker. Intl Arch Occup Envir Health 46:141–150

Kägi JHR, Nordberg M (1979) Metallothionein. Birkhäuser Verlag, Basel

McBride B, Merilees H, Cullen WR, Pickett W (1978) Anaerobic and aerobic alkylation of arsenic. In: Brinckman FE, Bellama JM (eds) Organometals and organometalloids: occurrence and fate in the environment. American Chemical Society, Washington DC, pp 94–115

Mertz W, Toepfer EW, Roginski EE, Polansky MM (1974) Present knowledge of the role of chromium. Fed Proc 33:2275–2280

Mushak P (1983) Mammalian biotransformation processes involving various toxic metalloids and metals. In: Brown SS, Savory J (eds) Chemical toxicology and clinical chemistry of metals. Academic Press, London, pp 227–245

Norin H, Christakopoulos A (1982) Evidence for the presence of arsenobetaine and another organoarsenical in shrimps. Chemosphere 11:287–298

Onosaka S, Cherian MG (1982) Comparison of metallothionein determination by polarographic and cadmium saturation methods. Toxicol Appl Pharmacol 63:270–274

Pershagen G, Lind B, Björklund N-E (1982) Lung retention and toxicity of some inorganic arsenic compounds. Envir Res 29:425–434

Petrilli FL, de Flora A (1978) Metabolic deactivation of hexavalent chromium mutagenicity. Mutat Res 54:137–147

Tam GKH, Charbonneau SM, Bryce F, Sandi E (1982) Excretion of a single oral dose of fish arsenic in man. Bull Envir Contam Toxicol 28:669–673

Tohyama C, Shaik ZA (1981) Metallothionein in plasma and urine of cadmium-exposed rats determined by a single-antibody radioimmunoassay. Fund Appl Toxicol 1:1–7

U.S. Environmental Protection Agency (1983 a) Health assessment document for inorganic arsenic. E.P.A.-600/8-83-021F. Office of Health and Environmental Assessment, Washington, DC

U.S. Environmental Protection Agency (1983 b) Health effects assessment document for chromium. Office of Health and Environmental Assessment, Washington, DC
Vahter M (1981) Biotransformation of trivalent and pentavalent inorganic arsenic in mice and rats. Envir Res 25:286–293
Vahter M, Envall J (1983) *In vivo* reduction of arsenate in mice and rabbits. Envir Res 32:14–24
Vahter M, Marafante E, Dencker L (1983) Metabolism of arsenobetaine in mice, rats and rabbits. Science Tot Envir 30:197–211
Vahter M, Marafante E, Lindgren A, Dencker L (1982) Tissue distribution and subcellular binding of arsenic in marmoset monkeys after injection of ^{74}As-arsenite. Arch Toxicol 51:65–77
WHO (1976) Environmental Health Criteria. 1. Mercury. World Health Organization, Geneva
WHO (1981) Environmental Health Criteria. 18. Arsenic. World Health Organization, Geneva
Yamamura Y, Yamauchi H (1980) Arsenic metabolites in hair, blood and urine in workers exposed to arsenic trioxide. Ind Health 18:203–210

The Importance of Chemical "Speciation" in Environmental Processes,
eds. M. Bernhard, F.E. Brinckman, and P.J. Sadler, pp. 71–83.
Dahlem Konferenzen 1986. Springer-Verlag Berlin, Heidelberg
© Dr. S. Bernhard, Dahlem Konferenzen

The Influence of Chemical Species on the Absorption and Physiological Utilization of Trace Elements from the Diet or Environment

C. F. Mills

Dept. of Inorganic Biochemistry
Rowett Research Institute
Aberdeen AB2 9SB, Scotland

Abstract. Differences in the physicochemical characteristics of complexes of trace elements influence trace element absorption, distribution within and between tissues, and pathological responses to deficient or excessive intakes from the diet or environment. Examples of the metabolic relevance of element "speciation" are considered.

Introduction

Chemical and physical characteristics of the trace elements influence their mobility at virtually all stages of their translocation from geochemical sources into water or food chains and, ultimately, into human or animal tissues. Variables influencing the partition of the elements between differing ligands become apparent from changes in their uptake, tissue utilization, or retention by the body and, in many cases, such effects are reflected by changed susceptibility to the pathological consequences of deficient or excessive supply.

Entry of Essential and Toxic Trace Elements into Food Chains

Although the principal theme of this paper is the influence of chemical species on mammalian trace element metabolism, it is important to indicate that factors affecting distribution of species in the external environment are often highly relevant. Thus the species in which elements are released from parent rocks and soils often determine the susceptibility of herbivores to deficiencies of copper, cobalt, and selenium and toxicities of molybdenum, fluoride, and iron, and may well influence the geographical distribution of iodine, selenium, and zinc deficiencies and possibly molybdenum and fluoride intoxication in human subjects. Thus, Mo uptake by crops is favored by neutral or high soil pH and low E_h, and such conditions are clearly relevant to the etiology of molybdenosis and Mo-induced Cu deficiency in ruminants *(see below)*. Increases in herbage Fe and Co content induced by low soil pH and low redox potential modify the incidence of Co deficiency and of Cu deficiency induced by Fe excess. Finally, the incidence of Se deficiency in ruminants is, in some instances, related to the sequestration of Se in high-Fe soils or to inhibition of Se uptake into crops arising ostensibly from the competitive effects of sulfate on selenate absorption.

Despite much evidence of the etiological significance of such relationships, little is known of the chemical species in which many trace elements are deposited in plant tissues used as food sources (Tinker, 1981).

Chemical Species in Mammalian Trace Element Absorption

The efficiency with which many essential and toxic trace elements are absorbed is governed both by the chemical forms in which they are ingested and by changes in species transformation occurring during digestion, uptake, and transport through the intestinal mucosa (Bremner and Mills, 1981).

Iron Utilization

The metabolic fate of ingested Fe is determined partly by physiological Fe status and by the species distribution of its sources. Little is known of the different Fe species involved in release of Fe from the food to the intestinal mucosa or in mediating its intramucosal transport and final transfer to the protein transferrin, the primary acceptor of Fe circulating within blood plasma. Uptake by the immature gut is achieved by pinocytosis and vesicular transport (Linder and Munro, 1973) and thus is likely to be modulated by the presence of Fe-carriers in the intestinal lumen which are immunochemically compatible with mucosal membrane receptors involved in vesicle formation.

Studies of the kinetics of Fe absorption in the mature gut and of the inhibitory effect of excess Zn on the absorption of the element from differing sources all suggest that heme-Fe is transported and its flux regulated by processes quite distinct from those involved in absorption of non-heme Fe sources. Absorption of non-heme Fe is less strongly inhibited in subjects with a normal or high Fe status than is that of soluble, non-heme Fe. Studies with isolated mucosal cells suggest that Fe entry is a passive process, inhibited by powerful Fe-binding ligands and regulated principally by the supply of soluble Fe species at the absorbing "brush border" surface (Savin and Cook, 1978). The site at which total Fe flux is regulated in proportion to demand is a matter of controversy. Studies with isolated cells suggest that control is achieved during exchange between ligands involved in transport from the mucosal cell to the blood system. However, this concept does not accord with evidence that despite a lower sensitivity to changes in the demand for Fe, absorption of heme-Fe occurs as an intact complex, the Fe of which is later released into the mucosal cell pool.

At the pH values encountered throughout virtually all of the absorptive tract except the stomach, the solubility of Fe(III) is virtually zero except when water- or lipid-soluble complexes are formed. Not surprisingly, the utilization of simple ionic sources of Fe is strongly influenced by the presence in the diet or gastrointestinal secretions of such ligands. There is limited evidence that a small proportion of ingested Fe(III) solubilized at gastric pH (ca. 2–3.5) is bound by a soluble ligand (ca. 25,000 daltons) present in intestinal secretions. This complex remains soluble at duodenal pH (ca. 6.5–7) and thus fulfills one important requirement of an effective Fe donor. While this ligand shows structural similarities to the Fe-binding protein, transferrin, its identity remains questionable since it retains Fe less avidly than transferrin when treated with the Fe-chelator, desferrioxamine. It is also evident that the total secretion and, by implication, the total Fe-binding capacity of this ligand is insufficient to preclude precipitation of the greater part of intralumenal Fe(III) as insoluble and unabsorbable polyhydroxy complexes within the intestinal tract unless other ligands capable of sequestering Fe(III) also play a role in Fe absorption (Huebers et al., 1976; Hungerford and Linder, 1983).

Ascorbate is pre-eminent among the exogenous (dietary) ligands for which there is strong presumptive evidence of involvement. While it has been speculated that its efficacy in promoting absorption of non-heme Fe may be exercised a) by reducing intraluminal pH or b) by promoting reducing conditions favoring uptake by Fe(II) receptors, it is now clear that the former concept is untenable (Hungerford and Linder, 1983). Suggestions that its locus of action lies at a rate-limiting reduction of Fe(III) to Fe(II) during Fe release from mucosal transferrins remain to be evaluated.

Clarification of these and other nutritionally significant aspects of Fe absorption will be delayed until techniques of adequate sensitivity become available to characterize kinetic relationships between the species of Fe involved at all stages. Aspects of this topic have remained controversial for more than 30 years. Diametrically opposed views are held on mechanisms of regulation of Fe supply in relation to demand, on the stimulatory or inhibitory role of the Fe-binding protein, lactoferrin, present in milk and on the beneficial or adverse effects of dietary phosphate and polyphosphate additives on Fe economy in human subjects.

Zinc Deficiency and Zinc Absorption

The manifestations of Zn deficiency include inhibition of growth, suppression of appetite, impaired immunity to infection, abnormalities in fetal development and, at the biochemical level, a wide range of metabolic lesions including defects in polynucleotide transcription, protein synthesis, and reduced activity of specific Zn-containing peptidases. Such consequences are most strongly evident in the rapidly growing subject or at stages of accelerated tissue proliferation such as repair following injury or infection.

Paradoxically, many of the tissues in which such metabolic defects develop suffer little or no apparent change in total Zn content. Furthermore, many are rectified rapidly if an external challenge, such as a brief period of food deprivation, initiates tissue catabolism and perhaps a minor redistribution of body Zn. Changes in the Zn concentration of tissues participating in such exchanges may be difficult to detect and, indeed, the only evidence of redistribution is usually a transient increase in plasma Zn content.

This and other evidence suggest that many tissues contain a small and labile pool of Zn, rapidly equilibrating with the Zn of functional centers of enzymes susceptible to Zn depletion but not readily equilibrating with a much larger, relatively immobile pool. Ignorance of the character of the Zn species in these pools and of variables controlling their rates of exchange currently limits our ability to predict the precise consequences of deficiency at differing stages of physiological development, during generalized malnutrition or during the stress of infection (Mills, 1982).

Other than by monitoring the development or remission of pathological defects, it is difficult to determine the species of the physiologically available pool of Zn. Unfortunately, plasma Zn concentration is an unreliable criterion, being influenced by transient changes in the balance between Zn liberated or utilized by tissues undergoing catabolic or anabolic change. The potentially greater diagnostic and discriminatory value of measuring plasma concentrations of a specific Zn-metalloprotein will be considered later.

Zinc Species and the Efficiency of Zinc Absorption

Absorption of Zn is mediated by at least two distinct carrier- or enzyme-mediated processes both exhibiting saturation kinetics. Although the nature of the Zn species involved in mucosal uptake in these rapid and slow pathways of mucosal translocation is not known, it is evident that excessive concentrations of intraluminal Zn result in nonspecific binding to the mucosa from which Zn is absorbed at a greatly reduced rate. Thus, variables influencing the intraluminal concentration of free Zn(II) markedly influence the efficiency of Zn absorption. Evidence that free amino acids within the intestinal lumen modify the efficiency of Zn absorption has promoted speculation that translocation of Zn as relatively stable complexes may also occur. Noteworthy in the context of the relevance of Zn speciation (distribution pattern of Zn species) to biological "availability" is evidence that the Zn of human milk is more readily utilized by human infants or laboratory rats than that of cow's milk. Low-molecular-weight ligands such as citrate have been suggested as Zn-carriers responsible for this effect. In fact, there is evidence that Zn-citrate complexes occur in higher concentrations in human than in bovine milk. For picolinate, the evidence presented is less convincing. Alternative postulates that macromolecular ligands such as the protein, lactoferrin, or an unidentified constituent of the casein fraction of milk promote Zn absorption (Lonnerdal et al., 1985) await confirmation.

Much clearer evidence exists of the nutritional relevance of phytate as a potential antagonist of Zn absorption. The biological effects of this compound (myo-inositol hexakis dihydrogen phosphate), a significant component of many diets rich in fiber or containing acid-precipitated soya proteins, are consistent with the concept that it forms insoluble complexes with Zn within the intestinal lumen. Thus, it restricts both the absorption of dietary Zn and the reutilization of Zn present in intestinal secretions.

Its role in the etiology of Zn deficiency in farm animals is well recognized and its effects are allowed for by increased provision of Zn in phytate-rich diets. However, its significance as a cause of Zn deficiency in human subjects remains controversial for reasons that relate directly to a failure to define the conditions under which its antagonistic action is most strongly potentiated.

Although the propensity of phytate to form complexes with a wide range of metals is well-known, the stabilities and solubilities of "mixed-metal phytates" that could form within the complex ionic environment of the digestive tract have not been studied adequately. It is recognized, however, that at pH 7 the solubility products of phytate complexes of Zn, Cu, Cd, or Pb are greatly reduced if Ca or Mg are incorporated. Furthermore, Ca increases the particle size (and thus decreases reactive surface area) of the amorphous Ca-Zn phytate precipitate and decreases the efficiency with which free amino acids in solution at pH 7 desorb Zn occluded therein (Wise, 1983).

The relevance of these phenomena has been evident from the earliest nutritional studies with laboratory and farm animals. These indicated that high dietary Ca concentration increased but high dietary protein or exogenous chelator concentration reduced the potency of phytate as a Zn antagonist. Such variables have frequently been ignored in the planning and interpretation of many recent studies with human subjects, and hence results are often inconsistent and difficult to interpret. In most human diets, the concentration of one important potentiating ion, Ca(II), is frequently lower than those used in experimental animal studies, demonstrating unequivocally the potency of phytate as a Zn inhibitor. Nevertheless, concentrations of Mg are frequently higher, but the solubilities and stability of Mg-Zn-phytate complexes are inadequately defined and thus their biological significance cannot readily be assessed. More effective appraisal of the nutritional significance of phytate will become possible once techniques become available for characterizing and quantifying the processes provoking Zn precipitation within the intestinal lumen.

Sulfur and Molybdenum-containing Species Influencing Copper Utilization

In many areas, the incidence of Cu deficiency in ruminants is associated with either geochemical backgrounds slightly enriched with Mo or soil conditions favoring uptake of Mo by forages. While this is attributed to inhibitory effects of Mo on Cu absorption, the mechanisms of this antagonism are complex and reflect the propensity of Mo to form a wide range of Mo/S complexes many of which have an affinity for Cu. Thus, the potency of Mo as a Cu antagonist varies substantially and is most strongly evident when the metabolic activity of microorganisms within the rumen yields significant quantities of sulfide from the reduction of ingested sulfate or by degradation of S-amino acids. Although methods of sufficient sensitivity to characterize the Mo- and S-containing ligands that are synthesized within the rumen are not yet available, it is suspected that these are either structurally related to the thio- or oxythiomolybdates or that such compounds are intermediates in their synthesis (Mills, 1982).

Several lines of evidence support this concept. First, sulfide generated within the rumen is normally removed rapidly by absorption and reoxidation in the rumen wall or incorporated into bacterial S-amino acids. Increased concentrations of Mo not only prolong the biological half-life of acid-labile sulfide but promote the appearance of fractions in rumen and duodenal digesta which contain Cu in a biologically unavailable species and probably have an unsatisfied Cu-binding capacity. Such conclusions are based on studies in which the bioavailability of Cu in the presence of digesta fractions or a range of model Mo/S complexes is monitored by following the recovery of (Cu-dependent) cytochrome oxidase activity in the intestinal mucosa of Cu-depleted rats. These show that, whereas many Mo-enriched digesta fractions contain unavailable Cu and inhibit utilization of added Cu, such effects are not replicated by addition of Mo to the diet in the form of molybdate. In contrast, tetrathiomolybdate (MoS_4^{2-}) is the most effective inhibitor of Cu absorption and utilization yet identified. Among the range of oxythiomolybdates and more complex Mo/S compounds tested, the most important determinant of antagonistic potency is the presence and frequency of terminal Mo-(sulfido) S groups. Terminal S-S groups or (bridging) μ-sulfido groups linked to Mo appear to be ineffective. However, none of the models yet tested exhibits as high an antagonistic potency as microbial and other particulate fractions derived from rumen and intestinal digesta of ruminants given diets enriched with Mo (e.g., 5 mg Mo/kg). Recent data remain consistent with the view that the action of Mo as a Cu antagonist is attributable to a) sequestration of reactive sulfide generated within the rumen, b) formation of Mo/S complexes less readily oxidized than S^{2-} or HS^- within the gut, and c) subsequent reaction of terminal sulfido groups of such complexes with Cu(I) or Cu(II) liberated at later stages of digestion in the acid (pH 3–5) environments of the abomasum and duodenum.

This role of Mo in sequestering endogenous sulfide and thus inhibiting Cu absorption may not be unique. Thus, high intakes of Fe(II) also reduce the Cu status of ruminants, its effects are also potentiated by increasing dietary S intake and accompanied by increased acid-labile S^{2-} concentrations within the rumen (Humphries et al., 1983; Suttle et al., 1984). Other *in vitro* studies have shown that precipitates of virtually all metallic sulfides with solubility products higher than those of Cu(I) or Cu(II) sulfides readily incorporate Cu from aqueous solutions at physiological pH ranges to form the corresponding insoluble Cu sulfides. This suggests the possibility that excessive ingestion of a wide range of transition elements or heavy metals may adversely affect Cu utilization and account for evidence that the incidence of Cu deficiency is more widespread than can be explained by either low concentrations of dietary Cu or high intakes of Mo or Fe. The frequency with which this disorder arises in the vicinity of residues derived from metalliferrous mineral workings exemplifies such situations.

Also fundamental to our understanding of variables influencing Cu utilization is how some dietary Cu, exposed at pH 6–7 and at 37 °C to an intraruminal environment rich in S^{2-}, ever avoids precipitation as physiologically unavailable CuS or Cu_2S. The answer may well lie in the defined structures of the (Cu-containing) leaf proteins, the plastocyanins (Lappin, 1981). Approximately 10–15% of the dietary Cu of herbivores may be in this form, in which bound Cu lies in an inaccessible hydrophobic environment within the protein "core" or the plastocyanin molecule. Whether such Cu, initially protected against the S-rich environment of the rumen, is liberated in available species by later proteolysis in regions of the gastrointestinal tract from which sulfide is normally absent is a possibility under investigation.

Trace Element Species and Mobility Within Tissues

Changes in the total concentration are often much less relevant to the development of metabolic defects initiated by deficiency or excess than changes in the intra- and extracellular partition, subcellular form, and species distribution of elements. Species transformation and distribution can arise from a) *de novo* induction of the synthesis of specific metal-binding proteins, b) sequestration of elements in polymeric, insoluble complexes, or c) metabolic defects inhibiting the efflux of elements from cells.

Sequestration of Metals by Metallothionein

Tissue accumulation of Zn, Cd, Cu, and Hg is frequently accompanied by marked increases in cytosolic concentrations of the 6,000 dalton, cysteine-rich metalloprotein, metallothionein [MT] (Vallee, 1979). Synthesis of its messenger RNA is very effectively induced by Cd and Hg. However, a more general physiological role is suggested for MT from evidence that its synthesis is also induced by virtually all situations which promote a rise in tissue Zn concentration or a redistribution of the physiologically available pool of tissue Zn. Such stimuli include transient periods of food deprivation, infection, and stress. Estimates of the biological half-lives of the protein moieties of the metal-MT complexes indicate that the nature of the bound metal influences metabolic stability. Although such estimates are complicated by the fact that concurrent incorporation of more than one of the above-mentioned elements frequently occurs *in vivo*, there is general agreement that in rats Cu-rich MT exhibits the shortest biological half-life (approx. 12 hr) and Cd-MT the longest (up to 5 d).

The *de novo* synthesis of such metal-binding proteins not only increases tissue metal content but may also interrupt metal flux through the cells in which such synthesis is taking place. Thus synthesis of MT in intestinal mucosal cells may well account for the antagonistic effects of high dietary Zn on absorption of Cu (Mills, 1980). Inhibition of Cu transport, accompanied by increases in mucosal Cu-MT content, probably arises first from the induction of MT synthesis in response to Zn exposure, followed by the intracellular substitution of Cu for Zn at MT-SH groups. Sequestration of up to 60% of mucosal Cu in the relatively stable Cu-protein, which with respect to the 24–28 hr half-life of mucosal cells has a slow rate of turnover, could account for a substantial diversion of Cu from its normal absorptive pathway.

An additional role for MT in the sequestration of potentially toxic tissue concentrations of Cu and thus in modifying Cu tolerance is suggested from several lines of evidence. Except when a preestablished Zn deficiency precludes its synthesis, Cu accumulation by the liver is followed first by the appearance of Cu-MT in the cytosol, followed, if Cu loading is excessive, by lysosomal accumulation of insoluble polymeric aggregates of this Cu-protein. Animal species particularly intolerant of Cu (e.g., sheep) have a restricted capacity to sequester Cu in either form, whereas in pigs given diets high in Zn or Cu, monomeric or polymeric MT accounts for half the liver content of these elements (Bremner, 1981).

Since changes in the concentration of MT in blood plasma reflect changes in the physiologically available reserves of tissue Zn required to initiate its synthesis, immunochemical assay of plasma MT holds the long-awaited promise of providing a highly specific indicator of Zn adequacy or deficiency (Sato et al., 1984).

Systemic Defects in Selenium, Iron, and Copper Metabolism

Selenium is an essential component of the enzyme, glutathione peroxidase, involved in protection of tissues against damage initiated by reactive metabolites of O_2. Utilization of tissue Se for glutathione peroxidase synthesis is interrupted if tissue Cu or Ag concentrations rise as a consequence of adventitious contamination of the diet or environment. Thus hepatic retention of Ag provokes a decline in glutathione peroxidase activity accompanied by an *increase* in hepatic Se content, whereas when Cu accumulates in tissues, a more specific inhibition of the turnover of a labile Se pool probably involved in glutathione peroxidase biosynthesis appears to occur (Al-Rahim, 1980). It has been suggested that such metal-induced inhibition of Se utilization results from intracellular sequestration of the element as metal selenides. If so, the acid-labile selenide fraction of tissues must have a much more significant role as an Se donor for glutathione peroxidase synthesis than is appreciated currently. Whatever mechanisms are involved, the inhibition of Se utilization demonstrated in such studies may well account for evidence from veterinary practice that the therapeutic response to Se is relatively poor if Cu is administered simultaneously.

The metabolic significance of modified intracellular forms of trace metals is also illustrated by defects in Fe metabolism arising during the development of Cu deficiency. These first become evident from a decline in plasma Fe despite marked increases in liver Fe content, initially in the form of ferritin but later as hemosiderin. A decreased efficiency of absorption of dietary Fe is frequently accompanied by histologically demonstrable mucosal deposits of Fe (Lee et al., 1976) and increased proportions of mucosal Fe extractable by perfusion of the intestinal vasculature with the Fe-binding protein, apotransferrin. Nevertheless, increases in dietary Fe supply ameliorate the anemia of Cu deficiency (Williams et al., 1983) and it has been claimed (but not yet substantiated) that such treatment also reduces the adverse effects of Cu deficiency upon growth but not on other Cu-dependent processes (Weisenberg et al., 1980). Such findings indicate that inadequately characterized changes in the content and forms of tissue Fe during Cu deficiency have a substantial influence on important pathological features of the Cu deficiency syndrome.

Influence of Molybdenum on Tissue Copper Distribution

A puzzling feature of Cu deficiency in ruminants has been that, when induced by excessively high intakes of Mo (e.g., <10 mkg Mo/kg diet), the development of clinical signs of deficiency is often accompanied by increases in plasma Cu concentration. Work with several animal species has shown that this effect results from a systemic inhibition of Cu utilization probably attributable to the absorption from the digestive tract of tetra- or trithiomolybdates. Both anions, associating with cationic sites of plasma albumin, react subsequently with Cu to form products which, although not yet identified, clearly retain Cu in unavailable species. Similar reactions probably account for marked decreases in the proportion of cellular Cu found in the cysotol of liver and kidney when tetra- or oxythiomolybdates are absorbed. The species in which Cu becomes retained in particulate fractions has not yet been identified, but it is believed to be associated with thiomolybdate residues for the following reasons: a) the Cu-enriched insoluble fraction also contains Mo, b) its Cu is unavailable for synthesis of the Cu enzymes superoxide dismutase and cytochrome oxidase, and c) the depressions in Cu enzyme activity and in the proportional solubility of cell Cu are related directly to the degree of thiosubstitution of the tetra- or oxythiomolybdates administered (Mills, 1982). Even though many aspects of this Mo/Cu interaction remain to be resolved, the fact that it results in the induction of pathological lesions of Cu deficiency often without concurrent evidence of marked decreases in total tissue copper illustrates the importance of determining the chemical species involved.

Conclusion

The topics considered above typify instances in which changes in trace element species in tissues have direct relevance to diseases induced by trace element deficiencies or excesses in human subjects or animals. Virtually all illustrate that inadequate understanding of the species in which the trace elements are translocated or utilized in tissues limits our ability to predict trace element availability or tolerance. In the absence of such information it is not only difficult to predict dose/response relationships but also to devise more effective criteria than changes in the total trace element content of fluids or tissues for the diagnosis of deficiency or toxicity. Until such uncertainties are resolved, progress in the study of the metabolism of the elements, particularly of the trace elements, will remain slow and existing knowledge will often remain inadequate to resolve the queries posed by the nutritionist, pathologist, or toxicologist.

With a few notable exceptions (e.g.,Williams 1981), little effort has been made to emphasize the contribution that studies of chemical species can make towards the resolution of many of the above questions. Technical developments permitting investigation, preferably *in situ*, of the partition of elements between differing species present at concentrations of parts per million or per billion are needed urgently. Until they are available, recourse to concentration and isolation procedures will continue (with all their associated uncertainties arising from redistribution artifacts) as will studies of metabolic responses to putative "model" species whose validity will remain questionable until more direct methods of species characterization become possible. Success in all the above approaches demands much closer liaison between physical chemists and biologists than is generally appreciated or encouraged.

References

Al-Rahim AG (1980) Dietary factors affecting selenium utilisation by animals. PhD Thesis, University of Aberdeen
Bremner I (1981) The nature and functions of metallothionein. In: Howell JM, Gawthorne JM, White CL (eds) Trace element metabolism in man and animals (TEMA-4). Australian Academy of Science, Canberra, pp 637–644
Bremner I, Mills CF (1981) Absorption, transport and tissue storage of essential trace elements. Phil Trans Roy Soc Lond B 249:75–89
Huebers H, Huebers E, Rummel W, Crichton RR (1976) Isolation and characterisation of iron-binding proteins from rat intestinal mucosa. Eur J Biochem 66:447–455
Humphries WR, Phillippo M, Young BW, Bremner I (1983) The influence of dietary iron and molybdenum on copper metabolism in calves. Br J Nutr 49:77–86
Hungerford DM, Linder MC (1983) Interactions of pH and ascorbate in intestinal iron absorption. J Nutr 113:2615–2622
Lappin AG (1981) Properties of copper "blue" proteins. In: Sigel H (ed) Metal ions in biological systems, vol 5, pp 15–71
Lee GM, Williams DR, Cartwright GE (1976) Role of copper in iron metabolism and heme biosynthesis. In: Prasad AS, Oberleas D (eds) Trace elements in human health and disease. Academic Press, New York, pp 373–390
Linder MC, Munro HN (1973) Iron and copper metabolism during development. Enzyme 15:111–138
Lonnerdal B, Keen CL, Ball JG, Hurley LS (1985) Zinc uptake and retention from chelates and milk fractions. In: Mills CF, Aggett PJ, Bremner I, Chesters JK (eds) Trace elements in man and animals (TEMA-5). C.A.B. Publications, Farnham Royal, in press
Mills CF (1980) Metabolic interactions of copper with other trace elements. In: Biological roles of copper. Excerpta Medica, Amsterdam, pp 49–69
Mills CF (1982) Molybdenum as a biochemical antagonist of copper. In: Barry HF, Mitchell PCH (eds) The chemistry and uses of molybdenum. Climax, Ann Arbor, pp 134–138
Mills CF (1982) Some outstanding problems in the detection of trace element deficiency diseases. Phil Trans Roy Soc Lond B 294:199–213

Sato M, Mehra RK, Bremner I (1984) Measurement of plasma metallothionein-I in the assessment of the zinc status of zinc deficient and stressed rats. J Nutr 114:1683–1689
Savin MA, Cook JD (1978) Iron transport by isolated rat mucosal cells. Gastroenterology 75:688–694
Suttle NF, Abrahams P, Thornton I (1984) The role of a soil x dietary sulphur interaction in the impairment of copper absorption by ingested soil in sheep. J Agric Sci Camb 103:81–86
Tinker PB (1981) Levels, distribution and chemical forms of trace elements in food plants. Phil Trans Roy Soc Lond B 249:41–55
Vallee BL (1979) Metallothionein: historical review and perspectives. In: Kägi JHR, Nordberg M (eds) Metallothionein. Birkhäuser, Basel, pp 19–46
Weisenberg E, Halbreich A, Mager J (1980) Biochemical lesions in copper-deficient rats caused by secondary iron deficiency. Biochem J 188:633–641
Williams DM, Scott-Kennedy F, Green BG (1983) Hepatic iron accumulation in copper-deficient rats. Br J Nutr 50:653–660
Williams RJP (1981) Physico-chemical aspects of inorganic element transfer through membranes. Phil Trans Roy Soc Lond B 294:57–74
Wise A (1983) Dietary factors determining the biological activities of phytate. Nutr Abstr Rev: Rev Clin Nutr 53:791–806

The Importance of Chemical "Speciation" in Environmental Processes,
eds. M. Bernhard, F.E. Brinckman, and P.J. Sadler, pp. 85–97.
Dahlem Konferenzen 1986. Springer-Verlag Berlin, Heidelberg
© Dr. S. Bernhard, Dahlem Konferenzen

The Interaction of Inorganic Species with Biomembranes

R. J. Mehlhorn

Membrane Bioenergetics Group
Lawrence Berkeley Laboratory
Berkeley, CA 94720, USA

Abstract: The interaction of inorganic species with membranes can play a significant role in their biological effects. To the extent that membranes prevent the entry of these species into cells, they will inhibit damaging effects. However, many interactions with membranes can result in direct damage or, more seriously, can lead to an amplification of toxicity. Direct damage can include structural alteration such as a disruption of vital lipid-protein associations or introduction of new surface charges. More substantial damage can occur when inorganic species impair membrane function catalytically, e.g., by providing ion conductance pathways or by promoting free-radical reactions which may cause massive oxidation of membrane lipids. Recognition of factors involved in the interaction of compounds with membranes may be useful in devising meaningful strategies for the determination of inorganic species in biological environments.

Introduction

Biological cells are surrounded by membranes which usually prevent the entry of toxic molecules into the cell and which control the flow of nutrients and ions between the external fluid and the cytoplasm. Other functions of membranes include providing an essentially two-dimensional matrix where collisions amongst molecules will occur more frequently than in aqueous environments, providing a nonpolar environment where organic chemical reactions can occur, providing a low dielectric region which allows membranes to act as electrical capacitors, and providing a macromolecular assembly which serves as an anchoring surface for structural elements of cells. Mammalian cells contain numerous subcellular compartments whose membranes usually account for the bulk of the total membrane content of the cell. The interior of a membrane is hydrophobic and this enables it to serve as a barrier to the passage of polar molecules. Among the toxic substances that cells seek to avoid are most heavy metal ions (e.g., mercuric ions), coordination complexes of metal ions (e.g., amino acid complexes with certain transition metals), and organometallic complexes (e.g., methylmercury). One purpose of this review is to consider the known permeability properties of membranes so that the potential toxicity of inorganic species can be assessed in terms of their ability to enter cells, and to provide a framework for identifying classes of compounds which are the most interesting candidates for speciation study.

Inorganic complexes can exert deleterious effects on cells through a variety of mechanisms. They can replace essential structural components because they mimic them, as in the case of arsenate which closely resembles phosphate and can therefore undergo similar metabolic transformations or selenium which can sometimes substitute for sulfur. Another toxicity mechanism is catalytic activity, particularly for certain transition metal ions and some of their complexes. If these can participate in electron transfer reactions with cellular reductants or oxidants, then destructive free-radical chain reactions are possible. A free-radical chain reaction with particularly destructive potential in cell membranes is lipid peroxidation. Hydrophobic or amphiphilic complexes can bind to membranes and alter structure-function relationships. Organotin compounds are known to transport anions across membranes and can cause transmembrane gradients to collapse; other organometallic species may exhibit similar activities. Many heavy metals and metalloids interact strongly with thiols and therefore have the capacity to block enzyme activities and deplete cells of glutathione and other vital sulfhydryl compounds. Metabolic processes involving organometallic complexes can liberate free-radical species into the cell and these free radicals can initiate chain reactions with considerable destructive potential. Although this list of toxicity mechanisms is not exhaustive, it points out the complexity of effects that must be considered in dealing with effects of inorganic species on biological membranes.

Membrane Permeability Properties

Membranes are usually not freely permeable to small cations and anions. These require transport systems to enter cells. Such transport systems are not always absolutely specific for a given ionic species and this lack of specificity enables toxic metal ions to enter cells. For example, in the bacterium *S. aureus*, Cd(II) enters the cell by the same transport system as does Mn(II) (Perry and Silver, 1982). Heavy metal toxicity mechanisms have been studied extensively in bacteria where mutations conferring resistance to these toxins can be conveniently characterized (Silver, 1984). Studies with cadmium-resistant and sensitive strains of *S. aureus* have shown that resistance is due to a pH gradient-driven and highly specific efflux system, which counteracts the uptake of cadmium ions through the manganese transporter (Tynecka et al., 1981). Another toxic species that enters cells by virtue of its resemblance to a required ion is arsenate, which enters bacterial cells by the phosphate pathway (Silver, 1981). Arsenate resistance in *S. aureus* and *E. coli* has also been ascribed to a specific efflux system (Silver and Keach, 1982). When transport systems cannot be utilized for the entry of toxic ions into cells, the ions may nevertheless be able to cross membranes by forming neutral complexes with organic ligands which surround the ion with a hydrophobic shell. Some ionic complexes with sufficiently large hydrophobic groups can cross the membranes as charged species, as exemplified by valinomycin.

For nonionic solutes the permeability is a function of their hydrophobicity. A reasonably quantitative relationship between permeabilities and the number of hydrogen bonds that a molecule can form as well as its content of hydrophobic residues (e.g., methylene groups) has been provided by Stein (1967). Small molecules such as glycerol cross membranes quite rapidly and such molecules cannot effectively be prevented from entering cells. Stein's formulation of permeabilities indicates that the permeability of small molecules decreases by a factor of about ten for every two hydrogen bonds that can be formed, e.g., for each OH residue. Thus, sugars, which can form a significantly larger number of hydrogen bonds than glycerol, do not enter cells at appreciable rates unless their uptake is facilitated by transport systems.

A useful technique for studying rates of solute permeability for molecules which can be spin-labeled, i.e., labeled with a paramagnetic nitroxide reporter group, is electron spin resonance (ESR) spectroscopy (Mehlhorn et al., 1982). With this technique it is possible to measure the concentrations of spin labels on both sides of a membrane interface simultaneously by perturbing the external signal with impermeable paramagnetic transition metal complexes. These complexes broaden the spectra of the nitroxides in a collision-dependent process, thus allowing a simultaneous determination of spectral intensities associated with both aqueous compartments. When this technique is applied to the measurement of uptake kinetics of a spin label, it is sufficient to add a large concentration of the paramagnetic transition metal complex to eliminate the external signal completely, and to monitor the time-dependent increase in the internal spin label signal. Manganese-EDTA has proven to be the most useful line broadening agent for these studies; it seems to be quite impermeable and broadens lines at low concentrations relative to other paramagnetic agents that have been tested.

The ESR method has been used to obtain the permeabilities of a large variety of available probes (Mehlhorn and Packer, 1983), and this information is useful for inferring some generalities about factors that control these permeabilities. Table 1 shows the uptake kinetics for a series of probes and a wide range of rates is evident. It is clear from this table and from Stein's (1967) analysis that uncharged molecules, e.g., neutral organometallic and metal ion complexes, will enter cells at a rapid rate. It is noteworthy that charged species can also enter cells provided that the charged sites are surrounded by lipophilic residues, as exemplified by the triphenylphosphonium nitroxide. The large bulk of the phosphonium ion does not prevent its entry into the vesicles. Smaller ions can also enter some membranes, perhaps by moving across as ion pairs. This seems to be the case for the CAT8 nitroxide which was observed to accumulate within cell envelope vesicles of halobacteria to the same extent in the presence and absence of a light-induced electrical transmembrane potential.

Table 1. Half-times for nitroxide uptake into envelope vesicles of *halobacterium halobium* (4 M NaCl, pH 7.0, 25 °C).

Structure	# H-bonds	Half-time
R_4-3-amino-1,2-propanediol	6	40 s
R_3-tashimik acid	8	30 min
R_4-glucosamine	10	15 hr
R_2-sulfate	–	70 min
R_2-N,N,N-trimethylammonium	–	280 d
R_2-N,N-dimethyl, N-octylammonium (CAT8)	–	10 min
R_1-triphenylphosphonium	–	2.5 min

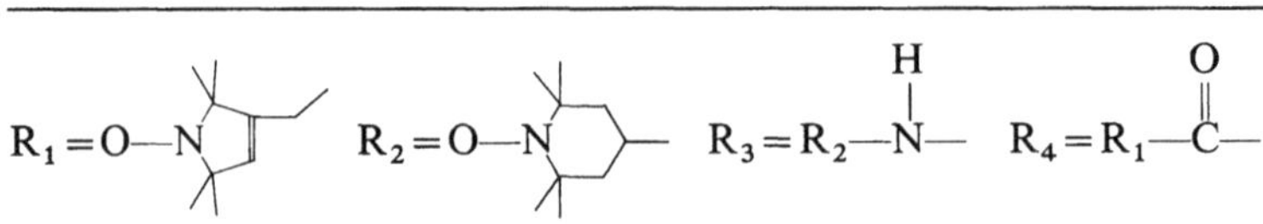

When inorganic species can cross membranes as charged complexes, then the possibility exists for substantial accumulation into compartments whose membranes maintain electrical potentials. Mitochondria generate large electrical potentials for the purpose of ATP synthesis, and these compartments can accumulate cationic inorganic species to perhaps a hundredfold higher concentration than the surrounding aqueous phase. In view of the vital role of this organelle in cell metabolism, such accumulation of a toxic substance could result in a substantial amplification of damage. In principle, metal complexes that are weak acids and amines can also be accumulated within alkaline and acidic subcellular compartments, respectively, provided that these residues are not strongly tied up as metal ligands and provided that the metal complex is sufficiently membrane-permeable in the uncharged state.

The ESR technique has also shown promise for the measurement of permeabilities of transition metal complexes which broaden nitroxides. In this approach, the leakage of the paramagnetic transition metal into a cell or vesicle is observed as a decrease of the internal nitroxide spectral line heights as the lines become broadened. In preliminary studies we have observed that amino acid complexes such as copper-valine cross the membranes of halobacterial membranes at a slow but easily measurable rate.

Interactions with Membranes

The interaction of a metal or organometallic complex with a biomembrane will depend on its hydrophobicity as outlined in the previous section. If the species is nonpolar it will become solubilized in the hydrophobic center of the membrane and usually disrupt the ordered structure of the lipids, possibly altering vital lipid-protein interactions. The effect of small hydrophobic molecules on membrane structure has been generally characterized as a fluidity increase. Many studies have shown that such fluidity increases alter membrane function, e.g., activities of membrane-bound enzymes are increased or decreased (Sauerheber et al., 1982). However, gross membrane integrity, as reflected in a capacity to serve as a permeability barrier, seems to be retained in the presence of significant concentrations of such structural perturbers.

The notion that perturbations of membranes by exogenous lipid-soluble solutes cause impaired excitability of membrane proteins forms the basis of the unitary hypothesis of general anesthesia. This hypothesis is consistent with a considerable body of evidence about the efficacy of small, structurally diverse molecules in causing anesthesia. The list of these molecules includes the gas xenon, halogenated hydrocarbons, simple alcohols, and ethers. Indeed, the only requirement for activity appears to be hydrophobicity. Thus, the potency of these anesthetics correlates quite well with their oil-water partitioning, except for some very hydrophobic compounds (e.g., tetradecanol) which are inactive. A hallmark of these general anesthetics is the relatively high concentrations that are required to elicit a physiological effect: at least millimolar concentrations in both lipids and water. Binding studies in lipid bilayer model membranes have shown that these aqueous concentrations correspond to a mole fraction of about 0.04 of the anesthetic relative to the lipid, a rather high membrane content of the perturbing molecule (Janoff and Miller, 1982). This can be taken to imply that comparably high concentrations of a perturbing inorganic species in a membrane would be required to cause significant effects on membrane function.

Large rigid hydrophobic metal compounds have the potential to decrease the fluidity of membranes. The most widely studied example of a molecule that decreases membrane fluidity is cholesterol. The addition of cholesterol to phospholipids in model membranes renders them more rigid, and high cholesterol concentrations eliminate phase transitions. Lateral diffusion of molecules within the plane of the membrane is also decreased upon cholesterol addition. Such fluidity decreases are unlikely to impair membrane integrity at low concentrations of a rigid inorganic species.

Polyvalent cations and amphiphilic ionic compounds will alter the surface charge and the fluidity characteristics of membranes when they become bound. The extent of binding of amphiphilic species will depend on the size of the hydrophobic groups. It has been shown that strong ionic detergents have a potent inhibitory effect on the respiratory function of mitochondria due to an alteration of membrane surface charge rather than to the fluidizing effects of the detergents (Mehlhorn and Packer, 1976). Most membrane surfaces bear a net negative charge which usually causes membranes to repel each other, to attract cations, and to bind peripheral proteins electrostatically. Thus, it is not surprising that compromised charge properties have serious effects on membrane integrity. This suggests that amphiphilic ionic metal complexes may exert a more destructive direct effect on membranes than neutral species. Polycations bind to membrane surfaces, including lipid headgroups, and also alter surface charge. In thus binding, they may also disturb regulatory functions of calcium or block critical transport systems for cations across membranes.

Another serious threat to cell survival is posed by complexes of inorganic species which profoundly alter membrane function, e.g., by stimulating ion exchange. An important case in point is the action of of triaryl- and trialkyltin complexes. These have been shown to be potent anion exchangers across membranes. As expected, the efficacy of these compounds increases with their lipid solubility (e.g., tributyltin is much more potent than triethyltin in red blood cells), mitochondria, and artificial membrane vesicles (Selwyn et al., 1970). The predominant anion in many biological environments is chloride, and this anion is readily transported across membranes as a neutral complex with the organotin cation. Hydroxide is also carried across membranes by this mechanism, and the presence of the organotin compound therefore causes a rapid dissipation of transmembrane pH gradients as hydroxide anions are exchanged for the much more abundant chloride anions. The results of this process can be disastrous for subcellular compartments which require the maintenance of acidic or alkaline conditions, e.g., the lysosomes, where acidic pH is required for activation of digestive enzymes. One assumes that this well characterized mode of action of trialkyl- and triaryltins will also occur with other organometallic compounds which will tend to form partially covalent bonds with halides, resulting in lipid-soluble neutral species. Data are needed on the effects of representative organometallic compounds, and speciation efforts should concentrate on those having particularly potent biological effects.

Oxidation of Membranes

Perhaps the greatest hazard posed to membranes by metal complexes is the initiation of oxidative damage via free-radical processes. Recent research on mechanisms of biological damage has suggested that free-radical processes, particularly those involving so-called "active oxygen molecules," play a major role in toxicology, disease, and aging (Cavalieri and Rogan, 1984; Pryor, 1984). There are many chemical species that can wreak havoc with living cells, but free radicals are considered to be especially damaging because of their unique chemistry, i.e., their ability to participate in chain reactions which, once initiated, have the potential to spread rapidly among membranes and macromolecules.

It has become clear that many cellular redox processes liberate superoxide radicals into the cellular environment and that these are rapidly dismutated to hydrogen peroxide. In principle, the existence of superoxide radicals and hydrogen peroxide is sufficient to lead to the production of the extremely destructive hydroxyl radicals, provided that traces of iron or copper ions are available for the Fenton reaction (Halliwell and Gutteridge, 1984).

The Fenton reaction has received considerable attention among biologists recently. In this process either Fe(II) or Cu(I) ions react with hydrogen peroxide or organic peroxides to generate extremely reactive hydroxyl or alkoxyl radicals (Eq. 1).

$$Fe^{2+} + H_2O_2 \rightarrow Fe^{3+} + OH^- + OH^\cdot \quad (1)$$

$$O_2^{\bar{\cdot}} + Fe^{3+} \rightarrow O_2 + Fe^{2+} \quad (2)$$

The regeneration of the reduced metal ion by superoxide radicals (Eq. 2) can be accomplished equally well by other reductants such as thiols and ascorbic acid which are abundant in many cells. The hydroxyl radical will react with virtually all biological targets, including DNA, proteins, and membrane lipids. It is well established that hydroperoxides are formed in aerobic cells, and attention has now shifted to the issue of whether transition metal ions are available in complexes which are compatible with the Fenton reaction. Considerable research is in progress to ascertain whether "loosely bound" iron or copper exists in cells under physiological conditions. A highly sensitive assay has been developed for the detection of iron in biological fluids (Gutteridge et al., 1981). This assay is based on the ability of bleomycin to form a complex with iron which can degrade DNA in the presence of a reductant. However, some iron complexes, such as the EDTA complex, are potent catalysts of the Fenton reaction and yet would not be expected to be detectable in this assay. Additional analytical schemes will be required to identify other transition metal complexes whose toxicity may reside in a capacity to promote free-radical reactions.

One consequence of oxidative processes in aerobic cells is lipid peroxidation. This process begins when some reactive free radical such as $OH^\cdot$ enters a biological membrane and strips a hydrogen atom from a lipid alkyl chain. Under aerobic conditions O_2 adds to the alkyl radical. The resulting peroxyl radical ($RO_2^\cdot$) abstracts a hydrogen atom from another lipid radical and a chain reaction ensues. Under normal conditions the resulting fatty acid hydroperoxides are probably reduced to alcohols by the protective enzyme glutathione peroxidase. However, if hydrophobic transition metal complexes are present in membranes, these could react with such lipid hydroperoxides to promote an intramembrane Fenton reaction, leading to the production of alkoxy radicals which could initiate new chain reactions. Such chain branching reactions can, in principle, greatly accelerate the process of lipid peroxidation. It is conceivable that organic complexes of other transition metal ions, e.g., nickel, could also catalyze intramembrane chain reactions. Therefore it is of interest to develop highly sensitive methods for the determination of metal species within membranes to identify other potentially damaging redox mediators.

Recently evidence has begun to accumulate showing that the microsomal enzyme cytochrome P450 acts on organometallic compounds as well as drugs, and that products of these reactions initiate lipid peroxidation (Prough et al., 1981; Wiebkin et al., 1982). The cytochrome P450 system oxidizes hydrophobic molecules to polar derivatives which are more readily excreted by the body, but occasionally this detoxification mechanism produces a more toxic chemical species rather than a less toxic one. Apparently this situation can occur when the enzyme reacts with organometallic species such as tetraethyllead or tetrabutyltin. These species are relatively innocuous until they are dealkylated by the cytochrome P450 enzyme. The products of dealkylation include species which can disrupt membrane ion gradients via anion exchange and species which have been shown to initiate lipid peroxidation.

The sequence of events resulting from the metabolism of trialkyltin complexes has been particularly well documented. Tetraalkyltins are converted to much more destructive trialkyltin and hydroxylated products. These species cause severe toxicity in animals, resulting primarily from damage to the central nervous system. The toxic effects have been ascribed largely to the inhibitory effects of trialkyltins on mitochondrial functions, including the anion-exchange phenomena discussed earlier, as well as a direct inhibition of the mitochondrial ATP synthetase complex (Aldridge et al., 1977). In addition, the metabolism of these compounds causes membrane lipid peroxidation in microsomes and in isolated hepatocytes (Prough et al., 1981; Wiebkin et al., 1982), and conversion of cytochrome P450 to the inactive P420 species (Prough et al., 1981; Rosenberg and Drummond, 1983). It has been suggested that this lipid peroxidation is initiated by alkyl radicals which arise in the process of reductive metabolism of the alkyltin species (Prough et al., 1981).

The occurrence of lipid peroxidation has been detected by an interesting noninvasive assay of hydrocarbons (e.g., ethane and pentane) in the expired breath of animals. These hydrocarbons are thought to result from the decomposition of fatty acid peroxyl radicals arising from lipid peroxidation. With this assay, the occurrence of lipid peroxidation has been inferred in rats that were injected with toxic doses of sodium selenite, but not with selenate (Dougherty and Hoekstra, 1982). When the animals' diets were supplemented with both the antioxidants vitamin E and selenium (at relatively low concentrations) the extent of lipid peroxidation was lower. A well-known effect of selenium is its reaction with thiols, resulting in their depletion from tissues. Such thiol depletion could weaken cellular antioxidant defenses. However, impaired protection cannot explain the lipid peroxidation, since it was found that lipid peroxidation could be demonstrated prior to the onset of significant losses of glutathione.

Ganther (1978) advanced the hypothesis that methylmercury toxicity is due primarily to free-radical processes. He cited the protective effects of vitamin E and selenium in support of this hypothesis. Since methyl radicals are produced in a simple reaction mixture containing methylmercury, ascorbate, and copper, he suggested that oxygen radicals arising in the course of aerobic metabolism may suffice to cause scission of the carbon-mercury bond in cells. This hypothesis explains the significant observation that much lower levels of selenium than those of the administered methylmercury can eliminate toxic symptoms and is also entirely consistent with a role of lipid peroxidation in methylmercury poisoning.

An example of lipid peroxidation is seen in mercuric chloride toxicity (Yonaha et al., 1982). In rats, $HgCl_2$ induces acute renal failure, which results from extensive membrane destruction following lipid peroxidation (Gstrauenthaler et al., 1983). However, in this case lipid peroxidation appears to be related to some effect of mercury on thiols. A major involvement of thiol destruction in the damage process is suggested by substantial losses of glutathione following $HgCl_2$ administration to rats (Gstrauenthaler et al., 1983). Thiol reactions are common for "soft" or class "b" metal ions such as mercury. Generally, thiol loss has been deemed to exert its destructive effect directly through the loss of crucial enzymes; however, it is now clear that free-radical oxidative processes must also be considered in seeking an understanding of metal toxicity mechanisms. A further effect of $HgCl_2$ in promoting oxidative damage is its inhibitory action on cellular antioxidant systems. In studies of $HgCl_2$ poisoning in rat kidneys, the enzymes which guard against superoxide and hydrogen peroxide accumulation, i.e., superoxide dismutase, catalase, and glutathione peroxidase, suffered activity losses of 48%, 32%, and 25%, respectively, 24 hours after subcutaneous administration of $HgCl_2$ (Gstrauenthaler et al., 1983).

Lipid peroxidation has also been implicated in the toxicity of lead. Measurements of pentane and ethane in the breath of rats which were fed lead-containing diets indicated that this metal promotes lipid peroxidation (Sifri and Hoekstra, 1978). Furthermore, dietary supplements of vitamin E and selenium decreased the concentrations of expired ethane.

Cadmium damage to testes also appears to involve lipid peroxidation (Omaye and Tappel, 1975). However, in experiments where the effects of cadmium on rat lung were studied, a protective response to selenium was clearly evident, yet no indications of lipid peroxidation could be found. In fact, the traditional assays of TBA-reactive materials revealed unexpected decreases in the cadmium-treated animals.

Silver, administered as Ag(I) salts, causes a decrease in glutathione peroxidase levels in animals (Wagner et al., 1975), and thus silver would be expected to cause increased susceptibility to oxidative damage. Indeed, animals deficient in vitamin E and in selenium exhibit greater susceptibility to silver, which can be counteracted by nutritional supplements with these and other antioxidants (Bunyan et al., 1968).

A commonly observed consequence of lipid peroxidation is protein polymerization (cross-linking). This can be a result of the combination of two free-radical residues to produce a covalent bond, but a more likely process is disulfide formation as thiols undergo oxidation, e.g., by reacting with lipid hydroperoxides. Such cross-links may be difficult to repair because the new bonds may be deeply imbedded within the hydrophobic membrane core where they may not be accessible to reduction by thiols. Thiols are also susceptible to cross-linking by more direct effects of inorganic species, as in the formation of selenotrisulfides (Ganther, 1968).

Conclusion

A great variety of destructive effects arising from the interaction of inorganic species with membranes can be envisioned. Of the many species that may interact with membranes and that are significant in the environment, only a few have been adequately characterized. However, the available data on toxicity mechanisms provide important clues about general characteristics of classes of inorganic species which may promote membrane destruction and which therefore warrant special scrutiny.

Acknowledgements. Supported by NIH AG-04818 and the Department of Energy.

References

Aldridge WN, Casida JE, Fish RH, Kimmel EC, Street BW (1977) Action on mitochondria and toxicity of metabolites of tri-n-butyltin derivatives. Biochem Pharmacol 26:1997–2000

Bunyan et al. (1968) Vitamin E and stress. I. Nutritional effects of dietary stress with silver in vitamin E deficient rats. Br J Nutr 22:165

Cavalieri EL, Rogan EG (1984) One and two electron oxidation in aromatic hydrocarbon carcinogenesis. In: Pryor WA (ed) Free radicals in biology, vol VI. Academic Press, New York, pp 323–369
Dougherty JJ, Hoekstra WG (1982) Stimulation of lipid peroxidation *in vivo* by injected selenite and lack of stimulation by selanate (41333). Proc Soc Exp Biol Med 169:209–215
Ganther HE (1968) Selenotrisulfides. Formation by the reaction of thiols with selenious acid. Biochemistry 7:2898–2905
Ganther HE (1978) Modification of methylmercury toxicity and metabolism by selenium and vitamin E: possible mechanisms. Envir Health Persp 25:71–76
Gstrauenthaler G, Pfaller W, Katanko P (1983) Glutathione depletion and *in vitro* lipid peroxidation in mercury or maleate induced acute renal failure. Biochem Pharmacol 32:2969–2972
Gutteridge JMC, Rowley DA, Halliwell B (1981) Superoxide-dependent formation of hydroxyl radicals in the presence of iron salts. Detection of "free" iron in biological systems by using bleomycin-dependent degradation of DNA. Biochem J 199:263–265
Halliwell B, Gutteridge JMC (1984) The role of transition metals in superoxide-mediated toxicity. In: Oberley LW (ed) Superoxide dismutase, vol III. CRC Press, Boca Raton
Janoff AS, Miller KW (1982) A critical assessment of the lipid theories of general anaesthetic action. In: Chapman D (ed) Biological membranes, vol IV. Academic Press, New York, pp 417–476
Mehlhorn RJ, Candau P, Packer L (1982) Measurements of volumes and electrochemical gradients with spin probes in membrane vesicles. Meth Enzymol 88:751–762
Mehlhorn RJ, Packer L (1976) Inactivation and reactivation of mitochondrial respiration by charged detergents. Biochim Biophys Acta 423:382–397
Mehlhorn RJ, Packer L (1983) Bioenergetic studies of cells with spin probes. Ann NY Acad Sci 416:180–189
Omaye ST, Tappel AL (1975) Effect of cadmium chloride on the rat testicular soluble selenoenzyme, glutathione peroxidase. Res Comm Chem Pathol Pharmacol 12:695–711
Perry RD, Silver S (1982) Cadmium and manganese transport in *Staphylococcus aureus* membrane vesicles. J Bacteriol 150:973–976
Prough RA, Stallmach MA, Wiebkin P, Bridges JA (1981) The microsomal metabolism of organometallic derivatives of the group IV elements, germanium, tin and lead. Biochem J 196:763–770
Pryor WA (1984) Free radicals in autooxidation and in aging. In: Armstong D (ed) Free radicals in biology and in aging. Raven Press, New York
Rosenberg DW, Drummond GS (1983) Direct *in vitro* effects of bis(tri-n-butyl)tinoxide on hepatic cytochrome P450. Biochem Pharmacol 32:3823–3829
Sauerheber RD, Esgate JA, Kuhn CE (1982) Alcohols inhibit adipocyte basal and insulin-stimulated glucose uptake and increase the membrane lipid fluidity. Biochim Biophys Acta 691:115–124
Selwyn MJ, Dawson AP, Stockdale M, Gains N (1970) Chloride-hydroxide exchange across mitochondrial, erythrocyte and artificial lipid membranes mediated by trialkyl and triphenyltin compounds. Eur J Biochem 14:120–126

Sifri M, Hoekstra WG (1978) Effect of lead on lipid peroxidation in rats deficient or adequate in selenium and vitamin E. Fed Proc Am Soc Exp Biol 37:757
Silver S (1981) Mechanisms of bacterial resistances to toxic heavy metals: arsenic, antimony, silver, cadmium, and mercury. In: Environmental speciation and monitoring needs for trace metal-containing substances from energy-related processes. National Bureau of Standards Special Publ 618:301–314
Silver S (1984) Bacterial transformations of and resistance to heavy metals. In: Nriagu JO (ed) Changing metal cycles and human health. Dahlem Konferenzen. Springer, Berlin Heidelberg New York Tokyo, pp 199–233
Silver S, Keach D (1982) Energy-dependent arsenate efflux: the mechanism of plasmid-mediated resistance. Proc Natl Acad Sci USA 79:6114–6118
Stein WD (1967) The movement of molecules across cell membranes, ch. 3. Academic Press, New York, pp 65–125
Tynecka Z, Gos Z, Zajak J (1981) Energy-dependent efflux of cadmium coded by a plasmid resistance determinant in *Staphylococcus aureus*. J Bacteriol 147:313–319
Wagner PA, Hoekstra WG, Ganther HE (1975) Alleviation of silver toxicity by selenite in the rat in relation to tissue glutathione peroxidase. Proc Soc Exp Biol Med 148:1106
Wiebkin P, Prough RA, Bridges JW (1982) The metabolism and toxicity of some organotin compounds in isolated rat hepatocytes. Toxicol Appl Pharmacol 62:409–420
Yonaha M, Ohbayashi Y, Ichinose T, Sagai M (1982) Lipid peroxidation stimulated by mercuric chloride and its relation to toxicity. Chem Pharm Bull 30:1437–1442

The Importance of Chemical "Speciation" in Environmental Processes,
eds. M. Bernhard, F. E. Brinckman, and P. J. Sadler, pp. 99–118.
Dahlem Konferenzen 1986. Springer-Verlag Berlin, Heidelberg
© Dr. S. Bernhard, Dahlem Konferenzen

Does the Coordination Environment Determine the Reactivity of Metals in Enzymes?

M. Zeppezauer and W. Maret

Fachbereich 15.2, Analytische und Biologische Chemie
Universität des Saarlandes
6600 Saarbrücken, F. R. Germany

Abstract. The control of the catalytic properties of metal ions in enzymes is discussed through three illustrations: the zinc enzymes carbonic anhydrase, carboxypeptidase, and alcohol dehydrogenase. They represent examples of the variation in reactivity of zinc ions placed in different biological environments leading to well-defined molecular species with different catalytic functions.

Two levels of speciation are discussed:

1) basic chemical characteristics of a species by choice of the number, the chemical nature and geometric arrangement of protein side chains binding the metal; and

2) modulation of the metal species reactivity by structural changes induced by substrate and/or coenzyme binding which, *inter alia*, may alter the geometry, accessibility, and polarity of the functional metal-binding site.

Thus, in general, the catalytic cycle is driven by a subtle interplay between all the constituents of the system, *viz.* protein metal, substrate, coenzyme, and water. Carbonic anhydrase and carboxypeptidase as reactive species catalyze reactions involving water and their catalytic metal centers are formed by hard ligands. Carbonic anhydrase is a rigid molecular system without any apparent tendency for changes in the protein structure upon binding of substrate. For carbonic anhydrase a combination of spectroscopic and kinetic methods has provided a detailed picture of coordination number, geometry, and ionization state of the metal-bound water and the location of the substrates. A mechanism consistent with most experimental observations has been derived.

Carboxypeptidase is a rigid protein which displays a selective mobility of functional side chains (e.g., tyrosine 248) upon substrate binding. Proposed catalytic functions of its metal are coordination and polarization of carbonyl groups of substrate, simultaneous binding of carbonyl and OH-groups, and concerted reaction. Metal substitution has provided a similar wealth of information as for carbonic anhydrase. There remain certain doubts concerning the coordination number of the free enzyme and the assignment of a pK_a around 9 to the metal-bound water molecule and, consequently, regarding details of the catalytic mechanism.

Alcohol dehydrogenase occurs as a dimeric or tetrameric protein with two zinc-binding sites per monomer, one being involved in catalysis. Two conformational states are observed, an open conformation without and a closed conformation with bound coenzyme. The metal environment and the electronic state of the catalytic metal ions are affected by this change of state. This has been visualized through the selective replacement of the catalytic zinc ions by transition metals.

For alcohol dehydrogenase, the problem of assigning exchanged metal ions, e.g., Co(II) and Cd(II), has only recently been solved. Both metal-binding sites offer particularly interesting examples of chemical species reactivity due to the somewhat unusual coordination of zinc to predominantly soft ligands. Transition metal substitution has clearly demonstrated the far-reaching analogy between the noncatalytic site and the iron-binding site of rubredoxin, whereas a convincing analogy has been established between the catalytic site and the copper-binding site of the so-called "blue" or type one copper proteins. Furthermore, the behavior of the catalytic metal ion as a Lewis acid has been supported by the spectroscopic properties of ternary complexes between enzyme, NADH, and the chromophoric substrate trans-4-n,n-dimethylaminocinnamaldehyde. By rapid scanning-stopped flow spectrophotometry the occurrence of at least four transient intermediates along the catalytic pathway has been demonstrated. Criteria have been developed to identify both the binding of anionic ligands to the catalytic metal ion and the effects of conformation change on it by optical spectroscopy. Thus both levels of chemical speciation have become accessible to spectroscopic analysis in the case of Co(II)-horse liver alcohol dehydrogenase.

Introduction

There is consensus among inorganic biochemists that a certain metal species (e.g., the Zn(II) ion) will exhibit a reactivity which depends markedly on its coordination environment in an enzyme or, more generally, in a metalloprotein. This holds true for all kinds of metal-binding sites in proteins, including both catalytic and noncatalytic (e.g., structural) sites. It is, however, much easier to assess a certain metal ion's reactivity in different *catalytic* sites, since these are loci of dynamic chemical processes linked to the presence and the particular chemical and physical properties of a metal ion. Both the chemical reactants and the metal itself – or other suitable metal ions replacing the native metal – may offer suitable probes for assessing the reactivity of the metal ion in its site. A *noncatalytic* metal site may be essential for the stability of the protein structure and thus for the proper function of the protein, e.g., as an enzyme. On the other hand, the coordination properties of a metal ion in a structural site are also dependent on the structure of the protein. In fact similar metal-binding sites in a particular metalloprotein class obtained from different biological species may differ in their chemical reactivity due to changes in the protein structure. A metalloprotein in which the native metal ion has been replaced artificially by another one represents a novel molecular species and a novel source of information by comparing its altered reactivity to that of the original (or native) species. While the question posed in the title of this paper may simply be answered yes, the difficult task remains to show *how* the coordination environment determines the reactivity of a metal in a given protein. The difficulty, in turn, may be pinpointed as the problem of *defining the term coordination environment:* Is it the set of ligands coordinating the metal ion or is it the entire protein molecule? In other words, in which kinds of chemical species do metal ions occur in metalloenzymes? This task is by no means easy and constitutes a primary focus of efforts in contemporary inorganic biochemistry.

Definitions and General Aspects

The subject of this review will be a discussion of the various factors controlling the catalytic properties of metal ions in enzymes. When discussing the influence of the coordination environment on a metal's reactivity, we must not forget that the interactions between metal ions and protein side chains are reciprocal ones, i.e., the metal ion is also able to modulate the reactivity of protein residues which may or may not participate in catalysis. This is a general principle irrespective of the strength of interactions between the metal ion, the protein ligands, the cofactors and/or substrates. With respect to the different strengths of binding, it is customary to distinguish between metalloenzymes with tightly bound metal ions and metal-activated enzymes with weakly bound metal ions ($K_d \geq 10^{-7}$ M).

Metalloenzyme Species

We shall discuss a few zinc metalloenzymes as examples of changes in metal species reactivities; among the more than hundred zinc proteins described in the literature, the enzymes carbonic anhydrase, carboxypeptidase, and alcohol dehydrogenase are particularly well suited to illustrate the intimate relationship between chemical characteristics of the species, its catalytic efficiency, and biological function of metals in living systems. Two aspects of speciation have to be discussed:

1. Basic chemical characteristics of a species are defined by the number, the chemical nature, and the gross geometric arrangement of the amino acid side chains liganding the metal (reactivity of the parent species – see Bernhard et al., this volume).
2. Functional characteristics of a species – During catalysis the metal's coordination environment, and hence its reactivity, is modulated by structural changes induced by substrate and/or coenzyme binding which, *inter alia*, may alter the geometry, accessibility, and polarity of the functional metal-binding site matrix or environmental effects on species reactivity (see Bernhard et al., this volume). Thus, in general the catalytic cycle of an enzyme is driven by a subtle interplay between all the constituents of the system, *viz.*, water, protein, metal, substrate, and coenzyme (if present). It is important to note that functional or matrix effects of a species, as compared to basic chemical reactivity, is related to a very different time scale since many catalytically important species are short-lived.

Identification of Chemical Species

Basic chemical characteristics of a species. For zinc proteins, X-ray structure analysis and perhaps extended X-ray absorption fine structure (EXAFS) spectroscopy seem to be the only direct methods available for obtaining structural information. It has to be kept in mind that the resolution rarely exceeds 2 Å, which means that on average the accuracy in determining the atomic positions is ±0.4 Å. In a few cases the errors have been reduced to 0.2 Å. In most cases, however, the elucidation of the three-dimensional structure of a protein by X-ray diffraction is critically dependent on the knowledge of its amino acid sequence.

Functional reactivity of a species. In a few cases, different states of an enzyme corresponding to different phases of the catalytic cycle (e.g., apoenzyme, binary or ternary enzyme complexes with bound coenzyme, substrate, or inhibitor) have been analyzed with X-ray diffraction techniques. However, most of the information regarding different functional states of metalloenzymes stems from the application of indirect methods. Metal ions, especially transition metal ions, are often valuable probes for studying structure-function relationships. Spectroscopy of equilibrium or transient species, fast kinetic studies and their pH dependencies have provided a wealth of data which render metalloenzymes the most thoroughly characterized catalysts known. In zinc enzymes, the replacement of Zn(II) by Co(II) and other transition metal ions has been most successful (Bertini and Luchinat, 1983).

Coordination Properties of Carbonic Anhydrase, Carboxypeptidase, and Alcohol Dehydrogenase

Table 1 gives a comparison of the basic chemical properties of the catalytic zinc ions in three well-known zinc enzymes: carbonic anhydrase, carboxypeptidase, and alcohol dehydrogenase. From a structural point of view, the first two proteins show similarities: they are monomeric proteins with a molecular weight of about 30 000, with one metal-binding site (i.e., the catalytic site). The metal ion is bound to hard ligands, i.e., N and O. The geometry is a distorted tetrahedron. Both catalyze reactions involving water as the reactant. Their turnover is very high (10^6 for carbonic anhydrase) or high (10^3 for carboxypeptidase). In addition to their biological function (namely, hydration of carbon dioxide and peptide hydrolysis, respectively) they hydrolyze esters *in vitro*. There is ample evidence for the participation of the metal-bound water molecule in catalysis. Alcohol dehydrogenase from liver is a dimeric enzyme built up from two identical polypeptide chains. Each subunit has one active center with one catalytic zinc ion; in addition each subunit has a noncatalytic, presumably structure-stabilizing zinc-binding site. The catalytic zinc ion binds a water molecule which can

Table 1. Basic chemical properties and catalytic function of the zinc enzymes carbonic anhydrase, carboxypeptidase A, and liver alcohol dehydrogenase.

Zinc enzymes	Molecular weight ($\times 10^{-3}$)	Number of subunits	Protein donor ligands	Ligand set[a]	Formal charge on the Zn^{2+}	k_{cat} (s^{-1})	Catalyzed reaction
Carbonic anhydrase	30	1	3 His	N_3O	2+	10^6	$CO_2 + H_2O \rightleftharpoons HCO_3^- + H^+$
Carboxypeptidase	30	1	2 His 1 Glu	N_2O_3	1+	10^3	Hydrolysis of peptides and esters
Alcohol dehydrogenase	40	2	2 Cys 1 His	S_2NO	0	10	$RCH_2OH + NAD^+ \rightleftharpoons RCHO + NADH + H^+$

[a] In the free enzymes, one water molecule in addition to the protein ligands was detected by X-ray crystallography. The carboxylate of glutamate 72 in carboxypeptidase behaves as a bidentate ligand.

be exchanged for a substrate. Alcohol dehydrogenase catalyzes the direct electron and hydrogen transfer ("hydride transfer") between organic substrate and organic cofactor molecules; the participation of water has not been determined. The catalytic metal-binding sites of this class of enzymes, which includes alcohol and sorbitol dehydrogenases, seem consistently to be composed of soft thiolate sulfur and hard metal-binding donor groups (imidazole nitrogen, water). The specific role of the metal in the catalytic mechanism cannot be easily deduced from a comparison of the basic chemical features of these three zinc enzymes. For example, the formal charge on the catalytic metal ion varies from 2+ (carbonic anhydrase) to 1+ (carboxypeptidase), to 0 (alcohol dehydrogenase). The latter two enzymes supposedly activate their substrates by direct coordination and polarization of carbonyl groups or alcohol C-0 bonds. An evaluation of this function would very much benefit from better knowledge of the *effective charge on the catalytic metal* ion in its coordination environment, which is not at present available. Similar arguments apply to the discussion of the properties of the metal-bound water molecule, in particular its acid-base properties. Thus a comparison of the structure (as far as it is available) and the basic chemical properties of an enzyme's metal-binding site is insufficient for the explanation of its catalytic properties. This becomes evident, *inter alia,* from the comparison of isoenzymes within one biological species, as in the case of carbonic anhydrases (Lindskog et al. 1971), or of the same enzyme from different biological species, such as for alcohol dehydrogenases (Brändén et al., 1975). In spite of identical coordination environments in the metal-binding site, they display greatly different reaction rates. This calls for a deeper examination of the whole system rather than of the nearest coordination environment and its time-dependent changes during the catalytic cycle.

Changes in the Metal Coordination Environment in the Active Sites of Carbonic Anhydrase, Carboxypeptidase, and Alcohol Dehydrogenase During the Catalytic Cycle

Carbonic Anhydrase

General mechanism. For carbonic anhydrase a combination of spectroscopic and kinetic methods has provided a detailed picture of coordination number and geometry of the catalytic zinc ion and ionization state of the

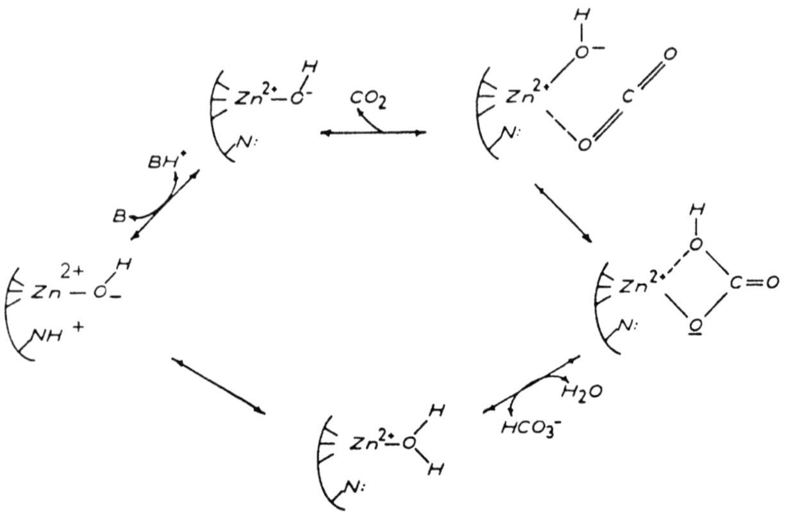

Fig. 1. The "zinc-hydroxide" mechanism of carbonic anhydrase that may dominate at high pH (From Lindskog et al., 1983, with permission)

metal-bound water molecule. A mechanism consistent with most experimental observations has been derived (Lindskog et al., 1983), based on the following sequence of elementary reactions: a) binding of H_2O; b) binding of CO_2; c) breaking of O-H bond in H_2O; d) formation of O-C bond: CO_2-HCO_3^--transformation; e) dissociation of HCO_3^-; f) dissociation of H^+. Figure 1 illustrates this pathway.

This "zinc-hydroxide" mechanism requires that the metal vary its coordination number during the catalytic cycle between four and five. No structural data are available which indicate significant changes in the protein structure upon binding of inhibitors (Lindskog et al., 1983). X-ray structure determinations of complexes between the enzyme and substrate molecules (i.e., CO_2 or HCO_3^-) have not yet been performed. Even in these species important conformation changes would not be expected. Altogether, at least five intermediate species account for the "zinc-hydroxide" mechanism. In all those species, hard acids and bases interact with the catalytic metal ion, and no gross changes in polarity and other chemical properties of the active site are expected. It has proven extremely difficult to trap and visualize all intermediate species.

Metal-substituted species of carbonic anhydrase. Carbonic anhydrase is easily depleted of its catalytic zinc ion by treatment with chelating agents at low pH. The resulting apocarbonic anhydrase is a stable protein with virtually unchanged tertiary structure, judging from various physical measurements. It can be reconstituted with a variety of divalent metal ions [e.g., Co(II), Ni(II), Cu(II), and Mn(II)] which all bind in the place of zinc in the active site. However, only Co(II) ions show activity comparable to that of Zn(II) (Bertini et al., 1982; Lindskog et al., 1983). The reasons for this behavior of the reconstituted enzyme are not well understood. Under the assumption that the protein is a very rigid matrix and that no significant conformational changes accompany the catalytic cycle, it must be concluded that the chemical properties of the catalytic metal ion in its fixed environment are decisive for the catalytic performance of the enzyme. Very few changes in these chemical properties are tolerated, which means that a high selective pressure has acted in the evolution of this particular catalyst. Nevertheless, the study of metal-substituted carbonic anhydrases of low activity has yielded much valuable information about the catalytic metal-binding site (Bertini et al., 1982).

Carboxypeptidase

Carboxypeptidase is an example of a metalloenzyme species where different functional states have been demonstrated by X-ray crystallographic (Christianson and Lipscomb, 1986) and spectroscopic methods. Binding of the slowly hydrolyzed substrate glycyl-L-tyrosine causes a movement of the side chain of tyrosine 248 such that an interaction with the bound substrate is made possible (Lipscomb, 1983). This is an example of the "induced fit" of a substrate molecule characterized by a local rearrangement of the active site without gross structural changes within the tertiary structure.

The replacement of the native zinc ion by other divalent metal ions has been a most valuable source of chemical, crystallographic (Rees et al., 1986), and spectroscopic information. In contrast to carbonic anhydrase, most metal-substituted species are enzymically active against both peptide and ester substrates (Table 2). This shows that the steric and electronic properties of the catalytic metal site are less rigorously limited for good catalytic performance than in carbonic anhydrase.

As in carbonic anhydrase, the Co(II)-derivative has been the most informative system. The catalytic mechanism has not yet been described unambiguously. There is more and more evidence accumulating which indicates that the enzyme forms a covalent intermediate between the acyl group of

Table 2. Hydrolysis of Bz-$(Gly)_2$-L-Phe and its ester analogue by various metallocarboxypeptidases at 25 °C, pH 7.5 (data from Hughes, 1981).

Metal Ion	Bz-$(Gly)_2$-L-Phe		Bz-$(Gly)_2$-L-OPhe	
	$10^2\,k_{cat}\,(s^{-1})$	$10^3\,K_m\,(mol\,dm^{-3})$	$10^{-2}k_{cat}\,(s^{-1})$	$10^4\,K_m\,(mol\,dm^{-3})$
Co (II)	1.0	0.66	6.5	3.03
Zn (II)	0.2	1.00	5.0	3.33
Mn (II)	0.04	0.36	6.0	15.2
Cd (II)	0.007	0.77	5.7	83.3

the substrate and glutamate 270 after the initial splitting of the ester bond. This is made possible by the polarizing effect of the catalytic metal ion on the carbonyl group of the ester bond. This acyl intermediate is then thought to be attacked by the metal-bound hydroxide ion. By rapid scan-stopped flow spectrophotometry at low temperatures using the Co(II)-enzyme and esters or peptides as substrates, Auld et al. detected two transient intermediates designated as ES_1 and ES_2 in a general scheme:

$$E+S \rightleftharpoons ES_1 \rightleftharpoons ES_2 \rightarrow E+P\,.$$

As shown by their optical spectra, these intermediates have different structures with respect to their catalytic metal sites; however, they are not believed to be covalent derivatives (Vallee et al., 1983). On the other hand, with 0-(trans-p-chlorocinnamoyl)-L-β-phenyllactate, spectroscopic and kinetic evidence was obtained for the formation of an acyl intermediate (Kuo et al., 1983). Finally, evidence has been presented for a chemically trapped intermediate, modified at glutamate 270 (Sander and Witzel, 1985). Thus, carboxypeptidase appears to be an enzyme with a limited flexibility of active site residues and with a metal-binding site which retains catalytic activity even after extensive chemical modification by metal replacement. Functional species reactivity is obvious both in stable and transient intermediates, but the nature of several transient species remains to be elucidated.

Alcohol Dehydrogenase

General aspects. Alcohol dehydrogenases are dimeric or tetrameric enzymes and therefore represent systems of a higher complexity than do the enzymes discussed above. In addition a coenzyme, NAD, is needed; it behaves kin-

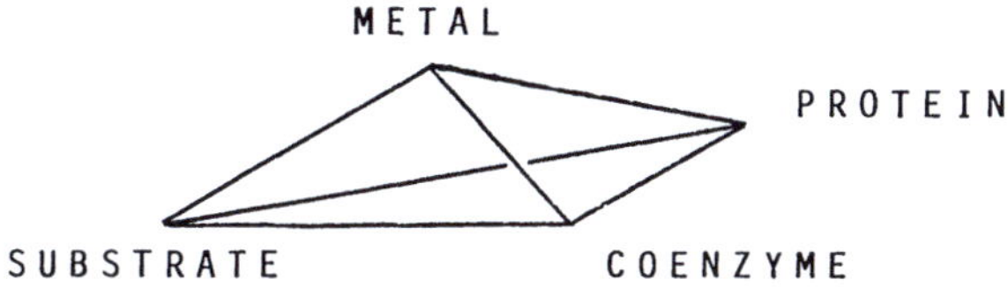

Fig. 2. Interactions in liver alcohol dehydrogenase

etically as a co-substrate. The four partners (substrate, coenzyme, metal, and protein) interact directly or indirectly in the way indicated by the tetrahedral relationship in Fig. 2.

The central step of the catalyzed reaction (cf. Table 1) is the transfer of two electrons and one proton ("hydride transfer") between primary or secondary alcohols and NAD^+. It requires the combination of enzyme, coenzyme, and substrate to a ternary complex. During the ternary complex interconversion, the redox step occurs without the participation of any protein side chain. The zinc ion is an essential but redox-inactive catalyst. It binds directly and activates the substrate (see below). According to the kinetic mechanism proposed by Theorell and Chance, the coenzyme binds first to the enzyme. In this step the protein is "conditioned" to bind more strongly and to activate the substrate. In the direction of alcohol oxidation, the formed aldehyde dissociates quickly from the ternary complex, and the dissociation of the reduced coenzyme NADH from the binary complex is rate-limiting for the enzymic turnover. A characteristic feature of the Theorell-Chance pathway is that the ternary complex interconversion is not rate-limiting, which means that the steady-state concentration of the ternary complex always remains low during catalysis. This is strictly valid only in the direction of alcohol oxidation. Concerning the various chemical interactions in the system (see Fig. 2), the Theorell-Chance pathway has several implications:

1) Only the metal and the protein remain chemically bound and interact during the entire catalytic cycle.
2) In the binary complexes enzyme/coenzyme or enzyme/substrate, only three partners interact, i.e., protein-metal-coenzyme or protein-metal-substrate.
3) In the ternary complexes the full set of interactions is operative, i.e., protein-metal-coenzyme-substrate.

In discussing the relationship between coordination environment and reactivity of the catalytic metal ion, it is useful to examine first the interactions between metal and protein in both the absence and presence of substrate or coenzyme, and subsequently the interactions between substrate, metal, and protein with and without the participation of the coenzyme.

Interactions between metal and protein. From X-ray crystallographic investigations a distorted tetrahedral coordination was established for the catalytic (c) zinc site in the enzyme from horse liver (HLADH) and for the species where the catalytic zinc ions have been replaced by Co(II)- or Cd(II)-ions (Zeppezauer et al., 1984). The noncatalytic (n) zinc ions are bound to four cysteine residues in a tetrahedral geometry. The tertiary structure of the enzyme depleted of its catalytic Zn(II)-ions, $H_4Zn(n)_2$-HLADH, is virtually unchanged, which shows that the protein provides a rigid structural framework imposing a tetrahedral geometry on all metal ions accommodated in the catalytic site (Schneider et al., 1983). This is further corroborated by the species $Ni(c)_2Zn(n)_2$-HLADH and $Cu(c)_2Zn(n)_2$-HLADH which show the optical and magnetic properties of tetrahedral Ni(II)- and Cu(II)-species with thiolate donor groups (Dietrich et al., 1981; Maret et al., 1980). The Cu(II)-substituted HLADH is therefore an excellent model for the type 1 or "blue" Cu(II)-proteins. From these studies a dominant influence by the protein on the coordination geometry of the catalytic metal site can be inferred. Little is known about the thermodynamics and kinetics of the interaction between the zinc ions and the protein structure in alcohol dehydrogenase. It has been observed that the noncatalytic Zn(II) ions, which seem to play a structure-stabilizing role, exchange faster with added Zn(II) ions (Drum et al., 1969), but neither the rate nor the activation parameters have hitherto been precisely determined. Treatment of crystal suspensions of HLADH with chelating agents, notably dipicolinic acid, selectively removes the catalytic Zn(II) ions whereas the structural ones remain *in situ* (Maret et al., 1979). Denaturation of the protein leads to a loss of all zinc, which can be removed by gel chromatography. Refolding and reassociation of the polypeptide chains requires Zn(II) ions within a critical range of low concentrations. The mode of participation of zinc in this process is not known. Stability constants have been determined neither for the structural nor the catalytic metal-binding sites. Rate and activation parameters have been determined for the recombination of the active site zinc-depleted protein with Co(II), Ni(II), and Zn(II) ions. It was found that the incoming metal ion first binds reversibly at some peripheral site from which it migrates to the catalytic binding site in a rate-limiting step. The latter step was shown to exhibit an exceptionally high negative entropy of activation as compared to the recombination of metal ions with other metal-depleted enzymes, possibly indicating a sterically restrictive transition state. The recombination process is also sensitive to the optimal metal concentration which is 20 μM for Zn(II) and 1 mM for Co(II). If $H_4Zn(n)_2$-HLADH is reconstituted with Zn(II) or Co(II) in crystal suspensions, higher metal concentrations are tolerated than in solution. Depleting the enzyme of metal in solution leads to irreversible loss of activity. This shows that the interaction of metal ions with the protein occurs at several sites which are important for the maintenance of the protein structure, and hence the function of the enzyme.

Interactions between metal, substrate, and protein. In the Theorell-Chance pathway this interaction is not kinetically dominating. In the absence of coenzyme, both alcohol and aldehyde substrates bind very weakly to the catalytic metal ion. In this state the catalytic metal ion is relatively inert. We recall from Table 1 that its formal charge is 0. Although the real charge is probably different, the catalytic metal ion binds negatively charged ligands only reluctantly. This is in striking contrast to the behavior of carbonic anhydrase. No binding to the catalytic metal ion of F^-, Cl^-, or Br^- ions was observed in either native HLADH or the species $Co(c)_2Zn(n)_2$-HLADH. In the latter enzyme, an absorption band around 575 nm characteristic for anion binding could be detected upon binding of strong donor anions (e.g., SH^-, N_3^-, CN^-, or acetate). The binding of acetate is supposedly enhanced by hydrophobic interactions. The origin of this spectral transition lies most probably in a local, geometric rearrangement of the catalytic metal-binding site as a consequence of charge compensation upon anion binding. The 575 nm band has proven to be of high diagnostic value for the characterization of transient chemical species during the catalytic cycle (see below). Although anions such as acetate may be regarded as substrate analogues, it is important to note that no spectroscopic observations of strong interactions between alcohol or aldehydes with the catalytic metal ion of HLADH or its metal-substituted species have been reported in the absence of coenzyme.

Interactions between coenzyme, protein, and metal. The binding of coenzyme has a profound influence on the protein structure, which in turn affects the properties and reactivity of the catalytic metal ion. The structural change induced by the binding of coenzyme is characterized by a domain rotation of the catalytic domain versus the coenzyme binding domain (Eklund and Brändén, 1983). The X-ray structures of this "closed" conformation have so far been obtained from complexes containing HLADH, NADH, or

NAD^+, and a suitable inhibitor such as dimethylsulfoxide with NADH and pyrazole with NAD^+. In the absence of inhibitor, molecules of the solvent used for the crystallization of the protein bind in the substrate binding cleft. Generally the coordination environment of the catalytic metal ion is much less polar in the "closed" conformation than it is in the "open" conformation. The majority of water molecules is expelled from the catalytic site and the substrate binding cleft. The latter is narrowed as a consequence of the domain rotation. The nicotinamide moiety is necessary for the induction of the domain rotation, since adenosine diphosphoribose, i.e., the coenzyme depleted of its nicotinamide part, binds to the "open" conformation of the enzyme. There are no bonding interactions between the nicotinamide moiety and the metal ion. Only Van der Waals contacts exist between the nicotinamide ring and the metal ligands, which has led to the impression that the metal ion is not important for the binding of coenzyme. In a structural sense this is true. $H_4Zn(n)_2$-HLADH, i.e., the enzyme deprived of its catalytic metal ion, binds coenzyme strongly in the "closed" protein conformation. The position and extended conformation of the bound NADH are the same as in the native complex HLADH/NADH (Schneider et al., 1983). However, the *dynamics* of coenzyme binding are affected strongly by the metal ion as demonstrated in Table 3, which shows the dependency of k_{off}, the rate constant of coenzyme dissociation, on the kind of metal ion in the catalytic site. The rates are inversely proportional to the ionic radius of the metal ions inserted into the catalytic site. It seems as though the polarity of the metal site is an important factor controlling the binding of the nicotinamide ring. The higher the ionic radius, and hence the lower the charge density at the metal, the lower the polarity around the metal-binding site and the stronger the attraction of the nicotinamide ring. Absence of the catalytic metal ion causes a dramatic, thousandfold reduction of the rate

Table 3. Comparison of the rates of dissociation of the complex Me(c)Zn(n)-HLADH/NADH for different metallo-alcohol dehydrogenases at 25 °C, pH 10, with the turnover for ethanol oxidation.

Species	Rate constant of NADH dissociation (s^{-1})	Enzymic turnover (s^{-1})
$H_4Zn(n)_2$-HLADH	5×10^{-3}	–
$Cu(c)_2Zn(n)_2$-HLADH	1×10^{-2}	Not detectable
$Cd(c)_2Zn(n)_2$-HLADH	5×10^{-1}	5×10^{-1}
Zn_4-HLADH	7	5.5
$Co(c)_2Zn(n)_2$-HLADH	18	7.0

constant, which is plausible in view of the greatly reduced polarity of the metal-depleted site.

An interesting case is provided by the insertion of Cu(II) ions in the catalytic site: the dissociation rate constant is reduced 100-fold as compared to Zn(II), which *per se* is sufficient to explain the absence of enzymic activity in this species. This is surprising in view of the ionic radius of tetrahedral Cu(II), which lies between those of Ni(II) and Zn(II). We therefore conclude that the catalytic metal-binding site in Cu(II)-HLADH is distorted similarly to the metal site in the blue copper proteins. This could create an abnormally large distance between the nicotinamide ring and the Cu(II) ion, and therefore a less polar environment for the former.

Interactions between metal, protein, coenzyme, and substrate. The direct coordination of substrate molecules to the catalytic zinc ion has been demonstrated by X-ray crystallography (Eklund and Brändén, 1983). p-Bromobenzyl alcohol was shown to bind to the zinc ion in the ternary complex HLADH/NAD^+/p-bromobenzyl alcohol and, similarly, the aldehyde substrate trans-4-N,N-dimethylaminocinnamaldehyde (DACA) was shown to form an inner-sphere bond in the complex HLADH/1,4,5,6-tetrahydronicotinamide adenine dinucleotide/DACA. The 1,4,5,6-tetrahydronicotinamide adenine dinucleotide (H_2NADH) acts as a coenzyme analogue, inducing the conformation change without participating in redox reactions. In the ternary complexes mentioned above, the protein exists in the "closed" conformation. Interestingly, the ternary complex HLADH/H_2NADH/2-methyl-2,4-pentanediol (MPD) crystallizes with the protein in the "open" conformation. MPD is a poor substrate and does not coordinate to the metal; rather it binds in the substrate binding cleft in hydrogen-bonding distance to the metal-bound water. Addition of an excess of DACA to a solution of the complex HLADH/H_2-NADH/MPD converts the complex into the ternary complex HLADH/H_2NADH/DACA. Thus it appears as if the interaction between the substrate and the metal as well as the protein does assist in the conformation change, and the probability of the protein changing from the "open" to the "closed" conformation depends to some extent on the strength with which the substrate interacts with the metal and the protein.

Aldehyde activation. As to the nature of the metal-substrate interaction, spectroscopic studies with chromophoric substrates and with metal-substituted alcohol dehydrogenases have been most illuminating. The visible spectra of DACA bound in the ternary complexes with HLADH and

NADH show a red shift from 398 nm in free DACA to 464 nm in the bound substrate which clearly illustrates the polarizing action of the metal. Further evidence for the Lewis acid function of the metal was derived from solution studies of the series of metallo-HLADHs with Cd(II), Ni(II), and Co(II) in the catalytic site (Zeppezauer et al., 1984). As shown in Table 4, the magnitude of the red shift for bound DACA is dependent on the kind of metal ion present in the catalytic site and increases in the series Cd < Zn < Ni < Co, which is consistent with decreasing ionic radii. It seems that the electrostatic interaction between the substrate molecule and the metal ion was the dominating factor in substrate activation and that ligand field effects play a less important role. A further illustration of the importance of electrostatic field effects by the catalytic metal ion and its surroundings has been given by Dahl and Dunn (1984). NAD^+ enhances the polarization of bound DACA, and hence its spectral red shift is 97 nm vs. 66 nm in the complex with NADH. Carboxymethylation of the metal ligand cysteine 46 lowers the red shift in the complexes with NADH/DACA (38 nm) and with NAD^+/DACA (60 nm). The carboxymethylated enzyme is particularly interesting since the *formal* charge at the metal increases by one unit which might be expected to enhance its polarizing strength. The observed effects are opposite and need explanation. Probably the negative charge introduced by the chemical modification counteracts the possible increase of charge density at the metal. In addition, this increase amounts to certainly less than one charge unit. These observations illustrate convincingly that the metal environment is highly important in regulating the reactivity of the catalytic metal ion and, in addition, that its electrostatic properties are modulated during the catalytic cycle.

Table 4. Dependence of absorption maxima of bound trans-4-(N,N-dimethylamino) cinnamaldehyde (DACA) on the catalytic metal ion in ternary complexes of alcohol dehydrogenase.

	Co^{2+}	Ni^{2+}	Zn^{2+}	Cd^{2+}
Me(c)Zn(n)-HLADH/NADH/DACA	478[a]	475	464	457
Red shift[b]	80	77	66	59
Me(c)Zn(n)-HLADH/NAD^+/DACA	499[c]	–	495[d]	490[c]
Red shift[b]	101	–	97	92

[a] In (nm)
[b] Bound DACA vs. free DACA ($\lambda_{max} = 398$ nm)
[c] C. Sartorius, unpublished results
[d] From Dahl and Dunn, 1984

Alcohol activation. The question as to whether the alcohol is bound as a neutral species or as a metal-coordinated alkoxide anion has been a matter of debate until now. In chemical terms, hydride transfer to NAD^+ would certainly be facilitated from a metal-bound alkoxide anion. Evidence in favor of this mechanism has been obtained from spectroscopic studies with Co(II)-HLADH. We have observed that binding to Co(II)-HLADH of chemically different, negatively charged ligands (e.g., acetate) causes a shift of an absorption band from 525 nm to 575 nm both in the free enzyme (see above) and in complexes with NAD^+. By ^{13}C-NMR spectroscopy it was shown unequivocally that acetate coordinates directly to the catalytic Co(II) ion. It is therefore conceivable that the same spectroscopic change is caused by directly coordinating anionic species. Most important, the spectroscopic features of coordinated anions have been detected both in stable complexes of Co(II)-HLADH with NAD^+ and 2,2,2-trifluoroethanol and in short-lived intermediates of the ternary complex interconversion. Rapid scan-stopped flow spectrophotometry of Co(II)-HLADH mixed with NAD^+ and alcohols or with NADH and aldehydes thus far have led to the detection of three intermediates, one of which has been identified as a complex containing NAD^+ and an anionic ligand. This species, which is observed in both directions of the redox reaction, most probably represents the complex of Co(II)-HLADH with NAD^+ and metal-bound alkoxide (Gerber et al., 1983).

Conformational changes in the ternary complex interconversion. The cobalt(II) ion probe offers the additional advantage of a red shift of the main visible absorption band at 650 to 680 nm due to the conformation change caused by the binding of reduced coenzyme. At low pH, binding of NAD^+ to Co(II)-HLADH causes only minor changes of this main band indicating that the protein stays largely in the "open" conformation. At high pH, the 650 nm band is red-shifted and split indicating that the conformation is closing. The apparent pK_a of this transition is 6.9.

In the alkaline form of this binary complex Co(II)-HLADH/NAD^+, the band at 575 nm is present suggesting the interaction of a negatively charged group with the metal ion. Whether this group is a hydroxide ion formed by ionization of the metal-bound water or another group remains to be established. It seems as though a negative group were necessary to stabilize the binding of NAD^+ in the "closed" conformation of the enzyme. Thus an important feature of the ternary complex interconversion is the concomitant formation of a metal-bound alkoxide anion and its interaction with the nicotinamide ring of NAD^+ to stabilize the productive ternary

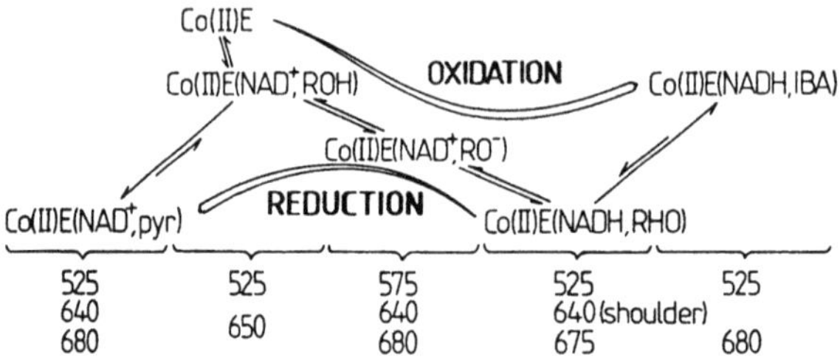

Fig. 3. Transient-spectroscopy of single-turnover runs of Co(II)-substituted horse liver alcohol dehydrogenase. Discernible species of the alcohol oxidation are Co(II)E or Co(II)E(NAD^+, ROH), Co(II)E(NAD^+, RO^-), and Co(II)E(NADH, IBA). The following species can be separated in the reduction of aldehydes: Co(II)E-(NADH, RHO), Co(II)E(NAD^+, RO^-), and Co(II)E(NAD^+, pyr). Numbers below the species indicate the wavelengths (in nm) of characteristic absorption maxima of the complexes. IBA = isobutyramide; pyr = pyrazole

complex in its "closed" protein conformation. It is important to note that most of the spectroscopic features which have been observed in equilibrium studies can be observed in the different transient intermediates by rapid scan-stopped flow spectrophotometry. The chemical characterization of these intermediates may be regarded as time-resolved species identification and prove a powerful means for the elucidation of the catalytic cycle. Figure 3 gives a scheme including the different short-lived intermediates detected hitherto by the absorption spectra of the Co(II) ion. Studies are now under way to characterize the structure of the metal ion in these intermediates which is being trapped by cryoenzymological techniques.

In conclusion, these examples demonstrate how the reactivity of the catalytic zinc sites in these enzymes is fine-tuned by interactions with the protein and effector molecules. Species identification in these systems requires a resolution of structures as a function of time.

Acknowledgement. This work was supported by the Deutsche Forschungsgemeinschaft, Bundesminister für Forschung und Technologie, and NATO.

References

Bertini I, Luchinat C (1983) An insight on the active site of zinc enzymes through metal substitution. In: Sigel H (ed) Metal ions in biological systems, vol 15, ch. 2. Marcel Dekker, New York

Bertini I, Luchinat C, Scozzafava A (1982) Carbonic anhydrase. An insight into the zinc binding site and into the active cavity through metal substitution. Struct Bonding 48:45–92

Brändén C-I, Jörnvall H, Eklund H, Furugren B (1975) Alcohol dehydrogenases. In: Boyer PD (ed) The enzymes. Academic Press, New York, pp 103–190

Christianson DW, Lipscomb WN (1986) Mechanistic inferences from the binding of ligands to the active site of carboxypeptidase A. In: Bertini I, Luchinat C, Maret W, Zeppezauer M (eds) Zinc enzymes, ch. 7. Birkhäuser, Basel, in preparation

Dahl KH, Dunn MF (1984) Reaction of 4-trans-(N,N-dimethylamino) cinnamaldehyde with the liver alcohol dehydrogenase – oxidized nicotinamide adenine dinucleotide complex. Biochemistry 23:4094–4100

Dietrich H, Maret W, Kozłowski H, Zeppezauer M (1981) Active-site-specifically reconstituted nickel(II) horse liver alcohol dehydrogenase: optical spectra of binary and ternary complexes with coenzymes, coenzyme analogues, substrates, and inhibitors. J Inorg Biochem 14:197–311

Drum DE, Li T-K, Vallee BL (1969) Zinc isotope exchange in horse liver alcohol dehydrogenase. Biochemistry 8:3792–3797

Eklund H, Brändén C-I (1983) The role of zinc in alcohol dehydrogenase. In: Spiro TG (ed) Zinc enzymes. J. Wiley and Sons, New York, pp 123–152

Gerber M, Zeppezauer M, Dunn MF (1983) Evidence for inner-sphere alkoxide ion intermediates in the catalytic mechanism of Co(II)-substituted liver alcohol dehydrogenase. Inorg Chim Acta 79:161–164

Hughes MN (1981) The inorganic chemistry of biological processes. J. Wiley and Sons, New York, p 108

Kuo LC, Fukuyama JM, Makinen MW (1983) Catalytic conformation of carboxypeptidase A. The structure of a true reaction intermediate stabilized at subzero temperatures. J Mol Biol 163:63–105

Lindskog S, Henderson LE, Kannan KK, Liljas A, Nyman PO, Strandberg B (1971) Carbonic anhydrase. In: Boyer PD (ed) The enzymes. Academic Press, New York, pp 587–665

Lindskog S, Ibrahim SA, Jonsson B-H, Simonsson I (1983) Carbonic anhydrase: structure, kinetics, and mechanism. In: Bertini I, Drago RS, Luchinat C (eds) The coordination chemistry of metalloenzymes. D. Reidel Publishing Company, Dordrecht/Holland, pp 49–64

Lipscomb WN (1983) Structure and catalysis of enzymes. Ann Rev Biochem 52:17–34

Maret W, Andersson I, Dietrich H, Schneider-Bernlöhr H, Einarsson R, Zeppezauer M (1979) Site-specific substituted cobalt(II) horse liver alcohol dehydrogenases. Eur J Biochem 98:501–512

Maret W, Dietrich H, Ruf H-H, Zeppezauer M (1980) Active site-specific reconstituted copper(II) horse liver alcohol dehydrogenase: a biological model for type 1 Cu^{2+} and its changes upon ligand binding and conformational transitions. J Inorg Biochem 12:241–252

Rees DC, Howard JB, Chakrabarti P, Yeates T, Hsu BT, Hardmann KD, Lipscomb WN (1986) Crystal structures of metallosubstituted carboxypeptidase A. In: Bertini I, Luchinat C, Maret W, Zeppezauer M (eds) Zinc enzymes, ch. 9. Birkhäuser, Basel, in preparation
Sander ME, Witzel H (1985) Carboxypeptidase A, evidence for an anhydride intermediate. Rev Port Quim 27:247
Schneider G, Eklund H, Cedergren-Zeppezauer E, Zeppezauer M (1983) Structure of the complex of active site metal-depleted horse liver alcohol dehydrogenase and NADH. EMBO J 2:685–689
Vallee BL, Galdes A, Auld DS, Riordan JF (1983) Carboxypeptidase A. In: Spiro TG (ed) Zinc enzymes. J. Wiley and Sons, New York, pp 25–75
Zeppezauer M, Andersson I, Dietrich H, Gerber M, Maret W, Schneider G, Schneider-Bernlöhr H (1984) Coordination chemistry and function of the catalytic metal ion in liver alcohol dehydrogenase. J Mol Catal 23:377–387

Standing left to right:
Donald Gamble, Fritz Prosi, Everett Jenne, Phil Kane, Jim Leckie

Seated (center), left to right:
Gerhard Brümmer, Heinz Tobschall, Alina Kabata-Pendias, German Müller, Fritz Frimmel

Seated (front), left to right:
Peter Baccini. Y. K. Chau, John Bauld

The Importance of Chemical "Speciation" in Environmental Processes,
eds. M. Bernhard, F. E. Brinckman, and P. J. Sadler, pp. 121–147.
Dahlem Konferenzen 1986. Springer-Verlag Berlin, Heidelberg
© Dr. S. Bernhard, Dahlem Konferenzen

Chemical Species in Freshwater and Terrestrial Systems

Group Report

E. A. Jenne, Rapporteur

P. Baccini
J. Bauld
G. W. Brümmer
Y. K. Chau
F. H. Frimmel
D. S. Gamble
A. Kabata-Pendias
P. F. Kane
J. O. Leckie
G. K. Müller
F. P. Prosi
H. J. Tobschall

Introduction

Group discussions were limited to metals and certain metalloids, except that N, P, and S were considered in the broader topic of biogeochemical cycles. In many instances, the statements concerning metals are also applicable to many, or even most, other elements. However, our discussions were focused on the metals and metalloids because of their major importance as environmental pollutants.

The distribution of a metal* between its oxidation states and among its various aqueous complexes in pore waters and surface waters exerts a very major influence upon its biogeochemical cycling. In particular, the distribution of the aqueous species of an element strongly influences its availability in pore waters, i.e., the extent to which it partitions among the solid, aqueous, and gaseous phases of soil or aquatic systems. The biouptake, i.e., the

* We found the common metal descriptors ("trace," "heavy," "toxic") inadequate for our purposes. By use of the term metal we infer first transition series elements as well as those in the following two rows of the periodic table excepting alkalis (IA), alkaline earths (IIA), and the last four columns. Metalloids include As, Se, and Sb.

tendency of that element to be accumulated by organisms, is similarly affected. Numerous biological, chemical, and physical processes are responsible for the observable distribution of a metal among its various gaseous and aqueous species as well as its solid forms. Various chemical, physical, and instrumental methods are used to determine a given species of a metal. Thus, the "speciation" of metals or metalloids or of organic compounds, as the term is used among scientists today, encompasses three aspects: a) the actual distribution among molecular level entities and forms; b) the processes responsible for an observable distribution; and c) the methods used to identify and quantify individual species and/or the distribution of species. It is noted that determination of the distribution of species of an element requires that both appropriate analytical methods and thermodynamic and/or kinetic calculations be used in parallel.

Up to now analysis of the species of metals, at the molecular level, in the solid phases of soils and sediments has not been carried out where these elements occur as minor constituents of minerals, amorphic inorganic solids, or as organic compounds. The difficulties in determining species at the molecular level in solids has led to the widespread use of "selective" extractants to estimate the quantity of a metal present in one of its forms (e.g., in amorphic iron oxides). Since interfacial characteristics are likely to be badly disrupted due to changes in oxidation state, dissolution, or precipitation/deposition during extraction, it is quite probable that observed metal releases during extraction sequences are derived from a variety of binding configurations that are each kinetically controlled. A sequential extraction procedure itself may cause changes in the solid phase forms and may lead to the formation of metals forms not originally present. As noted later, the selectivity of these methods for a particular form is rather low.

Extraction procedures may permit useful estimation of the capacity of the solid phase to resupply the aqueous phase, although in the past such methods have required extensive empirical calibration studies, e.g., paired field fertility and laboratory extraction studies.

In the longer term, the use of partially selective extractants to estimate the quantity of the more important sorption substrates (amorphic Fe and Mn oxides and reactive particulate organic carbon) may turn out to be a more valuable usage.

Since the methods applied to date rarely yield information at the molecular level in the solid phases, we use the more general term "form", rather than species, when referring to metals in soils and sediments. In contrast, available methods do yield molecular level information in plants and animals.

This panel was concerned with the current state of knowledge and with problems relating to the beneficial and adverse aspects of the availability and bioaccumulation of selected metals and metalloids (As, Se, and Sb) and to a lesser extent with certain other nutrient elements (e.g., N, P, and C). Since solute transport by water flow is one factor affecting the access of metals, metalloids, and organic compounds to organisms, this topic was included in the discussions. The discussions of the group were slightly structured within the following topics: chemical context (i.e., nature or characteristics) of selected systems (an ecosystem or an identified part thereof), transformations of chemical species and forms, transfer between phases (gas, aqueous, solid, and biological), transport, and other selected problems. We lacked the discussion time to reach a consensus on several points.

Chemical Context of a System

The aqueous species and solid forms of chemical elements in natural aqueous systems are typically established by the specific, local, chemical environment of the system. The chemical context that determines the specific distribution of species of an element is system-dependent. In addition, the biological, chemical, and physical characteristics of such natural systems vary in time and space. For example, natural systems such as soils or lakes can have chemical gradients with depth and can have significant seasonal alterations in such major master variables as pH and pE (negative logarithm of the theoretical activity of the electron). In some pore and surface waters, dissolved humic substances are sufficiently abundant to influence strongly and perhaps dominate the biogeochemistry of several metals. Scale is also important in such a discussion. A system may be defined as a whole lake or as a single soil particle.

The heterogeneous distribution of organic matter results in microenvironments within bulk oxic soil that in turn leads to steep gradients on a microscale of "master" parameters [e.g., pH, pE, dissolved oxygen, NO_3^-, SO_4^{2-}, S(-II), Fe(II), etc.]. In unsaturated soils, the lack of information on water thickness around particles makes prediction of water activity, hence of microbial processes, very uncertain. The lack of accurate *in situ* measurement of these critical chemical parameters (particularly in soils) and determination of their distribution and variability about the mean for subsystems at the level of individual pores is an important limitation on our understanding of the chemical context.

The mobilization, transport, and distribution of metals and metalloids within a natural system is a function of the aqueous species and solid forms of the element which, in turn, are controlled by the chemical context or condition established by the flux of energy and material involved in biogeochemical processes characteristic of the system. The extent to which these elements are found in the gas, aqueous, solid, and/or biological phase depends, in addition to the chemical properties of the element and of its aqueous species, on the chemical and biological characteristics of the system, the variation of these characteristics in time and space, and the available energy to drive the speciation reactions.

Transformation of Species

Electron Transfer Reactions

Electron exchange reactions are of major significance in establishing the distribution of the aqueous species of redox-sensitive metals and metalloids in many environments. There has not been major progress recently in our quantitative understanding of these processes, either thermodynamically or kinetically. The ubiquity of biochemically mediated processes suggests that the redox states of redox-sensitive elements are likely to be in disequilibrium with each other and that the occurrence of thermodynamically unstable redox species corresponds to a biogeochemically dynamic steady state. A consequence of this is the need for kinetic rate law expressions that account for the several variables normally encountered in natural waters.

From the point of view of the species transformation processes of redox-sensitive metals and metalloids and the interactions of these elements with other biogeochemical cycles, electron transfer reactions are important, particularly those producing or consuming soluble sulfide species. In the absence of volcanic sources, soluble sulfide S[II] species (e.g., H_2S) are predominantly the end product of the dissimilatory reduction of sulfate and elemental sulfur by microorganisms. Recent evidence indicates that the microbiological production of S[II] is not necessarily confined to anoxic environments (Cohen, 1984; Wood, personal communication, this workshop). Once formed, and in the absence of reactive chemicals, the distribution of S[II] species is determined by pH and the activity of complexing ligands. However, soluble sulfide species are highly reactive and may be subjected to a variety of biotic and abiotic processes such as complexation reactions with metals.

Chemical (abiotic) oxidation via dissolved oxygen to (usually) elemental sulfur and sulfate may take place. Biotic (microbial) oxidation to elemental sulfur and sulfate may occur under both oxic and anoxic conditions. The three most common reduced species (S^{2-}, HS^-, H_2S) are highly toxic, but the microorganisms that produce/consume S[II] can tolerate the ionic species, which dominate at neutral to alkaline pH values, at concentrations up to 10 to 20 millimole per liter.

Metals, if present, will compete with (some of) these processes for the dissolved sulfide species, abiotically forming soluble complexes first, then a solid phase of highly insoluble metal sulfides. The sulfides of Fe have received the most attention in aquatic habitats, a probable reflection of their relative abundance. The Fe sulfides formed during early diagenesis (e.g., FeS, $FeS_{0.9}$, Fe_3S_4) are thermodynamically metastable with respect to the final product, pyrite (FeS_2). Amorphic iron monosulfide (FeS) continues to be widely regarded as the first formed product, but in recent work Howarth (e.g., 1979) proposes that significant amounts of pyrite are formed rapidly upon production of biogenic S[II]. King (1983) has concluded that at least some of the "pyrite" reported to be rapidly formed was the result of an analytical artifact. However, the matter does not yet appear to be resolved and points to the requirement for unambiguous preparative and analytical techniques as a prerequisite to obtaining a clear understanding of the formation of these various iron sulfide species/phases.

In addition to transformations of metal species via linked biotic/abiotic processes, there is additional recent evidence implicating the involvement of microorganisms in the redox which causes a species transformation of Fe. The oxidation of Fe[II] to Fe[III] by microorganisms such as *Thiobacillus ferrooxidans* is well-known (e.g., see Olson, this volume). Recent work by Cohen and Gack (in press) provides evidence that Fe[II] to Fe[III] may be mediated by certain cyanobacteria (≡ blue-green algae) that can utilize Fe[II] as an alternative electron donor during photosynthetic CO_2 fixation. The reverse reaction has been reported to occur in coastal marine sediments. Sorensen (1982) provided evidence indicating that Fe[III] is used as an alternative electron acceptor for facultatively anaerobic NO_3^--reducing bacteria that switch to Fe[III] reduction when NO_3^- is exhausted. The sequence and rates of reactions under which the solid to aqueous phase transfer of Fe from oxidized coatings to soluble Fe[III] occurs, the stability of the latter, and the compatibility of these processes with the pE conditions required for microbially mediated reduction is not clear. Solubilization of Fe[III] by organic chelators (excreted by the bacteria?) is a possible mechanism (see also Wood et al., this volume).

The redox potential between two of the oxidation states of metals and metalloids have been used to predict the "ease" and "readiness" of an element for methylation (Ridley et al., 1977).

Ligand Exchange Reactions

Reversible labile equilibria occur between dissolved metals and humic substances in natural and polluted waters. These reactions are important because of the effect they have on availability and biouptake of dissolved metals. A wide range of conditional stability constants (5 to 12) have been reported for Cu, the most frequently studied metal (Baccini and Suter, 1979; Alberts and Giesy, 1983). It is our opinion that the majority of this variation results from the different experimental methods and conditions used with part of the total variation being due to inherent differences in dissolved organic matter present in pore and surface waters.

The functional groups out of which complexing or chelation sites are assembled may include combinations of carboxyl, phenolic, nitrogen-bearing, and sulfur-bearing functional groups. In principle, there is generally a theoretical upper limit for the complexing capacity of a mixture of organic ligands. In the case of dissolved humic substances, the upper limit may be deduced from a quantitative knowledge of the acidic functional groups. Complexation capacities (see Lund, this volume) have been measured as a bulk parameter to characterize the quantity of binding sites of dissolved organic matter with metals. Under real field or experimental conditions, three factors can cause the apparent complexing capacity actually observed to be less than the theoretical value. One is the degree of ionization. As the pH is lowered, less sites are ionized, hence fewer sites are available to complex metals. This accounts for the large decrease in conditional stability constants of metals with humic substances as the pH is decreased (Schnitzer and Skinner, 1966). A second factor can be the effect of dilution on the equilibria calculated via the law of mass action. The third factor is competition of other metals present on the equilibria involved. It should also be noted that complexation capacities alone are insufficient to calculate the distribution of species. A measure of the strengths of the complexes formed is also required.

Recently progress has been made in the development of the mathematical description of equilibria between dissolved metals and natural organic ligands. These theoretical descriptions are based directly on chemical reaction stoichiometry and account for the diversity of complexing sites and all necessary chemical composition variables. The theory provides guidelines that can be used in the design of effective experiments. This involves the progressive titration of the mixture of complexation sites. The resulting description may be used for predictive calculations. Although this approach has yet to be demonstrated for single-phase systems, multiple metal ion equilibria can readily be accommodated by this fundamental approach known as the "continuous stability function theory" (Gamble et al., 1980; Gamble et al., 1983). Other recently proposed continuous distribution models are the "affinity spectrum model" (Schuman et al., 1983) and the "normal distribution model" (Perdue and Lytle, 1983).

The influence of other metals present on the conditional stability constant is likely to have been one of the major errors in these values. For cases having labile equilibria, a theoretical treatment exists, however, which accounts rigorously for the effects of all the other metal ions on the complexing equilibrium of any given ion.

Organometallic Reactions

Biotic and abiotic transformations of elements are important reactions that may result in increased mobility and/or toxicity of several elements in aquatic systems. The processes involved in the transformation are degradation, which includes hydrolysis, photolysis, etc., and methylation, which involves the formation of organometals and organometalloids from their inorganic compounds. Several metallic elements (e.g., Hg, Sn, Pb, Pt, Pd) and metalloids (e.g., As, Se, Sb) can form element-carbon bonds that are hydrolytically stable. These organometallic and organic metalloidal compounds of metals such as Hg (Rubischung and Tobschall, 1980) may, therefore, be present in pore and surface waters. Examples of such compounds that have been detected are dimethylselenide, methylarsonic acid, dimethylarsenic acid, methyl- and dimethylmercury, and methyltin derivatives. However, compounds with methyl groups as the organic moieties are not the only organometallic derivatives found in nature. Organometals with more complex organic groups have also been identified, e.g., butyltin compounds (Maguire et al., 1982). The products of these reactions, especially those resulting from methylation reactions, have modified toxicological and mobility characteristics. Some organometals become more toxic (e.g., methylmercury, alkyllead, and alkyltin) than the antecedent inorganic species, whereas others have reduced toxicity (e.g., methylarsenic acids). The diverging toxic properties of organometallic compounds make the evaluation and regulation of environmental hazards purely in terms of total element concentrations quite unreasonable. Identification of organometallic compounds must be done on the molecular level.

Another aspect of organometallic reactions is derived from the anthropogenic input of organometallic compounds such as tetraalkyllead and alkyltin, as a result of their usage in industries and agricultural applications. These compounds degrade slowly in the environment, giving rise to a series of ionic organometallic species. Trialkyllead, dialkyllead (R = Me, Et), and butyltin species have been found in fish, sediment, and water (Chau et al., 1984; R. J. Maguire, personal communication, October, 1984).

The determination of species seems to be easier with organometallic compounds than with labile coordination compounds. The metal- or metalloid-carbon bond is in most cases sufficiently stable over time and in different chemical contexts to allow the separation of organometallic complexes from their matrix and, if necessary, to concentrate them to levels required by the techniques appropriate for identification and determination. Several concentration techniques and instrumental methods, including the coupling of chromatographic separation and element-specific detection, have been developed although the sensitivity of these methods is frequently inadequate.

A considerable effort has gone into deciphering the processes by which some metals and metalloids are methylated in nature, either biologically or abiologically. Yet much uncertainty persists: what is the relative importance of biotic and abiotic mediated reactions? Are the major methylating agents methylcobalamin and methyl iodide or betaine and/or silicones? Another important unanswered question concerns the possible methyl group transfer between metallic elements (e.g., Hg and Sn).

Organometallic and metalloid complexes occur both in biota and in the aqueous media in contact with the biota at very low concentrations, often at levels lower than nanomolar. It can be expected that at these low concentrations reaction pathways and rates are different from those taking place at higher concentrations. Thus, experimental studies of organometallic compounds should include concentrations at least as low as environmental levels.

Transfer Between Phases

Adsorption Reactions

The phenomena of adsorption on particulates (suspended or deposited) has undergone important conceptual development during the last decade. Present treatment of adsorption in terms of solution analogies (i.e., formation of surface complexes by reactions between solutes and functional groups attached to surfaces) has led to general treatment of interfacial phenomena in dominantly chemical terms (Leckie, this volume). Except for ion exchange at fixed charge sites (e.g., those resulting from isomorphic substitution), the binding of ions at surfaces can be adequately treated systematically as surface complex formation. With an additional set of reactions, ion exchange at fixed charge sites can be treated in a manner analogous to that of surface coordination sites. The surface coordination and classical ion exchange treatments both assume a stoichiometrically exact integer ratio.

Soils and sediments constitute complex mixtures of polyligand materials. A description of the interaction of dissolved metals at particle surfaces requires determination of reaction stoichiometries, including the proton stoichiometry. Because polyligand materials have distributed binding energies, the proton stoichiometry is a function of pH and degree of site occupancy (Altmann and Leckie, 1984). In addition, apparent binding constants can be affected by solid-to-solution ratios. The use of equilibrium surface complex formation models in general transport models awaits further work on the rates of both adsorption and desorption and a means of dealing with the solid-to-solution ratio problem.

Sorption of *Thiobacillus* onto the surface of pyrite crystals (Bennett and Tributsch, 1978) or colloidal sulfur particles (Baldensperger et al., 1974) could be mediated by polymers (e.g., polysaccharides) excreted by the organism (Marshall,1976). The initial stage of such sorption may be reversible but subsequently becomes irreversible (e.g., Marshall, 1976). The resulting changes of surface properties, presumably including ligand characteristics, may aid in the solubilization of Fe from pyrite crystals into soluble species and enable its biologically mediated oxidation to the ferric state.

In pore and overlying surface waters, multiple ion equilibria occur between aqueous, solid and swollen (highly hydrated), coagulated humic substances. These equilibria affect the availability (i.e., concentration over time in soil pore water) and potential biouptake (i.e., activity of the aqueous species) of individual metals. Mathematical descriptions are therefore necessary to predict the distribution between aqueous and coagulated humic substance phases. Calculations of these equilibria are complicated by a) the diverse assemblage of binding sites; b) the competition of cations for individual binding sites; and c) the competition of aqueous organic and inorganic ligands for the cations. An appropriate theoretical description has been demonstrated for a humic acid in a two-phase system (Gamble et al., 1983) and it is anticipated that it can be applied to other related systems.

Subsequent to surface adsorption onto solids, diffusion-limited absorption into the interior of aggregates and particles may occur. Diffusion of metals into iron oxides (hematite, goethite), manganese oxides (birnessite), and layer silicate minerals can reduce the availability (i.e., "immobilize" or "fix") of these elements in solid phases (Tiller and Hodgson, 1962; Gerth and Brümmer, 1983; Brümmer, this volume). The reduced availability is a result of the desorption rates of absorbed metals being very much slower than absorption rates. Diffusion into and out of such minerals may be a rate-determining step in reaching equilibrium conditions in soil solutions during transport through the soil profile. Since observed slow desorption rates may result in decreased concentrations of these metals in pore waters, diffusion-limited desorption rates may decrease the availability, hence toxicity responses of plants and benthic organisms in polluted soils and waters. That is, when metals are taken up by plant roots from soil pore waters or are diluted in surface waters, the dissolved metal concentration in these waters is less than it would be if it were not for the generally slow desorption. Thus, diffusion processes from solid phases to the solution phase may limit the rate of metal uptake by plants and animals from the pore water of soils and sediments as well as from surface waters.

These observations show a need for investigations to determine explicitly reaction kinetics of metals in soils and sediments.

Solubility Considerations

In this section we consider solid-liquid phase transformations due to dissolution and precipitation processes as well as coagulation and resolubilization of organic compounds.

At the most general level, those physical and chemical processes that occur in sediments are termed "diagenetic." Included in this term are recrystallization (aging) of metal hydrous oxides and the dissolution caused by a reduction in oxidation state of Fe oxides and occluded metals to form sulfides in an anaerobic environment. Energy to drive the systems arises ultimately from solar radiation, but in pores and sediments from decaying organic material. The formation of natural organic ligands may lead to an increased mobility and allow transport of metals to points of biouptake. Local, partial equilibria are often obtained in soil and surface-water systems. Solid phase equilibria are controlled by several factors, such as pH, redox potential, concentrations of elements in solution, microbiological effects, and temperature. In soils and sediments, solubility equilibria may change greatly within a few centimeters or even within a few millimeters at soil horizon and microenvironment boundaries, respectively. Thus, these equilibria are local in that the chemical context is often quite different in various parts (compartments) of the system. Transformation by recrystallization or by dissolution and reprecipitation are kinetic processes. Thus, the dynamic nature of soils causes dissolution and precipitation, ensuring that many of the minerals and amorphic solids will not be in equilibrium. Furthermore, some minerals equilibrate very slowly at ambient temperatures. This situation wherein only part of the minerals are in equilibrium with the aqueous phase is referred to as partial equilibrian.

Discrete solid phases are generally formed only by the major metals (e.g., Fe, Mn, Al, Ca) and nonmetals (C, S, P). The less abundant metals (Cd, Co, Cr, Cu, Pb, Zn) precipitate as discrete solid phases (such as $Cu_2(OH)_2CO_3$, $Pb_3(PO_4)_2$, $Pb_5(PO_4)_3Cl$, $Zn_5(OH)_6(CO_3)_2$, $Zn_3(PO_4)_2 \cdot H_2O$, $ZnSiO_4$) (Sposito, 1981; Brümmer et al., 1983; Herms and Brümmer, 1984) where they occur in elevated concentrations due to diagenetic processes or to pollution. The aqueous concentrations of these metals are believed to be limited by their tendency to co-precipitate with the major element solids and by adsorption.

Recently, analyses of soil pore water expressed with an immiscible liquid from samples of selected horizons of a number of soils were published. Distribution of aqueous species and solubility computations made on soil pore water (Kinniburgh and Miles, 1983) identified several minerals and amorphic solids that are in, or near to, equilibrium with the dissolved constituents present in these soil pore waters (E. A. Jenne, unpublished data). These solids include gibbsite ($Al[OH]_3$), chalcedony (cryptocrystalline SiO_2), allophane (an amorphic aluminosilicate whose Al-to-Si ratio is pH-dependent), laumonite (a calcium zeolite), phillipsite (a zeolite), Ca-montmorillonite, and amorphic zinc sulfide. The dynamic nature of mineral equilibria in these soils is indicated by the three solids (amorphic SiO_2, allophane, and ZnS), in permissible equilibria with the water, which are amorphic to cryptocrystalline.

The available information suggests that a) solids and natural water bodies locally obtain partial equilibria with a sufficient number of minerals and amorphic solids to make the use of thermodynamic calculation of species distribution and solubility models quite informative; and b) the available thermodynamic data, if critically reviewed, selected, and recalculated as necessary, are sufficiently accurate for these applications. These thermodynamic calculations indicate whether the water is oversaturated or undersaturated with particular solids that can thus be expected to be dissolving or precipitating, respectively. The quantification of sulfide complex formation for several trace metals (e.g., Cu[I], Zn[II], Ni[II], Pb[II]) may need to be redone since it is quite important and remains rather uncertain, even though there is good evidence for their existence from field data (Jacobs and Emerson, 1982).

Co-precipitation. As used in the technical literature, co-precipitation means the substitution of an element of suitable charge and radius for one of the principal constituents of a solid (isomorphic substitution), or it can imply the adsorption of an element onto a recent precipitate. We use the term here in the former sense. Inclusion of trace metals (e.g., Cd in solid $CaCO_3$) at low mole fractions of the Cd can reduce the concentration of Cd in pore and overlying surface waters to concentrations substantially below the saturation concentration with respect to a discrete $CdCO_3$ solid phase. Therefore, it is not an easy task to distinguish between the effects of adsorption and co-precipitation in limiting the concentration of less abundant metals in natural systems. Obvious examples where formation of co-precipitates are likely are Cd and Zn in calcareous sediments and soils (Sposito, 1981), and Ni, Zn, and Pb in Fe sulfides such as mackinawite (solid FeS[amorphic]) in anoxic pore and overlying surface waters of soils and sediments (Leckie and Nelson, 1975).

The solubility of the host mineral is generally increased by the presence of other elements as a solid solution. The solubility can be increased by an order of magnitude or more (Thorenstenson and Plummer, 1977).

Coagulation. Flocs of coagulated humic substances are commonly observed in the upper portion of the salinity gradient in estuaries. Similar mechanisms that cause coagulation in surface waters may also promote coagulation of dissolved organic compounds in soils, especially upon drying. The experimental investigation by Rayleigh light-scattering of fulvic acid aggregation has indicated two mechanisms (Underdown et al., 1984). One of these is hydrogen bonding at high ionic strength. The other is the formation of bridges between the polyelectrolyte molecules by divalent or trivalent metal ions. In suitable pH ranges, the presence of Fe[III] and Al[III] in sufficient concentrations in fulvic acid solutions can lead to the formation of mixed fulvic-hydroxide precipitates. The presence of organic coatings in at least partially continuous layers on soil peds suggests that organic compounds may irreversibly coagulate on ped surfaces in response to increased ionic strength and dehydration during the dry season.

Availability and Biouptake

Certain concepts are important for an overview of the importance of species in the uptake of metals to organisms. These concepts are also applicable to many other elements as well.

1) "Availability" (i.e., the quantity accessible for biouptake) is a function of the
   a) total dissolved concentration of the metal (note that this is an *intensity* factor);
   b) "resupply capacity" (i.e., the quantity of the various forms of the metal present in the solid phase that react rapidly enough to influence the dissolved concentration of the metal); and
   c) dissolved concentration of complexing ligands.
2) Biouptake (i.e., biouptake – efflux = bioaccumulation) from the aqueous phase is primarily dependent upon
   a) concentration (more correctly, the thermodynamic activity) of those aqueous species of the metal that are involved in biouptake; and
   b) physiological and ecological characteristics of the biota involved (e.g., swimming rate affects the volume of water contacting gill membranes).
3) Biouptake from the ingestion of solid materials is influenced by
   a) feeding habits,
   b) digestive processes, and
   c) form of the metals in the solid material.

It follows from the above concepts that an important effect of soils and sediments on biouptake is to regulate the dissolved concentration of metals and ligands.

The complexation of metals by dissolved organic ligands tends to decrease their co-precipitation with other solids as well as their sorption onto inorganic solid and coagulated organic phases. We assume that complexation of metals by dissolved organic ligands decreases their biouptake by virtue of the concomitant decrease in the thermodynamic activity of the chemical species involved in biouptake. This is, of course, only true if the amount of the metal released to solution from the solid phase in response to the complexation of the dissolved metal by the organic ligands is less than the amount complexed by the organic ligands and if none of the quantitatively important organically complexed metals have unique membrane-crossing abilities. In this regard, it may be appropriate to bear in mind that Cr absorption by the human digestive system is facilitated by presenting the Cr as an organic complex via a brewer's yeast extract (Mertz and Rozinski, 1971). Thus, some of the organic compounds excreted by organisms that form strong complexes with particular metals may be readsorbed along with their complexed metal.

It appears that biouptake of metals by prokaryotes and eukaryotes occurs by a) passive transport (diffusion) into cells through the outer membrane, b) carrier-facilitated transport into cells through the outer membrane, and c) adsorption onto outer surfaces of cell capsules (polymeric "exudates"), cell walls, and cell membranes.

The relative importance of these processes may vary depending on the distribution of chemical species in the aqueous phase (i.e., the species present and their concentration) and the molecular properties of the "barriers" such as the cell wall or capsule. For example, accumulation onto outer cell surfaces may not be harmful to the accumulating organisms at low environmental concentrations of the metal.

If biouptake is sufficiently in excess of efflux, i.e., if sufficient bioconcentration occurs, metabolic inhibition and death may result. Since microorganisms are quantitatively important components of sediments, even the metal-induced mortality may enhance higher trophic level bioaccumulation in that such bacteria are utilized by filter- and deposit-feeders. At fixed dissolved concentrations of a metal, the distribution of species may determine whether toxicity will occur. If a preponderance of the element is present in aqueous species that enhances biouptake, then the rate of resupply of the aqueous phase by desorption and dissolution may determine the extent of toxicity. Exceptions to this would occur where a significant portion of the metal intake is directly from ingested solids.

Competent models for the estimation of biouptake have to consider interelement competition effects and synergistic and antagonistic effects on physiological processes. The aqueous chemical species of Cd, Cu, etc., that are readily taken up by organisms and cause toxicity are partly known (Rawamoorthy and Kushner, 1975; Prosi, 1983; Cowan et al., 1984). Such data can only be gained by animal and plant bioassay procedures that include uptake kinetics from the system under investigation and lethal or sublethal tests with the organisms. However, traditional bioassays do not provide a definitive answer to the question of which species of a metal present in the systems is (are) responsible for biouptake, but do provide information on transfer rates of metals and equilibrium enrichment (bioaccumulation) in organisms. To determine which aqueous species are primarily involved in biouptake requires bioassays in which metals, ligands, and pH are varied over a useful range. For example, Cowan et al. (1984) have reanalyzed the freshwater (pH 7.4 to 8.2) study of Cu toxicity to *Daphnia* by Andrew et al. (1977) by recalculating the distribution of Cu inorganic species and determining the dependence of toxicity upon the activities of the various Cu species via the statistical methods of ridge regression, factor analysis, and principal component analysis. They concluded that the Cu hydroxy species, and a group of other aqueous Cu species that included Cu^{2+}, were responsible for the observed toxicity. In more acid systems, the uncomplexed metal is expected to be relatively more important than in these pH 7.4 to 8.2 waters.

Transport

Sedimentary particles are derived from the mechanical and chemical weathering of magmatic, metamorphic, and sedimentary rocks and to a lesser extent from the erosion of soils. Transportation (as bed load and as suspended load) in rivers leads to a grain size classification in relation to the current velocity: coarse material (gravel and sand) is mainly deposited in upstream areas, whereas silt and clay-sized materials (fractions <0.063 mm grain diameter) settle in downstream parts of rivers and in estuaries where mixing with marine suspended material occurs and leads to a preferential sedimentation in this transition zone.

The mechanical separation into different grain size classes parallels a mineralogical separation: sand fractions preferentially are made up of quartz, feldspar, rock fragments, and carbonates whereas in the clay fraction (<0.002 mm) layer silicate minerals dominate. The silt size class (0.063 to 0.002 mm) is intermediate between sand and clay.

Organic matter is added to the sediment by autochthonous production within an aquatic environment and by allochthonous supply from the drainage area (including soil organic material). The same holds true for hydrous oxides of Fe and Mn, which are mainly derived from soil-forming processes. Metal concentrations in sediments strongly depend on grain size: the finer the sediment, the larger the specific surface area and, in general, the higher the metal concentration. This leads to the assumption that a large portion of the heavy metal transport in rivers occurs via sorption by hydrous oxides and organic matter on the surfaces of suspended particles.

Ancient fine-grained sediments have a distinct element distribution pattern, which for a comparison with modern (polluted) sediments may be used as a reference value (geochemical background, "average shale standard").

Most modern riverine systems are strongly influenced by industrial and domestic sewage (the latter including highly polluted storm water runoff). As a result, metal concentrations in fine-grained riverine sediments have increased to such an extent during the past 100 years that in some rivers they now endanger aquatic organisms and limit the use of river water for drinking water purposes (Dissanayake et al., 1983, 1984).

The riverine and surface water transport of metals is enhanced when they are present as neutral or negatively charged chemical species (Robertson et al., 1984).

Biogeochemical Cycles

Environmental processes in aquatic and terrestrial systems are essentially biogeochemical processes. Soils and surficial sediments are complex assemblages of minerals, organic matter, and organisms and are natural bodies with properties that are not readily expressed by simply summing the constituent parts. Biogeochemical cycles in many ecosystems have already been altered by human activity frequently resulting in increased mobility of metals (Kabata-Pendias and Pendias, 1984; Kabata-Pendias and Wiacek, in press). Common causes of ecosystem alterations include changes in pH (caused by acidic rain, fertilizers, etc.), in pE (drainage, organic matter, microbiota), and in chemical composition (fertilizers, pollution). In order to elucidate the most important species of an element or compound in an ecosystem (i.e., with respect to their biogeochemical function), one must use two main approaches in combination:

1) characterization of chemical species on a molecular level and elucidation of their physicochemical properties with model compounds and/or model systems, and
2) analytical investigation of the ecosystem to elucidate those physicochemical and biochemical processes that are relevant for the whole system (Baccini, 1984).

The second approach begins with material balances of the ecosystem. First, information on the partitioning of an element can be obtained by

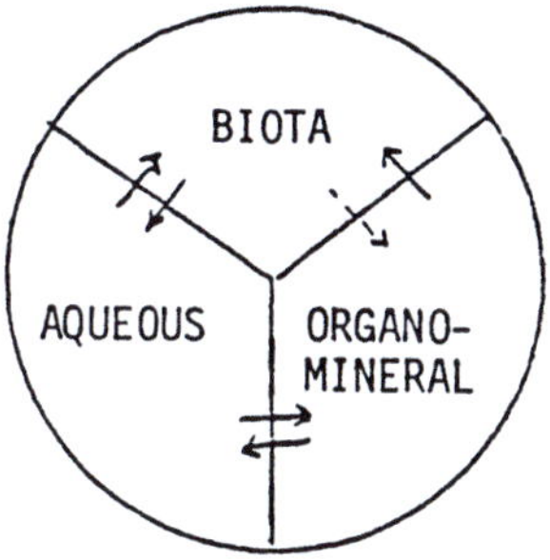

Fig. 1. Distribution among three phases (operational definition)

atmosphere

upper boundary layer

primary production

decomposition

lower boundary layer

Fig. 2. Stratification and subsystems (lakes, soils, etc.)

studying its distribution among the biota, aqueous organomineral compartments (Fig. 1), and in spatial distribution due to stratification (Fig. 2).

The system's balance of element and compound fluxes reflects the capacity of the "biological program" (under given physical and chemical boundary conditions) to retain chemical species for a longer or shorter time.

Examples

1) For phosphorus, a limiting nutrient element in a lake, the biological subsystem "sediment boundary layer" plays a crucial role in determining the predominant species, hence its fate. Within such an aerobic boundary layer an Fe[III] oxide barrier can contribute significantly to reduce the reflux of phosphorus into the hypolimnion.
2) In many terrestrial and aquatic systems, the total dissolved Cu[II] concentration would be toxic to most unicellular organisms if it were present exclusively in inorganic species. The presence of natural, dissolved organic matter in water containing a fixed quantity of dissolved Cu decreases the biouptake of dissolved Cu by formation of organic com-

plexes of Cu. Thus, the addition of dissolved organic matter to a water-sediment system generally results in desorption and/or dissolution of the metal increasing the dissolved concentration, i.e., the availability of the metal. In contrast, the addition of dissolved organic matter to a sediment-free system decreases the activity (i.e., effective concentration) of other chemical species that are more readily taken up by the organism.

System analysis also leads to the conclusion that the aqueous phase is the most dynamic and gives essential information with respect to the two neighboring phases. Its separation (pore water of soils and sediments, particle-free surface water) is still necessary because *in situ* measurements of these heterogeneous mixtures are not yet possible for most elements and compounds. Therefore, a sampling procedure has to be chosen that is apt to give the least perturbation (with respect to changes in pH, pE, T, ionic strength). However, it is not always necessary to get detailed information on a molecular level in order to qualify and quantify important processes.

Examples

1) It is sufficient to define operationally a group of phosphorus-containing species termed "soluble reactive phosphorus" in order to predict the approximate production of organic carbon by primary producers in lakes.
2) A distribution coefficient for NH_4^+ in a two-phase pore water/solid soil system might be sufficient to predict NH_4^+ concentrations in the aqueous phase of the rhizosphere in spite of a lack of sufficient knowledge with respect to the type of binding to the solid phase.
3) One can characterize organic complexes of Cu thermodynamically without elucidation of the structure of the ligands in order to predict, at equilibrium, the effective concentration of the species involved in biouptake. Of course, if these complexation rates are slow relative to biouptake or other reactions significantly affecting the dissolved concentration, the kinetics of the reaction(s) of the dissolved metal species and the organic matter need to be known in order to predict the activities of the various species of the metal.

At present there are no models available that predict quantitatively the element fluxes in ecosystems. This shortcoming is not on the level of adequate conceptual models but is the result of information deficiencies that lie mainly in the following three fields:

1) elucidation of the factors that govern the rates of decomposition of organic matter (and the nature and metal-complexing properties of the degradation products), authigenic mineral precipitation, and absorption desorption processes;
2) development of experimental boundary conditions very close to natural ones for a determination of the above-mentioned factors; and
3) development and/or refinement of analytical techniques for *in situ* determination of metal species at trace concentrations.

Pore Water Sampling

Questions concerning the representativeness of pore water sampling and the reliability of the concentration step often required prior to analysis arose numerous times. These discussions are summarized in this section.

The philosophy, techniques, and problems of operational separation of solid and liquid phases were discussed at length and rediscussed in a joint meeting with the group on chemical species in marine and estuarine systems (Kester et al., this volume). Several techniques for phase separation (centrifugation, suction, pressure, and dialysis bags) are available; each has problems associated with its use (Kabata-Pendias, 1972). Generally the problems can be classified in four groups: a) separation of solid and liquid phases, b) contamination, c) instability of constituents, and d) soil pore size and compositional heterogeneity. Several group members were concerned about the most desirable nominal pore size of the filter membranes used to obtain the phase separation and about comparability of data obtained using different pore size membranes. It was noted that this effect is most important in connection with the increasing clogging of the filter pores as the filtration proceeds, and often results in an additional uncertainty of the operational definition of the dissolved species by means of 0.45 μm membrane filtration. The large number of group participants (13) and shortage of time for discussion did not permit a consensus concerning the best available approaches to be reached. It was proposed that since the objective clearly is to achieve the separation of solid and liquid phases, the finest pore size consistent with reasonable flow rates should be used. Some members believed that pore sizes larger than 0.1 μm should not be used because sufficient solids have been shown to pass through the 0.45 μm membrane to result in significantly elevated levels of metals commonly associated with the clay size fraction (Kennedy et al., 1974). It was also reported that the removal of clay-sized particles is enhanced if initial aliquots of filtrate are discarded or used for purposes other than low-level metal analyses. This allows a sediment layer to build up on the membrane that is more effective in removing clay-sized particles than the membrane itself. The objection was raised that the accumulated sediment layer resulted in a localized solid-to-solution ratio much higher than the average value for the suspension and might cause desorption of a quantity of adsorbed metal. It is not clear that this represents a real problem. However, it is clear that fine-grained solids passing through the membrane are a problem in that spurious and variable metal concentrations result when the phase separation is inadequate. It was also pointed out that as filtration proceeds, the quantity of solids passing the membrane decreases with the development of a layer of sediment on the membrane. Thus, both more effective phase separation and a more consistent separation result by discarding the initial filtrate obtained before a sediment layer collects on the membrane.

The ceramic material in suction lysimeters may adsorb metals and may release Al and/or Si to the solid solution by dissolution of the porous ceramic. The extent of dissolution of the ceramic, releasing Al and Si, is quite dependent upon the chemistry of the pore water. This problem can be easily avoided by use of porous Teflon instead of ceramic. Unstable constituents include but are not restricted to CO_2(g), reduced valence states of redox-sensitive elements, and dissolved organic matter. Much concern was expressed concerning potential instability of water samples over time, i.e., to what extent does the distribution of species change between the time of sampling and that of analyses of the species? Suction lysimeters that are frequently used to sample soil solutions require small negative pressures to bring the pore water into the sampling device. This is likely to result in a partial degassing of CO_2 and a resultant rise in pH. Avoiding the entry of oxygen during the sampling of pore fluids and other waters containing the lower valence states of redox-sensitive elements is very difficult and special handling techniques are often required. The use of centrifugation to obtain a phase separation makes the complete exclusion of oxygen quite difficult. The use of dialysis bags may avoid some of these problems. Following filtration, reagents must generally be added to stabilize reduced valence states, e.g., a soluble salt of Zn or an alkaline EDTA buffer is added to fix S[II].

Soil heterogeneity exists at scales ranging from microns to hundreds of kilometers and may result in considerable variability between replicate samples. Soil pores may also cover a considerable range of effective sizes. This leads to problems in calculating the mass balances of solutes in pore water and drainage water since, under hydrostatic pressure, the larger pores drain first. Under conditions of frequent rain or irrigation, the mass movement of water downward may occur more rapidly than dissolution and desorption reactions. In this case the smaller pores, which have a higher surface area to volume ratio, will be closer to equilibrium than the larger pores. Thus, the composition of the throughput aliquots is sometimes observed to change (generally increasing concentrations) as more water is removed, and water is removed from smaller pores. It would appear to be advisable to use a maximum displacement pressure that just falls short of the pressure at which water within the electrical double layer is expelled (Manheim, 1966), to minimize this problem. The problem can be alleviated if adequate equilibration times are allowed. In some recent studies with glaciofluvial sediment (D. Girvin and E. A. Jenne, unpublished data, 1984), nearly constant dissolved concentrations were obtained within a week or two whereas months may be required for soil suspensions to reach a steady-state condition (E. A. Jenne and T. R. Garland, unpublished data, 1984).

Humic substances and other natural organic substances in solution sometimes coagulate and make the characterization of dissolved matter questionable. A fundamental understanding of this phenomena is needed. Then the appropriate analytical methods or separation techniques will need to be developed to permit the quantification of organic compounds, or classes of compounds, present in the original sample.

Available concentration techniques for metals and metalloids include evaporation, freeze-drying, adsorption, and gel permeation chromatography. These methods need to be reevaluated based on current analytical methods and on their impact on the determination of the distribution of species of individual metals and metalloids.

Research Needs

As in the preceding sections, much of the discussion of research needs apply to other elements of interest.

Instrumental Developments/Application Needs

1) Miniaturized *in situ* electrochemical measurement devices with a resolution down to 1 micron.
2) Continuous *in situ* chemical parameter measurement of increased reliability and stability.
3) Development of theory and Rayleigh light-scattering methods for determining the particle size continuum between dispersed colloidals and two-phase suspensions.

4) Improved characterization techniques for particulates in natural systems – size distribution, degree of particle porosity, surface coatings, binding site densities, specific surface area.
5) Identification and characterization of fine-grained solids passing through membrane filters, and review of physical and chemical concepts with a view to increasing efficiency of phase separation without adverse effects on aqueous samples.
6) Reliable chemical and coordination and/or crystallographic analyses of fresh precipitates [specifically of FeS and $Fe(OH)_3$].

Sampling and Analytical Methodology and Technique

1) Increased efficiency of aqueous-solid phase separation.
2) Synthesis of world literature on sample preservation techniques and evaluation of additional methods.
3) Analytical methods for small-volume aqueous sample:
   a) dissolved constituents
   b) microbiological processes.
4) Reference samples of fulvic and humic acids.
5) Reference samples for reduced valence-state metals and metalloids.
6) Forms of metals in solid phases. Improved ability to estimate solid phase reservoirs of metals with regard to capacity and kinetics of resupply reactions to aqueous phase.
7) Reevaluation of methods to measure quantity of amorphic oxides and reactive particulate organic carbon.
8) Develop improved understanding and quantitative description of the formation (rate, thermodynamics) of solid solution formation (e.g., carbonates, oxides, sulfides).

Ecosystem Modeling

1) Kinetic studies of organic matter decomposition in fresh water, with measurements for metals released and for classes or for key individual organic compounds released as a function of appropriate variables including seasonality.
2) Development of deterministic model to maximize usefulness of organic matter decomposition data.
3) Improvement of ecosystems/biogeochemical cycle models for metals and their species.
4) Linkage of microbiological biomass and processes model to geochemical cycling of elements with attention to mobile and potentially toxic species.

Organic Ligands

1) Metal complexation equilibria
   a) Further multiple metal ion equilibrium work for both one-phase and two-phase systems with experiments having more realistic collections of metal ions, and for the comparative chemistry of a variety of samples.
   b) The extension of equilibrium theory and methods developed for humic materials to clay-humic complexes and eventually to whole soils.
   c) The stoichiometrically exact investigation of proton dissociation and metal ion binding on the spectroscopic and photochemical properties of dissolved humic materials.
2) Chemical kinetics
   a) Rate constant studies of parallel reactions in mixed reactant systems.
   b) The stoichiometric investigation of the redox properties of calibrated fulvic acid samples, using potentiometric titrations.
3) Solubility
   a) Theory and methods for the calibration of the acidic properties of humic acids (two-phase systems) including pH-dependent solubility.
   b) The investigation of the physicochemical interactions of organic pesticides with humic materials, using humic samples for which the acidic functional groups have been stoichiometrically calibrated.

Adsorption Studies and Modeling

1) Rates and lability (reversibility) of surface reactions with soils and sediments.
2) Characterization of surfaces and surface reactions in anoxic systems.
3) The influence of particle/particle interactions on binding intensity and capacity.
4) Incorporation of surface reaction submodels into transport models.
5) Resolution of mathematical models to describe surface complex formation reactions over a broad range of concentrations and pH conditions.
6) Study and model adsorption of methylated metals.
7) Study and model adsorption of natural metal-organic complexes onto soils and sediments.

Organometallic Compound Studies

1) Establishment of key parameters for biotic and abiotic methylation reactions.
2) Extend chemical reaction studies of organometallics down to environmental levels.
3) Evaluate effect of solids in reactions of organometallic compounds in two-phase system.
4) Significance of environmental methylation of metals.
5) Significance of methylating agents such as methyliodide and cobalamine on the solubilization of compounds of redox-sensitive elements that are sinks for metals.
6) Methods of analysis of organometallics in biological samples.
7) Determination of mechanism of the biouptake of organometallics by biota.

Bioavailability and Toxicity

1) Evaluate effect of dissolved organic carbon levels for bioavailability of metals.
2) Determine structure-activity relationship for organometallic compounds.
3) Evaluate interelement competition for absorption sites and processes and interelement effects within plants.
4) Develop improved bioassay approaches with additional boundary conditions to circumvent the problem of modifying nutrient media to get close response curve.

References

Alberts JJ, Giesy JP (1983) Conditional stability constants of trace metal and naturally occurring humic materials: Chemical application in equilibrium models and verification with field data. In: Christman RF, Gessing ET (eds) Aquatic and terrestrial humic materials. Ann Arbor Press, Ann Arbor, MI, pp 333–348

Altmann RS, Leckie JO (1984) Metal binding in heterogeneous multicomponent systems: mathematical and experimental modeling. Proceedings, Fourth International Ocean Disposal Symposium, in press

Baccini P (1984) Regulation of trace metal concentrations in fresh water systems. In: Sigel H (ed) Metal ions in biological systems, vol 18, circulation of metals in the environment. Marcel Dekker, New York

Baccini P, Suter N (1979) MELINEX and experimental heavy metal pollution study: chemical speciation and biological availability of copper in lake water. Schweiz Z Hydrol 41:291–314

Baldensperger J, Guarraia LJ, Humphreys WJ (1974) Scanning electron microscopy of *Thiobacilli* grown on colloidal sulfur. Arch Microbiol 99:323–393
Bennett JC, Tributsch H (1978) Bacterial leaching patterns on pyrite crystal surfaces. J Bacteriol 134:310–317
Brümmer G, Tiller KG, Herms U, Clayton P (1983) Adsorption-desorption and/or precipitation-dissolution processes of zinc in soils. Geoderma 31:337–354
Chau YK, Wong PTS, Bengert GA, Dunn JL (1984) Determination of dialkyllead, trialkyllead, tetraalkyllead and lead(II) compounds in sediment and biological samples. Anal Chem 56:271–274
Cohen Y (1984) The solar lake cyanobacterial mats: strategies of photosynthetic life under sulfide. In: Cohen Y, Casterholz RW, Halvorson HO (eds) Microbial mats: Stromatolites. Alan R. Liss, Inc., New York, pp 133–148
Cohen Y, Gack E (1984) Fe'''' dependent photosynthesis in cyanobacteria. Nature, in press
Cowan CE, Jenne EA, Kinnison RR (1984) A methodology for determining the toxic chemical species of copper in toxicity experiments and natural waters, pp. 78–91. In Trace Substances in Environmental Health-XVIII, ed. D. D. Hemphill. Univ. of Missouri, Columbia
Dissanayake CB, Kritsotakis K, Tobschall HJ (1984) The abundance of Au, Pt, Pd, and the mode of heavy metal fixation in highly polluted sediments from the Rhine river near Mainz, West Germany. Intl J Envir Stud 22:109–119
Dissanayake CB, Tobschall HJ, Palme H, Rast U, Spettel B (1983) The abundance of some major and trace elements in highly polluted sediments from the Rhine River near Mainz, West Germany. Science Tot Envir 29:243–260
Frimmel FH, Innerz A, Niedermann H (1984) Complexation capacities of humic substances isolated from fresh water with respect to copper [II], mercury [II], and iron [II, III]. In: Kramer CJN, Buinger JC (eds) Complexation of trace metals in natural waters. Matenus Nijhoss-Dr. W. Junk, The Hague, pp 329–343
Gamble DS, Schnitzer M, Kerndorff H, Langford CH (1983) Multiple metal ion exchange equilibria with humic acids. Geochim Cosmochim Acta 47:1311–1323
Gamble DS, Underdown AW, Langford CH (1980) Copper [II] titration of fulvic acid ligand sites with theoretical, potentiometric and spectrophotometric analysis. Anal Chem 52:1901–1908
Gerth J, Brümmer G (1983) Adsorption und Festlegung von Nickel, Zink und Cadmium durch Goethit (α-FeOOH). Fresenius Z Anal Chem 316:616–620
Herms U, Brümmer G (1984) Einflußgrößen der Schwermetallöslichkeit und -bindung im Boden. Z Pflanzenernähr Bodenkd 147:400–424
Howarth RW (1979) Pyrite: its rapid formation in a salt marsh and its importance in ecosystem metabolism. Science 203:49–51
Jacobs L, Emerson S (1982) Trace metal solubility in an anoxic fjord. Earth Planet Sci Lett 60:237–252
Jenne EA, Kennedy VC, Burchard JM, Ball JW (1980) Sediment collection and processing for selective extraction and for total trace element analyses. In: Baker RA (ed) Contaminants and sediments, vol 2. Ann Arbor Science, Ann Arbor, MI, pp 169–190
Jenne EA, Wildung RE (eds) (1984) The biological availability of trace elements: chemical estimation, ecological and health implications. Science Tot Envir 28
Kabata-Pendias A (1972) Chemical composition of soil solution. T Roczniki Gleboznowcze 23:3–14 (in Polish)

Kabata-Pendias A, Pendias H (1984) Trace elements in soils and plants. Boca Raton FL, Chemical Rubber Company Press
Kabata-Pendias A, Wiacek K (1984) Excessive uptake of heavy metals by plants from contaminated soils. T Roczniki Gleboznowcze, in press
Kennedy VC, Zellweger GW, Jones BF (1974) Filter pore-size effects on the analyses of Al, Mn and Ti in water. Water Resources Res 10:785–790
King GM (1983) Sulfate reduction in Georgia salt marsh soils. An evaluation of pyrite formation by use of ^{35}S and ^{55}Fe tracers. Limnol Oceanog 28:987–995
Kinniburgh DG, Miles DL (1983) Extraction and chemical analysis of interstitial water from soils and rocks. Envir Sci Technol 17:362–368
Leckie JO, Nelson MB (1975) Role of natural heterogeneous sulfide systems in controlling the concentration and distribution of heavy metals. Paper presented at the 2nd International Symposium on Environmental Biogeochemistry, Ontario, Canada
Lindsay WL (1979) Chemical Equilibria in Soils. Wiley, New York
Maguire RJ, Chau YK, Bengert GA, Hales EJ (1982) Occurrence of organotin compounds in Ontario lakes and rivers. Envir Sci Technol 16:698–702
Manheim FT (1966) A hydraulic squeezer for obtaining interstitial water from consolidated and unconsolidated sediments. US Geol Surv Prof Paper 550-C:256–261
Marshall KC (1976) Interfaces in microbial ecology. Harvard University Press, Cambridge, MA
Mertz W, Rozinski EE (1971) Chromium metabolism: glucose tolerance factor. In: Mertz W, Cornatzer W. F. (eds) Newer trace elements in nutrition. Dekker, New York, pp 123–153
Perdue EM, Lytle CR (1983) Distribution model for binding of protons and metal ions by humic substances. Envir Sci Technol 17:654–660
Prosi F (1983) Storage of heavy metals in organics of limnic and terestric invertebrates and their effects on the cellular level. In: Management and Control of Heavy Metals in the Environment. Heidelberg, pp 459–462
Rawamoorthy S, Kushner DJ (1975) Binding of mercuric and other heavy metal ions by microbial growth media. Microb Ecol 2:162–176
Ridley WP, Dizikes LJ, Wood JM (1977) Biomethylation of toxic elements in the environment. Science 197:329–332
Robertson DE, Toste AP, Abel KH, Cowan CE, Jenne EA, Thomas CW (1984) Speciation and transport of radionuclides in groundwater. In: Alexander DH, Birchard GF (eds) NRC nuclear waste geochemistry '83. NUREG /CP-0052, pp 297–325
Rubischung P, Tobschall HJ (1980) Identification and determination of some environmental organomercurials in recent fluviatile sediments by means of thin-layer chromatography. Chem Erde 39:239–275
Schnitzer M, Skinner SIM (1966) Organo-metallic interactions in soils: 5. Stability constants of Cu^{+2}-, Fe^{+2}- and Zn^{2+}-fulvic acid complexes. Soil Sci 102:361–365
Schuman MS, Collins BJ, Fitzgerald PJ, Olsen DL (1983) Distribution of stability constants and dissociation rate constants among binding sites on estuarine copper-organic complexes: Rotated disk electrode studies and an affinity spectrum analysis of iron-selective electrode and photometric data. In: Christman RF, Gessing ET (eds) Aquatic and terrestrial humic materials. Ann Arbor Science, Ann Arbor, MI, pp 349–370

Sorensen J (1982) Reduction of ferric iron in anaerobic, marine sediment and interaction with reduction of nitrate and sulfate. Appl Envir Microbiol 43:319–324
Sposito G (1981) The thermodynamics of soil solution. Claredon Press, London
Sposito G (1983) The chemical forms of trace metals in soils. In: Thornton I (ed) Applied environmental geochemistry. Academic Press, London
Thorenstenson DC, Plummer LN (1977) Equilibrium criteria for two-component solids reacting with fixed composition in an aqueous phase – example: the magnesian calcites. Am J Sci 277:1203–1223
Tiller KG, Hodgson JF (1962) The specific sorption of cobalt and zinc by layer silicates. Clays Clay Min 9:404–411
Underdown AW, Langford CH, Gamble DS (1984) Light scattering studies of the relationship between cation binding and aggregation of a fulvic acid. Envir Sci Technol, in press

The Importance of Chemical "Speciation" in Environmental Processes,
eds. M. Bernhard, F.E. Brinckman, and P.J. Sadler, pp. 149–167.
Dahlem Konferenzen 1986. Springer-Verlag Berlin, Heidelberg
© Dr. S. Bernhard, Dahlem Konferenzen

Analytical Aspects of Organometallic Species Determination in Freshwater Systems

Y.K. Chau

National Water Research Institute
Canada Center for Inland Waters
Burlington, Ontario L7R 4A6, Canada

Abstract. The recent techniques and developments applicable to the determination of molecular and ionic organometallic compounds in the aquatic environment are reviewed. Techniques for quantitative digestion of biological samples prior to species analysis are discussed. Chromatographic separation coupled to an element-specific detector is currently the best technique for the analysis of organometallic species.

Introduction

The study of organometallic compounds in the environment is a relatively new subject, in spite of the fact that organometals have long been used in many industries and in agriculture. After the catastrophic mercury poisonings in Japan, intensive research led to the discovery of biological methylation of inorganic mercury by microorganisms in sediment. During the last decade, reports about the environmental methylation of many other metals, such as Pb, Sn, As, and Se, and the toxicity of the methylated compounds continued to appear in the literature. Organometals such as alkyltin and alkyllead compounds were found in air, water, sediments, and biota. A special symposium on organometals and organometalloids and their occurrence and fate in the environment was held under the auspices of the American Chemical Society in 1978 to discuss this new area of chemistry. All these studies during the last decade have given impetus to the development of a new type of analytical technique useful for the determination of organometals at environmental concentrations. Credit must be given to analytical chemists who design new hybrid analytical systems for the investigations of the fate of organometallics, which not long ago were considered not even to exist in aquatic systems.

Techniques for Species Analysis

Many techniques are available for the determination of a particular species of an organometal, but not too many techniques are suitable for the simultaneous determination of several species in the same sample at environmental concentrations. In the latter case, sensitivity as well as specificity is of prime importance. The method should directly identify and quantitate a species in the form in which it is present in the sample with as little chemical modification as possible. Thus, many methods that determine one species and obtain the concentrations of other species by difference are not desirable because of the possibility of accumulation of errors and the absence of affirmative identification. Of all the available methods, those using combinations of a separation technique and an element-specific detector are currently most suitable for the determination of organometals. The systems combine mainly chromatographs, such as gas chromatographs (GC) and liquid chromatographs (LC), with atomic spectrometers in the absorption (AAS), emission (AES), or fluorescence (AFS) modes as the detectors. The following are the most common and widely used systems.

Gas Chromatography-Atomic Absorption Spectrometry (GC-AAS)

The interfacing of a GC to an AAS detection system is simple, requiring very little or no structural modification of the individual instrument. The instruments can be easily decoupled as required. The column exit of the GC at the detector base is connected with a transfer line to the atomization device of the AAS. Both flame and furnace modes have been successfully used.

GC-Flame AAS

To combine a GC with a flame AAS, the transfer line can be either connected to the nebulizer or directly fed into the flame at the burner head. The latter arrangement gives better sensitivity because the analyte is not diluted by the nebulizer gases. The sensitivity of flame AAS is generally poorer than that of the graphite furnace. Recently, Ebdon et al. (1982) increased the

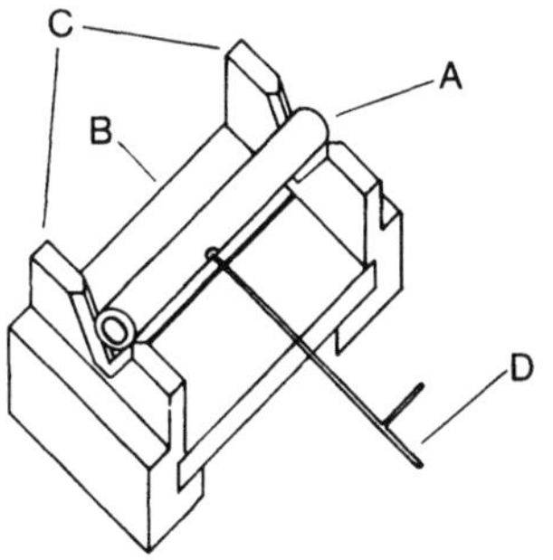

Fig. 1. Flame-heated ceramic tube atomizer. A: ceramic tube; B: air-acetylene burner head; C: stainless steel supporters; D: glass-lined T-piece. (Ebdon, Ward, and Leathard, 1982, with permission)

residence time of the atoms in the absorption path by feeding the GC effluent, mixed with hydrogen, into a flame-heated ceramic tube. This system has excellent sensitivity of 17 pg for tetramethyllead. Figure 1 illustrates the configuration of the atomization device.

The height of the ceramic tube above the burner is critical. The device is simple, sensitive, and low in cost. However, some operational inconveniences may be encountered in its use. These include the optimization of many gas flow rates and the continuous operation of a dry flame for several hours a day. The effect of long operation times on the burner and the nebulizer coatings has yet to be assessed.

GC-Graphite Furnace AAS (GC-GFAAS)

Furnace AAS provides much better sensitivity than flame AAS. Several designs are available for GC-GFAAS systems. Segar (1974) used a tungsten tube to deliver the GC effluent into a heated GF tube at ca. 1700 °C. In the analysis of mixed tetraalkyllead compounds in gasoline he achieved detection limits of 10 ng Pb. The sensitivity was improved by the use of a specially designed carbon rod atomizer operated at a constant temperature of 2000 °C. The detection limit for tetramethyllead under these conditions was 0.1 ng Pb.

Other designs used the inert gas purge tubes leading to the furnaces. The GC effluent was introduced into the furnace either through a T-fitting from both ends or through the central injection opening of the furnace tube. The sensitivity was much enhanced and a detection limit of 0.04 ng Pb was reported by De Jonghe et al. (1980) for tetraalkyllead. Commercial furnaces are not designed for continuous operation. The graphite furnace in a GC-GFAAS must be kept at the atomization temperature (1500–2000 °C) during the course of chromatography. The average life of a graphite furnace tube is about 10–15 hrs. Thus, the operation of a GC-GFAAS system could become very expensive. Furthermore, graphite tubes deteriorate on prolonged heating over a period of hours as indicated by a gradual loss of sensitivity.

Electrothermal Silica Tube Furnace

A simple, electrothermally heated, open-ended furnace made of a silica tube (Fig. 2a) was used (Chau and Wong, 1977; Chau et al., 1982) for the determination of organolead and organotin compounds. The silica tube (7 mm, i.d.; 4 cm long) is wrapped with 26 gauge resistance wire (5 ohms) and insulated with asbestos tape. The unit is housed in a block of preshaped fire brick that is mounted on top of the burner assembly. Hydrogen, air, or mixtures of both can be introduced to the furnace through a side arm to enhance atomization as required. Twenty volts a.c. applied to the furnace

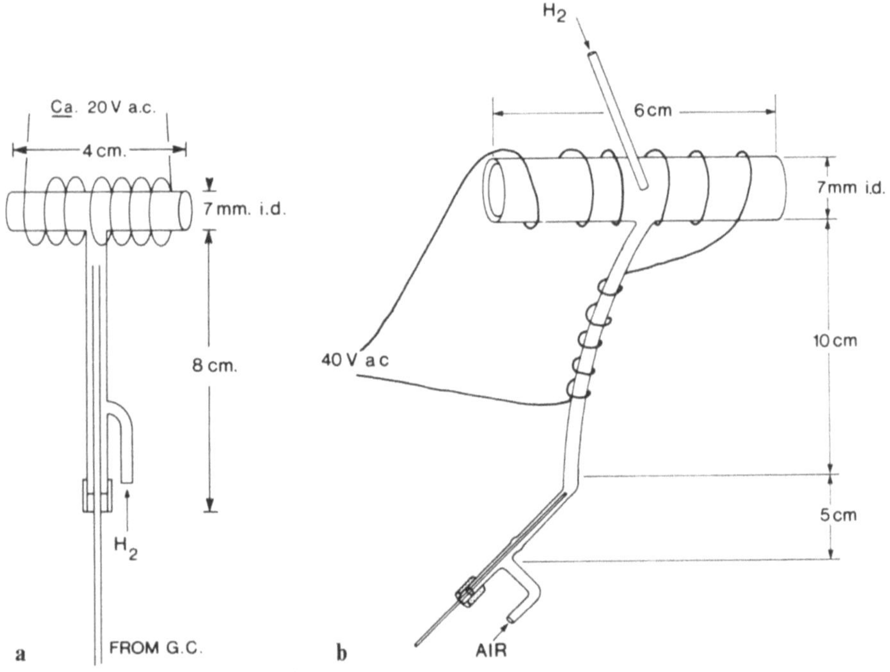

Fig. 2. a: Silica furnace; **b:** Silica furnace with precombustion section. (Chau and Wong, 1977, with permission)

assisted by the hydrogen flame produce temperatures as high as 900 °C. Such a temperature is sufficient for the atomization of organometals such as alkyltin, alkyllead, and alkyl selenides. The furnace can be operated at its maximal temperature continuously for several hours per day for up to one month without deterioration. A layer of metal oxide may form in the interior walls of the furnace after long use, causing loss of sensitivity. At that time the furnace should be replaced.

Solvents, particularly chlorinated hydrocarbons that are trapped with the sample, may interfere with the determination of elements with absorption lines in the far UV region. Under certain conditions, the broad band absorbance may be so high as to be unmanageable by deuterium background correction. A precombustion section made from a silica tube with an air inlet can be installed as part of the furnace to burn off the contaminants (Fig. 2b).

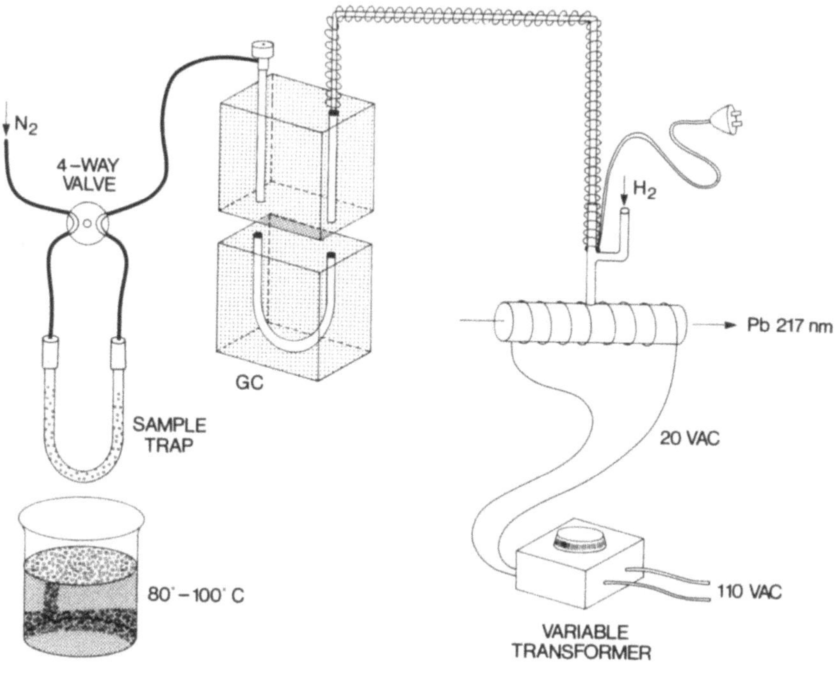

Fig. 3. Schematic diagram showing the interface of GC-AAS (Chau and Wong, 1983, with permission)

The transfer line is 1.6 mm stainless steel tubing, i.e., joining the furnace and the GC column exit. Several kinds of tubing materials can be used, such as Teflon-lined aluminum, glass-lined stainless steel, and Teflon. The transfer line is wrapped with heating tapes to control its temperature and prevent condensation of the analytes. The GC electrothermally heated silica furnace system is schematically shown in Fig. 3. The four-way valve installed between the carrier gas inlet and the injection port makes it possible to use a cold trap for volatile samples, whereas samples contained in a solvent can still be injected through the injection port.

Other Systems

The other commonly used spectrometric instruments for element-specific detection are atomic emission spectrometers. Braman et al. (1977) used a d.c. arc discharge spectrometer for the determination of methylarsines at the nanogram level. Higher sensitivity was achieved by the same research group for methyltin compounds using a hydrogen-rich flame in the emission excitation (Braman and Tompkin, 1979). Recently plasmas were used as excitation sources (Reamer et al., 1978; Talmi and Bostick, 1975) to detect picogram quantities of arsenic and lead. Similarly, microwave-excited helium plasma was used to attain the same sensitivity in the determination of alkyllead compounds (Estes et al., 1982). Plasma is a more effective source of excitation for emission.

Gas Chromatography-Mass Spectrometry (GC-MS) systems do not receive wide application in the analysis of organometallic compounds, probably because these instruments have high installation costs and require special skills for operation. The sensitivity of MS is not particularly superior to any of the spectrometric methods. Only a few reports on GC-MS determination of organometals appear in the literature.

Liquid Chromatography-Atomic Absorption Spectrometry (LC-AAS)

Certain organometals are polar and may exist in solution as hydrated cations. These compounds may not have adequate volatility for GC separation. Liquid chromatography can separate these species without chemical modification. The latitude in choosing solvent systems and elution modes has made LC a very versatile technique for the separation of organic compounds. However, the application of LC to organometallics is still in its infant stage.

LC-AAS Systems

An LC-AAS system is more complicated than a GC-AAS system because a continuous liquid flow is not always compatible with a high temperature furnace. Different modes of AAS have been interfaced with LC. When flame AAS is used, the LC effluent can easily be routed to the nebulizer of the burner. The LC flow rate should be slightly higher than that of the nebulizer in order to achieve better sensitivity. The overall sensitivity of the IC-flame AAS system is in the μg range, which limits its extensive use in environmental analysis.

Much sensitivity can be gained with furnace AAS. In coupling LC with GFAAS, the most difficult part is the introduction of the effluent into the furnace. Several designs that are still far from perfection are available. One design uses an automatic sampler to take aliquots from the effluent stream. A sequencer-controlled multiport sampling valve was used by Cantillo and Segar (1975). A commercial AA sampler was employed by Brinckman et al. (1977) to sample the effluent from a micro flow-through well. Similarly, Stockton and Irgolic (1979) sampled the LC effluent with a sequencer-controlled slider injector. All these techniques give chromatograms in pulse form. The sum of the signals in a chromatographic peak quantifies the analyte. The precision of the measurement will depend on the number of signals in a peak, which is determined by the cycle time of the furnace. The commercial furnaces generally operate in three cycles: drying, ashing, and atomizing. The minimal time to complete these cycles and cool the furnace is 40 seconds. Thus, if a band has a width of only 1–2 min, the chromatographic peak is not well-defined. The precision of the results is also influenced by the synchronization of the autosampler with the GFAAS cycle.

A second approach is the peak storage technique suggested by Vickrey et al. (1983). Fractions of the effluent containing the components and separated by plugs of an inert gas were stored in capillary tubes. The fractions were analyzed off-line by GFAAS. More aliquots can be analyzed in this manner to quantify a peak. This method has better precision and accuracy than the pulse and the on-line analysis methods but is quite cumbersome. A detection limit of 0.048 ng was obtained for tetraphenyllead with a 20 μL sample injection onto the LC column.

The third approach does not involve discrete sampling of the effluent, but isolation of the analytes from the effluent stream in volatile forms. This technique can be applied to organometals that readily form covalent hydrides, such as those of As, Ge, Sn, Pb, and Sb. A post-column hydride generator is installed on-line and the volatile hydrides are continuously fed to

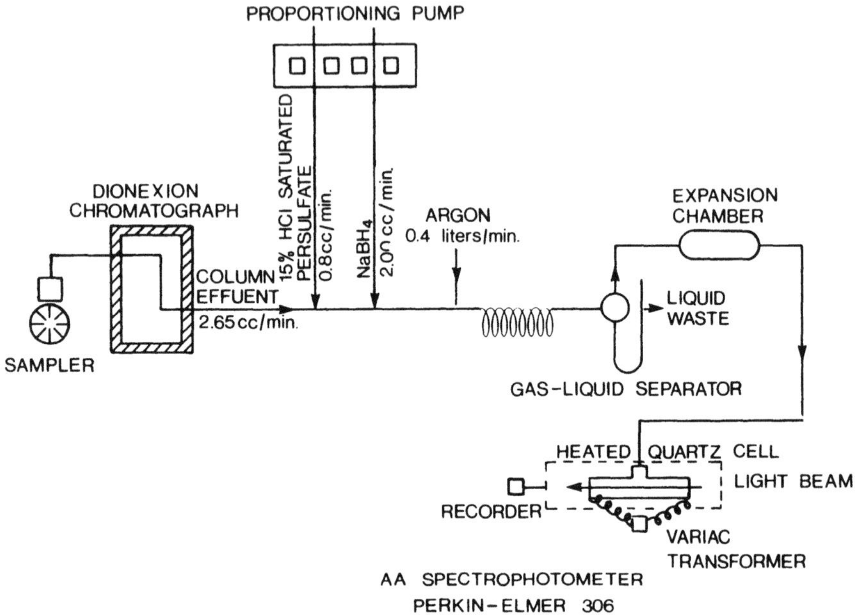

Fig. 4. Dionex ion chromatograph coupled with automatic arsine generator and AAS detector [reprinted with permission from Ricci et al., 1981. Copyright (1981) American Chemical Society]

a heated silica furnace. A recent report by Burns et al. (1981) described the combination of an HPLC and an electrothermally heated silica furnace AAS via an automatic hydride generator. A mixture of methyltin compounds was analyzed with detection limits of 2–20 pg. An automatic system has also been developed by Ricci et al. (1981) for the analysis of As(III), As(V), monomethylarsonic, dimethylarsinic, and p-aminophenylarsonic acids using ion chromatographic separation and determination of the hydrides by AAS. Figure 4 shows the interfacing of the two systems. Detection limits were less than 10 ng mL^{-1} for each species. Similarly, a conventional gravity-flow column chromatograph coupled to an automatic hydride system was used for the determination of methylarsenic acids and inorganic As (Chau and Wong, 1983). The post-column hydride formation has certain advantages over the batch hydride generation. The molecular rearrangement (Talmi and Bostick, 1975) during the hydride reaction can be avoided. In addition, on-line automatic hydride generation gives more uniform and reproducible results than the manual batch system. The on-line hydride generation is at present the best interface for LC and furnace AAS but is applicable only to compounds reducible to volatile derivatives. Consumption of the whole sample during analysis provides high sensitivity.

Other LC Element-Specific Systems

A d.c. argon plasma emission detector was used with an HPLC (Uden et al., 1978) for determination of several transition metal complexes. The use of inductively coupled plasma emission spectrometers (ICP) as element-specific detectors is described in the analytical background papers of this volume (Irgolic and Brinckman, this volume). An on-line, post-column, and continuous hydride generation technique coupled to an ICP has been developed for the analysis of methyltin species (Krull and Panaro, 1985). A review on the use of ion chromatography and liquid chromatography coupled to element-specific detectors such as atomic emission and atomic fluorescent spectrometry for analysis of inorganic species has also been provided (Krull, this volume).

Electrochemical detectors based on oxidation and reduction of the analytes have been used for organic analysis. These detectors have good sensitivities but generally lack element specificity and versatility in comparison to the spectrometric detectors. A cyclic voltammetric detector was used with HPLC (MacCrehan et al., 1977) for the determination of alkyl and phenyl derivatives of Hg, Sn, and Pb with submicrogram sensitivity. Using the differential pulse mode of the amperometric detector further improved the sensitivity and selectivity (MacCrehan and Durst, 1978). The method was applied to the determination of methylmercury in fish with a detection limit of 40 pg Hg. Reviews on the use of electrochemical detectors in liquid chromatography are available (Buchta and Papa, 1976; Pungor et al., 1975). Although electrochemical detectors have not been used very frequently with LC, they are expected to find wider application in the future.

Electrochemical Methods

Electrochemical methods have been used for the determination of organometals in natural waters. These methods are sensitive, relatively specific if the sample matrix is known, and require very little or no sample preparation. Most methods developed earlier were based on the determination of one of the organometals and obtained the concentrations of other compounds by difference. For example, Plazzogna and Pilloni (1967) determined the R_2Sn^{2+} and R_3Sn^+ species in solution by titrating their total amounts potentiometrically with alkali and then determining the R_2Sn^{2+} in another aliquot by amperometric titration with standard 8-hydroxyquinoline solution. In mixtures of R_2Pb^{2+} and R_3Pb^+ compounds, dialkyllead ion can be titrated amperometrically with ferrocyanide solution, and trialkyllead ion with tetraphenylboron solution. Hodges and Noden (1979), using two different plating potentials in anodic stripping voltammetry, were able to determine the concentrations of dialkyl- and trialkyllead in solution. These methods, however, differentiate only the dialkyl and trialkyl metals as a class without identifying the alkyl groups. At the same time, a similar stripping voltammetric method was developed by Columbini et al. (1981) for consecutive determinations of Me_4Pb, Et_4Pb, Me_3Pb^+, Et_3Pb^+, Me_2Pb^{2+}, Et_2Pb^{2+} and Pb(II) in natural waters with detection limits in the nanomolar range. The species were separated by selective solvent extraction. This method is sensitive and relatively specific but is tedious and complex, involving many steps of calculation by difference. Despite these disadvantages, this method represents the current state of the art for species analysis by electrochemical techniques.

Analysis of Molecular and Ionic Organometals

Molecular Organometals

Volatile, molecular organometallic compounds such as tetraalkyllead, tetraalkyltin, methylarsines, and methylselenides generated either in experimental systems or in the ambient atmosphere can be collected cryogenically at ca. -150 °C in a trap containing nonpolar chromatographic column materials (OV-1, OV-101, glass beads). For very volatile compounds such as methylarsines and dimethyl sulfide the cold trap can serve as GC, allowing fractional volatilization of the components into the detector in the order of their boiling points. A GC would give more accurate control of temperature and gas flow and could produce more accurate retention times.

For the determination of volatile organometals in natural waters, the purge and trap techniques with an inert gas and cryogenic trap are often used. Andreae and Barnard (1983) determined dimethyl sulfide in seawater with a flame photometric detector. Jackson et al. (1982) determined the extremely volatile methyltin hydrides in an estuary with a tin-selective photometric detector. An alternate method is to extract the molecular organometals such as tetraalkyllead from water or sediment with a nonpolar solvent; the extract is then analyzed by GC-AAS.

Ionic Organometals

Ionic organometal species behave like aquated metal ions or salts in solution. Typical examples of such compounds are the dialkyllead, trialkyllead, and a variety of alkyltin cations. These organometals are difficult to determine because they are not easily extracted from aqueous media or biological tissues. In addition, the thermal instability of these compounds complicates the chromatographic separation. Both problems have to be resolved before reliable analyses can be performed. Previous methods were based on the salting-out extraction of Bolanowska (1967) for trialkyllead compounds followed by spectrometric determination with dithizone or gas chromatographic determination with an electron capture detector.

Several methods were published that used GC-microwave helium plasma, GC with flame ionization detector, and GFAAS after sequential extraction of samples. The results were not satisfactory. The main difficulty lies with the incomplete extraction of these compounds from aqueous medium. Recently, the method of quantitative extraction and determination of dialkyl- and trialkyllead compounds was improved (Chau et al., 1983; Forsyth and Marshall, 1983). This improved method made it possible for the first time to establish the presence of dialkyllead and trialkyllead in a variety of environmental materials (Chau et al., 1984).

Extraction of Ionic Alkyllead Compounds

The extraction of dialkyl- and trialkyllead compounds has been difficult because of the highly polar nature of these species. Triethyllead species can be extracted almost quantitatively into benzene in the presence of saturated sodium chloride (Bolanowska, 1967), but only 20 percent of trimethyllead species were recovered under the same conditions. The addition of potassium chloride made the quantitative extraction of trimethyllead possible (Noden, 1980). In spite of these improvements, only 30–40 percent of dialkyllead, and especially of dimethyllead, species could be extracted. Many attempts were made in enhancing the recovery of dialkyllead. Additions of inorganic lead salt, sodium benzoate, and other compounds were largely ineffective (Birnie and Hodges, 1981). Recently, sodium diethyl-dithiocarbamate was found to extract quantitatively all dialkyl- and trialkyllead species from an aqueous medium into benzene (Chau et al., 1983). Tetraalkyllead and lead(II) species were also extracted. Similarly, dithizone was used for the quantitative extraction of the dialkyl- and trialkyllead from a variety of media including water, buffer solutions, and eggs (Forsyth and Marshall, 1983).

Ionic butyltin species were extracted from water into benzene by tropolone solutions (Meinema et al., 1978). The more polar methyltin species are not quantitatively recovered unless sodium chloride is added to the medium. Once the ionic alkyl metals are isolated from the matrix they can be derivatized for final determination.

Methods of Derivatization

For GC-AAS analysis, the polar and high-boiling ionic organometals must be converted to more volatile yet stable derivatives. Two methods are currently used.

Conversion to Hydrides

Compounds readily converted to covalent hydrides can be isolated as volatile derivatives and then determined. Hydride techniques were used to analyze mixtures of methylarsenic acids (Andreae, 1977; Braman et al., 1977). In a similar manner methyltin, dimethyltin, and trimethyltin were converted to form the corresponding hydrides and Sn(IV) to stannane (Braman and Tompkins, 1979; Hodge et al., 1979). The hydrides were collected in a trap cooled by liquid nitrogen and were then vaporized into the detector. The hydride derivatization can be extended to alkyl derivatives of metals such as Pb(IV), Ge(IV), and Sb(V). This method has advantages and disadvantages. Isolation of the analytes from the sample matrix in volatile forms is a great advantage. All of the analyte is collected and the sensitivity is therefore improved. A disadvantage is the possibility of molecular rearrangement during the reduction. Such rearrangements were observed during the reduction of organoarsenic compounds and methyltin compounds (Chau, unpublished results). The cause of molecular rearrangement is not fully known. Talmi (1975) suggested the use of sodium borohydride tablets instead of borohydride solution to minimize rearrangements. Another disadvantage observed in the author's laboratory is a mysterious contamination which occurs now and then, even when the most stringent precautions are exercised. The hydride might be such a reactive agent that it acts on anything it contacts. Laboratories not equipped with clean-room facilities may find it difficult to use successfully the hydride generation technique.

Alkylation

Alkylated metals of group IV, $R_nM^{(4-n)+}$, readily accept further alkyl groups from a Grignard reagent to form stable tetraalkyl derivatives, $R_nMR'_{(4-n)}$, that are more volatile than their parent compounds. Methylation was used (Meinema et al., 1978) to determine butyltin compounds. Pentyl Grignard was employed to derivatize butyltin species (Maguire and Huneault, 1981), and butyl Grignard for the determination of methyltin species (Chau et al., 1982). The butylation reactions (Eq. 1) were also used to prepare volatile derivatives of dialkyl- and trialkyllead compounds:

$$R_nPb^{(4-n)+} + (4\text{-}n)\,BuMgCl \rightarrow R_nPbBu_{(4-n)} + (4\text{-}n)Mg^{++} + (4\text{-}n)Cl^- \quad (1)$$

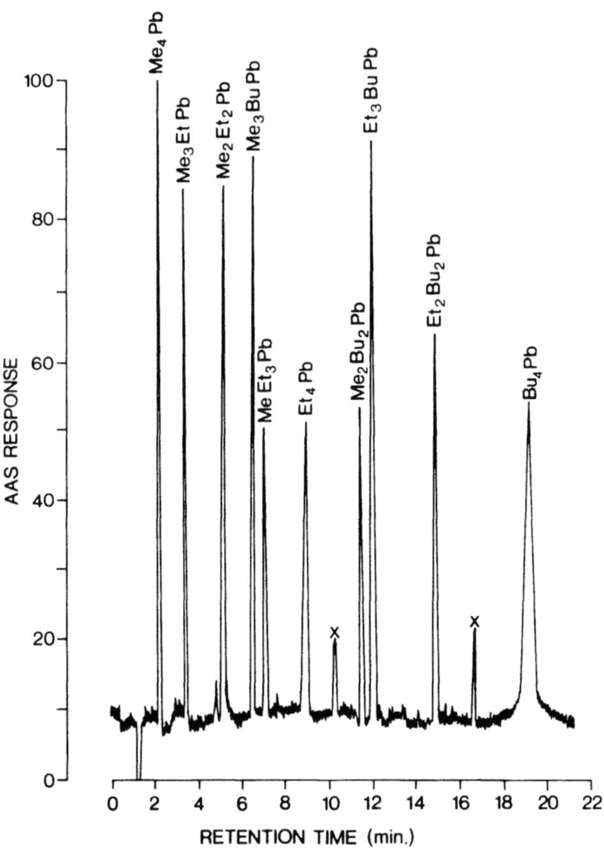

Fig. 5. GC-AAS chromatograms of five R Pb (10 ng each); four butyl derivatives of R Pb^{++} and R Pb$^+$ (8 ng each) and Pb(II) (15 ng) expressed as Pb. x: unidentified Pb compound. (Chau et al., 1983, with permission)

The butyl derivatives, $R_nPbBu_{(4-n)}$ (R = Me or Et), have relatively low boiling points and high thermal stability. The GC-AAS chromatograms of the butyl-derivatized dialkyl-, trialkyllead and Pb(II) and tetraalkyllead species are shown in Fig. 5.

Phenylmagnesium bromide was used for the determinations of alkyllead compounds extracted by dithizone (Forsyth and Marshall, 1983). Alkylation is easy to control and gives uniform and consistent results. The Grignard reagents can also be contaminated but are generally purer than the hydride reagent. This technique does not use all of the analyte for the analysis. The derivatization is carried out in a non-aqueous medium of which only a small aliquot is injected. To the analytical system. However, the extraction used to isolate the analytes for alkylation generally concentrates the analytes.

Sampling, Sample Preservation, and Sample Preparation

Sampling of Surface Microlayers

Surface microlayers are important strata at which water and air meet. This natural interface consists of a thin hydrophobic layer that has been known to concentrate metals and organic species through various processes. Hydrophobic organometals, such as tetraalkyllead and tetraalkyltin, and the naturally occurring metal hydrides are believed to be concentrated in this layer. We observed high concentrations of mixed alkyllead species in the microlayer that were not found in the adjoining water (Chau et al., unpublished results).

Special samplers are available to collect surface microlayers. The most convenient and simple collector is a glass plate, 40 × 40 cm, that is manually dipped into the water. The adhering film is scraped into a container with a neoprene blade. Though cumbersome, this sampler is simple and economical. A more elaborate sampler (designed by the National Water Research Institute, Burlington, Ontario, Canada) consists of a motor-driven rotating ceramic drum of ca. 1 m length, 40 cm diameter mounted in the front of a small boat. While the boat is trolling at about 1–2 knots, the drum is rotating at ca. 10 rpm. The depth of immersion of the drum in the water is kept at 2–3 cm with adjustment cranks. The surface film adhering to the drum is scraped by a blade and collected in a bottle. The drum is far more efficient than the glass plate. A four-liter sample can be collected in ca. 10 minutes.

A regular trace metal sampler can be used for taking water samples. Contamination in sampling for organometals is not as serious as for heavy metals because organometals are not universal contaminants. However, precautions must be taken to obtain representative samples. For example, plastic ware should not be used for organotin sampling because of the alkyltin stabilizers they contain. Boat paints normally contain butyltin antifouling agents. Tetraalkyllead in gasoline used in boat motors can be a source of alkyllead contamination.

Sample Preservation

Some organometals are unstable in the presence of light. Alkyllead compounds were observed to degrade slowly in water to inorganic lead. However, lake water samples enriched with dimethyllead and trimethyllead chlorides are stable for at least one month if refrigerated and stored in glass bottles in the dark (Chau et al., 1983). Alkyltin compounds are stable in water. Because extraction of methyltin compounds required saturated sodium chloride, Chau et al. (1982) recommended the addition of sodium chloride to samples at the time of sampling to reduce the risk of adsorption of inorganic tin and alkyltin onto the container walls.

Acidification of water samples with hydrochloric acid has been proposed to reduce adsorption loss. Maguire (personal communication) also found it desirable to acidify water samples to pH 1 to preserve butyltin and inorganic tin species during short-term storage. In general, storage in glass containers and in the dark are essential. If volatile species are to be determined, the samples should be immediately extracted. It is convenient to add a layer of hexane to seal off the surface for subsequent processing in the laboratory.

Digestion of Biological Samples

Digestion of samples without altering the chemical nature of the analytes is indeed a most challenging analytical problem. For some samples that contain the organometals in a physical mixture, straight extraction may be sufficient to separate the analytes. In biological samples the organometals could be part of the tissue. Under these circumstances extraction would not recover them efficiently. This is particularly the case for ionic organometals such as dialkyl- and trialkyllead compounds. The addition of various salts during extraction can recover the trialkyllead species but cannot adequately recover dialkyllead compounds. Recently two digestion procedures were published. Chau et al. (1984) digested fish, algae, and macrophytes in tetramethylammonium hydroxide, and Forsyth and Marshall (1983) used a mixture of lipases and proteases to hydrolyze egg homogenates. Both methods were effective in releasing the alkyllead compounds chemically unchanged from biological tissues.

Extraction of Sediments

Organometals in sediments are either derived from anthropogenic sources or formed by the biological and chemical processes of alkylation or dealkylation. For the determination of organometals in sediments, total digestion is not necessary. An effective extraction procedure can often recover the organometals.

Tetraalkyllead species can be recovered by extraction with hexane. The ionic alkyllead species can be extracted with sodium diethyldithiocarbamate after addition of sodium chloride, sodium iodide, and sodium benzoate (Chau et al., 1984). Similarly, methyltin compounds can be extracted from sediments with a solution of tropolone in benzene in the presence of sodium chloride (Y. K. Chau, unpublished results). For complete extraction of butyltin compounds, the sediment has to be refluxed for two hours with a benzene solution of tropolone (R. J. Maguire, personal communication).

Difficulties and Research Needs

Difficulties encountered in the analysis of organometal species in environmental samples arise mainly from sample preparation which includes incomplete recovery of individual analytes and interferences caused by sample matrix. Polarity of the organometallic molecules varies according to the nature of the organic component parts and the degree of the organic substitution which controls the efficiency of recovery. It is not unusual to experience variations of extraction efficiency among a series of organometals of the same homolog. Matrix interferences can be serious with biological samples such as eggs and fish tissues during the extraction and final determination.

Data interpretation is another difficult area in dealing with transient and labile species whose half-lives may not be in the same scale as that of the analytical processes. For example, the measurement of volatile tetraalkyllead species in natural water may not have any significance if their degradation products were not simultaneously measured.

Further research needs in the chromatography element-specific detection systems include several aspects. Liquid chromatography and furnace atomic absorption spectrometry are not fully compatible and further improvements are desirable. Electrochemical detection systems for chromatography have not yet been fully developed. Atomic fluorescence detectors have not been widely applied in spite of their multi-element capability and sensitivity. Complexation with a fluorogenic ligand after column separation can be a sensitive technique for organometals and its selectivity should be further investigated. The feasibility of interfacing a micropore HPLC directly into a micro emission flame or a graphite furnace should also be explored.

In the sample treatment aspect, recovery of organometals from biological matrices without destruction of their chemical forms has been a challenging problem for analytical chemists. Many organometallic compounds are involved in biogeochemical reactions under environmental conditions; their interactions with other organometals and their environmental pathways and metabolism require further investigations and assessment.

References

Andreae MO (1977) Determination of arsenic in natural waters. Anal Chem 49:820–823

Andreae MO, Barnard WR (1983) Determination of trace quantities of dimethyl sulfide in aqueous solution. Anal Chem 55:608–612

Birnie SE, Hodges DJ (1981) Determination of ionic alkyl lead species in marine fauna. Envir Tech Lett 2:433–442

Bolanowska W (1967) A method for the determination of triethyllead in blood and urine. Chemia Analityczna (Warsaw) 12:121–129

Braman RS, Johnson DL, Foreback CC, Ammons JM, Bricker JL (1977) Separation and determination of nanogram amounts of inorganic arsenic and methylarsenic compounds. Anal Chem 49:621–625

Braman RS, Tompkins MA (1979) Separation and determination of nanogram amounts of inorganic tin and methyltin compounds in the environment. Anal Chem 51:12–19

Brinckman FE, Blair WR, Jewett KL, Iverson WP (1977) Application of a liquid chromatograph coupled with a flameless atomic absorption detector for speciation of trace organometallic compounds. J Chrom Sci 15:493–503

Buchta RC, Papa LJ (1976) Electrochemical detector for liquid chromatography. J Chrom Sci 14:213–219

Burns DT, Glockling F, Harriott M (1981) Investigation of the determination of tin tetraalkyls and alkyltin chloride by atomic absorption spectrometry after separation by gas-liquid chromatography or high pressure liquid-liquid chromatography. Analyst (London) 106:921–930

Cantillo AY, Segar DA (1975) Metal species identification in the environment a major challenge for the analyst. Proceedings of the International Conference on Heavy Metals in the Environment, Toronto, pp 183–204
Chau YK, Wong PTS (1977) An element- and speciation-specific technique for the determination of organometallic compounds. In: Ewing GW (ed) Environmental analysis. Academic Press, New York, pp 215–225
Chau YK, Wong PTS (1983) Direct speciation analysis of molecular and ionic organometals. In: Leppard GG (ed) Trace element speciation in surface waters and its ecological implications. Plenum Publishing Corp, New York, pp 87–103
Chau YK, Wong PTS, Bengert GA (1982) Determination of methyltin(IV) and tin(IV) species in water by gas chromatography-atomic absorption spectrophotometry. Anal Chem 54:246–249
Chau YK, Wong PTS, Bengert GA, Dunn JL (1984) Determination of dialkyllead, trialkyllead, tetraalkyllead and lead(II) compounds in sediment and biological samples. Anal Chem 56:271–274
Chau YK, Wong PTS, Kramar O (1983) The determination of dialkyllead, trialkyllead, tetraalkyllead and lead(II) ion in water by chelation/extraction and gas chromatography/atomic absorption spectrometry. Anal Chim Acta 146:211–217
Columbini MP, Corbini G, Fuoco R, Papoff P (1981) Speciation of tetra-, tri-, dialkyllead compounds and inorganic lead at nanomolar levels (sub ppb) in water samples by differential pulse electrochemical techniques. Ann Chim (Rome) 71:609–629
De Jonghe W, Chakraborti D, Adams F (1980) Graphite furnace atomic absorption spectrometry as a metal specific detection system for tetraalkyllead compounds separated by gas-liquid chromatography. Anal Chim Acta 115:89–101
Ebdon L, Ward RW, Leathard DA (1982) Development and optimization of atom cells for sensitive coupled gas chromatography-flame atomic absorption spectrometry. Analyst 107:129–143
Estes SA, Uden PC, Barnes RM (1982) Determination of n-butylated trialkyllead compounds by gas chromatography with microwave plasma detection. Anal Chem 54:2402–2405
Forsyth DS, Marshall WD (1983) Determination of alkyllead salts in water and whole eggs by capillary gas chromatography with electron capture detection. Anal Chem 55:2132–2137
Hodge VF, Seidal SL, Goldberg ED (1979) Determination of tin(IV) and organotin compounds in natural waters, coastal sediments and macro algae by atomic absorption spectrometry. Anal Chem 51:1256–1259
Hodges DJ, Noden FG (1979) The determination of alkyllead species in natural waters by polarographic techniques. In: Proceedings of the Conference on Management and Control of Heavy Metals in the Environment. CEP Consultants, Edinburgh, pp 408–411
Jackson JA, Blair WR, Brinckmann FE, Iverson WP (1982) Gas chromatographic separation of methylstannanes in the Chesapeake Bay using purge and trap sampling with a tin selective detector. Envir Sci Technol 16:100–119
Krull IS, Panaro KW (1985) Trace analysis and speciation for methylated organotins by HPLC-hydride generation-direct current plasma emission spectroscopy. Anal Chem, in press

MacCrehan WA, Durst RA (1978) Measurement of organomercury species in biological samples by liquid chromatography with differential pulse electrochemical detection. Anal Chem 50:2108–2112

MacCrehan WA, Durst RA, Bellama JM (1977) Electro-chemical detection in liquid chromatography: applications to organometallic speciation. Anal Lett 10:1175–1188

Maguire RJ, Huneault H (1981) Determination of butyltin species in water by gas chromatography with flame photometric detection. J Chrom 209:458–462

Meinema HA, Burger-Wiersma T, Versluis-de Haan G, Giver EC (1978) Determination of trace amounts of butyltin compounds in aqueous systems by gas chromatography/mass spectrometry. Envir Sci Technol 12:288–293

Noden GF (1980) The determination of tetraalkyllead compounds and their degradation products in natural water. In: Branica M, Konrad Z (eds) Lead in the marine environment. Pergamon Press, New York, pp 83–91

Plazzogna G, Pilloni G (1967) Amperometric titration of organolead and organotin ions. Anal Chim Acta 37:260–266

Pungor E, Toth K, Feher Z, Nagy G, Varadi M (1975) Application of electroanalytical detectors in chromatography. Anal Lett 8:ix–xxiii

Reamer DC, Zoller WH, O'Haver TC (1978) Gas chromatography-microwave plasma detector for the determination of tetraalkyllead species in the atmosphere. Anal Chem 50:1449–1453

Ricci GR, Colovos G, Hester NE (1981) Ion chromatography with atomic absorption spectrometric detector for determination of organic and inorganic arsenic species. Anal Chem 53:610–613

Segar DA (1974) Flameless atomic absorption gas chromatography. Anal Lett 7:89–95

Stockton RA, Irgolic KJ (1979) The Hitachi graphite furnace-Zeeman atomic absorption spectrometer as an automated element-specific detector for high pressure liquid chromatography. Intl J Envir Chem 6:313–319

Talmi Y, Bostick DT (1975) Determination of alkylarsenic acids in pesticide and environmental samples by gas chromatography with a microwave emission spectrometric detection system. Anal Chem 47:2145–2150

Uden PC, Quimby BD, Barnes RM, Elliott WG (1978) Interfaced D.C. argon-plasma emission spectroscopic detection for high-pressure liquid chromatography of metal compounds. Anal Chim Acta 101:99–109

Vickrey TM, Howell HE, Paradise MT (1983) Liquid chromatogram peak storage and analysis by atomic absorption spectrometry. Anal Chem 51:1880–1883

The Importance of Chemical "Speciation" in Environmental Processes,
eds. M. Bernhard, F. E. Brinckman, and P. J. Sadler, pp. 169–192.
Dahlem Konferenzen 1986. Springer-Verlag Berlin, Heidelberg
© Dr. S. Bernhard, Dahlem Konferenzen

Heavy Metal Species, Mobility and Availability in Soils

G. W. Brümmer

Institut für Bodenkunde
Rheinische Friedrich-Wilhelms-Universität Bonn
5300 Bonn 1, F. R. Germany

Abstract. The ecological effects of heavy metals in soils are closely related to their content and species distribution in the solid and liquid phases of soils. Methods for the determination of metal species in both phases are described and critically evaluated. In connection with the possible binding mechanisms of heavy metals in soils the concept of specific and nonspecific adsorption is discussed. Recent results indicate that the adsorption of heavy metals on soil particles is not only restricted to the formation of surface complexes but can also take place in the structure of minerals. Diffusion processes of heavy metals into soil minerals are described for goethite. In order to assess metal mobility and availability in soils, the concentration of the species in the soil solution or in comparable aqueous equilibrium solutions of soil samples, and the transfer of metals from solid pools to the liquid phase, and *vice versa,* have to be assessed. Methods for the determination of the amount of plant-available heavy metals are described and the associated problems discussed.

Introduction

During the last decade numerous investigations connected with trace and toxic metals in soils have been conducted in order to assess micronutrient availability or to point out the continuing pollution of soils by heavy metals. The recognition of the associated environmental hazards connected with either trace-element deficiency of living organisms or enrichment of toxic metals in the food chain and groundwater is justification for many studies of the reactions of metals such as Mn, Co, Ni, Cu, Zn, Cd, Hg, and Pb with soils or different soil components and model substances thought to behave like soil materials (Kabata-Pendias and Pendias, 1984; Scheffer and Schachtschabel, 1982).

As soils consist of heterogeneous mixtures of different organic and organomineral substances, clay minerals, oxides of iron, aluminum, and manganese, and other solid components as well as a variety of soluble substances, the binding mechanisms for heavy metals in soils are manifold and vary with the composition of the soils, the soil reaction, and redox conditions (Fig. 1). Thus, a metal may form different species according to whether it is bound to various soil compounds, reacting surfaces, or to external or internal binding sites with different bonding energies.

The composition of soils changes with depth from surface to subsurface horizons. Especially under humid climatic conditions precipitation percolates through soils and may transport soluble heavy metals with the soil solution. The concentration of heavy metals in the solution phase is of major importance for all ecological considerations because plants take up the

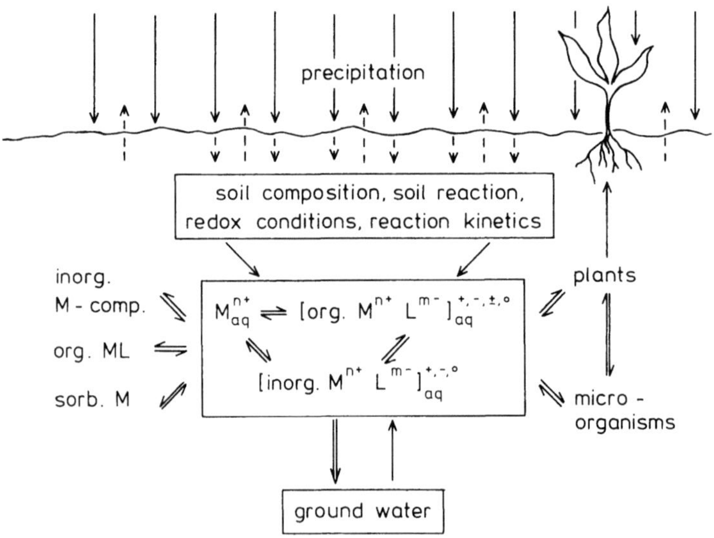

Fig. 1. Schematic presentation of heavy metal reactions in soils. Relationship between solid metal phases [inorganic metal compounds, organic metal compounds (org. ML), sorbed metals], soluble metal species (metal ions, organic and inorganic metal complexes), metal uptake by plants and microorganisms, and metal transfer from decomposing plants and microorganisms to the soil solution

available metals from the soil solution, and the transport of metals within the soil or even to the groundwater also depends on the metal concentration of the solution phase. However, the processes which determine the concentration of heavy metals in the soil solution are not yet fully understood. Changes in soil water content, departures from equilibrium or steady-state to nonequilibrium conditions, and changes in the activity of microorganisms which influence redox conditions, contents of soluble chelating agents, and the composition of soil atmosphere require simultaneous consideration. These processes moderate the reactions of heavy metals such as precipitation-dissolution, adsorption-desorption, and complex and ion-pair formation in relation to the pH and thus influence the distribution of the various metal species in the solid phases of the soil and in the soil solution (Fig. 1).

The complexity of possible reactions and often unknown reaction kinetics in natural soil systems restrict studies of metal species distribution in solid phases mainly to operationally defined analytical procedures. Several authors have suggested methods for the fractionation of heavy metals in solid phases (Elsokkary and Lag, 1978; Fischer and Fechter, 1982; Kuo et al., 1983; Shuman, 1982, 1983; Sposito et al., 1982; see also Förstner, this volume). For the determination of metal species in soil solutions methods are used which are based on a more profound analytical background (Camerlynck and Kiekens, 1982; Gerritse and Salomons, 1983; Gregson and Alloway, 1984; Jeffery and Uren, 1983; McBride and Bouldin, 1984; Tills and Alloway, 1983). Also, successful computer programs have been developed for modeling the distribution of metal species in soil solutions (Jenne 1979; Sposito, 1981; Sposito and Bingham, 1981) and metal transport through the soil profile (Mayer, 1978; Poelstra et al., 1979).

Soil Sampling and Sample Preparation

For analysis of heavy metals in solid soil material, samples should be taken from surface to subsurface horizons in order to determine the metal distribution, total metal concentrations, and operationally defined metal species within the whole soil profile.

If the soil surface is substantially polluted by atmospheric deposition, the surface few centimeters and especially surface organic materials may contain high total concentrations of heavy metals compared with the original background contents of materials at greater depth. Especially in forest

Table 1. Content (mg kg^{-1} air-dried soil) and distribution of different heavy metals in a forest soil near Hamburg (soil type: Gleyic Podzol; location: Sachsenwald, F.R. Germany)

Depth (cm)	Horizon*	pH ($CaCl_2$)	Pb_T	Cd_T	Zn_T	Cu_T
10– 7	L	2.8	135	0.8	67	91
7– 4	Of	2.7	220	1.6	85	104
4– 0	Oh	2.6	240	3.0	99	57
0– 8	Aeh	2.7	24	1.0	17	4
8–18	Ahe	2.8	12	0.32	9	2
18–32	Ae	3.2	5	0.26	8	1
32–44	Bsh	3.5	7	0.19	13	2
44–50	Go	4.0	6	0.17	18	3
Several Oh horizons from the same area		2.6 –2.9	135 –475	0.18 –3.0	72 –99	57 –114

* The main soil horizons are designated by capital letters. Small letters indicate special properties of the soil horizons
L, O: organic layers on the surface of the mineral soil horizons; L (litter); Of (fermentation layer; organic residues, partly decomposed by the soil fauna and flora); Oh (humified organic residues)
A horizon: upper mineral soil horizon. B horizon: subsurface mineral soil horizon. C horizon: parent rock material, relatively little affected by pedogenic processes. G horizon: mineral horizon affected by groundwater
h: horizon enriched with humic substances; e: eluviated (leached) horizon because of downward movement of iron, manganese, and aluminum. s: horizon enriched with sesquioxides of Fe, Mn, and Al. o: oxidized part of groundwater horizon

soils the organic surface layers (L, Of, Oh in Table 1) often show an enrichment of heavy metals as a result of the air-filtering effect of the forst vegetation (Mayer, 1981; Schwertmann et al., 1982; Ulrich et al., 1979). Thus, the content of lead and other heavy metals in the organic surface layers of forest soils can be taken as an indication of the degree of atmospheric deposition. But a translocation of metals from subsurface horizons to organic surface layers by metal uptake of the plant roots, transport to the overground parts of the plants, and accumulation of organic material on the soil surface may also cause an increase of metal contents in the organic layers.

Representative samples from different soil horizons are normally air-dried for several days or dried for 24 hours at 60 °C. Then the soil samples are ground to pass through a 2 mm sieve and mixed thoroughly. If the analysis requires small subsamples (<1 g), the samples should be crushed to pass through a 40-mesh screen. Careful treatment of the samples is necessary to avoid accumulation or loss of heavy metals during sampling, sample preparation, and analysis. If losses are expected from the fresh material, e.g., as volatile Hg or heavy metal alkyls, the samples should be transported to the laboratory in a sealed container, preferably frozen, and analyzed with their natural water content (Page, 1982). In the latter case, the moisture content is determined on a separate subsample.

For analysis of the heavy metal species in natural soil solutions, the solution phase can be collected in different soil horizons by installation of ceramic or Teflon cells or lysimeters in different depths of the soil. The soil solution can then be sampled by applying a vacuum (Hetsch et al., 1979; Horrison, 1982). The advantage of this technique is that changes in the composition of the solution phase during the transport through the soil and associated transport processes of metals can be estimated. Also, input-output analysis can be performed in conjunction with the total metal budget of the soil. However, if ceramic filters are used, possible losses or gains of heavy metals by adsorption or desorption processes on the ceramic material have to be evaluated.

Other techniques for the separation of solid and liquid phases of soils require centrifugation of moist soil samples or replacement of the soil solution by organic solvents.

Also, laboratory experiments with aqueous leaching of soil columns or equilibration of soil samples with aqueous solution and separation of the solution phase by centrifugation or filtration can be used for the determination of metal species in soil solutions. Such experiments have the advantage that the experimental and analytical conditions are generally better defined than in field investigations but the results may not be as representative as those obtained under field conditions.

The analytical methods for the determination of metal species in liquid phases have been described by several authors (Camerlynck and Kiekens, 1982; Gerritse and Salomons, 1983; Jeffery and Uren, 1983; McBride and Bouldin, 1984; Page, 1982; see also Chau, this volume).

Chemical Extraction of Different Metal Species from Solid Phases

In order to assess the reactivity of the species or binding forms of heavy metals in solid materials, different procedures involving successive extractions have been developed which assume that the following heavy metal species exist in soils:

1. Water-soluble (i.e., in the soil solution);
2. Exchangeable;
3. Organically bound;
4. Occluded in iron and manganese oxides;
5. Definite compounds (e.g., heavy metal carbonates, phosphates, sulfides); and
6. Structurally bound in silicates (i.e., the residual fraction).

The heavy metals in the soluble fraction can be determined from the saturation extract of soil samples or by extracting soil material with deionized water at a certain soil-water ratio (Biddappa et al., 1982; Horak, 1979; Sposito et al., 1982).

The exchangeable fraction can be extracted by neutral salt solutions, e.g., KNO_3, $Ca(NO_3)_2$, $CaCl_2$, $MgCl_2$ (Biddappa et al., 1982; Elsokkary and Lag, 1978; Kuo et al., 1983; Lion et al., 1982; Sposito et al., 1982; see also Förstner, this volume). For most soil samples an extraction with 0.1 M $Ca(NO_3)_2$ or $CaCl_2$ solution gives an appropriate estimation of the amount of this fraction.

For the extraction of the organically bound heavy metals, several procedures have been proposed. The destruction of the organic matter by H_2O_2 or NaOCl treatment and subsequent extraction of heavy metals by dilute acid were suggested (Kuo et al., 1983; Lion et al., 1982; Shuman, 1983; see also Förstner, this volume). Other authors use NaOH, $Na_4P_2O_7$, $K_4P_2O_7$, or Na_2EDTA for the extraction of the organically bound fraction (Elsokkary and Lag, 1978; Fischer and Fechter, 1982; Sposito et al., 1982). Cu-salt solutions can also be used because Cu in excess displaces other metals in metal-organic complexes (Juo and Kamprath, 1979).

For the separation of metals occluded by manganese oxides from the iron oxide-bound fraction, an extraction with NH_2OH and dilute HCl has been proposed. In a successive step, the metal fraction occluded in amorphous and weakly crystalline iron oxides can be extracted with oxalate in the dark, and afterwards the crystalline iron oxide fraction with ascorbic acid and oxalate (Shuman, 1982). However, the analytical differentiation between these metal fractions is still a problem, because the reagents suggested are not very selective. Especially, besides manganese oxides, the NH_2OH-HCl extraction also dissolves part of the amorphic iron oxides (Lion et al., 1982). When the preceding steps for the separation of the different oxide fractions are left out, a treatment with boiling oxalate (Fischer and Fechter, 1982; Schwertmann et al., 1982) or dithionate-citrate-bicarbonate (DCB) (Fischer and Fechter, 1982; Kuo et al., 1983) extracts both the manganese oxide and the total iron oxide-bound metal fraction. However, several analytical problems are connected with the DCB method. Therefore, the boiling oxalate extraction is favored by some authors (Fischer and Fechter, 1982).

Heavy metals associated with carbonates can be dissolved by sodium acetate adjusted to pH 5 (Förstner, this volume) or by Na_2EDTA at pH 4.5 (Sposito et al., 1982). For the extraction of heavy metal (phosphates and) sulfides, strong acids are suggested (Sposito et al., 1982). However, in the sequence of extractions the position for the dissolution of heavy metal carbonates, phosphates, and sulfides has to be well considered because several other solvents may also dissolve these compounds.

The residual fraction, which mainly consists of silicate-bound metals, can be dissolved by an HF-$HClO_4$ digestion.

The soluble plus exchangeable fraction characterizes the mobile fraction of heavy metals in soils. This fraction also allows an estimation of the amount of bioavailable metals. The other metal fractions are more or less immobile. A mobilization of metals from these fractions or transformation of mobile to immobile metal fractions is often a very slow process which is strongly controlled by reaction kinetics (Gerth, 1985; Gerth and Brümmer, 1983; Brümmer et al., 1987).

The methods suggested for the extraction of different heavy metal species are based on the assumption that the above-mentioned binding forms exist. However, in multi-component systems such as soils it is difficult to assess all possibilities of heavy metal binding. For example, organomineral binding forms also exist in soils, which are not considered in the described extraction procedures and can hardly be determined at all. Therefore, a separation of soil material into fractions of different density such as organic, organomineral, and mineral fractions (Lichtfuß and Brümmer, 1981) and/or into different grain-size fractions (Shuman, 1979) with subsequent extraction of the bound metals is suggested.

Furthermore, the question is still open as to whether, and under which conditions, definite heavy metal compounds such as carbonates, phosphates, silicates, and others exist in soils. It seems more likely that heavy metals are bound in soil minerals (e.g., iron and manganese oxides, carbonates, and clay minerals), by isomorphic substitution, or by fixation at free structural places. The results of several authors also show that the adsorption capacity of several soil components for heavy metals can be very high (Brümmer et al., 1983; Chubin and Street, 1981; McKenzie, 1980; McLaren et al., 1983). Therefore, a considerable amount of heavy metals can be bound by adsorption before a formation of definite heavy metal compounds takes place (Brümmer et al., 1983).

Because it is uncertain whether the expected binding forms or metal species in soils really exist, other chemical extraction procedures have been proposed. Here the soil material is sequentially extracted with organic complexing agents, which form metal complexes of increasing stability and thus allow conclusions to be drawn about the binding strength of heavy metal species present in soils (Lindsay, 1979).

Direct measurements of the metal content of different soil components by using an electron microprobe or a scanning electron microscope attached to an energy-dispersive system (EDS) also allow the analysis of heavy metal binding forms in soils (Norrish, 1975) and probably give the most reliable information on species distribution of metals.

Adsorption Capacity of Different Soil Components and Adsorption Mechanisms

In order to assess the adsorption capacity of soil components for heavy metals, model experiments were carried out in simple systems under controlled conditions (Brümmer et al., 1983; Chubin and Street, 1981; McKenzie, 1980; McLaren et al., 1983). As an example of this kind of work the results of zinc adsorption by different model substances in carbonate-buffered systems are shown in Fig. 2 and Table 2 (Brümmer et al., 1983). Even pure $CaCO_3$ adsorbs zinc (Fig. 2, lower curve). Up to a solution concentration of about 10^2 µmol dm^{-3} Zn the relation between adsorbed and dissolved zinc can be described by adsorption isotherms. Then follows a near-vertical solubility line, which indicates the precipitation of zinc carbonate. In a system of pure $CaCO_3$ under the given experimental conditions, the precipitation of zinc carbonates begins at about 0.5 µmol Zn g^{-1} $CaCO_3$ (~30 ppm Zn). If $CaCO_3$ is mixed with 1% MnO_2 (birnessite), then nearly one hundred times the concentration of Zn in the solid phase (50 µmol g^{-1} solid, ~3000 ppm) is necessary for the formation of zinc carbonate (Fig. 2, upper curve). With 10% amorphous iron oxide or humic acid in the $CaCO_3$ system (not shown) $ZnCO_3$ is not formed, and up to 50 µmol Zn g^{-1} of mixture, in the whole range changes in concentrations are only due to adsorption. Amorphous aluminum oxide gives results similar to those of amorphous iron oxide. Additions of clay minerals (bentonite) also considerably increase the quantity of zinc adsorption. These results indicate that in the presence of soil components with strongly adsorbing surfaces, a relatively large input of zinc into soils is required (the amounts depending on the composition of the soil) to reach the zinc concentration necessary for

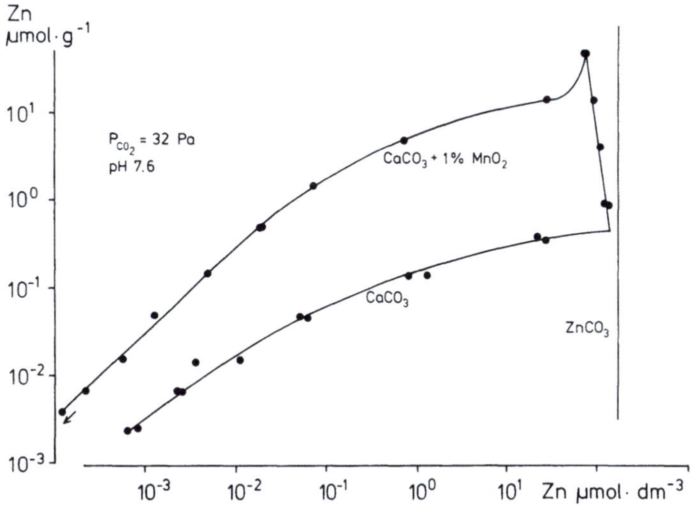

Fig. 2. Zinc concentrations in equilibrium solutions of pure calcite and a mixture of calcite with δ-MnO_2 (1%) in relation to the amount of adsorbed or precipitated zinc. The vertical line gives the calculated solubility of $ZnCO_3$ for the conditions of the experiment (Brümmer et al., 1983). $CaCO_3$ content was 4 g in 20 mL solution, reaction time was seven days

the formation of $ZnCO_3$. As the zinc content of most soils is below 1–2 µmol Zn g^{-1} (65–130 ppm Zn), these results also predict that $ZnCO_3$ is not common under natural conditions.

In Table 2 the zinc adsorption capacity of different soil components in a $CaCO_3$-containing, pH-buffered system is given and compared with the cation-exchange capacity (CEC) of these substances. The high zinc adsorption capacities of manganese, iron, and aluminum oxides clearly show the importance of these substances for the adsorption of zinc and other heavy metals at neutral to slightly alkaline pH values. The selective accumulation of heavy metals by the oxides under natural conditions has been demonstrated by several authors (Jenne, 1968; Kalbasi and Racz, 1978; Lion et al., 1982; McKenzie, 1975; Norrish, 1975; Tiller et al., 1984).

The data in Table 2 also show that the zinc adsorption capacity of bentonite and humic acid is only 0.1 and 0.5 times that of the CECs of these substances, respectively, but 6.7 to 26.7 times that of the oxides. The high values for the oxides indicate to which extent zinc is adsorbed at specific sites, i.e., associated with the hydroxyl surfaces of the oxides which are not

Table 2. Cation-exchange capacity (CEC) and zinc adsorption capacity of different substances in $CaCO_3$-buffered systems (Brümmer et al., 1983)

Substances	CEC at pH 7.6 µmol M^{2+} g^{-1} [a] (a)	Zn-ads. cap. µmol Zn g^{-1} [b] (b)	b/a
$CaCO_3$	–	0.44	–
Bentonite	450	44	0.10
Humic acid	1700	842	0.50
Amorphous Fe-oxide	160	1190	7.4
Amorphous Al-oxide	50	1310	26.2
δ-MnO_2	230	1540	6.7

[a] CEC measurements with Ba^{2+}
[b] Zn adsorption capacity (=adsorption maximum) calculated from Langmuir adsorption isotherms for zinc concentrations in the solution up to the equilibrium concentration of zinc carbonate after a reaction time of seven days (Fig. 2)

available for normal cation exchange. The mechanisms and binding sites of cation exchange of alkaline and alkaline-earth cations are quite different from those of heavy metal adsorption. In order to distinguish between both binding mechanisms the concept of specific and nonspecific adsorption of heavy metals was developed (Forbes et al., 1976; Hodgson et al., 1964; Kinnigurgh et al., 1976; Tiller and Hodgson, 1962; Tiller et al., 1984). If this concept is applied to the above-described chemical extraction procedure for the separation of different heavy metal species, then the fraction of the exchangeable heavy metals is more or less identical with the nonspecifically adsorbed metal fraction, whereas the other chemical treatments with stronger solvents extract specifically adsorbed heavy metals of different bonding strenghts.

A generally accepted view is that the specific adsorption of heavy metals by soil minerals is mainly related to metal-ion hydrolysis (Forbes et al., 1976; Grimme, 1968; Hodgson et al., 1964). With the increasing ability of the metals to form hydroxy complexes, the specific adsorption of the metals also generally increases (Benjamin and Leckie, 1982; Herms and Brümmer, 1984; Kinniburgh et al., 1976; McKenzie, 1980; Tiller et al., 1984). Thus the pK values of the reaction

$$M^{2+} + H_2O = MOH^+ + H^+$$

characterize the general adsorption behavior of the different metals. With decreasing pK values (in brackets), and at equal pK values with increasing ionic size, heavy metal adsorption increases in the sequence Cd (10.1) < Ni (9.9) < Co (9.7) < Zn (9.0) < < Cu (7.7) < Pb (7.7) < Hg (3.4).

The close correlation between metal-ion hydrolysis and specific adsorption is especially well established for goethite. However, in the presence of anions such as phosphates and others, an extension of this concept is necessary (Benjamin and Leckie, 1982). For other soil minerals considerable modifications of the concept may also be unavoidable, taking into consideration structural and spatial peculiarities of the different minerals which can lead to specific key-keyhole relations between heavy metals and soil minerals (Gerth, 1985).

Diffusion of Metals into Soil Minerals

The chemical separation of heavy metal species becomes even more complicated when specific adsorption is not only restricted to surface sites of minerals but also takes place within the surface layer or inside the structure of the minerals. Recent results show that a slow diffusion of heavy metals into the goethite structure is possible (Gerth, 1985; Gerth and Brümmer, 1983; Brümmer et al.,1987). In Fig. 3 the relative diffusion rate (RDR) of Zn, Ni, and Cd into goethite particles is shown in relation to pH. The RDR values of the three metals increase with pH to a maximum (RDRmax) and then decrease again. The pH values of RDRmax of Zn, Ni, and Cd (5.1, 6.1, and 6.0, respectively) and the pK values of the first constant of metal-ion hydrolysis (9.0, 9.9, and 10.1, respectively) show relatively constant differences of nearly four units (3.9, 3.8, and 4.1, respectively). From this observation it can be concluded that RDR increases with pH up to a value which is equal to the pK for $a_{M^{2+}} = a_{MOH^+}$ on the goethite surface. At higher pH values and $a_{MOH^+} > a_{M^{2+}}$ on the goethite surface the RDR values decrease again.

The maximum relative diffusion rates decrease in the order Ni < Zn < Cd (Fig. 3) and can be related to the ionic diameters of these metals (0.69, 0.74, and 0.97 Å, respectively).

The pathways along which metal ions diffuse into the goethite particles are still unknown, but it is very likely that diffusion takes place in micropores or at positions with defects or dislocations in the goethite structure. Thus, adsorption of heavy metals by goethite is determined by three different steps: surface adsorption, diffusion into the goethite particles, and ad-

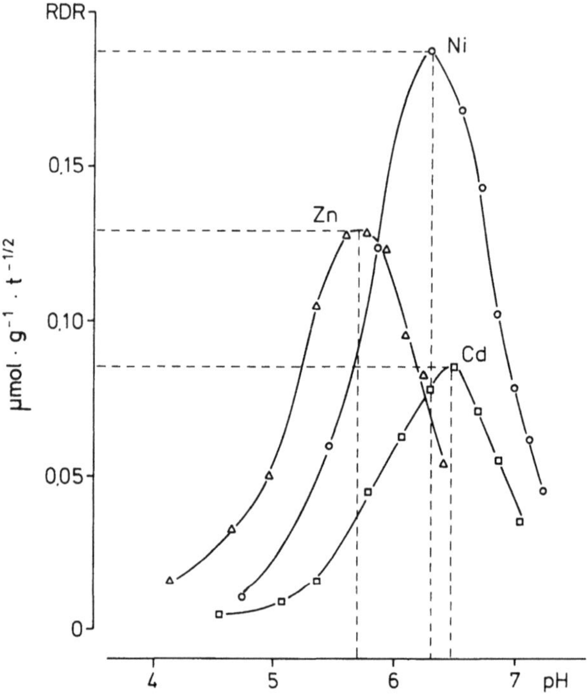

Fig. 3. Relative diffusion rates of zinc, nickel, and cadmium into the structure of goethite in relation to pH (surface area of the goethite: 73 m^2 g^{-1}, thickness of the needles: about 100 Å) (Gerth, 1985; Brümmer et al., 1987)

sorption and fixation at positions inside the particles (Gerth, 1985; Gerth and Brümmer, 1983; Brümmer et al., 1987).

An adsorption of metals within the structure is also observed for manganese oxides (Gerth, 1985; McKenzie, 1980) and for illites and smectites (Gerth, 1985) and seems also to be likely for several other minerals.

The slow diffusion of trace or toxic heavy metals into soil minerals under natural conditions may in part explain why heavy metals added to soils as soluble salts or deposited by pollution processes become immobilized and less available with time.

Adsorption-Desorption and/or Precipitation-Dissolution Processes

Generally it can be stated that the lower the content of heavy metals in the soil and the more sites there are available for high-affinity adsorption, the more likely it is that adsorption-desorption processes will determine the concentration of the elements in soil solution (Brümmer et al., 1983). It is a stringent criterion for the efficiency of adsorption-desorption processes that the concentrations of heavy metals in solution rise with increasing metal content of the solid soil material. In Fig. 4 the concentrations of Zn and Cd in the equilibrium solution of three soil samples containing different amounts of heavy metals are shown in relation to adjusted pH values of 3 to 8 (Herms and Brümmer, 1984). Two of the soils were highly contaminated by industrial pollution. The concentrations of zinc and cadmium in the equilibrium solutions strongly increased with decreasing pH and in-

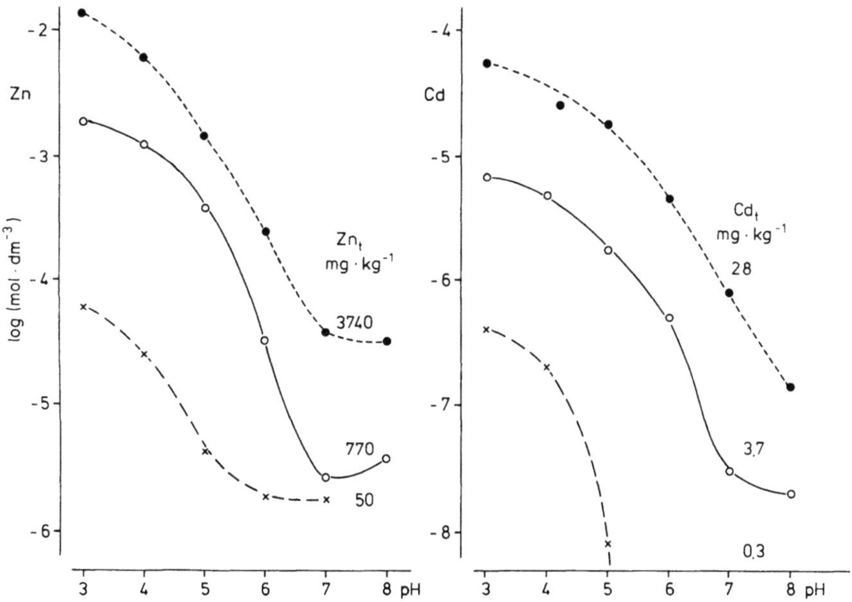

Fig. 4. Concentrations of zinc and cadmium in equilibrium solutions of three soil samples (A horizons) containing different amounts of total Zn and Cd in relation to pH. (×) loamy Luvisol, (o) eutric Gleysol, (●) calcaric Fluvisol (Herms and Brümmer, 1984)

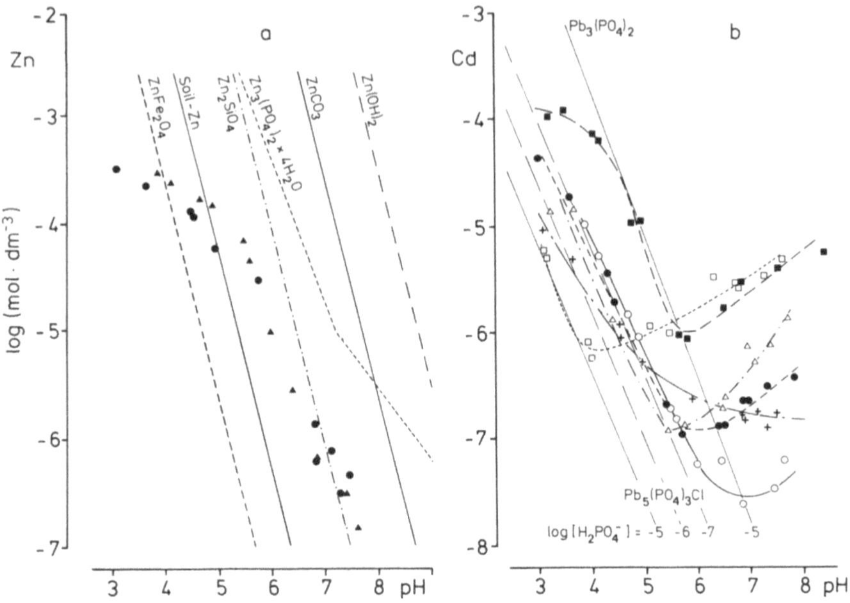

Fig. 5. Solution concentration of **a** zinc and **b** lead of different subsurface horizons in relation to pH (Herms and Brümmer, 1984) and calculated solubility isotherms. (For details of the calculated solubility isotherms see Herms, 1982; Lindsay, 1979)

creasing total content of these metals in the soils. In the same way the concentrations of Cu and Pb in the solution were influenced by both factors (Herms, 1982). The shape of these concentration curves versus pH, namely, of slopes that vary with pH, the total amount of metals present, as well as the kind of soil, are characteristic of adsorption reactions (Brümmer et al., 1983; McBride and Blasiak, 1979). These contrast with precipitation-dissolution reactions of heavy metal species for which theoretically a single relationship should result at a given pH. This relationship is independent of the amount of reactant or of the different mineral surfaces present. In Figs. 5

Fig. 6 a–d. Concentrations of Zn, Cd, Cu, and Pb in equilibrium solutions of samples from A horizons of different soils and a marine mud in relation to pH (Herms and Brümmer, 1984) and calculated solubility isotherms of definite metal species. (For details of the calculated solubility isotherms see Herms 1982; Lindsay, 1979). (●) marine mud; (○) calcaric Fluvisol; (+) loamy Luvisol; (×) eutric Cambisol; (□) leptic Podzol; (△) humic Podzol

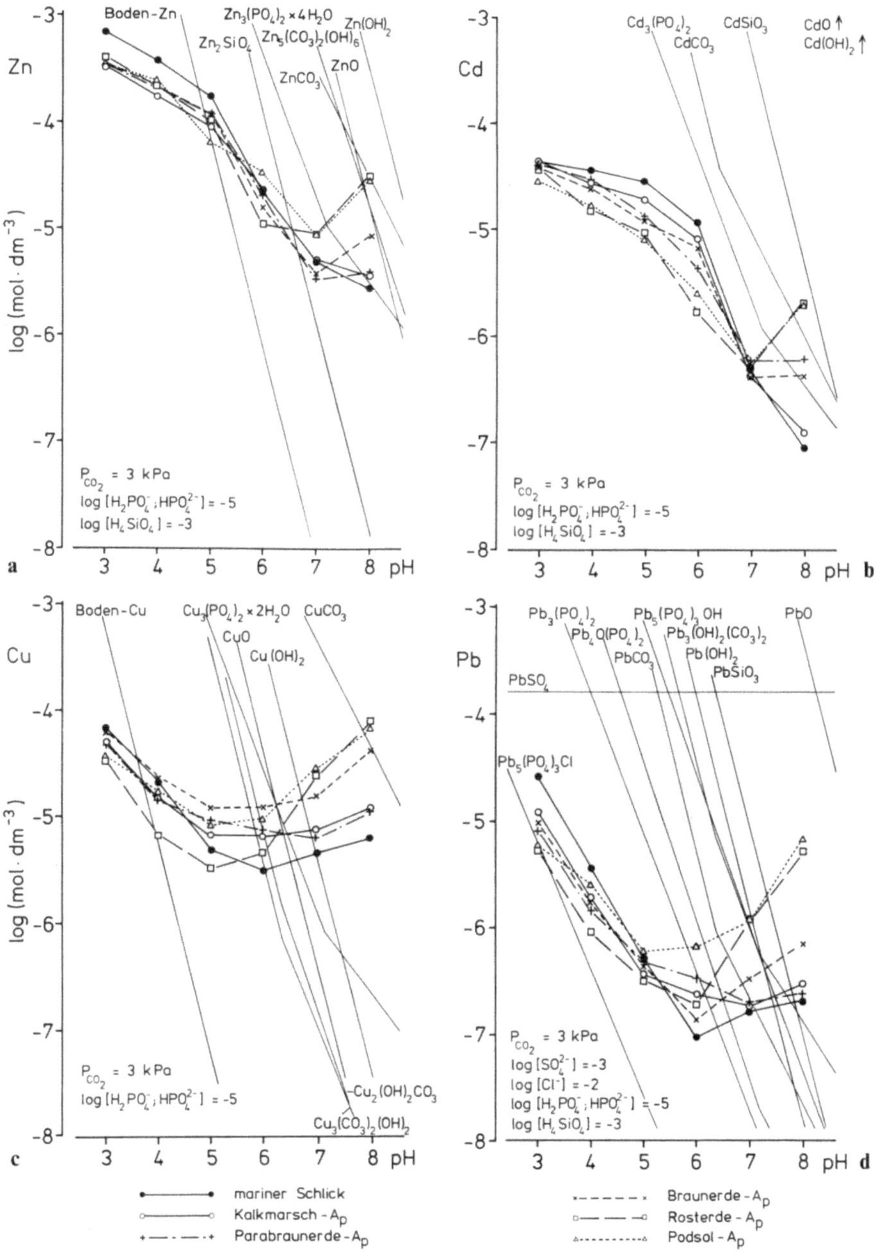
Boden-Zn
$Zn_3(PO_4)_2 \times 4H_2O$
Zn_2SiO_4
$Zn_5(CO_3)_2(OH)_6$
$Zn(OH)_2$
$ZnCO_3$
ZnO
Zn
log (mol · dm⁻³)
P_{CO_2} = 3 kPa
log [$H_2PO_4^-$; HPO_4^{2-}] = -5
log [H_4SiO_4] = -3
a
pH
$Cd_3(PO_4)_2$
$CdSiO_3$
$CdCO_3$
CdO ↑
$Cd(OH)_2$ ↑
Cd
P_{CO_2} = 3 kPa
log [$H_2PO_4^-$; HPO_4^{2-}] = -5
log [H_4SiO_4] = -3
b
Boden-Cu
$Cu_3(PO_4)_2 \times 2H_2O$
$CuCO_3$
CuO
$Cu(OH)_2$
Cu
$Cu_2(OH)_2CO_3$
$Cu_3(CO_3)_2(OH)_2$
P_{CO_2} = 3 kPa
log [$H_2PO_4^-$; HPO_4^{2-}] = -5
c
$Pb_3(PO_4)_2$
$Pb_5(PO_4)_3OH$
PbO
$Pb_4O(PO_4)_2$
$Pb_3(OH)_2(CO_3)_2$
$PbCO_3$
$Pb(OH)_2$
$PbSiO_3$
Pb
$PbSO_4$
$Pb_5(PO_4)_3Cl$
P_{CO_2} = 3 kPa
log [SO_4^{2-}] = -3
log [Cl^-] = -2
log [$H_2PO_4^-$; HPO_4^{2-}] = -5
log [H_4SiO_4] = -3
d
mariner Schlick
Kalkmarsch - A_p
Parabraunerde - A_p
Braunerde - A_p
Rosterde - A_p
Podsol - A_p

and 6, such solubility isotherms are presented in relation to pH for several heavy metal species (Lindsay, 1979).

The results presented in Fig. 6 also indicate that the curves of the pH-dependent solution concentrations do not correspond to solubility isotherms of definite metal compounds. Although in soils (especially at acidic soil reaction) adsorption-desorption processes of heavy metals combined with complexation processes of organic and inorganic ligands (see below) mainly determine the composition of the soil solution, a formation of definite heavy metal compounds is also possible under specific conditions. These conditions include a relatively high content of the heavy metal concerned, a very low solubility of the metal species, a sufficient amount of anions and cations needed for the formation of metal species, and a low content of specific adsorption sites as well as substances (such as organic matter) which may prevent the precipitation of definite compounds (Brümmer et al., 1983).

The necessary low solubility of metal compounds is often associated with a neutral-to-alkaline soil reaction (cf. solubility isotherms in Fig. 6). In this pH range several authors observed or expected the formation of definite metal species such as $Cu_2(OH)_2CO_3$ (McBride and Bouldin, 1984), $Zn_5(OH)_6(CO_3)_2$, $Zn_3(PO_4)_2 \cdot 4H_2O$ (Sposito, 1981), and $CdCO_3$ (Soon, 1981). There are also some indications that in soil samples from subsurface horizons with a very low content of organic matter and a zinc content of more than 100 mg kg^{-1} a formation of Zn_2SiO_4 could be possible at pH values above 7.0 (Fig. 5) (Brümmer et al., 1983). The formation of lead phosphates [$Pb_3(PO_4)_2$, $Pb_5(PO_4)_3Cl$, and others] in polluted soils also seems to be very likely (Fig. 5) (Herms and Brümmer, 1984; Santillan-Medrano and Jurinak, 1975). In pot experiments with increasing Pb additions, phosphate deficiency of the plants was observed as a result of phosphate fixation by Pb in the soil (von Hodenberg, unpublished results, 1974). Under reducing conditions and at high sulfide contents the existence of heavy metal sulfides (PbS, ZnS, CuS, and complex sulfides of different metals) was also observed (Norrish and Brümmer, unpublished results, 1985).

Heavy Metals in the Liquid Phase of Soils

For ecological considerations the concentration and kind of heavy metal species in the soil solution is of primary importance, because metal mobility and availability are closely related to the composition of the liquid phase.

Solubility data also allow conclusions to be drawn about the binding forms of metal species in the solid phases of soils. Several authors analyzed the solution concentration of metals in relation to pH (and redox conditions) in model experiments under controlled conditions (Cottenie et al., 1982, 1983; Herms, 1982; Herms and Brümmer, 1984; McBride and Blasiak, 1979; Sims and Patrick, 1978). As an example of this kind of work some results are presented in Fig. 6.

Five samples from A horizons of different uncontaminated soils were adjusted to comparable total contents of metals by adding to each sample 100 mg kg^{-1} Zn, Cu, and Pb and 15 mg kg^{-1} Cd. The added metals were equilibrated with the soil samples by alternating drying and wetting during eight weeks. Then the pH of soil suspensions was adjusted to values between 3 and 8. After equilibration times of about an additional three weeks the concentrations of heavy metals were measured in the equilibrium solutions (Herms, 1982).

The solution concentrations of Zn and Cd are generally low at pH 7 to 8 but strongly increase below pH 6. The Cu and Pb concentrations show distinct minima between pH 5 and 6 and increase below pH 4 to 5 and above pH 6 to 7. The pH-dependence of the solution concentration of these elements shows that the mobility and availability of heavy metals is high at strongly to extremely acidic soil reaction. Therefore, a high input of heavy metals, especially in strongly acidic forest soils, may lead to toxic environmental effects due to the accumulated elements.

The relative mobility of the different metals can be roughly expressed by the ratio of the dissolved to bound amounts of each metal (Herms and Brümmer, 1984) or by the percentage of the total content of each metal which is dissolved at pH values below 6 (Cottenie et al., 1982). According to this interpretation and in close agreement with the concept of metal-ion hydrolysis (see above), the mobility of heavy metals in acidic soils decreases in the order Cd > Zn > Ni > Mn > Cu > Pb > Hg.

The influence of pH on the concentration of heavy metals in solution is modified to some extent by the content and kind of organic matter (OM) in soils. The results in Fig. 6 show that in some soil samples with a higher content of OM (humic Podzol: A horizon, 4.4% OM; leptic Podzol: A horizon, 8.0% OM), the Cu and Pb concentrations as well as the Zn and Cd concentrations increase at pH values above 6 or 7 because soluble metal-organic complexes are formed in this pH range. In soil samples with a low content of OM (calcaric Fluvisol: A horizon, 2.6% OM; loamy Luvisol: A horizon, 2.8% OM), the concentrations of Cu and Pb in solution show only a slight increase at neutral to alkaline soil reaction (Brümmer and Herms,

1983; Herms and Brümmer, 1984). In this pH range the major proportion of the heavy metals in solution is present as metal-organic complexes (Bloomfield et al., 1976; Jeffery and Uren, 1983; McBride and Blasiak, 1979; McBride and Bouldin, 1984; Sims and Patrick, 1978). Between the content of Cu and organic matter in solution a close correlation exists (Jeffery and Uren, 1983). At pH 7, 99.5% of the dissolved Cu is complexed by soluble organic substances. With decreasing pH, the proportion of complexed Cu also decreases, whereas the Cu^{2+} concentration rises. Below pH 4 the amount of complexed Cu reaches very low values (McBride and Blasiak, 1979; McBride and Bouldin, 1984). Pb forms metal-organic complexes of similar stability but probably higher molecular weight (Herms, 1982), whereas the Cd and especially the Zn complexes possess a lower stability. At pH values below 6 the proportion of complexed Zn decreases rapidly (Bloomfield et al., 1976; McBride and Blasiak, 1979; Sims and Patrick, 1978).

The metal-organic complexes of the liquid phases are very likely dominated by negatively charged species (Bloomfield et al., 1976; Sims and Patrick, 1978). At pH values below the point of zero charge (PZC) of pedogenetic iron oxides ($\sim$pH 6), an adsorption and probably a stabilization of these complexes on sites with positive charge takes place. With decreasing pH the solubility of humic acids also decreases. Thus it is probable that solid metal-organic complexes can exist even at extremely acidic soil reaction. There are several indications that considerable proportions of Pb and Cu are bound by solid organic substances even at pH values around 3, and those of Cd still at pH values between 3 and 4, whereas solid metal-organic complexes of Zn are not stable below pH 5 (Herms and Brümmer, 1984). The data of Fig. 6 also show that the soil samples with a relatively high content of OM (A horizons of humic and leptic Podzol) are characterized at pH 3 to 5 by lower solution concentrations of Cd, Cu, and Pb than are soil samples with a low content of OM and high content of clay minerals and iron oxides (A horizons of calcaric Fluvisol and loamy Luvisol, marine mud). Thus, OM obviously possesses a much higher binding capacity for heavy metals at strongly acidic soil reaction than do mineral substances (Brümmer and Herms, 1983; Herms and Brümmer, 1984; Tiller et al., 1984), and therefore soil OM can reduce toxic effects of heavy metals in contaminated acidic soils. The accumulation of metals in the humic layer of acidic forest soils (Table 1) (see also Mayer, 1981; Schwertmann et al., 1982; Ulrich et al., 1979) is explained in part by this fact.

Besides the complexation of metals by organic substances, inorganic anions can also decrease adsorption and increase the concentration of metals in solution by forming soluble inorganic metal complexes. Several authors investigated the importance of inorganic complexation for heavy metal mobility and availability in soils. Especially chloride can increase the concentrations of Cd (Bingham et al., 1983; Gerth et al., 1981; Hahne and Kroontje, 1973) and Hg (Hahne and Kroontje, 1973; Kinniburgh and Jackson, 1978) in solution, and sulfate those of Ni (Bowman and O'Connor, 1982) and Cd (Garcia-Miragaya and Page, 1976) by forming chloro- and sulfato-complexes, respectively.

Characterization of Heavy Metal Availability for Plants

In order to characterize the availability of nutrients as well as toxic elements for plants, at least three parameters have to be considered: the total amount of potentially available elements (quantity factor), the concentration or activity as well as the ionic ratios of elements in the soil solution (intensity factor), and the rate of element transfer from solid to liquid phases and to plant roots (rate factor or reaction kinetics) (Fig. 7).

The total pool of potentially available metals comprises those fractions which can deliver metals from solid phases to the soil solution within relatively short times, e.g., in one vegetation period. Probably the amount of

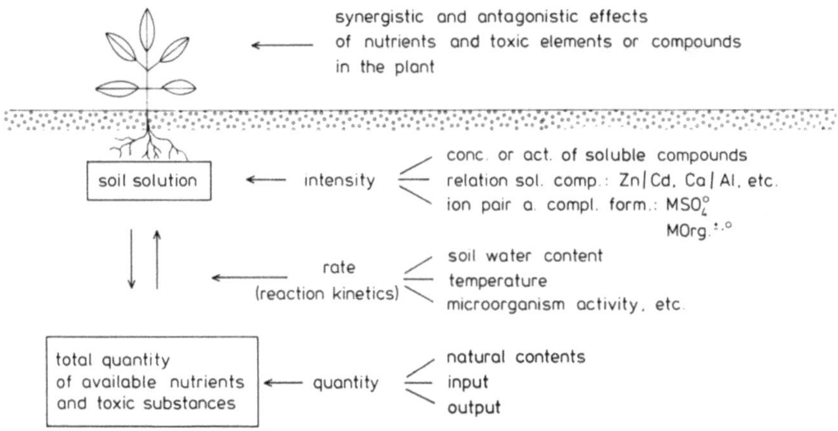

Fig. 7. Schematic presentation of the parameters which determine the availability of heavy metals for plants

EDTA- or DTPA-extractable heavy metals (Köster and Merkel, 1983; Lindsay, 1979) gives satisfactory estimates of this pool. Methods which also extract metals occluded by iron, aluminum, and manganese oxides or bound in the structure of silicates are not suitable for this purpose.

Because the uptake of elements by the plant roots is restricted to the liquid phase, the content of heavy metals in the soil solution is of primary importance. This fraction can be determined in the water saturation extract of soil samples (Bingham et al., 1983; Horak, 1979) or in soil extracts obtained with dilute salt solutions such as 0.1 M $CaCl_2$ or $Ca(NO_3)_2$ (Köster and Merkel, 1983). Both fractions are often satisfactorily correlated with the uptake of heavy metals by plants.

Recently published results indicate that metal-ion activity in the water saturation extract probably allows the best estimation of the intensity factor (Bingham et al., 1983; Sparks, 1984; Sposito, 1984; Sposito and Bingham, 1981). For the calculation of metal-ion activity, corrections were made for ion-pair and complex formation. Although this approach seems to be quite successful, the significance of single-ion activity is limited to its use, through a self-consistent set of conventions, for predicting activities of electrolytes (Sposito, 1984). More information is also necessary concerning multiple-ion activity ratios (Sparks, 1984) and the antagonistic or synergistic effects of accompanying cations and anions in the soil solution in relation to uptake and physiological effects of heavy metals in plants (Fig. 7). Only insufficient information exists regarding the availability of organic heavy metal complexes in the solution phase (Baccini and Suter, 1979).

While an accurate measurement of the intensity factor is difficult, there are hardly any methods developed and standardized for the estimation of the rate factor. Detailed information concerning the reaction kinetics of the mobilization and immobilization processes of heavy metals in soils as well as the importance of mass flow and diffusion for the availability of heavy metals is urgently necessary. Thus, many fundamental investigations still have to be carried out before we can sufficiently well characterize heavy metal availability in soils.

References

Baccini P, Suter U (1979) MELIMEX, an experimental heavy metal pollution study: Chemical speciation and biological availability of copper in lake water. Schweiz Z Hydrol 41:291–314

Benjamin MM, Leckie JO (1981) Multi-site adsorption of Cd, Cu, Zn, and Pb on amorphous iron oxyhydroxide. J Colloid Interf Sci 79:209–221

Benjamin MM, Leckie JO (1982) Effects of complexation by Cl, SO_4, and S_2O_3 on adsorption behaviour of Cd on oxide surfaces. Envir Sci Technol 16:162–170
Biddappa CC, Chino M, Kumazawa K (1982) Migration of heavy metals in two Japanese soils. Plant Soil 66:299–316
Bingham FT, Strong JE, Sposito G (1983) Influence of chloride salinity on cadmium uptake by Swiss chard. Soil Sci 135:160–165
Bloomfield C, Kelso WJ, Pruden G (1976) Reactions between metals and humified organic matter. J Soil Sci 27:16–31
Bowman RS, O'Connor GA (1982) Control of nickel and strontium sorption by free metal ion activity. Soil Sci Soc Am J 46:933–936
Blume H-P, Hellriegel T (1981) Blei- und Cadmium-Status Berliner Böden. Z Pflanzenernähr Bodenkd 144:181–196
Brümmer G, Gerth J, Tiller KG (1987) Reaction kinetics of the adsorption ans desorption of nickel, zinc, and cadmium by goethite. I. Adsorption and diffusion of metals. J Soil Sci, in press
Brümmer G, Herms U (1983) Influence of soil reaction and organic matter on the solubility of heavy metals in soils. In: Ulrich B, Pankrath J (eds) Effects of Accumulation of Air Pollutants in Forest Ecosystems. D. Reidel Publishing Co, pp 233–243
Brümmer G, Tiller KG, Herms U, Clayton PM (1983) Adsorption-desorption and/or precipitation-dissolution processes of zinc in soils. Geoderma 31:337–354
Camerlynck R, Kiekens L (1982) Speciation of heavy metals in soils based on charge separation. Plant Soil 68:331–339
Chubin RG, Street JJ (1981) Adsorption of cadmium on soil constituents in the presence of complexing ligands. J Envir Qual 10:225–228
Cottenie A, Velghe G, Verloo M, Kiekens L (1982) Biological and Analytical Aspects of Soil Pollution. Ghent, Belgium: Laboratory of Analytical and Agrochemistry, State University of Ghent
Cottenie A, Verloo M, Kiekens L, Camerlynck R, Velghe G, Dhaese A (1983) Essential and Nonessential Trace Elements in the System Soil-Water-Plant. Ghent, Belgium: Laboratory of Analytical and Agrochemistry, State University of Ghent
Elsokkary JH, Lag J (1978) Distribution of different fractions of Cd, Pb, Zn, and Cu in industrially polluted and non-polluted soils of Odda region, Norway. Acta Agric Scand 28:262–268
Fischer WR, Fechter H (1982) Analytische Bestimmung und Fraktionierung von Cu, Zn, Pb, Cd, Ni und Co in Böden und Unterwasserböden. Z Pflanzenernähr Bodenkd 145:151–160
Forbes EA, Posner AM, Quirk JP (1976) The specific adsorption of divalent Cd, Co, Cu, Pb, and Zn on goethite. J Soil Sci 27:154–166
Garcia-Miragaya J, Page AL (1976) Influence of ionic strength and inorganic complex formation on the sorption of trace amounts of Cd by montmorillonite. Soil Sci Soc Am J 40:658–663
Gerritse RG, Salomons W (1983) The chromatographic behaviour of heavy metal species in sephadex gel columns. Envir Technol Lett 4:463–468
Gerth J (1985) Untersuchungen zur Adsorption von Nickel, Zink und Cadmium durch Bodentonfraktionen unterschiedlichen Stoffbestandes und verschiedene Bodenkomponenten. Dissertation, University of Kiel

Gerth J, Brümmer G (1983) Adsorption und Festlegung von Nickel, Zink und Cadmium durch Goethit (α-FeOOH). Fresenius Z Anal Chem 316:616–620
Gerth J, Schimming CG, Brümmer G (1981) Einfluß der Chloro-Komplexbildung auf Löslichkeit und Adsorption von Nickel, Zink und Cadmium. Mitteilgn Dtsch Bodenkundl Gesellsch 30:19–30
Gregson SK, Alloway BJ (1984) Gel permeation chromatography studies on the speciation of lead in solutions of heavily polluted soils. J Soil Sci 35:55–61
Grimme H (1968) Die Adsorption von Mn, Co, Cu and Zn durch Goethit aus verdünnten Lösungen. Z Pflanzenernähr Düng Bodenkd 121:58–65
Hahne HCH, Kroontje W (1973) Significance of pH and chloride concentration on behaviour of heavy metal pollutants: Mercury(II), cadmium(II), zinc(II) and lead(II). J Envir Qual 2:444–450
Herms U (1982) Untersuchungen zur Schwermetallöslichkeit in kontaminierten Böden und kompostierten Siedlungsabfällen in Abhängigkeit von Bodenrekation, Redoxbedingungen und Stoffbestand. Dissertation, University of Kiel
Herms U, Brümmer G (1984) Einflußgrößen der Schwermetall-Löslichkeit und -Bindung in Böden. Z Pflanzenernähr Bodenkd 147:400–424
Hetsch W, Beese F, Ulrich B (1979) Die Beeinflussung der Bodenlösung durch Saugkerzen aus Ni-Sintermetall und Keramik. Z Pflanzenernähr Bodenkd 142:29–38
Hodgson JF, Tiller KG, Fellows M (1964) The role of hydrolysis in the reaction of heavy metals with soil-forming materials. Soil Sci Soc Am Proc 28:42–46
Horak O (1979) Untersuchungen zur Bleiaufnahme der Pflanze. Bodenkultur 30:120–126
Horrison RD (1982) A modified vacuum-pressure lysimeter for soil water sampling. Soil Sci 134:206–210
Jeffery JJ, Uren NC (1983) Copper and zinc species in the soil solution and the effects of soil pH. Aust J Soil Res 21:479–488
Jenne EA (1968) Controls of Mn, Fe, Co, Ni, Cu, and Zn concentrations in soils and water: the significant role of hydrous Mn and Fe oxides. Adv Chem Ser No 73:337–387
Jenne EA (ed) (1979) Chemical Modelling in Aqueous Systems. ACS Symposium Series No 93. American Chemical Society, Washington, DC
Juo ASR, Kamprath EJ (1979) Copper chloride as an extractant for estimating the potentially reactive aluminium pool in acid soils. Soil Sci Soc Am J 43:35–38
Kabata-Pendias A, Pendias H (1984) Trace Elements in Soils and Plants. Chemical Rubber Company
Kalbasi M, Racz GJ (1978) Association of zinc with oxides of iron and aluminium in some Manitoba soils. Can J Soil Sci 58:61–68
Kinniburgh DG, Jackson ML (1978) Adsorption of mercury(II) by iron hydrous oxide gel. Soil Sci Soc Am J 42:45–47
Kinniburgh DG, Jackson ML, Syers JK (1976) Adsorption of alkaline earth, transition and heavy metal cations by hydrous oxide gels of iron and aluminium. Soil Sci Soc Am J 40:796–799
Köster W, Merkel D (1983) Beziehungen zwischen den Gehalten an Zink, Cadmium, Blei und Kupfer in Böden und Pflanzen bei Anwendung unterschiedlicher Bodenuntersuchungsmethoden. Landwirtsch Forsch Sonderh 39:245–254

Kuo S, Heilman PE, Baker AS (1983) Distribution and forms of copper, zinc, cadmium, iron, and manganese in soils near a copper smelter. Soil Sci 135:101–109

Lichtfuß R, Brümmer G (1981) Gehalte an organischer Substanz, Schwermetallen und Phosphor in Dichtefraktionen von fluvialen Unterwasserböden. Geoderma 25:245–265

Lindsay WL (1979) Chemical Equilibria in Soils. John Wiley and Sons, New York

Lion LW, Altmann RS, Leckie JO (1982) Trace-metal adsorption characteristics of estuarine particulate matter: Evaluation of contributions of Fe/Mn oxide and organic surface coatings. Envir Sci Technol 16:660–666

Mayer R (1978) Adsorptionsisothermen als Regelgrößen beim Transport von Schwermetallen in Böden. Z Pflanzenernähr Bodenkd 141:11–28

Mayer R (1981) Natürliche und anthropogene Komponenten des Schwermetallhaushalts von Waldökosystemen. Göttinger Bodenkdl Ber 70:1–292

McBride MB, Blasiak JJ (1979) Zinc and copper solubility as a function of pH in an acid soil. Soil Sci Soc Am J 43:866–870

McBride MB, Bouldin DR (1984) Long-term reactions of copper(II) in a contaminated calcareous soil. Soil Sci Soc Am J 48:56–59

McKenzie RM (1975) An electron microprobe study of the relationship between heavy metals and manganese and iron in soils and ocean floor nodules. Aust J Soil Res 13:177–188

McKenzie RM (1980) The adsorption of lead and other heavy metals on oxides of manganese and iron. Aust J Soil Res 18:61–73

McLaren RG, Williams JG, Swift RS (1983) The adsorption of copper by soil samples from Scotland at low equilibrium solution concentration. Geoderma 31:97–106

Norrish K (1975) Geochemistry and mineralogy of trace elements. In: Trace Elements in Soil-Plant-Animal Systems. Academic Press, New York, pp 55–81

Page AL (ed) (1982) Methods of Soil Analysis. Part 2. Chemical and Microbiological Properties. Agronomy No 9. American Society of Agronomy, Madison, WI

Poelstra P, Frissel MJ, El-Bassam N (1979) Transport and accumulation of Cd ions in soils and plants. Z Pflanzenernähr Bodenkd 142:848–864

Santillan-Medrano J, Jurinak JJ (1975) The chemistry of lead and cadmium in soils: Solid phase formation. Soil Sci Soc Am Proc 39:851–856

Scheffer F, Schachtschabel P (1982) Lehrbuch der Bodenkunde. F. Enke-Verlag, Stuttgart

Schwertmann U, Fischer WR, Fechter H (1982) Spurenelemente in Pedosequenzen. I. Zwei Braunerde-Podsol-Sequenzen aus Tonschieferschutt. II. Zwei Pararendzina-Pseudogley-Sequenzen aus Löß. Z Pflanzenernähr Bodenkd 145:161–180; 180–196

Shuman LM (1979) Zinc, manganese, and copper in soil fractions. Soil Sci 127:10–17

Shuman LM (1982) Separating soil iron- and manganese-oxide fractions for microelement analysis. Soil Sci Soc Am J 46:1099–1102

Shuman LM (1983) Sodium hypochlorite methods for extracting microelements associated with soil organic matter. Soil Sci Soc Am J 47:656–660

Sims JL, Patrick WH (1978) The distribution of micronutrient cations in soil under conditions of varying redox potential and pH. Soil Sci Soc Am J 42:258–262

Soon YK (1981) Solubility and sorption of cadmium in soils amended with sewage sludge. J Soil Sci 32:85–95
Sparks DL (1984) Ion activities: An historical and theoretical overview. Soil Sci Soc Am J 48:514–518
Sposito G (1981) The Thermodynamics of Soil Solutions. Oxford: Clarendon Press
Sposito G (1984) The future of an illusion: Ion activities in soil solutions. Soil Sci Soc Am J 48:531–536
Sposito G, Bingham FT (1981) Computer modeling of trace metal speciation in soil solutions: Correlation with trace metal uptake by higher plants. J Plant Nutr 3:35–49
Sposito G, Lund LJ, Chang AL (1982) Trace metal chemistry in arid-zone field soils amended with sewage sludge. I. Fractionation of Ni, Cu, Zn, Cd, and Pb in solid phases. Soil Sci Soc Am J 46:260–264
Tiller KG, Gerth J, Brümmer G (1984) The relative affinities of Cd, Ni, and Zn for different soil clay fractions and goethite. Geoderma 34:17–36
Tiller KG, Hodgson JF (1962) The specific sorption of cobalt and zinc by layer silicates. Clays Clay Miner 9:393–403
Tills AR, Alloway BJ (1983) The use of liquid chromatography in the study of cadmium speciation in soil solutions from polluted soils. J Soil Sci 34:769–781
Ulrich B, Mayer R, Khanna PK (1979) Deposition von Luftverunreinigungen und ihre Auswirkungen in Waldökosystemen im Solling. Sauerländer's Verlag, Frankfurt a.M.

The Importance of Chemical "Speciation" in Environmental Processes,
eds. M. Bernhard, F. E. Brinckman, and P. J. Sadler, pp. 193–216.
Dahlem Konferenzen 1986. Springer-Verlag Berlin, Heidelberg
© Dr. S. Bernhard, Dahlem Konferenzen

Chemical Species and Metal Transport in Lakes

W. Salomons * and P. Baccini **

* Delft Hydraulics Laboratory-Haren Branch
c/o Institute for Soil Fertility
9750 RA Haren (Gr), The Netherlands

** Institute for Water Resources and Water Pollution Control
Swiss Federal Institute of Technology
8600 Duebendorf, Switzerland

Abstract. The processes affecting distribution of trace metal species in lakes can be divided in two categories:

1) Species transformation at system boundaries: river-lake, atmosphere-lake, and sediment-water.
2) Biochemical processes of assimilation and respiration in the system which can transform species and alter transport processes by increasing or decreasing proton activity, by producing ligands, and by increasing or decreasing electron activity. Descriptive data are only available for lakes in temperate climates

Introduction

Freshwaters such as lakes and rivers represent less than one per mill of the total mass of water which is continually recycled on the surface of our planet. Although their quantitative contribution on a global scale seems to be negligible, their qualitative properties are indispensable for life.

Trace metal sources for lakes are the atmosphere, riverine inputs, and waste discharges (Fig. 1). Metals are introduced in lakes in both dissolved and particulate forms. The allochthonous particles will partially settle to the bottom, whereas the dissolved trace metals are subject to removal processes such as adsorption, uptake by biota, and incorporation in authigenous phases. In the oxic zone there is a strong interaction between the carbon

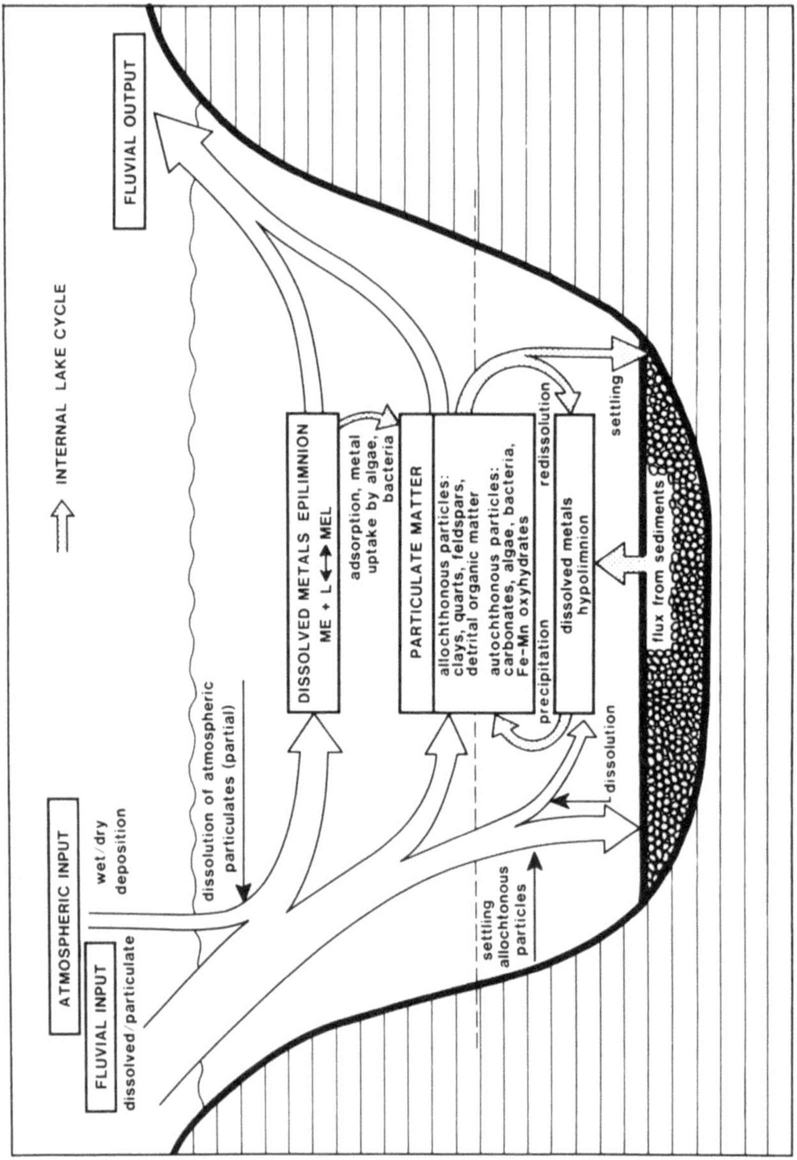

Fig. 1. Metal cycling in lakes (Salomons and Förstner, 1984)

cycle and the various metal cycles. The growth of algae causes changes in pH and an excretion of organic products. A high pH might result in an additional precipitation of calcium carbonate. The various autochthonous particles provide additional surfaces for adsorption and subsequent removal of trace metals from the water column by settling. While falling through the water column, in the same way as in oceans, a partial dissolution of the trace metals may take place. In lakes with an anoxic hypolimnion there is a strong interaction between the metals and the redox cycle. In the anoxic waters a strong redissolution of iron and manganese takes place (in the water column and in the sediments). The two metals diffuse upwards and are precipitated at the oxic/anoxic interface in the lake. The resulting precipitates are able to remove dissolved metals.

The pathways/transport processes of trace metals in lakes can be distinguished as a) the processes at the systems boundaries (the river-lake boundary, the atmosphere-lake boundary, and the sediment-lake boundary), and b) the processes inside the system.

In addition, human perturbations may change the metal fluxes in lakes. Important parameters in this respect are a) nutrient inputs, causing eutrophication; b) acid input, causing a drop in pH in weakly buffered systems; and c) input of complexing agents.

It should be noted that we know much more about lakes in moderate climates than about cycling of metals in tropical lakes.

Interactions at the System's Boundaries

The River-Lake Boundary

Metals enter lakes from rivers in both dissolved and particulate forms (including metal species associated with the suspended matter). In a first approximation a distinction between these two *forms* is sufficient for describing the processes. The dissolved and particulate forms are operationally defined by the use of a 0.45 μm membrane filter. The ratio between the concentration of dissolved and particulate forms of metals depends on chemical conditions in the river system, input from waste, and natural weathering processes. The dissolved transport index (percentage of dissolved transport over total transport) for a number of elements is presented in Table 1.

Physical processes predominate in fast-flowing rivers. At low water discharge there is little suspended matter and no bed-load transport. At high water discharge, surface runoff contributes **significantly:** earlier deposited

Table 1. Dissolved transport index (DTI) in rivers (Martin and Meybeck, 1979). Percentage of dissolved transport over total transport (ranked in decreasing order)

90–50%	Br, I, Sa, Cl, Ca, Na, Sr
50–10%	Li, N, Sb, As, Mg, B, Mo, F, Cu, Zn, Ba, K
10–1%	P, Ni, Si, Rb, U, Co, Mn, Cr, Mn, Th, Pb, V, Cs
1–0.1%	Ga, Tm, Lu, Gd, Ti, Er, Nd, Ho, La, Sm, Tb, Yb, Fe, Eu, Ce, Pr, Al

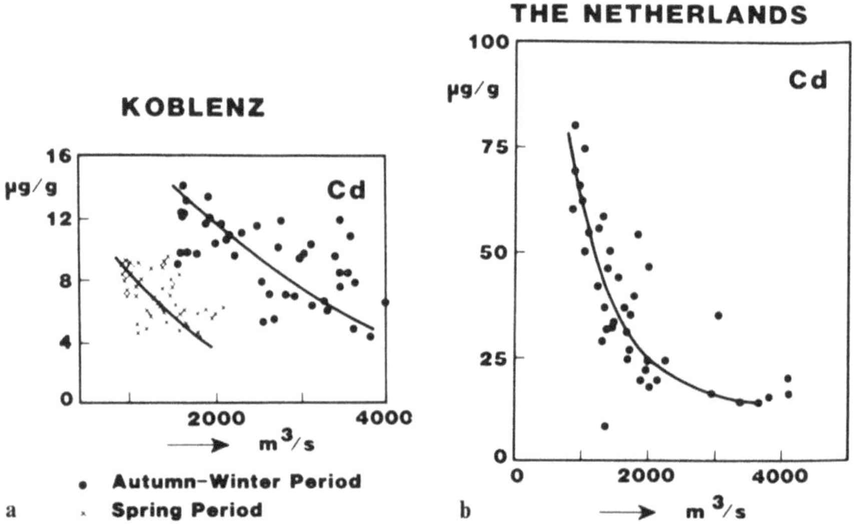

Fig. 2. The relationship between cadmium concentrations (mg Cd g^{-1} suspended matter) and water velocity (m^3 s^{-1}) in the river Rhine at Koblenz (Germany) and in the Netherlands (Salomons and Eysink, 1981)

sediments from the main river and its tributaries are brought into suspension, increasing the concentration of particles in the river (Salomons and Eysink, 1981). For example, the two distinct relationships in cadmium concentrations as a function of water discharge in the Rhine at Koblenz are caused by flushing of its tributary Neckar during high rainfall periods (Fig. 2a). Downstream, the contribution of other pollutant sources masks the influence of the Neckar (Fig. 2b).

For a number of rivers it has been found that the particulate cadmium concentration (in fact, that of most trace metals) depends on the discharge.

This phenomenon is probably caused by the following effects:

1) Proportional dilution as the discharge increases, assuming a constant load of cadmium into the river system;
2) Dilution as the increased erosion during high discharge (surface runoff) causes a mixing of contaminated fluvial particulates with soil particles;
3) Difference in grain size composition. At low discharge suspended matter is relatively finer. The coarser particles are settled on the river floor. Cadmium (and most contaminants) is preferentially associated to the finer particles; and
4) Difference in residence time. The residence time of the particles in the river during high discharge will be low in contrast to periods of low discharge. As the amount of cadmium which can adsorb to the particles depends on their residence time, relatively small amounts will be available for adsorption during periods of high discharge.

In rivers containing locks such as sluices, the relation between particulate cadmium concentration and discharge looks somewhat different. During low discharge the suspended matter with relatively high Cd concentrations settles in front of these locks. This sediment is eroded at higher discharge, the result being that the Cd concentration does not differ much from that at low discharge. Only after longer periods of high discharge, when all the sediment is removed, does the same relation exist as in nonlock-regulated rivers. An example is the river Meuse.

The distribution of trace metal species on and in suspended particles has been studied by sequential chemical extraction techniques (Salomons and Förstner, 1980).

This scheme consists of an extraction with ammonium-acetate (exchangeable fraction), followed by an extraction with hydroxylamine-HCl (pH 2) presumably releasing metals associated with manganese (and partly iron) hydroxides and carbonates. The third step is an extraction with hydrogen peroxide (30%, pH 2.5) releasing metals associated with organic solids. The amount of metal not released by these three extractions is called the resistant fraction and is presumably locked up in silicates and unavailable for short-term geochemical processes. The results show that in more polluted rivers a larger part of the trace metals is associated with mobile forms (the first three fractions) when compared with less- or nonpolluted rivers. In nearly all cases copper was associated with organic solids more than the other metals.

Data on species composition in the dissolved form are scarce. Mississippi and Minnesota river waters contain dissolved organics which can complex copper, cadmium, and lead (Eisenreich et al., 1980). In general the composition of lake water differs from that of river water, e.g., with respect to pH and dissolved organic matter concentrations. In addition, the flow velocity of the river water is higher. As a consequence, upon entering the lake environment a large number of changes affect the distribution of trace metals between the particulate and dissolved phases and the species distribution within these two phases. Settling of the coarser particles will cause an increase in metal concentrations in the remaining suspended matter, because the coarse-grained particles contain fewer trace metals. Estuarine research has addressed this aspect in detail (Duinker, 1983). A characteristic dependence is found on metal concentrations in suspended matter as a function of suspended matter concentration in the water (Fig. 3). This can be explained by a model that considers suspended matter to be composed of small low-density particles that are permanently in suspension and large high-density particles belonging to suspension in bottom sediments. These fractions have different settling properties as well as different metal concentrations. The relative contribution of the permanently suspended fraction to the total suspended matter is low at high suspended matter concentrations (eroded bottom particles dominate) but is high at low suspended matter concentrations.

Whether similar effects occur in the mixing zone of the river (entry into a lake) is not known, but this seems highly probable. The quantification of these processes should make it possible to determine metal input into the main body of the lake. The differences in chemical composition encountered when river water enters the lake environment should cause a redistribution of the dissolved metal species, a change in the surface properties of the particulates, and consequently a change in the distribution of trace metals between the dissolved and particulate phases. The chemical processes associated with the mixing of river and lake water are not well studied in comparison with estuarine studies. For example, in the IJsselmeer a de-

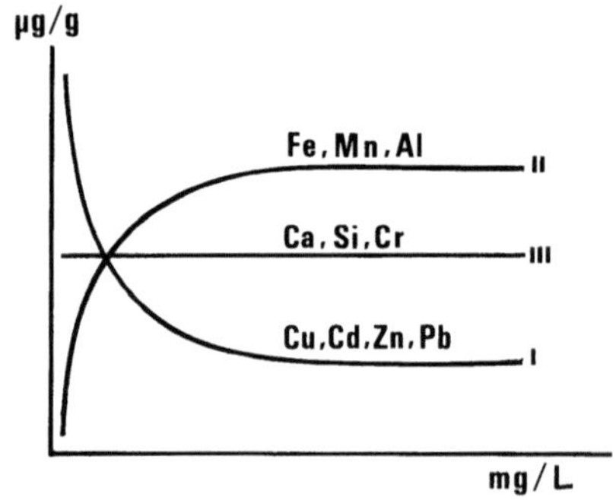

Fig. 3. Relationship between (mg L^{-1}) suspended matter concentrations and metal concentrations in suspended matter (mg metal g^{-1} suspended matter) (Duinker, 1983)

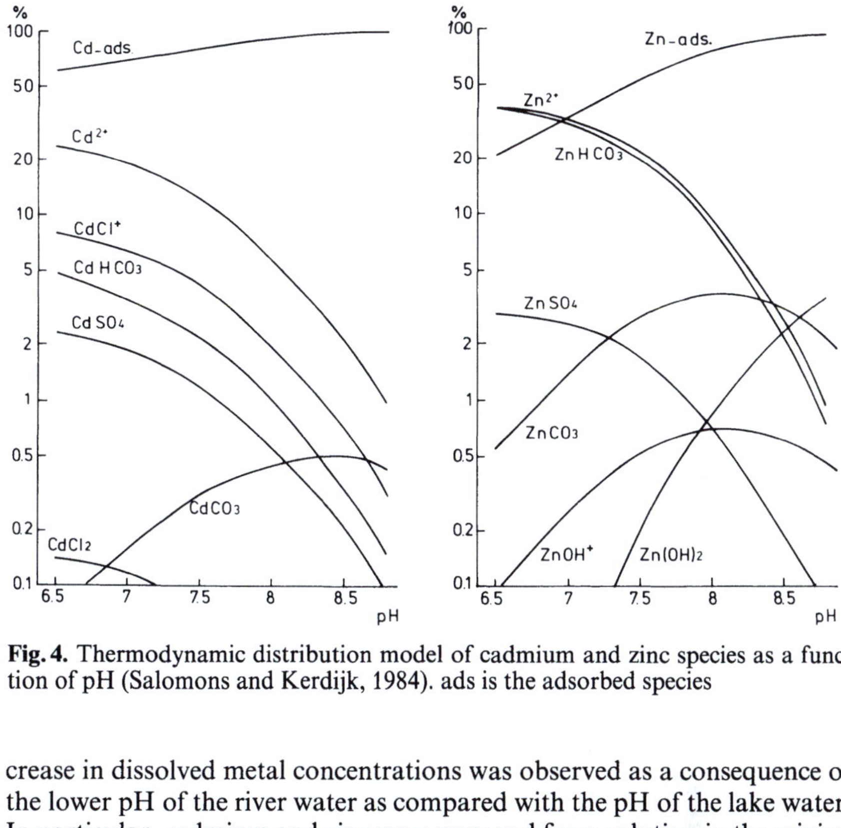

Fig. 4. Thermodynamic distribution model of cadmium and zinc species as a function of pH (Salomons and Kerdijk, 1984). ads is the adsorbed species

crease in dissolved metal concentrations was observed as a consequence of the lower pH of the river water as compared with the pH of the lake water. In particular, cadmium and zinc were removed from solution in the mixing zone (Salomons, 1983). Examples of the species transformation predicted on the basis of thermodynamic calculation of the dissolved forms of Cd and Zn as a function of pH are shown in Fig. 4. These equilibrium predictions were successful in predicting seasonal cycles of zinc concentrations in the IJsselmeer (see Fig. 9, below).

The Atmosphere-Lake Boundary

The atmosphere can be a major source for the input of trace metals into the freshwater system. Its relative importance depends on the contribution from other sources and on the residence time of the water in the lake. A high residence time in and a high surface-area-to-depth ratio of the lake causes a relatively large atmospheric contribution. For some lakes, the atmospheric input is an especially important anthropogenic source where other pollutant sources are small or absent. In some cases, it was possible to attribute the metal concentrations along profiles in the deposited sediments to an increased atmospheric input (Salomons, 1984a). For lakes close to industrial areas, the atmosphere can be an important source. For Lake Ontario and other Great Lakes, for example, about 30%–60% of the annual trace metal input is atmospheric (Salomons, 1984b). The relative importance for each element varies to a considerable extent. In Lake Michigan (Fig. 5) atmospheric loading is especially important for lead (60%) and is significant for zinc (33%). For copper it contributes 13% and for cadmium and manganese 11% (Eisenreich, 1980).

The input from the atmosphere is by both wet and dry deposition. In Lake Michigan the dry deposition contributes 75% of the total loading for lead and 50% for zinc (Sievering et al., 1981).

Atmospheric trace metals enter the surface microlayer which may be operationally defined as extending from 3 Å (the diameter of a water molecule) to about 3 mm, which is the extreme limit of nonturbulent kinetics with no wind as well as the penetration depth of jets from small bubbles (MacIntyre, 1974). Most information on the surface microlayer has been obtained from the marine environment (Salomons, 1984b; Hardy, 1982). The surface layer is highly enriched in organic matter as compared with the bulk water and has a higher microbial activity. The organic material is predominantly of proteinaceous or carbohydrate nature (Lion and Leckie, 1981).

The residence time of atmospheric particles in the surface layer may be up to 30 minutes, sufficiently long to allow significant interactions (Elzer-

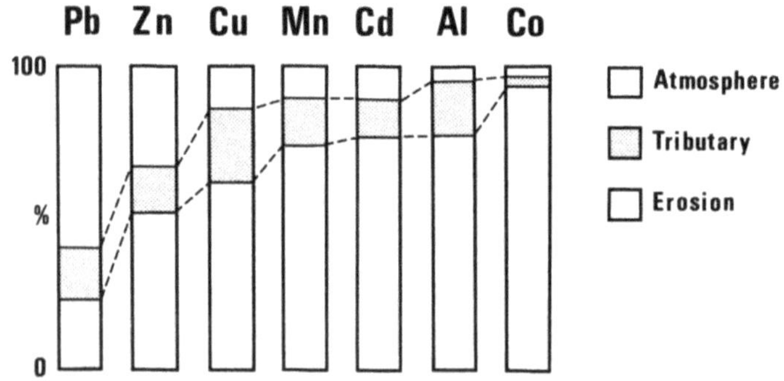

Fig. 5. Relative importance (in %) of the input of trace metals in Lake Michigan from the atmosphere, tributaries, and erosion (Eisenreich, 1980)

man et al., 1979). The species distribution of trace metals in atmospheric particulate matter will determine their reactivity at the lake-air interface. Information of the mode of occurrence of trace metals in urban particulates has been obtained with selective chemical leaching techniques (Lum, 1982). The results show high proportions of soluble metals, especially Zn, Pb, and Cd. Co and Cr are more tightly bound in the residual fraction. Lindberg and Harris (1983) studied in detail the relationship between solubility and particle size (Fig. 6).

All elements exhibit a trend of increasing mean solubility as particle size decreases. Differences in origin of the various particles are probably responsible for the different solubilities found. The high solubility of trace metals contained in atmospheric particles is a general phenomenon (Salomons, 1984 b) and shows that dry deposition in lakes will result in a release of trace metals to the surface layer and the bulk water.

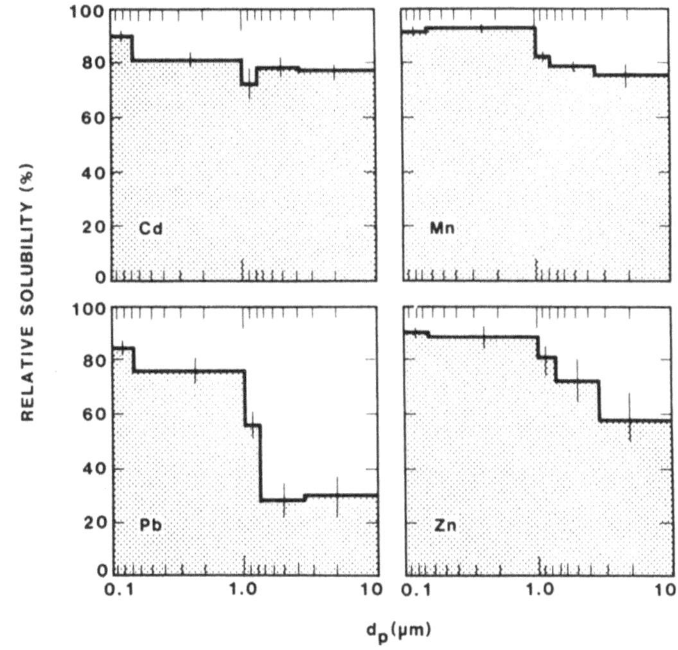

Fig. 6. Relationship between relative solubility (water-soluble concentrations x 100/water-soluble plus dilute acid leachable concentrations) and particle diameters (d_p μm) for cadmium, manganese, lead, and zinc (Lindberg and Harris, 1983)

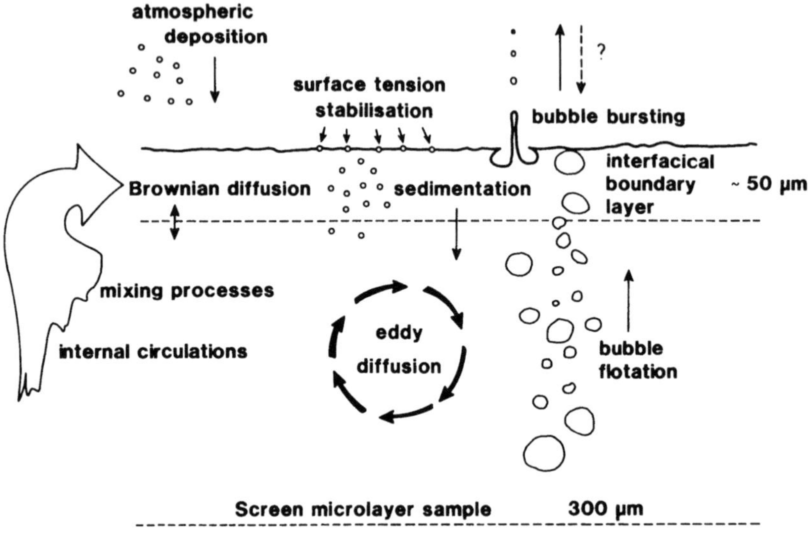

Fig. 7. Schematic presentation of the processes affecting trace metal species in the surface microlayer (Hunter, 1980)

The surface microlayer receives metals not only from "above" but also from "below" (see Fig. 7). Rising bubbles [derived from waves breaking or from gases produced in the sediments (methane)] scavenge the water column. Organic matter and particles become attached at the water-air interface of the bubble and are transported in this way to the surface. Experiments (Wallace and Duce, 1975) showed an aggregation of particles due to the bubble flux; the increase in particle size favors sedimentation from the surface layer. Model calculation (Lion and Leckie, 1981) from the marine environment have shown that if transport to the surface layer takes place by organics, then the transport of copper and lead is favored over cadmium and mercury due to their stronger tendency to complex with organics (cadmium and mercury are complexed mainly by chloride ions). The situation in the lake environment will be different due to low concentrations of chloride. However, similar calculations have not been carried out for the freshwater environment.

Results obtained on freshwater systems (Barnes et al., 1982; Elzerman and Armstrong, 1979; Maquire et al., 1982; Meyers et al., 1981; Meyers et al., 1983) all show significant enrichment in trace metals (and organic micropollutants) in the surface microlayer. Processes in this dynamic system are far from understood and need more research.

Sediment-Water Boundary

The sediment/water interface may act as either a source or sink for elements and compounds. In deep lakes having aerobic hypolimnions all year round, the sediment is a sink for most elements. Mass balance studies of metals in Swiss lakes (Baccini, 1976) lead to the following conclusions:

1) Allochthonous particulate metals are sedimented almost completely;
2) Within the lake assimilation of metals by biota leads to an additional transfer to the sediment; and
3) In early diagenesis microbial decomposition of organic matter leads to a partial reflux of metals into the hypolimnion.

In shallow lakes such as the IJsselmeer in the Netherlands the erosion of sediments and the release of pore waters during storm periods lead to a net transport of trace metals such as manganese and arsenic (both highly enriched in the pore waters) from the system (Salomons and Kerdijk, 1983).

For a better understanding of metal cycling in lakes one has to model the subsystem "sediment/water boundary layer." At present there are no models available to predict correctly the metal fluxes from the sediment despite the fact that several conceptual models have been developed. Good mathematical tools are available to treat the chemical dynamics of early diagenesis (Berner, 1980). The difficulties lay mainly in a) describing the factors which govern the rates of decomposition of organic matter and the nature and complexing properties of the degradation products, the authigenic mineral precipitation, and the adsorption processes; b) developing experimental boundary conditions very close to natural ones for a determination of the above-mentioned factors; and c) refining chemical operational schemes for the determination of metal species in the subsystem (pore waters and sediment). Especially for the species composition of pore water, very little information is available (Salomons, 1984a).

Recent investigations in different Swiss lakes (Baccini, 1985b) have led to a simple two-box model for the sediment boundary layer (Fig. 8). The crucial zone is the oxic/anoxic interface where sharp redox gradients are observed. The transport of redox-sensitive species can be governed by these redox gradients. The classical examples are iron and manganese for which a simple conceptual model can well describe the qualitative aspects of transport phenomenology (Davison, 1985). For other metals (e.g., Cu, Zn, Cd, Pb) analogous models are still lacking. However, there is much experimental evidence (Baccini and Joller, 1981; Laxen and Sholkovitz, 1981; Leckie et al., 1980; Tipping et al., 1981) that the newly formed iron(III) and manganese(IV) oxides offer new adsorbing surfaces to transfer trace metals from the dissolved to the particulate phase. Upwardly diffusing trace metals encounter this freshly precipitated material and are adsorbed onto it, whereby the reflux of trace metals is reduced (see iron oxide barrier in Fig. 8). This iron barrier moves upward with continued sedimentation (Hallberg, 1978). The chemical nature of these newly formed inorganic polymers is not yet fully understood (Schneider, personal communication, 1984). The presence of dissolved organic matter (concentration of dissolved organic carbon in interstitial waters of the sediment boundary layer: 10^{-3} M) influences strongly the size and stability of these polymers. Moreover, the distribution of the dissolved metal species in the pore waters, and hence their adsorbing properties, will be influenced.

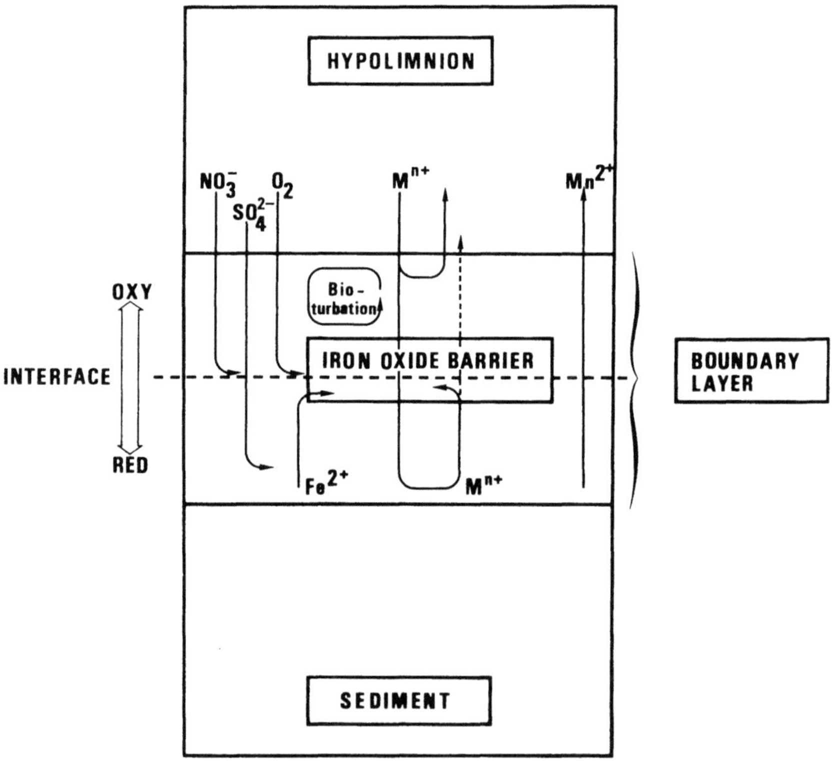

Fig. 8. Two-Box model for the sediment-water interface (Baccini, 1985 b)

Very often the important role of microbial activity is not taken into account. There exist microbial strategies to transform trace metal species directly to more mobile and often more toxic forms in order to remove them from the system (Thayer and Brinckman, 1982; Wood, 1974). Quantitative information on their importance for the lacustrine environment is lacking.

In conclusion it can be said that the distribution of chemicals and the transport processes of metal species in lacustrine sediments are still poorly understood in spite of the rich collection of geochemical data on the solid phase (Förstner and Wittmann, 1979).

Interactions within the System

The total trace metal loading of lakes is generally much higher compared with that of the ocean; despite this fact, the concentrations observed in lakes with a similar pH are within an order of magnitude of that of the ocean. This shows that efficient removal processes are taking place in the lacustrine environment (Sigg et al., 1982). As in the ocean, the removal takes place through the sedimentation of particulates. Particulates from outside the system settle close to the rivers' mouths and exert very little control on metal concentrations within the lake. Only in some cases where the chemical gradients are strong (e.g., a strong pH rise) do the allochthonous particles play a role in the removal of dissolved metals (Salomons, 1983). Various species of the dissolved metals play an important role in the adsorption processes. For a first approach to the distribution of inorganic trace metal species several mathematical models are available (Salomons and Förstner, 1984). However, incorporating the role of organics into the models of trace metal species is very difficult. There are two main sources of organic ligands in aquatic ecosystems: a) freshly excreted products from primary and secondary producers, e.g., carbohydrates, peptides, and lipides, which are degraded within days to weeks; and b) refractory organic matter to which 70%–80% of the total dissolved organic molecules belong and which is the result of enzymatic degradation processes in the drainage area or in the system itself (Buffle, 1984).

Freshwater phytoplankton can produce extracellular organic complexing agents which are partly capable of complexing copper and other trace metals under lacustrine conditions.

Several methods are in use for determining the conditional equilibrium constant of trace metal species [mainly Cu(II) ions] with dissolved ligands (Salomons and Förstner, 1984). In general the values vary between 8 and 12 (log K). Often two distinct sites for complexation are observed, one at low concentrations with a high K-value and a second site at higher concentrations with a lower K-value. However, observed differences in the K-values can also be due to different metal-to-ligand ratios (Buffle, 1984). Another major problem is the equilibration time in the experimental procedure (Hart, 1981).

In addition, convincing evidence from the marine field shows that quite large amounts of unstable (at least during preservation time) organic compounds are found which may escape the procedures used for the determination of the equilibrium constants (Piotrowicz et al., 1983), especially when pre-concentration steps for the DOC are used.

Analytical schemes (Salomons and Förstner, 1984) have been used in a few cases to study the distribution of trace metal species in lakes (Del Castilho and Salomons, 1984; Hart and Davies, 1981).

Particulate matter in deep lakes originates directly (organic matter) or indirectly (e.g., through the precipitation of calcium carbonate) from biota. The particularization of dissolved trace metal species can be attributed to the processes of assimilation by biota. It is mainly a transfer from the dissolved metal reservoir to the particulate biomass. Due to internal lake decomposition processes of organic matter, the total particulate concentration of trace metals is lowest in the hypolimnion (Baccini, 1984). It follows that the net biogenic flux of trace metals to the sediment depends mainly on the lake's capacity to produce organic particulate matter and to decompose it before it is buried definitely in the sediment, i.e., after early diagenesis (see below).

There are distinct differences between aquatic biota in particularizing trace metal species. If one compares the concentration factors, defined as the ratio of trace metal concentrations in the phytoplankton (mole per kg dry weight) to the concentrations in the substrate (mole per L), the following sequence of lakes' capacity to keep trace metals in solution is observed:

$$Cu > Zn > Cd > Pb > Fe.$$

With the exception of cadmium, the same sequence is observed in river waters where the differences are attributed to abiotic interactions between trace metal species and mineral surfaces. It is therefore concluded that biotic transfers in an aerobic lake do not alter the characteristic pattern of metal transport already observed in rivers. However, biota can contribute quantitatively in a significant way. Not only is their direct influence on the particularization of trace metals a significant factor, but sometimes their indirect influence by increasing the proton activity (pH) is even more important. In fresh waters with a carbonate buffer the proton activity can vary between 7 and 9.5 (i.e., more than two orders of magnitude). Adsorption studies have shown that the amounts of cadmium, zinc, and chromium adsorbed depend very much on the pH in this range. In fact, a relatively small shift in the pH can cause a large decrease in metal concentrations. On the other hand, copper and lead are not very sensitive to changes in adsorption over this pH range. In the IJsselmeer, for example, the high pH of the lake in summer ($pH \geq 9$) compared to that of the river water (pH 7.2) entering the lake causes an extensive removal of cadmium and zinc and almost no removal of lead and copper by adsorption processes (Salomons, 1983), as well as a decrease in their bioavailability. The seasonal changes in pH cause seasonal cycles in the dissolved zinc concentrations (Fig. 9). As a consequence of the pH increase a large retention of dissolved cadmium and zinc is observed in the IJsselmeer. However, in the Haringvliet (Netherlands) the pH increases less, and consequently less cadmium and zinc are removed by adsorption.

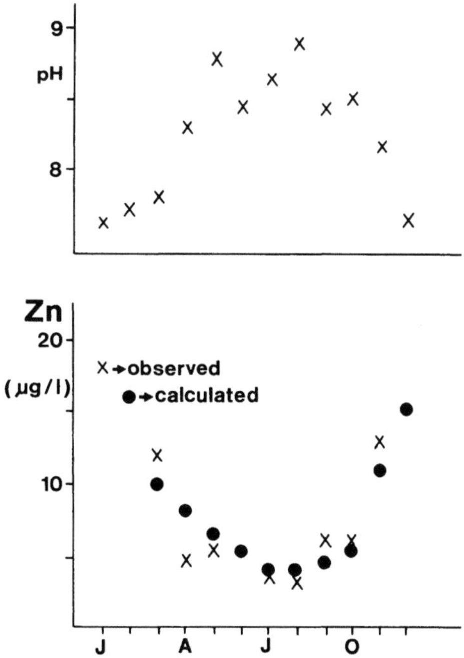

Fig. 9. Seasonal cycles in pH and in the dissolved zinc concentrations in a polluted lake (IJsselmeer) (Salomons, 1983)

An increase in proton activity occurs in lakes with a low buffering capacity and an acidic input, as has been shown for Scandinavian and Canadian lakes. In these cases an increase in dissolved metal concentrations is observed. Again relatively small changes in the pH are able to cause large changes in dissolved trace metal levels. This decrease in pH causes more metal to remain in solution; in addition, metals may thereby also be desorbed from the sediments. Determining species distribution along sediment cores showed that, in particular, the easily reducible fractions become solubilized (Salomons and Förstner, 1984).

The concentration of dissolved trace metal species may also increase if complexing agents are introduced into the system by man. Important in this respect are NTA (Salomons and Kerdijk, 1984) and chloride from salt mines (Salomons and Kerdijk, 1984) or from road deicing (Sutcliffe and Carrick, 1983). Laboratory experiments have shown that NTA is capable of releasing trace metals from sediments or of preventing adsorption onto suspended matter (Salomons and Kerdijk, 1984). An example of the influence of NTA on the speciation of trace metals in sediment-water systems (modelled after laboratory data) is given in Fig. 10.

In lakes with an anaerobic hypolimnion new adsorbing surfaces are formed at the oxic/anoxic interface in the water column. The flux of iron(II) and manganese(II) from the reduced deeper layers into an oxic environment is stopped by biotic or abiotic oxidation of these two cations because particulate hydroxides and oxides are formed which are again sedimented. Lake Baldegg is such a system which shows an additional precipitation of copper at the Fe^{3+}/Fe^{2+} interface. However, an analogous phenomenon was not observed for zinc (Baccini and Joller, 1981). These findings are in agreement with measurements on freshly precipitated iron hydroxides from an anaerobic English lake (Laxen and Sholkovitz, 1981). Hydrous ferric oxide can co-precipitate significant amounts of cadmium, copper, and lead at conditions very close to natural ones. According to laboratory experiments zinc stays in solution (Baccini and Joller, 1981). Neither freshly precipitated calcium carbonate nor manganese oxides have any influence. The finding that co-precipitation of trace metals with calcium carbonate is negligible from a quantitative point of view has been experimentally shown under natural conditions (Baccini et al., 1979). In fact, the production of calcium carbonate containing low metal concentrations causes an "apparent" decrease in particulate metal concentrations [dilution effect (Salomons, 1983)]. However, biotic manganese oxidation under natural conditions might contribute significantly to the hypolimnic transport of zinc, cadmium, and lead. This supports the hypothesis that the manganese and iron cycles operate

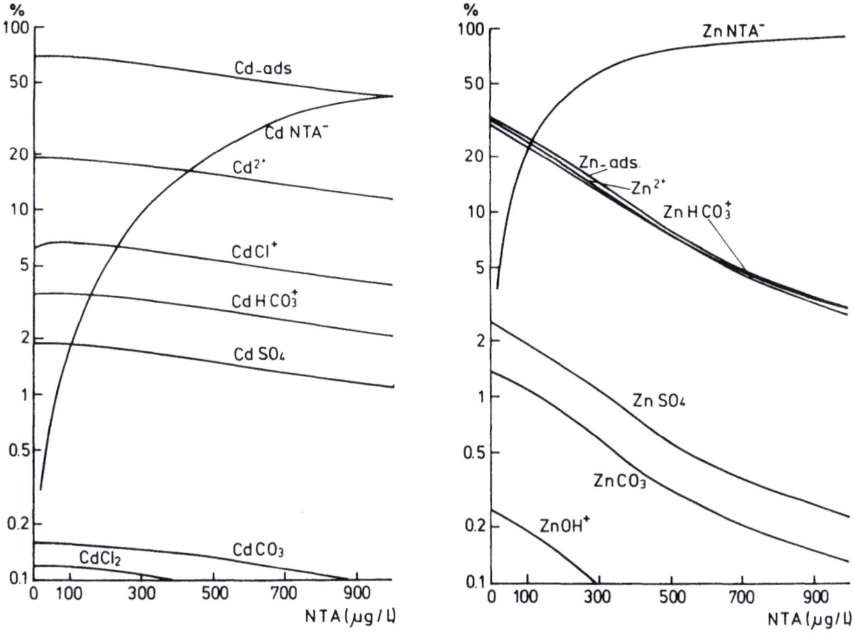

Fig. 10. Changes in the species distribution of cadmium and zinc as a function of the NTA concentration (Salomons and Kerdijk, 1984). ads is the adsorbed species

practically independently. Copper seems to be influenced mainly by the "iron wheel" whereas zinc, cadmium, and lead can be transported preferably on manganese oxides. Sulfides, although present at high concentrations in deeper layers of Lake Baldegg (approx. 10^{-4} M), have no observable effects on copper and zinc transport (Baccini and Joller, 1981).

Dissolved organic matter can influence the adsorption properties of mineral particles (Salomons and Förstner, 1984). Under lake water conditions, i.e., with particle concentrations of 2–16 mg L^{-1} and dissolved organic carbon concentrations of 1–4 mg L^{-1} at pH = 8, the adsorption of copper and zinc is reduced significantly by the presence of natural organic matter. The organic molecules are competing with copper, zinc, and other ions for the surface-binding sites on the particles. These findings can give an explanation as to why incoming mineral particles are relatively poor scavengers for trace metals unless the pH of the lake water is much higher than that of the river water.

Both in oxic lakes and in lakes with an anaerobic hypolimnion the settling particles will partly release their adsorbed trace metals after entering the anaerobic zone. The extent of the process depends on the reversibility of the adsorption process and the kinetics of decomposition, an area in need of further research.

Transport Processes and Species Distribution: Biogeochemical Principles

As discussed, a large number of processes have been observed which control and alter transport processes to the lakes and in the lakes. Metal fluxes to lakes depend on the biogeochemical processes taking place in the drainage area and on the interactions of trace metals with particles.

From a chemical point of view the material fluxes which alter significantly the boundary conditions (flux to the lake, Fig. 1) consist of four types:

1) Protons (e.g., acidification of soils);
2) Nutrients (e.g., eutrophication by increasing input of phosphorus);
3) Metals (e.g., the nonessential elements cadmium, mercury, and lead); and
4) Organics (e.g., xenobiotic chlorinated hydrocarbons).

Each of these groups can influence the biochemical processes in the aquatic ecosystem and thereby alter the transport of metals to the lake system.

In the aquatic ecosystem of a lake (Fig. 1) the biochemical processes of assimilation and respiration alter the chemical boundary conditions for trace metals in time and space by a) increasing or decreasing proton activity, b) producing new ligands, and c) increasing or decreasing electron activity. These three processess affect the metal concentrations and their species distribution.

It has been postulated that primary producers and the microbial decomposers (which are the most sensitive species within the aquatic biota to increased concentrations of metals) aim to control the concentrations of their dissolved and biologically available forms. Supposing a constant input, they have to prevent an accumulation of trace metals in the system, more precisely in the water column and at the sediment/water interface, in order to maintain an optimal concentration.

The following biochemical mechanisms are applied to reduce the transfer of metals into the biota (Fig. 11).

1) Trophogenic control of trace metals – Decrease of biological uptake by reduction of proton activity in the upper part of the lake (up to the depth of effective light penetration, the trophogenic zone):

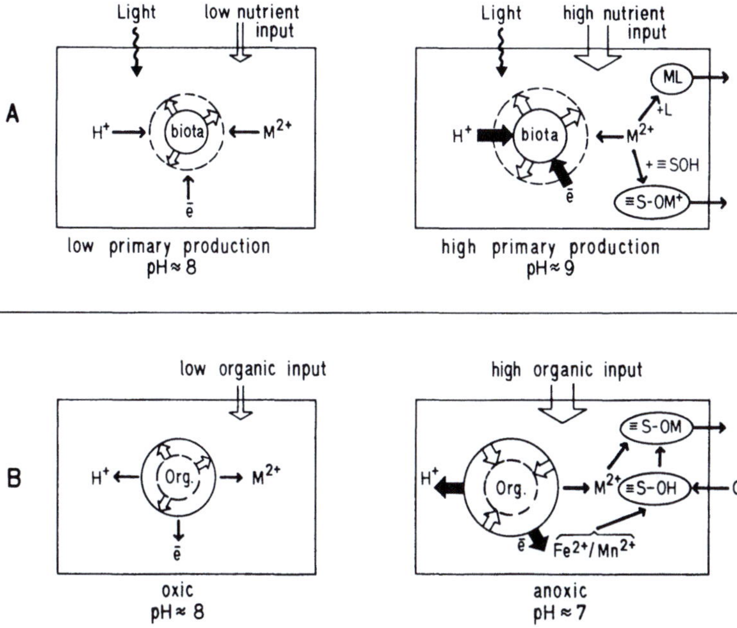

Fig. 11. The light arrows indicate the direction of the quantitative change of biomass in A growing, in B decreasing. The dark arrows indicate the net transfer of protons, electrons, and metals between the biomass and its environment. In oligotrophic systems (low nutrient input in **A**, low organic input in **B**) these transfers to and from the biota/org. cannot change significantly the chemical composition of the aquatic environment. In eutrophic systems proton and electron activity is altered significantly. The assimilation of metals can be reduced due to the formation of dissolved organic complexes (ML) and/or surface complexes of particles (≡S-OM) which are both biologically unavailable

A higher primary production rate increases the pH value. For a number of trace metals (especially cadmium and zinc) whose adsorption edge is within the range observed for nonacidifying lakes (pH 7–9), this causes a strong increase in the adsorption:

$$\{\equiv\text{S-OH}\}\,(s) + M^{2+}\,(aq) \rightarrow \{\equiv\text{S-OM}^{+}\} + H^{+}(aq). \quad (1)$$

This change in species composition – the adsorption of the metal ion on suspended matter surfaces (symbolized by $\{\equiv\text{S-OH}\}$) – lowers the concentration of the available form M^{2+} (aq). This reduction of proton activity thus favors the formation of biologically unavailable species (adsorbed metals). This mechanism of trace metal control is proposed for zinc if

$$\{\equiv \text{S-OH}\}_{\text{Tot}} >> (\text{Zn}^{2+})_{\text{Tot}}.$$

The adsorption of copper is not sensitive to pH change over the range observed in most lakes, and a different control mechanism causes a reduction in the concentrations of biologically available species (e.g., Cu^{2+}). Unlike zinc, copper forms strong complexes with organic ligands:

$$\text{LH(aq)} + \text{M}^{2+}\text{(aq)} \rightarrow \text{ML}^{+}\text{(aq)} + \text{H}^{+}\text{(aq)}. \qquad (2)$$

An increase in pH favors the formation of CuL^{+}(aq), thereby lowering that of the biologically available form Cu^{2+}(aq).

In both cases the proton activity becomes a master variable for the transfer of zinc and copper from available to unavailable biological species (Baccini, 1985a). The consequences of these two different mechanisms are found in the resulting transport paths. In spite of a reduced biological uptake, zinc is sedimented more strongly than copper.

2) Tropholytic control of trace metals – Decrease of trace metal concentration by formation of new adsorbing surfaces in that part of the lake below the depth of effective light penetration (tropholytic zone):

In a tropholytic system an increased flux of particulate organic matter leads subsequently to an increase in proton activity. This would favor the formation of available trace metal forms as indicated above. However, the increased decomposition of organic matter increases simultaneously the electron activity in the water. At the boundary layer of a reduced tropholytic system the electron acceptors iron and manganese are oxidized and precipitated as efficient scavengers for trace metals (see also Fig. 1). In other words, new $\{\equiv \text{SO-H}\}$ is produced to apply the mechanism formulated in Eq. 1. It is possible that ternary surface complexes are also formed:

$$\{\equiv \text{S-OH}\}\text{ (s)} + \text{ML}^{+}\text{ (aq)} \rightarrow \{\equiv \text{S-OML}\} + \text{H}^{+}\text{ (aq)} \qquad (3)$$

In both cases (1) and (2) dissolved organic matter plays a crucial role. By competing with trace metals for surface-binding sites on mineral surfaces, it can reduce a transfer to the hypolimnion:

$$\{\equiv \text{-OM}^{+}\text{(s)} + \text{LH(aq)} \rightarrow \{\equiv \text{S-O-LH}\}\text{(s)} + \text{M}^{2+}. \qquad (4)$$

However, in the tropholytic zone the opposite effect is possible. In the abiotic formation of new adsorbing surfaces the dissolved organic matter can enlarge the specific surface area by stabilizing relatively small particles.

It goes without saying that different populations on different trophic levels have their individual strategies for controlling trace metal uptake. The above-mentioned mechanisms have been chosen to serve the ecosystem as a whole.

3) Selection of a new biocenose by metal stress – An increased metal flux leading to phytotoxic concentrations reduces primary production of the established biocenose. Within weeks or months, changes in the phytoplankton variety towards species are observed which assimilate fewer metals and which are less susceptible to the increased metal concentration. This adaptation results in a biota which reduces the lake's capacity to retain metals.

4) Selection of a new biocenose by acidification – An increased proton flux into a lake with a relatively small buffer capacity also alters the biocenose and reduces primary production. Analogous to case (3), the transfer of metals to the sediment is reduced. An increase of proton concentration in the tributaries also enhances metal transfer from the particulate to the dissolved phase, thereby increasing the dissolved metal flux from the drainage area.

These examples may serve to show the strong interactions (direct and indirect) between the major element cycles and trace metal cycles in lakes and the influence that biota has on the speciation of trace metals and, subsequently, on their transport pathways.

References

Baccini P (1976) Untersuchungen über den Schwermetallhaushalt in Seen. Schweiz Z Hydrol 38:121–158

Baccini P (1984) Regulation of trace metal concentrations in fresh water systems. In: Siegel H (ed) Metal ions in biological systems, vol. 18. Circulation of trace metals in the environment. Marcel Dekker, New York

Baccini P (1985a) Metal transport and metal/biota interactions in lakes. Env Technol Lett 6:327–334

Baccini P (1985b) Phosphate interactions at the sediment-water interface. In: Stumm W (ed) Chemical processes in lakes. Wiley, London New York

Baccini P, Joller T (1981) Transport processes of copper and zinc in a high eutrophic and meromictic lake. Schweiz Z Hydrol 43:176–199

Baccini P, Ruchti J, Warner O, Grieder E (1979) Regulation of trace metal concentrations in limno-corals. Schweiz Z Hydrol 41:202–227
Barnes RK, Batley GE, Sharp JH (1982) Heavy metal enrichment in the surface microlayer of the Nepean-Hawkesbury river system. Aust J Mar Freshw Res 33:414–430
Berner RA (1980) Early diagenesis: a theoretical approach. Princeton University Press, Princeton
Buffle J (1984) Natural organic matter and metal-organic interactions in aquatic systems. In: Siegel H (ed) Metal ions in biological systems, vol 18. Circulation of trace metals in the environment. Marcel Dekker, New York
Davison W (1985) Conceptual models for transport at a redox boundary, implications for lacustrine geochemistry. In: Stumm W (ed) Chemical processes in lakes. Wiley, London New York
Del Castilho P, Salomons W (1984) Biological activity and its influence on metal complexation in fresh water systems. International Conference on Environmental Contamination, London
Duinker JC (1983) Effects of particle size and density on the transport of metals to the oceans. In: Wong CS, Boyle E, Bruland KW, Burton JD, Goldberg ED (eds) Trace metals in sea water. Plenum, New York London, pp 209–226
Eisenreich SJ (1980) Atmospheric input of trace metals to Lake Michigan. Water Air Soil Poll 13:287–301
Eisenreich SJ, Hoffmann RM, Rastetter D, Yost E, Maier WJ (1980) Metal transport phases in the upper Mississippi River. In: Kavanaugh MC, Leckie JO (eds) Particulates in water. Am Chem Soc Adv Chem Ser 189:135–176
Elzerman AW, Armstrong DE (1979) Enrichment of Zn, Cd, Pb and Cu in the surface microlayer of Lakes Michigan, Ontario and Mendota. Limnol Oceanog 24:133–144
Elzerman AW, Armstrong DE, Andren AW (1979) Particulate zinc, cadmium, lead, and copper in the surface layer of Southern Lake Michigan. Envir Sci Technol 13:720–725
Förstner U, Wittmann G (1979) Metal pollution in the aquatic environment. Springer-Verlag, Berlin Heidelberg New York
Hallberg RO (1978) Metal-organic interaction at the redox-cline. In: Krumbeing WE (ed) Environmental biogeochemistry and geomicrobiology, vol 3. Ann Arbor Science Publishers, Ann Arbor, pp 947–954
Hardy JT (1982) The sea surface micro layer, biology, chemistry and anthropogenic enrichment. Prog Oceanog 11:307–328
Hart BT (1981) Trace metal complexing capacity of natural waters: a review. Envir Technol Lett 2:95–110
Hart BT, Davies SHR (1981) Trace metal speciation in three Vistorian lakes. Aust J Mar Freshw Res 32:175–189
Hunter KA (1980) Processes affecting particulate trace metals in the sea surface micro-layer. Mar Chem 9:49–70
Laxen DPH, Sholkovitz ER (1981) Adsorption (co-precipitation) of trace metals at natural concentrations on hydrous ferric oxide in lake water samples. Envir Technol Lett 2:561–569
Leckie JO, Benjamin M, Hayes K, Kaufman G, Altmann S (1980) Adsorption/coprecipitation of trace elements from water with iron hydroxide. Final Report, EPRI RP-910, Electric Power. Palo Alto, CA: Research Institute

Lindberg SE, Harris RC (1983) Water and acid soluble trace metals in atmospheric particles. J Geophys Res 88:5091–5100
Lion LW, Leckie JO (1981) Chemical speciation of trace metals at the air-sea interface: The application of an equilibrium model. Envir Geol 3:293–314
Lum KR (1982) The potential availability of P, Al, Cd, Co, Cr, Cu, Fe, Mn, Ni, Pb and Zn in urban particulate matter. Envir Technol Lett 3:57–62
MacIntyre F (1974) Chemical fractionation and sea surface micro layer processes in the sea. In: Goldberg ED (ed) Marine chemistry, vol 5. Wiley, New York, pp 245–299
Maquire RJ, Chau YK, Bengert GA, Hale EJ, Wong PTS, Chay YK (1982) Occurrence of organo-tin compounds in Ontario lakes and rivers. Envir Sci Technol 16:698–702
Martin JM, Meybeck M (1979) Elemental mass balance of material carried by major world rivers. Mar Chem 7:173–206
Meyers PA, Owen RM, Mackin JE (1981) Organic matter and heavy metal concentrations in the particulate phase of Lake Michigan surface microlayers. In: Eisenreich SJ (ed) Atmospheric inputs of pollutants to natural waters. Ann Arbor Science Publishers, Ann Arbor, MI, pp 129–141
Meyer PA, Rice CP, Owen RM (1983) Input and removal of natural and pollutant materials in the surface microlayer on Lake Michigan. Ecol Bull 35:519–532
Piotrowicz SR, Harvey GR, Springer-Young M, Courant RA, Boren DA (1983) Studies of cadmium, copper and zinc complexation by marine fulvic and humic materials in sea water using anodic stripping voltammetry. In: Wong CS, Boyle E, Bruland KW, Burton JD, Goldberg ED (eds) Trace metals in sea water. Plenum, New York London, pp 699–717
Salomons W (1983) Trace metal cycling in a polluted lake (IJsselmeer, the Netherlands). Delft Hydraulics Laboratory Report S 357/EEC, Contract No. 199-7-1-ENV N
Salomons W (1984a) Contaminants in sediments: out of sight out of mind? International Conference on Environmental Contamination, London, June 1984
Salomons W (1984b) Impact of atmospheric trace metals on the hydrological cycle. In: Nriagu J, Davidson C (eds) Toxic metals in the air. Wiley, New York
Salomons W, Eysink W (1981) Pathways of mud and particulate trace metals from rivers to the Southern North Sea. In: Nio SD, Schuttenhelm RTE, van Weering TCE (eds) Holocene marine sedimentation in the North Sea Basin. Spec Publ Assoc Sedimentol 5:429–450
Salomons W, Förstner U (1980) Trace metal analysis on polluted sediments. II. Evaluation of environmental impact. Envir Technol Lett 1:506–517
Salomons W, Förstner U (1984) Metals in the hydrocycle. Springer-Verlag, Berlin Heidelberg New York Tokyo
Salomons W, Kerdijk H (1983) Trace metal cycling in lakes (IJsselmeer, The Netherlands). Proceedings of the International Conference on Heavy Metals in the Environment, Heidelberg, September 1983
Salomons W, Kerdijk H (1984) Influence of natural and man-made complexing agents on adsorption processes in river systems. In preparation
Sievering H, Dave M, Dolske DA, McCoy P (1981) Transport and deposition of trace metals over southern Lake Michigan. In: Eisenreich SJ (ed) Atmospheric pollutants in natural waters. Ann Arbor Science Publishers, Ann Arbor, MI

Sigg L, Sturm M, Stumm W, Mart L, Nurnberg HW (1982) Schwermetalle im Bodensee, Mechnismen der Konzentrationsregulierung. Naturwiss 69:546–548
Sutcliffe DW, Carrick TR (1983) Relationship between chloride and major cations in precipitation and streamwaters in the Windermere catchment (English Lake District) Freshw Biol 13:415–441
Thayer JS, Brinckman FE (1982) The biological methylation of metals and metalloids. Adv Organomet Chem 20:313–356
Tipping E, Woof C, Cooke D (1981) Iron oxide from a seasonally anoxic lake. Geochim Cosmochim Acta 45:191–199
Wallace GT, Duce RA (1975) Concentration of particulate organic carbon in marine waters by a bubble flotation mechanism. Mar Chem 3:157–181
Wood JM (1974) Biological cycles for toxic elements in the environment. Science 15:1049–1052

The Importance of Chemical "Speciation" in Environmental Processes,
eds. M. Bernhard, F.E. Brinckman, and P.J. Sadler, pp. 217–236.
Dahlem Konferenzen 1986. Springer-Verlag Berlin, Heidelberg
© Dr. S. Bernhard, Dahlem Konferenzen

Interactions Between Natural Organic Polymers and Metals in Soil and Freshwater Systems: Equilibria

D.S. Gamble

Chemistry and Biology Research Institute
Agriculture Canada, Ottawa, Ontario, K1A 0C6, and
Dept. of Chemistry, Concordia University
Montreal, Québec, Canada

Abstract. The equilibrium metal ion-binding reactions of natural organic polymers in soils and surface waters can be understood and predicted with mathematical descriptions that are based on chemical stoichiometry. The reactivity of individual species depends on physical states and chemical composition variables. Equilibria are described by a formalism that applies the concept of weighted averages to the law of mass action and accounts for chemical composition variables with mole fraction relationships. The chemically blind "black box" empirical descriptions lack predictive abilities and are no longer necessary. The analytical chemistry and the physical chemistry of this class of systems have become indistinguishable, because of the intimate interrelationship between measurements of stoichiometry and the theoretical basis for predictive equations. Future advances depend on analytical chemical research.

Introduction

The application of analytical chemistry to a pure, monomeric reactant produces a set of constants which uniquely characterize the sample. In contrast, the application of analytical chemistry to a natural product that consists of a mixture of polymeric reactants requires a somewhat more sophisticated mode of thought. The rigorous chemical characterization of such a sample must have a conceptual basis, otherwise the analytical chemical description will lack both scientific rigor and practical usefulness.

For systems with pure monomeric reagents, it is conventional practice to describe the equilibrium between the reactant and a dissolved metal with a mathematical constant defined in accordance with the law of mass action. This is not applicable to a system having a mixture of polymeric reactants. In this case it becomes necessary to replace the concept of a mathematical constant with the concept of a mathematical function. The function is also defined in accordance with the law of mass action, and its independent variables are temperature and various chemical composition variables. The physical chemistry of the system must have a rigorous mathematical description that takes into account the effects of the independent variables on the equilibrium function. This leads at once to equally important implications for the analytical chemistry of the system. This type of system presents analytical chemistry with a much greater challenge than does a simpler reactant system. There are at least four reasons for this:

1) A greater number of variables must be measured.
2) The existence of an equilibrium function instead of an equilibrium constant requires that each variable has to be either controlled or measured over a range of numerical values. In other words, the equilibrium properties of the system must be characterized by experimental scans of its variables. This contrasts fundamentally with the "single point" measurements with which a mathematical constant can be characterized. The chemical analyses will tend to resemble equilibrium titrations.
3) The physical chemical calculations may become complex, because the natural system has a number of variables instead of a single constant as obtains with a single metal and reactant. This puts heavier demands on the quality and quantity of chemical analyses.
4) There is an intimate feedback relationship between the analytical chemistry and the physical chemistry of the system. Chemical analysis identifies variables that must be accounted for in the mathematical description. The physical chemistry elucidated in this way then determines the experimental designs which provide specifications for subsequent chemical analyses.

The purpose of this combination of analytical and physical chemistry is to make *predictive* calculations possible. Practical applications may then be investigated. For this to be possible, two long-standing difficulties must be removed. One of these is that much of the earlier work has tended to produce chemically blind empirical descriptions. These lack predictive capacity. The other difficulty is that, until recently, the distinction between the concepts of mathematical functions and mathematical constants in the application of the law of mass action had not been very widely recognized. Years of confusion and lack of progress with practical applications have been the consequences.

Polyelectrolyte-metal ion reaction equilibrium systems constitute a large generic class of chemical systems. It includes two categories. One of these consists of *biochemical* examples, such as proteins, pectins, and histones. The other consists of *geochemical systems* such as fulvic acids, humic acids, and polysaccharides of soils and natural waters. Extensive experimental work has been done on the metal ion equilibria of both categories. The mathematical description of equilibria in mixed polymeric reactants applies equally well to both. To date, however, it has been most successfully developed and experimentally demonstrated for the geochemical category. While this is the one under discussion here, it could with great profit be adapted to the biochemical systems. (A much more extensive paper on which the present paper is based is now in press in the journal *Chemie der Erde,* Jena, GDR, Gustav Fischer Verlag.)

The Definition of Physical States of Natural Systems

The physical states in which natural organic polymers may exist in aqueous systems range between two limiting cases, with a conceptually identifiable state somewhere between them. The reason that this question of physical states is important is that it is the basis of the distinction between solution phase complexing and two-phase (i.e., solid and solution) cation exchange. This leads to direct implications for the mobility and bioavailability of metal ions. In the investigation of this problem, analytical chemistry and physical chemistry become almost indistinguishable. The two limiting cases are a) true solutions, and b) undissolved water-swollen gels.

Colloidal sols are the important intermediate type of state. The distinction between the two limiting cases is especially important for insight into the theoretical description and practical behavior of metal ion reaction equilibria. If a single true solution state exists, then there is a metal ion-complexing equilibrium with bound and free metal ions both randomly distributed and mobile throughout the whole system. A practical consequence of this is that bound and free metal ions can both be leached or transported in soils and surface waters. Analytical chemical methods which distinguish between free and bound metal ions must be developed or adapted. The concentrations of these ions must then be measured with sufficient accuracy so that physicochemical calculations may be properly carried out. The second limiting case provides a sharp contrast in terms of an undissolved water-swollen gel in an aqueous system: there are now two physical states in the system. The metal ions have an equilibrium distribution between two phases. The bound metal ions are confined to ion exchange sites within the polymer gel by the Donnan potential. In examples with complexation, covalent chemical bonding exists together with the Donnan potential. Only cation exchange elution can release the "trapped" (absorbed) metal ions into the external (bulk) solution (Helfferich, 1962). Elution by other cations or mechanical transport of the natural organic polymer are therefore the only means by which the bound metal ions can move in soils or natural waters. The metal ion analysis is much less difficult for a two-phase system. With the separation of phases by filtration or centrifugation, the analytical chemical method does not have to distinguish between free and bound species. This gives a freer choice of spectroscopic and electrochemical methods.

The physical chemistry of the third physical state introduces some quite difficult questions for analytical chemistry. The concept of colloidal sols is defined by two boundary regions. These two boundary regions are not themselves well-defined, however. There can exist a continuum of solute particle sizes and system properties between a true solution and a colloidal sol. It will probably never be possible to define a sharp boundary between the two physical states. In the first place, no theoretical concepts are available with which to formulate the necessary criteria. What is more important, however, is the operational interrelationship between particle size and metal ion-complexing equilibria. It has now become evident that analytical chemical research can make substantial progress here. In particular, the application of Rayleigh light-scattering methods correlated with electroanalytical chemical and filtration methods should reveal the particle size – metal ion-complexing phenomena (Underdown, 1982). The second boundary region lies between colloidal sols and suspensions of undissolved organic polymer in an aqueous solution. The undissolved particles consist of water-swollen gel. The serious, unresolved difficulty that confronts both theoretical physical chemistry and operational analytical chemistry is that, within this boundary region, a continuum of particle sizes and system properties also exists. No theoretical or logical criteria are available with which to distinguish between colloidal sols and two-phase suspensions. Because some kind of decision is necessary for practical purposes, it has become common practice to use 0.45 μm filters for the approximate distinction between dissolved and undissolved fractions of samples (Reuter and Perdue,

1977). While this arbitrary practice may be unsatisfactory, theoretical chemistry has not yet produced an alternative. Analytical chemistry research will have to turn to such experimental methods as Rayleigh light-scattering, ultra- and micro-filtration, and centrifugation. There have already been some experimental beginnings (Orlov and Gorshkova, 1965; Pagenkopf and Whitworth, 1981; Saar and Weber, 1979a).

Reactive Functional Groups: The Source of all the Sites and Stoichiometry for Metal Ion Reactions

Reactive functional groups are the key to the metal ion reactions, and the equilibria must be defined in terms of them. This immediately raises questions about what functional groups are present and how they react with metal ions. Stumm and Morgan emphasized this (1981) when they said, "The literature abounds with empirical evaluations of the complex-forming tendency of fulvic or humic acids. Much of this information cannot be generalized because the acid-base characteristics of the ligand were not described." It has in fact been possible to deduce the essential points from half a century of experimental evidence published by many authors (Gamble et al., 1983). For any particular sample of natural organic polymer, the first step is an examination of elemental analysis. This is because all subsequent chemical analyses related to functional groups must be consistent with the functional group analysis which provides clues that can guide subsequent research. Analytical chemical methods used, for example, by Schnitzer and his co-workers have included reaction with calcium acetate followed by back titration, preparation of organic derivatives, potentiometric acid-base titrations, and infrared spectrophotometry. Since then, NMR spectrometry has come into general use. The functional groups that have been reported include carboxyl groups, phenolic groups, alcoholic-OH, ketonic, and quinone groups (Gamble, 1972; Gamble and Schnitzer, 1973; Schnitzer and Khan, 1978). Tables 1 and 2 show elemental and functional group analyses for the particular example of Armadale Bh horizon fulvic acid. Figure 1 shows its postulated generic type of structure. In Table 2, the values for Type A carboxyl groups and total titratable acid groups have been obtained by potentiometric acid-base titration. The other results in Tables 1 and 2 have been published by Schnitzer and Kodama (1972).

Because of their well-known ability to complex metal ions, it is necessary to consider whether or not amino groups are also present. First

Table 1. Armadale Bh horizon fulvic acid (Gamble and Schnitzer, 1973; Schnitzer and Khan, 1978)

Elemental analysis	
C	50.90 wt. %
H	3.35 wt. %
N	0.75 wt. %
S	0.25 wt. %
O	44.75 wt. %

Table 2. Fulvic acid functional groups (Gamble, 1972; Gamble and Schnitzer, 1973)

Functional group	Amount (m equiv.) fulvic acid (g)
Type A carboxyl	4.99 ± 0.25
Titratable groups	7.71 ± 0.25
Phenolic OH	3.0
Alcoholic OH	3.6
Ketonic C=O	2.5
Quinoid C–O	0.6

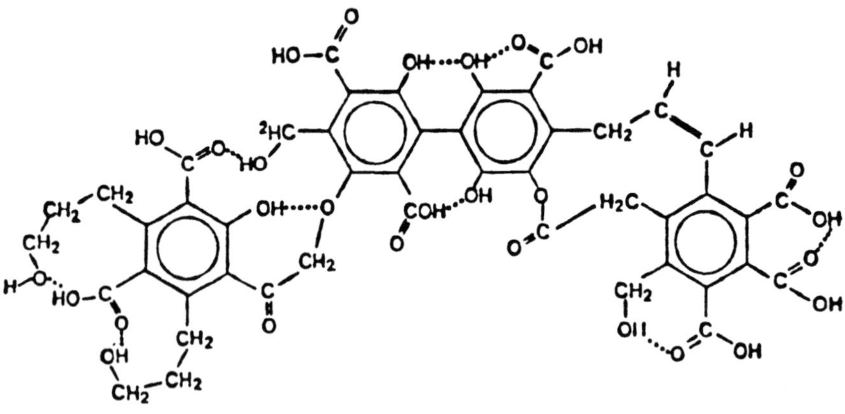

Fig. 1. A postulated generic type of structure in Armadale fulvic acid

Table 3. Bidentate chelation of metal ions M(II) by fulvic acid: postulated general case

$$SH^- + M^{++} \rightleftharpoons SM + H^+$$

Type of chelation site	Chemical Form of Chelation Site: chelate		chelation site, singly ionized		chelation site, fully protonated	
	symbol	structure	symbol	structure	symbol	structure
salicylic	LM	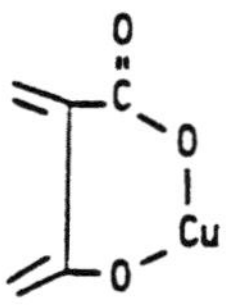	LH^-	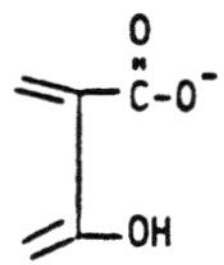	LH_2^-	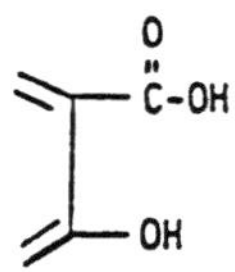
dicarboxylic	DM	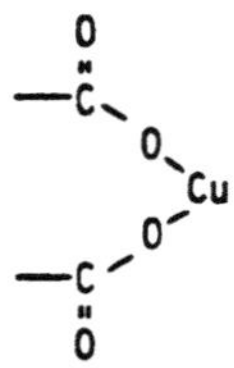	DH^-	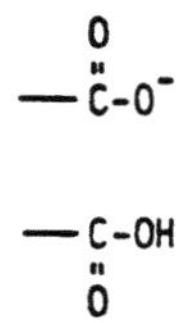	DH_2	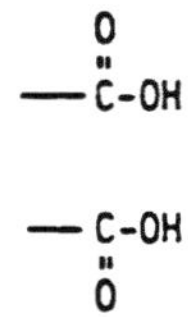
total, (salicylic + dicarboxylic)	SM	both of the above structures	SH^-	both of the above structures	SH_2	both of the above structures

Khanna and Stevenson (1962) and later Khan and Sowden (1971) reported experimental evidence indicating that they do not normally make substantial contributions to complexing capacity (Bremner, 1957; Khan and Sowden, 1971; Khanna and Stevenson, 1962; Sowden, 1957; Sowden et al., 1976). A considerable amount of published work leads instead to the conclusion that metal ions generally interact with carboxyl groups, and some of the metal ions, such as Cu(III) and Fe(III), can react simultaneously with both carboxyl and phenolic groups (Beckwith, 1959; Broadbent and Bradford, 1952; Gamble et al., 1970; Gamble et al. 1983; Griffith and Schnitzer, 1975; Wagner and Stevenson, 1965). The types of bidentate chelation sites for which the literature provides evidence have been classified previously (Gamble et al., 1980). They are illustrated in Table 3. There are, however, some other possibilities that merit future investigation. The picolinic acid and catechol types could be important, for example, at the trace level of metal ion loadings.

The acid-base characteristics to which Stumm and Morgan referred include two phenomena. One is simply the number of acidic functional groups per g of natural organic polymer. The other is the spectrum of weak acid K_A values which exists for the collection of chemically nonidentical acidic groups in a natural product mixture. Correspondingly, there are two separate but related analytical chemical problems. The two analytical chemical problems are so totally interconnected that they must be solved together. Potentiometric equilibrium scan titration is the analytical chemical method by which this is done. The first of the two problems is that stoichiometrically defined equivalence points must be evaluated instead of chemically blind end points obtained by strictly mathematical procedures. First and second derivatives of pH vs. mL curves, for example, are not defined in terms of numbers of acidic functional groups. The importance of Gran's functions to the analytical chemistry of natural organic polymers

Table 4. Analytical characteristics of the HA (on a dry, ash-free basis)

%C	53.5
%H	6.2
%N	5.5
%S	0.8
%O	34.0
meq total acidity g^{-1}	6.4
meq CO_2H g^{-1}	3.4
meq phenolic OH g^{-1}	3.4

depends on the fact that they mathematically relate the number of acidic functional groups to measurable quantities. For titrations of natural organic polymers, the calculations must account for the spread of K_A values, strong buffering, and even the dissociation of H_2O. The required theory has been computerized and published (Burch et al., 1978; Gamble, 1970, 1972).

Table 4 gives analytical data for the humic acid from the surface layers of the Bainsvill clay loam. This humic acid was used by Kerndorff and Schnitzer for their multiple metal ion exchange experiments (1980).

Published Research

Solution Phase equilibria: Cu(II) chelation

It has always been general practice to apply the law of mass action to the whole system of cation-binding sites. For the mixture of Cu(II) chelation sites in fulvic acid, this gives Eq. 1:

$$SH^- + Cu^{2+} \underset{}{\overset{\bar{K}_4}{\rightleftharpoons}} SCu + H^+ \tag{1}$$

and the equilibria for the general case is

$$\bar{K}_4 = m_c m_H / m_{SH} m_M .$$

The evaluation of $\bar{K}_4$ requires the use of material balance equations together with an experimental value for complexing capacity, such as that shown in Fig. 2.

Equation 2 is defined for the ith small portion of the mixture of sites. $(\Delta G_i/RT)$ is the electrostatic work term which results from the electrical charge on the polymer.

$$S_iH^- + Cu^{2+} \overset{\bar{K}_{4i}}{\rightleftharpoons} S_iCu + H^+$$

$$K_{4i} = \frac{m_{ci} m_H}{m_{S_iH} m_M} = K_{4i}^0 e^{-(\Delta G_i^0/RT)} \tag{2}$$

The behavior of $\bar{K}_4$ for the whole mixture is deduced from the properties of the components, by combining Eqs. 1 and 2 with some material balance relationships (Gamble et al., 1980). This produces the general working Eq. 3:

$$\bar{K}_4 = -\frac{1}{(1-\chi_c-\chi_{SH_2})}\int_0^{\chi_c} K_4 d\chi_c - \frac{1}{(1-\chi_c-\chi_{SH_2})}\int_0^{\chi_{SH_2}} K_4 d\chi_{SH_2} . \tag{3}$$

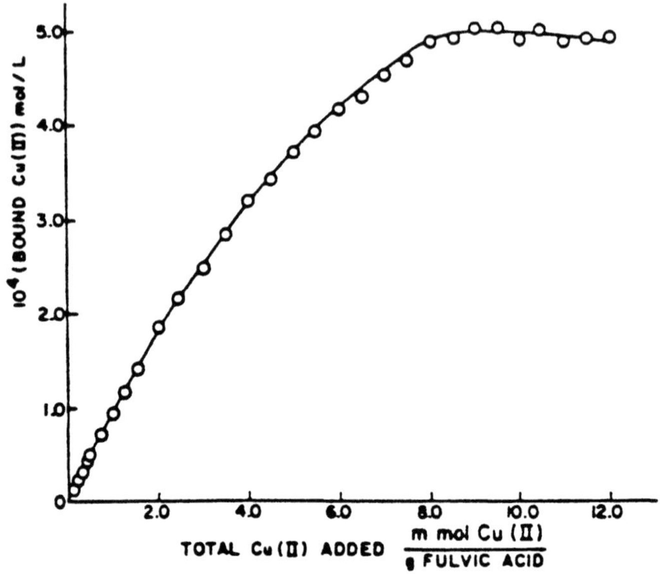

Fig. 2. Titration of fulvic acid with Cu(II) at 25.0 °C. Determination of bound Cu(II) and the total chelation capacity of the fulvic acid

Table 5 lists a few typical Cu(II) results for Armadale Bh Horizon fulvic acid

Table 5. Calculation of K_4, the differential equilibrium $^x_{SH} \simeq 0.37$, function for Cu(II)-fulvic acid chelation[a].

K_4	K_4	$-(\Delta G_4^0 + RT \ln K_4)$ (kJ/mol)	χ_c
1.53×10^3	5.90×10^4	27.3	0.0181
3.92×10^2	1.39×10^4	23.6	0.0363
1.97×10^2	4.30×10^3	20.7	0.0540
9.78×10^1	1.80×10^3	18.6	0.0717
6.75×10^1	7.00×10^2	16.2	0.0896

[a] T = 25.0 °C

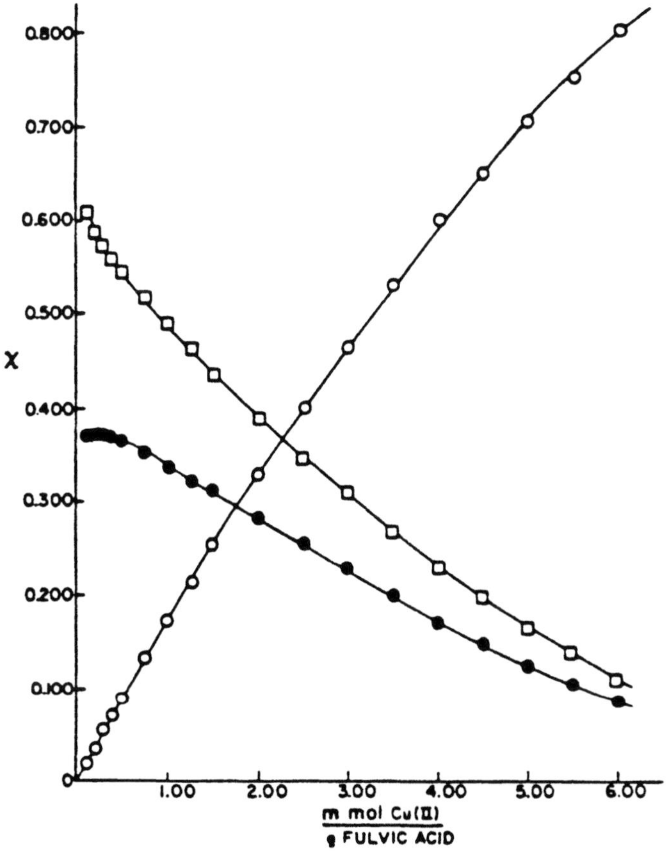

Fig. 3. Mole fractions of un-ionized, ionized, and complexed chelation sites. (●)χ_{SH2} = mole fraction of un-ionized bidentate chelation sites. (□)χ_{SH} = mole fraction of bidentate chelation sites free from 1st proton and from Cu(II). (○)χ_c = mole fraction of bidentate sites occupied by chelated Cu(II)

It gives $\bar{K}_4$ as a mathematical function, in terms of the chemical composition of the whole system. The importance of the chemical composition variables is seen in Fig. 3, while Fig. 4 emphasizes the point that $\bar{K}_4$ is a mathematical function. Figure 3 explains the independent variables in Eq. 3. The differential function K_4 observed in Eq. 3 may be evaluated from experimental data. ΔG_4 estimates obtained from Eq. 4 may then be used for characterizing the whole system. This is done by experimentally scanning

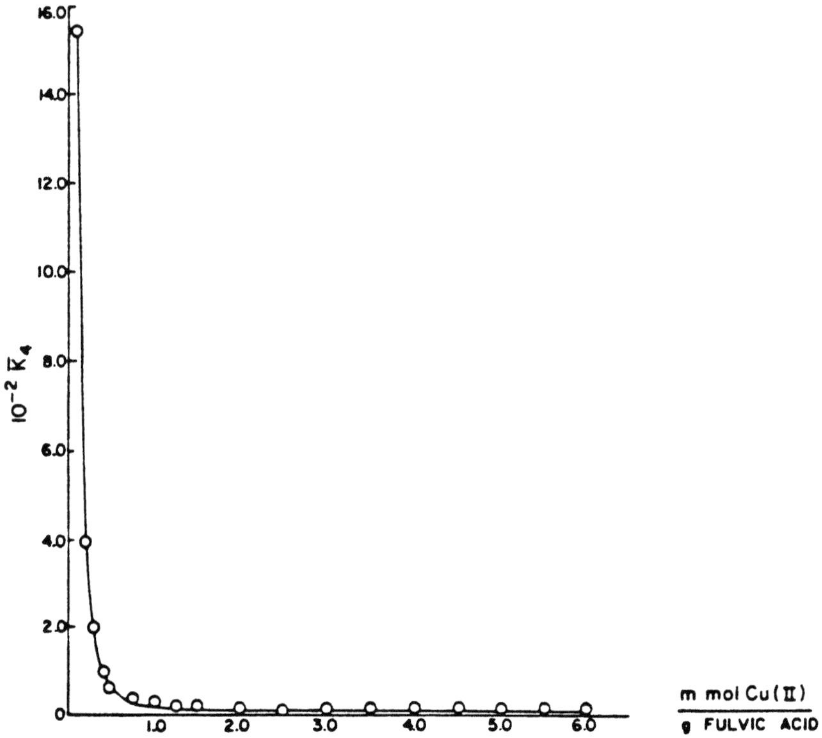

Fig. 4. $\bar{K}_4$, the weighted average mass action quotient for Cu(II)-fulvic acid chelation equilibrium

$(\Delta G_4 + RT \ln \Gamma)$ across the mixture of components.

$$RT \ln K_4 = -(\Delta G_4^0 + RT \ln \Gamma_4), \tag{4}$$

where $\Gamma_4 = \dfrac{\gamma_c \gamma_H}{\gamma_{SH} \gamma_M}$.

Two-phase equilibria: multiple metal ion exchange

Using the humic acid described in Table 4, Kerndorff and Schnitzer carried out the important multiple metal ion experiments outlined in Table 6.

The effect of pH on metal ion uptake by the humic acid is shown in Fig. 5.

Table 6. Effect of pH on sorption of metals on HA[a]

	pH							
	2.4			3.7		4.7	5.8	5.8
	HA (mg)							
Metal	1000	500	200	500	200	250	250	0
	% sorption							
Hg	80	88	99	98	98	100	98	24[b]
Fe	88	87	81	96	96	100	100	88[b]
Pb	65	33	19	96	80	97	98	24[b]
Cu	66	31	12	91	59	95	97	52[b]
Al	59	29	7	98	86	100	100	38[b]
Ni	14	7	5	41	6	29	61	0
Cr	10	4	0	83	70	100	100	53[b]
Cd	8	3	0	43	7	41	77	0
Zn	10	5	0	33	8	29	64	0
Co	4	2	0	23	2	22	45	0
Mn	5	2	0	14	3	13	36	2[b]

[a] Each system contained 0.5×10^{-4} mol of each metal.
[b] Hydroxide precipitates.

In order that these experiments may become useful for practical predictive calculations, it was necessary to analyze them according to the same kind of weighted average equilibrium function theory that has been outlined for the solution phase Cu(II) chelation. The theoretical study was based on a stoichiometric description of cation exchange equilibria as illustrated in Fig. 6. For each of the eleven metal ions listed in Table 6, a weighted average equilibrium function was obtained in this way. Each of the eleven metal ions therefore has a $\bar{K}_M$ definition and general working equation of the types indicated by Eqs. 5 and 6:

$$SH_2 + M^{n+} \xrightleftharpoons{\bar{K}_M} SM^{(n-2)+} + 2H^+$$

$$\bar{K}_M = \frac{m_c m_H^2}{m_{SH_2} m_M} \tag{5}$$

$$\bar{K}_m = -\frac{1}{\chi_{SH_2}} \int_0^{\chi_{c1}} K_M \, d\chi_{c_1} - \sum_{j=2}^{k} \frac{1}{\chi_{SH_2}} \int_0^{\chi_{c1}} K_M \, d\chi_{c_1} . \tag{6}$$

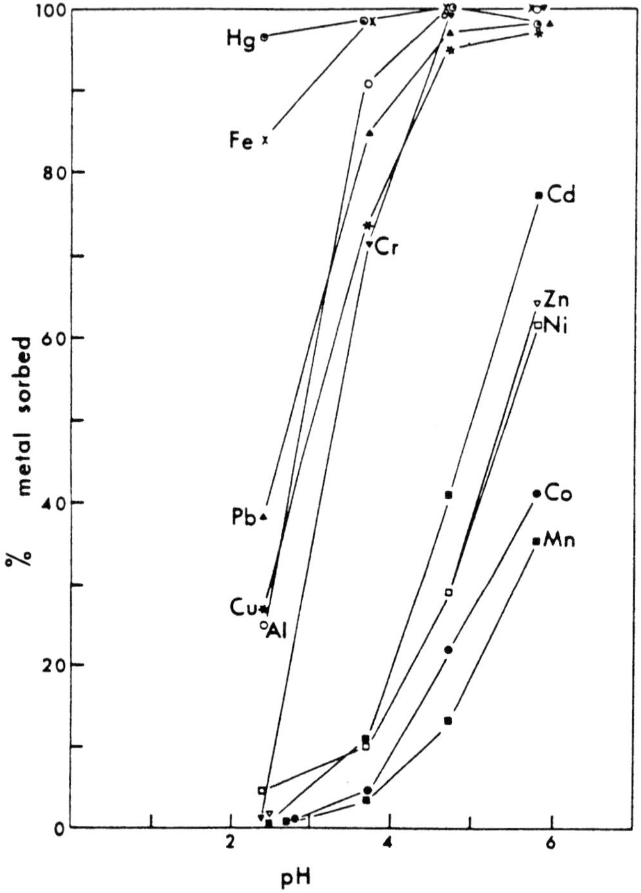

Fig. 5. Proportions of metals sorbed by 250 mg of HA at different pH levels. Initial metal content was 0.5×10^{-4} mol of each metal in 100 mL of solution

In Eq. 6, the first integral describes the effects on the $\bar{K}_M$ of a given metal ion of the loading of ion exchange sites with this same metal ion. Each of the other terms accounts for the influence of one of the other metal ions. Figures 7 and 8 show Cu(II) and Cd(II) scans across the mixture of ion exchange sites, obtained from the Kerndorff-Schnitzer experiments.

Together with the related $\bar{K}_M$ and ΔG° calculations, a full set of such scans can be used to characterize fully even a quite complicated natural ion exchange system. Final ion exchange results for most of the metal ions are summarized in Table 1.

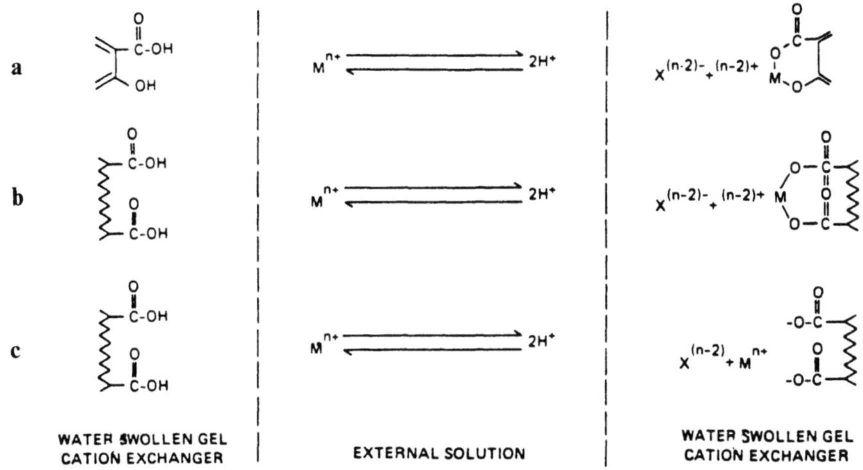

Fig. 6. The humic acid cation exchange system in Fig. 1, with divalent metal ions

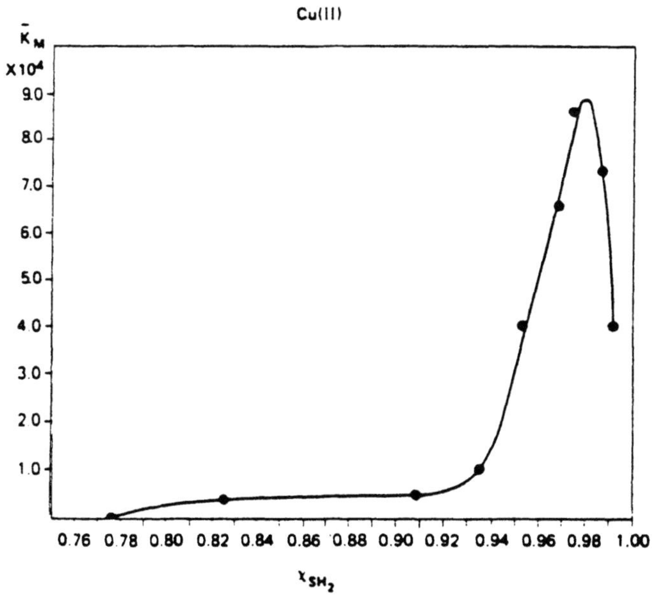

Fig. 7. Cu^{2+}-$2H^{+}$ exchange, in the humic acid-multiple metal ion system at 25 °C. The weighted average equilibrium function vs. the mole fraction of protonated exchange sites

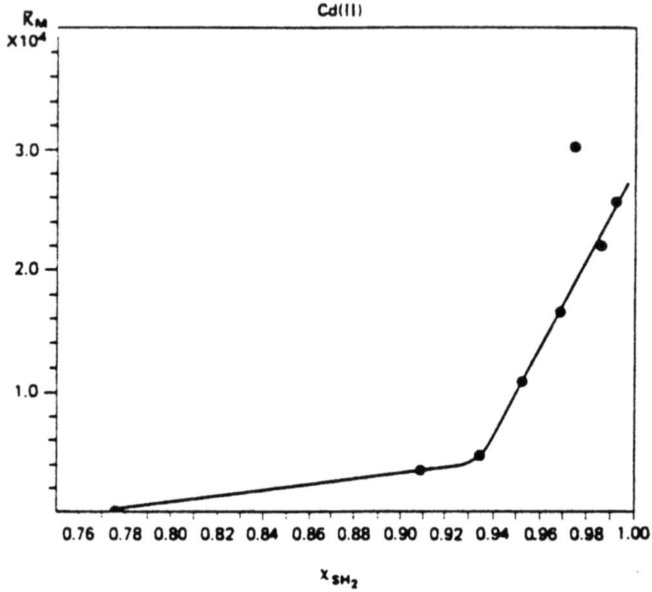

Fig. 8. Cd^{2+}-$2H^+$ exchange, in the humic acid-multiple metal ion system at 25 °C. The weighted average equilibrium function vs. the mole fraction of protonated exchange sites

Future Research Requirements

The ultimate research objective is to produce conclusions and predictive equations that have generality. The theoretical concepts, calculation techniques, and analytical chemical methods now available have been developed in research on only a very few samples of natural organic polymer. For these results to be exploited properly their generality must be tested and a broader scientific perspective attained. The most urgent needs therefore are the comparative analytical chemistry and metal ion reaction studies for samples of several types and from many geographical locations. Samples should represent soluble and insoluble fractions from soil horizons, water columns, and sediments of fresh waters, and samples from marine water columns and sediments. The chemical analyses and physicochemical measurement employed should include those described here, in order that the comparative chemistry be done according to the principles of stoichiometry. This is essential for practical purposes, because the chemically blind "black box" work found in the earlier literature lacks predictive capacity. Because natural geochemical systems contain mixtures of metal ions, the multiple metal ion equilibrium work should be extended in at least three ways. Future experiments should use metal ion compositions that resemble more closely those of natural soils and surface waters. Samples should be experimentally scanned over the ranges of their metal ion loadings, so that the partial differential calculus methods indicated in the mixed sample equilibria theory may be employed. This will permit the equilibria to be described as mathematical functions of important chemical composition variables. Finally, as mentioned above, the scope of the research should encompass a wide variety of samples. With extensions of these types, the research will almost automatically become the kind of mathematical modeling that is needed for practical predictions.

Two new fields of analytical chemistry are indicated for future progress. Rayleigh light-scattering methods show promise and should be developed for the investigation of the aggregation – metal ion-complexing aspects of species analysis. The other field of activity in need of expansion is chemical computing. The titration, data logging, data processing, and stoichiometric calculation methods that must be used are all labor-intensive. Most, if not all, of the technology with which this difficulty can be solved is now available. The appropriate combination of integrated instrumentation, computers, and software has, however, yet to be assembled and applied to the research. Such an enterprise falls clearly within the realm of analytical chemistry.

References

Arnold R, Overbeek JTG (1950) The Dissociation and Specific Viscosity of Polymethacrylic Acid Rec Trav Chim 69:192

Beckwith RS (1959) Titration Curves of Soil Organic Matter. Nature 184:745

Bremner JM (1957) II. Observations on the Estimation of Free Amino Groups. Reactions of Humic Acid and Lignin Preparations with Nitrous Acid. J Agr Sci 48:352

Bresnahan WT, Grant CL, Weber JH (1978) Stability Constants for the Complexation of Copper (II) Ions with Water and Soil-derived Fulvic Acids: Effect of pH and Fulvic Acid Concentration

Broadbent FE, Bradford GK (1952) Cation Exchange Groupings in the Soil Organic Fraction. Soil Sci 74:447

Burch RD, Langford CH, Gamble DS (1978) Methods for the Comparison of Fulvic Acid Samples: The Effects of Origin and Concentration on Acidic Properties 56:1196

Gamble DS (1970) Titration Curves of Fulvic Acid: The Analytical Chemistry of a Weak Acid Polyelectrolyte. Can J Chem 48:2662
Gamble DS (1972) Potentiometric titration of Fulvic Acid: Equivalence Point Calculations and Acidic Functional Groups. Can J Chem 50:2680
Gamble DS, Schnitzer M (1973) Trace Metals and Metal and Organic Interactions in Natural Waters. In: Singer PC (ed) Trace Metals and Metal Organic Interactions in Natural Waters, ch. 9. Ann Arbor Science Publishers, Ann Arbor, MI
Gamble DS, Schnitzer M, Hoffmann I (1970) Cu^{2+} – Fulvic Acid Chelation Equilibrium in 0.1 im KCl at 25.0 °C. Can J Chem 48:3197
Gamble DS, Schnitzer M, Kerndorff H, Langford CH (1983). Multiple Metal Ion Equilibria with Humic Acid. Geochim Cosmochim Acta 47:1311
Gamble DS, Underdown AW, Langford CH (1980) Copper (II) Titration of Fulvic Acid Ligand Sites with Theoretical, Potentiometric, and Spectrophotometric Analysis. Anal Chem 52:1901
Ghosh K, Schnitzer M (1981) Fluorescence Excitation Spectra and Viscosity Behaviour of a Fulvic Acid and its Copper and Iron Complexes. Soil Sci Soc Amer Proc 21:368
Griffith SM, Schnitzer M (1975) The Isolation and Characterization of Metal Organic Complexes from Tropical Volcanic Soils. Soil Sci 120:126
Hanzlik RP (1976) Inorganic Aspects of Biological and Organic Chemistry. Academic Press
Helfferich F (1962) Ion exchange. McGraw-Hill, New York
Himes FL, Barber SA (1957) Chelating Ability of Soil Organic Matter. Soil Sci Amer Proc 21:368
Huaglin MB (1972) Light Scattering from Polymer Solution. Academic Press, New York
Irving H, Williams RJP (1953) The Stability of Transition-Metal Complexes. J Chem Soc 3192
Katchalsky A, Gillis J (1949) Theory of the Potentiometric Titration of Polymeric Acids. Rec Trav Chim 68:879
Katchalsky A, Spitnik P (1947) Potentiometric Titrations of Polymethacrylic Acid. J Polymer Sci 2:432
Kennedy VC, Jellwiger GW, Jones BF (1974) Filter pore-size effects on the analysis of Al, Fe, Mn, and Ti in water. Wat Lat Resources Res 10(4):785–790
Kerndorff H, Schnitzer M (1980) Sorption of Metals on Humic Acid. Geochim Cosmochim Acta 28:1701
Khan SU, Sowden FJ (1971) Distribution of Nitrogen in the Black Solonetzic and Black Chernozemic Soils of Alberta. Can J Soil Sci 51:185
Khanna SS, Stevenson FJ (1962) Metalloorganic Matter Isolates in the Presence of Transition Metals. Soil Sci 93:298
Kodama H, Schnitzer M (1972) Differential Thermal Analysis of Metal-Fulvic Acid Salts and Complexes. Geoderma 7:93
Lee SLF (1982) CU^{2+} Complexing in Fulvic Acid Solutions at Different pH and Ionic Strength Conditions. Ph D Thesis, Dept of Chemistry, Carleton University, Ottawa, Ontario
Marinsky JA, Wolf A, Bunzl K (1980) The Binding of Trace Amounts of Lead(II), Copper(II), Cadmium(II), Zinc and Calcium to Soil Organic Matter. Talanta 27:461

Morel F, McDuff RE, Morgan JJ (1973) Interactions and Chemostasis in Aquatic Systems: Role of pH, pE, Solubility and Complexation. In Trace Metals and Metal Organic Interactions in Natural Waters. Ed Singer PC, ch 6. Ann Arbor Sci Publ
Orlov DS, Gorshkova El (1965) Dimensions and Shape of Humic Acid Particles from Chernozem and Sod-Podzolic Soil. Dokl Vyssh Skl Biol Nauk 1:207
Pagenkopf GK, Whitworth C (1981) Precipitation of Metal-Humate Complexes. J Inor Nucl Chem 43:1219
Reuter JH, Perdue EM (1977) The Importance of Heavy Metal-Organic Matter Interactions in Natural Waters. Geochim Cosmochim Acta 41:325
Sarr RA, Weber JH (1979) Lead(II) Complexation by Fulvic Acid: How it Differs from Fulvic Acid Complexation of Copper(II) and Cadmium(II). Geochim Cosmochim Acta 44:1381
Sarr RA, Weber JH (1979) Complexation of Cadmium(II) with Water- and Soil-derived Fulvic Acids: Effect of pH and Fulvic Acid Concentrations. Can J Chem 57:1263
Schnitzer M, Khan SU (1978) Soil Organic Matter. Elsevier
Schnitzer M, Skinner SIM (1963) Organo-Metallic Interactions in Soils: 1, Reactions Between a Number of Metal Ions and the Organic Matter of a Podzol Bh Horizon. Soil Sci 96:86
Schnitzer M, Skinner SIM (1965) Organo-Metallic Interactions in Soils: 4, Carboxyl and Hydroxyl Groups in Organic Matter and Metal Retention. Soil Sci 99:278
Senesi N, Chen Y, Schnitzer M (1977) Aggregation Dispersion Phenomena in Humic Substances. In: Soil Matter Studies, vol 2. International Atomic Energy Agency, Vienna, pp 143–155
Sillen LG, Martell AE (1964) Stability Constants of Metal Ion Complexes. Special Publication No 17. The Chemical Society, London
Sillen LG, Martell AE (1971) Stability Constants of Metal Ion Complexes. Special Publication No 25. The Chemical Society, London
Swoden FJ (1957) Note on the Occurrence of Amino Groups in Soil Organic Matter. Can J Soil Sci 37:143
Sowden FJ, Griffith SM, Schnitzer M (1976) The Distribution of Nitrogen in Some Highly Organic Tropical Volcanic Soils. Soil Biol Biochem 8:55
Stevenson FJ (1976) Binding of Metal Ions by Humic Acids. In: Nriagu JO (ed) Environmental Biogeochemistry, vol 2. Ann Arbor Science Publishers, Ann Arbor, MI, pp 519–540
Stumm W, Morgan JJ (1981) Aquatic Chemistry, 2nd ed Wiley Interscience. John Wiley & Sons, New York
Underdown AW (1982) Polyelectrolyte, Aggregation and Light Scattering Properties of Fulvic Acid-Metal Ion Systems. Ph D Thesis, Dept of Chemistry, Carleton University, Ottawa, Ontario
Underdown AW, Langford CH, Gamble DS (1981) Light Scattering of a Polydisperse Fulvic Acid. Anal Chem 53:2139
Wagner GH, Stevenson FJ (1965) Structural Arrangements of Functional Groups in Soil Humic Acid as Revealed by Infrared Analysis. Soil Sci Soc Amer Proc 29:43
Wershaw RL, Burcar PJ, Sutula CL, Wiginton BJ (1967) Sodium Humate Solution Studied with Small-angle X-Ray Scattering. Science 157:1429

Wershaw RL, Heller SJ, Pinckney DJ (1970) Measurement of the Molecular Size of a Sodium Humate Fraction. J Adv X-Ray Anal 13:609
Wershaw RL, Pinckney DJ (1971) Association and Dissociation of a Humic Acid Fraction as a Function of pH. Geological Surv Prof Pap 0750-D: D 216
Wershaw RL, Pinckney DJ (1973 a) The Fractionation of Humic Acids from Natural Water Systems. J Res US Geol Surv 1, 361
Wershaw RL, Pinckney DJ (1973 b) Determination of the Association and Dissociation of Humic Acid Fractions by Small-angle X-Ray Scattering. J Res US Geol Surv 1, 701

The Importance of Chemical "Speciation" in Environmental Processes,
eds. M. Bernhard, F. E. Brinckman, and P. J. Sadler, pp. 237–254.
Dahlem Konferenzen 1986. Springer-Verlag Berlin, Heidelberg
© Dr. S. Bernhard, Dahlem Konferenzen

Adsorption and Transformation of Trace Element Species at Sediment/Water Interfaces

J. O. Leckie

Environmental Engineering and Science Dept.
Stanford University
Stanford, CA 94305, USA

Abstract. The transformation of trace element species in aqueous systems includes the formation of surface complexes at the solid/solution interface. The reactions describing the formation of these surface coordination species follow all the concepts and mathematical formalism of coordination chemistry. The development of an electric field in the solid/solution interphase region requires either an explicit accounting for electrostatic effects or incorporation of field effects in rational activity coefficients. Reactions at sediment/water interfaces can be understood best when treated as polyligand systems. Both organic and inorganic materials are observed to exhibit variable binding intensities with trace metals. These systems can be modeled as polyligand systems with variable proton stoichiometry. Real soils and sediments exhibit similar variable binding intensities and can be modeled as polyligand materials.

Introduction

Chemical species in aqueous systems containing solids inevitably includes surface species formed between surface functional groups and solutes from bulk solution. The influence of species on the reactions at solid/solution interfaces in natural aquatic systems becomes apparent when we consider the type and nature of particulates in natural waters (suspended or deposited). The particulates in natural waters consist predominantly of detrital organic colloidal matter, living cells (bacteria and algae), and inorganic solids such as metal oxides and hydroxides, carbonates, and clays.

Reactions between solutes and functional groups on solid surface are essentially coordination reactions which conform to the concepts and mathematical formalism of coordination processes. Forces acting at the solid/solution interface are extensions of those forces predominant within the solid and solution phases. Because atoms at solid surfaces have an imbalance of chemical forces they are quite reactive. Reaction between the aqueous solvent and the solid surface (e.g., metal oxide) leads to the development of a charged interface. The importance of the electrostatic interactions between adsorbing solutes and the charged surface groups has led to the overall interaction being separated into a chemical term and an electrostatic term. In addition, the surface charge depends strongly on the composition of the electrolyte solution.

Because direct, unambiguous measurements of surface complexes are presently not possible, a large number of conceptual models and theories of surface processes have been proposed. Despite all the research activity devoted to adsorption phenomena over the past two decades, not all observed adsorption reactions can be accounted for within a single framework.

General Considerations

The species distribution of a particular element in an aqueous environment is determined through resolution of a wide array of competing reactions including complexation with both inorganic and organic soluble ligands, changes in oxidation state, precipitation/dissolution, solid-phase adsorption/desorption, and ion exchange. The importance of metal or metalloid-particulate association in aquatic environments has been thoroughly demonstrated (Förstner and Wittman, 1979). Surficial sediments are likely sinks for trace metals through complexation with particulate-associated organics (e.g., humics) and/or adsorption to mineral solids (e.g., hydrous metal oxides).

Experimental observations of the adsorption/complexation characteristics of trace metals with both humic-type substances (Gamble et al., 1980) and hydrous metal oxides (Benjamin and Leckie, 1981) have revealed variation in binding intensity as a function of degree of site occupancy and a hierarchy of binding intensities for solutes with the same solid and for different solids with the same solute. The naturally occurring solid phases found in oxidized environments can be conveniently classified as either organic or mineral. Important mineral constituents include various amorphous to crystalline hydrous metal oxides (e.g., SiO_2, MnO_2, FeOOH, Al_2O_3, TiO_2), the layered aluminosilicate clays (e.g., montmorillonite), and metal carbonates. Organic materials are represented by the variably oxidized, refractory, moderate-to-very-high molecular weight, complex polymers known as humic and fulvic acids, macrophyte exudates, and bacterial and algal cell surfaces and associated exopolymers. Although diverse, this melange does possess certain common characteristics which are expected to influence the overall nature of their interaction with trace metal species. Of particular note are the near universality of bound oxygen as an available electron donor, the high degree of polyfunctionality, and the structural complexity and heterogeneity with respect to ligand site microscopic physical and chemical environments. Experimental observations indicate that humic materials and the hydrous oxides of iron and manganese are of particular importance vis-à-vis trace metal binding in environments ranging from seawater to soils (Balistrieri and Murray, 1982a, 1982b; Jenne, 1976; Luoma and Bryant, 1981; Sposito, 1983a). Adsorption behavior of trace metals on estuarine sediments exhibits qualitative similarities to trace metal binding on oxides (Lion et al., 1982).

The adsorption/complexation intensity and capacity characteristics of humic substances have been examined in considerable detail for binding of trace metals (Buffle et al., 1977; Gamble et al., 1980; Mantoura and Riley, 1975; Stevenson, 1977). Although we will draw comparisons between polyligand humics and hydrous metal oxides below, emphasis will be placed on details of adsorption models for hydrous oxide surfaces.

Concepts of Adsorption

Surface Functional Groups

Because metal oxides and hydroxides represent a major fraction of reactive inorganic particulate matter in natural aquatic systems, a description of the oxide/water interface will be useful. Upon contact with water, metal oxide surfaces coordinate with water molecules thus satisfying the coordination requirements of the surface metal ions leading to a hydroxylated surface (Fig. 1). The development of the amphoteric surface hydroxyls imparts acid-base properties to the hydrous oxide surface. This surface acidity is an important property of hydrous oxides; the presence of surface functional groups is closely related to the type and extent of interfacial reactions such

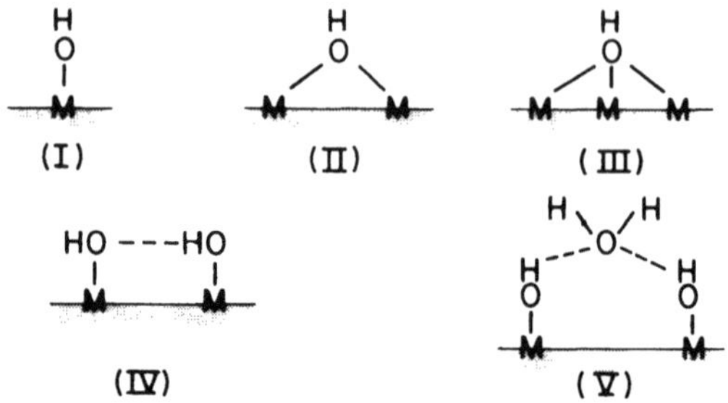

Fig. 1. Different types of surface hydroxyl groups found on hydrous oxide surfaces: "free" hydroxyls bound to one (I), two (II), or three (III) surface metals; hydrogen-bonded surface hydroxyls (IV), i.e., two surface hydroxyl groups in close proximity; adsorbed water (V)

as adsorption and coagulation. The acid sites can initially be considered to be of two types: a) Lewis acid sites such as Fe^{3+} ions in a crystal edge, which are capable of accepting electrons from the adsorbate molecules, and b) Brønsted acid sites such as Si-OH, Fe-OH, Al-OH, or H_3O^+ which can donate a proton. These amphoteric surface hydroxyls can fulfill the same coordination role with metal ions as other solution ligands. The currently accepted formalism represents surface hydroxyl groups of metal oxides by a generalized hydrolyzed species, SOH, and acid/base reactions are characterized by the gain or loss of a proton rather than the exchange of a hydroxyl ion:

$$SOH = SO^- + H^+$$

$$SOH + H^+ = SOH_2^+$$

Thus reactions in which the proton is replaced by a metal ion or a ligand exchanges for the hydroxyl are the equivalent to the formation of surface inner-sphere complexes (Table 1), and similar reactions where the adsorbing species bridges across a chemisorbed water molecule represent surface ion pair complexes. Clearly, several types of surface species are possible, including inner-sphere and outer-sphere (ion pairs) complexes, multidentate, and ternary or mixed ligand complexes (Table 1). Unfortunately, it is not currently possible to unambiguously examine the configuration of these adsorbed species. Since thermodynamic arguments do not require a specific model at the molecular level, we can still analyze equilibrium data within a thermodynamic context.

Table 1. Possible surface species which may form between oxide functional groups and solutes. Current analytical techniques do not allow direct confirmation of the presence of species on the surface

Surface reactions	Surface species
Proton/Surface Reactions	$\vdash O\langle^{H}_{H^+}$
$SOH_2^+ = SOH + H^+$	$\vdash O-H$
$SOH = SO^- + H^+$	$\vdash O^-$
Metal/Surface Reactions (Inner-Sphere)	
$SOH_2^+ + M = SOM + 2H^+$	
$SOH + M = SOM + H^+$	$\vdash O-M$
$SO^- + M = SOM$	
$SOH + H_2O + M = SOMOH + 2H^+$	$\vdash O-M-OH$
$SOH + 2H_2O + M = SOM(OH)_2 + 3H^+$	$\vdash O-M\langle^{OH}_{OH}$
$(SOH)_2 + M = (SO)_2M + 2H^+$	$^{\vdash O}_{\vdash O}\rangle M$
$(SOH)_2 + M = S_2OOHM + H^+$	$\vdash O\langle^{H}$, $^{\vdash O}_{\vdash O}\rangle M$
Ligand/Surface Reactions (Inner-Sphere)	
$SOH_2^+ + L = SL + H_2O$	
$SOH + H^+ + L = SL + H_2O$	$\vdash L$
$SOH + 2H^+ + L = SLH + H_2O$	$\vdash LH$
$(SOH)_2 + 2H^+ + L = S_2L + 2H_2O$	$\vert\!>\!L$
$(SOH)_3 + 3H^+ + L = S_3L + 3H_2O$	$\models\!L$
Mixed Ligand (Ternary) Reactions (Inner-Sphere)	
$SOH + ML = SOML + H^+$	$\vdash O-M-L$
$SOH + ML = SLM + OH^-$	$\vdash L-M$
$(SOH)_2 + ML = SOMLS + H_2O$	$^{\vdash O-M}_{\vdash L\diagup}$
Ion Pair Reactions (Outer-Sphere)	
$SOH + M = SO^- - M + H^+$	$\vdash O^- - \; - M$
$SOH + H_2O + M = SO^- - MOH + 2H^+$	$\vdash O^- - \; - MOH$
$SOH + H^+ + L = SOH_2^+ - L$	$\vdash OH_2^+ - \; - L$
$SOH + 2H^+ + L = SOH_2^+ - LH$	$\vdash OH_2^+ - \; - LH$
Exchange Reactions	
$SO^- - Na^+ + M = SO^- - M + Na^+$	$\vdash O^- - \; - M$
$SO^- - Na^+ + M = SOM + Na^+$	$\vdash O-M$
$SOH_2^+ - SO_4^{2-} + L = SOH_2^+ - L + SO_4^{2-}$	$\vdash OH_2^+ - \; - L$
$SOH_2^+ - SO_4^{2-} + L = SL + SO_4^{2-} + H_2O$	$\vdash L$

Most of the surface reactions are inferred from experimental observations of reaction stoichiometries. Thus the choice of specific surface species such as SOMOH, S_3L, $(SO)_2M$, or SOH_2^+-L, given as examples in Table 1, is dependent upon two considerations: a) a need to depict reaction stoichiometry with observed experimental measurements, and b) the selection of a model which allows representation of the binding intensity. Since we cannot directly observe the surface species we cannot verify the existence of either ion pairs or coordinate complexes at the surface, although surely both types must exist. Since the type and concentration of background electrolyte is observed to affect adsorption phenomena at oxide surfaces, these observations can be accounted for either by explicitly incorporating the electrolyte/surface interactions or by lumping effects into activity coefficients. These details are discussed subsequently.

Historically, the description of adsorption or partitioning of solutes to the solid/solution interface has been accomplished in a manner that assumes only two states for the surface sites: occupied (SM) or unoccupied (S). This approach leads directly to the Langmuir isotherm where the assumption is made that the free energy of adsorption is independent of surface coverage. This assumption is often taken to mean that adsorbed molecules do not interact with one another. This is the equivalent of an ideal surface solution. In the mathematical derivation of the Langmuir isotherm, the system is assumed to have a fixed concentration of sites to which adsorbate can attach, S_T (moles of sites/*l*). The adsorption reaction is modeled analogous to solution-phase reactions:

$$M + S = SM \tag{1}$$

$$^{c}K = \frac{[SM]}{[S]\,[M]}, \tag{2}$$

and the mass balance on surface sites is given by

$$S_T = [S] + [SM], \tag{3}$$

where we assume that activity coefficients on the surface and in solution are constant and c_K is a conditional equilibrium constant. Combining the mass action expression and mass balance yields the Langmuir isotherm

$$[SM] = \frac{S_T\,{}^{c}K[M]}{1 + {}^{c}K[M]}, \tag{4}$$

where brackets denote concentration. The assumption that adsorption energy is independent of surface coverage means that c_K is independent of [SM]. The simple stoichiometry portrayed in Eq. 1 does not allow for multiple surface species distribution nor does it account for dependence on solution composition. To reflect more fully the complexity of the system, models must include concepts of solution analogy for surface reactions, complex adsorption stoichiometries, and variable trace metal-binding energies with polyligand substrates.

Surface Complexation

Here we consider that hydrosols containing functional groups with specific, well-defined surface sites can be treated in direct analogy with solution coordination chemistry. That is, for each type of site with a total concentration of sites S_T, a standard free energy of adsorption (ΔG^o_{ads}) can be defined for each specific adsorption reaction. Chemical reactions at metal oxide surfaces can be described solely in thermodynamic terms if experimental methods exist for determining the reaction stoichiometry (mole numbers) of the reactants and products at equilibrium. Since these chemical equilibria are described by thermodynamic arguments formulated independently of specific mechanisms, the detailed nature of surface complexes (e.g., inner-sphere or outer-sphere) has no part in the thermodynamic description. Furthermore, the conditional equilibrium constants for these surface reactions and the rational activity coefficients of the surface species can be determined solely from data on the composition of the solid-phase interfacial region and the aqueous solution at equilibrium (Sposito, 1981). Although it is not necessary to invoke molecular hypotheses (e.g., diffuse double-layer theory) to render chemical significance to the interphase region (Sposito, 1983 b), currently used models express the electric field effects explicitly in terms of the electrical double-layer model rather than incorporating these effects into the rational activity coefficients.

Westall and Hohl (1980) have shown that five different models for the electrostatics of the oxide/water interface can be made to fit experimental data equally well. The surface complexation models invoke specific molecular configurations in the interphase region (i.e., inner- or outer-sphere complexes) as a means of developing surface charge-surface potential relationships to allow mathematical description of the electric field portion of the rational activity coefficients for the surface species.

Table 2. Surface complexation reactions with surface hydroxyls represented as outer-sphere complexes. (Data taken from Balistrieri and Murray, 1982a; Davis and Leckie, 1978; Davis and Leckie, 1980)

Reactions	pK (25 °C)
Goethite (α-FeOOH)	
$FeOH_2^+ = FeOH + H^+$	4.9
$FeOH = FeO^- + H^+$	10.4
$FeOH + Na^+ = FeO^-Na^+ + H^+$	9.6
$FeOH + K^+ = FeO^-K^+ + H^+$	9.6
$(FeO)_2H_2 + Mg^{2+} = (FeO^-)_2Mg^{2+} + 2H^+$	11.9
$FeOH + Mg^{2+} + H_2O = FeO^-MgOH^+ + 2H^+$	12.2
$(FeO)_2H_2 + Ca^{2+} = (FeO^-)_2Ca^{2+} + 2H^+$	15.8
$FeOH + Ca^{2+} + H_2O = FeO^-CaOH^+ + 2H^+$	16.5
$FeOH + Zn^{2+} + H_2O = FeO^-ZnOH^+ + 2H^+$	9.15
$FeOH + Pb^{2+} = FeO^-Pb^{2+} + H^+$	1.8
$FeOH + Pb^{2+} + H_2O = FeO^-PbOH^+ + 2H^+$	5.0
$FeOH + Cu^{2+} = FeO^-Cu^{2+} + H^+$	3.0
$FeOH + Cu^{2+} + H_2O = FeO^-CuOH^+ + 2H^+$	7.0
$FeOH + Cd^{2+} = FeO^-Cd^{2+} + H^+$	1.3
$FeOH + Cd^{2+} + H_2O = FeO^-CdOH^+ + 2H^+$	9.35
$(FeO)_2H_2 + SO_4^{2-} + 2H^+ = (FeOH_2^+)_2SO_4^{2-}$	−14.1
$FeOH + SO_4^{2-} + 2H^+ = FeOH_2^+SO_4H^-$	−14.4
$FeOH + Cl^- + H^+ = FeOH_2^+Cl^-$	− 5.5
Alumina (γ-Al_2O_3)	
$AlOH_2^+ = AlOH + H^+$	5.7
$AlOH = AlO^- + H^+$	11.5
$AlOH + Pb^{2+} = AlO^-Pb^{2+} + H^+$	5.0
$AlOH + Pb^{2+} + H_2O = AlO^-PbOH^+ + 2H^+$	10.3
Titania (TiO_2)	
$TiOH_2^+ = TiOH + H^+$	2.6
$TiOH = TiO^- + H^+$	9.0
$TiOH + Cd^{2+} = TiO^-Cd^{2+} + H^+$	1.8
$TiOH + Cd^{2+} + H_2O = TiO^-CdOH^+ + 2H^+$	8.7
Quartz (α-SiO_2)	
$SiOH = SiO^- + H^+$	7.2
$SiOH + Cu^{2+} = SiO^-Cu^{2+} + H^+$	4.4
$SiOH + Cu^{2+} + H_2O = SiO^-CuOH^+ + 2H^+$	8.6

Reaction Stoichiometry

A general overall reaction stoichiometry can be formulated for the reaction of a metal cation and ligand with a surface hydroxyl group. This general reaction accounts for the range of possibilities including adsorption of the aquo metal ion and the first hydrolysis product, or the adsorption of a complex thus forming ternary surface-complexed species. The general overall reaction stoichiometry is given by

$$SO_aH_a + bM^{m+} + cL^{l-} + dH^+ + eH_2O = (SO)_aM_b(OH)_eH_dL_c^{\delta} + (a+e)H^+, \tag{5}$$

and

$$K = \frac{((SO)_aM_b(OH)_eH_dL_c^{\delta})\,(H^+)^{(a+e)}}{(SO_aH_a)\,(M^{m+})^b(L^{l-})^c(H^+)^d(H_2O)^e}, \tag{6}$$

where SOH represents a mole of reactive hydroxyls on the hydrous oxide surface in contact with solution, δ represents the valence of the solid-phase product, and parenthesis ( ) designate activities of species. M and L represent the metal cation and ligand, respectively. Given the normal constraints of mass balance and electroneutrality, Eqs. 5 and 6 represent conventional chemical equilibria. Table 2 lists several examples of reactions expressed in terms of the general reaction stoichiometry given in Eq. 5. Although the anion-complexed formation reactions in Table 2 are depicted in terms of outer-sphere complexes, they could as well have been constructed according to the alternative stoichiometry depicting exchange with the surface hydroxyl:

$$S_a(OH)_a + bM^{m+} + cL^{l-} + dH^+ + eH_2O = S_aH_dL_cM_b(OH)_e^{\delta} + (a+e)OH^- \tag{7}$$

$$K = \frac{(S_aH_dL_cM_b(OH)_e^{\delta})\,(H^+)^{(a+e)}}{(S_a(OH)_a)\,(M^{m+})^b(L^{l-})^c(H^+)^d(H_2O)^e}, \tag{8}$$

where notation is defined as before. Examples of ligand exchange reactions (Stumm et al., 1980) of organic and inorganic anions for surface hydrolysis are given in Table 3.

Table 3. Ligand exchange reaction with surface hydroxyls represented as inner-sphere complexes. Data taken from Stumm et al. (1980). Constants valid at constant ionic strength $I=0.1$

Reaction	pK (22 °C)
Goethite (α-FeOOH)	
$FeOH_2^+ = FeOH + H^+$	6.4
$FeOH = FeO^- + H^+$	9.25
$FeOH + F^- = FeF + OH^-$	4.8
$FeOH + SO_4^{2-} = FeSO_4 + OH^-$	5.8
$Fe_2(OH)_2 + SO_4^{2-} = Fe_2SO_4 + 2OH^-$	13.5
$FeOH + HAc = FeAc + H_2O$ (acetate)	− 2.9
$FeOH + H_4SiO_4 + FeSiO_4H_3 + H_2O$	− 4.1
$FeOH + H_3PO_4 = FePO_4H_2 + H_2O$	− 9.5
$FeOH + H_3PO_4 = FePO_4H^- + H_2O + H^+$	− 5.1
Alumina (γ-Al_2O_3)	
$AlOH_2^+ = AlOH + H^+$	7.4
$AlOH = AlO^- + H^+$	10.0
$AlOH + HA = AlA + H_2O$ (benzoate)	− 3.7
$AlOH + H_2A = AlAH + H_2O$ (catechol)	− 3.7
$AlOH + H_2A = AlAH + H_2O$ (phthalate)	− 7.3
$AlOH + H_2A = AlAH + H_2O$ (salicylate)	− 6.0

Proton Stoichiometry

Experimental determination of adsorption reaction stoichiometry involves observation of macroscopic solution composition. This observed reaction stoichiometry represents overall, average values and can be described for cation binding by

$$SH_\chi + M \overset{^cK_M}{=} SM + \chi H\,, \tag{9}$$

and for anion binding by

$$SH + \chi H + A \overset{^cK_A}{=} SH_{(1+\chi)}A\,, \tag{10}$$

where the conditional equilibrium constants ^cK_M and ^cK_A are found to be quite sensitive to the uncertainty in χ. No assumptions are made concerning interphase configuration.

Surface complexation reactions involve either an increase or decrease in hydronium ions, here given by the stoichiometric coefficient χ. Seldom is

Table 4. Proton stoichiometric coefficients for reactions at the hydrous oxide surface

Adsorbate	Adsorbent	χ	pH	−log Γ	Ref.
Pb^{2+}	am-$Fe_2O_3 \cdot H_2O$	1.98	4.5–5.0	2.05	Perona and Leckie, 1985
	"FeOOH"	1.18	5.0	–	Loganathan and Burau, 1973
		1.59	6.0	–	Loganathan and Burau, 1973
	δ-MnO_2	1.4	6.0	–	Gadde and Laitenen, 1974
	γ-Al_2O_3	1.5	4.0–7.0	–	Vydra and Galba, 1969
Cd^{2+}	am-$Fe_2O_3 \cdot H_2O$	1.67	6.6–7.2	2.2	Perona and Leckie, 1985
	δ-MnO_2	1.3	6.0	–	Gadde and Laitenen, 1974
Cu^{2+}	am-$Fe_2O_3 \cdot H_2O$	1.9	5.1–5.8	2.17	Perona and Leckie, 1985
Zn^{2+}	am-$Fe_2O_3 \cdot H_2O$	2.3	6.0–6.5	2.36	Perona and Leckie, 1985
	δ-MnO_2	2.1	4.0	–	Gadde and Laitenen, 1973
		1.1	6.0	–	Gadde and Laitenen, 1974
	SiO_2	1.19	6.0	–	Vydra and Galba, 1969
Co^{2+}	δ-MnO_2	2.1	4.0	–	Gadde and Laitenen, 1973
	SiO_2	1.18	6.3	–	Vydra and Galba, 1969
		1.34	7.6	–	Vydra and Galba, 1969
Ag^{+}	am-$Fe_2O_3 \cdot H_2O$	0.86	6.5–8.0	2.0	Perona and Leckie, 1985
CrO_4^{2-}	am-$Fe_2O_3 \cdot H_2O$	0.89	6.0–7.0	1.5	Perona and Leckie, 1985
SeO_4^{2-}	am-$Fe_2O_3 \cdot H_2O$	0.85	4.5–6.5	1.5	Perona and Leckie, 1985
AsO_4^{2-}	am-$Fe_2O_3 \cdot H_2O$	1.2	9.5–10.5	1.5	Perona and Leckie, 1985
$Sb(OH)_3$	am-$Fe_2O_3 \cdot H_2O$	0.4	7.0–9.0	3.0	Perona and Leckie, 1985

the stoichiometric coefficient an integer (Table 4), suggesting multiple simultaneous surface complexation reactions of the type given in Tables 3 and 4. Thus, the application of simple two-state surface models (i.e., Langmuir isotherm) cannot easily be applied if we are to have a more complete description of the surface reactions. In addition, the species reactivity on the surface is strongly dependent upon the concentration of solid in the system as demonstrated by Fig. 2 (where $\chi = 1$). The proton stoichiometry can be either measured experimentally or calculated from adsorption data by the method of Kurbatov et al. (1951), where the mass action expression for Eq. 9 is linearized to give

$$\log\left[\frac{[SM]}{[SH_\chi][M]}\right] = \chi pH + \log {}^cK_M . \tag{11}$$

Plotting log $[SM]/[SH_\chi][M]$ as a function of pH yields a line with slope χ and intercept $\log {}^cK_M$. An alternative method uses the adsorption isotherms at different pH values (Perona and Leckie, 1985)

$$\chi = \left(\frac{\partial \bar{n}_H}{\partial \bar{n}_M}\right)_{\mu_i} = \left[\frac{\Delta \log[M]}{\Delta(pH)}\right]_{\Gamma,\mu_i} , \tag{12}$$

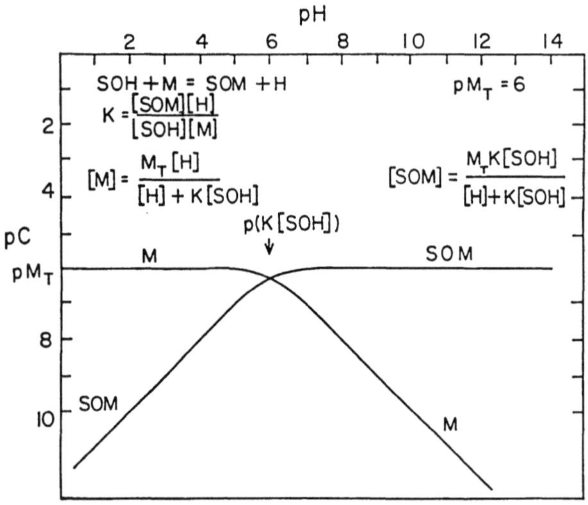

Fig. 2. pH-log concentration showing dependence of species reactivity on available surface-binding sites ([SOH]) and on the intensity of binding $(K)_o$. One proton exchange is assumed for simplicity

where $\bar{n}_H$ and $\bar{n}_M$ are the moles of H^+ and M adsorbed/desorbed per mole of adsorbent, and μ_i indicates that the chemical potential of all other species remains constant. The proton coefficient is estimated at constant surface coverage Γ. The proton stoichiometry is a function of both pH and degree of surface coverage. Because the proton stoichiometry depends on both pH and surface coverage and the conditional equilibrium constants depend markedly on χ, it is difficult to compare conditional equilibrium constants from different researchers without first placing the data in the same framework.

The general cation-adsorption reaction (Eq. 9) can be reformulated in terms of fractional adsorption by defining $f = [SM]/M_T$, making appropriate substitutions in Eq. 9, and rearranging

$$^cK_M \frac{[SM]}{[H]^\chi} = \frac{f}{1-f}. \tag{13}$$

For conditions where ^cK_M is constant, fractional adsorption, f, is independent of total metal, M_T, and dependent on available surface sites $[SH_\chi]$ and pH, and the slope of the Langmuir plot of the data should be one. At high site occupancy (large Γ), the Langmuir model predicts increasing fractional adsorption dependence on M_T. However, fractional adsorption becomes dependent on metal/site ratio (Benjamin and Leckie, 1981) at surface coverages far below site saturation (Fig. 3). The interpretation was that a range of site-binding energies existed. Although all sites are being filled simultaneously, higher energy sites fill preferentially and, since they are fewer in number than lower energy sites, they become limiting first. As lower energy sites are gradually filled, the overall binding constant decreases.

The apparent equilibrium constant, K_e, is conditional because the electrical double-layer term is considered relatively invariant in the swamping electrolyte used and was included in the constant. It is not a true equilibrium constant for another reason. It is an average binding constant because the adsorbed metal is the consequence of several different reactions, as is χ. K_e has the value of the equilibrium constant that would be computed if one assumed that all surface sites were identical and had a binding strength equal to the average binding strength at the given adsorption density. At low surface coverage, K_e is independent of Γ (Fig. 3). Above a critical surface coverage, $\Gamma^\star$, K_e decreases with increasing Γ. Because below $\Gamma^\star$, K_e is constant, the position of the pH/fractional adsorption edge does not move in response to changing Me_T. However, above $\Gamma^\star$, K_e is a function of Γ [or at fixed SOH_T, $K_e = f(Me_T)$] and the position of the fractional ad-

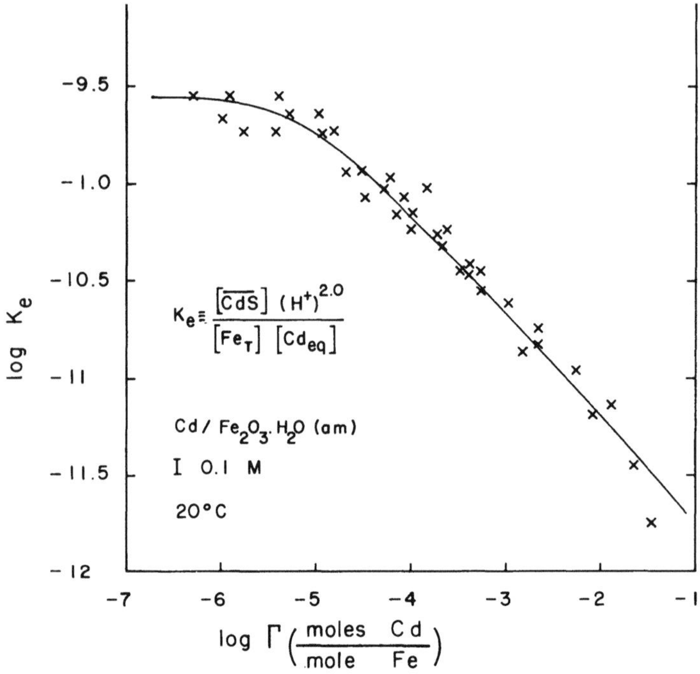

Fig. 3. Apparent overall adsorption equilibrium constant K_e as a function of adsorption density Γ in the system Cd/Am-$Fe_2O_3 \cdot H_2O$

sorption edge moves in response to the ratio Me_T/SOH_T. Thus the position of the fractional adsorption edge is not unique for a particular metal and oxide but changes as a function of the solid/solution ratio.

Adsorption in Multicomponent Systems

The interpretation of metal binding by oxides and humics has generally been accomplished by assuming single or several reactions with specific thermodynamic properties. The specified reactions and associated equilibrium constants are considered to be the same for all ligand sites present on the polyligand. In most instances, for both organic and inorganic polyligands, experimental observations, when made for conditions of low surface coverage and over a relatively narrow range of metal-to-solid ratio, conform to the homogeneous polyligand assumption suggested above. The homogeneous polyligand assumption does not hold, however, when systems are examined over an extensive range of surface coverage. Variable binding intensity has been reported (Benjamin and Leckie, 1981) for Cd, Zn, Cu, and Pb for oxide binding and for fulvic acid and Cu (Gamble et al., 1980), and has been observed by others (Kinneberg et al., 1983).

In summary, it has been shown that under appropriate (but limited) system conditions, monomeric solution chemistry analogies provide useful representations of metal complexation by polyligand solids. A reasonable extension of the single-polyligand, homogeneous-energetics approach toward understanding the complexation behavior of natural materials is to consider known mixtures of polyligands under similar conditions. This approach has been taken by Honeyman (1984) in examining metal and metalloid binding by well-defined combinations of oxides and a clay where adsorption sites were treated as monomeric using surface area and site density as normalizing parameters. Others (Luoma and Davis, 1983; Oakley et al., 1981) have also considered that whole system behavior might be predictable from consideration of the components but did not subject their hypothesis to experimental verification.

The results of Honeyman (1984) indicate behavior in accordance with assumed simple polyligand/metal-binding additivity to be the exception rather than the rule under low loading conditions. Composite adsorption behavior was found variously to be either enhanced or decreased relative to strict additivity depending on system conditions (i.e., solid types and concentrations, adsorbing species). Apparently changes in the net system surface properties resulting from varying degrees of particle/particle interactions (e.g., coagulation, heterocoagulation) must be considered if adsorption in mixed oxide systems is to be understood. Such considerations will be necessary in any summed behavior hypothesis regardless of the approach taken in modeling single polyligand behavior.

Metal binding by polyfunctional organic macromolecules exhibit characteristics inconsistent with the above simple models under certain conditions. Gamble et al. (1980) and Perdue and Lytle (1983) have observed and formulated models for this type of behavior in metal and proton binding by humic substances, a natural heteropolymer. A similar variable binding behavior for hydrous metal oxide systems has been experimentally investigated and theoretically considered (Benjamin and Leckie, 1981).

Numerous approaches are possible in conceptualizing and mathematically modeling variable binding behavior. The approach taken depends upon the level of detail available on the structure and reaction chemistry of the polyligand. Where differing site types are identifiable, a model with discrete mechanistic aspects may be appropriate. When a binding intensity continuum is observed and distinctions concerning site types are not easily made, a model utilizing general system parameters is preferred. Many of the polyligands found in natural environments (e.g., clays, oxides, humics) exhibit a continuously changing macroscopic metal partitioning when metal-to-ligand ratio is varied over a wide range. The conceptual approach taken to describe these systems determines the model structure. The predominant approach taken to describe metal complexation by humic substances has been that of identifiable site classes. This predisposition derives from structural information indicating a predominance of two bidentate site structures, salicylate (ortho carboxylate/phenolic) and phthalate (ortho carboxylate).

Some Questions and Problems

Key questions concerning the formation of surface complexes on sediment and soil materials are related to the problem of relating multi-component polyligand binding constants from system to system. The dependence of binding constants for surface complexes on particle-particle interactions and, therefore, solid/solution ratio invites additional study. The strong dependence of the binding constants on the proton stoichiometry also requires additional clarification. A general methodology for estimating rational activity coefficients for surface complexes would allow more general applications. The question of kinetics of both adsorption and desorption, as well as lability vis-à-vis real soils and sediments, needs to be resolved before adsorption models can be generally used in transport models.

Acknowledgements. I thank B. D. Honeyman, R. S. Altmann, and K. F. Hayes for their stimulating discussions and useful advice. The work and interest of W. Stumm has been a continuing stimulus.

References

Balistrieri L, Murray J (1982a) Surface of goethite (α-FeOOH) in seawater. ACS Symp Ser 93:275–298

Balistrieri L, Murray J (1982b) The adsorption of Cu, Pb, Zn and Cd on goethite from major ion seawater. Geochim Cosmochim Acta 46:1253–1265

Benjamin MM, Leckie JO (1981) Multiple site adsorption of Cd, Cu, Zn, and Pb on amorphous iron oxyhydroxide. J Colloid Interf Sci 79:209–221

Buffle J, Gaeter FL, Haerdi W (1977) Measurements of complexation properties of humic and fulvic acids in natural waters with lead and copper ion-selective electrode. Anal Chem 49:216–221
Davis JA, Leckie JO (1978) Surface ionization and complexation at the oxide/water interface. 2. Surface properties of amorphous iron oxyhydroxide and adsorption of metal ions. J Colloid Interf Sci 67(1):90–107
Davis JA, Leckie JO (1980) Surface ionization and complexation at the oxide/water interface. 3. Adsorption of anions. J Colloid Interf Sci 74:32–43
Förstner U, Wittman GTW (1979) Metal pollution in the aquatic environment. Springer-Verlag, Berlin
Gadde RR, Laitinen HA (1973) Study of the sorption of lead by hydrous ferric oxide. Envir Lett 5:223–235
Gadde RR, Laitinen HA (1974) Studies of heavy metal adsorption by hydrous iron and manganese oxides. Anal Chem 46:2022–2026
Gamble DS, Underwood AW, Langford CH (1980) Copper(II) titration of fulvic acid ligand sites with theoretical potentiometric, and spectrophotometric analysis. Anal Chem 52:1901–1908
Honeyman BD (1984) Metal and metalloid adsorption at the oxide/water interface in systems containing mixtures of adsorbents. Ph.D. Thesis, Stanford University, Stanford, CA
Jenne EA (1976) Trace element adsorption by sediments and soils – sites and processes. In: Chappell WR, Peterson KK (eds) Molybdenum in the environment, vol 2. Marcel Dekker, Inc., New York, pp 425–553
Kinneberg DG, Barker JA, Whitfield M (1983) A comparison of some simple adsorption isotherms for describing divalent cation adsorption by ferrihydrite. J Colloid Interf Sci 95:370–384
Kurbatov MH, Wood GB, Kurbatov JD (1951) Isothermal adsorption of cobalt from dilute solutions. J Phys Chem 55:1170–1182
Lion LW, Altmann RS, Leckie JO (1982) Trace-metal adsorption characteristics of estuarine particulate matter: evaluation of contributions of Fe, Mn oxide and organic surface coatings. Envir Sci Technol 16:660–666
Loganathan P, Burau R (1973) Sorption of heavy metal ions by a hydrous manganese oxide. Geochim Cosmochim Acta 37:1277–1293
Luoma SN, Bryant GW (1981) A statistical assessment of the form of trace metals in oxidized estuarine sediments employing chemical extractants. Sci Tot Envir 17:165–196
Luoma SN, Davis JA (1983) Requirements for modeling trace metal partitioning in oxidized estuarine sediments. Mar Chem 12:159–181
Mantoura RFC, Riley JP (1975) The use of gel filtration in the study of metal binding by humic acids and related compounds. Anal Chim Acta 78:193–200
Oakley SM, Nelson PO, Williamson KJ (1981) Model of trace metal partitioning in marine sediments. Envir Sci Technol 4:474–480
Perdue EM, Lytle CR (1983) Distribution model for binding of protons and metal ions by humic substances. Envir Sci Technol 17:654–660
Perona MJ, Leckie JO (1985) Proton stoichiometry for the adsorption of anions and cations on oxide surfaces. J Colloid Interf Sci, submitted
Sposito G (1981) The thermodynamics of soil solutions. Clarendon Press, Oxford
Sposito G (1983 a) The chemical forms of trace metals in soil. In: Thornton I (ed) Applied environmental geochemistry. Academic Press, London

Sposito G (1983 b) On the surface complexation model of the oxide-aqueous solution interface. J Colloid Interf Sci 91:239–340
Stevenson FJ (1977) Nature of divalent transition metal complexes of humic acids as revealed by a modified potentiometric method. Soil Sci 123:10–16
Stumm W, Kummert R, Sigg L (1980) A ligand exchange model for the adsorption of inorganic and organic ligands at hydrous oxide interfaces. Croatica Chem Acta 53:291–312
Westall J, Hohl H (1980) A comparison of electrostatic models for the oxide/solution interface. Adv Colloid Interf Sci 12:265–294
Vydra F, Galba J (1969) Sorption von Metal-Komplexen an Silikagel. V. Sorption von Hydrolyseprodukten des CO^{2+}, Mn^{2+}, Cu^{2+}, Ni^{2+}, Zn^{2+} an Silikagel. Coll Czech Chem 34:3471–3478

The Importance of Chemical "Speciation" in Environmental Processes,
eds. M. Bernhard, F.E. Brinckman, and P.J. Sadler, pp. 255–273.
Dahlem Konferenzen 1986. Springer-Verlag Berlin, Heidelberg
© *Dr. S. Bernhard, Dahlem Konferenzen*

Transformation of Sulfur Species by Phototrophic and Chemotrophic Microbes

J. Bauld

Baas Becking Geobiological Laboratory
Canberra City, ACT 2601, Australia

Abstract. Transformations between the various oxidation states of sulfur occurring within the biosphere are dominantly microbially mediated. Species transformations may occur during the biosynthesis of cell biomass (e.g., the assimilatory reduction of sulfate) or its subsequent decomposition. Dissimilatory processes may be oxidative or reductive in character. The former commonly provide a source of electrons for energy-generating metabolism and/or a source of reducing power for autotrophic CO_2 fixation. The latter employs oxidized S species as terminal electron acceptors for anaerobic respirations. In contrast to the diversity of microbes which selectively assimilate species, dissimilatory processes are mediated by distinct and specialized physiological groups having widely differing responses to parameters such as oxygen, light, and sulfide. These restrictions effectively partition certain sulfur species transformations into separate, though often adjacent, habitats. Some species transformations can only proceed at the interface between mutually exclusive habitats. Under anoxic conditions dissimilatory sulfur species transformations are exclusively microbial.

Recent investigations have been marked by an increasing awareness of the quantitative importance of certain organic S species in aquatic sediments and the wide variation of spatial and temporal scales across oxic-anoxic interfaces. The ability of microbially mediated oxidations to compete with abiotic reactions under oxic conditions appears to be variable and habitat-dependent. Under anoxic conditions iron may act as a significant, though possibly transitory, sink for sulfide produced during dissimilatory sulfate reduction.

Introduction

Sulfur is an essential element for all plants, animals, and microorganisms. It usually accounts for <1% of the dry weight biomass (e.g., Anderson, 1978; Schlegel, 1981) and is commonly in the range 0.4%–0.8% for microorganisms (Postgate, 1968). Much of this sulfur is present as the sulfur-containing amino acids, methionine (-S-) and cysteine (-SH), which occur in both structural proteins and enzymes (Siegel, 1975), while lesser amounts occur in S-containing molecules such as coenzyme A, vitamins, ferredoxins, and major cellular metabolites such as glutathione (Siegel, 1975). Cellular sulfur may also be present in structural components (algal cell walls) and storage compounds (fungi) in the form of sulfate esters (Roy and Trudinger, 1970; Siegel, 1975).

Bacteria dominate the transformation of sulfur species in the biosphere, particularly transformations of inorganic species (see Fig. 1) (e.g., Schlegel, 1981). Decomposition (mineralization) of organic S compounds to inorganic S species is the preserve of microbes, including fungi. For some microbes, reduced forms of inorganic sulfur can serve as energy sources and/or as sources of reducing power (Kelly and Kuenen, 1985). The oxidized forms, principally sulfate, can serve as electron acceptors in energy-generating, anaerobic respiration (Pfennig and Widdel, 1982). The transformation of sulfur species during such dissimilatory oxidations and reductions may cause dramatic changes in the surrounding environment by producing me-

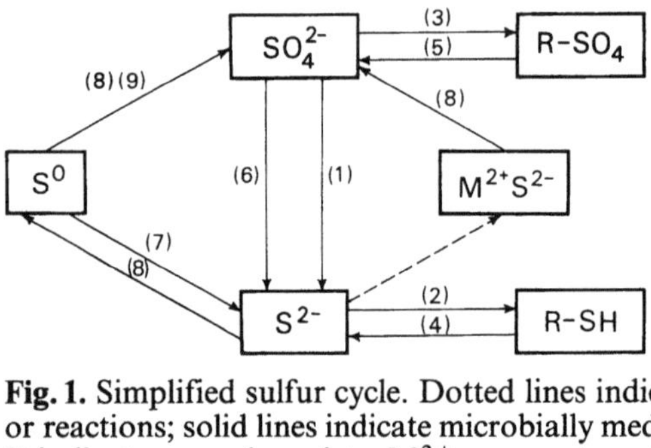

Fig. 1. Simplified sulfur cycle. Dotted lines indicate abiotic species transformations or reactions; solid lines indicate microbially mediated sulfur species transformation; R indicates organic moiety; M^{2+} represents metals (e.g., Fe).
1 – Assimilatory sulfate reduction; 2 – Sulfide assimilation; 3 – Formation of ester sulfates; 4 – Desulfuration; 5 – Ester sulfate hydrolysis; 6 – Dissimilatory sulfate reduction; 7 – Dissimilatory sulfur reduction; 8 – Sulfide oxidation; 9 – Sulfur oxidation. For simplicity, some intermediates – e.g. sulfite, thiosulfate, tetrathionate – are omitted from this representation

tabolites toxic to other life forms (Postgate, 1982; Schlegel, 1981). Dissimilatory oxidation of reduced sulfur species may be accomplished either under anoxic conditions in the light, by phototrophic bacteria, or under oxic conditions by chemotrophic bacteria. Some members of the latter group are facultative anaerobes able to use nitrate as an alternative electron acceptor in the absence of oxygen. Dissimilatory sulfate and sulfur reduction only occur under anoxic conditions. Microbes play a unique and quantitatively important role in these aspects of sulfur cycling within the biosphere (e.g., Postgate, 1982).

It should be noted that the species transformation of sulfur is inextricably linked to the biogeochemical cycles of other elements, particularly carbon, iron, and oxygen (Trudinger, 1982). Dissimilatory oxidations are frequently linked to CO_2 fixation (and hence primary production) by chemoautotrophic and photoautotrophic microbes, while dissimilatory reductions are associated with decomposition of organic carbon (Pfennig, 1978; Trüper, 1982).

Sulfur Species and Reservoirs

Sulfur exists in a large number of oxidation states (see Table 1). Of those species which are soluble in aqueous media at ambient temperature and pressure, sulfate (+6, most oxidized) and sulfide (−2, most reduced) are the most chemically stable (Nriagu and Hem, 1978). Anoxia is a prerequisite for stability of the latter. Although they may be widely distributed in

Table 1. Some inorganic sulfur species and their oxidation states; after Ehrlich (1981)

Species	Formula	Oxidation state(s)
Sulfate	SO_4^{2-}	+6
Tetrathionate	$S_4O_6^{2-}$	0, +4[a]
Dithionate	$S_2O_6^{2-}$	+4
Thiosulfate	$S_2O_3^{2-}$	0, +4
Sulfite	SO_3^{2-}	+4
Dithionite	$S_2O_4^{2-}$	+2
Elemental sulfur	S_8^0[b]	0
Sulfide	S^{2-}	−2

[a] Outer S, 0; inner S, +4 [b] Orthorhombic form

low concentrations, species of intermediate oxidation state (e.g., sulfite, dithionite, thiosulfate, polysulfide) are not generally dominant (Nriagu and Hem, 1978) under the near neutral pH and the oxidizing conditions which predominate at the Earth's surface.

Sulfate is one of the most stable species (Anderson, 1978) and is the predominant form of dissolved inorganic S in most freshwater and marine environments (e.g., Wetzel, 1983). In anoxic aquatic ecosystems such as waterlogged soils or sediments both sulfate and sulfide are stable, but the latter may become the dominant species. At ambient temperature and pressure and at neutral/alkaline pH, H_2S is very soluble in water – 100 times more soluble than O_2 (Nriagu and Hem, 1978; Wetzel, 1983) – and is toxic to many aerobic organisms (Postgate, 1968). H_2S is a weak acid and a strong reducing agent which, when oxidized, normally yields S^o and various ionic species, including thiosulfate (Jørgensen, 1982; Nriagu and Hem, 1978). Predictions based on thermodynamic considerations (Nriagu and Hem, 1978) indicate an increase in polysulfides beyond pH 9, with the more highly polymerized forms being favored in sulfur-rich systems. It has been shown that polysulfides occur in reduced marine sediments at pH between 7.5–8.3 (Boulege and Michard, 1973). Calculations suggested that at pH <7 only small amounts of thiosulfate would occur at equilibrium (Nriagu and Hem, 1978), but the high concentrations reported for Lake Ontario sediments suggest that thiosulfate, and polythionates, may persist in micro-oxic sediments (Nriagu et al., 1979). Elemental sulfur is chemically stable in most environments in the presence of O_2.

Early studies of aquatic ecosystems emphasized microbially mediated species transformations of inorganic S species, the consensus generally being that, while significant, the pool size of organic S species in detrital and living biomass is small relative to the inorganic S pool (e.g., see Wetzel, 1983). However, such a preponderance of inorganic S species did not appear to extend to soil ecosystems. Fitzgerald (1978), in a comprehensive review, noted that on average no more than about 8% of total soil S occurred as $[SO_4^{2-} + S^{2-} + S^o]$ and that organic S species constitute the majority of the sulfur pool in soils worldwide. Results obtained from agricultural (Freney et al., 1971) and forest (Swank et al., 1984) soils indicate that the microbial conversion of sulfate to organic S species may be a major pathway in the S cycle of these ecosystems.

Fitzgerald (1976) has also emphasized the potential importance of sulfate esters as a constituent of soil organic S. King and Klug (1980, 1982), as part of a long-term study of the dynamics of sulfur cycling in the shallow hypereutrophic Wintergreen Lake (Michigan), estimated that ester sulfates comprised ca. 30% of total sediment S and were up to two to three orders of magnitude greater than total inorganic species. It is possible that sulfate esters may be present in significant amounts in some marine sediments where red algae are numerous since the latter are reported to contain heavily sulfated polysaccharides within their cell walls (Percival and McDowell, 1967). Capsular polysaccharides of certain unicellular red algae are reported to consist of sulfated heteroglycans having a sulfate content of 7%–8% (McCandless and Craigie, 1979).

Generally the amount of S in the biosphere is small when compared to annual consumption by living organisms (Anderson, 1978). Localized concentrations of sulfur species may occur in the geosphere as e.g. S^o, metal sulfides, petroleum, and gypsum, but these are released to and withdrawn from the biosphere at very slow rates compared to cycling by microbes within the biosphere (Anderson, 1978). Within the biosphere, interconversions of S species occur on a spatial scale ranging from oceans/large lakes ($\sim 10 - 10^4$ km^2) to small springs ($\sim 10^{-2} - 10^{-4}$ m^2). Within each ecosystem, cycling of species may be largely or wholly internal over short periods. Some processes (for example, photosynthetic oxidation of reduced inorganic S species) may be spatially restricted but, within a confined environment, quantitatively dominant.

Microbially Mediated Interconversions of Sulfur Species

Biosynthetic S Assimilation

In contrast to animals, which require preformed cysteine and methionine, plants and most microorganisms (including fungi) can utilize sulfate by a process known as assimilatory sulfate reduction (Fig. 1) as their sole source of sulfur in order to meet biosynthetic requirements (Anderson, 1978; Trüper, 1982). Sulfate assimilation is a reductive, energy-requiring process and is accomplished by both aerobic and anaerobic chemotrophs and phototrophs. However, certain anaerobic microbes (e.g., the phototrophic green S bacteria) are not capable of sulfate assimilation and must rely on sulfide assimilation for their biosynthetic needs, a process which requires less energy than sulfate assimilation (Trüper, 1982). A few microbes exhibit an obligate biosynthetic requirement for the S-containing amino acids cysteine or methionine. For some microorganisms, assimilatory sulfate reduction is repressed in the presence of cysteine, which is preferentially assimilated (Siegel, 1975) no doubt because the latter, like sulfide, requires less energy for assimilation than does sulfate. A considerable proportion of the sulfate esters in soil habitats originates from plants and animals, and the quantitative role of microbial synthesis is not yet clearly established (Fitzgerald, 1978).

Degradation (Mineralization) of Organic S Compounds

H_2S may be liberated during (an)aerobic decomposition of organic S compounds containing –SH (thiol) groups, e.g., from cystine, cysteine, and methionine present in proteinaceous material. There is accumulating evidence that degradation products also include a variety of volatile organic S species such as methyl mercaptan (CH_3SH = MeSH) and dimethyl sulfide (CH_3SCH_3 = DMS), and dimethyl disulfide (CH_3SSCH = DMDS) may be liberated by microbial decomposition of, e.g., methionine (Bremner and Steele, 1978). DMS, DMDS, and MeSH are also formed by axenic cultures of both freshwater cyanobacteria (Bechard and Rayburn,1979) and marine phytoplankton (Andreae, 1980). There is evidence that, in anoxic lake sediments, MeSH can be metabolized to CH_4, CO_2, and H_2S (Zinder and Brock, 1978 b, 1978 c). Bremner and Steele (1978), who comprehensively reviewed the microbial production of volatile sulfur compounds in terrestrial and aquatic habitats, noted that other such compounds may be produced – for example, carbon disulfide (CS_2) from cysteine (and cystine), and carbonyl sulfide (COS) which may be derived from (iso)thiocyanates present in plant residues. Phototrophic bacteria do not play a role in these processes.

The decomposition of sulfate esters is catalyzed by sulfhydrolase enzymes which release sulfate and an alcohol. Sulfhydrolase activity appears to be primarily due to heterotrophic bacteria, but mammalian enzymes are also known (Fitzgerald, 1978). Sulfhydrolase activity has been detected in soils (Fitzgerald, 1978) and salt marsh sediments (Oshrain and Weibe, 1979) and may be a major source of sulfate in lacustrine sediments (King and Klug, 1980). The isolation from soil of arylsulfatase-producing microbes capable of releasing sulfate ester groups from lignosulfonates (Steinitz, 1981) suggests that sulfate bound to "recalcitrant" organic matter may in fact be released in significant amounts.

Dissimilatory Reduction of Inorganic S Species

While mechanistically resembling assimilatory sulfate reduction, this process, performed by the obligately anaerobic sulfate- and sulfur-reducing bacteria, employs the oxidized S species as terminal electron acceptors for an anaerobic respiration. Most species of sulfate-reducing bacteria can also reduce sulfite to sulfide (Trudinger and Loughlin, 1981). The resultant copious production of H_2S maintains and enlarges the anoxic environment and low redox required by these microorganisms. The toxicity of H_2S effectively excludes many aerobes (but see below). Anaerobes are also affected, although the sulfate-reducing bacteria are reported to be active at H_2S concentrations as high as 25–30 mM (Jørgensen, 1983). Other microbes may well carry out dissimilatory sulfate reduction. The alga *Cyanidium caldarium* appears to miminize the toxic effect of the metals by sulfide precipitation on its cell surface. Wood and Wang (1983) suggested that *C. caldarium* produces H_2S by means of a membrane-associated sulfate reductase.

Recent investigations by Pfennig and his collaborators have provided a clearer understanding of the physiological capability and ecological role of the dissimilatory sulfate and sulfur reducers, particularly with respect to the wide range of organic compounds which can be oxidized, and their role in the complete oxidation of organic matter to CO_2 in "sulfate-sufficient" anoxic environments. Of particular note was the discovery and elucidation of microbes (*Desulfuromonas* spp.) capable of using S^o as a terminal electron acceptor (Biebl and Pfennig, 1978; Pfennig and Biebl, 1976). As a consequence of consortial associations with other microbes, the dissimilatory S-reducing bacteria are cosmopolitan in anoxic habitats and not excluded from low sulfate environments (Peck and LeGall, 1982). Dissimilatory reduction of dimethyl sulfoxide (DMSO) to dimethyl sulfide (DMS) by a bacterium isolated from lake sediment has been reported (Zinder and Brock, 1978 a). DMSO performed a role analogous to that of sulfate in acting as an electron acceptor during anaerobic growth on lactate of succinate (Zinder and Brock, 1978 a). Other S species utilized as electron acceptors included elemental sulfur, sulfite, and thiosulfate, but not sulfate. Useful reviews of the physiology, ecology, and biochemistry of dissimilatory sulfate- and sulfur-reducing bacteria are those by Peck and LeGall (1982), Pfennig and Widdel (1981, 1982), Pfennig et al. (1981), and Postgate (1984).

Oxidation of Sulfide and Other S Species

Hydrogen sulfide may be chemically or biologically oxidized in the presence of oxygen or, in the absence of O_2, oxidation to sulfate may be accomplished by the anoxygenic phototrophic bacteria and cyanobacteria. H_2S may also be precipitated by heavy metals (see below). The abiotic (chemical) oxidation of sulfide by molecular oxygen is discussed below. Microbially mediated oxidation may proceed under either oxic or anoxic conditions. For some chemotrophic bacteria, sulfide (or S^o) serves a dual function: as an electron donor for energy-generating metabolism and as a source of reducing power for autotrophic CO_2 fixation. Since phototrophic microbes utilize light to generate ATP, sulfide is required by them only as a reductant for anoxygenic CO_2 fixation. Some chemotrophic bacteria (obligate chemolithotrophs) require the simultaneous presence of H_2S and O_2 for their energy-generating metabolism and growth.

Anaerobic Oxidation of Reduced S Species

Sulfide arising from dissimilatory sulfate reduction or other sources can be oxidized under anoxic conditions only by the phototrophic bacteria and by those cyanobacteria capable of anoxygenic photosynthesis. Table 2, adapted from Hansen (1983), gives some idea of the range of reduced inorganic S species which can be utilized as phototrophic electron donors, together with intermediates and end products. Trüper (1981, Tables 1–4) has summarized such data for about 70 species of phototrophic bacteria. Points to note include the fact that all Chlorobiaceae (green S bacteria), Chromatiaceae (purple S bacteria), and facultative anoxygenic cyanobacteria can use sulfide as an electron donor. Those species of the Rhodospirillaceae (purple non-S bacteria) which utilize sulfide do so only at concentrations

Table 2. Inorganic sulfur electron donors utilized by various phototrophic bacteria; after Hansen (1983)

	H_2S	$S_2O_3^{2-}$
Cyanobacteria	Strain-dependent	Few strains studied
Rhodospirillaceae	Strain-dependent; (1) $S^{2-} \rightarrow S^0$ ext (2) $S^{2-} \rightarrow SO_4^{2-}$	 *Rhodopseudomonas palustris* *R. sulfidophila* (no S^0 accum.)
Chromatiaceae	All strains; $S^{2-} \rightarrow S^0$ int $\rightarrow SO_4^{2-}$ except *Ectothiorhodospira* (S^0 ext)	Several strains (S^0 accum.)
Chlorobiaceae	All strains; $S^{2-} \rightarrow S^0$ ext $\rightarrow SO_4^{2-}$	Some strains
Chloroflexaceae	One strain $S^{2-} \rightarrow S^0$ ext	No growth

considerably lower than those routinely tolerated by green (~4–8 mM) and purple (~0.8–4 mM) sulfur bacteria (Pfennig, 1978). The end product of these anaerobic oxidations is commonly but not universally sulfate (Chromatiaceae, Chlorobiaceae). However, S^o is the end product for *Chloroflexus, Oscillatoria limnetica* (and other anoxygenic cyanobacteria), and a number of Rhodospirillaceae (Trüper, 1981).

Elemental sulfur is an intermediate which may accumulate as globules either internally (all Chromatiaceae except *Ectothiorhodospira*) or externally (Chlorobiaceae, *Ectothiorhodospira, Chloroflexus, O. limnetica*) to the cells. All species of Chromatiaceae and Chlorobiaceae can utilize externally supplied S^o (Trüper, 1978). According to Hageage et al. (1970), the S^o globules occur as "liquid" or "plastic" sulfur and are stable until dried, when they form an unstable allotrope which converts to the more stable orthorhombic crystalline forms. The oxidation of internal or external sulfur globules to sulfate appears to be inhibited in the presence of external sulfide (Trüper, 1982). *Ectothiorhodospira* spp. converts sulfide to polysulfides which subsequently form external S^o globules. However, polysulfides are unstable at less alkaline pH favorable to all other Chromatiaceae (and Chlorobiaceae), and polysulfide formation is not observed for these phototrophs (Trüper, 1982).

Many of the Chromatiaceae can utilize thiosulfate as an electron donor. Thiosulfate utilization is reported to proceed with S^o as an intermediate. The sulfane-S gives rise to S^o, while the sulfone-S is converted to sulfate and excreted (Trüper, 1978). Thiosulfate utilization is reported to be constitutive in *Chromatium vinosum* (Hashwa, 1975) but may be inducible in other genera (e.g., Rolls and Lindstrom, 1967). Sulfite and tetrathionate are utilized by very few species. Sulfite is considered to be an intermediate of sulfur oxidation (Trüper, 1978) but is both toxic and unstable, and only a few species are reported to use it (Trüper, 1981). Biochemical and ecological investigations have been recently reviewed by van Gemerden and Beeftink (1983) and Trüper and Fischer (1982).

Aerobic Oxidation of Reduced S Species

The colorless sulfur bacteria comprise a diverse group of microbes exhibiting a broad spectrum of physiological attributes and ecological niches. They have the potential to oxidize reduced inorganic sulfur species using molecular oxygen as the terminal electron acceptor (see Fig. 1). Some species are facultative anaerobes and can utilize nitrate as an alternative electron acceptor in an anaerobic respiration. Some members of the group may be obligate or facultative chemolithotrophs while others are heterotrophs. Detailed compilations of sulfur and carbon species transformations carried out by this group are given by Kuenen and Beudeker (1982, Table 1) and Kelly and Kuenen (1985, Tables 1 and 2).

The physiology of the thiobacilli is perhaps the best understood at the moment and illustrates the range of metabolic capability which characterizes the colorless sulfur bacteria. The thiobacilli include obligate chemolithoautotrophs (deriving energy from oxidation of reduced S species; fix CO_2) and facultative chemolithotrophs (can grow as heterotrophs on organic compounds; also may derive energy from S oxidation). The properties of other colorless sulfur bacteria are presently less well understood but are currently under investigation from both physiological (Kuenen and Beudeker, 1982) and ecological (Jørgensen, 1982) aspects. Biochemistry of S transformations by the thiobacilli has been reviewed by Kelly (1982). *Thiobacillis ferrooxidans* can also obtain metabolic energy from the oxidation of Fe(II) and U(IV) to Fe(III) and U(VI), respectively, at acid pH (Olson, this volume).

The thiobacilli utilize sulfide, S°, thiosulfate, and other intermediates en route to sulfate. Some heterotrophs produce tetrathionate or other polythionates, usually from thiosulfate (Kelly and Kuenen, 1985). The obligately aerobic chemolithotrophs among the colorless sulfur bacteria require the simultaneous presence of H_2S and O_2, and these organisms are closely associated with the oxycline (oxic-anoxic transition zone) between oxidizing and reducing environments. Consequently, they may face competition for available sulfide and/or oxygen from abiotic oxidation (see below). Competition may also arise from certain phototrophic microbes. It has been known for some time that *Chromatium vinosum* strain D remains viable during prolonged exposure to O_2 (Hurlbert, 1967). Recently, seven species of Chromatiaceae (of 17 tested) were found to be capable of microaerobic chemotrophic growth in the dark using thiosulfate as an electron donor for both energy and CO_2 fixation (Kämpf and Pfennig, 1980). However, none of the Chlorobiaceae strains tested were able to do so.

The Transformation of Sulfur Species in Nature (and Interaction with Other Biogeochemical Cycles)

Under natural conditions the biogeochemical cycle of sulfur is inextricably linked with other elemental cycles through the biosynthetic requirements of, and the dissimilatory utilization by, microbial communities. The interactions between C and S cycling are evident from the preceding discussion. Microbially catalyzed species transformations of S may require specialized, even quite restrictive, environmental conditions, and the end products of some microbial species transformations may modify the surrounding (micro)environment and thus limit or promote other transformation pathways and/or chemical stability of other species. The interaction between physicochemical characteristics (e.g., O_2, Eh, H_2S, light intensity, pH) and microbial communities produces a continuum or gradient exemplified in aquatic ecosystems such as stratified lakes or sulfureta (Fig. 2). While such ecosystems are now reasonably well understood, they are less widespread than, for example, coastal marine sediments or soils, both of which may exhibit a more heterogeneous spatial distribution of reduced microenvironments within more oxidized sediments (Jørgensen, 1977; Nriagu et al., 1979).

The establishment of a stratified ecosystem (Fig. 2) is driven by the primary production of organic matter (which may be of allochthonous or autochthonous derivation). Subsequent aerobic decomposition depletes O_2, particularly at lower light intensities, and some H_2S is produced from the anaerobic decomposition of organic thiol (-SH) groups. Dissimilatory reduction of oxidized S species can now proceed under such anoxic, reducing conditions. H_2S toxicity excludes all but a few special groups of aerobic (O_2-requiring) microbes. H_2S removal in the anoxic, aphotic zone may occur via precipitation with Fe^{2+} released from anoxic sediments (see below). Microbial transformation of species will occur only when H_2S diffuses up-

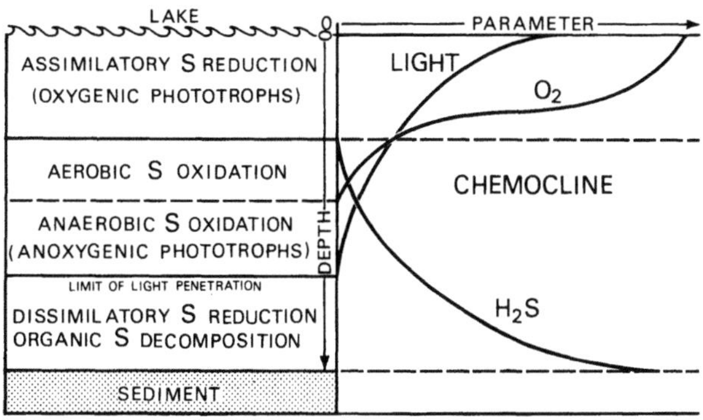

Fig. 2. Schematic of stratified aquatic ecosystem illustrating vertical profile of microbes involved in sulfur species transformation and associated physicochemical conditions

ward into the photic zone, where anaerobic phototrophic S oxidation takes place, and resultant S° and SO_4^{2-} species may reenter the reductive pathways.

In stratified water bodies the phototrophic bacteria may form successive vertical layers or "plates" whose location is determined by their H_2S tolerance and light intensity requirements – the Chlorobiaceae operating at low light intensity and high sulfide concentrations, the Chromatiaceae above them in higher light intensity but lower sulfide (e.g., van Gemerden and Beeftink, 1983). Located above the plate(s) lies the transition zone (metalimnion), or interface, between the lower anoxic zone (hypolimnion) and the upper oxic zone (epilimnion), where both O_2 and H_2S may coexist. It is in this zone of coexistence, which may be diurnally transient, that oxidation by aerobic, colorless sulfur bacteria can occur. These activities, together with abiotic H_2S oxidation (via O_2 moving down from the air-water interface or formed by oxygenic planktonic photosynthesis), remove the last vestiges of the H_2S generated below.

The metalimnion – which may also be referred to as (or coincide with) the oxycline, redoxcline, chemocline, or thermocline – does not occupy a vertically fixed position in the water body. Long-term variation may be determined, for example, by SO_4^{2-} concentration in the anoxic zone, light penetration, and H_2S input of groundwaters. Diurnal fluctuations, which allow H_2S penetration higher into the water column at night when phototrophic anaerobic oxidation stops, will tend to favor microbes that can tolerate if not utilize intermittent H_2S: those that are evasively motile by flagella or passively motile by gas vesicles (van Gemerden and Beeftink, 1983), and those that transiently retain S° for later oxidation (Jørgensen, 1982). The vertical extent of the interface may vary considerably – from tens of meters (e.g., Black Sea) to tens of microns (e.g., *Beggiatoa* mats) (Jørgensen, 1982).

The transformation of sulfur species may be sustained at high turnover rates under anoxic conditions, despite low extracellular concentrations, by interaction between phototrophic green S bacteria and heterotrophic S-dissimilating bacteria (Pfennig, 1978; Trüper, 1982) such that S° (or SO_4^{2-}) reduced to H_2S by the heterotroph is retained after H_2S oxidation (to extracellular S° or SO_4^{2-}) by the phototroph. Such syntrophy is demonstrable in both laboratory mixed cultures (Biebl and Pfennig, 1978) and naturally occurring consortia (Pfennig, 1978). Geothermal H_2S appears to provide the major energy input to the dense invertebrate communities associated with deep-sea hydrothermal vents (Jannasch and Wirsen, 1979). This is considered to be accomplished via nonphotosynthetic primary production, and the high concentration of H_2S relative to other electron donors suggests that chemoautotrophic S-oxidizing bacteria are the dominant CO_2-fixing microbes in this aphotic environment (Ruby et al., 1981). Isolates from hydrothermal vent habitats include obligately heterotrophic, mixotrophic, and obligately autotrophic S oxidizers (Jannasch and Nelson, 1984). In addition, there is now good evidence to support the hypothesis that organic carbon derived from CO_2 fixation by symbiotic, S-oxidizing microbes is the nutrient source for some members of the vent invertebrate community (Cavanaugh, 1983; Tuttle et al.,1983).

Microbial Versus Abiotic Sulfide Oxidation

Investigators of abiotic (chemical; spontaneous) sulfide oxidation report half-times ranging from several minutes to several hours (see Nriagu and Hem, 1978, Table 4). This disparity, together with the conflicting views concerning proposed reaction mechanisms (Almgren and Hagström, 1974), may be attributable to methodological differences and to the dependence of oxidation rates on parameters such as salinity and temperature, the presence of trace metals, organic compounds, polysulfides (which may have a catalytic effect), and the initial H_2S and O_2 concentrations (Nriagu and Hem, 1978, and references therein). For example, when O_2 or H_2S concentrations are low, thiosulfate and sulfate are the major species formed, but S^o may assume quantitative importance when H_2S levels are initially in the millimolar range (Jørgensen, 1982). Other S species resulting from abiotic oxidation include sulfite, polysulfides, and polythionates (Nriagu and Hem, 1978). Oxygen and H_2S are reported to coexist at concentrations between 0.1–0.001 mM (Kelly and Kuenen, 1985). The aerobic, colorless sulfur bacteria occupying the interface between oxic and anoxic zones must compete with abiotic oxidation for the metabolically necessary electrons and reducing power stored in the H_2S (presumably they must also compete with H_2S for O_2?).

Both field and laboratory studies indicate that microbial transformation of species can compete successfully with abiotic transformations. However, the degree to which they are successful appears to be variable and is habitat-dependent. Investigations using sulfide-limited chemostats have demonstrated that measured growth yields of thiobacilli are those that would be predicted if there were no concomitant abiotic H_2S oxidation (Kelly and Kuenen, 1985). One of the organisms studied, *Thiobacillus* strain A2, is incapable of oxidizing S^o. The latter did not accumulate in the me-

Table 3. Comparison of three H_2S–O_2 interfaces from stratified aquatic habitats; after Jørgensen (1982)

Parameter	Black sea	Solar lake	*Beggiatoa* mat
Vertical extent of interface	35 m	10 cm	50 μm
Calculated H_2S residence time	5 d	10–20 min	0.6 s
Microbial oxidation of H_2S (% total)	0	30–50	100
Concentration (μM) at which $C_{O_2} = C_{H_2S}$	1–3	2–4	3–5
Intermediate products	SO_4^{2-}, $S_2O_3^{2-}$	S^0	S^0

dium, consistent with the absence of abiotic oxidation. Such data indicate that thiobacilli can compete successfully with abiotic oxidation.

Jørgensen and co-workers (1982, and references therein) have investigated the H_2S-O_2 interfaces in several stratified ecosystems – including Solar Lake (water column and microbial mats), *Beggiatoa* mats (coastal marine sediments), and *Thiovulum* veils – and have compared these with data available for the Black Sea (Table 3). Calculated H_2S residence times, together with experimental data, indicate that microbial oxidation of these environments ranges from 0%–100% of the total oxidation. Jørgensen (1982) highlights the tremendous variations in scale characterizing these environments, which may be spatial (vertical extent of the H_2S-O_2 interface) or temporal (H_2S residence time), and suggests that the Black Sea S oxidizers may be too dispersed to compete effectively with physical mixing and abiotic oxidation. In more "compressed" systems, residence times fall dramatically (Table 3) as the effectiveness of microbial transformations increases and surpasses abiotic transformations, and this may be related to the increased importance of S^o as a nondiffusible intermediate. In some environments, facultatively anoxygenic cyanobacteria capable of oxidizing H_2S to S^o reside at the H_2S-O_2 interface (Paden, 1979). After overnight H_2S accumulation they oxidize the H_2S while fixing CO_2 photosynthetically. After H_2S depletion they switch to O_2-evolving photosynthesis, thus enhancing the extent of the oxic zone.

Iron as a Sulfur Sink

Insoluble sulfides form readily in those aquatic and soil habitats subject to anoxia and low redox conditions. The relative abundance of Fe in most natural habitats compared to other metals such as Zn, Pb, Ni, and Cu ensures that only iron sulfide precipitates can be a major S sink to sediments. The proportion of biogenic H_2S precipitated in this manner will depend upon factors such as the rate of H_2S production, diffusion losses, degree of bioturbation (Jørgensen, 1983), and the availability of Fe as affected by, for example, pH and redox potential. Iron (Fe^{2+}) diffuses readily from freshwater sediments when Eh values fall to ca. 200–250 mV. Iron and other metal sulfides are highly insoluble at the neutral-alkaline pH common in natural waters, and the Fe^{2+} released will react with H_2S precipitating iron sulfides. In lakes, this may lead to a substantial reduction in Fe^{2+} levels, particularly in late summer (e.g., Wetzel, 1983). If the water is alkaline, H_2S (HS^-, S^{2-}) will accumulate only after most of the Fe^{2+} has been precipitated; otherwise, appreciable H_2S accumulation will only be evident if the anoxic hypolimnion is somewhat acidic (Wetzel, 1983). The formation of other metal sulfides (for example, Cu, Pb, and Zn, which are even less soluble than those of Fe) may provide an important buffer system which controls the level of dissolved heavy metals in sulfide-rich water and sediments (Nriagu and Hem, 1978). In addition, organic sulfhydryl groups may be important ligands for metal complexation and thus participate in the transformation of metal species (Nriagu and Hem, 1978).

In coastal sediments, much of the sulfide formed would appear to undergo rapid reoxidation. For example, $<10\%$ of biogenic sulfide is fixed as metal sulfides in the sediments of Limfjorden (Jørgensen, 1977). Chambers (1982) deduced from stable isotope data that all of the sulfide formed in microbial mat-associated sediments was recycled within a relatively short period (weeks to months). While the FeS sink may be transitory in nature, a major consequence is to reduce H_2S concentrations such that the oxic zone remains a significant proportion of the water mass and that the aquatic ecosystem remains relatively stable or that the anoxic episode is diminished. Jørgensen (1982) has speculated that in coastal sediments where the O_2 and H_2S layers do not overlap but may be separated by several cm, iron might be involved as an intermediate electron carrier which is reduced by H_2S and subsequently oxidized by O_2.

In addition to amorphous ferrous sulfide ($Fe_{1-x}S$), other metastable iron sulfides formed in aquatic sediments and soils include mackinawite ($Fe_{1-x}S$) and greigite (Fe_3S_4). The thermodynamically stable form, pyrite (FeS_2), is generally considered to form from amorphous FeS by reaction with S^o in a relatively slow process (Berner, 1971). While some investigators report that amorphous FeS is the predominant product of dissimilatory sulfate reduction in Fe-rich sediments (e.g., Skyring et al., 1983), others (e.g., Howarth, 1979; Howarth and Merkel, 1984) consider pyrite itself to be a major end product which forms rapidly without intermediate FeS formation. The contradictory conclusions reached may be attributable to, perhaps even characteristic of, the different sediment types examined (e.g., Howarth and Jørgensen, 1984) and to the assay procedures used to determine pyrite in the presence of other sedimentary sulfur species. The latter aspect is considered in some detail by King (1983) and Howarth and Merkel (1984).

Acknowledgements. The Baas Becking Geobiological Laboratory is supported by the Bureau of Mineral Resources, the Commonwealth Scientific and Industrial Research Organization, and the Australian Mineral Industries Research Association, Ltd. This paper was prepared while the author was carrying out research at Indiana University (Research Program in Biogeochemistry) where financial support was provided by NSF Grant No. PCM 79-10747 (H. Gest, Biology) and NASA Grant No. NGR-15-003-118 (J. M. Hayes, Chemistry). Travel support was provided by the Ian Potter Foundation. I thank G. J. Olson and P. A. Trudinger for helpful criticism and discussion.

References

Almgren T, Hagström I (1974) The oxidation rate of sulfide in seawater. Water Res 8:395–400

Anderson JW (1978) Sulphur in biology. Edward Arnold, London

Andreae MO (1980) The production of methylated sulfur compounds by marine phytoplankton. In: Trudinger PA, Walter MR, Ralph BJ (eds) Biogeochemistry of ancient and modern environments. Australian Academy of Science, Canberra, pp 253–259

Bechard MJ, Rayburn WR (1979) Volatile organic sulfides from freshwater algae. J Phycol 15:379–383

Berner RA (1971) Principles of chemical sedimentology. McGraw-Hill, New York

Biebl H, Pfennig N (1978) Growth yields of green sulfur bacteria in mixed cultures with sulfur- and sulfate-reducing bacteria. Arch Microbiol 117:9–16

Boulege J, Michard G (1973) Formation de polysulfures dans les conditions physico-chimiques de l'eau de mer. CR Acad Sci (Paris) 277:2613–2616

Bremner JM, Steele CG (1978) Role of microorganisms in the atmospheric sulfur cycle. In: Alexander M (ed) Advances in microbial ecology. Plenum Press, New York, pp 155–201

Cavanaugh CM (1983) Symbiotic chemoautotrophic bacteria in marine invertebrates from sulphide-rich habitats. Nature 302:58–61

Chambers LA (1982) Sulfur isotope study of a modern intertidal environment and interpretation of ancient sulfides. Geochim Cosmochim Acta 46:721–728
Ehrlich HL (1981) Geomicrobiology. Marcel Dekker, New York
Fitzgerald JW (1976) Sulfate ester formation and hydrolysis: a potentially important yet often ignored aspect of the sulfur cycle of aerobic soils. Bacteriol Rev 40:698–721
Fitzgerald JW (1978) Naturally occurring organosulfur compounds in soil. In: Nriagu JO (ed) Sulfur in the environment. II. Ecological impacts. John Wiley, New York, pp 391–443
Freney JR, Melville GE, Williams CH (1971) Organic sulphur fractions labelled by addition of ^{35}S-sulphate to soil. Soil Biol Biochem 3:133–141
Hageage GJ, Eanes ED, Gherna RL (1970) X-ray diffraction studies of the sulfur globules accumulated by *Chromatium* species. J Bacteriol 101:464–469
Hansen TA (1983) Electron donor metabolism in phototrophic bacteria. In: Ormerod JG (ed) The phototrophic bacteria: anaerobic life in the light. Blackwell Scientific Publications, Oxford, pp 76–99
Hashwa F (1975) Thiosulfate metabolism in some red phototrophic bacteria. Pl Soil 43:41–47
Howarth RW (1979) Pyrite: its rapid formation in a salt marsh and its importance in ecosystem metabolism. Science 203:49–51
Howarth RW, Jørgensen BB (1984) Formation of ^{35}S-labelled elemental sulfur and pyrite in coastal marine sediments (Limfjorden and Kysing Fjord, Denmark) during short-term $^{35}SO^{2-}$-reduction measurements. Geochim Cosmochim. Acta 48:1807–1818
Howarth RW, Merkel S (1984) Pyrite formation and the measurement of sulfate reduction in salt marsh sediments. Limnol Oceanog 29:598–608
Hurlbert RE (1967) Effect of oxygen on viability and substrate utilization in *Chromatium.* J Bacteriol 93:1346–1352
Jannasch HW, Nelson DC (1984) Recent progress in the microbiology of hydrothermal vents. In: Klug MJ, Reddy CA (eds) Current perspectives in microbial ecology. American Society for Microbiology, Washington, DC, pp 170–176
Jannasch HW, Wirsen CO (1979) Chemosynthetic primary production at East Pacific sea floor spreading centers. Bioscience 29:592–598
Jørgensen BB (1977) The sulfur cycle of a coastal marine sediment (Limfjorden, Denmark). Limnol Oceanog 22:814–832
Jørgensen BB (1982) Ecology of the bacteria of the sulphur cycle with special reference to anoxic-oxic interface environments. Phil Trans Roy Soc Lond B 298:543–561
Jørgensen BB (1983) The microbial sulfur cycle. In: Krumbein WE (ed) Microbial geochemistry. Blackwell, Oxford, pp 91–124
Kämpf C, Pfennig N (1980) Capacity of Chromatiaceae for chemotrophic growth. Specific respiration rates of *Thiocystis violacea* and *Chromatium vinosum.* Arch Microbiol 127:125–137
Kelly DP (1982) Biochemistry of the chemolithotrophic oxidation of inorganic sulphur. Phil Trans Roy Soc Lond B 298:499–528
Kelly DP, Kuenen JG (1985) Ecology of the colourless sulfur bacteria. In: Codd GA (ed) Aspects of microbial metabolism and ecology. Academic Press, San Diego, pp 211–240

King GM (1983) Sulfate reduction in Georgia salt marsh soils: an evaluation of pyrite formation by use of ^{35}S and ^{55}Fe tracers. Limnol Oceanog 28:987–995
King GM, Klug MJ (1980) Sulfhydrolase activity in sediments of Wintergreen Lake, Kalamazoo County, Michigan. Appl Envir Microbiol 39:950–956
King GM, Klug MJ (1982) Comparative aspects of sulfur mineralization in sediments of a eutrophic lake basin. Appl Envir Microbiol 43:1406–1412
Kuenen JG, Beudeker RF (1982) Microbiology of thiobacilli and other sulphur-oxidizing autotrophs, mixotrophs, and heterotrophs. Phil Trans Roy Soc Lond B 298:473–497
McCandless EL, Craigie JS (1979) Sulfated polysaccharides in red and brown algae. Ann Rev Plant Physiol 30:41–53
Nriagu JO, Coker RD, Kemp ALW (1979) Thiosulfate, polythionates, and rhodanese activity in Lake Erie and Ontario sediments. Limnol Oceanog 24:383–389
Nriagu JO, Hem JD (1978) Chemistry of pollutant sulfur in natural waters. In: Nriagu JO (ed) Sulfur in the environment. II. Ecological impacts. John Wiley, New York, pp 211–270
Oshrain RL, Wiebe WJ (1979) Arylsulfatase activity in salt marsh soils. Appl Envir Microbiol 38:337–340
Padan E (1979) Impact of facultatively anaerobic photoautotrophic metabolism on ecology of cyanobacteria (blue-green algae). Adv Microb Ecol 3:1–48
Peck HD, LeGall J (1982) Biochemistry of dissimilatory sulphate reduction. Phil Trans Roy Soc Lond B 298:443–466
Percival E, McDowell RH (1967) Chemistry and enzymology of marine algal polysaccharides. Academic Press, London
Pfennig N (1978) General physiology and ecology of photosynthetic bacteria. In: Clayton RK, Sistrom WR (eds) The photosynthetic bacteria. Plenum Press, New York, pp 3–18
Pfennig N, Biebl H (1976) *Desulfuromonas acetoxidans* gen. nov. and sp. nov., a new anaerobic,sulfur-reducing, acetate-oxidizing bacterium. Arch Microbiol 110:3–12
Pfennig N, Widdel F (1981) Ecology and physiology of some anaerobic bacteria from the microbial sulfur cycle. In: Bothe H, Trebst A (eds) Biology of inorganic nitrogen and sulfur. Springer-Verlag, Heidelberg, pp 169–177
Pfennig N, Widdel F (1982) The bacteria of the sulphur cycle. Phil Trans Roy Soc Lond B 298:433–441
Pfennig N, Widdel F, Trüper HG (1981) The dissimilatory sulfate-reducing bacteria. In: Starr MP, Stolp H, Trüper HG, Balows A, Schlegel HG (eds) The prokaryotes. Springer-Verlag, New York, pp 926–940
Postgate JR (1968) The sulfur cycle. In: Nickless G (ed) Inorganic sulfur chemistry. Elsevier, Amsterdam, pp 259–279
Postgate JR (1982) Economic importance of sulphur bacteria. Phil Trans Roy Soc Lond B 298:583–600
Postgate JR (1984) The sulphate-reducing bacteria. Cambridge University Press, Cambridge
Rolls JP, Lindstrom ES (1967) Effect of thiosulfate on the photosynthetic growth of *Rhodopseudomonas palustris*. J Bacteriol 94:860–866
Roy AB, Trudinger PA (1970) The biochemistry of inorganic compounds of sulfur. Cambridge University Press, London

Ruby EG, Wirsen CO, Jannasch HW (1981) Chemolithotrophic sulfur-oxidizing bacteria from the Galápagos Rift hydrothermal vents. Appl Envir Microbiol 42:317–324

Schlegel HG (1981) Microorganisms involved in the nitrogen and sulfur cycles. In: Bothe H, Trebst A (eds) Biology of inorganic nitrogen and sulfur. Springer-Verlag, Berlin, pp 3–12

Siegel LM (1975) Biochemistry of the sulfur cycle. In: Greenberg DM (ed) Metabolic pathways. Metabolism of sulfur compounds. Academic Press, New York, pp 217–286

Skyring GW, Chambers LA, Bauld J (1983) Sulfate reduction in sediments colonized by cyanobacteria, Spencer Gulf, South Australia. Aust J Mar Freshw Res 34:359–374

Steinitz YL (1981) Microbial desulfonation of lignosulfonate – a new approach. Eur J Appl Microbiol Biotechnol 13:216–221

Swank WT, Fitzgerald JW, Ash JT (1984) Microbial transformation of sulfate in forest soils. Science 223:182–184

Trudinger PA (1982) Geological significance of sulphur oxidoreduction by bacteria. Phil Trans Roy Soc Lond B 298:563–581

Trudinger PA, Loughlin RE (1981) Metabolism of simple sulfur compounds. In: Florkin M, Neuberger A, van Deemem LLM (eds) Comprehensive biochemistry. 19A. Amino acid metabolism and sulfur metabolism. Elsevier, Amsterdam, pp 165–256

Trüper HG (1978) Sulfur metabolism. In: Clayton RK, Sistrom WR (eds) The photosynthetic bacteria. Plenum Press, New York, pp 677–690

Trüper HG (1981) Photolithotrophic sulfur oxidation. In: Bothe H, Trebst A (eds) Biology in inorganic nitrogen and sulfur. Springer-Verlag, Berlin, pp 199–211

Trüper HG (1982) Microbial processes in the sulfur cycle through time. In: Holland HD, Schidlowski M (eds) Mineral deposits and the evolution of the biosphere. Dahlem Konferenzen. Springer-Verlag, Berlin Heidelberg New York

Trüper HG, Fischer U (1982) Anaerobic oxidation of sulphur compounds as electron donors for bacterial photosynthesis. Phil Trans Roy Soc Lond B 298:529–542

Tuttle JH, Wirsen CO, Jannasch HW (1983) Microbial activities in the emitted hydrothermal waters of the Galápagos Rift vents. Mar Biol 73:293–299

van Gemerden H, Beeftink HH (1983) Ecology of phototrophic bacteria: In: Ormerod JG (ed) The phototrophic bacteria: anaerobic life in the light. Blackwell Scientific Publications, Oxford, pp 146–185

Wetzel RG (1983) Limnology. Saunders College Publishing, Philadelphia

Wood JM, Wang HK (1983) Microbial resistance to heavy metals. Envir Sci Technol 17:582A–590A

Zinder SH, Brock TD (1978a) Dimethyl sulfoxide as an electron acceptor for anaerobic growth. Arch Microbiol 116:35–40

Zinder SH, Brock TD (1978b) Methane, carbon dioxide and hydrogen sulfide production from the terminal methiol group of methionine by anaerobic lake sediments. Appl Envir Microbiol 35:344–352

Zinder SH, Brock TD (1978c) Production of methane and carbon dioxide from methane thiol and dimethyl sulfide by anaerobic lake sediments. Nature 273:226–228

Standing, left to right:
Marko Brancia, Marvin Vestal, Steve George, Wolfgang Calmano, Jan Duinker

Seated (center), left to right:
Michael Bernhard, Samuel Luoma, Rance Velapoldi, Dana Kester

Seated (front), left to right:
Walter Lund, Andi Andreae, Bernard Tramier

The Importance of Chemical "Speciation" in Environmental Processes, eds. M. Bernhard, F. E. Brinckman, and P. J. Sadler, pp. 275–299. Dahlem Konferenzen 1986. Springer-Verlag Berlin, Heidelberg
© Dr. S. Bernhard, Dahlem Konferenzen

Chemical Species in Marine and Estuarine Systems

Group Report

D. R. Kester, Rapporteur

M. O. Andreae
M. Bernhard
M. Branica
W. Calmano
J. C. Duinker
S. G. George
W. Lund
S. N. Luoma
B. Tramier
R. A. Velapoldi
M. L. Vestal

Characteristics of Marine and Estuarine Environments for Chemical Species

It is essential to define at the outset those aspects of a system that influence chemical species. In marine and estuarine environments we recognize a variety of important phases such as the aqueous, colloidal, particulate, and biological components of the system. Each of these will have distinctive properties that influence the molecular form of its constituents. A variety of criteria are available to identify these phases in oceans, coastal regions, and estuaries.

Salinity (S) is one of the master variables for distinguishing fresh, estuarine, and sea waters. Conceptually, salinity is the grams of dissolved inorganic solutes per kilogram of solution, though its formal definitions over the past 80 years have been expressed in terms of operational procedures (Lewis, 1980). Any attempt to establish specific criteria to distinguish fresh, estuarine, and ocean waters will be inherently arbitrary. One of the characteristics of waters from the sea is that the relative composition of the major ions is nearly constant, so that changes in salinity reflect changes in water content rather than salt components (Kester, this volume). Ocean waters have ranges in salinity of $S = 30$ to $S = 38$ (parts per thousand or grams per kg). Fresh waters do not exhibit constancy of composition; their major constituents are present at much lower concentration than in ocean waters, and their relative abundances depend on the rates of geochemical processes such as weathering and erosion (Meybeck, 1979). We will consider fresh waters to be those that have not been influenced by mixing with seawater. Estuarine waters then complete the continuum between fresh and ocean waters and have salinities typically in the range of S = 0.1 to S = 30. Over most of this range the relative composition of major ions in estuaries is nearly the same as in the ocean. A body of water with S = 3 will be about 10% ocean water, but 98% of its salt content will be derived from the ocean water.

One of the important distinctions between the waters of the ocean and of the estuaries is the dynamic range of salinities. In the ocean the variation amounts to about 25%, whereas in estuaries the range is a factor of 100. This difference has two implications for considerations of chemical species. For many chemical processes the chemical potential (or activity) of a constituent is the important variable, rather than its concentration. The relationship between concentration and activity varies with the ionic strength of the medium, which in estuaries ranges from 0.006 to 0.6 and in the ocean varies from 0.60 to 0.76. This latter range is one in which ionic activity coefficients are only slowly varying functions of ionic strength. The second factor is that changes in salinity result in concentration changes in major ions which may interact in specific ways with the constituents whose species composition is considered.

We have taken salinity as the master variable to distinguish fresh, estuarine, and ocean waters, but there are several other chemical properties of the aqueous phase that differ among these three classes of natural water. The pH of seawater is buffered in the range 7.6–8.3, whereas the lower buffer capacities of estuarine and fresh waters allow a wider range of pH, perhaps 6–9 in estuaries, and 3–8 in fresh and groundwaters.

Estuaries and oceans have different amounts and types of suspended particulate material, which are important factors in determining the distribution of chemical species of the elements. Flocculation processes in estuaries may increase the concentration and reactivity of particulate matter, and erosional processes transport large amounts of continental particulate phases to estuaries and oceans. The major sources of particulate matter in the open ocean are biological processes and deposition from the atmosphere. The factors of particulate matter that influence the distribution of species include a) their primary chemical composition (silicate, carbonate, metal oxides, aluminosilicates, and organic detritus); b) their size, and c) their surface properties (electrical charge, hydrophobic or hydrophilic nature). The surface properties of natural particles in the ocean may not be directly related to the bulk composition of the particles when organic films coat the surface (Hunter, 1980).

Living organisms are a special class of particulate material in estuarine and ocean waters because in addition to their surface chemical reactivity, their metabolic processes may produce localized and large-scale changes in the chemical properties of the medium. These "particles" include bacteria, phytoplankton, zooplankton, and higher trophic organisms such as fish, shellfish, and marine mammals. Biological activity leads to the production of a complex milieu of organic substances in the dissolved and particulate phases and as films on surfaces.

The presence of organic substances and microbial activity is also an important factor in formation and transformation of chemical species. Changes in the redox state of an element can completely change its species distribution and consequently its chemical behavior in the environment. In simple systems it is possible to define a redox potential, Eh, that will characterize the oxidation state of the constituents in a manner similar to the use of pH for acid-base equilibria. In natural systems, however, there are many redox reactions that do not achieve equilibrium. One approach to this problem is first to identify those redox reactions with kinetics that enable equilibrium to be approached on the time scales that are important in the environment (minutes, hours, and days) and then restrict the application of the concept of Eh to those chemical systems.

In waters containing dissolved oxygen it appears that almost each redox couple with favorable kinetics leads to different Eh values. If these differences cannot be resolved, Eh will not be a useful parameter for oxic systems. In spite of the ambiguity of what establishes the redox condition of oxygenated waters, we generally recognize that these waters favor the oxidized forms of many elements such as Fe(III), Mn(IV), Cu(II), Cr(VI), nitrate, and iodate instead of Fe(II), Mn(II), Cu(I), Cr(III), ammonia, and iodide. In oxygenated systems it is perhaps best to consider that the redox states of elements are set not by a chemical redox potential, but rather by the extent of microbial activity that can catalyze or overcome the kinetic barrier between different valence states.

The behavior of redox processes in marine and estuarine environments that are depleted of dissolved oxygen is somewhat more amenable to the use of an Eh parameter. The sulfate-sulfide couple regulates the availability of electrons for redox reactions in many anoxic marine environments, and there is good correspondence between thermodynamic predictions and analytical measurements.

This discussion of the formation and transformation of chemical species in marine and estuarine systems has thus far focused mainly on the aqueous phase, but the role and nature of particles – especially important for estuarine systems – should not be overlooked. The above considerations of the aqueous phase, such as its salinity, pH, Eh, and dissolved organic matter content, also apply to the pore waters of sediments. An additional variable in these systems is the very large surface-area-to-volume ratio which amplifies the role of surface chemistry.

One final consideration in the factors that determine the distribution of chemical species in oceans and estuaries is the activity of humans. For the most part human activity has not yet changed the basic properties of seawater on a global scale, and in most instances this is not likely due to the immense volume of ocean waters. Nevertheless, we can identify global-scale changes of element concentrations in the environment due to the combustion of fossil fuels, the atmospheric testing of nuclear weapons, and the production of persistent synthetic organic substances. However, an evaluation of the potential magnitudes of global chemical and physical changes due to human activity is still very difficult.

There are numerous examples in which human activity has modified distribution of chemical species on local scales (i.e., in harbors, estuaries, and coastal regions). Eutrophication resulting from the anthropogenic input of organic substances and nutrients has led to changes in redox states of many chemical species. The ability of human activity to alter the rates of geochemical cycles of the chemical species can have a major effect on their distribution.

The interaction of parameters which influence the distribution of chemical species is very complex in estuarine and seawater. However, many parameters are reasonably well characterized and the following can be identified as being of special importance: salinity, temperature, pressure, pH, redox state, inorganic solids and particles, organic substances, organic detritus, and living organisms.

The Meaning of Species Identification in Marine and Estuarine Systems

The term speciation has been applied to chemical systems in the environment with increasing frequency in the past decade. This term has been used, and perhaps misused, through conventional practice; it may be worthwhile to consider limits to its meaning in the interest of precise communication.

The concepts of species and speciation have been transferred from biology to chemistry, to refer to distinguishable forms of a chemical substance that are unique to some degree. We will adopt the view that a chemical species is a specific form of an element that can be represented in molecular terms. This eliminates such generic chemical forms as dissolved and particulate, inorganic and organic, or labile and inert from designation as species. These and similar descriptions of chemical forms are based on operationally defined procedures and represent classes of chemical species. Investigations that provide information on the generic forms of an element should be termed as partitioning or fractionating the element into identifiable forms, which is not "speciation" (i.e., species identification) because it lacks molecular representations.

Even when it is agreed that the term chemical species means a form of an element that can be represented in molecular terms, there is a dual usage of the term speciation. The phrase "chemical speciation in environmental processes" refers to the existence of a multitude of chemical species, whereas "speciation methods" refers to the act of determining and identifying chemical species (both of these phrases appeared in the goal of this Dahlem Workshop). The ambiguity in this dual usage can be eliminated if we state explicitly what is meant and avoid using the term speciation, specifying whether we refer to the analytic identification and quantification of chemical species, their distribution pattern in the environment or in a sample, or the transformation of species (see Bernhard and George, this volume).

As we consider the chemical species in marine and estuarine systems, the concept of phases is encountered. In the simplest sense one can identify three types of phase: gas, liquid, and solid. A criterion for the identification of a phase is the presence of a phase boundary or interface. Interfacial regions are the locations of many important chemical processes in natural systems. In marine and estuarine systems the principal liquid phase is aqueous and the distinction between gas and solution phases is generally evident. The presence of biogenic materials leads to distinctions between aqueous and organic phases. There can be a great variety of solid phases, and their distinction from the liquid must often be made using operational procedures rather than a fundamental criterion. Conventionally, studies in environmental systems have identified particulate forms of an element as those that are retained on a 0.45 μm membrane filter. Consequently, those that pass through the filter are operationally regarded as dissolved. Technological advances have provided alternatives to the 0.45 μm fibrous matrix filters, such as thin polycarbonate or fluorocarbon membranes punched by holes of different sizes down to 0.1 μm. It is not clear whether we should maintain consistency with the previous convention or utilize progressively smaller pore-sized materials as they become available, to obtain a better approximation of dissolved and particulate phases. Probably the best approach consists of using smaller pore size but preserving consistency by also filtering a representative number of samples through 0.45 μm filters. In practice there have been criteria other than the pore size that have determined the choice of filters. These have included the chemical properties of the filter such as plastic membranes for trace-metal analyses and glass fibers for particulate organic substances. The flow rate through the filter and the volume that can be filtered without substantial clogging are also factors in selecting a procedure to distinguish dissolved and particulate phases by filtration. The retention of particles by a filter is likely to change during filtration, rendering the distinction between dissolved and particulate even more ambiguous.

A dissolved species is one that is solvated. In aqueous systems solvation may be viewed as leading to a hydrophilic association between the solute and solvent, or to a hydrophobic relationship as in the formation of clathrate structures. One may also have substances dissolved in organic phases. In terms of physical sizes it is possible to consider a continuum of forms from individually dissolved solute molecules to polymeric solutes to colloids to particles. Such a continuum appears to lead to an arbitrary ultimate distinction between dissolved and particulate forms, but this is not the case. A dissolved species must be solvated, and a particulate-phase species must be separated from the solvent by an interface. Thus, molecular monomers and polymers are species, but colloids and particles are not species, they are types of solid phases. Colloids may contain occluded solvent, but this does not exchange readily with the liquid phase. While solid phases are not species *per se,* they contain species, and one may contemplate the characterization of species associated with solid phases.

The transformations among solute species involve considerations of several types of chemical process such as phase transfer, ligand exchange, and electron transfer. It is important to recognize that bare ions or free ions are not quantitatively important in natural waters. If ions are designated as being "free", it is understood that they are in fact hydrated. This is important when one considers complexation reactions, because they can lead to outer-sphere complexes when the first hydration sphere is retained or to inner-sphere complexes when a ligand replaces a water molecule in the coordination sphere of the central ion. The exchange between a ligand and a water molecule of hydration can lead to species of different kinetic and spectroscopic characteristics.

One of the factors that may be more appreciated in determination of the species of elements in environmental systems than in that involving laboratory systems is the role of photochemical processes. Life processes driven by photosynthesis, an obvious example, lead to changes in chemical species. There is increasing evidence that other photochemical processes are important in the redox states of manganese, copper, and iron (Hong and Kester, 1986; Miles and Brezonik, 1981; Moffet and Zika, 1983; Waite and Morel, 1984).

Chemical species encompass a broad range of molecular forms in environmental systems. They include dissolved, gaseous, and solid-phase species. It is necessary to determine the oxidation state and complexation of the elements and to characterize the equilibrium and kinetic aspects of the system. The solid phases within marine and estuarine systems are especially diverse, including colloidal phases, inorganic particulate phases, hydrophobic phases, and living organisms which in themselves are multi-phase systems.

Role of Kinetics in Transformation of Chemical Species

Chemical species in environmental systems undergo a continuous series of transformations. The rates of these transformations are an essential factor in characterizing species and their role in environmental processes. It is important to identify the principal time scales of processes in marine and estuarine systems: the rates of specific types of reactions can then be related to environmental processes. For biochemical processes such as the uptake of species by cells, it is important to determine the rate-determining step and its relationship to particular chemical species.

Basic Time Scales for Estuarine and Oceanic Processes

Environmental processes occur over the complete range of time scales from those of molecular vibrations (10^{-12} s) to billions of years. Within this range, one may identify specific time scales that are important for transformation of chemical species in oceanic and estuarine regions. Biological processes establish one set of important time scales. For photosynthetic activity, time scales of hours are important. Tidal oscillations have periods of tens of hours, with higher-order harmonics of monthly and seasonal periods. Phytoplankton have generation times on the order of days. Meteorological events such as storms have durations on the order of hours to days.

Seasonal cycles are important in variations of temperature, vertical stability of natural waters, and rainfall. The time scales for human perturbations to the environment (and the recovery of the environment from such perturbations) appear to be on the order of years, decades, and centuries.

Kinetic information about chemical species will be of special importance in environmental processes when the rates of their reactions are in the range of minutes to years. For reactions that reach equilibrium on much shorter time scales, the equilibrium condition provides an adequate description. Very slow reactions may be important for geological processes, but their kinetics are not required for the understanding of many marine processes and their variability.

Time Scales of Species Reactions

The bioavailability, geochemical reactivity, and transport of trace metals in natural waters are determined mainly by their particular chemical species. The kinetics of transformation among these species due to physical and biological processes span a broad range of time scales from very fast ($<10^{-10}$ s) to very slow (>1 yr), depending on the nature of exchanged ligands as well as on the central metal ion. Some species reactions are generally fast, providing equilibrium conditions in times of a minute or less: examples include weak acid proton equilibria and inorganic outer-sphere complexation equilibria. Some types of organic ligands have rapid complexation kinetics with metal ions, whereas others have slow reaction kinetics. These differences lead to operationally defined labile and inert forms of metals in environmental systems. Adsorption reactions tend to occur more rapidly than desorption reactions for some systems, leading to hysteresis in the exchange between dissolved and particulate phases.

Important time scales for estuarine processes are determined by the residence times of water and particles in the system. In general, the residence time of water decreases as the freshwater discharge to an estuary increases (Pilson, 1985); residence time increases with increasing vertical mixing, ranging from days in a salt-wedge type estuary to months in a well-mixed estuary. The non-steady state conditions induced by variations in freshwater discharge, in tidal amplitude, and in wind stress make it difficult to calculate transport rates from the salinity distribution. The residence times of particles in estuaries are usually longer and more variable than those in the open ocean (Wollast and Duinker, 1982).

Role of Particles

Particles play an important role in the formation and distribution of metal species. The most important processes which control metal concentrations are sorption reactions, including the following:

1) Nonspecific adsorption of species on a particle surface, which is a physical process originating from electrostatic forces. Bonding energies are on the order of $\sim 1\ \mathrm{kcal\ M^{-1}}$.
2) Chemisorption of species, which takes place in hydrolytic adsorption or condensation reactions, including co-precipitation with iron and manganese hydroxides. Bonding energies are on the order of about $10\ \mathrm{kcal\ M^{-1}}$.
3) Ion exchange, which is also a chemical process.

Metals or species are more or less strongly bound by sorption reactions with particles in natural waters. These particles are transported and eventually deposited as sediment or dissolved particles. Very important matrices for binding metal species in sediments are carbonates, Mn-oxides, amorphous, hydrous, and crystalline Fe-oxides, organic substances such as humic and fulvic acids, bio-organic detritus such as cell-wall material, and bacteria. Solids such as SiO_2 or clay minerals seem to play a less important role in metal sorption reactions in natural waters, although they are nucleation centers for coatings of the other phases.

Processes which control the physicochemical forms of these phases as well as the surface metal species in sediments include redox reactions, pH, bacterial activity, pressure, and temperature. Redox conditions seem to have the greatest importance for physicochemical forms of many metal species in sediments. Anoxic conditions cause changes in the oxidation state of species of metal and other elements: e.g., Mn(IV) is converted to Mn(II), Fe(III) to Fe(II), or NO_3^- to NH_4^+ and SO_4^{2-} to S^{2-}. These changes have different consequences: if Mn-oxides or amorphous Fe-oxides are solubilized, metal species which are bound during these phases can go into solution. However, at very low redox conditions (sulfidic conditions) they may be precipitated as sulfides. However, metal concentrations have been found that exceed the solubility of their metal-sulfide solid phase, suggesting the importance of polysulfide or organic complexes in these environments. Under anoxic conditions a certain part of the Fe and Mn in sediments may exist as Fe- and Mn-carbonates, but if the sediment becomes oxic, this phase will not occur. For example, an extraction sequence was performed on an anoxic sediment of the Rotterdam harbor, giving different results for two methods (Kersten, personal communication). One part had been freeze-dried and then extracted, the other was extracted without drying or exposure to oxygen. Under the original anoxic conditions the Fe was less strongly bound than would have been the case had the sediment been oxidized. On the other hand, Zn was less strongly bound under oxic conditions. This has practical consequences for the dredging of anoxic sediments and subsequent disposal on land. It may be possible that some other elements, especially Cd, which show a similar extraction behavior to Zn are relatively easily mobilized under conditions which change from anoxic to oxic. Because of these effects, sediment samples for the analysis of chemical species should be taken, stored, and processed under the redox conditions from which they were obtained.

Particles in Estuaries

Particles in estuaries are agglomerates of various amorphous and crystalline, authigenic and detrital mineral components and organic matter (living and dead organisms, coatings of organic material). Fe and Mn hydrous oxides can sorb significant amounts of trace elements from solution. Freshly formed particulate material, because of its high surface-area-to-volume ratio, is a more effective sorbent than aged material. Distribution coefficients for trace elements on freshly formed MnO_2 exceed those for alkali and alkaline-earth elements. Clay minerals contribute relatively little to the exchange capacity of estuarine and marine particulates for trace elements, as their distribution coefficients are similar to those for Ca, Mg, and Na. Their main role is in their ability to act as nucleation centers for Mn and Fe hydrous oxides and organic matter. These modify the solubility, redox potential, and precipitation behavior of trace elements by ion exchange, surface adsorption, and chelation. They may be precipitated in estuaries, most likely by sorption to minerals through physical adsorption, cation exchange, and chemical reactions. Interactions between particles and solutes include electrostatic interaction (rapid exchange).

Many estuaries have a turbidity maximum, in the low salinity regime, in connection with residual (nontidal) current distribution characteristics. The positions of the turbidity maximum and the associated mud maximum vary within the tidal and neap-spring cycles and, to a larger extent, with variations in freshwater discharge and meteorological conditions (wind). Thus, more than one turbidity maximum can be generated and may exist over long periods. The particles may consist of relatively large flocs with high contents of water and organic matter; they are characterized by a narrow range of settling velocities (v_s). Particles with higher v_s can be deposited to the bottom, whereas those with lower v_s can escape estuarine trapping mechanisms and be transported out into the sea. These particles have higher contents of Cd, Zn, Cu, and Hg and lower contents of Fe, Mn, and Al. The particles in a turbidity maximum at increased concentrations may be enriched in Fe and Mn, hydrous oxides, and organic substances and thus have a large capacity for exchange of species between solution and particles. Such transitions are favored by minima in pH and O_2 distributions, often occurring near the region where the turbidity maximum is found.

Oceanic Particles

Most of the particles present in the open ocean are of biogenic origin. They consist of a mixture of mineral components ($CaCO_3$, SiO_2) and organic matter. Only the bulk chemical composition of these particles has been studied so far (with the exception of some leaching studies discussed below). The molecular basis for the scavenging of trace substances from seawater (and their redissolution) remains unknown.

Specific mechanisms that lead to the formation of marine organic particles include a) photochemical production of humic-type substances (Harvey et al., 1983), b) production by animals of 1–3 μm diameter particles through intracellular peroxidation of lipoprotein polymers, c) condensation of soluble excretion products from animals, d) production of organic detritus such as chitinous exoskeletal material and phytoplankton cell walls, and e) fecal pellets from plankton organisms, etc.

Marine particles are modified and distributed through the ocean by a complex series of biological and physical processes. The diverse physicochemical and biochemical environments encountered by the particles lead to species transformations. The packaging of particles into fecal pellets, their microbial decomposition and subsequent repackaging will alter portions of an element in solid-phase species.

There is evidence that particles in the sea are covered by organic matter independent of the composition of the solid substrate. Techniques are needed which will identify the nature of this organic layer and which can characterize the binding environment of the metals bound to marine particles.

Importance of Particle Size

The binding of substances on the surfaces of particles is an important process in marine and estuarine systems. Surface area per unit mass increases with decreasing particle size; thus the importance of surface reactions is highly dependent upon particle size distribution. All the methods employed to assess particle size distribution or surface areas of particles in estuarine and marine systems (e.g., "Coulter" counting, settling velocity, nitrogen adsorption) are limited by the complexities in the shape, porosity, and composition of the particles, variations in particle density, and the tendency of particles to agglomerate or dissociate under different conditions. Particle size and density can also be a factor in compositional variations of organic carbon, trace elements, and major constituents. The reactivity of particulate elements as reflected by a mild acid leach (0.1 N HCl) varies systematically with particle size and density.

Selective Chemical Leaching

Characterizing the distribution of trace elements among the components of the particulate phase (partitioning) is a critical chemical and biological problem. Mathematical models of trace-element partitioning in sediments (analogous to models of element distribution among species in solution) are not possible at present because of limitations in our understanding of the processes involved (Luoma and Davis, 1983). Selective chemical leaching theoretically should yield some operational separation of metals bound to different components of sediments. In reality, however, these methods have a number of limitations. For example, the metal released from the dissolution of one component of the solid phase may resorb to the remaining solid phases, low pH leaching solutions may remove several phases simultaneously, and high pH extractants may result in the precipitation of metals. It should be evident that there is not likely to be a one-to-one correspondence between metals released by a leaching procedure and the different solid phases present in natural particles.

When the results of leaching techniques yield operationally defined fractions of the solid phase, they provide useful information. A leaching procedure can identify the amounts of various elements which can be released from natural particles upon changes in ionic strength, pH, or redox conditions. There is evidence that particulate metal species from anthropogenic sources are more easily leached, and therefore potentially more mobile in the marine environment, than species of the same metals in the natural geochemical cycle.

When a selective leaching procedure is applied to natural particles an effort should be made to verify its performance by empirical means. For example, it is useful to demonstrate that metal species present in organic phases are released during an oxidation leach and not during a weak acid leach. Mixtures of several types of particles such as carbonates, metal oxides, organic material, and clay minerals, which have been tested separately, can be used to evaluate the selectivity of the leaching procedure.

There has not been a convincing demonstration of the relationship between a leaching procedure and the bioavailable fraction of metals in sediment. Correlations between weak acid leaches and the benthic organism content of metals indicate that the weak acid leach is not an adequate indication of metal mobility in the digestive tract of the organisms. The stronger 1 N HCl leach has provided a better correlation with bioavailability. Evidently simple weak acid solutions do not adequately simulate digestive enzymatic processes.

Kinetics: Metals in Sediment

When sediment chemistry changes, the kinetics of species redistribution will greatly influence whether metal species remain bound to a single component of the sediment or redistribute among components. It has been well established that the rate of metal desorption from particles to solution is often quite slow. However, when two different types of surface with different binding constants are present, desorption from the weaker surface and adsorption to the stronger surface may be accelerated by continual depletion of the solution phase by the stronger surface. Recently Kuwabara et al. (in preparation) showed an accelerated transfer of Zn from titanium dioxide particles to algal cells, compared to the rate of Zn desorption from titanium dioxide and transformation into soluble Zn^{2+}.

Chemical Species and Biological Systems

For a better understanding of the biological processes occurring in marine and estuarine systems, a good knowledge of chemical species involved and their transformation kinetics is of fundamental importance (Luoma, 1983). Biological processes are a source of organic ligands. The organic substances present in seawater, most of which are either poorly characterized or completely unknown, are the ligands for complexation of a number of elements in the marine environment. Copper has been the most intensively studied example for this type of complexation, but nevertheless the nature of these ligands, in terms of both functionality and steric characteristics, remains largely unknown. As a consequence, there are many paradoxes. The assumption that only one organic copper species exists is inadequate to account for the behavior of copper in the ocean. There may be organic Cu species that are in equilibrium with the inorganic Cu species or there may be others that are kinetically inert over substantial periods of time. There may be some ligands that exhibit the relative affinities for divalent transition metal and alkaline-earth cations shown by simple functional groups such as carboxylic acids, phenols, and amines; others may involve specific chelation geometries for a metal. Some species of organically bound metal may be hydrophobic, whereas others may be hydrophilic or amphiphilic. Experiments studying the distribution of species should be designed and interpreted with this diversity of metal organic species in mind. Further investigations should be made of the properties of natural organic ligands in marine systems.

The interior of biological cells is characterized by high electron activity (reducing conditions). These electrons, which have been extracted from the $H_2O \rightarrow O_2$ reaction during photosynthesis, are the source of the free energy which drives biological processes. A fraction of this reducing power becomes available for the reduction of trace elements (e.g., arsenic and antimony) resulting in the presence of reduced, nonequilibrium species in the marine and estuarine environments. In addition, species with metal-carbon bonds (e.g., methylarsenic and methylgermanium compounds) are biologically synthesized and released into the aquatic environment (Andreae, 1985).

The cell membrane of an organism presents both a phase boundary and a chemically reactive barrier between the organism and the external medium. Thus interfacial reactions are critically important in determining transport across a cell membrane, and membrane permeability is species-dependent. The various processes of specific element transport through membranes – carrier-mediated, active, and passive transport (pores, channels, pumps) – are considered in detail by Wood et al. (this volume). Microorganisms have much more specific mechanisms for metal uptake than eukaryotes because their intracellular machinery and genetic material are unprotected from a reactive species which might enter the cell, due to their lack of the elegant intracellular membrane system present in higher organisms. The latter enables specialization of organs for discrete and different roles.

Attention should be given to the unique properties of the marine environment (high salinity) and the marine biota, such as the presence of high Cl^- and the tendency to form neutral chloro-complexes of many metals. In the marine environment many uncharged species are present and these may diffuse into the membrane. Membrane structure must be considered a three-dimensional structure with defined orientations of charged, amphiphilic and hydrophobic regions (Mehlhorn, this volume). Metal-organic complexes, even if they are charged, can partition in the membrane. The charged complexes form ion pairs. These complexes can cross the membrane by removal of the metal species from the external medium at the intracellular boundary through complexation of the metal with soluble macromolecules in the cell cytoplasm. This transport process is a diffusion process driven by the intracellular complexation. Ultimately, control of the concentration of a species is exerted via control of synthesis of the intracellular ligand. The rate-limiting steps are the partitioning into the membrane (interface reactions) and the synthesis of the intracellular binding ligands. Ag, Hg, and Tl (which form planar chloro-complexes) appear to be passively transported.

For many elements transport occurs via active transport mechanisms (requiring energy expenditure for transmembrane movements) – examples include Na^+, K^+, Ca^{2+}, Mg^{2+}, and HCO_3^- – evidence suggests that these are not entirely specific and that other metals (Cd^{2+}, Ni^{2+}) are transported by carriers designed for other metals (e.g., Ni^{2+} via Mg^{2+}-carrier). Transmembrane and transepithelial transport must be distinguished when uptake is considered. Many anions are transported across epithelia through pores between cells: these include chloride, orthovanadate, arsenate, phosphate, sulfate, selenate, and chromate.

Marine invertebrates appear unable to control uptake of many metals into their cells. It is uncertain whether this is due to the extracellular species differences (cf. freshwater animals) or to a difference in excretory fluxes due to co-transport with water (due to the osmotic difference in seawater). There is a balance of pumps and leaks of most elements. However, most studies to date have only measured the net uptake, and in the future both influx and efflux must be measured.

There are highly developed mechanisms for regulation of intracellular concentrations in the cytoplasm of many of the cells. In general, concentrations of uncomplexed ions are very low: around 10^{-8} M – 10^{-9} M. This enables their use in organ-specific processes in multicellular organisms (transfer of chemical signals, movement, catalysis). The most widely studied heavy-metal regulatory protein is metallothionein. This protein binds Cu, Zn, Cd, Mg, and Ag (and possibly Ni), and therefore the metabolic interactions between these elements are complex (George and Viarrengo, 1984). A study of marine invertebrates has led to the proposal of a model (fitting the data for mammals as well) for control of intracellular concentrations and distribution of these elements. The hypothesis depends upon lysosomal digestion and excretion of the different metal-metallothionein species as the key process. The residue from lysosomal digestion of cellular macromolecules cannot be further degraded because divalent metal catalysts (Cu^{2+}, Fe^{2+}) promote free-radical generation, causing peroxidation and cross-linking of unsaturated lipids which cannot be enzymatically hydrolyzed (their steric form is wrong). These compounds have definable metal-binding properties (George, 1984) and characteristics similar to humic substances. Metallothionein is excreted by organisms and is a major component of invertebrate fecal pellets. Due to its resistance to enzyme attack, it has a long environmental residence time.

There is also a mechanism which produces inorganic phosphorites containing high concentrations of Zn, Mn, and other divalent cations (review by George, 1982). This process may be responsible for the production of recent marine phosphorite sediments.

The identification of the chemical species involved is fundamental to understanding toxicity mechanisms in the marine environment. In a given family of products (organotin compounds, for example), some species are toxic at low concentration and others are nontoxic. A discharge of a nontoxic species into the environment might have harmful consequences because of the interactions with suspended solids, light, bacteria, and other species, which might transform a nontoxic form into a toxic one. But the reverse is also true: a toxic species can be transformed into a nontoxic one.

Antifouling paints containing tributyltin oxide have recently been held responsible for abnormal growth of the oyster *Crassostrea gigas*. Experimental tests have shown an impact for these paints, but the same phenomenon also occurs with paints containing no tin compounds. Tin is emitted from the paint in the form of tributyltin oxide, but what is its fate in the marine environment? The problem is difficult to solve as organotin compounds are light-sensitive and can be biodegraded, and the species must be determined at concentrations lower than a microgram per liter in an environment where other metallic (Zn, Pb, Cu, Cd, etc.) and organic species are also present.

Distinctive Types of Marine Environments for Studies of Chemical Species

For some investigations it is useful to recognize that particular types of marine environments may enhance the role of specific types of chemical species. Estuaries, for example, provide regions with large salinity gradients and high concentrations of suspended particles. By selecting estuaries in which physical processes lead to differences in residence time between fresh and ocean waters, it is possible to find conditions where rapid chemical processes influence strongly the distribution of chemical species or where biological activity and slower geochemical processes have important influence.

Oligotrophic regions of the open ocean are areas where highly efficient use of nutrients and their extensive recycling may lead to unusual chemical species. Highly productive regions such as those associated with coastal upwelling and water mass divergence provide organic-rich particulate material, low oxygen concentrations, low pH, and potentially high concentrations of chemically active organic ligands.

The chemical species of an element may show different features in the polar seas where temperatures are low and seasonal extremes in light and in biological activity are great.

Oceanic rings are bodies of water that may be tracked over periods of months and resampled to provide a time series of changes in physical, chemical, and biological properties. These rings provide an opportunity for examining changes in the distribution of species with time in an oceanic ecosystem.

Hydrothermal vents provide an opportunity to examine how chemical species respond to high-temperature and high-pressure conditions, and in a chemosynthetic, rather than photosynthetic, ecosystem.

The Use of Models in Studies of Chemical Species

There are many types of models that have been applied to chemical studies in marine and estuarine environments. Conceptual models are summary formulations of existing knowledge with the projection of hypotheses that can guide further research. It is important that the role of feedback mechanisms in environmental processes be considered. These mechanisms may help to stabilize the environment or, in some instances, they may amplify perturbations. Models can be helpful in identifying feedback mechanisms.

Box models are often used in geochemical investigations. The environment is partitioned into various reservoirs, and one examines the pathways and fluxes of chemical exchange among the reservoirs. The reservoirs are treated as well-mixed systems, and no attempt is made to account for the processes that occur within a single reservoir. Key parameters of box models are the identification of sources and sinks (i.e., the pathways among the reservoirs), the quantity of a substance within each reservoir, and the rate of transfer of a substance between reservoirs (i.e., the flux). From these parameters it is possible to estimate the residence time of the substance within a reservoir.

Box models have generally not been applied to chemicals in the environment at the species level. They are commonly formulated for a chemical element and can be extended to operationally defined fractions of an element. An example of a box model for the partitioning of copper among various forms is shown in Fig. 1. At the stage of formulation presented in Fig. 1, a box model serves to organize the knowledge of chemical forms and to identify the principal pathways of exchange among different chemical forms. By carrying out an analytical program of a specific environmental system, it would be possible to determine the mass of copper in each box. If rate constants can be determined for the exchange of copper among the boxes, values could be assigned to fluxes. In some cases the fluxes between boxes might be determined from environmental measurements.

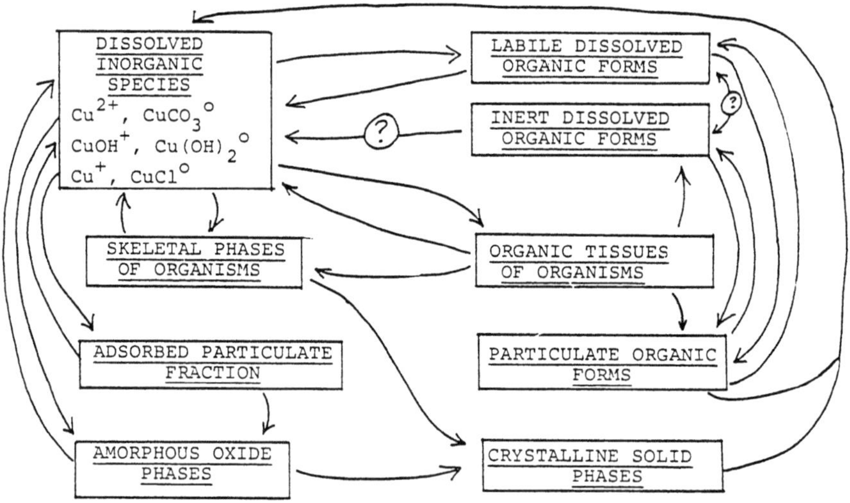

Fig. 1. Example of a box model applied to various forms of copper

Equilibrium models of chemical species have been extensively developed in marine and estuarine systems. They have provided quantitative predictions for the marine chemistry of inorganic species (see Kester, this volume). Equilibrium models are less able to account for the role of organic species in seawater, because organic ligands have not yet been adequately characterized and many organic species are not in equilibrium with inorganic chemical systems. Organic substances are inherently unstable in oxidizing environments, yet many of them can persist for years, centuries, or millenia.

When equilibrium models are combined with the knowledge that their species have rapid kinetics, they provide a reliable description of the chemical species they comprise. They should, however, be regarded as a submodel of the species in a system. An equilibrium model cannot account for species that persist for kinetic reasons nor for those that exist in a steady state of production and decomposition processes.

In a kinetic model, the rates of reactions within the system are assembled into a set of differential equations and initialized with a set of boundary conditions. Numerical solutions result in either time-invariant (steady-state) or time-dependent descriptions of the system. These models can be applied to an analysis of the response of the system to a perturbation or to the tendency for the system to approach equilibrium. The quasi-steady-state condition, which may be far from equilibrium, is a very important one in marine environments where biological and photochemical processes are a continual source of perturbation.

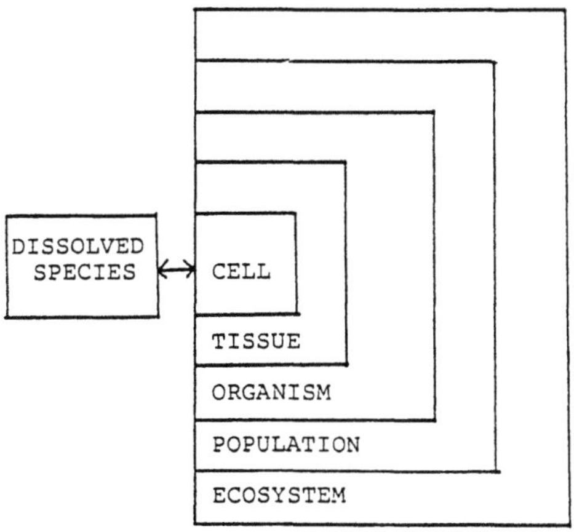

Fig. 2. Schematic representation of the propagation of a dissolved species-biological cell interaction through several stages to influence an ecosystem

One may envision a series of models that is nested to represent various levels of detail in the role of chemical species within a system. Figure 2 illustrates such a nested set in which chemical species propagate from the solution phase through several levels of biological organization to influence the ecosystem.

While the ultimate objective of a chemical species model might be to obtain a predictive understanding of the dynamics of species in the environment, the complexity of natural systems makes limited goals more realistic. Models can integrate the information available and guide future research. They can point out gaps in existing data. They can provide an indication of the sensitivity of the system to specific variables. Even for equilibrium models of inorganic species, there is a need to develop a more reliable data base of chemical constants under marine and estuarine conditions. Models should be developed in conjunction with investigations designed to test them.

Analytical Considerations of Chemical Species

The most important concept in analyzing environmental samples for chemical species is the development of and adherence to a critical analytical strategy that incorporates quality assurance concepts and leads to reliable concentration values of specific chemical species with stated estimates of uncertainties. The analytical strategy should follow accepted principles (Kieth et al., 1983) which encompass interdisciplinary interactions in the following steps: problem identification, planning, sampling, sample handling (processing, preservation, storage, etc.), measurement process (transformation of the sample into a measurable form, separation, detection, and measurement), and data handling and interpretation. Although all steps in the total analytical process should be considered carefully, special consideration has to be given in an operational sense to sampling, sample processing, and the measurement process, all of which are discussed below.

Sampling

In situ analysis is possible for only a small number of marine constituents, therefore sampling, pretreating, and storing of the materials are necessary. These are essential operations, as there is no point in making very accurate and sophisticated analyses if the samples are not representative of the environment from which they were collected. In many cases sampling will modify the species. Sampling operations must be planned *according to the analytical objective.* This could imply taking multiple samples and treating them differently according to the species of interest. Sampling might also correspond to a unique situation that cannot be reproduced. In most cases variability in the ocean and estuaries is undersampled: there is a need for higher resolution sampling and analysis in time and space.

Sampling devices may produce artifacts in several ways. Adsorption and desorption on surfaces are always possible. In some instances, preconditioning a sampler to the environment may be better than having it chemically clean (i.e., completely void of a particular constituent).

Analysis of Chemical Species

For open ocean waters the total concentration of most trace metals, such as Cu, Pb, Hg, Zn, and Cd, is so low that it becomes very difficult to distinguish experimentally among different chemical species. In many cases it has been only within recent years that reliable measurements have been made of the total metal concentrations. The prospect of identifying and quantifying species at these low concentrations is a substantial challenge for an analyst.

For trace metals present at the picomolar level the introduction of a preconcentration step becomes essential. For stable molecular species such preconcentrations can probably be performed, but for more labile species preconcentration methods will alter the distribution of species. Careful evaluation and validation procedures will be required wherever preconcentration is used in connection with the determination of chemical species. The effect of the preconcentration step is not easily predicted and must be investigated in each case.

Chemical species are also a factor in preconcentration methods intended to yield total element analyses. It is inevitable that a preconcentration method will be somewhat selective. It is very difficult to prove that total element recovery has been achieved when unknown species of the element may exist in the sample.

Electrochemical methods are applicable for determination of some chemical species at very low concentrations in natural systems. Additional analytical techniques are available or are being developed which allow direct determination of organic and inorganic molecular species at concentrations as low as about 10^{-6} ML^{-1}, even when they are present in relatively complex matrices. While some species exist in this concentration range in ocean waters, many species of interest occur at concentrations of 10^{-9} ML^{-1} and less. With the exception of electrochemical methods, there remains a considerable gap between the sensitivity of analytical technology and the concentration of environmentally important chemical species.

The instability of many chemical species makes sample storage very difficult, even impossible. The analysis of samples in the field (e.g., on shipboard) can often at least partially mitigate this problem.

In addition, on-site analysis allows the instant feedback of the results into the planning of sample collection and processing. This can help to eliminate sampling artifacts and contamination, and substantially improve sampling strategies. This approach is now possible for a wide range of instrumental measurements including UV-visible spectroscopy, gas chromatography, liquid chromatography, electrochemical analysis, and atomic absorption spectroscopy. As marine chemists continue to extend their laboratory capabilities in the field, it will become possible not only to improve sampling procedures, but also to reveal the chemical dynamics of marine systems through experimental studies of unstored samples.

To achive reliable data for the concentration values of chemical species in complex samples, the measurement process must be in statistical control and must be validated. Statistical control is accomplished by determining the precision of the measurement process (agreement among repeated measurements of the same quantity). Method validation is accomplished by inferring the accuracy of the values obtained by the measurement process (discrepancy between the true value and the measured value). Method validation is more complex than statistical control and includes calibration of the instrument scales (absorbance, wavelength, mass, etc.) as well as testing the measurement process using certified reference materials *or* exhaustive research on each step in the process. It is only through methods that are in statistical control and are validated that data interpretation can be performed by accepted statistical practices and final uncertainties assigned to measured quantities.

Methods for Characterizing Particles

Instrument methodology for defining specific associations between elements and their associated ligands on particle surfaces is developing rapidly (e.g., ESCA). The compositional and morphological complexity of the estuarine and marine particles appears to create significant difficulty in unambiguous interpretations using these methods. A great deal of information can be obtained from pure solid-phase species, but mixed phases with small sizes do not provide as quantitative results. Methods that require subjecting the particulate material to a high vacuum are of limited use in investigating chemical species on hydrous surfaces.

Methods for Characterizing Organic Substances

Detailed characterization of organic species present in marine and estuarine systems is an enormous undertaking. While detailed characterization may well be impractical, analytical instrumentation is now available which can provide identification and quantification of the major organic ligands present, as well as information on the binding of the ligands to metal ions. Recent advances in high-performance liquid chromatography detectors, together with the successful development of on-line combinations of the techniques provide important new tools for molecular characterization of nonvolatile samples. Interdisciplinary research is necessary to develop and evaluate the new techniques for application to the determination of organic species in environmental systems.

Recommendations for Future Research

It is evident from the composition of this discussion group that there was a bias towards inorganic chemistry and an emphasis on the chemistry of metals. To some extent this is a reflection of the present state of the art and areas of recent research on chemical species in oceanic and estuarine environments. A knowledge of chemical species is important for understanding marine organic chemistry, and there are recent reviews that address this area of work (Mantoura, 1981; Williams, 1975). Our considerations of chemical species in biochemical processes focused on the cellular level, but chemical species play important roles at higher levels of biological complexity where differentiation of tissues and organ function are factors. Within these limitations on the scope of our discussions, we summarize our deliberations with the following recommendations:

1) Chemists and biologists should work jointly to determine which species are available to organisms, which species are taken up by organisms, and which species alter rates of biological activity. In considerations of toxicity, particular attention should be given to redox-sensitive species because they may lead to the production of free radicals that are damaging to biological systems. Substantial contributions can result from broadly interdisciplinary research.
2) Improvements should be made in the methods for determining particle size, surface properties, and composition. For some problems it will be useful to classify particles in terms of settling velocity (which depends on size, shape, and density) rather than by filtration or size alone. Future

studies of adsorption should emphasize natural rather than ideal solid phases. The binding strengths of coordination sites on solids and the valence state of metals in solids should be determined.

3) There is a need to improve the quality of data for chemical transport to the ocean by rivers. Has contamination been a factor in riverine trace-element analyses, as it has been for oceanic measurements? Considerations of geochemical behavior should be examined at the level of chemical species rather than merely that of the chemical elements.
4) The kinetics of species transformation processes should be emphasized in future research. Photochemical reactions deserve special attention.
5) Studies of chemical species in estuaries should include investigation of their spatial and temporal distribution. Auxiliary measurements of salinity, current velocity, temperature, pH, dissolved oxygen, and turbidity are required for meaningful interpretations of the data obtained.
6) There should be continued development of *in situ* measurement systems and of real-time analytical investigations in the field.
7) Certified Reference Materials (CRMs) are required for a variety of marine matrices, including seawater (for trace elements), biota (for algal, fish, and crustacean specimens), and sediment (for hemipelagic, calcareous ooze, siliceous ooze, and anoxic organic sediments). These CRMs will aid in intercalibration, standardization, and validation of analytical methods and should be specified for total element content. The present state of knowledge of chemical species in marine matrices is too limited to specify the demand for individual chemical species, but major efforts should be made to arrive at CRMs for the toxicologically most important species.

References

Ahlf W, Calmano W, Förstner U (1984) Correlations between chemical and biological evaluation procedures for the determination of trace metal availability from suspended solids. Third International Symposium on Interactions Between Sediments and Water, Geneva

Andreae MO (1985) Organoarsenic compounds in the environment. In: Craig MPJ (ed) Organometallic compounds in the environment. London: Longmans in press

Calmano W, Förstner U (1983) Chemical extraction of heavy metals in polluted river sediments in Central Europe. Sci Tot Envir 28:77–90

George SG (1982) Subcellular accumulation and detoxication of metals in aquatic animals. In: Vernberg WB, Calabrese A, Thurberg FP, Vernberg FJ (eds) Physiological mechanisms of marine pollutant transfer. Academic Press, New York, pp 3–52

George SG (1984) Heavy metal detoxication in *Mytilus* kidney: an *in vitro* study of Cd- and Zn-binding to isolated tertiary lysosomes. Comp Biochem Physiol 76C:53–57
George SG, Viarrengo A (1984) A model for metal homeostasis and detoxication in mussels. In: Vernberg WB, Calabrese A, Thurberg FP, Vernberg FJ (eds) Physiological effects of pollutants on marine animals. University of South Carolina Press, in press
Harvey GR, Boran DA, Chesal LA, Tokar JM (1983) The structure of marine fulvic and humic acids. Mar Chem 12:119–132
Hong H-S, Kester DR (1986) Redox state of iron in the offshore waters of Peru. Limnol Oceanog, 31(3):512–524
Hunter KA (1980) Microelectrophoretic properties of natural surface-active organic matter in coastal seawater. Limnol Oceanog 25(5):807–822
Kieth LH, Crummett W, Deegan J Jr, Libbey RA, Taylor JK, Wentler G (1983) Principles of environmental analysis. Anal Chem 55:2210–2218
Kuwabara J, Davis JA, Chang C (1985) Nutrient particulate interactions with plant growth. In preparation
Lewis EL (1980) The practical salinity scale 1978 and its antecedents. IEEE J Ocean Eng 5(1):3–8
Luoma SN (1983) Bioavailability of trace metals to aquatic organisms – A review. Sci Tot Envir 28:1–22
Luoma SN, Davis JA (1983) Requirements for modeling trace metal partitioning in oxidized estuarine sediment. Mar Chem 12:159–181
Mantoura RFC (1981) Organo-metallic interactions in natural waters. In: Duursma EK, Dawson R (eds) Marine organic chemistry. Elsevier, London, pp 179–223
Meybeck M (1979) Concentration des eaux fluviales en elements majeurs et apports en solution aux oceans. Rev Geol dynam Geograph phys 21(3):215–246
Miles CJ, Brezonik PL (1981) Oxygen consumption in humic-colored waters by a photochemical ferrous-ferric catalytic cycle. Envir Sci Technol 15(9):1089–1095
Moffett JW, Zika RG (1983) Oxidation kinetics of Cu(I) in seawater: implications for its existence in the marine environment. Mar Chem 13(3):239–251
Pilson MEQ (1985) On the residence time of water in Narragansett Bay. Estuaries 8(1):2–14
Waite TD, Morel FMM (1984) Coulombic study of the redox dynamics of iron in seawater. Anal Chem 56(4):787–792
Williams PJleB (1975) Biological and chemical aspects of dissolved organic matter in sea water. In: Riley JP, Skirrow G (eds) Chemical oceanography, 2nd ed. vol 2. Academic Press, London, pp 301–363
Wollast R, Duinker JC (1982) General methodology and sampling strategy for studies on the behavior of chemicals in estuaries. Thallasia Yugoslavica 18:471–491

The Importance of Chemical "Speciation" in Environmental Processes,
eds. M. Bernhard, F.E. Brinckman, and P.J. Sadler, pp. 301–335.
Dahlem Konferenzen 1986. Springer-Verlag Berlin, Heidelberg
© Dr. S. Bernhard, Dahlem Konferenzen

Chemical Species in Seawater and Marine Particulates

M.O. Andreae

Dept. of Oceanography
Florida State University
Tallahassee, FL 32306, USA

Abstract. In recent years analytical techniques have become available which make it possible to identify individual chemical species in seawater and marine particulates, or at least to obtain operational information on the structural environment of some elements. These methods include various extraction and chromatographic techniques, mass spectrometry, hydride generation, and electrochemical techniques.

The major processes affecting chemical species in seawater are solid/aqueous phase exchange, electron exchange (redox chemistry), proton exchange (acid-base chemistry), and ligand exchange (complex chemistry). For a few elements, mostly the nonmetallic and metalloid elements, the formation of covalent bonds is of importance in controlling species distribution in the marine environment. Kinetic factors are important in the persistence of nonequilibrium species found for a number of elements.

Introduction

Knowledge of the chemical species of the elements in seawater and in marine particulates is essential for our understanding of their biogeochemical behavior. While this has always been obvious for elements such as carbon and nitrogen, it has only recently been appreciated for metals and metalloid elements. Initially, the chemical species present in seawater were derived on the basis of chemical equilibrium models (discussed by Kester, this volume). The development of analytical methods which are able to differentiate between chemical species of an element (at least in a more or less operational manner) showed that many elements were present in nonequilibrium species distributions or in forms not represented in the thermodynamic models. This led in recent years to an emphasis on empirical rather than theoretical studies on chemical species in seawater. This paper is limited to observational aspects of chemical species in seawater and mentions modelling results only to contrast them with empirical data.

Chemical Determinants of Species Distribution

The processes which determine the species distribution of an element can be understood in the context of the fundamental interactions in the bonding environment around the atom: the formation of covalent bonds, the electron exchange in redox reactions, and various types of ligand exchanges (complex formation, acid-base chemistry, and surface interactions). Covalent bonding dominates the species distribution of the nonmetals and remains important for the metalloids. Coordination chemistry, on the other hand, determines the species distribution of the metallic elements. In this context we consider acid-base reactions, precipitation-dissolution, and complex formation all as aspects of coordination chemistry (Stumm and Morgan, 1981). Redox reactions lead to changes in the electron configuration of the elements, which is then reflected in both covalent and coordinative bonding characteristics. These concepts provide an organizing principle for the discussion of chemical species in the sea.

Different bonding interactions control the kinetics of species transformation reactions. Acid-base reactions (coordination change of the proton) are typically very fast and consequently equilibrium is approached rapidly (time scales of fractions of seconds). The acid-base aspects of chemical species distribution in seawater can therefore be most easily understood on the basis of equilibrium thermodynamic models. Coordination reactions involving large, organic ligands are considerably slower, and disequilibrium may persist over geochemically relevant times (hours to weeks). Reactions involving change of the covalent bonding structure tend to have much higher activation energies and consequently often cannot proceed at appreciable rates without catalysis (e.g., enzymes, light, or surfaces) at the temperatures found in marine systems. Species transformation kinetics also limit the types of analytical approaches we can use for species identification. The species has to be stable on the time scale characteristic of the analytical method. The kinetically most inert species can be isolated from the sample in pure form and analyzed, e.g., by X-ray crystallography. Labile species, on the other hand, can usually only be unequivocally identified by spectrographic methods, which rely on interactions at very short time scales. The problem here is that these analytical methods are usually not sensitive enough to be directly applied to seawater, so that preconcentration or model studies at elevated concentrations are necessary. This approach, however, brings with it the necessity to alter and disturb the systems to such an extent that the validity of the results for *in situ* conditions becomes questionable. The determination of species which are not stable enough to be isolated should ideally be done by *in situ* techniques. With very few exceptions (e.g., the hydrogen ion), concentrations in seawater are too low to make this possible with current analytical technology.

Analytical Approaches to the Determination of Chemical Species

Since most of the analytical techniques applicable to species determination in seawater are discussed in detail elsewhere in this volume as well as in recent reviews (Florence, 1982; Florence and Batley, 1980; Turner, 1984), I focus here on the application of these methods to marine systems rather than to descriptions of the analytical methodology. A summary of the best current information on element concentrations and our best estimate of the most important species is given in the Appendix.

In addition to the problems usually encountered in the determination of trace elements in seawater, especially the problems of sample contamination and/or analyte loss, the determination of chemical species adds the problems of species transformation during sampling and storage. No general rules can be given regarding how to handle these problems; sample collection and storage techniques must be validated for each individual species.

The first analytical step is usually the separation of dissolved and particulate phases. Conventionally, this is done with filters having pore sizes near 0.45 μm. It must be emphasized that this represents an arbitrary cutpoint in a continuum of particle sizes (Fig. 1). Some particle groups overlap this size cut between "dissolved" and "particulate" states, e.g., bacterioplankton and fine clay particles. The separation of true dissolved compounds from colloidal forms is possible using gel and membrane filtration and dialysis (e.g., Dawson and Duursma, 1974; Smith, 1976; Sugai and Healy, 1978). These physical separation techniques have recently been reviewed by de Mora and Harrison (1983).

← filterable →

← membrane filterable →

← dialyzable →

← in true solution →

Free metal ions	Inorganic ion pairs, inorganic complexes, organo-metallic compounds	Organic complexes, chelates	Metal species bound to high-molecular-weight organic material	Metal species in the form of highly dispersed colloids	Metal species sorbed on colloids	Precipitates, organic particles, remains of living organisms

Diameter range: ← 1 nm — 10 nm — 100 nm 0.45 µm

Examples:

Cu^{3+}aq.	$Cu_2(OH)_2^{2+}$	Me-SR	Me-lipids	FeOOH	$Me_x(OH)_y$	Metals in:
Fe^{3+}aq.	$Pb(CO_3)^0$	Me-OOCR	Me-humic-acid polymers	$Fe(OH)_3$	$MeCO_3$, MeS etc. on clays	Zooplankton
Pb^{2+}aq.	$CuCO_3$	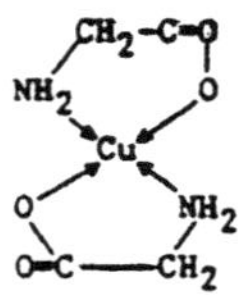	"lakes"	Mn(IV) oxides	FeOOH or Mn(IV) on oxides	Phytoplankton
	AgSH		"Gelbstoffe"	$Mn_7O_{13} \cdot 5H_2O$		Bacterioplankton
	$CdCl^+$		Me-polysaccharides	$Na_4Mn_{14}O_{27}$		$CaCO_3$
	$CoOH^+$			Ag_2S		SiO_2
	$Zn(OH)_3^-$					Fe_3O_4
	$(CH_3)_3Ge^+$					Organically bound metals
	$(CH_3)_2AsOOH$					

Fig. 1. Forms of occurrence of metal species (adapted from Stumm and Morgan, 1981)

The ultimate objective in species analysis is always the identification of specific molecular entities. Classically, spectroscopy in one of its many forms (e.g., IR, Raman, UV-VIS, NMR, mass spectroscopy) has been used for this purpose. In marine systems the species concentrations are too low to be directly detectable with spectroscopic means. The necessity to include enrichment and separation steps interferes significantly with the *in situ* species distribution, so that the results of essentially all species determinations in seawater have a more or less operational validity. An exception is provided in the analysis of marine particulate materials, where phases and oxidation states have been identified by the application of spectroscopic techniques. An interesting recent example is the identification of the Cu(I) oxidation state in hydrothermal manganese precipitates by electron spectroscopy (ESCA) and Extended X-ray Absorption Fine Structure (EXAFS) (Stouff and Boulegue, 1983). The application of electron spectroscopy to the study of marine particles has been considered a promising tool for quite some time, but little practical progress has been made.

Methods of optical spectrometry are used in classical procedures for the determination of several seawater ions, e.g., Mn^{2+}, Fe^{2+}, sulfur species, and several nutrient species, e.g., NH_4^+, NO_2^-, and NO_3^- (Grasshoff et al., 1983; Jacobs and Emerson, 1982, and references therein). It should be emphasized that these ions are measured not directly, but rather after reaction with various reagents to produce strongly absorbing compounds. The species selectivity of these determinations is dependent upon the selectivity of the color-forming reactions. Iodate has been measured in seawater by iodometric titration in the cuvette of a spectrophotometer (Wong and Brewer, 1974). Due to the relatively long time characteristic of spectrophotometric measurements (minutes to hours) and the need to add various reagents, these measurements are only useful for distinguishing between different redox states of an element – they cannot identify individual molecular species.

Mass spectroscopy has not yet been applied to the identification of dissolved species in seawater, most likely due to the problem of sample preparation. Its use has been suggested for the determination of butyltin species by GC/MS, but its applicability will be restricted to extremely polluted systems due to the high limits of detection (Meinema et al., 1978). On the other hand, MS is a very powerful tool for the identification of organometallic compounds in solid phases, e.g., the determination of copper and nickel porphyrins in marine sediments (Palmer and Baker, 1978) and of arsenobetaine in marine biota (Luten et al., 1983).

Precipitation and absorption techniques which make use of species-selective interactions between solid phases and dissolved species have been used for the determination of the dissolved species of a number of elements in seawater. These interactions can occur either during the process of precipitation of a solid (co-precipitation) or through adsorption to a preexisting surface. Selective co-precipitation was used for the determination of chromium species in seawater by Murray et al. (1983). Total chromium [Cr(VI) + Cr(III) + $Cr_{particulate}$] was precipitated with Fe(II)-hydroxide, and reduced chromium [Cr(III) + Cr_p] was co-precipitated with Fe(III)-hydroxide. Cr_p was determined by filtration. Nakayama et al. (1981) have used a similar co-precipitation method but added a persulfate oxidation step to determine organically bound Cr. Precipitation of low-solubility compounds has often been used to determine the formation constants of metal complexes (e.g., Cosovic et al., 1982, for Co complexes; Washburne, 1981, for Pb complexes). These studies involve the use of high metal concentrations in artificial systems, however, and are more related to the modelling approach discussed by Kester (this volume) rather than to the observational approach discussed here. The adsorption of Cu^{2+} to an MnO_2 surface competes with organic ligands in solution for free cupric ion, and the distribution between the two phases has been used by Van den Berg (1982 a, 1982 b) to characterize organometallic complexation of Cu^{2+} in seawater.

Adsorption of organic complexes of a number of metals to nonpolar substrates, e.g., the Amberlite XAD resins, or to C_{18}-bonded silica (Sep-Pak), has been used to separate the organic fractions of a number of metals from seawater. After elution with organic solvents, the organic fraction can then be either analyzed for total metal content or further fractionated, e.g., by HPLC. Lyons et al. (1979) separated iron from sediment pore waters into inorganic and organic fractions using adsorption on Amberlite XAD-2. It is possible, however, that colloidal materials are also adsorbed to the macroreticular Amberlite resin and thus lead to false identification of inorganic colloids as organic complexes. Kremling et al. (1983) collected the organic fraction of dissolved copper in seawater using a silanized, porous glass substrate. Mills et al. (1982) and Mackey (1983) collected organically complexed metals on Sep-Pak cartridges. The opposite approach, adsorption to chelating resins, has been used to collect selectively the component of the metal present as free ions or as kinetically labile complexes (Bruland et al., 1979; Figura and McDuffie, 1980).

The application of reversed-phase HPLC on C_{18}-bonded silica columns is a logical extension of the collection of organic complexes by cartridges containing the same material (Sep-Pak). In its most elegant form, the

HPLC unit is directly interfaced with an element-specific detector, e.g., an atomic absorption or emission system (Jewett and Brinckman, 1983). Using an atomic fluorescence detector, Mackey (1983) demonstrated evidence for the existence of organic complexes of Mg, Cu, Zn, Fe, Ni, and Mn in seawater. Mills et al. (1982) used a simplified version of this concept by eluting Sep-Pak cartridges sequentially with eluent mixtures of decreasing polarity.

Liquid-liquid extraction techniques have classically been used for the separation of redox states of a number of elements. Gohda (1975) investigated the species distribution of arsenic and antimony in seawater by extraction of the diethyldithiocarbamate complexes of the trivalent species into chloroform. Bertrand and Choppin (1982) developed an extraction scheme for the separation of the redox states of plutonium in seawater, which was used by Roberts (1984) for the investigation of the redox kinetics of Pu. Measures et al. (1980) separated Se(IV) and Se(VI) by the extraction of the tetravalent species as the complex with 4-nitro-o-phenylenediamine into toluene, followed by gas chromatographic separation of the complex and detection with an electron capture detector.

The hydride generation methods made use of the relative chemical stability and the volatility of the hydrogen compounds of a number of metals and metalloids (Ge, Sn, Pb, As, Sb, Se, Te). The use of this method for the determination of metal and metalloid species in seawater has been reviewed recently (Andreae, 1983c). The procedure consists of a volatilization step, where the elements are reduced with sodium borohydride to the hydrides. Organometallic species (those with carbon-metal bonds) are reduced to the corresponding organically substituted hydrides. The conditions under which the reduction is performed afford redox-state selectivity, and the inorganic and organic hydrides are separated by gas chromatography. Detection is accomplished by a variety of element-specific detectors, e.g., electron capture (organoarsenicals: Andreae, 1977), atomic absorption (As: Andreae, 1977; Se: Cutter, 1978; Ge: Andreae and Froelich, 1981; Hambrick et al., 1984; Sb: Andreae et al., 1981; Sn: Andreae, 1983c; Hodge et al, 1979, flame emission (Sn: Andreae and Byrd, 1984; Braman and Tompkins, 1979), and dc-arc emission (Braman and Foreback, 1973; Feldman and Batistoni, 1977).

The complexation of Zn, Cd, Pb, and Bi with marine humic acid has been investigated using changes in electrophoretic mobility as a function of the addition of humic acids to seawater containing radioactive isotopes of these elements (Musani et al., 1980, 1981). Hunter and Liss (1979) determined the electrokinetic charge of marine particles using measurements of electrophoretic mobility.

A number of electrochemical methods have received widespread use in species distribution studies of seawater. This subject has recently been reviewed (Nürnberg and Valenta, 1983) and is discussed elsewhere by Lund (this volume). Three modes of electrochemical analysis have been most commonly applied: anodic stripping voltammetry (ASV), polarography, and ion-specific electrodes. Of these methods, only ASV has sufficient sensitivity to detect elements at their natural concentrations in seawater. The ion-specific electrode has been used for the determination of reduced inorganic sulfur species in marine pore waters (Boulegue et al., 1982). Its use for the determination of trace-metal species has been limited to free ion detection in complexometric titration studies at elevated metal levels (e.g., Ramamoorthy and Kushner, 1975). The most commonly used electrochemical technique is ASV, which is used to measure the concentration of kinetically labile, "electroactive" species in seawater. Variation in metal and ligand concentration is used to characterize the complexation constants and capacities of organic ligands in seawater (e.g., Figura and McDuffie, 1980; Fisher and Fabris, 1982; Piotrowicz et al., 1982, 1983; Plavsic et al., 1982; Sugai and Healy, 1978). The determination of free ion concentrations by ASV, however, suffers from a number of problems (Lund, this volume; also see Turner, 1984, and references therein): a) Labile inorganic and organic complexes contribute to the electrochemical response; b) depending on the deposition potential applied, irreversible reduction of some complexes can occur, which will result in an increased response; c) natural organic matter adsorbed on the electrode causes interference in the response; and d) slow kinetics of complex formation following the addition of metal in complexometric titrations can make the interpretation of the titration curves problematic. The variation of the ASV response as a function of the applied deposition potential results in a "pseudo-polarogram" (Nürnberg and Valenta, 1983, and references therein) from which the complexation constants of certain labile complexes can be deduced. This approach has been used to study the inorganic complexes of Pb(II) and Cd(II) in seawater (Sipos et al., 1980).

For some elements, in particular Cu^{2+}, it has been shown that the free ionic species (present in the hydrated state) is the only form toxic to marine biota (Gillespie and Vaccaro, 1978; Sunda and Ferguson, 1983). This phenomenon has been used for free ion concentration measurements using bioassay techniques. Using the uptake of Am and Pu by a marine diatom as an indicator of free ion concentration, Fisher et al. (1983) have suggested that little or no complexation of these elements with organic ligands occurs in seawater.

The foregoing discussion shows that operational procedures dominate the analytical approaches to species determination in seawater. For organic complexes this situation appears difficult to improve, mostly due to the fact that the ligands are not present as a few distinct compounds but rather as a continuum of structures. The kinetic stability of some of the organometallic compounds (i.e., compounds with covalent metal-carbon bonds) should allow their characterization at the molecular level, however. The efforts of trace species analysts in seawater will have to be directed towards this goal.

Chemical Species in Seawater: Observational Aspects

The focus of this paper on observational aspects makes it necessary to leave out important information which relates more closely to the modelling aspect. The review by Mantoura (1981) and the paper by Kester (this volume) should complement this section.

Alkali and Alkaline-Earth Elements

This group contains the major seawater cations which determine the fundamental electrolyte characteristics of seawater (together with the corresponding major anions). The high concentrations of these cations relative to trace-metal ions would suggest that they compete with the trace metals for ligand sites, even if they have much lower complex stabilities. The outer-sphere complexes (ion pairs) which are thought to be present in seawater are not amenable to direct analysis, so that their presence has to be inferred on the basis of the observed activity coefficients (with the exception of $MgSO_4^0$, which has been observed by sound absorption measurements (Fisher, 1967). The evidence for the formation of organic complexes of these metals is either negative or rather circumstantial (cf. Mantoura, 1981). It should be remembered, however, that the concentrations of Ca and Mg are high enough that even a relatively small fraction of the total concentration of these metals could bind a major fraction of the organic ligands present in seawater. The only direct measurement of possible organic complexes of a group I or II element in seawater has been made by Mackey (1983) who observed Mg by an element-specific detector in the eluate from an HPLC system. Within algal cells, Mg is present as a chelate in chlorophyll, but the fate of this complex during the evolution of marine particulates from phytoplankton through fecal pellets to the sediments has not yet been studied.

Chromium and Manganese

The chemical behavior of chromium and manganese in seawater is largely determined by the properties of the different redox states of these elements. Both elements have one highly soluble redox state [Cr(VI) and Mn(II)] and one very insoluble state [Cr(III) and Mn(IV)]. Redox transformations between these oxidation states determine the distribution of these elements between the dissolved and particulate fractions in seawater.

At this time there is no evidence for the existence of significant organic complexation of Mn (Mantoura, 1981, and references therein), with the exception of the HPLC data of Mackey (1983) which are not interpretable in a quantitative fashion. The detection limit stated by Mackey (10 $\mu g\ L^{-1}$) is much higher than the total Mn concentration found by other authors (e.g., Landing and Bruland, 1980), which makes the source of the Mn observed by Mackey questionable. Manganese chemistry is thus largely restricted to the inorganic species in the three redox states, Mn(II), Mn(IV), and possibly Mn(III). These oxidation states have been identified in seawater not on the basis of direct measurement, but rather on the basis of the inference that dissolved Mn is Mn(II) and that particulate Mn is either Mn(III) or Mn(IV). Only in estuarine environments and in anoxic basins has Mn(II) been determined directly by a spectrophotometric method (e.g., Jacobs and Emerson, 1982). The high solubility of Mn(II) and the low solubility of Mn(IV) lead to the "classical" cycling of Mn between oxidizing and reducing environments, characterized by the precipitation of Mn-oxides in the oxic zone (and consequently very low Mn concentrations there) and the dissolution of these oxides in the anoxic zone, leading to high concentrations of Mn(II) in this zone. Diffusion of dissolved Mn(II) into the oxic zone from the underlying anoxic waters closes the cycle (Emerson et al., 1979, and references therein). This process occurs both in anoxic basins and within sedimentary pore waters. The solubilization of Mn by reduction of particulate Mn(IV)-oxides in shelf sediments and within the oxygen minimum zone in the marine water column has been suggested as the mechanism leading to the Mn maxima observed in near-surface waters and in the oxygen minimum zone (Klinkhammer and Bender, 1980; Landing and Bruland, 1980; Murray et al., 1983). Klinkhammer and Bender (1980) suggested that the maximum of Mn concentration in the oxygen minimum zone was due to the presence of the Mn(III) solid phase, hausmannite, but this hypothesis was not supported by solid-phase characterization or by the solid/solution partitioning found there by Murray et al. (1983).

The oxidation and reduction reactions between Mn(II), Mn(III), and Mn(IV) are subject to microbial and photochemical catalysis. Emerson et al. (1982) have suggested that the oxidation of Mn(II) is bacterially catalyzed at the oxic/anoxic interface. Sunda et al. (1983) have shown that the reduction of MnO_2 to Mn(II) can occur even in the presence of dissolved oxygen in marine surface waters. They observed some reduction of MnO_2 by humic acids even in the dark, but a substantial increase in the reduction rate was produced by illumination with sunlight. Stone (1983) determined the rate of MnO_2 reduction with a large variety of organic compounds.

In contrast to manganese, chromium is solubilized by oxidation and precipitated by reduction. Consequently, Cr(VI) shows a pronounced minimum and Cr(III) a maximum in the oxygen minimum waters of the eastern tropical Pacific (Fig. 2), where an Mn maximum had been observed (Murray et al., 1983). In Saanich Inlet, the reduction of Cr(VI) to Cr(III) at the O_2/H_2S interface leads to a steep decrease of Cr(VI) from the oxic into the anoxic zone, accompanied by an increase in Cr(III) and particulate Cr (presumably also in the +3 oxidation state), but with only a minor change in total Cr throughout the water column (Emerson et al., 1979). Nakayama et al. (1981) have presented evidence that organically bound Cr, probably organic complexes of Cr(III), represents the most abundant fraction of dissolved chromium in seawater. Their conclusion, which is based on persulfate digestion and co-precipitation experiments, should be verified by alternative methods.

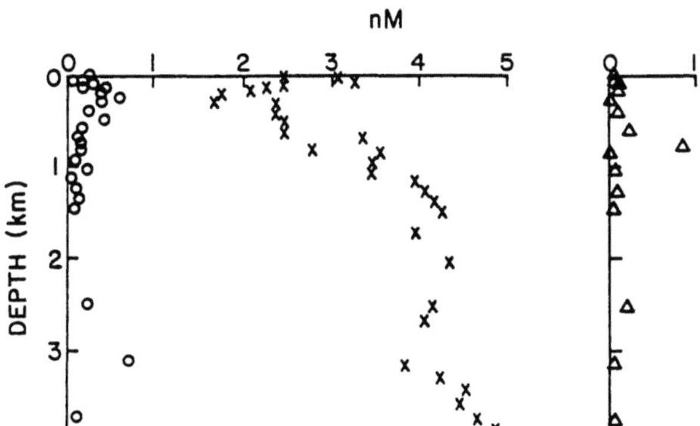

Fig. 2. Chromium species distribution at Station TT145/6 off Baja California (adapted from Murray et al., 1983). Cr(III): O; Cr(VI): X; particulate chromium: Δ

Iron, Cobalt, and Nickel

A large body of information exists on the "organic interactions" of iron in natural waters (reviewed by Mantoura, 1981). Unfortunately, most of this work is either so operational that its interpretation is very questionable, or is conducted only in freshwater systems and due to the large differences in iron and ligand concentrations is not applicable to seawater. At this time few reliable data on the concentration of iron in seawater exist (Gordon et al., 1982; Landing and Bruland, 1981), and its chemical species remains unknown. It must be assumed that iron is present predominantly in the thermodynamically stable +3 oxidation state, since the oxidation rate of Fe(II) in seawater is rapid (Roekens and Van Grieken, 1983). There is some evidence for the existence of organically bound iron in seawater (Sugimura et al., 1978 b) and in sediment pore fluids (Lyons et al., 1979) based on adsorption to Amberlite XAD-2, but Mantoura (1981) has suggested that the uptake of Fe by the macroreticular resin was due to the trapping of colloidal Fe particles.

In anoxic basins and sediment pore waters, Fe(II) is formed and can reach high concentrations (micromolar) due to the solubility of the +2 oxidation state. Fe(II) can therefore be directly determined by spectrophotometry in these waters (Emerson et al., 1979). The solubility of iron in anoxic waters appears to be limited by the formation of iron sulfides, in particular greigite (Boulegue et al., 1982; Jacobs and Emerson, 1982). In the water column above the Peru shelf, Fe(II) was detected near the bottom (due to fluxes from the sediments), and also near the sea surface, where it is probably formed by photochemical processes (Kester, 1984, personal communication).

Cobalt is present in biological systems as part of the vitamin B_{12} molecule, and it may occur as this complex in seawater at low concentrations (Williams, 1975). The data of Sugimura et al. (1978 a), who observed a large fraction of the Co in seawater being adsorbed to Amberlite XAD resin and concluded from this a high degree of complexation, are in contrast to the work of Bruland (1983), who found Chelex to extract Co quantitatively from seawater. The existence of significant amounts of organic complexation for this element does not appear very likely on the basis of determined stability constants for such complexes. Cosovic et al. (1982) have determined the species distribution of cobalt in seawater by model experiments using polarographic and precipitation techniques and found free ion, $CoSo_4^0$, and $CoCl^+$ to be dominant.

Little is known about the chemical species of nickel in seawater. Sugimura et al. (1978a) and Morris (1974) suggested the existence of organic complexes, but the total concentrations observed by these workers are so much higher than those of Bruland (1983, and references therein) that sample contamination must be suspected. On the other hand, Bruland et al. (1979) found that some 10–20% of dissolved Ni was not retained by a Chelex column and therefore may be in organically complexed form.

Copper, Zinc, and Cadmium

The organic complexation of copper in seawater has been the subject of intense study for quite some time (for a review of the older literature the reader is referred to Mantoura, 1981). This element best exemplifies the problems encountered with operational approaches to species determination. Practically all of the operational methods known have been applied to this element: anodic stripping, Amberlite adsorption, Chelex extraction, MnO_2 adsorption, EDTA ligand exchange, reversed-phase HPLC, and bioassays. All of these methods suggest that there is a significant fraction of Cu in seawater which is "organically bound". There exists, however, wide disagreement about the percentage of copper present in the form of organic complexes. This disagreement is in part due to the fact that different researchers have applied different methods to samples of widely diverse origin. Much of the older literature must be considered useless due to sample contamination. A comparison of the results of recent work on determination of Cu species by a variety of techniques is presented in Table 1. It is apparent that good agreement exists about the magnitude of the complex stability constant for Cu with dissolved organic matter (DOM), while a broad range in the percentage of Cu complexed has been reported. This may be due at least in part to differences in DOM concentrations in the samples studied by the various authors. Comparison studies appear essential in this field. The interpretation of all these data is complicated by the fact that the Cu-ligand interactions appear to represent a continuum both in terms of binding constants and in terms of ligand-exchange rates, rather than discrete properties of individual compounds. Different analytical probes have implicated different time constants for the interactions to which they respond, and thus will give different results (cf. Figura and McDuffie, 1980). The work by Piotrowicz et al. (1983) and by Kremling et al. (1983) shows that temporal variations in the degree of apparent Cu- and Zn-binding occur over time spans of days or even hours. Longer-term temporal variations have been shown by Hanson et al. (1986) by tracking the apparent Cu complexation in warm core rings in the North Atlantic.

Table 1. Summary of some recent investigations of dissolved Cu species in coastal and oceanic waters (from Hanson et al., 1984)

Region	Dissolved Cu (nmol kg^{-1})			% Cu-organic			Complexing organic ligands (nmol kg^{-1})		Conditional stability constants			Methodology[a]	Reference
	$\bar{x}$	Range	n	$\bar{x}$	Range	n	C_{L1}	C_{L2}	pH	log K_{CuL}	log K_{CuL2}		
N.W. Pacific Ocean	–[b]	15–16	8	–[b]	83–90	8	ND			ND		XAD, AAS	Sugimura et al. (1978)
N.W. Pacific Ocean	24	–	1	97[c]			21	7	8.1	11.8	13.8	EDTA-LER, XAD, AAS	Hirose et al. (1982)
Irish Sea		ND			ND		ND		8.0	9.37 8.45	(I=0.02) (I=0.65)	XAD, GFC	Mantoura et al. (1978)
Gulf of Mexico	4.5	2.4–8.8	6		82–>99[c]		15–130	5–20	8.1	8.9–9.8	11–>12	Bioassay	Sunda and Ferguson (1983)
Gulf of Mexico	4.5	2.4–8.8	6		83–>90		ND			ND		pH, ASV	Sunda and Ferguson (1983) (Data of Piotrowicz)
Atlantic Ocean		ND			ND		110		8.1	9.86±0.23		MnO_2-A	Van den Berg (1982)
Irish Sea		ND			ND		174		8.01	9.84±0.23		MnO_2-A	Van den Berg (1982)
Middle Atlantic Bight	4	0.8–9	24	27	9–60	24	ND			ND		C_{18}-RPLC, GFAAS	Hanson and Quinn (1983)
N.W. Atlantic Ocean Warm Core Ring							4.7		8.2	9.87±0.13		C_{18}-RPLC, Cu titration, GFAAS	Hanson et al. (unpublished)
Euphotic waters							8.3		8.2	9.95±0.04		ASV Cu titration	Hanson et al. (unpublished)
	1.41	1.3–1.8	11	–	61–>90	11	–			–		pH, ASV	
	1.05	0.83–1.34	26	57	38–92	26	–			–		C_{18}-RPLC, GFAAS	

[a] AAS: atomic absorption spectrophotometry; GFAAS: graphite furnace AAS; ASV: anodic stripping voltammetry; C_{18}-RPLC: C_{18} reversed-phase liquid chromatography; pH: pH adjustment; XAD: Amberlite XAD resin chromatography; LER: ligand-exchange reaction; MnO_2-A: MnO_2 adsorption equilibration; GFC: gel filtration chromatography.
[b] Individual sample results not reported.
[c] %Cu-organic based on model calculations using C_L and K values determined.

Despite all the work on Cu-organic binding, the ligands are characterized only in a very indirect way. Most studies have not even attempted to identify the chemical nature of the ligands but simply treat them as a "black box." Probably the most significant recent progress has been made by the groups of Piotrowicz and Harvey, who have proposed a model structure of marine humic and fulvic acids (Harvey et al., 1983) and who investigated the binding properties of Cu, Zn, and Cd with marine humic and fulvic acids which were characterized by NMR (Piotrowicz et al., 1984). They found significant differences in the binding properties between the different organic fractions and for the different metals. While the identification of the ligands as individual chemical compounds may be impossible due to the fact that the mechanisms of their formation lead to a continuum of structures which does not allow any individual compounds to be isolated, it appears imperative that more effort be devoted towards the characterization of the binding environment in metallo-organic complexes at the molecular level.

In addition to the +2 oxidation state, which up until recently has been considered the only one present in oxygenated seawater, there may be significant amounts of Cu(I) present in both oxic and anoxic waters. Moffet and Zika (1983) have studied the redox kinetics of Cu(I) in seawater and have proposed that Cu(II) may be reduced to Cu(I) by the photocatalytic reduction of Cu(II) with organic matter in surface seawater. Their work suggests that significant amounts of Cu(I) may be present at steady state in surface seawater. In anoxic waters Boulegue et al. (1982) have suggested that Cu is present as Cu(I)-organosulfur complexes. The observed concentrations of Cu and sulfur species in anoxic pore waters and anoxic basins are close to the solubilities of either covellite (CuS) or chalcolite (Cu_2S) (Jacobs and Emerson, 1982). Stouff and Boulegue (1983) have applied ESCA and EXAFS to copper-rich particles from the vicinity of submarine hydrothermal vents and have concluded that the copper is mostly present as Cu(I) in these materials.

The availability and toxicity of trace metals to phytoplankton appears to be influenced by the presence of "marine humus," supporting the validity of the organic complex model for Cu (Ortner et al., 1983). Since it has been shown that complexation of Cu, Zn, and Cd can be effected by materials excreted from phytoplankton (e.g., Fisher and Fabris, 1982), it is possible that the complexation of trace metals represents an evolutionary strategy for controlling trace-metal availability in the aquatic environment.

Davis and Leckie (1978) have studied the effect of organic ligands adsorbed to hydrous oxides on the uptake of Cu(II) and Ag(I) by these oxides.

They found an important effect on the solid/liquid distribution of these ions as a consequence of variations in the presence of organic ligands. These phenomena remain to be investigated in the natural seawater system. The fact that most of the older literature on Cd and Zn species or forms present in seawater gives concentrations which can only be due to sample contamination severely limits our understanding of the species distribution of these elements. Most of the evidence available at this time suggests that zinc and cadmium are not significantly complexed by organic ligands at the DOM concentrations typical for seawater (Bruland et al., 1979; Mantoura, 1981; Musani et al., 1980, 1981). However, the data of Piotrowicz et al. (1983, 1984) suggest that there may be some binding capacity present in marine humic and fulvic acid and that this complexation capacity may be subject to rapid temporal changes. The importance of this complexation at ligand concentrations typical for seawater remains to be established.

Mercury

Besides forming a variety of inorganic complexes, mercury has very high stability constants with organic ligands and can form stable bonds to carbon and, therefore, true organometallic compounds. The presence of organic Hg-complexes in seawater is supported both by the high complexing constants determined for Hg with fulvic and humic acids (Andren and Harriss, 1975; Kerndorff and Schnitzer, 1980; Kritsotakis et al., 1979; Lindberg et al., 1975) and by operational determinations in seawater. Andren and Harriss (1975) showed a large fraction of the mercury in seawater to be associated with intermediate molecular weights in ultrafiltration experiments, and other investigators found that a significant fraction of mercury was released by UV irradiation (Fitzgerald, 1975; Jewett et al., 1975). Complexation of Hg with organic materials appears to be slow and essentially irreversible (Lindberg et al., 1975; Schnitzer and Kahn, 1972). Methylated mercury compounds, dimethylmercury and methylmercury cation, are the dominant mercury species in fish (National Academy of Sciences, 1978; Nriagu, 1979, and references therein). It is generally assumed that fish take up methylmercury either directly from seawater or via their food intake (Bernhard and Andreae, 1984, and references therein), but the concentrations of methylmercury in open ocean seawater are still uncertain. Measurements of methylmercury concentrations in waters near Japan have given higher values than expected: near 1 pM (picomole per liter), or about 1%–10% of total Hg (Fujita and Iwashima, 1981; Yamamoto et al., 1983). Since the species identification in these studies was based on complex extraction schemes, some uncertainty also persists about the possibility of other organic mercury compounds being included with the methylmercury determination. In model ecosystems, Topping and Davies (1981) have observed the formation of methylmercury in the marine water column, but the observed methylmercury was measured in the settled particulates only. Knauer and Martin (1972) have measured methylmercury in natural marine phytoplankton, but the source of these compounds still remains unclear: Was it methylation by phytoplankton, by bacterioplankton, or by abiotic reactions?

Germanium, Tin, and Lead

The stability of the metal-carbon bond in aqueous solution decreases from germanium to lead. Consequently, while the methylated species of germanium are the most abundant fraction of germanium in seawater (Hambrick et al., 1984), methyllead compounds are only found in the direct vicinity of anthropogenic sources.

Inorganic germanium in seawater, probably present as $Ge(OH)_4^0$, is present in an essentially constant ratio to dissolved silica (Froelich and Andreae, 1981). On the other hand, the methylated forms of germanium are present in seawater at a constant ratio to salinity, at levels near 300 pM for monomethylgermanium and 120 pM for dimethylgermanium. These compounds also show completely conservative behavior in estuaries (Andreae and Froelich, 1984) (Fig. 3). Their ultimate source remains unknown at this time; preliminary data point towards biomethylation by bacteria and/or fungi in the marine environment. These compounds are extremely stable against chemical attack by acids, comparable to the hydrocarbons. Their persistence in the marine environment is most likely due to their chemical stability as well as to the fact that they are present in seawater at concentration levels at which they are not an effective energy source for microorganisms.

Methylated tin species have been observed in nearshore polluted waters (Brinckman et al., 1983; Byrd and Andreae, 1982) but are at or below the limits for detection (<0.2 pM) in the open ocean. Butyltin compounds were also observed in the marine environment (Seidel et al., 1980). Both methyl- and butyltin compounds are produced industrially, and their use in plastics and antifouling agents is likely to be a major source to the coastal marine environment. The methylation of inorganic tin by estuarine microorganisms, albeit at extremely elevated concentrations of inorganic tin (millimolar), has been observed by Hallas et al. (1982). Manders et al. (1984) have

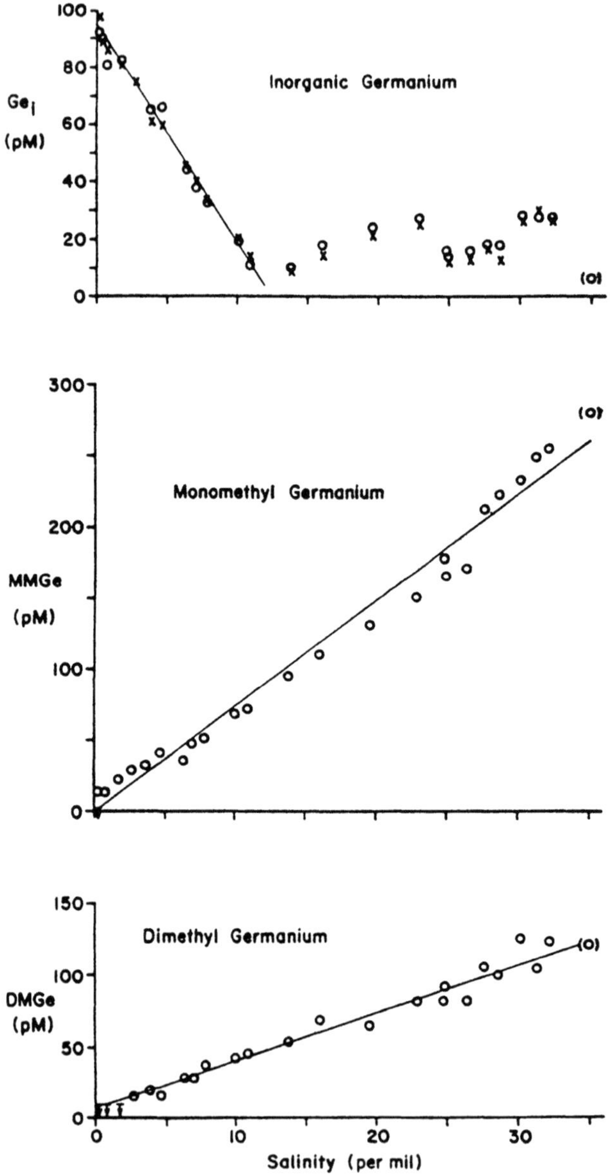

Fig. 3. Germanium species as a function of salinity in the Charlotte Harbor Estuary, Florida (adapted from Froelich et al., 1984)

suggested the abiotic reaction between methyliodide and tin(II)sulfide as a source of methyltin species.

The redox state and coordination environment of inorganic tin has not yet been experimentally determined in seawater. On the basis of thermodynamic calculations and some model experiments at elevated concentrations, the species $SnO(OH)_3^-$ has been predicted (Macci and Pettine, 1980), but the uncharged species $Sn(OH)_4^0$ would also be possible on the basis of other model calculations (Baes and Mesmer, 1976).

Experimental investigations with lead at elevated concentrations in seawater point towards the carbonato- and chloro-complexes of Pb(II) as the dominant inorganic species (Byrne, 1981; Nürnberg and Valenta, 1983; Washburne, 1981). There are a number of studies on the stability constants of lead-organic complexes (cf. Mantoura, 1981) but no evidence for the existence of organic complexes of Pb in seawater. There is still a controversy about the biological methylation of lead in the environment (*viz.*, Craig, 1980). While some laboratory studies in both biological systems (Huber et al., 1978; Wong et al., 1975) and abiotic systems (Ahmad et al., 1980) have shown the production of methyllead, other studies did not observe this process under similar conditions (Reisinger et al., 1981; Snyder and Bentz, 1982). On the other hand, the possibility of biomethylation in actual aquatic systems may be a moot question, since the demethylation reactions for methyllead compounds are in any case fast enough to result in negligible concentrations at steady state (De Jonghe, 1983; Huber et al., 1978). There is no analytical evidence for the existence of methylleads in seawater at this time.

Arsenic and Antimony

The first reliable data on arsenic species in the marine environment were presented by Braman and Foreback (1973). They demonstrated the presence of As(III), As(V), and methylated arsenic species in seawater. They obtained their data using selective borohydride reduction to the arsines, a method which has been used with some modifications in later, more comprehensive studies (Andreae, 1977, 1978; Waslenchuk, 1978). Since this technique is essentially a derivatization method, it cannot rigorously identify the actual species present in solution before derivatization, e.g., the species determined as As(V) could be present as a labile organic complex of arsenate as well as free hydrogen-arsenate ion, or the species identified as dimethylarsinic acid could in fact be an unknown compound with two methyl groups on arsenic, which is also reduced to dimethylarsine with sodium borohydride. While no evidence exists for the presence of such an organoarsenic compound, Waslenchuk (1977) has presented some evidence for an organic association of As(V) in estuaries from the southeastern U.S. The hydride method therefore characterizes the redox state and number of bonds to carbon rather than the actual solution species. The data on arsenic species (and that of the other hydride elements) must be viewed with this caveat.

Measurements on seawater show arsenate to be the most abundant species in most environments, especially the deep ocean. As(III) reaches maxima near the surface, where it is probably produced by phytoplankton, and in anoxic basins (Andreae, 1977, 1978; Andreae and Froelich, 1984). Only rarely does As(III) reach levels above 10%–20% of the total arsenic in surface waters; the highest concentrations were observed in the waters of the subtropical gyres. In some samples from the Sargasso Sea, As(III) accounted for over 95% of the total inorganic arsenic (Byrd and Andreae, unpublished data). This is most likely due to rapid internal cycling of the phosphate pool in these waters and, consequently, to a high rate of arsenate uptake and excretion after reduction by phytoplankton. The biogeochemical cycling of As has recently been reviewed (Andreae, 1983b, 1985). Phytoplankton take up arsenate in surface water together with phosphate and excrete arsenite and the methylated species, methanearsonate and dimethylarsinate (Andreae and Klumpp, 1979). This leads to a nutrient-like distribution of As(V) in the marine water column and to the presence of typically about 10% of the total arsenic as methylated species in the marine euphotic zone (Fig. 4). In marine particulates, a substantial if not dominant fraction of arsenic is in various organic forms. This is shown in Table 2 for the example of marine phytoplankton collected off the coast of California.

The species distribution of antimony resembles that of arsenic in some respects: Sb(V) is quantitatively the most important species, and Sb(III) and methylated antimony species are present at levels typically about 10% of the total Sb (Andreae, 1983c; Andreae and Froelich, 1984; Bertine and Lee, 1983; Gohda, 1975). Again the highest concentrations of the reduced species are typically found near the sea surface. In the case of antimony, however, the monomethylated species, methanestibonic acid, is more abundant than dimethylstibinic acid, whereas in the case of arsenic the dimethyl species predominates. The methylantimony species are found throughout the water column, whereas the methylarsenic species are usually below the limit of detection at depths greater than 200 m. This may be due to the lower concentrations of the methylantimony species which may make them unavailable to biological demethylation, similar to the methylgermanium

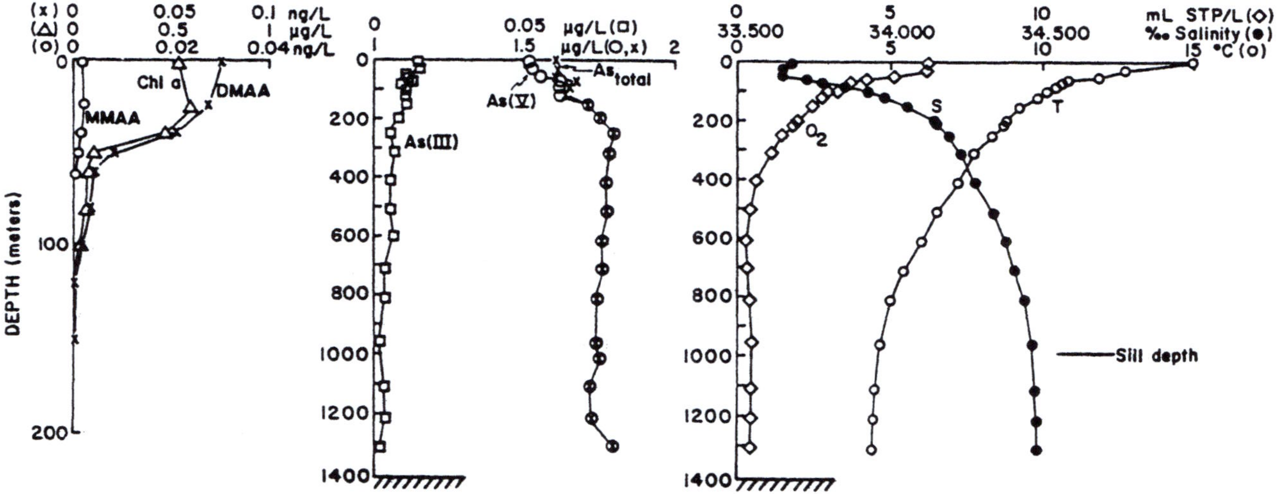

Fig. 4. Arsenic species distribution at Station AS3-2 in the Santa Catalina Basin off California (33°23.9'N, 118°49.8'W) (from Andreae, 1983 b)

Table 2. Species distribution of arsenic and antimony in phytoplankton from the eastern North Pacific (ng g^{-1} dry weight) (from Andreae and Froelich, 1984)

Sample	As(III)	As(V)	MeAs[a]	Me_2As[b]	As_t[c]	% org	Sb(III)	Sb(V)	Sb_t[d]
001	63	30	5.9	490	2010	95	43	11	54
018	26	94	20	1290	3520	96	ND	220	220
033	27	429	27	740	3400	87	18	46	64

[a] As in the form of methylarsonic acid, $CH_3AsO(OH)_2$
[b] As in the form of dimethylarsinic acid, $(CH_3)_2AsO(OH)$
[c] Total arsenic by $MgO/Mg(NO_3)_2$-dry ashing
[d] Total soluble antimony (in 2 N HCl)

species. Methylantimony species do not appear to be produced by marine algae; they are not present in phytoplankton (Table 2) and have not been observed in marine macroalgae (Kantin, 1983).

Selenium and Tellurium

Sugimura et al. (1977) observed the presence of both Se(IV) and Se(VI) in seawater. The species distribution of Se in seawater has been systematically studied by two groups, Cutter and co-workers (1978, and personal communication, 1984) and Measures and co-workers (1980). Both groups agree on the Se species concentrations in deep water: 800–900 pM Se(IV) and ca. 1400 pM Se(VI). Since only Se(VI) should be thermodynamically stable under oxic conditions, this represents a kinetically controlled redox species distribution, as has already been observed for other elements (e.g., As). In surface waters the total Se concentrations observed by both groups are also comparable, but the work by Cutter and Bruland (1984) suggests that most of the Se is in the form of organic selenium compounds, e.g., Se-amino acids, while Measures et al. (1980) argue that most of the surface water Se is in the form of Se(VI). Since both groups use operational definitions of the critical species [photoreduction of Se(VI) in Measures' work and persulfate reduction of organic Se in Cutter's method], it is essential that these methods be compared. The distribution of the Se species in the oceans appears to be controlled by uptake of Se(VI) and possibly of some Se(VI) by plankton, followed by regeneration of Se as either Se(IV) or organic Se, which are eventually reoxidized to Se(VI). In marine particulates, most of the Se appears to be present as organic Se (Cutter, personal communication, 1984).

There are few published data available on the concentration or species distribution of Te in the oceans (Andreae, 1984). Preliminary work in our laboratory suggests concentrations on the order of 0.5–6 pM Te(IV) in the water column in the North Atlantic, with a silicate-like vertical structure. No Te(VI) was detectable.

Iodine

Iodine occurs in seawater in a number of chemical species: iodide (I^-), iodate (IO_3^-), methyliodide (CH_3I), and possibly elemental iodine (I^0) (Emerson et al., 1979; Lovelock et al., 1973; Su and Goldberg, 1976; Wong, 1977; Wong and Brewer, 1977). In oxygenated waters iodate predominates, but iodide is present at levels of some 10–20% of the total iodine in open ocean surface waters, and it predominates completely in anoxic waters. In the deep ocean iodide is near or below the limits of detection. It is assumed, but has not been proven, that the presence of iodide in surface water is a result of biological uptake and reduction of iodate. The vertical distribution of methyliodide in the oceans is not known; it is assumed that this compound is released by phytoplankton.

Actinides and Rare-Earth Elements

In view of the obvious importance of the chemical species distribution for the fate of the transuranics (especially plutonium and americium) in the marine environment, it is surprising how little is known about both their equilibrium species in seawater and their species transformation kinetics. Nelson and Lovett (1978) found that over 75% of the dissolved Pu in the Irish Sea was Pu(V+VI) and over 90% of the particulate Pu was present as Pu(III+IV). They could not distinguish between Pu(V) and Pu(VI) or between Pu(III) and Pu(IV). The complexation of Pu by seawater DOM is also controversial: Fisher et al. (1983) have observed no evidence of complexation using a bioassay method, while earlier work (Dahlman et al., 1976) had suggested complexation by marine humic substances. In model studies using the addition of known Pu oxidation states to seawater, Roberts (1984) found Pu(VI) to be kinetically very labile and rapidly reducible to Pu(V), which was found to be the most stable Pu species in solution. It was, however, slowly reduced to Pu(IV), which was removed from solution by adsorption. The presence of marine humic acids accelerated the reduction of Pu(VI) and Pu(V) to Pu(IV).

Thorium should only be stable in the +4 oxidation state as $Th(OH)_4^0$. There is no evidence available for organic complexation of this element. Bacon and Anderson (1982) found about 96% of the Th in seawater in the dissolved form (< 1.0 μm) and concluded that an equilibrium exists between dissolved Th and the Th on marine particulates.

The rare-earth element (REE) species in seawater have not been experimentally investigated. The stable oxidation state for most REEs is +3, with the exception of Ce, which can also occur in the +4 state, and Eu, which can occur in the +2 state. de Baar (1983) has presented oceanographic REE distributions which are consistent with these assumed oxidation states. He has observed negative Ce anomalies which can be related to scavenging of Ce(VI) in addition to Ce(III) in oxygenated deep waters. He found positive Ce anomalies relative to the other rare-earth elements in Atlantic surface waters and in the oxygen minimum in the Pacific. These positive anomalies may be related to regeneration of particulate Ce by reduction. Gadolinium anomalies observed by de Baar (1983) are consistent with the anomalous chemical properties of Gd(III) due to the exactly half-filled 4f electron shell. It is clear, however, that experimental work both in model systems and by actual species determination is needed for the rare-earth elements.

Chemical Species in Marine Particulates

With the few exceptions mentioned above, the chemical species of the elements in marine particulates have not been experimentally investigated. Elements may be present in these complex materials either as defined solid phases (e.g., opal in diatom frustules), as adsorbed compounds, or as part of biological cells. Various digestion and leaching procedures have been applied to separate the different fractions (e.g., Cutter, 1978). Collier and Edmond (1983) determined a number of trace elements in different leaches of marine plankton and found most of them directly associated with the non-skeletal organic phases of the plankton. They differentiated between a labile fraction and a more refractory component bound to metal-specific sites. Calcium carbonate and opal did not contain significant amounts of trace metals; neither did the refractory terrigenous aluminosilicate fraction.

Conclusions

In this review the information currently available on the chemical species of a number of elements in seawater has been discussed. A few general observations emerge from this overview.

We have reasonably good information for some elements on those species which are kinetically either very labile or very inert. The very labile spe-

Appendix. Species, concentrations, and distribution types of trace elements in ocean water (adapted from Bruland, 1983).

Element	Probable main species in oxygenated seawater	Range and average concentration at S = 35[a]	Type of distribution
Li	Li^+	25 µmol kg^{-1}	Conservative
Be	$BeOH^+$, $Be(OH)_2^0$	4–30 pmol kg^{-1}; 20 pmol kg^{-1}	Nutrient-type and scavenging
B	H_3BO_3	0.416 mmol kg^{-1}	Conservative
C	HCO_3^-, CO_3^{2-}	2.0–2.5 mmol kg^{-1}; 2.3 mmol kg^{-1}	Nutrient-type
N	NO_3^- (also as N_2)	<0.1–45 µmol kg^{-1}; 30 µmol kg^{-1}	Nutrient-type
O	O_2 (also as H_2O)	0–300 µmol kg^{-1}	Mirror image of nutrient-type
F	F^-, MgF^+	68 µmol kg^{-1}	Conservative
Na	Na^+	0.468 mol kg^{-1}	Conservative
Mg	Mg^{2+}	53.2 mmol kg^{-1}	Conservative
Al	$Al(OH)_4^-$, $Al(OH)_3^0$	(5–40 nmol kg^{-1}; 20 nmol kg^{-1})	Mid-depth minima
Si	H_4SiO_4	<1–180 µmol kg^{-1}; 100 µmol kg^{-1}	Nutrient-type
P	HPO_4^{2-}, $NaHPO_4^-$, $MgHPO_4^0$	<1–3.5 µmol kg^{-1}; 2.3 µmol kg^{-1}	Nutrient-type
S	SO_4^{2-}, $NaSO_4^-$ · $MgSO_4^0$	28.2 mmol kg^{-1}	Conservative
Cl	Cl^-	0.546 mol kg^{-1}	Conservative
K	K^+	10.2 mmol kg^{-1}	Conservative
Ca	Ca^{2+}	10.3 mmol kg^{-1}	Slight surface depletion
Sc	$Sc(OH)_3^0$	8–20 pmol kg^{-1}; 15 pmol kg^{-1}	Surface depletion
Ti	$Ti(OH)_4^0$	(<20 nmol kg^{-1})	?
V	HVO_4^{2-}, $H_2VO_4^-$, $NaHVO_4^-$	20–35 nmol kg^{-1}; 30 nmol kg^{-1}	Slight surface depletion
Cr	CrO_4^{2-}, $NaCrO_4^-$	2–5 nmol kg^{-1}; 4 nmol kg^{-1}	Nutrient-type
Mn	Mn^{2+}, $MnCl^+$	0.2–3 nmol kg^{-1}; 0.5 nmol kg^{-1}	Depletion at depth
Fe	$Fe(OH)_3^0$	0.1–2.5 nmol kg^{-1}; 1 nmol kg^{-1}	Surface depletion, depletion at depth
Co	Co^{2+}, $CoCO_3^0$, $CoCl^+$	(0.01–0.1 nmol kg^{-1}; 0.02 nmol kg^{-1})	Surface depletion, depletion at depth
Ni	Ni^{2+}, $NiCO_3^0$, $NiCl^+$	2–12 nmol kg^{-1}; 8 nmol kg^{-1}	Nutrient-type
Cu	$CuCO_3^0$, $CiOH^+$, Cu^{2+}	0.5–6 nmol kg^{-1}; 4 nmol kg^{-1}	Nutrient-type and scavenging
Zn	Zn^{2+}, $ZnOH^+$, $ZnCO_3^0$, $ZnCl^+$	0.05–9 nmol kg^{-1}; 6 nmol kg^{-1}	Nutrient-type
Ga	$Ga(OH)_4^-$	(0.3 nmol kg^{-1})	?

Appendix (continued)

Element	Probable main species in oxygenated seawater	Range and average concentration at S=35[a]	Type of distribution
Ge	H_4GeO_4, $H_3GeO_4^-$	$\leqq$7–115 pmol kg^{-1}; 70 pmol kg^{-1}	Nutrient-type
As	$HAsO_4^{2-}$	15–25 nmol kg^{-1}; 23 nmol kg^{-1}	Nutrient-type
Se	SeO_4^{2-}, SeO_3^{2-}, $HSeO_3^-$	0.5–2.3 nmol kg^{-1}; 1.7 nmol kg^{-1}	Nutrient-type
Br	Br^-	0.84 mmol kg^{-1}	Conservative
Rb	Rb^+	1.4 µmol kg^{-1}	Conservative
Sr	Sr^{2+}	90 µmol kg^{-1}	Slight surface depletion
Y	YCO_3^+, YOH^{2+}, Y^{3+}	(0.15 nmol kg^{-1})	?
Zr	$Zr(OH)_4^0$, $Zr(OH)_5^-$	(0.3 nmol kg^{-1})	?
Nb	$Nb(OH)_6^-$, $Nb(OH)_5^0$	($\leqq$50 pmol kg^{-1})	?
Mo	MoO_4^{2-}	0.11 µmol kg^{-1}	Conservative
(Tc)	TcO_4^-	No stable isotope	–
Ru	?	?	?
Rh	?	?	?
Pd	?	0.2–0.6 pmol kg^{-1}	?
Ag	$AgCl_2^-$	(0.5–35 pmol kg^{-1}; 25 pmol kg^{-1}	Nutrient-type
Cd	$CdCl_2^0$	0.001–1.1 nmol kg^{-1}; 0.7 nmol kg^{-1}	Nutrient-type
In	$In(OH)_3^0$	(1 pmol kg^{-1})	?
Sn	$SnO(OH)_3^-$	0.5–12; ~2 pmol kg^{-1})	High in surface waters
Sb	$Sb(OH)_6^-$	(1.2 nmol kg^{-1})	?
Te	TeO_3^{2-}, $HTeO_3^-$	0.6–6 pmol kg^{-1}	?
I	IO_3^-	0.2–0.5 µmol kg^{-1}; 0.4 µmol kg^{-1}	Nutrient-type
Cs	Cs^+	2.2 nmol kg^{-1}	Conservative
Ba	Ba^{2+}	32–150 nmol kg^{-1}; 100 nmol kg^{-1}	Nutrient-type
La	La^{3+}, $LaCO_3^+$, $LaCl^{2+}$	13–37 pmol kg^{-1}; 30 pmol kg^{-1}	Surface depletion
Ce	$CeCO_3^+$, Ce^{3+}, $CeCl^{2+}$	16–26 pmol kg^{-1}; 20 pmol kg^{-1}	Surface depletion
Pr	$PrCO_3^+$, Pr^{3+}, $PrSO_4^+$	(4 pmol kg^{-1})	Surface depletion
Nd	$NdCO_3^+$, Nd^{3+}, $NdSO_4^+$	12–25 pmol kg^{-1}; 20 pmol kg^{-1}	Surface depletion

Sm	$SmCO_3^+$, Sm^{3+}, $SmSO_4^+$	2.7–4.8 pmol kg^{-1}; 4 pmol kg^{-1}	Surface depletion
Eu	$EuCO_3^+$, Eu^{3+}, $EuOH^{2+}$	0.6–1.0 pmol kg^{-1}; 0.9 pmol kg^{-1}	Surface depletion
Gd	$GdCO_3^+$, Gd^{3+}	3.4–7.2 pmol kg^{-1}; 6 pmol kg^{-1}	Surface depletion
Tb	$TbCO_3^+$, Tb^{3+}, $TbOH^{2+}$	(0.9 pmol kg^{-1})	Surface depletion
Dy	$DyCO_3^+$, Dy^{3+}, $DyOH^{2+}$	(4.8–6.1 pmol kg^{-1}; 6 pmol kg^{-1}	Surface depletion
Ho	$HoCO_3^+$, Ho^{3+}, $HoOH^{2+}$	(1.9 pmol kg^{-1})	Surface depletion
Er	$ErCO_3^+$, $ErOH^{2+}$, Er^{3+}	4.1–5.8 pmol kg^{-1}; 5 pmol kg^{-1}	Surface depletion
Tm	$TmCO_3^+$, $TmOH^{2+}$, Tm^{3+}	(0.8 pmol kg^{-1})	Surface depletion
Yb	$YbCO_3^+$, $YbOH^{2+}$	3.5–5.4 pmol kg^{-1}; 5 pmol kg^{-1}	Surface depletion
Lu	$LuCO_3^+$, $LuOH^{2+}$	(0.9 pmol kg^{-1})	Surface depletion
Hf	$Hf(OH)_4^0$, $Hf(OH)_5^-$	(< 40 pmol kg^{-1})	?
Ta	$Ta(OH)_5^0$	(< 14 pmol kg^{-1})	?
W	WO_4^{2-}	0.5 nmol kg^{-1}	?
Re	ReO_4^-	(14–30 pmol kg^{-1}; 20 pmol kg^{-1})	?
Os	?	?	?
Ir	?	?	?
Pt	?	?	?
Au	$AuCl_2^-$	(25 pmol kg^{-1})	?
Hg	$HgCl_4^{2-}$	(2–10 pmol kg^{-1}; 5 pmol kg^{-1})	?
Tl	Tl^+, $TlCl^0$; or $Tl(OH)_3^0$	60 pmol kg^{-1}	Conservative
Pb	$PbCO_3^0$, $Pb(CO_3)_2^{2-}$, $PbCl^+$	5–175 pmol kg^{-1}; 10 pmol kg^{-1}	High in surface waters, depleted at depth
Bi	BiO^+, $Bi(OH)_2^+$	≦0.015–0.24 pmol kg^{-1}	Depletion at depth

[a] Parentheses indicate uncertainty

cies can be considered to be in thermodynamic equilibrium, which allows the prediction of species distribution from mathematical models (*if* all relevant species are known). The very inert species can be isolated and subjected to spectrometric and chromatographic techniques. The biggest problem exists with those species which are inert enough to persist in nonequilibrium but are too labile to allow isolation.

We can usually identify covalently bonded molecules (or parts of molecules) reasonably well but have difficulties with coordinatively bound ones. This is due either to difficulty of measuring the concentration of the complex in seawater because of limitations of analytical sensitivity, or to the lack of knowledge about the characteristics of the ligands. For some molecules, e.g., the methylarsenic compounds in seawater, we have characterized the covalent aspects of the molecule (e.g., two methyl groups bound to As) but not the coordinative ones.

We know very little about the chemical species or forms of the elements in marine particulate matter. This is largely due to the difficulty of characterizing the extremely complicated binding environment in impure, mixed organic-inorganic solids. Applications of newly developing analytical technologies for surface characterization and single-particle analysis may offer some promise in this area.

References

Ahmad I, Chau YK, Wong PTS, Carty AJ, Taylor L (1980) Chemical alkylation of lead(II) salts to tetraalkyllead(IV) in aqueous solution. Nature 287:716–717

Andreae MO (1977) Determination of arsenic species in natural waters. Anal Chem 49:820–823

Andreae MO (1978) Distribution and speciation of arsenic in natural waters and marine algae. Deep-Sea Res 25:391–402

Andreae MO (1983a) Arsenic (by hydride generation/AAS); antimony (by hydride generation/AAS); germanium (by hydride generation/AAS). In: Grasshoff K, Ehrhardt M, Kremling K (eds) Methods of seawater analysis. Verlag Chemie, Weinheim, pp 218–236

Andreae MO (1983b) Biotransformation of arsenic in the marine environment. In: Lederer WH, Fensterheim RJ (eds) Industrial, biomedical, environmental perspectives. Van Nostrand Reinhold, New York, pp 378–392

Andreae MO (1983c) The determination of the chemical species of some of the "hydride elements" (arsenic, antimony, tin, and germanium) in seawater: methodology and results. In: Wong CS, Boyle E, Bruland KW, Burton JD, Goldberg ED (eds) Trace metals in sea water. Plenum Press, New York, pp 1–19

Andreae MO (1984) Determination of inorganic tellurium species in natural waters. Anal Chem 56:2064–2066

Andreae MO (1985) Organoarsenic compounds in the environment. In: Craig PJ (ed) Organometallic compounds in the environment. Longman, London, pp 198–228

Andreae MO, Byrd JT (1984) Determination of tin and methyltin species by hydride generation and detection with graphite-furnace atomic absorption or flame emission spectrometry. Anal Chim Acta 156:147–157

Andreae MO, Froelich PN (1981) Determination of germanium in natural waters by graphite furnace atomic absorption spectrometry with hydride generation. Anal Chem 53:287–291

Andreae MO, Froelich PN (1984) Arsenic, antimony, and germanium biogeochemistry in the Baltic Sea. Tellus 36B:101–117

Andreae MO, Klumpp DW (1979) Biosynthesis and release of organoarsenic compounds by marine algae. Envir Sci Technol 13:738–741

Andreae MO, Asmodé JF, Foster P, Van't dack L (1981) Determination of antimony(III), antimony(V), and methylantimony species in natural waters by atomic absorption spectrometry with hydride generation. Anal Chem 53:1766–1771

Andren AW, Harriss RC (1975) Observations on the association between mercury and organic matter dissolved in natural waters. Geochim Cosmochim Acta 39:1253–1257

Bacon MP, Anderson RF (1982) Distribution of thorium isotopes between dissolved and particulate forms in the deep sea. J Geophys Res 87:2045–2056

Baes, CF, Mesmer RE (1976) The hydrolysis of cations. Wiley, New York

Bernhard M, Andreae MO (1984) Transport of trace metals in marine food chains. In: Nriagu JO (ed) Changing metal cycles and human health. Dahlem Konferenzen. Springer-Verlag, Berlin Heidelberg New York Tokyo, pp 143–167

Bertine KK, Lee DS (1983) Antimony content and speciation in the water column and interstitial waters of Saanich Inlet. In: Wong CS, Boyle E, Bruland KW, Burton JD, Goldberg ED (eds) Trace metals in sea water. Plenum Press, New York, pp 21–38

Bertrand PA, Choppin GR (1982) Separation of actinides in different oxidation states by solvent extraction. Radiochim Acta 31:135–137

Boulegue J, Lord CJ III, Church TM (1982) Sulfur speciation and associated trace metals (Fe, Cu) in the pore waters of Great Marsh, Delaware. Geochim Cosmochim Acta 46:453–464

Braman RS, Foreback CC (1973) Methylated forms of arsenic in the environment. Science 182:1247–1249

Braman RS, Tompkins MA (1979) Separation and determination of nanogram amounts of inorganic tin and methyltin compounds in the environment. Anal Chem 51:12–19

Brinckman FE, Jackson JA, Blair WR, Olson GJ, Iverson WP (1983) Ultratrace speciation and biogenesis of methyltin transport species in estuarine waters. In: Wong CS, Boyle E, Bruland KW, Burton JD, Goldberg ED (eds) Trace metals in sea water. Plenum Press, New York, pp 39–72

Bruland KW (1983) Trace elements in sea-water. In: Chemical oceanography, vol 8. Academic Press, London, pp 157–220

Bruland KW, Franks RP, Knauer GA, Martin JH (1979) Sampling and analytical methods for the determination of copper, cadmium, zinc, and nickel at the nanogram per liter level in sea water. Anal Chim Acta 105:233–245

Byrd JT, Andreae MO (1982) Tin and methyltin species in seawater: concentrations and fluxes. Science 218:565–569
Byrne RH (1981) Inorganic lead complexation in natural seawater determined by UV spectroscopy. Nature 290:487–489
Collier RW, Edmond JM (1983) Plankton compositions and trace element fluxes from the surface ocean. In: Wong CS, Boyle E, Bruland KW, Burton JD, Goldberg ED (eds) Trace metals in sea water. Plenum Press, New York, pp 789–809
Cosovic B, Degobbis D, Bilinski H, Branica M (1982) Inorganic cobalt species in seawater. Geochim Cosmochim Acta 46:151–158
Craig PJ (1980) Metal cycles and biological methylation. In: Hutzinger O (ed) Handbook of environmental chemistry, vol 1, part A. Springer-Verlag, Berlin, pp 169–227
Cutter GA (1978) Species determination of selenium in natural waters. Anal Chim Acta 98:59–66
Cutter GA (1982) Selenium in reducing water. Science 217:829–831
Cutter GA, Bruland KW (1984) The marine biogeochemistry of selenium: a reevaluation. Limnol Oceanog, 29:1179–1192
Dahlman RC, Bondietti EA, Eyman LD (1976) Biological pathways and chemical behavior of plutonium and other actinides in the environment. ACS Symp Ser 35:47–80
Davis JA, Leckie JO (1978) Effect of adsorbed complexing ligands on trace metal uptake by hydrous oxides. Envir Sci Technol 12:1309–1315
Dawson R, Duursma EK (1974) Distribution of radioisotopes between phytoplankton, sediment and sea water in a dialysis compartment system. Netherlands J Sea Res 8:339–353
de Baar HJW (1983) The Marine Geochemistry of the Rare Earth Elements. Doctoral Dissertation, Woods Hole Oceanographic Institution/M.I.T
De Jonghe WRA (1983) Selective determination and concentration levels of organic lead in the environment. Doctoral Dissertation, Universitaire Instelling Antwerpen
de Mora SJ, Harrison RM (1983) The use of physical separation techniques in trace metal speciation studies. Water Res 17:723–733
Emerson S, Cranston RE, Liss PS (1979) Redox species in a reducing fjord: equilibrium and kinetic considerations. Deep-Sea Res 26:859–878
Emerson S, Kalhorn S, Jacobs L, Tebo BM, Nealson KH, Rosson RA (1982) Environmental oxidation rate of manganese(II): bacterial catalysis. Geochim Cosmochim Acta 46:1073–1079
Feldman C, Batistoni DA (1977) Spectroscopic element detector for gas chromatography. Anal Chem 49:2215–2221
Figura P, McDuffie B (1980) Determination of labilities of soluble trace metal species in aqueous environmental samples by anodic stripping voltammetry and Chelex column and batch methods. Anal Chem 52:1433–1439
Fisher FH (1967) Ion pairing of magnesium sulfate in seawater: determined by ultrasonic absorption. Science 157:823
Fisher NS, Bjerregard P, Huynh-Ngoc L, Harvey GR (1983) Interactions of marine plankton with transuranic elements. II. Influence of dissolved organic compounds on americium and plutonium accumulation in a diatom. Mar Chem 13:45–56

Fisher NS, Fabris JG (1982) Complexation of Cu, Zn and Cd by metabolites excreted from marine diatoms. Mar Chem 11:245–255
Fitzgerald WF (1975) Mercury analysis in seawater using cold-trap preconcentration and gas phase detection. Adv Chem Ser 147:99–109
Florence TM (1982) The speciation of trace elements in waters. Talanta 29:345–364
Florence TM, Batley GE (1980) Chemical speciation in natural waters. CRC Crit Rev 9:219–296
Froelich PN, Andreae MO (1981) The marine geochemistry of germanium: ekasilicon. Science 213:205–207
Froelich PN, Kaul LW, Byrd JT, Andreae MO, Roe KK (1984) Arsenic, barium, germanium tin, dimethylsulfide and nutrient biogeochemistry in Charlotte Harbor, Florida, a phosphorus-enriched estuary. Est Coast Shelf Sci, 20:239–264
Fujita M, Iwashima K (1981) Estimation of organic and total mercury in seawater around the Japanese archipelago. Envir Sci Technol 15:929–933
Gillespie PA, Vaccaro RF (1978) A bacterial bioassay for measuring the copper chelation capacity of sea water. Limnol Oceanog 23:543–548
Gohda S (1975) Valence states of arsenic and antimony in sea water. Bull Chem Soc Japan 48:1213–1216
Gordon RM, Martin JH, Knauer GA (1982) Iron in north-east Pacific waters. Nature 299:611–612
Grasshoff K, Ehrhardt M, Kremling K (eds) (1983) Methods of seawater analysis, 2nd ed. Verlag Chemie, Weinheim
Hallas LE, Means JC, Cooney JJ (1982) Methylation of tin by estuarine microorganisms. Science 215:1505–1507
Hambrick GA, Froelich PN, Andreae MO, Lewis BL (1984) Determination of methylgermanium species in natural waters by graphite furnace atomic absorption spectrometry with hydride generation. Anal Chem 56:421–424
Hanson AK Jr, Quinn JG (1983) The distribution of dissolved and organically complexed copper and nickel in the Middle Atlantic Bight. Can J Fish Aquat Sci, 40:151–161
Hanson AK Jr, Sakamoto-Arnold CM, Huizenga DL, Kester DR (1986) The distribution and speciation of copper in a Gulf Stream warm core ring. Mar Chem, in press
Harvey GR, Boran DA, Chesal LA, Tokar JM (1983) The structure of marine fulvic and humic acids. Mar Chem 12:119–132
Hirose K, Dokiya Y, Sugimura Y (1982) Determination of conditional stability constants of organic copper and zinc complexes dissolved in seawater using ligand exchange method with EDTA. Mar Chem 11:343–354
Hodge VF, Seidel SL, Goldberg ED (1979) Determination of tin(IV) and organotin compounds in natural waters, coastal sediments and macro algae by atomic absorption spectrometry. Anal Chem 51:1256–1259
Huber F, Schmidt U, Kirchman H (1978) Aqueous chemistry of organolead and organothallium compounds in the presence of microorganisms. ACS Symp Ser 82:65–81
Hunter KA, Liss PS (1979) The surface charge of suspended particles in estuarine and coastal waters. Nature 282:823–825
Jacobs L, Emerson S (1982) Trace metal solubility in an anoxic fjord. Earth Planet Sci Lett 60:237–252

Jewett KL, Brinckman FE (1983) The use of element-specific detectors coupled with high-performance liquid chromatographs. In: Vickrey TM (ed) Liquid chromatography detectors. Marcel Dekker, New York, pp 205–241
Jewett KL, Brinckman FE, Bellama JM (1975) Chemical factors influencing metal alkylation in water. ACS Symp Ser 18:304–318
Kantin R (1983) Chemical speciation of antimony in marine algae. Limnol Oceanog 28:165–168
Kerndorff H, Schnitzer M (1980) Sorption of metals on humic acid. Geochim Cosmochim Acta 44:1701–1708
Klinkhammer GP, Bender ML (1980) The distribution of manganese in the Pacific Ocean. Earth Planet Sci Lett 46:361–384
Knauer GA, Martin JH (1972) Mercury in a marine pelagic food chain. Limnol Oceanog 17:868–876
Kremling K, Wenck A, Osterroht C (1983) Variations of dissolved organic copper in marine waters. In: Wong CS, Boyle E, Bruland KW, Burton JD, Goldberg ED (eds) Trace metals in sea water. Plenum Press, New York, pp 609–620
Kritsotakis K, Rubischung P, Tobshall HJ (1979) Invers-voltammetrische Untersuchung zur Speziation des Quecksilbers in Flußwasser. Fresenius Z Anal Chem 296:358–364
Landing WM, Bruland KW (1980) Manganese in the North Pacific. Earth Planet Sci Lett 49:45–56
Landing WM, Bruland KW (1981) The vertical distribution of iron in the northeast Pacific. EOS 62:906
Lindberg SE, Andren AW, Harriss RC (1975) Geochemistry of mercury in the estuarine environment. In: Cronin L (ed) Estuarine research, vol 1. Wiley, New York, pp 64–107
Lovelock JE, Maggs RJ, Wade RJ (1973) Halogenated hydrocarbons in and over the Atlantic. Nature 241:194–196
Luten JB, Riekwel-Booy G, Greef J vd, ten Noever de Brauw MC (1983) Identification of arsenobetaine in sole, lemon sole, flounder, dab, crab and shrimps by field desorption and fast atom bombardment mass spectrometry. Chemosphere 12:131–141
Lyons WB, Gaudette HE, Armstrong PB (1979) Evidence for organically associated iron in nearshore pore fluids. Nature 282:202–203
Macchi G, Pettine M (1980) Voltammetric characterization and chemical behavior of inorganic tin in natural waters. Envir Sci Technol 14:815–818
Mackey DJ (1983) Metal-organic complexes in seawater: an investigation of naturally occurring complexes of Cu, Zn, Fe, Mg, Ni, Cr, Mn and Cs using high-performance liquid chromatography with atomic fluorescence detection. Mar Chem 13:169–180
Manders WF, Olson GJ, Brinckman FE, Bellama JM (1984) A novel synthesis of methyltin triiodide with environmental implications. J Chem Soc Chem Comm, 538–540
Mantoura RFC (1981) Organo-metallic interactions in natural waters. In: Duursma EK, Dawson R (eds) Marine organic chemistry: evolution, composition, interactions and chemistry of organic matter in seawater. Elsevier, Amsterdam, pp 179–223
Mantoura RFC, Dickson A, Riley JP (1978) The complexation of metals with humic materials in natural waters. Est Coast Mar Sci 6:387–408

Measures CI, Burton JD (1980) The vertical distribution and oxidation states of dissolved selenium in the northeast Atlantic Ocean and their relationship to biological processes. Earth Planet Sci Lett 46:385–396

Measures CI, McDuff RE, Edmond JM (1980) Selenium redox chemistry at GEOSECS I re-occupation. Earth Planet Sci Lett 49:102–108

Meinema HA, Burger-Wiersma T, Versluis-de Haan G, Gevers EC (1978) Determination of trace amounts of butyltin compounds in aqueous systems by gas chromatography/mass spectrometry. Envir Sci Technol 12:288–293

Mills GL, Hanson AK Jr, Quinn JG, Lammela WR, Chasten ND (1982) Chemical studies of copper-organic complexes isolated from estuarine waters using C_{18} reverse-phase liquid chromatography. Mar Chem 11:355–377

Moffett JW, Zika RG (1983) Oxidation kinetics of Cu(I) in seawater: implications for its existence in the marine environment. Mar Chem 13:239–251

Morris AW (1974) Seasonal variation of dissolved metals in inshore waters of the Menai Straits. Mar Poll Bull 5:54–59

Murray JW, Spell B, Paul P (1983) The contrasting geochemistry of manganese and chromium in the eastern tropical Pacific Ocean. In: Wong CS, Boyle E, Bruland KW, Burton JD, Goldberg ED (eds) Trace metals in sea water. Plenum Press, New York, pp 643–669

Musani L, Nürnberg HW, Valenta P, Konrad Z, Branica M (1981) Interaction of ^{65}Zn and humic acid in seawater. Thalassia Jugoslavica 17:71–81

Musani L, Valenta P, Nürnberg HW, Konrad Z, Branica M (1980) On the chelation of toxic trace metals by humic acid of marine origin. Est Coast Mar Sci 11:639–649

Nakayama E, Tokoro H, Kuwamoto T, Fujinaga T (1981) Dissolved state of chromium in seawater. Nature 290:768–770

National Academy of Sciences (1978) An Assessment of Mercury in the Environment. Washington, DC: National Academy of Sciences

Nelson DM, Lovett MB (1978) Oxidation state of plutonium in the Irish Sea. Nature 276:599–601

Nriagu JO (ed) (1979) The biogeochemistry of mercury in the environment. Elsevier/North Holland Biomedical Press, New York

Nürnberg HW, Valenta P (1983) Potentialities and applications of voltammetry in chemical speciation of trace metals in the sea. In: Wong CS, Boyle E, Bruland KW, Burton JD, Goldberg ED (eds) Trace metals in sea water. Plenum Press, New York, pp 671–697

Ortner PB, Kreader C, Jarvey GR (1983) Interactive effects of metals and humus on marine phytoplankton carbon uptake. Nature 301:57–59

Palmer SE, Baker EW (1978) Copper porphyrins in deep-sea sediments: a possible indicator of oxidized terrestrial organic matter. Science 201:49–51

Piotrowicz SR, Harvey GR, Boran DA, Weisel CP, Springer-Young M (1984) Cadmium, copper and zinc interactions with marine humus as a function of ligand structure. Mar Chem 14:333–346

Piotrowicz SR, Harvey GR, Springer-Young M, Courant RA, Boran DA (1983) Studies of cadmium, copper and zinc interactions with marine fulvic and humic materials in seawater using anodic stripping voltammetry. In: Wong CS, Boyle E, Bruland KW, Burton JD, Goldberg ED (eds) Trace metals in sea water. Plenum Press, New York, pp 699–717

Piotrowicz SR, Springer-Young M, Pulg JA, Spencer MJ (1982) Anodic stripping voltammetry for evaluation of organic-metal interactions in seawater. Anal Chem 54:1367–1371
Plavsic M, Krznaric D, Branica M (1982) Determination of the apparent copper complexing capacity of sea-water by anodic stripping voltammetry. Mar Chem 11:17–31
Ramamoorthy S, Kushner DJ (1975) Heavy metal binding sites in river water. Nature 256:399–401
Reisinger K, Stoeppler M, Nürnberg HW (1981) Evidence for the absence of biological methylation of lead in the environment. Nature 291:228–230
Roberts RA (1984) Speciation of Actinides in Marine Waters. PhD Dissertation, Florida State University
Roekens EJ, Van Grieken RE (1983) Kinetics of iron(II) oxidation in seawater of various pH. Mar Chem 13:195–202
Schnitzer M, Kahn SU (1972) Humic substances in the environment. Dekker, New York
Seidel SL, Hodge VF, Goldberg ED (1980) Tin as an environmental pollutant. Thalassia Jugoslavica 16:209–233
Sipos L, Valenta P, Nürnberg HW, Branica M (1980) Voltammetric determination of the stability constants of the predominant labile lead complexes in sea water. In: Branica M, Konrad Z (eds) Lead in the marine environment. Pergamon Press, Oxford, pp 61–76
Smith RG Jr (1976) Evaluation of combined applications of ultrafiltration and complexation capacity techniques to natural waters. Anal Chem 48:74–76
Snyder LJ, Bentz JM (1982) Alkylation of lead(II) salts to tetraalkyllead in aqueous solution. Nature 296:228–229
Stone AT (1983) The Reduction and Dissolution of Mn(III) and Mn(IV) Oxides by Organics. Report No AC-1-83, California Institute of Technology
Stouff P, Boulegue J (1983) Control of copper speciation in sea water in the vicinity of hydrothermal vents. In: Proceedings of the International Conference on Heavy Metals in the Environment, vol 2. Heidelberg, September 1983
Stumm W, Morgan JJ (1981) Aquatic chemistry, 2nd ed. Wiley, New York
Su C, Goldberg ED (1976) Environmental concentrations and fluxes of some halocarbons. In: Windom HL, Duce RA (eds) Marine pollutant transfer. Lexington Books, Lexington MA: D.C. Heath and Co, pp 353–374
Sugai SF, Healy ML (1978) Voltammetric studies of the organic association of copper and lead in two Canadian inlets. Mar Chem 6:291–308
Sugimura Y, Suzuki Y, Miyake Y (1977) The content of selenium and its chemical form in seawater. J Oceanog Soc Japan 32:235–241
Sugimura Y, Suzuki Y, Miyake Y (1978a) On the chemical forms of minor metallic elements in the ocean. J Oceanog Soc Japan 34:93–96
Sugimura Y, Suzuki Y, Miyake Y (1978b) The dissolved organic iron in seawater. Deep-Sea Res 25:309–314
Sunda WG, Ferguson RL (1983) Sensitivity of natural bacterial communities to additions of copper and to cupric ion activity: a bioassay of copper complexation in seawater. In: Wong CS, Boyle E, Bruland KW, Burton JD, Goldberg ED (eds) Trace metals in sea water. Plenum Press, New York, pp 871–891
Sunda WG, Huntsman SA, Harvey GR (1983) Photoreduction of manganese oxides in seawater and its geochemical and biological implications. Nature 301:234–236

Topping G, Davies IM (1981) Methylmercury production in the marine water column. Nature 290:243–244
Turner DR (1984) Relationships between biological availability and chemical measurements. In: Sigel H (ed) Metal ions in biological systems. Marcel Dekker, New York
Van den Berg CMG (1982a) Determination of copper complexation with natural organic ligands in seawater by equilibration with MnO_2. I. Theory. Mar Chem 11:307–322
Van den Berg CMG (1982b) Determination of copper complexation with natural organic ligands in seawater by equilibration with MnO_2. II. Experimental procedures and application to surface seawater. Mar Chem 11:323–342
Washburne CL (1981) Thermodynamics and Speciation of Lead in Seawater. MS Thesis, University of Delaware
Waslenchuk DG (1977) The Geochemistry of Arsenic in the Continental Shelf Environment. PhD Dissertation, Georgia Institute of Technology
Waslenchuk DG (1978) The budget and geochemistry of arsenic in a continental shelf environment. Mar Chem 7:39–52
Williams PJ LeB (1975) Biological and chemical aspects of dissolved organic material in sea water. In: Riley JP, Skirrow G (eds) Chemical Oceanography 2, 2nd ed. Academic Press, London, pp 301–363
Wong GTF (1977) The distribution of iodine in the upper layers of the equatorial Atlantic. Deep-Sea Res 24:115–125
Wong GTF, Brewer PG (1974) The determination and distribution of iodate in South Atlantic waters. J Mar Res 32:25–36
Wong GTF, Brewer PG (1977) The marine chemistry of iodine in anoxic basins. Geochim Cosmochim Acta 41:151–159
Wong GTF, Chau YK, Luxon PL (1975) Methylation of lead in the environment. Nature 253:263–264
Yamamoto J, Kaneda Y, Hikasa Y (1983) Picogram determination of methylmercury in seawater by gold amalgamation and atomic absorption spectrophotometry. Intl J Envir Anal Chem 16:1–16

The Importance of Chemical "Speciation" in Environmental Processes,
eds. M. Bernhard, F. E. Brinckman, and P. J. Sadler, pp. 337–363.
Dahlem Konferenzen 1986. Springer-Verlag Berlin, Heidelberg
© Dr. S. Bernhard, Dahlem Konferenzen

Equilibrium Models in Seawater: Applications and Limitations

D. R. Kester

Graduate School of Oceanography
University of Rhode Island
Kingston, RI 02881, USA

Abstract. A knowledge of chemical species is fundamental to understanding the behavior of substances in environmental processes. Thermodynamic data provide a basis for predicting the equilibrium state of a chemical system. There are three basic steps in formulating an equilibrium model for trace constituents in seawater. The composition of the medium with respect to the major ions must be specified on a conventional concentration scale such as molality. The interactions among the major ions must be considered in order to determine the free-ligand concentrations or activities which are available to interact with the trace constituents. Finally, the complexation of a trace element by the ligands in seawater can be calculated. This procedure is illustrated using an ion-pair model for the major ions and considering copper, iron(II), iron(III), cadmium, and lead as examples of trace elements. Equilibrium models have been particularly successful in accounting for the chemical behavior of many solution-phase reactions such as ionic interactions, weak acid dissociation, and some aspects of trace-metal complexation. Caution must be used in applying this approach to redox reactions and heterogeneous phase reactions because kinetic factors often limit the attainment of equilibrium. Biological systems are inherently far from equilibrium. Nevertheless, considerations of the equilibrium state can be useful even in systems that are not in equilibrium; departures from equilibrium indicate the important role of kinetic factors. Substantial advances have been made in recent years in obtaining a suitable base of thermodynamic data for marine systems, but limitations exist because data are lacking for many trace-element reactions involving some of the important oxyanions in seawater. The existing equilibrium data base is inadequate to predict the effects of temperature and pressure on most reactions in the marine environment. A variety of analytical methods are available to test the predictions of equilibrium models.

Introduction

The reactivity of substances in the ocean is greatly influenced by variations in their chemical species. Whether an element will be toxic to organisms, remain dissolved to be transported by currents and mixing, or become associated with particles and concentrated in sediments are all determined by its distribution among chemical species. The consequences and nature of chemical species have been explored most extensively in marine systems for solution phases, but the same concepts are equally important in considerations of solid- and gas-phase chemistry in the environment. Geochemical approaches to the environment have tended to emphasize elemental sources, sinks, and fluxes; these studies have provided a large amount of valuable information. But it is clear that one must go beyond considerations of chemical elements and examine specific chemical species in order to link biogeochemical behavior to predictable mechanisms. For example, it is not particularly useful to consider the atmospheric flux of gaseous carbon without distinguishing among its chemical species – carbon dioxide, carbon monoxide, and methane. The distinct molecular forms of gases are readily recognized and there is increasing emphasis on distinguishing molecular forms of dissolved elements, but the chemical species comprising environmental solid phases have not been examined extensively. A great deal of information has been obtained on the elemental composition of marine particulate material, but understanding the chemical reactivity of these phases requires a knowledge of the species associated with the solid phases.

This paper will examine equilibrium considerations of several solution-phase systems that are of interest in seawater. It is important to distinguish between distribution of chemical species and the partitioning of an element among a number of operationally defined fractions. There has been a tendency to use the term species interchangeably to refer to particular molecular entities of an element and to operationally determined fractions of an element. Inasmuch as a one-to-one correspondence does not generally exist between specific molecular forms and operational fractionation, inconsistencies are likely to occur between the two descriptions of a system, which are only obscured by considering both of them as chemical species. This paper will use the term species to refer to forms of an element that can be represented in molecular terms.

Chemical models provide a means of linking together analytical, thermodynamic, and kinetic data to make predictions of chemical behavior, stability, and reactivity. Because of the chemical complexity of marine systems and limitations in existing data, it is most useful if analytical or experimental tests can be made for specific models. The application of equilibrium methods to nonequilibrium natural systems must take into account the rates and time scales of processes within the system. It is obvious that an equilibrium model will be inadequate if chemical quantities change with time. But constancy of chemical properties with time in multicomponent natural systems is an insufficient condition for assuring the reliability of an equilibrium model. A set of processes within a system can result in a steady-state condition in which fluxes occur through the system but chemical concentrations remain invariant. Such a system is not at equilibrium and thermodynamic predictions will not correspond to the steady state. Consequently, in applying equilibrium models to chemical species it is necessary to identify the time scales of the reactions that can alter the distribution of the species. Many processes such as acid-base and ion-pairing reactions have time scales of less than a second, so that equilibrium models are very useful for the important time scales in natural systems that span the range of hours, days, months, and years. Other types of reactions are kinetically limited, such as a) the precipitation of some solid phases including those of calcium carbonate and b) redox processes including the oxidation of nitrogen by oxygen. While predictions based on equilibrium considerations will not be valid for these kinetically limited systems, knowledge of departures from equilibrium establishes the need for kinetic data.

Construction of Equilibrium Models

There are three main steps in constructing an equilibrium model for natural water systems. First, the composition of the medium must be established from analytical information. The medium will determine the ionic strength of the system and the ligands that are available to interact with a particular constituent. Second, the interactions among the major components of the medium must be established before the chemical behavior of minor and trace constituents can be considered. Third, the chemical reactions involving the constituent of interest must be identified. These reactions can include complexation by inorganic and organic ligands, oxidation-reduction processes, the incorporation into solid phases, or the formation of gaseous species. These three steps will be examined for the variations of inorganic species of several metals in the medium of seawater.

Composition of Seawater

There are twelve major ionic constituents in seawater (Fig. 1) with concentrations greater than about 0.0001 molal (moles/kg H_2O). These major ions tend to cluster at the edges of the Periodic Table, where the gain or loss of one or two electrons leads to complete electron subshells. The formula weight for these ions ranges from about 20 to 100. Because of their relatively high concentrations and slow rates of geochemical activity, the oceans are well mixed with regard to these major ions. Worldwide ocean circulation results in an oceanic mixing time on the order of 10^3 years; the time scales for processes that vary the relative concentrations of the major ions are greater than 10^5 years. Thus the major ions in the ocean maintain a constant composition relative to each other. A single parameter such as salinity or chlorinity can be used as an index of the composition of seawater, and the amounts of the major ions will be very nearly proportional to that index. Salinity is conceptually the grams of dissolved salts per kg of seawater; chlorinity is conceptually the grams of chloride plus bromide (ex-

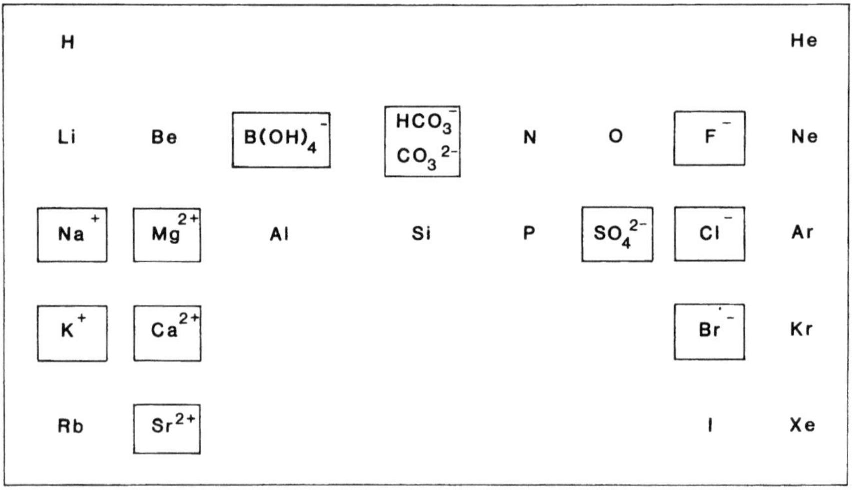

Fig. 1. Major ions of seawater (enclosed in rectangles) positioned according to a simplified Periodic Table

pressed as an equivalent weight of chloride) per kg of seawater. The variations of major ion concentrations in seawater over the salinity range of S = 5 to S = 40 are almost entirely due to variations in the water content of seawater caused by evaporation, rainfall, and mixing with rivers.

The concentrations of major ions are usually expressed as mass ratios to chlorinity, from which one may calculate more conventional concentrations such as molality at a particular salinity. The pH of ocean waters is generally in the range from 7.6 to 8.4, and under these conditions the carbonic and boric acids in seawater dissociate to produce three of the major anions – bicarbonate, carbonate, and borate. The molality of these anions will depend on the pH of seawater. Given the temperature, pressure, salinity, and pH, the molality of the twelve major ions in seawater can be calculated, thereby defining the composition of the medium for considerations of equilibrium models (Table 1).

This approach to specifying the composition of seawater includes equilibrium considerations in distributing the carbonic and boric acids into the major anions that comprise the titration alkalinity. The equilibrium constants required for this calculation are well known from a number of experimental studies over the ranges of temperature, salinity, and pressure found in marine systems (Skirrow, 1975). There have been many demonstrations that an equilibrium model is essential for understanding the

Table 1. Molality of major ions in seawater at 25 °C, 1 atm pressure, and several salinities and pH. These values were computed from major ion chlorinity ratios and dissociation constants of carbonic and boric acid

Salinity	5	15	35		
pH	8	8	7	8	9
Na^+	0.06735	0.20410	0.48617	0.48617	0.48617
Mg^{2+}	0.00759	0.02299	0.05477	0.05477	0.05477
Ca^{2+}	0.00148	0.00447	0.01065	0.01065	0.01065
K^+	0.00147	0.00444	0.01058	0.01058	0.01058
Sr^{2+}	0.00001	0.00004	0.00009	0.00009	0.00009
Cl^-	0.07838	0.23753	0.56578	0.56578	0.56578
SO_4^{2-}	0.00405	0.01229	0.02927	0.02927	0.02927
HCO_3^-	0.00033	0.00096	0.00247	0.00211	0.00087
CO_3^{2-}	0.00001	0.00004	0.00002	0.00016	0.00067
$B(OH)_4^-$	0.00001	0.00002	0.00001	0.00008	0.00030
Br^-	0.00012	0.00037	0.00087	0.00087	0.00087
F^-	0.00001	0.00003	0.00007	0.00007	0.00007

chemical behavior of weak acids in seawater. Without this knowledge we would be unable to explain the exchange of CO_2 between the atmosphere and ocean, or the formation and dissolution of carbonate sediments on the seafloor.

Major Ion Interactions

The effects of ionic interactions among the major constituents can be taken into account in either of two ways. Activities and activity coefficients can be defined relative to a common reference state, such as infinite dilution of all solutes, or the equilibrium data can be determined for the ionic medium of interest, in which case activity coefficients become unity for all ionic medium solutes. Seawater is particularly well suited for the ionic medium approach because of its constancy of composition and because its range of ionic strengths (0.5 to 0.8, for salinities between 25 and 40 is a region where activity coefficients, based on an infinite dilution scale, change only slowly with ionic strength. An infinite-dilution activity scale is a better approach for fresh and brackish waters because of variability in their composition and because there are well-developed theoretical methods for estimating activity coefficients at low ionic strengths.

Millero and Schreiber (1982) compiled a selected set of parameters for describing the major ionic interactions in seawater based on both the infinite-dilution and ionic medium approaches. These parameters consist of empirical coefficients for the variation of activity coefficients with ionic strength. The equations do not have mechanistic significance, but they cover a broad range of ionic strengths from infinite dilution to beyond the range of most marine waters, and they have internal consistency with additivity relationships for the underlying free energy of the ionic constituents.

In this consideration of equilibrium models for seawater the ionic medium approach will be used. This approach can be directly related to experimental data in electrolyte solutions related to seawater, and it does not require the estimation of unknown activity coefficients. This method is readily extended from considerations of major ionic interactions to the behavior of other chemical systems such as weak acid equilibria, redox reactions, and trace-element species. While the ionic medium approach expresses ionic interactions in terms of the law of mass action and derives the necessary equilibrium constants from free-energy data, it can lead to predictions of properties of the system that go beyond the free-energy measurements on which it is based. For example, species concentrations derived from ionic interactions can elucidate kinetic phenomena such as ultrasonic absorption (Fisher, 1967) and spectral properties including Raman (Daley et al., 1972) and optical (Brown and Kester, 1980) absorption.

The specific interactions among major ions can be represented by the law of mass action with the formation of ion-pair species. This approach can be illustrated for the interactions between magnesium and sulfate ions by the following equilibrium:

$$Mg^{2+} + SO_4^{2-} \rightleftharpoons MgSO_4^0 , \tag{1}$$

with the stoichiometric association constant given by:

$$K^*_{MgSO_4} = \frac{[MgSO_4^0]}{[Mg^{2+}]\,[SO_4^{2-}]} . \tag{2}$$

The brackets designate the molality of the enclosed species (free ion or ion-pair). The values of the stoichiometric association constants for the major ion-pairs in seawater have been determined using the ionic medium approach. These values are based on the premise that a reference electrolyte solution with minimal specific interactions can be used to represent the nonspecific electrostatic interactions that vary with ionic strength. Common choices for these reference solutions include those with chloride, nitrate, or perchlorate as the principal anion because these anions have only weak interactions with the major cations in seawater.

Several studies have examined the various alternatives for selecting a reference solution for ion pairing. Johnson and Pytkowicz (1979) demonstrated that if perchlorate media are used a set of internally consistent ion-pairing constants can be obtained for the interactions of alkali and alkaline-earth cations with chloride. Byrne and Miller (1982) applied the concept of alkali and alkaline-earth cation ion-pairs with chloride to studies of lead-chloride complexes in various media. They arrived at a self-consistent model, but unfortunately they were forced to treat the alkali/alkaline-earth chloride ion-pairs as adjustable constants, rather than using the values obtained by Johnson and Pytkowicz. It will take considerable experimental data for marine chemists to alter the prevailing view of most electrolyte solution chemists that NaCl, $MgCl_2$, and $CaCl_2$ are fully dissociated electrolytes.

Nearly all studies of transition and heavy-metal complexation have been based on the premise that NaCl does not form ion-pairs. Table 2 lists the major ion-pair equilibria in seawater taking chloride media as the reference solution for distinguishing specific and nonspecific interactions.

Table 2. Major ion-pair equilibria in seawater based on chloride media as the reference solution

$Na^+ \; SO_4^{2-} \rightleftharpoons NaSO_4^-$	$Mg^{2+} + CO_3^{2-} \rightleftharpoons MgCO_3^0$
$Mg^{2+} + SO_4^{2-} \rightleftharpoons MgSO_4^0$	$Ca^{2+} + CO_3^{2-} \rightleftharpoons CaCO_3^0$
$Ca^{2+} + SO_4^{2-} \rightleftharpoons CaSO_4^0$	$Na^+ + B(OH)_4^- \rightleftharpoons NaB(OH)_4^0$
$K^+ + SO_4^{2-} \rightleftharpoons KSO_4^-$	$Mg^{2+} + B(OH)_4^- \rightleftharpoons MgB(OH)_4^+$
$Na^+ + HCO_3^- \rightleftharpoons NaHCO_3^0$	$Ca^{2+} + B(OH)_4^- \rightleftharpoons CaB(OH)_4^+$
$Mg^{2+} + HCO_3^- \rightleftharpoons MgHCO_3^+$	$Na^+ + F^- \rightleftharpoons NaF^0$
$Ca^{2+} + HCO_3^- \rightleftharpoons CaHCO_3^+$	$Mg^{2+} + F^- \rightleftharpoons MgF^+$
$Na^+ + CO_3^{2-} \rightleftharpoons NaCO_3^-$	$Ca^{2+} + F^- \rightleftharpoons CaF^+$

Table 3. Stoichiometric ion-pair constants for the major ions in seawater. The values have been fit to a Pitzer-type equation (given below) over the range 0 to 1 ionic strength by Millero and Schreiber (1982)

$$\ln K^*_{MA} = \ln K_{MA} + z^2_{MA} f^\gamma + B^0_{MA} I + B'_{MA} f' + C_{MA} I^2$$

$$\text{with: } f^\gamma = -0.392\left(\frac{\sqrt{I}}{1+1.2\sqrt{I}} + \frac{2}{1.2}\ln\left(1+1.2\sqrt{I}\right)\right)$$

$$\text{and } f' = 1 - e^{-2\sqrt{I}}\left(1 + 2\sqrt{I} - 2\,I\right)$$

Ion-Pair	$\ln K_{MA}$	z^2_{MA}	B^0_{MA}	B'_{MA}	C_{MA}
$NaSO_4^-$	2.17	4	−0.9626	1.7874	0.76042
$MgSO_4^0$	5.09	8	−0.2948	2.6510	0.75863
$CaSO_4^0$	5.25	8	−0.3920	2.6060	0.75341
KSO_4^-	2.03	4	−1.0752	1.7332	0.75409
$NaHCO_3^0$	−0.43	2	0.1438	0.3876	0.00819
$MgHCO_3^+$	2.19	4	0.4618	0.9553	0.00199
$CaHCO_3^+$	2.30	4	0.3646	0.9103	−0.00323
$NaCO_3^-$	2.35	4	0.3706	0.9463	0.01911
$MgCO_3^0$	6.68	8	1.7682	1.8099	0.01732
$CaCO_3^0$	7.37	8	1.6710	1.7649	0.01210
$NaB(OH)_4^0$	0.51	2	0.1449	0.5260	−0.31334
$MgB(OH)_4^+$	3.54	4	0.4629	1.0937	−0.31955
$CaB(OH)_4^+$	4.14	4	0.3657	1.0487	−0.32476
NaF^0	−1.47	2	0.3000	0.2563	0.00912
MgF^+	4.38	4	0.6180	0.8240	0.00292
CaF^+	3.02	4	0.5208	0.7790	0.00230

Table 3 gives the functional form and coefficients used to calculate the stoichiometric association constants from infinite dilution to an ionic strength up to 1.0; these values are also based on chloride media being fully dissociated.

The ion-pairing constants can be combined with mass-balance equations for each major constituent to form two sets of equations of the form:

$$T(M) = [M] + \Sigma K^*_{MA}[M][A], \tag{3}$$

and

$$T(A) = [A] + \Sigma K^*_{MA}[M][A]. \tag{4}$$

M represents the cations Na^+, Mg^{2+}, Ca^{2+}, and K^+; A represents the anions SO_4^{2-}, HCO_3^-, CO_3^{2-}, $B(OH)_4^-$, and F^-. T(M) and T(A) are the total molality of each cation and anion in seawater. Equations 3 and 4 provide a set of nine relationships with the free-ion molalities as the nine unknowns. The stoichiometric association constants K^*_{MA} are given as functions of ionic strength (Table 3). The effective ionic strength of the medium, I_e, depends on the extent of ion pairing:

$$I_e = \tfrac{1}{2}\Sigma\{z_M^2[M] + z_A^2[A] + z_{MA}^2[MA]\}, \tag{5}$$

with z being the charge on each species. An iterative procedure can be used to calculate the ion-pair equilibria in seawater. The ionic strength initially can be estimated from salinity, S, to be $I = 0.02$ S; this estimate of I provides a first approximation of the values for the K^*'s. The set of five equations for the anions (with the form of Eq. 4) can be solved for a first approximation of the free-anion molalities, [A], by using T(M) as an estimate of [M]. These values of [A] can be substituted into the four cation equations (with the form of Eq. 3) to obtain an estimate of the free-cation molalities, [M]. The molalities of the ion-pairs are obtained from:

$$[MA] = K^*_{MA}[M][A]. \tag{6}$$

A refined estimate of I_e can be made with Eq. 5 and the calculation can be repeated with revised values of K^* and with improved estimates of the free-cation molalities. This procedure requires about four iterations to converge to a solution in which the effective ionic strength and the species

Table 4. Ion-pair calculations for seawater at 25 °C, 1 atm pressure, and several salinities and pH expressed as a percentage of the total anion molality of each species based on the data in Tables 1 and 3

Salinity	5	15	35		
pH	8	8	7	8	9
SO_4^{2-}	67%	51%	38%	38%	38%
$NaSO_4^-$	17%	28%	38%	38%	38%
$MgSO_4^0$	13%	17%	19%	19%	19%
$CaSO_4^0$	3%	4%	4%	4%	4%
KSO_4^-	0.3%	0.5%	1%	1%	1%
HCO_3^-	95%	88%	79%	79%	79%
$NaHCO_3^0$	2%	6%	11%	11%	11%
$MgHCO_3^+$	2%	5%	8%	8%	8%
$CaHCO_3^+$	1%	1%	2%	2%	2%
CO_3^{2-}	38%	25%	14%	14%	14%
$NaCO_3^-$	11%	15%	16%	16%	16%
$MgCO_3^-$	37%	44%	52%	52%	52%
$CaCO_3^0$	14%	16%	18%	18%	18%
$B(OH)_4^-$	83%	68%	54%	54%	54%
$NaB(OH)_4^0$	6%	12%	18%	18%	18%
$MgB(OH)_4^+$	8%	15%	21%	21%	21%
$CaB(OH)_4^+$	3%	5%	7%	7%	7%
F^-	80%	65%	49%	49%	49%
NaF^0	1%	1%	2%	2%	2%
MgF^+	18%	32%	47%	47%	47%
CaF^+	1%	2%	2%	2%	2%

molalities do not vary by more than 1 part in 10^4. A simple microcomputer program, *SWSPECIES,* has been written in *BASIC* to perform these calculations; it requires about 30 seconds to complete the iterative calculation. The results of several calculations are shown in Table 4. The percentage of each total anion molality that exists as each species varies considerably with salinity but not with pH. The percentage of each total cation molality at $S = 35$ salinity that exists as each species varies by less than 0.3% over the pH range of 7 to 9.

Distribution of Trace-element Species

The free-ion molalities of the major constituents provide the ligand concentration values needed to calculate the species of trace elements. These calculations can be performed more easily than those for the ion-pairs because the results do not alter the ionic strength or the ligand concentrations significantly, so that iterations are not necessary. The same ionic medium approach can be used for the trace elements, in which the basic data set required is the stoichiometric stability constants for the species at the ionic strength, temperature, and pressure of interest. Turner, Whitfield, and Dickson (1981), TWD, compiled a set of stability-constant values as functions of ionic strength based on the critically selected values of Smith and Martell (1976) and of Baes and Mesmer (1976). These values can be updated and extended as more data become available, especially for media related to seawater. The TWD equilibrium models for trace metals include up to fourth-order inorganic complexes with the ligands OH^-, F^-, Cl^-, SO_4^{2-}, and CO_3^{2-}.

The equilibrium species of a metal with a specific oxidation state such as Fe(III) can be obtained from:

$$[Fe^{3+}]/T(Fe(III)) = 1/(1 + \Sigma \beta_{FeL_n}[L]^n), \tag{7}$$

where n is the number of ligands bound to the Fe(III) in a complex. The ratios of each complexed species to the total Fe(III) is given by:

$$[FeL_n]/T(Fe(III)) = \beta_{FeL_n}[L]/(1 + \Sigma \beta_{FeL_n}[L]^n). \tag{8}$$

The values of the β's can be obtained from a source such as TWD and the free-ligand concentrations [L] can be obtained from the ion-pairing calculations for the medium.

For some metals it is important to consider the formation of complexes with organic ligands in seawater. There is undoubtedly a wide range of organic substances in natural waters, only some of which have been characterized to date. While some specific types of organic compounds such as amino acids, carbohydrates, aldehydes, and fatty acids have been measured in the ocean, most of the dissolved organic carbon has not been molecularly characterized, including humic-like substances and fulvic acid (Williams, 1975). The acid-base properties of this material suggest that it contains carboxylic, phenolic, and possibly amino functional groups that dissociate progressively over the pH range of 4 to 8 (Huizenga and Kester, 1979; Varney et al., 1983). The charged sites resulting from this acid dissociation provide potential locations for the binding of trace elements. In addition to this bulk organic matter that may account for 70%–80% of the organic carbon in seawater, there may be organic ligands with specific metal-chelating properties such as siderophores, but their concentrations in natural marine environments have not been established (Morel and Morel-Laurens, 1983; Trick et al., 1983).

At present there is no satisfactory method for quantitatively accounting for organically complexed metals in an equilibrium model. But there have been many attempts to place some limits on the relative importance of organic and inorganic ligands for specific metals, and there are a number of analytical approaches to gain information about the presence of organically bound metals. If one assumes that the organic ligands in seawater have chemical properties similar to some generic types of organic functional groups, such as those associated with carboxylic acids or EDTA, it is possible to examine the potential importance of equilibrium metal-organic species. When such an analysis was done for Fe(III) the results produced two important conclusions (Kester et al., 1975). In many cases the presence of high concentrations of magnesium and calcium ions in seawater binds a large portion of the organic ligands so that few sites are available to metals which have low concentrations. With Fe(III) the formation of very stable inorganic complexes (primarily with OH^-) overwhelmed the tendency to form organic complexes at pH values greater than 7. Equilibrium models can be used to identify those metals that are most likely to be influenced by organic ligands, but direct analytical observations are necessary to verify and quantify these predictions.

One final important aspect of trace-element species in an equilibrium model is consideration of the redox state of the element. Many elements can have multiple oxidation states under marine conditions leading to major changes in their chemical behavior; examples include Fe(II)/Fe(III), Cr(III)/Cr(VI), Mn(II)/Mn(IV), H_2S/SO_4^{2-}, As(III)/As(V), Hg(0)/Hg(I)/Hg(II). Other trace elements are believed to occur with only one oxidation state in marine systems – Cd(II), Al(III), Pb(II), and Zn(II). The ionic medium equilibrium model that has been used for the major constituents and the trace-element complexes can be extended to include redox equilibria among oxidation states (Kester et al., 1975). One of the major limitations with these redox considerations is the lack of a definitive value for the redox potential of marine waters. The equivalent of a reliable pH electrode is lacking for redox potentials, and it has not been possible to predict unambiguously these potentials from chemical data (Breck, 1974; Parsons, 1975). Some redox reactions do not readily achieve equilibrium, and some geochemical data suggest that no single redox potential exists for the mixture of redox couples in the environment (Lindberg and Runnells, 1984). Nevertheless, a number of analytical studies have shown a progression of redox couples that are consistent with systematic variations in redox potential (Emerson et al., 1979; Landing et al., 1984; Liss et al., 1973).

Examples of Equilibrium Models for Trace Metals

Predicting inorganic species based on an equilibrium model provides a means of examining the chemical behavior of metals in natural systems. Copper, iron, cadmium, and lead will be used as examples, but the general approach can be applied to a wide range of elements (Turner et al., 1981). A complete equilibrium model for metals in seawater must consider organic as well as inorganic ligands. While some progress has been made in characterizing the organic ligands for a metal such as copper (Zuehlke and Kester, 1983), the present knowledge of organic ligands in marine systems is not adequate for a systematic and quantitative model for metals in general. It will be necessary to have more data on the functional groups that complex with metals, as well as their protonation equilibria and their interaction with the major cations in seawater in order to formulate a complete model of metal species.

Copper

The marine chemistry of copper has received a great deal of attention in recent years because of its active geochemistry and its importance for biological systems. In the ocean copper is depleted in surface waters, its concentration reaches maximum values near the seafloor, and there is evidence of scavenging from the water column at mid-depths. Copper is an economically important constituent in the manganese nodules that form on the seafloor. Biologically, copper is an essential element, but at high concentrations it is toxic to marine organisms.

The variations in copper species in seawater is an important factor in understanding the marine chemistry of this metal. The solubility of solid phases composed of copper, such as malachite, is modified as the distribution of species changes (Symes and Kester, 1984 a). The adsorption of copper onto suspended particles in marine systems depends on its species (O'Connor and Kester, 1975). The response of phytoplankton to copper depends not on its total concentration but on the amount of copper that is uncomplexed (Sunda and Lewis, 1978).

The first step in examining an equilibrium model for a metal such as copper can be to draw upon the available data for its complexes with the ligands in seawater and determine which species are most important (Turner et al., 1981). The reliability of such a model can be improved if an experimental program is then undertaken to measure a consistent set of equilibrium constants for the major species in a medium designed to simulate the ionic-medium effects of seawater. Such an investigation has recently been completed for copper in which its hydrolysis products, $CuOH^+$ and $Cu(OH)_2^0$ (Paulson and Kester, 1980), its carbonate and bicarbonate complexes (Zuehlke and Kester, 1983), and its chloride complexes (Byrne et al., 1983) were investigated. These studies have eliminated much of the disparity that existed in earlier models based on data extracted from the chemical literature and then applied to seawater without experimental verification. Previous models had reported the major inorganic species of copper in seawater to be $Cu(OH)_2^0$, $CuCO_3^0$, or $CuOHCl^0$ (Ahrland, 1975; Dyrssen and Wedborg, 1974; Long and Angino, 1977; Zirino and Yamamoto, 1972). The species that result from the recently determined stability constants at the ionic strength of seawater (0.6 to 0.7) are given in Fig. 2.

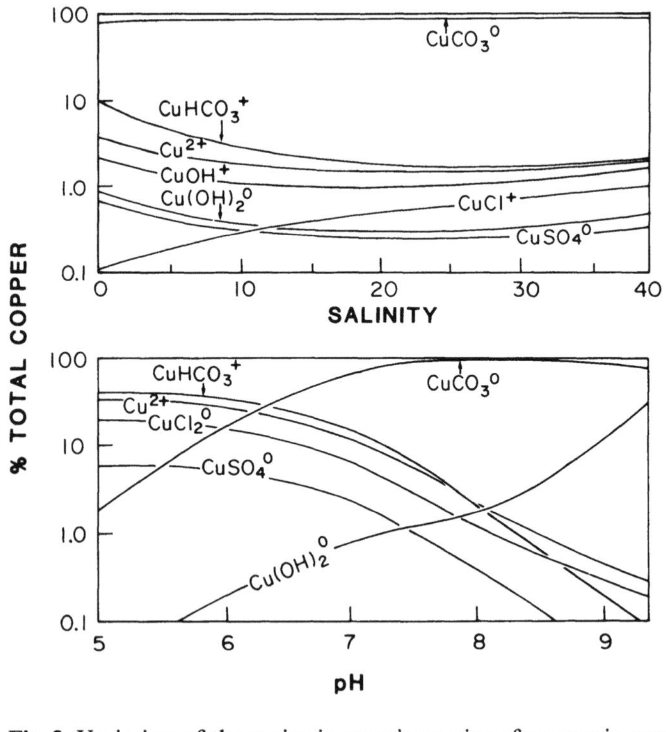

Fig. 2. Variation of the major inorganic species of copper in seawater over a range of salinity and pH. The salinity variation is at pH 8.0 and the pH variation is at 35.0 salinity. These calculations are based on stability constants determined in our laboratory

The importance of carbonate and bicarbonate ligands in the formation of inorganic copper species means that the environmental variables pH and alkalinity will have first-order effects on the marine chemistry of copper. Salinity will have an effect inasmuch as it reflects alkalinity changes and because the stability constants vary with ionic strength, but the salinity-dependent ligands chloride and sulfate are not of major importance with copper.

Iron

The species of iron(III) in seawater are controlled by hydrolysis products such as the three complexes $FeOH^{2+}$, $Fe(OH)_2^+$, and $Fe(OH)_3^0$ (Table 5). A major difficulty in relating a chemical model for iron in seawater with analytical data is the distinction between solution and solid phases when making measurements of iron. In many chemical systems iron is known to form colloidal species that cannot be separated from the dissolved phase by normal filtration methods. These colloidal species can be charge-stabilized in solutions of low ionic strength, and one might expect that they would flocculate when the surface charge is neutralized by the ionic medium of seawater. On the other hand, colloidal species can be stabilized by adsorbed organic substances, which are important in marine systems.

The stability constants for the hydrolysis species of iron have been determined under conditions related to seawater using potentiometric, spectroscopic, and solubility methods (Byrne and Kester, 1976a, 1981). One may debate the exact molecular nature of the $Fe(OH)_3^0$ species – whether it is monomolecular as implied by the chemical formula, or whether it could be polymeric up to colloidal dimensions (Zafiriou and True, 1980). Its stability constant was derived from a series of experiments that provided in-

Table 5. Percentage of dissolved iron (III) in seawater as various species over a range of salinity and pH calculated in this study

Salinity	5	15	35		
pH	8	8	7	8	9
Fe^{3+}	1.6×10^{-9}	2.5×10^{-9}	3.8×10^{-6}	3.5×10^{-9}	9.3×10^{-13}
$FeOH^{2+}$	< 0.1	< 0.1	< 0.1	< 0.1	< 0.1
$Fe(OH)_2^+$	2.2	2.8	22.6	2.1	< 0.1
$Fe(OH)_3^0$	86.5	80.8	74.0	67.2	17.9
$Fe(OH)_4^-$	11.3	16.4	3.4	30.7	82.0

formation on the chemical behavior of this form of iron. The ratio of this species' concentration to that of free Fe varies with the cube of the hydroxyl-ion activity; when the pH is reduced this species dissociates very rapidly compared to iron that is retained on a filter; this species migrated through a dialysis membrane implying molecular dimensions of less than 3 nm (Byrne and Kester, 1976b). It is likely that these operationally defined characteristics of the $Fe(OH)_3^0$ species are of primary importance in the marine chemistry of iron.

The redox behavior of iron is also important. Under reducing conditions in the marine environment Fe(II) will be the stable form of iron. Fe(II) forms complexes with chloride and hydroxide in seawater with a significant portion remaining as the free divalent cation (Fig. 3). The chemical model for iron indicates that pH and Eh are the major environmental variables affecting the behavior of iron in ocean water.

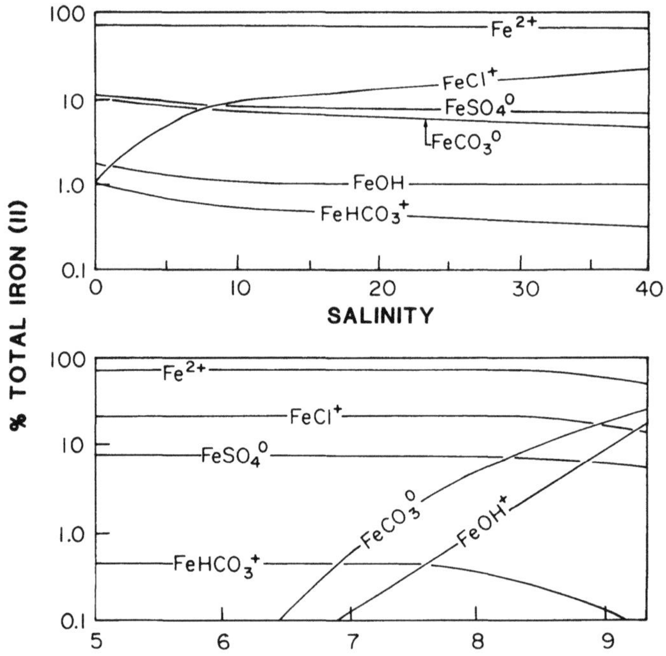

Fig. 3. Variation of iron (II) species in seawater with salinity and pH. The salinity variation is at pH 8.0 and the pH variation is at 35.0 salinity

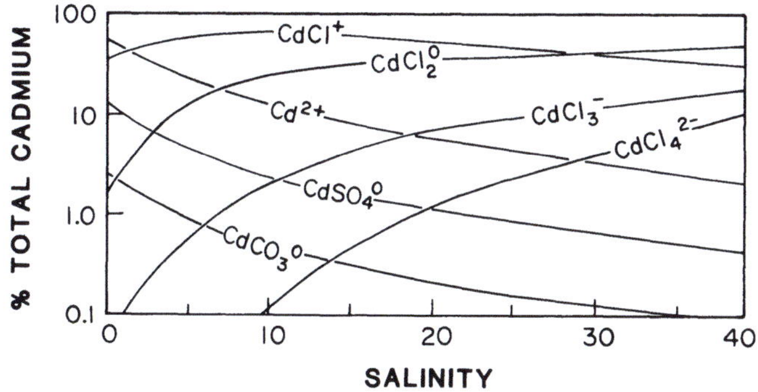

Fig. 4. Variation of cadmium species in seawater with salinity at pH 8.0 based on the constants compiled by Turner et al. (1981)

Cadmium

The chemistry of cadmium in marine systems is controlled by the formation of chloro-complexes (Turner et al., 1981). Consequently, salinity is the major environmental variable influencing the species of cadmium. Figure 4 shows the variations in cadmium species over the range of salinities found in estuaries. The pH-dependent species of cadmium never exceed a few percent of the total cadmium. Conditions exist in which cadmium may be predominantly a divalent or a univalent cation species, a neutral species, or an anionic species. These differences in net electrical charge may be important in the tendency of cadmium to adsorb onto particles with surface-charge sites.

Lead

An initial assessment of the lead species in marine systems indicates competitive equilibria between the pH-dependent ligands, carbonate and hydroxide, and the salinity-dependent ligands, chloride and sulfate (Fig. 5; Turner et al., 1981). The importance of carbonate species for lead suggests that bicarbonate complexes of lead should also be considered in an equilibrium model related to seawater. Electrolyte-solution chemists have not given a great deal of attention to systems containing carbonate, perhaps because these systems involve a complex set of acid-base, ion-pair, and gas-exchange equilibria making it difficult to test or apply simple chemical models. Many of the studies that have been the basis for determining metal-carbonate stability constants have ignored the presence of competing

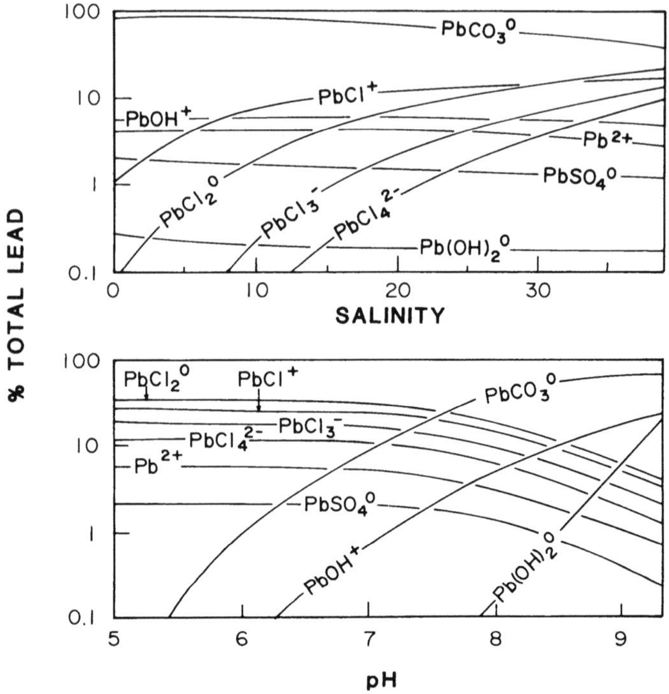

Fig. 5. Variation of lead species in seawater with salinity at pH 8.0 and with pH at 35.0 salinity. Calculations are based on the stability constants compiled by Turner et al. (1981)

equilibria between the metal and bicarbonate or hydroxide ions. Marine chemists may have a greater appreciation for the possible importance of these additional species because of the key role of the carbonate system in establishing the chemical properties of natural waters.

Recent experimental studies have provided better data for the complexation of lead in seawater. Byrne and Miller (1984) reported new values for the stability constants of $PbCl^+$, $PbCl_2^0$, and $PbCl_3^-$ in media of various ionic strengths. Byrne (1981) also examined the ultraviolet absorption of lead species in seawater media of various pH and analyzed the results in terms of $PbCO_3^0$, $PbOH^+$, $PbHCO_3^+$, and $Pb(CO_3)_2^{2-}$ species. It was not possible to obtain unique values for each of the pH-dependent species, but the experiments provided a consistent estimate of the stability constant for $PbCO_3^0$ and upper limits for the stability constants of the $PbOH^+$,

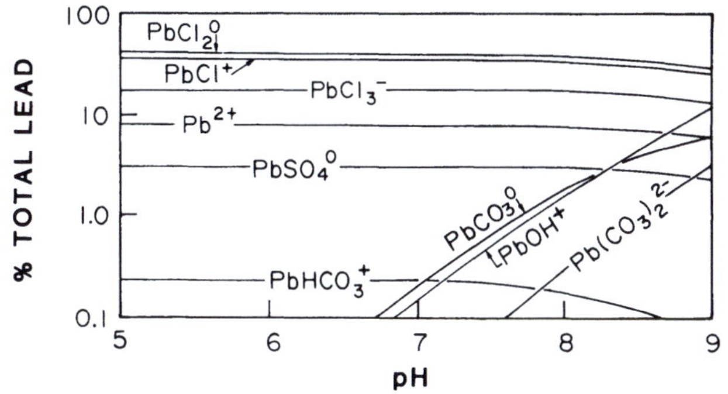

Fig. 6. Lead species in seawater as a function of pH based on the stability constants of Byrne (1981) and Byrne and Miller (1984). The diagram is for 35.0 salinity

$PbHCO_3^+$, and $Pb(CO_3)_2^{2-}$ species. A revised species model for lead based on these new data is shown in Fig. 6. The new data indicate that the pH-dependent lead species in seawater are less important than would be inferred from literature compilations (compare Figs. 5 and 6). The uncomplexed lead in seawater, based on inorganic species, comprises 6%–8% of the total inorganic lead. Chloride complexes of lead are the major species. The amounts of $PbOH^+$ and $Pb(CO_3)_2^{2-}$ indicated in Fig. 6 are maximum estimates; they are based on upper limits for both stability constants. The ultraviolet spectra can be accounted for by either of these species along with the $PbCO_3^0$. The comparison of an equilibrium model based on compiled constants and self-consistent experimental observations illustrates the need for experiments in media related to seawater in order to obtain quantitatively reliable models.

Influence of Environmental Variables

Several generalizations can be drawn from these examples of trace-metal species. A consideration of the major species of a metal in solution provides information about its net electrical charge, which will determine its electrostatic interaction in the medium and with surfaces. By identifying the ligands that form complexes with the major portion of a metal, the environmental variables that are most important in the chemistry of the metal can be determined. When the chloride species dominate as in the case of Cd(II), Cu(I), Ag(I), Hg(II), and Au(II), one can expect that the chemical behavior and biological effects of the metal will change substantially along the mixing gradient between fresh and ocean waters in an estuary. When hydrolysis products are the major species as for Fe(III) and Al(III), variations in pH from surface waters to the oxygen-minimum zone of the ocean are likely to be important. If carbonate (or bicarbonate) are among the most important ligands for a metal as for Cu(II) and Pb(II), then titration alkalinity and pH will be important variables. For some metals (Cu^{2+}, Mn^{2+}, and Zn^{2+}) experiments have suggested that phytoplankton growth responds to the uncomplexed metal-ion concentration or activity. Equilibrium models can be used to identify the environmental conditions under which the uncomplexed metal will be among the most and least important species.

Limitations of Equilibrium Models

In the preceding discussion allusion was made to the limitations of equilibrium models. There are three principal limitations: chemical systems may not be at equilibrium, equilibrium data compiled for a model from independent sources may not be internally consistent, and the data may be incomplete for the conditions of natural systems.

There are numerous examples of chemical systems in the environment that are not at equilibrium. Generalizations can be made about which types of reactions are most likely, or are least likely, to be at equilibrium. Ionic reactions such as acid-base, ion-pair, and inorganic-metal complexation tend to achieve equilibrium within minutes. Some reactions between solution and solid phases, redox reactions, and some metal-organic species do not achieve equilibrium on environmentally important time scales of days, months, and years. Comparisons between analytical data and equilibrium models can be used to identify the nonequilibrium systems. Increasing attention is being given to characterizing the role of kinetics in natural systems, including the effects of photochemical processes that lead to nonequilibrium states. As this work continues efforts should be made to develop kinetic models of natural systems that complement the equilibrium models.

An equilibrium model will yield erroneous results if the stability constants are not internally consistent. When separate studies are used to derive the equilibrium constants, the resulting model may overaccount for the inorganic complexes due to inconsistency in the interpretation of the experiments on which the complexes are based. For example, one investigator may attribute an observed phenomenon entirely to the $CuOH^+$ species; another investigator may make similar observations but attribute them to the $Cu(OH)_2^0$ species (unless corrections are applied for the $CuOH^+$); and a third experiment may observe greater complexing effects due to the presence of carbonate, without taking into account the $CuOH^+$ and $Cu(OH)_2^0$ species or the possible formation of a $CuHCO_3^+$ species. It appears that the recently completed studies of copper hydrolysis, bicarbonate, and carbonate complexes relative to seawater are the first to provide an internally consistent set of data for these complexes (Byrne et al., 1983; Paulson and Kester, 1980; Symes and Kester, 1985; Zuehlke and Kester, 1983). It is uncertain whether attempts were made to screen data for consistency in accounting for side-reaction equilibria when tabulations of critically selected stability constants were provided (Baes and Mesmer, 1976; Smith and Martell, 1976).

The final limitation in formulating equilibrium models is that the existing data base is inadequate for many purposes. This is especially evident when one tries to examine the possible importance of complexes formed by the anions of weak acids in seawater. More attention should be given to carbonate, bicarbonate, borate, and phosphate species. It is also difficult to consider the effects of low temperature and high pressure in an equilibrium model due to the lack of data. Variations in chemical behavior through the water column may be induced by the gradients in temperature and pressure. The chemical processes occurring in sedimentary pore waters can not be predicted reliably using data obtained at room temperature and one atmosphere pressure. One general class of metal complexes that may be important in marine systems but which have not been studied extensively by chemists are mixed-ligand species. Byrne has considered mixed-ligand species using theoretical and experimental approaches (Byrne, 1980; Byrne, 1983; Byrne and Young, 1982) and these species can be an important factor in metal complexation.

Analytical Tests of Equilibrium Models

An important aspect of dealing with equilibrium models is to consider analytical methods that can provide verification of the major classes of species. Analytical techniques can often be devised to provide operationally defined partitioning of an element into fractions that can be related to the results of an equilibrium species model. Such tests can identify whether or not the system is at equilibrium and they can indicate if a major species has been neglected.

Anodic stripping voltammetry (ASV) has been a widely used analytical method that depends on the species present in solution. In marine systems this technique can be used to investigate cadmium, copper, lead, and zinc.

The metal is initially electrochemically reduced from its oxidation state in solution to form an elemental amalgam with mercury; this step serves to preconcentrate the metal from the solution. The quantity of metal incorporated into the electrode is then measured by varying the voltage on the electrode and measuring the current which flows as the metal is oxidized back into solution. This technique depends on the kinetic characteristics of the metal species in solution; if a portion of a metal such as copper is bound to organic ligands that have slow dissociation kinetics, that fraction of the metal will not be detected (for more complete considerations see Lund, this volume). Whereas ASV methods have been usefully applied to natural waters (Branica et al., 1976; Huizenga and Kester, 1983; Nurnberg and Valenta, 1983; Piotrowicz et al., 1983), the interpretation of the results in terms of specific species is not always straightforward. The species inferred may sometimes depend on the electrochemical reactions occurring at the surface of the electrode. For example, there is evidence that in media containing chloride, such as seawater, the reduction of copper(II) involves an intermediate step that is stabilized by the copper(I)-chloride complex (Huizenga and Kester, 1984).

Ultraviolet spectroscopy of copper and lead in seawater media has been used to demonstrate the important inorganic complexes formed by these metals (Byrne, 1981; Byrne and Miller, 1985). This technique requires experiments at higher concentrations of metals than occur naturally in seawater, but it has the advantage of being able to identify the major complexes and their response to such seawater variables as pH, alkalinity, or salinity with a minimum of assumptions and auxiliary data.

Another method that has proved helpful in characterizing metal-organic species has been reversed-phase liquid chromatography, RPLC (Mills et al., 1982; Hanson and Quinn, 1983). With this technique seawater is passed through a cartridge that selectively retains those species of a metal that have a particular chemical property, such as being hydrophobic. This method has been used to examine the chemical forms of copper in seawater, as it can distinguish between organically bound and ionic forms. Hanson et al. (1986) recently completed a study in an oceanic ecosystem (a warm-core Gulf Stream ring) in which an equilibrium model, ASV, RPLC, and atomic absorption spectroscopy were used concurrently to characterize the copper species. We found good agreement among these methods that provided information on the uncomplexed, electrochemically labile, and hydrophobic forms of copper. One striking result was that the ASV-labile copper reached a minimum concentration and the hydrophobic copper reached a maximum concentration in the chlorophyll-maximum layer of the water column. This observation demonstrates the ability of oceanic organisms to alter the species of copper, possibly to reduce its toxicity.

One more approach that can be useful for testing an equilibrium model is to measure variations in the solubility of a relatively insoluble solid phase as the composition of the medium is changed. A study was recently completed that examined the solubility of the copper mineral, malachite ($Cu_2(OH)_2CO_3$), in various solutions including seawater (Symes and Kester, 1984 b). These measurements provided an overall test of the equilibrium model for inorganic complexes of copper(II) in seawater, and they contributed new information indicating the possible importance of a $CuOHCO_3^0$ mixed-ligand species.

Conclusions

There are several stages in the application of equilibrium models to seawater. The first is the use of available equilibrium-constant data with whatever assumptions might be required to make an initial set of predictions about the major species in the system. The second is to conduct experimental investigations that will remove as many of the initial assumptions as possible and provide data as directly related to the medium of seawater as possible. The third step is to seek analytical verification of the model.

This progression can be identified in studies of marine chemistry that span the past six decades. It was soon after the development of electrolyte-solution theory (with the ionic dissociation of electrolytes and the Debye-Hückel theory) that marine chemists realized that the carbon dioxide system in seawater did not conform to simple predictions. This led Buch et al. (1932) to define and measure apparent dissociation constants for carbonic acid in seawater. Subsequent to their work many analytical studies have confirmed the equilibrium relationships that exist among such properties of seawater as the total CO_2 concentration, alkalinity, pH, the partial pressure of carbon dioxide gas, and the concentration of carbonate ions. Studies of other weak-acid systems including phosphate, arsenate, borate, ammonia, and hydrogen sulfide have followed the same approach. Another example is the model prediction of Garrels and Thompson (1962) concerning major ion interactions in seawater in terms of ion-pair formation, which have subsequently been refined with the experimental studies of Pytkowicz and coworkers (Johnson and Pytkowicz, 1979; Kester and Pytkowicz, 1969; Pytkowicz and Hawley, 1974). These models have been tested and substantially verified by measurements of ultrasonic absorption and Raman spectroscopy. Investigations of trace metals are yielding a similar pattern in which initial model predictions have provided the basis for experimental investigations. Trace-metal models have become sufficiently well developed and analytical methods have been refined to the point that it is now becoming possible to seek analytical verification of model predictions. It will be important that future applications of these analytical methods be conducted in the field to reveal the ways in which marine biogeochemical processes respond to and modify chemical species.

The distinction between equilibrium and nonequilibrium systems in the environment will continue to be an important topic. In addition to kinetic barriers such as large activation energies that can prevent equilibrium, natural systems are perturbed from equilibrium by photochemical and biological processes. It is important to improve our understanding of these factors in marine chemistry. There is a need to enhance kinetically based models of marine processes to complement equilibrium models.

The data for equilibrium calculations need to be expanded to include the effects of pressure and temperature. There are chemical systems that are important in seawater which have not received adequate attention by inorganic chemists. The interactions of carbonate and bicarbonate with cationic species in seawater is of major importance in marine chemistry.

Acknowledgments. The development of the equilibrium models and microcomputer programs used in this work was conducted at The Laboratory of the Marine Biological Association of the U.K., Plymouth, England, while on a sabbatical leave from the University of Rhode Island. During that time I greatly appreciated many stimulating discussions with M. Whitfield. I appreciate the comments provided on a draft of the paper by R. W. Zuehlke, M. Bernhard, and M. O. Andreae. This research has been supported by Office of Naval Research contract no. N000-14-81-C-0062 and by National Oceanic and Atmospheric Administration contract no. M-04-8-MO1-192.

References

Ahrland W (1975) Metal complexes present in seawater. In: Goldberg ED (ed) The Nature of Seawater. Dahlem Konferenzen, Berlin, pp 219–244

Baes CF Jr, Mesmer RE (1976) The Hydrolysis of Cations. John Wiley, New York

Branica M, Sipos L, Bubic S, Kozar S (1976) Electroanalytical determination and characterization of some heavy metals in seawater. In: LaFleur PD (ed) Accuracy in Trace Analysis: Sampling, Sample Handling, Analysis, National Bureau of Standards Special Publication 422, vol 2. U.S. Government Printing Office, Washington, DC, pp 917–928

Breck WG (1974) Redox levels in the sea. In: Goldberg ED (ed) The Sea: Ideas and Observations on Progress in the Study of the Sea, vol 5. John Wiley, New York, pp 153–179

Brown MF, Kester DR (1980) Ultraviolet spectroscopic studies related to iron complexes in marine systems. Thalassia Jugoslavica 16:191–201

Buch K, Harvey HW, Wattenberg H, Gripenberg S (1932) Über das Kohlensäuresystem im Meerwasser. Conseil Permanent pour l'Exploration de la Mer. Rapport et Proces-Verbaux 79:1–70

Byrne RH (1981) Inorganic lead complexation in natural seawater determined by UV spectroscopy. Nature 290(5806):487–489

Byrne RH (1983) Trace metal complexation in high ligand variety natural media. Mar Chem 12:15–24

Byrne RH, Kester DR (1976a) A potentiometric study of ferric ion complexes in synthetic media and seawater. Mar Chem 4:275–287

Byrne RH, Kester DR (1976b) Solubility of hydrous ferric oxide and iron speciation in seawater. Mar Chem 4:255–274

Byrne RH, Kester DR (1981) Ultraviolet spectroscopic study of ferric equilibria at high chloride concentrations. J Sol Chem 10:51–67

Byrne RH, Miller WL (1984) Medium composition dependence of lead(II) complexation by chloride ion. Am J Sci 284:79–94

Byrne RH, Miller WL (1985) Copper(II) carbonate complexation in seawater. Geochim Cosmochim Acta 49(8):1837–1844

Byrne RH, Young RW (1982) Mixed halide complexes of lead. A comparison with theoretical predictions. J Sol Chem 11(2):127–136

Byrne RH, van der Weijden CH, Kester DR, Zuehlke RW (1983) Evaluation of the $CuCl^+$ stability constant and molar absorptivity in aqueous media. J Sol Chem 12:581–595

Daly FJ, Brown CW, Kester DR (1972) Sodium and magnesium sulfate ion-pairing. Evidence from Raman spectroscopy. J Phys Chem 76:3664–3668

Dyrssen D, Wedborg M (1974) Equilibrium calculations of the speciation of elements in seawater. In: Goldberg ED (ed) The Sea: Ideas and Observations on Progress in the Study of the Sea, vol 5. John Wiley, New York, pp 181–195

Emerson S, Cranston RE, Liss PS (1979) Redox species in a reducing fjord: equilibrium and kinetic considerations. Deep-Sea Res 26(8):859–878

Fisher F (1967) Ion pairing of magnesium sulfate in seawater: determined by ultrasonic absorption. Science 157:823

Garrels RM, Thompson ME (1962) A chemical model for seawater at 25 °C and one atmosphere total pressure. Am J Sci 260:57–66

Hanson AK Jr, Quinn JG (1983) The distribution of dissolved and organically complexed copper and nickel in the Middle Atlantic Bight. Can J Fish Aquat Sci 40:151–161

Hanson AK Jr, Sakamoto-Arnold CM, Huizenga DL, Kester DR (1986) Copper speciation in oceanic waters. Mar Chem, in press

Huizenga DL, Kester DR (1979) Protonation equilibria of marine dissolved organic matter. Limnol Oceanog 24:145–150

Huizenga DL, Kester DR (1983) The distribution of total and electrochemically available copper in the northwestern Atlantic Ocean. Mar Chem 13:281–291

Huizenga DL, Kester DR (1984) Stripping polarograms for film electrodes. J Electroan Chem 164:229–236

Johnson KS, Pytkowicz RM (1979) Ion association of chloride and sulphate with sodium, potassium, magnesium, and calcium in seawater at 25 °C. Mar Chem 8:87–93

Kester DR, Byrne RH (1972) Chemical forms of iron in seawater. In: Horn DR (ed) Ferromanganese Deposits on the Ocean Floor. Lamont-Doherty Geological Observatory, Palisades, NY, pp 107–116

Kester DR, Byrne RH Jr, Liang Y-J (1975) Redox reactions and solution complexes of iron in marine systems. In: Church TM (ed) Chemistry in the Coastal Environment, ACS Symposium Series No. 18. American Chemical Society, Washington, DC, pp 56–79

Kester DR, Pytkowicz RM (1969) Sodium, magnesium, and calcium ion-pairs in seawater at 25 °C. Limnol Oceanog 14:686–692

Landing WM, Cutter GA, Smith GJ, Bruland KW (1984) Suboxic redox chemistry at VERTEX-II and -III. EOS 65(45):924

Lindberg RD, Runnells DD (1984) Ground water redox reactions: an analysis of equilibrium state applied to Eh measurements and geochemical modeling. Science 225:925–927

Liss PS, Herring JR, Goldberg ED (1978) The iodide/iodate system in seawater as a possible measure of redox potential. Nature Phys Sci 242:108–109

Long DT, Angino EE (1977) Chemical speciation of Cd, Cu, Pb, and Zn in mixed freshwater, seawater, and brine solutions. Geochim Cosmochim Acta 41:1183–1191

Millero FJ, Schreiber DR (1982) Use of the ion pairing model to estimate activity coefficients of the ionic components of natural waters. Am J Sci 282:1508–1540

Mills GL, Hanson AK Jr, Quinn JG, Lammela WR, Chasteen ND (1982) Chemical studies of copper-organic complexes isolated from estuarine waters using C-18 reverse-phase liquid chromatography. Mar Chem 11:355–377

Morel FMM, Morel-Laurens MML (1983) Trace metals and plankton in the oceans: facts and speculations. In: Wong CS, Boyle E, Bruland KW, Burton JD, Goldberg ED (eds) Trace Metals in Seawater. Plenum Press, New York, pp 841–869

Nürnberg HW, Valenta P (1983) Potentialities and applications of voltammetry in chemical speciation of trace metals in the sea. In: Wong CS, Boyle E, Bruland KW, Burton JD, Goldberg ED (eds) Trace Metals in Seawater. Plenum Press, New York, pp 671–697

O'Connor TP, Kester DR (1975) Adsorption of copper and cobalt from fresh and marine systems. Geochim Cosmochim Acta 39:1531–1543

Parsons R (1975) The role of oxygen in redox processes in aqueous solutions. In: Goldberg ED (ed) The Nature of Seawater. Dahlem Konferenzen, Berlin, pp 505–522

Paulson AJ, Kester DR (1980) Copper(II) ion hydrolysis in aqueous solution. J Sol Chem 9:269–277

Piotrowicz SR, Harvey GR, Springer-Young M, Courant RA, Boran DA (1983) Studies of cadmium, copper, and zinc interactions with marine fulvic and humic materials in seawater using anodic stripping voltammetry. In: Wong CS, Boyle E, Bruland KW, Burton JD, Goldberg ED (eds) Trace Metals in Seawater. Plenum Press, New York, pp 699–717

Pytkowicz RM, Hawley JE (1974) Bicarbonate and carbonate ion-pairs and a model of seawater at 25 °C. Limnol Oceanog 19:223–234
Skirrow G (1975) The dissolved gases – carbon dioxide. In: Riley JP, Skirrow G (eds) Chemical Oceanography, 2nd edn, vol 2. Academic Press, London, pp 1–192
Smith RM, Martell AE (1976) Critical Stability Constants, vol 4: Inorganic Complexes. Plenum Press, New York
Sunda WG, Lewis JAM (1978) Effect of complexation by natural ligands on the toxicity of copper to a unicellular alga, *Monochrysis lutheri*. Limnol Oceanog 23:870–876
Symes JL, Kester DR (1984a) Thermodynamic stability studies of the basic copper carbonate mineral, malachite. Geochim Cosmochim Acta 48:2219–2229
Symes JL, Kester DR (1984b) Copper(II) interaction with carbonate species based on malachite solubility in perchlorate medium at the ionic strength of seawater. Mar Chem 16:189–211
Trick CG, Andersen RJ, Gillain A, Harrison PJ (1983) Prorocentrum: an extracellular siderophore produced by the marine dinoflagellate *Prorocentrum minimum*. Science 219:306–308
Turner DR, Whitfield M, Dickson AG (1981) The equilibrium speciation of dissolved components in freshwater and seawater at 25 °C and 1 atm pressure. Geochim Cosmochim Acta 45:855–881
Varney MJ, Mantoura RFC, Whitfield M, Turner DR, Riley JP (1983) Potentiometric and conformational studies of the acid-base properties of fulvic acid from natural waters. In: Wong CS, Boyle E, Bruland KW, Burton JD, Goldberg ED (eds) Trace Metals in Seawater. Plenum Press, New York, pp 751–772
Williams PJ leB (1975) Biological and chemical aspects of dissolved organic matter in sea water. In: Riley JP, Skirrow G (eds) Chemical Oceanography, 2nd edn, vol 2. Academic Press, London, pp 301–363
Zafiriou OC, True MB (1980) Interconversion of iron(III) hydroxy complexes in seawater. Mar Chem 8:281–288
Zirino A, Yamamoto S (1972) A pH-dependent model for the chemical speciation of copper, zinc, cadmium, and lead in seawater. Limnol Oceanog 17:661–671
Zuehlke RW, Kester DR (1983) Ultraviolet spectroscopic determination of the stability constants for copper carbonate and bicarbonate complexes up to the ionic strength of seawater. Mar Chem 13:203–226

The Importance of Chemical "Speciation" in Environmental Processes, eds. M. Bernhard, F. E. Brinckman, and P. J. Sadler, pp. 365–384. Dahlem Konferenzen 1986. Springer-Verlag Berlin, Heidelberg
© Dr. S. Bernhard, Dahlem Konferenzen

Formation and Transformation of Element Species in Estuaries

J. C. Duinker

Institut für Meereskunde
2300 Kiel 1, F. R. Germany

Abstract. The major mechanisms involved in the transport, distribution, and transformation of element species in estuaries are summarized. The importance of transport and mixing characteristics of water and particulates and their effects on master variables (such as pH, O_2) is emphasized. The interpretation of observed distribution patterns of non-conservative trace-element species in estuaries is in most cases complicated because of lack of chemical equilibrium conditions. A rather unique situation is found in the Scheldt Estuary, where (quasi-)equilibrium conditions may exist for several trace-element species. The behavior of operationally defined species of Cd, Zn, Mn, Ni, As, and V is described in terms of the various chemical properties and the water and sediment transport characteristics. This may assist the understanding of species distribution patterns of these elements under different conditions, i.e., in other estuaries.

Introduction

An estuary is an area where a river meets an inlet of the sea, or more precisely in geomorphological terms, a river valley that is open to the ocean. The estuary may include a freshwater section of the river, above the limit of maximum salt intrusion. Rather than on the basis of the presence of seawater (Pritchard, 1967), the definition of the upper boundary of an estuary is taken from the definition of an estuary as "an inlet of the sea reaching into a river valley as far as the upper limit of tidal rise" (Fairbridge, 1980). This definition allows the distinction of three sections: a) a marine or lower estuary, in free connection with the open sea; b) a middle estuary, subject to strong salt- and freshwater mixing; and c) an upper (or fluvial) estuary, characterized by freshwater but subject to daily tidal action.

Fresh- and seawater have very different characteristics. Mixing of these waters in estuaries results in large gradients of physicochemical properties such as total salt concentration (salinity*), major ion and trace-element species composition, pH, turbidity (seston concentration), dissolved oxygen concentration, etc. Dissolved and particulate suspended element species react in response to the variations in chemical environment to which they are exposed in an estuary.

Dissolved species are readily available for chemical reactions, e.g., complexation and exchange processes. These reactions may also involve surfaces of solid particles and, as a consequence, elements may be removed from or released into solution during estuarine mixing. Slower reaction rates tend to be characteristic for element species associated with particles. A distinction has to be made between element species bound in forms from which they may become available under certain environmental conditions, e.g., in iron and manganese oxides and organic matter, and forms which are firmly bound in inert lattice positions of heavy minerals or clastic fragments, e.g., Al-silicates. An important factor is also the time available for reaction during their residence in the estuary. If conditions and available time for reactions are favorable, certain particulate forms of an element, which may be stable in the fresh or marine regions, may become involved in physicochemical transformations in the estuary. However, no analytical methods are available which go beyond an investigation of operationally defined species, and this has been a drawback in estuarine as well as fresh- and seawater research.

The distribution pattern of any particular element between dissolved and particulate forms in an estuary is a complex function of a number of physical, chemical, and biological processes which occur simultaneously and partly independently. They depend on water and sediment transport characteristics, which vary within one tidal cycle and within the neap-spring cycle. They may vary even more strongly between seasons and between periods of extreme meteorological conditions. Fundamental problems and approaches have been described by Lauff (1967), Burton and Liss (1976), Aston (1978), and Olausson and Cato (1980).

---

* For a new definition of salinity see footnote under **The Scheldt Estuary: A Case Study**.

It has become clear that the often unexpected distribution patterns of chemicals (including contaminants) in estuaries cannot be interpreted without detailed knowledge of water and particulate matter transport characteristics. The popular plots of dissolved concentrations vs. salinity used to decide whether a given element behaves conservatively (i.e., whether its concentration in solution in the estuary can be accounted for by mere mixing of fresh- and seawater) are usually insufficient to allow conclusions on processes and mechanisms without additional data on hydrology and dynamics as well as on the physicochemical master variables which govern the distribution of element species and their transformation in an estuary. Some general major aspects of water and sediment characteristics in estuaries are summarized below.

Water Transport and Mixing

The circulation pattern of water in estuaries determines the residence time of water and transfer mechanisms and accumulation of particles. The interaction between river flow, tidal action, and topography results in a range of circulation types (Fig. 1).

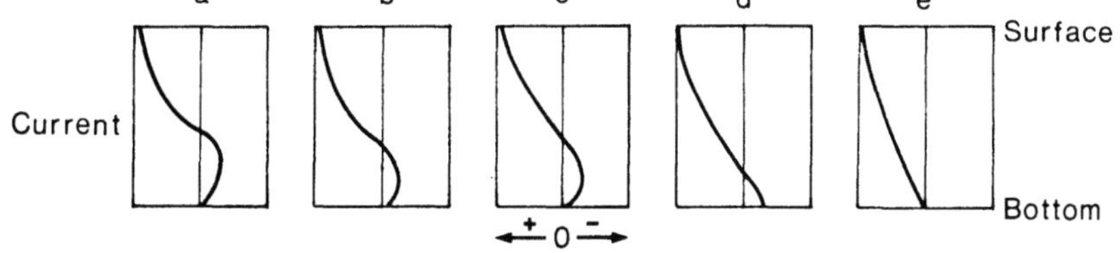

Fig. 1. Approximate representation of salinity (S) versus depth (upper) and residual current (lower: - = upstream, + = downstream) in typical estuary types. Each profile represents a specific position in the estuary where both fresh- (S_o) and seawater (S_s) contribute. The forms change in the upstream and downstream direction. **a** salt wedge without entrainment or mixing; **b** salt wedge with entrainment and some mixing with halocline; **c** stratified with halocline and mixing into the lower layer; **d** partially mixed; **e** well mixed. (Modified after Peters and Sterling, 1976)

In a salt-wedge estuary (highly stratified), the circulation is dominated by river flow (Pritchard, 1967; Bowden, 1980). Above certain velocities of the outward flowing layer, salt water is entrained in the upper layer. With increasing disturbances at the interface, a halocline with large salinity gradients may be formed, separating two layers with only minor internal gradients. Increased tidal action results in a partially mixed estuary, where salinity decreases gradually from bottom to surface. When the tidal action relative to the river flow is very strong, a well-mixed estuary will result, with only small or zero vertical gradients of salinity. If, in addition, the vertical velocity gradients would vanish as well, the estuary would be classified as homogeneous.

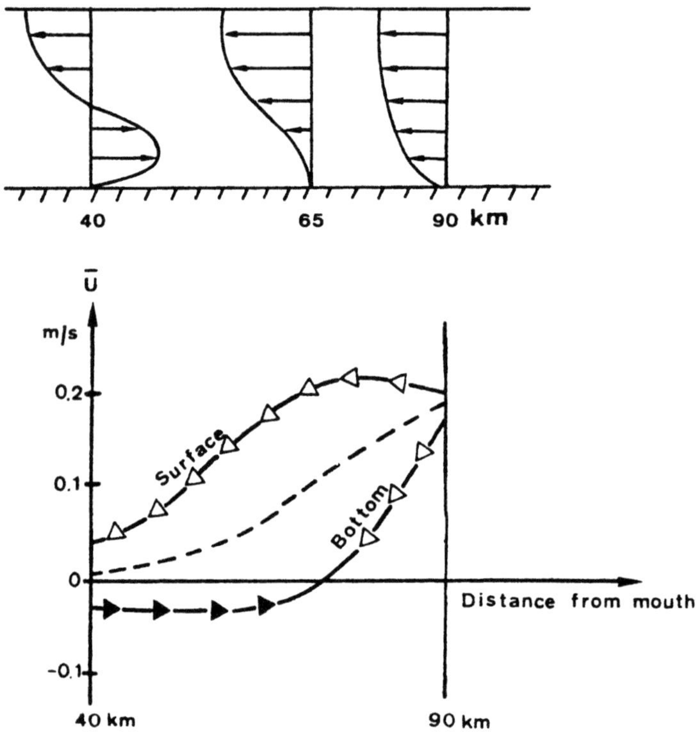

Fig. 2. Mean residual velocities in relation to depth at three positions in the Scheldt Estuary (top) and mean residual velocities in the surface layer and the bottom layer at various distances from the mouth of the estuary (bottom) (from Peters and Sterling, 1976)

The salt intrusion has an important effect on the vertical distribution of velocities because of the salinity and thus density gradients. The denser water masses have a tendency to flow upstream near the bottom (Postma, 1980). The mean velocities integrated over a tidal period (i.e., residual or nontidal velocities) show a two-layer system with an upstream flow in the lower layer. The distortion of the vertical distribution, depending on the horizontal density gradient, will change along the estuary. At positions sufficiently upstream, the residual current will be seaward from surface to bottom. Thus, there must be a region where the residual current near the bottom equals zero (Fig. 2). This is the region where particles may accumulate (turbidity maximum).

Residence Time of Water

The residence times of water vary among estuaries from one or a few days to several months. In perfectly stratified estuaries, freshwater is removed rapidly in a restricted surface layer, without mixing with the underlying salt wedge. Generally the residence time increases with increasing vertical mixing, due to the dilution of freshwater in a large body of seawater. Qualitatively, it thus increases from a salt-wedge type to a well-mixed estuary.

The circulation type of an estuary can, at first approximation, be estimated from the ratio between the freshwater discharge through a cross section during a tidal cycle and the amount of water entering the estuary through the section during flood tide. However, morphological aspects have to be taken into account as well. For instance, decreasing depth within an estuary tends to favor mixing (Bowden, 1980; Wollast and Duinker, 1982).

Sediment Transport and Accumulation

The residence time of particulates in an estuary is usually considerably larger than for water, except for small and low-density particles which are transported like dissolved components. In many estuaries, a region exists in the low salinity range where suspended particulate matter (SPM) concentrations largely exceed those in the source waters. Colloidal fractions which are stable in the freshwater region can be destabilized in the low salinity range of the estuary, and they may form a significant fraction of the material in suspension and in sediments (Wollast and Peters, 1978). This area is an exclusive site in an estuary where particles can collect and accumulate despite prevailing strong currents. The particles – mainly silt- and clay-sized fractions – belong to a narrow range of settling velocities. The agglomeration of coherent particles with large water contents in the presence of organics may be a dominant factor in these processes (Wellershaus, 1981). Also, discrete flocs of clays, e.g., may be produced due to increased total salt concentrations in the early stages of mixing. In general, however, it is difficult actually to observe the latter process *in situ* (Meade, 1972).

Collection of particles occurs near the head of salt intrusion or, more accurately, in the region where the nontidal near-bottom current vanishes. A part of these particles can be deposited on the bottom, depending on vertical settling velocities and typical estuarine vertical mixing velocities (Festa and Hansen, 1978). From there they can return into the overlying water during those phases of the tidal cycle when increased turbulent dispersion occurs after rapid changes in current velocities. During high and low water slacks, particles are deposited on the bottom if their settling velocities exceed the mean vertical mixing velocity. This can measurably affect the size distribution and the mineral composition (particle sorting) and trace-metal content of the material in suspension (Duinker et al., 1982a).

Distribution and Transformation of Chemicals in the Estuarine Environment

The strong gradients of various properties in estuaries favor species transformation between different physicochemical forms, including dissolution, precipitation, redox and exchange reactions and flocculation/deflocculation processes.

Theoretical Models

The behavior of components in solution can, in principle, be dealt with by theoretical models, under the assumption that chemical equilibrium conditions prevail and that activity coefficients can be obtained experimentally and theoretically (Dyrssen and Wedborg, 1980). A complicating factor is the role of dissolved organic ligands for which available qualitative and quantitative information is rather limited, both experimentally and theoretically. The role of humic material derived from terrestrial sources gradually diminishes from river to sea; inorganic ligands become progressively more important. Rapid processes involving particles and dissolved components are known to exist, e.g., the replacement of Ca on sorption sites of clays by Na in waters of increased salinities (Russell, 1970; Sayles and Mangelsdorf, 1977, 1979).

Conflicting evidence has been presented in the literature on the relative role of particulate organic matter and mineral fragments such as iron and manganese oxides and clay minerals in estuaries, with respect to the exchange of certain trace elements between solution and particulates (Duinker, 1980). The scatter of opinions is partly due to the fact that these mineral and organic fractions usually occur as extremely complex associations of variable compositions.

Plots of dissolved concentrations against salinity (more accurately, chlorinity) can be used to evaluate the relative importance of removal and production processes. The possibilities and limitations of this approach have been extensively discussed (Boyle et al., 1974; Liss, 1976).

Laboratory Studies

The evolution in an estuary of the master variables pH, dissolved oxygen concentration, and turbidity is usually not linearly related to salinity. Moreover, the relations with salinity are different between these variables. For instance, a minimum can occur in the distribution of both pH (Mook and Koene, 1975) and dissolved oxygen concentration and a maximum in the distribution of turbidity, often at different salinities. These relations vary with river discharge, tidal action, meteorological conditions, etc.

In principle, it should be possible to determine experimentally or theoretically how the different species of a given element would behave when exposed to a particular well-defined condition (in terms of all relevant variables) for a given period of time. A realistic physicochemical model for the behavior of the species of an element within an estuary would require simulation of all likely combinations of conditions in succession, during different periods of time. Laboratory experiments, designed to represent conditions such as occur in estuaries, have therefore only limited value for understanding the complexity of estuarine systems. However, they are useful in analyzing specific processes of mechanisms and evaluating their kinetics.

Field Studies

The general lack of equilibrium or steady-state conditions in estuaries is a major problem for any interpretation of observed distribution patterns of the species of trace elements. Both temporal and regional variability have been observed in several estuaries (e.g., see Duinker, 1980). A rather unique situation exists in the Scheldt Estuary. Although its upper, middle, and lower estuarine sections differ considerably in physicochemical properties due to their morphological and hydrodynamic characteristics, (quasi-)equilibrium conditions exist in each region. This allows a detailed study of the effects of exposure of various elements to the different conditions in succession, during (slow) transport downstream.

The Scheldt Estuary: A Case Study

Apart from the occurrence of local small salinity gradients, the Scheldt Estuary is well mixed. It is characterized by a long residence time for freshwater (1–3 months, depending on river flow), and even more so for particulate matter accumulating in the turbidity zone with a large organic matter content. The long residence time of water is a consequence of the low freshwater discharge (average 80 $m^3\ s^{-1}$) and a strong tidal action, affecting the vertical water movement as far as 150 km upstream from the mouth, far above the limit of salt intrusion (Peters and Sterling, 1976). Vertical and lateral salinity differences are small (S = 1–2) *. Nevertheless, they are important for water and sediment transport. A region of zero nontidal current near the bottom is found at about km 80 near Antwerpen (Fig. 3). This is the site of the turbidity maximum, collecting particles with high organic C contents from the river, the lower estuary, the city of Antwerpen, and bordering agricultural land. The entire water column in this area is permanently undersaturated and often totally depleted of dissolved oxygen.

The distribution of dissolved components of several redox pairs (nitrogen, manganese, iron, and sulfur species) have been approximated by thermodynamical equilibrium calculations (Billen et al., 1976; Wollast,

* Salinity has recently been redefined (IAPSO, 1985). The term salinity is now defined as "practical salinity" and has replaced the term salinity defined as a function of chlorinity. This new definition was necessary because salinity is now obtained by conductivity measurements instead of by chemical titration. Practical salinity (S) has no dimension and is based on conductivity ratios. Symbols such as ‰ and 10^{-3} have been abandoned. Due to the constant relationships between major elements in seawater the equations defining practical salinity are valid between S = 2 and S = 42. However, to obtain some idea of the significance of the values in a first approximation the old definition of salinity may be helpful: the salinity of a sample of seawater divided by the mass of this sample when all carbonate has been converted into oxide, the bromine and iodine replaced by chlorine, and all organic matter completely oxidized.

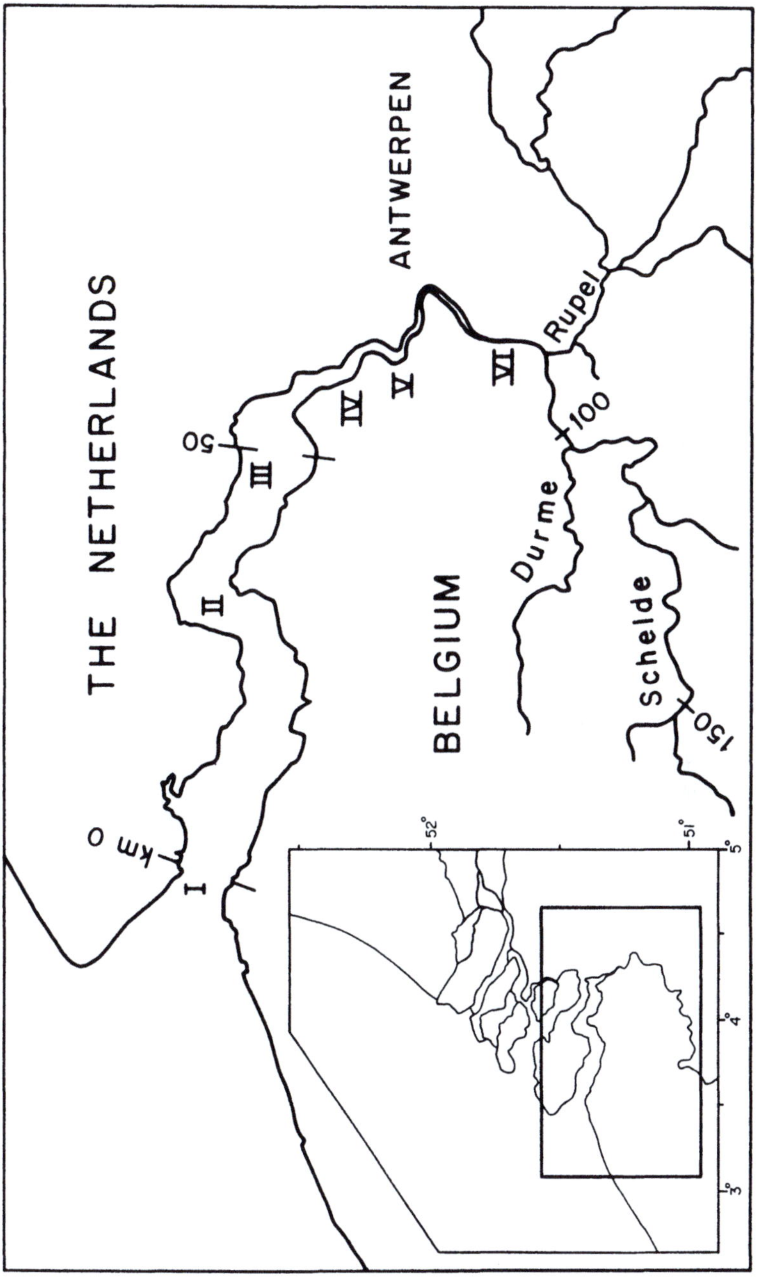

Fig. 3. The Scheldt Estuary and the location of the six positions (I-VI) where observations were made during tidal cycles. Distance from the mouth in km measured from Vlissingen (km 0)

1982; Wollast et al., 1979; Wollast and Duinker, 1982). We shall summarize here some recent findings on physicochemical transitions involving Cd, Zn, Ni, Mn, V, and As (Duinker et al., 1982b; van der Sloot et al., 1985) in connection with water and sediment transport characteristics.

Interpretation of observations made during complete tidal cycles at six fixed positions (I–VI, Fig. 3) in different, partly overlapping salinity regimes of the estuary is facilitated by the small nontidal water excursions (on the average 0.08 m s^{-1} at km 90 and 0.01 m s^{-1} at the mouth of the Scheldt Estuary (Peters and Sterling, 1976), causing essentially the same water body, i.e., with the same salinity, to return at each of the positions in the same phase of consecutive tidal cycles. The dependence of each parameter (pH, dissolved O_2 concentration, turbidity) on salinity thus reflects the changes of this parameter occurring during the slow residual water transport downstream where mixing with higher salinity water as well as exposure to other factors typical of another geographical region occur (of particular importance in the maximum turbidity zone).

The evolution of pH and dissolved oxygen concentration in the longitudinal profile reflect the effects of mixing and interaction with sediments (Fig. 4). pH showed large gradients at stations VI and III and small gradients (and lower values) at stations V and IV (the range of salinities at the stations is represented as horizontal bars below the pH and oxygen profiles). Oxygen contents were low, with minor variations, at stations VI and V while values increased rapidly, with large variations, at stations IV and III. Low dissolved O_2 concentrations resulted from microbial degradation of organic matter in the high turbidity zone. Re-aeration was insufficient to balance the rate of disappearance of this oxidant (Billen et al., 1976). From station IV onwards, less oxygen was required for microbial activity and concentrations in solution were restored by supply from higher salinity water and by increased re-aeration in the broadening estuary. The seston concentration shows, in general, a decrease with salinity.

Any chemical transported downstream in a water parcel has been exposed to a continuous series of slowly changing conditions. If the movements were slow enough, conditions might have been favorable for (quasi-)equilibria to be established. This is supported by the following find-

Fig. 4. Evolution of pH and dissolved O_2 concentration (% of saturation) and turbidity (mgdm^{-3}), dissolved V and As (nM), and dissolved Mn, Cd, Zn, and Ni (μM) in the Scheldt Estuary. Data obtained at the six fixed positions I–VI (see Fig. 3). The salinity range at each position during a tidal cycle is indicated below the O_2/pH graph. (From Duinker et al., 1982b; van der Sloot et al., 1985

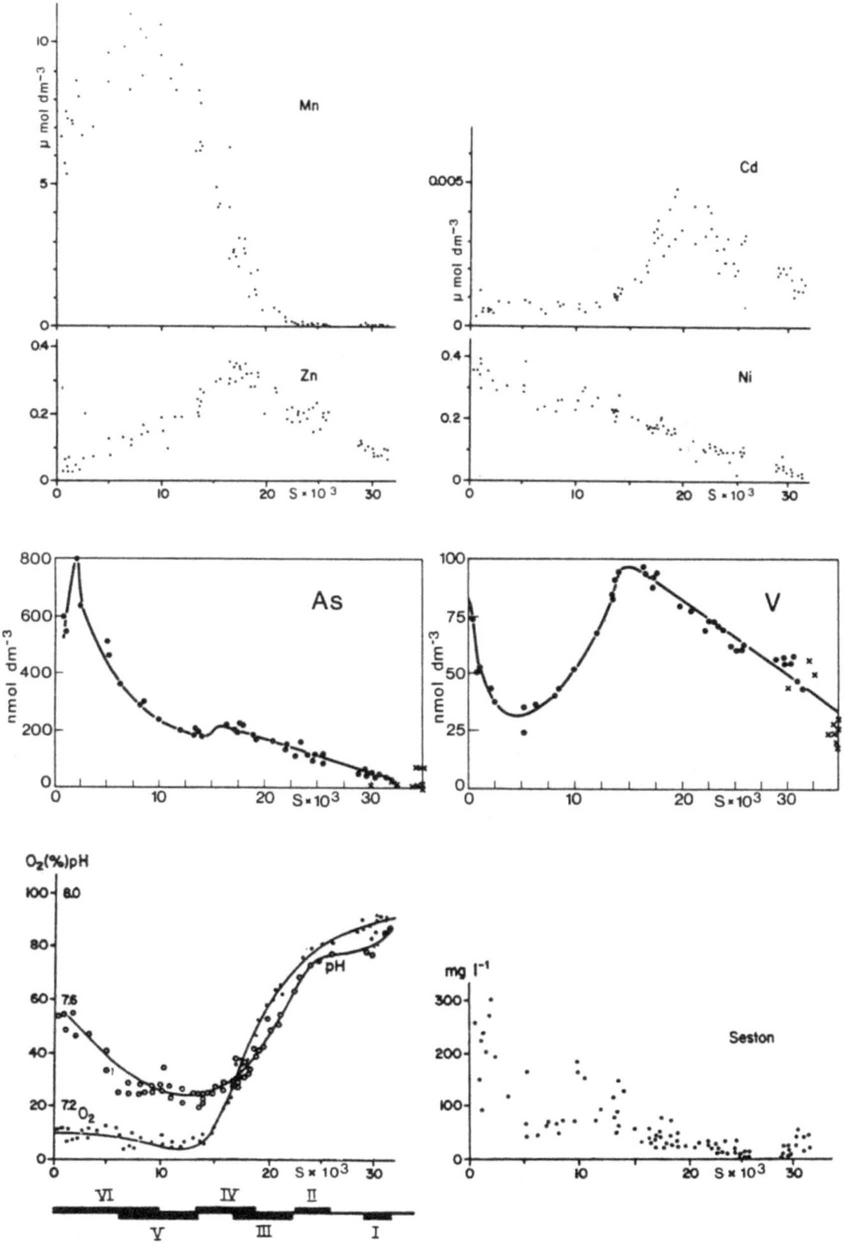
Mn
µ mol dm⁻³
Cd
µ mol dm⁻³
Zn
Ni
S × 10³
As
nmol dm⁻³
V
nmol dm⁻³
S×10³
O₂(%)pH
pH
O₂
VI
IV
II
V
III
I
mg l⁻¹
Seston
S × 10³

ings. Dissolved concentrations of several trace elements and the contents of these elements in suspended matter (w/w) demonstrated all the same characteristics: each water parcel with a particular salinity was characterized by a specific concentration and content of Mn, Cd, Ni, Zn, V, and As. The variability of these values at any salinity value, irrespective of location, is considerably smaller than that found in practically all other estuaries studied. Their variations with salinity (in other words, during net downstream transport) were considerable; in line with the latter observation, the relations between dissolved concentrations and salinity are remarkably smooth (Fig. 4). For example, the relation is linear for Ni and clear-cut maxima are found for Cd, Zn, V, and Mn. These maxima occur at different salinities.

In some cases, these extremes can be related to the evolution of contents of the corresponding element in suspension (e.g., Mn and As). The observed profiles of concentration in solution and contents in suspension therefore have element-specific characteristics; we shall discuss the findings for each element in turn. The samples were filtered on 0.45 μm millipore filters on board ship. Mn, Cd, Zn, and Ni were analyzed by AAS, and As and V by NAA. Details are given by van der Sloot et al. 1985.

Manganese

Maximum concentrations of dissolved manganese occurred in the relatively low salinity range, i.e., around S = 8 (Fig. 4). On the other hand, strongly increased manganese contents were present in suspended particulate matter (SPM) at higher salinities (around S = 21). It has been demonstrated that the master variables for the species distribution of Mn are the concentration of dissolved oxygen and the pH (Wollast et al., 1979). Both of these parameters have minimum values at salinities between S = 8 and S = 12 (Fig. 4). It has been shown with the help of thermodynamic models that under these conditions $MnCO_3$ is the thermodynamically stable solid species. At the strongly increased concentration of dissolved O_2 and the higher pH values downstream, Mn(II) is oxidized into higher oxidation states (MnOOH and MnO_2) with lower solubilities (Fig. 5). A significant part of both the dissolved and particulate Mn carried by the river is recycled between the lower and upper estuary. An important part of the particulate Mn accumulates in the sediments and the fraction that escapes the estuarine trapping mechanisms enters the sea mainly in particulate forms (Duinker et al., 1979).

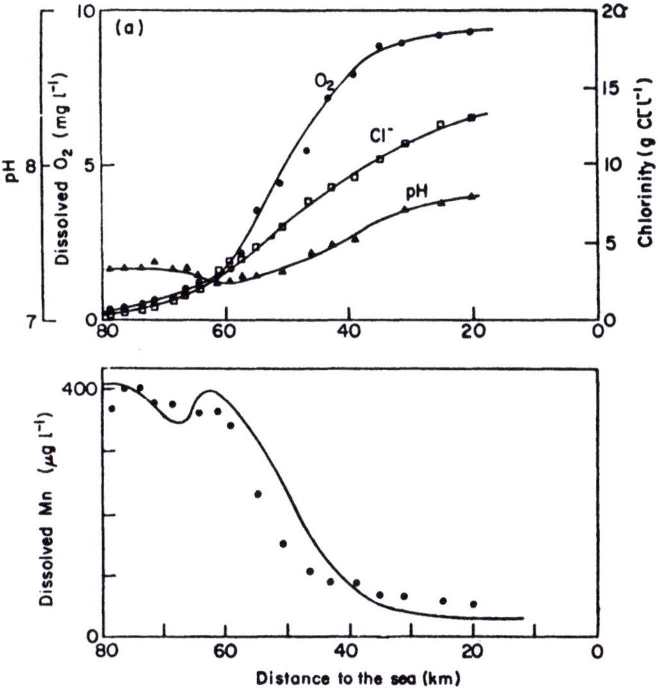

Fig. 5. Comparison between experimental profiles of dissolved manganese concentration in the Scheldt Estuary (black dots, lower diagram) and profiles computed with the thermodynamic equilibrium model using the experimental profiles of chlorinity, oxygen concentration, and pH smoothed as shown in the upper diagram. (April–May 1974, from Wollast et al., 1979)

Cadmium

The maximum in the dissolved Cd concentration occurs at much higher salinities (around S = 20) than for Mn (Fig. 4). Low concentrations of dissolved Cd occurred in the high turbidity zone (where seston concentrations are high; Fig. 4). If the levels of dissolved Cd would be governed by equilibrium with solid cadmium sulfide, still lower values should be expected in the low O_2 zone (i.e., below S = 15). Sulfate reduction is inhibited in the presence of oxygen and nitrate. In the low oxygen zone the levels of nitrate were as low as 2 µM. Also, Eh values were not low enough (minimum values were 30 mV) to account for sulfate reduction during our measurements. However, this is a common phenomenon during summer in the

Scheldt Estuary (Billen et al., 1976), and sulfides (or rather, polysulfides with increased solubilities) remaining from this period might be responsible for the relatively high dissolved Cd values observed in the low oxygen zone. Oxidation of sulfides is known to be a rapid process and this may have caused the strong increase in dissolved Cd in the reoxygenated zone (i.e., below S = 5). An important factor likely contributing to the dissolution process is the tendency for Cd to form soluble chloro-complexes with increased salinities.

The mixing of water with S = 20 having maximum dissolved cadmium concentrations with more saline water showed conservative mixing, while the mixing with lower salinity water showed a negative deviation from an ideal dilution relation. This difference may be related to the difference in concentrations of suspended matter in the lower and higher salinity waters, as increased concentrations of SPM (suspended particulate matter) tend to favor Cd adsorption onto particles, especially if the SPM contains a high

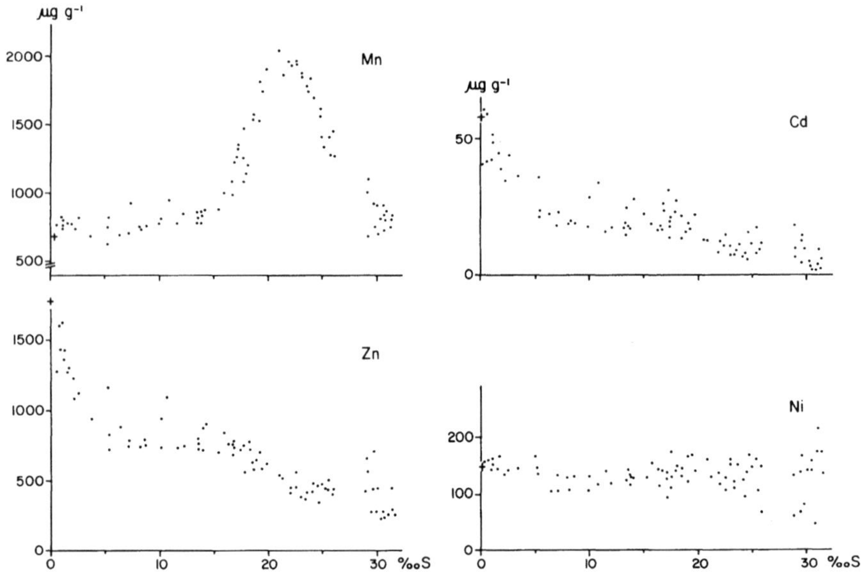

Fig. 6. Data on contents of Mn, Zn, Cd, and Ni (in $\mu g\ g^{-1}$) measured in all suspended particulate matter (SPM) samples collected at fixed positions I–VI during tidal cycles, in relation to salinity. The concentrations of the trace metals in SPM collected further upstream at km 130–150 are indicated on the vertical axis by the symbol +. (From Duinker et al., 1982b)

fraction of small particles and carries a significant fraction of freshly deposited iron (hydr)oxides.

The concentrations of particulate Cd in suspension on a volume basis (μM dm^{-3}) were more than sufficient to account for the observed maxima of dissolved Cd at $S=20$. As a consequence, the concentrations as well as the contents (i.e., concentrations per unit weight of SPM) of Cd contained in suspended matter were not affected measurably by the dissolution process (Fig. 6).

Zinc

The maximum for the dissolved zinc concentration occurred at a salinity of $S=17$ (Fig. 4). Mixing of this water body with waters both upstream and downstream resulted in two linear relations with salinity: positive upstream and negative downstream. Most likely, in the absence of a local land-based source, the dissolved zinc maximum was produced by release from particulates in the reoxygenated zone (i.e., above $S=15$). The explanation suggested for Cd, that removal of dissolved Zn from solution might occur by increased concentration of SPM in the turbidity zone, thus does not apply to the dissolved zinc profile. The more pronounced tendency for dissolved Zn to resist removal during estuarine mixing has been observed in other estuaries as well (Duinker, 1980). The concentrations of particulate suspended zinc on a volume basis were one to two orders of magnitude above those in solution. As in the case of Cd, the Zn content in SPM was also not affected measurably (Fig. 6).

Nickel

The concentration profile of Ni in solution in relation to salinity suggests conservative behavior. The content of Ni in SPM is remarkably constant (150 μg g^{-1}). The concentrations of Ni in suspension (0.25 μM particulate Ni per dm^3 at an SPM concentration of 100 mg SPM dm^{-3}) are similar to the concentrations in solution. It is therefore to be expected that significant phase transitions of Ni between solution and particulates would show up in concentration/content profiles in the longitudinal section. It thus appears that the reactivity of Ni in the Scheldt Estuary is relatively low, despite the existence of large chemical gradients.

Vanadium

A minimum in the dissolved concentration of V is present in the low salinity range, where pH and O_2 content were near minimum values (Fig. 4). In the analytical method, involving adsorption on active carbon, both V(IV) and V(V) are included. Eh-pH conditions [with Eh calculated according to Breck (1972)] at salinities $S = <15$ and $S = >20$ are in the stability field of H_2VO_4 and those in the intermediate region in the stability field of $VOOH^+$ (Pourbaix, 1963). According to Hem (1970), V species have a minimum solubility near or below 20 nM at neutral pH and under mildly reducing conditions. Thus, $VOOH^+$ would be less soluble than H_2VO_4. The dissolved V distribution then satisfies, at least qualitatively, a thermodynamic equilibrium model involving H_2VO_4 and $VOOH^+$.

The average V content in SPM in 53 samples over the entire salinity range was 2.02 ± 0.29 μmol g^{-1}. The relation with salinity is therefore not depicted here. The concentration of particulate suspended V (200 nM particulate suspended V at an SPM concentration of 100 mg dm^{-3}) would be more than sufficient to account for the observed change in dissolved V concentration (about 80 nmol dm^{-3}; Fig. 4) without showing an observable change in the concentration or content of V in particulate form, as is in fact shown by the data.

Arsenic

High (total) dissolved concentrations of As were observed at low salinities, where local industries are known to contribute significantly (Fig. 4). The relation with salinity would suggest removal from solution up to $S = 15$ and physical dilution with saline waters from there onwards. The removal would involve about 400 nM; this is very similar to the average concentration of suspended particulate As, i.e., 260 nM particulate As at an SPM concentration of 100 mg SPM dm^{-3} with 2600 nmol As g^{-1} SPM. This removal mechanism was expected to result in measurably increased As contents in SPM at low salinities. This was actually observed. Moreover, the strong correlation between particulate As and Fe contents (Fig. 7) suggests that As was removed primarily by $Fe(OH)_3$ co-precipitation. Similar observations were reported for other estuaries (Crecelius et al., 1975; Leckie, 1978). However, linear As-salinity relations have also been found. In these cases the inefficiency of iron hydroxide removed by co-precipitation was attributed to the interference of organics in the Fe co-precipitation (Waslenchuk and Windom, 1978).

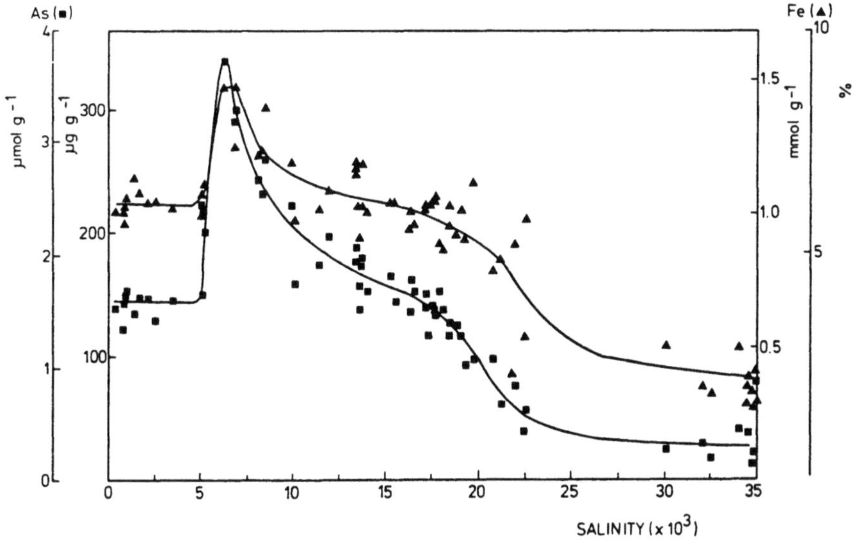

Fig. 7. Fe ($mmol\,g^{-1}$ and %) and As ($\mu mol\,g^{-1}$) contents in SPM sampled at the six stations (Fig. 3) in the Scheldt Estuary, in relation to salinity (van der Sloot et al. 1985)

Conclusions

It can be concluded that in estuaries the species distribution of several river-borne trace elements can be modified in response to the large typically estuarine gradients in master variables, in particular total salt concentration (salinity), major element composition, pH, O_2 content, and turbidity (particulate matter, seston concentration).

For the Scheldt Estuary, the following processes were found to be involved: redox reactions involving various solid species for Mn and V, solubility changes under different dissolved oxygen regimes, possibly involving sulfides for Cd and Zn and adsorption onto existing particles for Cd and As. Reactivity of Ni was found to be low; no evidence for solution-particulate transitions could be detected.

In general, such species transformations and transitions cannot be studied by measurements of concentrations of the species involved alone, but the distribution patterns of the species involved should be studied in connection with the water and sediment transport processes and other master variables involved in the species transformation.

Components surviving estuarine processes in dissolved, colloidal, or small/low density particulates will be transported out of the estuary. Other particulates may have a long residence time for sorption processes involving solutes; they have been identified both as source and sink for trace elements in solution, depending on conditions. Every estuary is unique with respect to the complex interplay between the factors responsible for these conditions.

For future work one must keep in mind that important problems in species distribution and transformation are related to a) hydrodynamics: evolution of salinity and currents and their effects on sediment transport and accumulation, residence time of water, and the distribution of pH and O_2 content; b) composition of particulate matter in terms of individual minerals and organic matter; the formation of aggregates, identification of sorption sites and their quantitative roles; c) identification of organic matter components and the role of individual compounds as potential ligands for complexation of trace-element species; and d) the extent to which and rates at which degradation of organic matter affect the transitions of trace-element species in bottom sediments.

Such problems can be resolved successfully only by interdisciplinary research groups. Field work should consider a variety of estuaries with different discharges and different meteorological and tidal conditions.

References

Aston SR (1978) Estuarine chemistry. In: Riley JP, Chester R (eds) Chemical oceanography, vol 7, pp 361–440

Billen G, Smitz J, Somville M, Wollast R (1976) Dégradation de la matière organique et processus d'oxydo-réduction dans l'estuaire de l'Escaut. In: Nihoul JC, Wollast R (eds) Modèle mathematique de la Mer du Nord, vol 10, L'Estuaire de l'Escaut. Service du Premier Ministre, Programme de la Politique Scientifique, Bruxelles, pp 101–152

Bowden KF (1980) Physical factors: salinity, temperature, circulation and mixing processes. In: Olausson E, Cato I (eds) Chemistry and biogeochemistry of estuaries. J Wiley and Sons, New York, pp 37–70

Boyle EA, Collier R, Dengler AT, Edmond JM, Ng AC, Stallard RF (1974) On the chemical mass balance in estuaries. Geochim Cosmochim Acta 38:1719–1728

Breck WG (1972) Redox potentials by equilibration. J Mar Res 30:121–139

Burton JD, Liss PS (eds) (1976) Estuarine chemistry. Academic Press, London New York San Francisco

Crecelius EA, Bothner MH, Carpenter R (1975) Geochemistries of As, Sb, Hg and related elements in sediments of Puget Sound. Envir Sci Technol 9:325–333

Duinker JC (1980) Suspended matter in estuaries: Adsorption and desorption processes. In: Olausson E, Cato I (eds) Chemistry and Biogeochemistry of Estuaries. J Wiley and Sons, New York, pp 121–151
Duinker JC, Wollast R, Billen G (1979) Behaviour of manganese in the Rhine and Scheldt estuaries. Part II: Geochemical cycling. Est Coast Mar Sci 9:727–738
Duinker JC, Hillebrand MTJ, Nolting RF, Wellershaus S (1982a) The river Elbe: processes affecting the behaviour of metals and organochlorines during estuarine mixing. Neth J Sea Res 15:141–169
Duinker JC, Nolting RF, Michel D (1982b) Effects of salinity, pH, and redox conditions on the behaviour of Cd, Zn, Ni and Mn in the Scheldt estuary. Thalassia Jugoslavica 18:191–202
Dyrssen D, Wedborg M (1980) Major and minor elements, chemical speciation in estuarine waters. In: Olausson E, Cato I (eds) Chemistry and biogeochemistry of estuaries. J Wiley and Sons, New York, pp 71–119
Fairbridge RW (1980) The estuary: its definition and geodynamical cycle. In: Olausson E, Cato I (eds) Chemistry and biogeochemistry of estuaries. J Wiley and Sons, New York, pp 1–35
Festa JF, Hansen DV (1978) Turbidity maxima in partially mixed estuaries: a two-dimensional model. Est Coast Mar Sci 7:347–359
Hem JD (1971) Study and interpretation of the chemical characteristics of natural water. Geological survey water – Special Paper 1473. U.S. Government Printing Office, Washington, DC
IAPSO (1985) The International System of Units (SI) in Oceanography. IAPSO Pulication Scientifique No 32 (UNESCO Technical Papers in Marine Science No 45, pp 40–42) Paris, UNESCO
Lauff GH (ed) (1967) Estuaries. American Association of Advanced Science Publication Number 83
Leckie JO (1978) Trace element removal from water by coprecipitation with iron hydroxide. EPRI-Report RP-910
Liss PS (1976) Conservative and non-conservative behaviour of dissolved constituents during estuarine mixing. In: Burton JD, Liss PS (eds) Estuarine chemistry. Associated Press, London, pp 93–130
Meade RH (1972) Transport and deposition of sediments in estuaries. Geol Soc Mem 133:91–120
Mook WG, Koene BKS (1975) Chemistry of dissolved inorganic carbon in estuarine and coastal brackish waters. Est Coast Mar Sci 31:325–336
Olausson E, Cato I (eds) (1980) Chemistry and biogeochemistry of estuaries. J Wiley and Sons, New York
Peters JJ, Sterling A (1976) Hydrodynamique et transport de sédiments de l'estuaire de l'Escaut. In: Nihoul CJ, Wollast R (eds) L'Estuaire de l'Escaut, Project Mer, vol 10. Programme de la Politique Scientifique
Postma H (1980) Sediment transport and sedimentation. In: Olausson E, Cato I (eds) Chemistry and biogeochemistry of estuaries. J Wiley and Sons, New York, pp 153–186
Pourbaix M (1963) Atlas d'équilibres électroniques à 25 °C. Grauthier-Villars, Paris
Pritchard DW (1967) Observations of circulation in coastal plain estuaries. In: Lauff GH (ed) Estuaries, American Association of Advanced Science Publication Number 83, pp 37–44

Russell KL (1970) Geochemistry and halmyrolysis of clay minerals Rio Ameca – Mexico. Geochim Cosmochim Acta 34:893–907
Sayles FL, Mangelsdorf PC (1977) The equilibria of clay minerals with sea water: exchange reactions. Geochim Cosmochim Acta 41:951–960
Sayles FL, Mangelsdorf PC (1979) Cation-exchange characteristics of Amazon River suspended sediment and its reaction with seawater. Geochim Cosmochim Acta 43:767–779
van der Sloot HA, Wijkstra J, Duinker JC, Nolting RF (1985) Anionic species of V, As, Se, Mo, Sb, Te and W in the Scheldt and Rhine estuaries and the Southern Bight. Est Coast Shelf Sci 21, 533–655
Waslenchuk DF, Windom HL (1978) Factors controlling the estuarine chemistry of arsenic. Est Coast Mar Sci 7:455–464
Wellershaus S (1981) Turbidity maximum and mud shoaling in the Weser estuary. Arch Hydrobiol 29:161–198
Wollast R (1982) Behaviour of organic carbon, nitrogen and phosphorus in the Scheldt estuary. Thalassia Yugoslavica 18:11–34
Wollast R, Peters J (1978) Biogeochemical properties of an estuarine system: The River Scheldt. In: Biogeochemistry of estuarine sediments. Proceedings of a UNESCO/SCOR Workshop, Melreux, 1976. UNESCO, Paris
Wollast R, Billen G, Duinker JC (1979) Manganese in the Scheldt and Rhine estuaries. Part I: Physico-chemical aspects. Est Coast Mar Sci 9:161–169
Wollast R, Duinker JC (1982) General methodology and sampling strategy for studies on the behaviour of chemicals in estuaries. Thalassia Yugoslavica 18:471–491

The Importance of Chemical "Speciation" in Environmental Processes,
eds. M. Bernhard, F. E. Brinckman, and P. J. Sadler, pp. 385–422.
Dahlem Konferenzen 1986. Springer-Verlag Berlin, Heidelberg
© Dr. S. Bernhard, Dahlem Konferenzen

Importance of Chemical Species in Uptake, Loss, and Toxicity of Elements for Marine Organisms

M. Bernhard * and S. G. George **

* Centro Studi Ambiente Marino, ENEA
19100 La Spezia, Italy
** Unit for Aquatic Biochemistry University of Stirling
Stirling FK9 4LA, Scotland

Abstract. The influence of chemical species on the uptake and the toxicity of As, Cd, Cr, Cu, Fe, I, Hg, Ru, and Zn is reviewed in this paper. Experiments in which trace metals were complexed with natural and artificial ligands show that with very few exceptions the complexed forms are accumulated at lower rates, thus leading to a lowering of toxicity. Although definitive proof is still lacking, for Cu, Cd, and Zn uptake and toxicity appear to be related to the free metal-ion activity and the species labile to anodic stripping voltametry. At present the latter technique appears to provide the best measure of soluble, bioavailable metal species. Many organisms are capable of taking up microparticulate, colloidal metal hydroxides such as those of Fe, Au, Th, and Pu directly by pinocystosis. Marine organisms can excrete a number of organic substances which can chelate trace metals; whether these are responsible for the chelation capacity of seawater is not clear. Elements which are capable of forming stable, covalent organic compounds, such as As and Hg, are predominantly accumulated with size and higher trophic levels in this form as opposed to inorganic species. In many animals several metals are bound and detoxified by specialized metal-binding proteins such as metallothionein and ferritin. Metals are also stored intracellularly in insoluble forms in granules: these are either inorganic, predominantly based on Ca phosphates, or organics such as peroxidized pigments. The former may account for recent biogenic phosphorites in marine sediments and the latter for detrital material. Most evidence concerning the influence of chemical species on toxicity and uptake is derived from experiments using conditions far removed from those in the natural environment, and this should be rectified. The definition of the molecular mechanisms of uptake and toxicity of biochemical and physiological techniques may prove to be the most fruitful approach to the definition of the chemical species involved in these processes.

Introduction

During the last decade the importance of chemical species for the interpretation of uptake, loss, and toxicity of elements has received wide recognition. Indirect evidence for the occurrence of different trace metal species could be deduced much earlier from the use of soil extracts in algal cultures (Schreiber, 1935). The same effect could later be obtained by employing artificial chelating agents (e.g., EDTA) and it could be shown that these chelating agents reduce the toxicity of "heavy metals" in the culture medium (see below).

The development of new chemical techniques sensitive to different species and the recognition that many chemical methods will not determine the total concentration of an element have further increased the awareness in environmental studies that different chemical species must be distinguished. Social concern has resulted in an uneven distribution of the amount of information available on the distribution of chemical species of the various elements. For example, much more information is available on species of heavy metals and certain radioisotopes that on many other elements.

Alkaline and Alkaline Earth Metals

The only published evidence on the possible influence of chemical species on the uptake, loss, and toxicity of alkaline and alkaline-earth metals is the data of Chambers and Whiteley (1966) and Harrison (1979). Chambers and Whiteley presented data which indicated that the Ca and Mg ion-pair of phosphate may be preferentially assimilated relative to unassociated phosphate (Chambers and Whiteley, 1966). Harrison (1979) reported that during summer, oysters placed in the outfall canal of a nuclear plant accumulated more Cs^{134} and Cs^{137} from unfiltered water than from filtered seawater (Table 1) indicating that most of the bioavailable Cs was in particulate form. However, the differences measured in winter were insignificant. There is also some indication that Cs^{137} was in a different physicochemical

Table 1. Radionuclides in oysters accumulated from filtered and unfiltered seawater before and after a release of radioactive waste (modified from Harrison, 1979)

	Cs^{134}	Cs^{137}
Nonfiltered seawater (July 1973)		
pre-release (pCi/kg)	ND	ND
post-release (pCi/kg)	160	340
concentration factor	0.62	0.8
Filtered seawater (July 1973)		
pre-release (pCi/kg)	1	8
post-release (pCi/kg)	99	210
concentration factor	0.38	0.5
Nonfiltered seawater (Dec. 1973)		
pre-release (pCi/kg)	ND	ND
post-release (pCi/kg)	324	500
concentration factor	0.9	1.1
Filtered seawater (Dec. 1973)		
pre-release (pCi/kg)	ND	ND
post-release (pCi/kg)	310	498
concentration factor	0.9	1.1

ND = not detected

form than Cs^{134}. In fact, in both experiments (summer and winter) and for both conditions (i.e., filtered and unfiltered water) the concentration factors (CF) of Cs^{137} was higher than the CF of CS^{134}. This difference may be due to the fact that a part of the Cs^{137} came from fallout and hence has a different origin. De Bartoli et al. (1968) also found that the CF of Cs^{137} was different from that of stable cesium in fishes from four Italian lakes because the Cs^{137} came from fallout (Table 2).

Table 2. Concentration factors of stable and radioactive cesium in fishes from northern Italian lakes (de Bartoli et al., 1968)

	Lake Maggiore	Lake Varese	Lake Comabbio	Lake Monate
Perca fluvialis				
ratio $CFCs^{137}$/CFCs stable	1.9	9.8	2.6	3.9
Scardinius erythrophtamus				
ratio $CFCs^{137}$/CFCs stable	1.9	5.7	1.1	3.1
Lepomis gibbosus				
ratio $CFCs^{137}$/CFCs stable	1.1	3.3	1.5	3.1

This lack of distribution into different chemical species by the alkaline and alkaline-earth metals is undoubtedly determined by the chemical properties of their interactions with organic ligands, i.e., the formation of weak ionic interactions and preferences of 0-liganding. At high pO_2 and pCO_2 levels, however, metals may be stripped from seawater as insoluble precipitates and, for example, produce Mg deficiency in fish (Cowey et al., 1977).

In some organisms, particularly invertebrates with a calcified exoskeleton which have to rapidly mobilize large amounts of calcium at molt, the calcium is stored in extracellular granules (0.5–2 μm diameter) of calcium carbonate (Simkiss, 1981). These granules have been described in some algae (Coccolithophores), crustaceans, and molluscs. There is also evidence to suggest that Pu may accumulate in the same granules, and for the occurrence of extracellular ferric-phosphate granules. Intercellular Ca-based granules are also found in most animal species but are particularly common in invertebrates (Simkiss, 1976; George, 1982). These insoluble, spherical granules (1–100 μm diameter) are comprised mainly of Ca and Mg phosphates, but other metals such as Al, Be, Co, Fe, Mn, Pb, Sn, and Zn may also contain pyrophosphate which increases their insolubility.

Arsenic

Total As concentrations in unpolluted seawater: 15–33 nM.

Arsenite (As(III)), arsenate (As(V)), monomethylarsonate (MMA), and dimethylarsinate (DMA) have been detected in seawater. In open-sea samples from the euphotic zone As(III) is usually only a few percent of the total As. DMA is usually between 10 and 20% and MMA $<1\%$ (see Andreae, this workshop). Positive correlations have been observed between primary production and both As(III) and organoarsenicals, indicating biological modification of the distribution of As species. In both unpolluted and polluted coastal areas the relative amounts of arsenite can be higher, especially in estuaries with arsenite inflow from rivers (Klumpp and Peterson, 1979).

Several laboratory experiments and observations in large *in situ* enclosures (mesocosms) have demonstrated that marine unicellular algae and bacteria can markedly influence the distribution of As species. As(V) taken up (together with phosphate) by phytoplankton species and benthic algae is transformed to As(III) and organic arsenicals. However, large differences in accumulation and transformation rates have been observed in different laboratory studies. Andreae and Klumpp (1979) found in batch cultures that the nine phytoplankton species examined accumulated up to 85% of the As present in the medium as As(V). Two species excreted 30% and 66%, respectively, as As(III) but the others excreted only very small amounts of As(III) (not detected to 3%). Somewhat higher amounts of organoarsenicals (not detected to 18%) were excreted into the culture medium. *Cricosphera carteri,* which for 14 days grew only in the stationary phase, accumulated only very small amounts of As but converted 29% of the As(V) present in the medium into As(III). On the other hand, Sanders and Windom (1980) found that the As(III) increase due to As(V) reduction was about twice the concentration of the released DMA in the culture medium. Cultures enriched with As(V) showed a marked increase of DMA in the medium and slow increases of As(III) and As(V). No significant changes occurred in the lag phase (stationary-growth phase) of growth.

Sanders and Windom (1980) also examined the As transformations in large *in situ* containers (mesocosms). Here, where conditions are more similar to natural ones, the changes observed were much more moderate. When an *in situ* mesocosm was enriched with As(V), the concentration of As(V) slowly decreased and the concentration of As(III) increased at about the same rate (Fig. 1 a). When, on the other hand, a container was enriched with As(III) (Fig. 1 b), first As(III) decreased relatively rapidly and As(V) increased, but later there was a slight increase following a pattern similar to that of the As(V) enrichment (Fig. 1 a). The initial decrease of As(III) is in part due to the chemical oxidation (see dotted line) of As(III) to As(V). In both enrichments DMA increased in the containers to about 15% of the total As in the containers. The initial decrease of the As(III) concentration and increasing As(V) concentration in Fig. 1 b of Sanders and Windom (1980) is very likely due to the oxidation of As(III) by marine heterotrophic microorganisms. Scudlark and Johnson (1982) showed that the oxidation rate of As(III) to As(V) is constant in distilled water (2.6 nMd^{-2}) and in sterile seawater samples (3.5 nMd^{-1}). In nonsterile samples the oxidation rate is exponential and follows first-order kinetics. Experiments with filtered seawater samples showed that the arsenite-oxidizing organisms are in the 0.1–0.8 µm size range. Since filtration of the biological population through 0.1 µm did not show any increase in the chemical oxidation rate, it can be concluded that the biomediated oxidation is not due to extracellular products.

There are few experiments published on uptake of arsenic by marine animals. Fowler and Unlu (1978) found that radioactive arsenate is slowly accumulated by the shrimp *Lysmata seticaudata* from seawater. They observed that the steady-state concentration factor reached with the radioac-

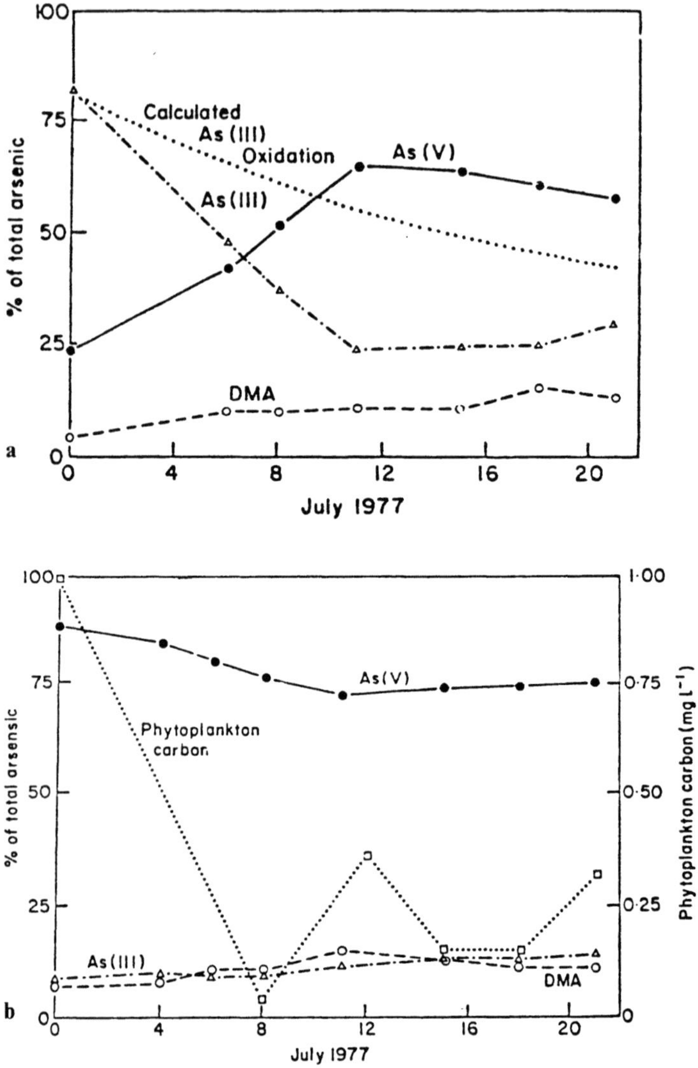

Fig. 1a, b. Changes in As species in *in situ* enclosures spiked with 67 nM As(III) **a** or As(V) **b,** respectively (from Sanders and Windom, 1980)

tive As was much below corresponding stable concentration factors, which led them to suggest that the pathway from water is of little significance. They also found that As accumulated from one ration of radioactive Artemia was eliminated from the shrimp, with a half-time about seven times longer (18 d) than that of the As accumulated from seawater only. Although stable As data were not obtained during the experiments and, therefore, fluxes cannot be estimated, these observations seem nevertheless to indicate that the As accumulated from water is in either a different form or a different compartment than the As accumulated from food.

Klumpp (1979) investigated As uptake from water and food by the gastropod molluscs *Littorina* and *Nucella* and also concluded that the diet was the most important route; indeed, in *Littorina* there is a direct correlation between tissue As levels and dietary As concentrations up to 80 μg/g fresh weight (FW) food.

The results of Penrose (1975) are most probably also valid for marine fish. He administered inorganic radioactive arsenic to brown trout (*S. trutta*) orally and by intramuscular injections. Only 4.5 h after oral administration, 82 to 93% of the inorganic As was found in blood, muscle, or liver as nonanionic ("organic") As. On the other hand, at 4.5 h after injection only 0.8 to 7.5% was found in the same tissues as nonanionic As. After three days about 50% and after ten days about 85% of the injected inorganic As had been transformed into organic As in the liver. Incubation of gut content with As(V)74 and incubation of isolated, ligated gastrointestinal tract did not convert inorganic As into organic As. These experiments show that inorganic As taken up via the gastrointestinal tract was much more efficiently converted to the nonanionic (organic) form than when injected. Since it could be shown that the inorganic and organic forms are excreted in the bile, a cycle between the body and lumen of the fish exists allowing the preferential reabsorption of nonanionic As. Lunde (1972) obtained very similar results.

Several authors have analyzed marine organisms distinguishing only between inorganic and organic forms of As (Brooks and Evans, 1981; Flanjak, 1982; Lunde, 1973; Reinke et al., 1975; Sanders, 1979b; Shinagawa et al., 1983). Their data show that seaweed contain 25 to 80% organic As. Shrimps, molluscs, and fish contain typically more than 90% organic As.

In macroalgae such as the brown kelp (*Eklonia*) arsenosugars are the major form of As (Edmonds and Francesconi, 1981; Klumpp and Peterson, 1981; Sanders, 1979b). Analyses of the kidneys of the giant clam, *Tridacna,* which contain a symbiotic alga have revealed the presence of trimethylarsonium lactate and arsenosugars (Benson and Summons, 1981; Conney et al., 1978; Edmonds and Francesconi, 1982). In the few higher organisms so far investigated it has been shown (Klumpp and Peterson, 1981; Shiomi et al., 1983, and references therein) that in snails many different organoarsenicals exist but no arsenobetaine can be found. In shrimp, octopus, and fish the predominate organic arsenic compound (>60%) is arsenobetaine.

Combining the results from laboratory experiments and chemical analysis of natural marine organisms, the following As pathways in the sea may occur. Marine plants (micro- and macrophytes) accumulate As(V) from the seawater and transform it into arsenite and organic arsenic compounds. A part of the As(III) and organic arsenic is released into the surrounding seawater and part is retained within the plants. The released As compounds are probably oxidized by microorganisms present in the seawater into arsenate, and also the organic arsenic compounds are demethylated. Animals feeding on plants will take up the organic As contained in the plants. Animals situated in lower levels of the food chain seem to accumulate many different water-soluble As organic substances which will be converted by still unknown processes into arsenobetaine, which is the dominant form of As in marine organisms belonging to higher trophic levels. Some marine organisms (especially benthic ones) have very high As levels which suggest that the organic substances containing As (as with Hg) are accumulated quickly and released slowly by marine animals. Observing that different forms of arsenic are metabolized differently, it is not surprising that a distinction between chemical species also is important in toxicity. Exposing phytoplankton species to 40 nM to 1.3 μM arsenate showed that three out of seven species terminated growth at the range tested, while others were affected but continued growth at low rates of cell division (Sanders and Vermersch, 1982). As(III) seems to be slightly more toxic (>13 nM As(III) than As(V) (>160 nM (As(V)) to phytoplankton. A phosphate enrichment (>300 nM) can counteract this toxicity (Sanders, 1979a): arsenate and arsenite were both toxic to the marine diatom *Skeletonema costatum* at concentrations as low as 67 nM As, while dimethylarsonic acid showed no toxic effect even at 333 μM As.

Cadmium

Total Cd in open-ocean seawater: 0.01–1 nM.

In coastal waters 15–20% and in estuarian waters (up to 85%) of the cadmium present is associated with particulate matter (review by Phillips,

1980). Calculations based on thermodynamic principles (equilibrium) predict that in seawater the soluble inorganic cadmium species occur mainly as chloro-complexes. Anodic stripping voltammetry shows likewise that in the soluble phase electroactive species predominate.

Experimental studies have shown that for both algae and animals Cd uptake is reduced in the presence of chelating agents. Thus Cd uptake by the diatom *Phaeodactilum tricornutum* is negligible when the Cd is complexed with EDTA. Cd-contaminated *P. tricornutum* cells lose nearly all previously accumulated Cd when transferred into a medium containing 21 mM cysteine (Cossa, 1976). Similar observations were made by Foster and Morel (1982) who found that EDTA (10 to 100 μM) reduced the cadmium toxicity to another diatom (*Thalassiosira weissflogii*) of otherwise toxic concentrations. However, it was not possible to reverse the Cd toxicity by increasing only the EDTA concentration (to 100 μM); the Fe concentration also had to be increased. The authors suggest that the interaction between Cd and Fe does not occur in the medium but is due to a physiological interaction within the cell. In higher organisms a reduction of Cd tissue levels was also observed in the presence of chelating agents. For example, Cd uptake was decreased in the barnacle *Semibalanus balanoides* in the presence of humate, alginate, and EDTA (Rainbow et al., 1980), and in the clam *Macoma balthica* in the presence of EDTA (McLeese and Ray, 1984). When the marine worm *Nereis virens* or the marine shrimp *Pandalus montagui* were exposed to Cd-EDTA the Cd levels were reduced by 40% and 20%, respectively (Ray et al., 1979). In the American oyster (*Crassostrea virginica*) Cd complexed with EDTA, NTA, or humic acid was accumulated by as much as 70% less than when only inorganic, unchelated Cd was present at the same concentration (Hung, 1982). Experimenting with the grass shrimp *Palaemonetes pugio* Sunda et al. (1978) showed that the observed Cd toxicity is correlated with the calculated free cadmium ion. Only George and Coombs (1977) reported that chelating agents (EDTA, alginate, humus, and pectin) doubled the amount of Cd taken up by mussels compared with the uptake from unchelated Cd (900 nM). However, further experiments have cast doubts on the enhancement of Cd uptake in the presence of EDTA observed previously, since the uptake of $Cd^{115m}Cl_2$ (the standard reference in this study) varied between experiments. This is thought to have been caused by the use of different ages of stock isotopes (George, personal communication, 1984) and the method of preparing the spiked seawater. Furthermore, unpublished experiments with isolated gills by the same author showed that the rate of Cd^{109} influx was very much lower when EDTA was present. These recent observations are in accordance with those of other authors who found a decrease in the Cd tissue levels in the presence of EDTA.

As noted elsewhere (Piscator, this volume), vertebrate cells synthesize a unique protein, metallothionein, in response to exposure to a number of heavy metals including Ag, Cd, Cu, Hg, and Zn (Kägi and Nordberg, 1979). Metallothioneins are a group of low-molecular-weight proteins (about 6100 daltons) which bind up to seven g-atoms of metal per mol of thionein, have a high content of cysteine (30 to 35%) and virtually no aromatic amino acids. Cd-thioneins have been isolated and characterized from a marine flatfish, a finfish, and a crab and have been shown to have a high degree of homology with mammalian metallothionein. Heavy metal-binding proteins with many of the characteristics of metallothionein have also been reported in several other marine phyla, including marine molluscs and mammals, after these animals had been exposed to Cd, Cu, Hg, and Zn (George, 1982; Roesijadi, 1980). Many organisms also contain high concentrations of non-metallothionein Cd-binding proteins and their occurrence has been recently reviewed (Stone and Overnell, 1985). In vertebrates metallothionein accounts for over 90% of the intracellular Cd. However, in molluscs and many other invertebrates up to 50% of the intracellular Cd may be associated with particulate subcellular structures, the lysosomes. The characteristics of these lysosomes will be described in the section on iron (see below).

Chromium

Total chromium in open-ocean seawater: 2–5 nM.

In the open sea soluble Cr(VI) predominates over particulate Cr(III) (see Andreae, this volume). In estuaries and coastal zones a few percent of the dissolved Cr occur as Cr(III). While passing from rivers into the estuaries, through the salinity* range $S=2$ to $S=10$, 40–70% of the dissolved Cr may be converted into particulate Cr (Cranston and Murray, 1980).

The influence of different physicochemical forms of chromium on the bioavailability to organisms with different feeding behavior was demonstrated by comparing the accumulation of soluble Cr(VI) and particulate Cr(III) by a filter-feeding clam and a worm. The filter feeder preferentially accumulated the particulate Cr(III) while the worm showed enhanced Cr(VI) uptake (Chipman, 1966, 1967). Similar results were obtained on three crustaceans, a polychaete worm, and a bivalve mollusc, which accumulated Cr during short-term exposure to drilling mud-seawater mixtures.

* For adefinition of salinity see Duinker (this volume).

Cr uptake and release were species-dependent. The amount of Cr taken up was relatively low and appeared to be associated with the soluble or fine particulate phase of the drilling mud (Carr et al., 1982).

Cr(III) is virtually nonabsorbed by most animals (CrO_3 is used as a marker for fecal excretion rates in nutrient studies) and is nontoxic. There have, however, been reports that Cr(VI) is highly toxic to marine animals (Oshida et al., 1976). This is probably due to Cr(VI) being a very strong oxidizing agent. The most extensive work on uptake of Cr(VI) has been conducted with freshwater fish in which uptake and toxicity are both increased at lower pH. Bell et al. (1983) showed that uptake across the intestinal epithelium is uncontrolled.

Copper

Total Cu in open-ocean seawater: 0.1–3.7 nM.

Large amounts of copper are usually bound to inorganic particles (Batley and Gardner, 1978; Van den Berg, 1984). Model calculations (equilibrium) suggest that at salinity $S = 35$ the generally small amount of soluble inorganic copper is 82% $Cu(OH)_2$ and about 5% $CuCO_3$. The other inorganic Cu species contribute less than 1% each (Mantoura, 1981). However, more recent models indicate that $CuCO_3$ is by far the most frequent dissolved species (Millero, 1981; Van den Berg, 1984). In seawater from certain areas, up to >98% of the total copper can be associated with organic matter. This is predominately humic material. It has been suggested that with increasing salinity, copper associated with humus will decrease from 100% in fresh water to about 10% at $S = 35$ (Mantoura, 1981). This may be due to the mass action effects of increased Na^+ competition for COO^--sites since these are poor co-coordination sites for Cu.

Natural and synthetic chelators such as amino acids, humic acids, soil or sediment extract, detergents, EDTA, etc., reduce the toxicity of copper salts to sea-urchin eggs and larvae (e.g., Bernhard, 1955; Tyler, 1953), to unicellular algae (e.g., Morris and Russell, 1973; Steemann-Nielsen and Wium-Anderson, 1970), and to clams and oysters (Harrison, 1979; Stephenson and Taylor, 1975). Recently experiments with different marine organisms have shown that toxicity is related to the calculated cupric-ion activity and not to the total Cu concentration. The addition of TRIS (trihydroxymethyl-aminomethane) to culture media of marine unicellular algae containing otherwise toxic Cu concentrations reduced the Cu toxicity (e.g., Sunda and Guillard, 1976). Similarly, partial destruction of natural organic matter with UV irradiation increased Cu toxicity. This inhibition in UV-treated seawater is then again annulled by the addition of 10 μM EDTA. A reduction in toxicity of Cu to a marine bacterium and the American oyster (*C. virginica*) by NTA (nitrolotriacetic acid) showed that toxicity is also related to calculated cupric-ion activity in these organisms (Sunda and Gillespie, 1979; Zamuda and Sunda, 1982). Florence et al. (1983) tried to overcome the difficulties inherent in relying on calculated cupric-ion activ-

Table 3. Species analysis and toxicity of copper complexes to *Nitzschia closterium*

Ligand[a]	Ligand conc. (10^{-6} M)	Percent of copper added[b]			
		ASV-labile			Chelex-100 labile
		pH 8.2		pH 4.7	
		−0.6 V	−1.3 V	−0.6 V	
None	–	100	100	100	100
Fulvic acid	10	1.5	29	41	66
Tannic acid	0.59	5.5	10.5	75	100
Iron–humic colloid	–[e]	70	74	65	100
Desferal	2.0	100	95	100	95
Alginic acid	10 mg L^{-1}	100	100	100	95
Lecithin	2.0	32	32	32	95
LAS	0.5 mg L^{-1}	65	100	NQ	95
NTA	20	100	100	100	100
Acetylacetone	2.0	100	100	100	100
Linoleic acid	2.0	77	100	100	100
Oxine	2.0	64	100	90	98
2-Methyloxine	2.0	68	63	91	96
Oxine-5-sulfonate	2.0	100	100	100	77
1,10-Phenanthroline	2.0	35	100	59	100
DMP	2.0	2.5	NQ	52	100
α-Benzoinoxime	2.0	35	50	59	100
DDTC	2.0	< 0.5	< 1.0	< 0.5	70
APDC	2.0	9.0	8.0	15.5	53
Ethyl xanthogenate	2.0	10.5	48	48	63

[a] Abbreviations: LAS, linear alkylbenzene sulphonate; NTA, nitrilotriacetic acid; oxine, 8-quinolinol; DMP, 2,9-dimethyl-1,10-phenanthroline; DDTC, diethyldithiocarbamate; APDC, ammonium pyrrolidinedithiocarbamate

[b] Total of 20 μg Cu L^{-1} (3.15×10^{-7} M) added. An acetate buffer (0.06 M) was used for pH 4.7

ity by determining the concentration of copper species chemically and with a bioassay. They investigated the distribution of Cu species in the presence of nineteen different natural and artificial complexing agents in order to establish a correlation between the copper-species concentrations determined with chemical techniques and the toxic effects observed on the growth rate of a marine unicellular alga and the copper associated with the algae. From Table 3 it can be seen that most ligands decreased the Cu toxicity. However, no consistent correlation was observed between the fraction of the total copper measured by algal bioassay and the labile copper determined by the

(Florence et al., 1983)

Thiol labile	Solvent extractable[c]	Labile copper from growth curve[d]	Copper on algae	Copper adsorbed on algae (10^{-13} g Cu/cell)	Depression of algal growth rate by ligand alone (%)
100	< 10	100	42	0.75	–
100	< 10	7.5	< 5	< 0.07	< 10
100	< 10	12.5	77	0.66	< 10
100	13.0	60	53	0.80	< 10
100	< 10	55	60	0.68	< 10
100	27	155	58	0.90	< 10
100	22	15	55	0.73	< 10
100	13.0	25	65	1.06	15
100	< 10	20	40	0.25	23
100	25	95	56	0.72	12
100	< 10	>200	69	0.84	21
98	70	–	95	>20	>100
98	61	–	98	>20	>100
100	< 10	35	23	0.37	27
100	< 10	120	46	1.64	< 10
100	< 10	–	91	>20	90
98	< 10	35	20	0.33	32
95	85	190	80	2.8	49
100	100	110	93	3.7	14
94	86	>200	80	6.4	22

[c] n-Hexane/n-butanol (9:1)
[d] Corrected for depression by ligand alone
[e] 1.0 mg Fe L^{-1} and 5.3 mg humic acid L^{-1}

different methods, showing that all techniques "look" at different chemical species or groups of species. It is worthwhile noting that four ligands increased the toxicity (see the values much higher than 100 in the column "labile copper from growth curve" in Table 3).

Iodine

Total dissolved iodine in seawater: 0.44–0.48 μM.

The dominant form of iodine in rocks and soils occurs as iodide. The thermodynamically stable iodine species in oxygenated river water and seawater is iodate. However, chemical determinations have shown that in river water iodide dominates. Iodide is not oxidized in the estuary but is only oxidized to iodate after arriving in the sea. Because the half-time of the iodide-iodate conversion is supposed to be tens of years (Brewer, 1975), the residence time of the iodide in the estuary is too short for a detectable transformation. Smith and Butler (1979) also found an organo-iodate (15% of the total iodine). In the ocean below 200 m depth iodine can exist only as iodate, but above in the euphotic zone up to 50% of the total iodine can occur as iodide. In anoxic water only iodide is found (see Andreae, this volume). Iodide may be released from sediments when the system changes from oxic to anoxic (Chapman, 1983). Oxidation of iodide is slow, whereas the reduction of iodate at the anoxic boundary is generally considered to be rapid.

Biological activity could be a source for thermodynamically unstable iodide. Bacteria can reduce iodate when oxygen concentrations are low and nitrate concentrations are high (Hickett, 1971, cited by Chapman, 1983). Also, Tsunogai and Sase (1969) observed that aerobic organisms can reduce iodate to iodide. However, experiments with phytoplankton cultured in a medium containing iodide yielded conflicting observations. Sugawara and Terada (1967) found that diatoms cultured in 126 nM iodide converted 24 nM iodide to iodate in four weeks. Also Truesdale (1978) observed a similar conversion. But Butler et al. (1981), on the basis of experiments of only 10 days duration, maintain that phytoplankton alone could not influence the relative abundance of iodine species.

There were also differences observed between the uptake and loss of iodate and iodide by seaweed and marine fish. The organisms accumulated about ten times more iodide than iodate and released iodate at faster rates than iodide (Hirano et al., 1983). Kuenzler (1969), however, suggested that the iodine released from zooplankton labeled under natural conditions in a fallout area was mostly iodide and little was iodate. Some of this iodine was also organically bound.

Iron

Reactive iron in open-ocean seawater: 1–10 nM; Hydrolyzable iron: 7–60 nM.

Most iron is present in seawater as aggregates of highly insoluble Fe(III)-hydroxide or as Fe(III) associated with organic chelators. At pH >6 Fe(III) is the thermodynamically most stable species.

Extensive investigations on the iron uptake by a coastal diatom indicate that iron is transported across the membrane by a membrane-bound metal-binding complex which in analogy to transferrin has been called phytotransferrin (Anderson and Morel, 1982). A technique which differentiates iron uptake from the formation of adsorbed or co-filtered colloidal ferric hydroxide was used to measure Fe uptake in the presence and absence of chelation (EDTA, CDTA, DTPA, NTA). Differences in uptake with and without light were observed. In the dark a comparison of the uptake rates of different chelators showed that Fe uptake was determined by the (calculated) free Fe^{3+} activity. In the light EDTA and CDTA increased the uptake rate by a factor of 2 and 6, respectively, at the same total iron concentration due to photodegradation of the Fe(III)-complex, making more free Fe^{2+} available. Ascorbic acid (liberating Fe(II) from Fe(III)-aminocarboxylate complexes) enhanced the uptake rate even more. In the absence of chelators, ascorbate inhibits the iron uptake by 50%. When very strong iron-chelating agents such as EDDHA or desferri-ferrioxamine B were added to the medium, no iron uptake could be detected. Also, Murphy et al. (1976) found that hydroxamate chelators (siderophores excreted by blue-green freshwater algae and bacteria) inhibit the growth of eukaryotic phytoplankton by rendering iron unavailable to them. Total Fe (ferric and ferrous) concentration determined uptake rates in the absence of chelators because the equilibrium between Fe^{3+} and the Fe^{2+} taken up by the algae is established sufficiently quickly. In nature, photooxidation of naturally occurring organic complexes or dissolution of inorganic complexes could also supply the essential Fe^{2+}.

Wells et al. (1983) studied the bioavailability of different colloidal sources of iron to the marine diatom *Thalassiosira pseudonana*. They found that iron concentrations below 2 nM would decrease the cell production of the diatom and that freshly prepared colloidal ferric hydroxide allowed a higher cell production than aged or heated ferric hydroxide or the more crystalline goethite or hematite. Their results suggest that the reduced bioavailability is due to greater thermodynamic stability of goethite, hematite, or treated ferric hydroxide. This shows that the thermodynamic stability of colloidal iron affects its bioavailability. Kuenzler (1969) observed that naturally labeled zooplankton lost Fe^{55} in both particulate and soluble forms.

Comparing the uptake and internal body distribution of radioactive ferric hydroxide in mussels with that of ferritin and transferrin containing Fe-59, George et al. (1976) observed that 20% of the inorganic Fe ingested was absorbed through the gills and 80% via the gut, but that 30% of the ingested Fe was not absorbed but eliminated with the feces. Ferric-hydroxide particles and ferritin are absorbed by pinocytosis both in the gills and in the gut. Although ferritin and ferric hydroxide were absorbed very similarly, ferritin was accumulated about 20% faster than the inorganic Fe. No excretion nor any degradation of the ferritin could be observed. The accumulation rate of the Fe-protein transferrin was even about 35% higher than that of the inorganic Fe. Ferritin was only absorbed in the digestive diverticulum by pinocystosis of mucus-bound transferrin and not as free, unbound protein.

Although these investigations suggest that algae can take up iron in the dissolved form while mussels absorb iron as particulates, it must be remembered that unicellular algae have also been observed to ingest particles directly.

The major storage form of iron for metabolic use is in the protein ferritin. In cases of iron overload, such as occur in man after multiple whole-blood transfusion or in benthic organisms ingesting large amounts of sediment, the excess Fe is immobilized in the tertiary lysosomes of the cells. These lysosomes contain the residual, tractable remnants of intercellular digestion. They are mainly composed of cross-linked and highly peroxidized lipoproteins (George et al., 1982). This peroxidation process is catalyzed by divalent metals such as Fe(II), and during the process the metal becomes physically trapped and immobilized. These structures also act as chelation sinks within the cell and sequester many other metal ions, including Cd, Cu, Pb, Pu, and Zn (George, 1982). An additional route of metals is via metalloproteins such as metallothionein, ferritin, ceruloplasmin, etc., which are degraded in the lysosomes. The peroxidation process is a free-radical reaction and is autocatalytic; consequently, the structures become more highly polymerized with age and the metals become trapped: they are totally insoluble and inert (George et al., 1982). Vertebrates tertiary lysosomes tend not to be excreted and therefore accumulate within cells in many invertebrates (hence they are often termed age pigments or lipofuschins), particularly in bivalves. They are excreted in the urine and feces. It should be noted that this material has many characteristics of marine humic material and may form a large proportion of marine detritus.

Mercury

Total mercury concentration in open-ocean seawater: 5–50 pM; Organic mercury: 0.5–6% of Hg_T.

The total Hg (Hg_T) concentrations have continuously been lowered in recent years, mainly because more attention has been paid to avoiding sample contamination. Computer models based on thermodynamic principles (equilibria) predict that soluble inorganic mercury (Hg_I) species are present in seawater exclusively as chloro-complexes with $HgCl_4^-$ as the dominant species. Since methylmercury (MeHg) predominates in marine organisms, it is the most important Hg compound from the biological and health-protection point of view. The emphasis on MeHg in the biogeochemical cycle of mercury has most probably distracted attention from the fact that MeHg is not the dominant form of organic mercury in natural waters. Unfortunately, only very few MeHg data for seawater exist (Egawa et al., 1982; Fujita and Iwashima, 1981; Yamamoto et al., 1983). Their values range from <0.03 to 6% of the Hg_T present. Also, Andren and Harriss (1975) observed that the dissolved mono- and di-MeHg concentration in water samples from Mississippi and the Everglades, ranging in salinity from $S=1$ to $S=33$ were less than 2% and 1%, respectively, of the Hg_T present in the samples, whereas the rest of the organic Hg was associated with organic matter of size fractions ranging from <500 to >300000. The highest percentages of Hg were associated with the <500 molecular-size fraction. In estuarine water of $S=3$, 82% of the total dissolved Hg was associated with the <500 fraction. Wallace (1982) found that 4 to 50% of the Hg_T in coastal seawater was associated with surface-active, dissolved colloidal organic matter isolated from the water column in a controlled experimental system. Most recently Suzuki and Sugimura (1985) found Hg associated with organic matter with a molecular size of 9000. These authors could not detect alkyl-Hg and even suggest that the alkyl forms of Hg do not play an important role in the biogeochemical cycle of Hg.

Many heterotrophic microorganisms (bacteria), both aerobes and anaerobes, can methylate mercury (reviews by Beijer and Jerneloev, 1979; Windom and Kendall, 1979). At the pH 8 of seawater both mono-MeHg and di-MeHg are formed, but chemical methylation of mercury will also occur in seawater if suitable methyl donors (e.g., tin and lead alkyls, humic substances, methylcobalamine) are available. Degradation of MeHg by microorganisms also occurs. A summary of the mercury cycle is shown in Fig. 2. Two problems arise from this scheme: MeHg is only formed in sediments, and the formation of MeHg is a detoxification process. If MeHg is only formed in sediments, then it is difficult to explain how marine pelagic fishes living several hundred meters above the sea bottom obtain MeHg, especially in summer when the thermocline hinders vertical transport. All bacteria which can methylate mercury can also transform Hg(II) into metallic mercury which is much less toxic than MeHg. Why should they, therefore, produce the highly toxic MeHg instead of the much less toxic metallic Hg to reduce toxic environmental levels? Are the extremely high Hg(II) concentrations used in the methylation experiments

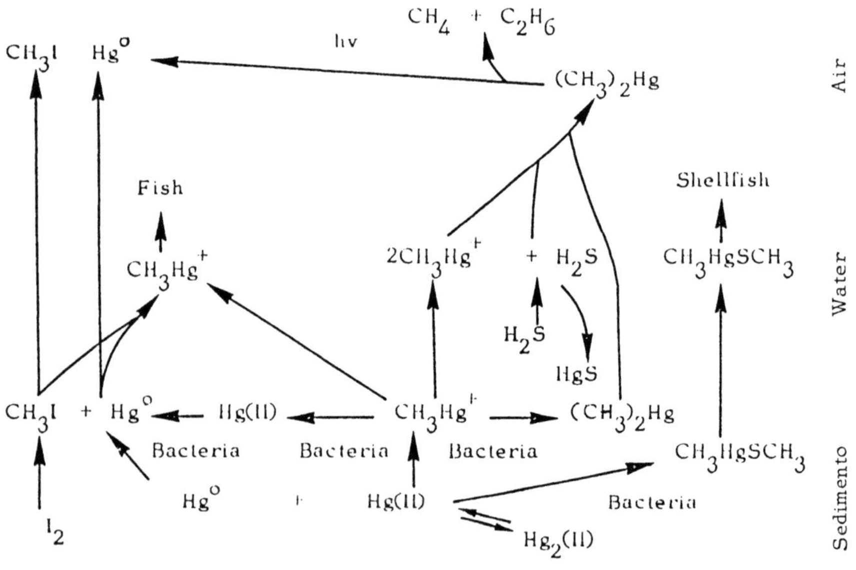

Fig. 2. The mercury cycle (Wood and Wang, 1983), showing the reactions catalyzed by bacteria, chemical disproportionation by H_2S, and the photochemistry of organomercury compounds

responsible for the MeHg formation, and hence is the methylation of Hg in sediments an artifact? On the other hand, direct methylation by marine organisms or freshwater organisms higher in the trophic chain than bacteria could not be demonstrated (Huckabee et al., 1978; Pentreath, 1976 b); only indirect mercury methylation by intestinal microorganisms occurs (e.g., Rowland et al., 1980). However, this seems not always to be the case. Pan-Hou and Imura (1981) did not observe any methylation by microorganisms isolated from the intestine of yellowfin tuna. This observation is very important because in marine organisms the percentage of MeHg is much higher then the percentage of MeHg in seawater (0.5 to 6%). From the few data available, it seems that in mixed phyto- and zooplankton typically 20 to 50% of the mercury is MeHg (Knauer and Martin, 1972). At higher levels in the food chain, for example, in plankton-feeding fishes such as sardines and anchovies, 60 to 80% MeHg has been found, while in their predators (bonito and tuna) MeHg comprises 70 to 90% of the Hg_T. For an understanding of the processes leading to the high accumulation of mercury it is also important to remember that the mercury concentration increases with the size of the organisms.

A relatively simple model of a pelagic food chain (seawater, plankton, sardine, tuna) based on general data available on Hg metabolism and specific mercury concentration in pelagic marine organisms from the Mediterranean (Bernhard, 1985) shows how the Hg_T in the organisms increases with their size and with increasing level int he food chain, and how at the same time the percentage of MeHg in organisms increases.

Fish accumulate Hg_I and MeHg directly from seawater, but MeHg is taken up much faster than $HgCl_2$ (Pentreath, 1973, 1976 a–c). Unfortunately, good uptake data for MeHg by marine fish are not available, but the uptake efficiency is very high (about 90%) compared with that of Hg_I (10%). On the other hand, MeHg is much more slowly released (biological half-time ($Tb_{1/2}$) = 275 days) than Hg_I ($Tb_{1/2}$ = 60 days. Pelagic bluefin tuna caught in the Western Mediterranean can be divided into two distinct populations according to their mercury concentration, one "low-mercury population" and one "high-mercury population". Fishery biology observations suggest that the low-mercury population migrates from the Atlantic into the Mediterranean for spawning, whereas the high-mercury population is indigenous to the Mediterranean. Sardines and anchovies (food of tuna) caught in the Strait of Gibraltar have lower concentrations than specimens of the same size caught in Italian coastal waters, although the published mercury concentrations in seawater from the Mediterranean and the Atlantic show no significant difference. The best fit to the observed Hg con-

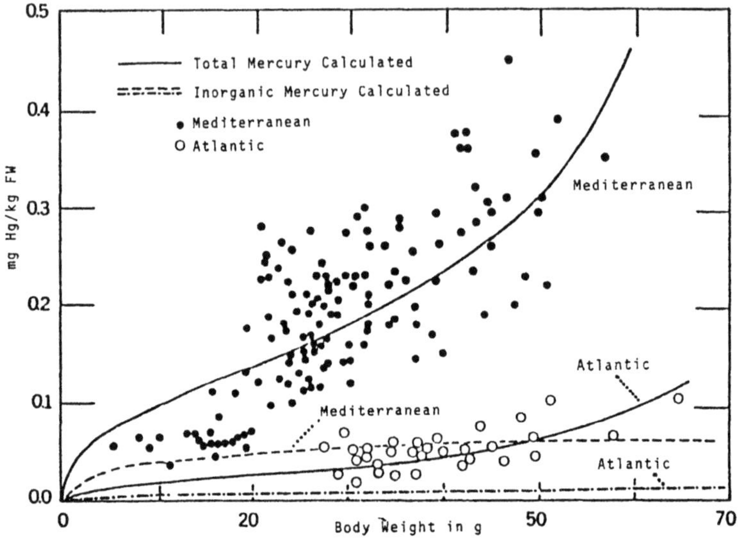

Fig. 3. Model calculation of total mercury (continuous line) and inorganic mercury (dotted line) versus body weight in *Sardinia pilchardus* from the Strait of Gibraltar (A) and the Tyrhennian Sea (M) (Bernhard, 1985)

centrations in sardines and tuna were obtained assuming that in the Atlantic the concentration of mercury in seawater is five times lower than in the Western Mediterranean, even though such differences could not be established from the literature due to the wide spread of the data. In Fig. 3 the predictions of the model are compared with Hg concentration in sardines from the above two locations. As can be seen, the Hg_I concentration increases with increasing size (e.g., age) in the sardine. Above about 5 grams body weight the Hg_I concentration remains constant (uptake and loss have come to a dynamic equilibrium) while the MeHg concentration steadily increases. At 10 grams body weight MeHg comprises 60% and at 60 grams 87.5%. If we now take the tuna model (Fig. 4) we see that the Hg_I is only a small fraction of the Hg_T. This model shows that, starting with 2% MeHg in seawater, sardines will reach 60 to 90% according to age. Tuna feeding on sardines will reach nearly 100% MeHg at a very early age. This example may be an extreme case since tunas can grow very large (400 kg) and live to a relatively old age (> 15 years). In organisms with shorter life spans the *relative enrichment of MeHg* will not be so great, but the model predicts that even for very small tuna the MeHg percentage is high.

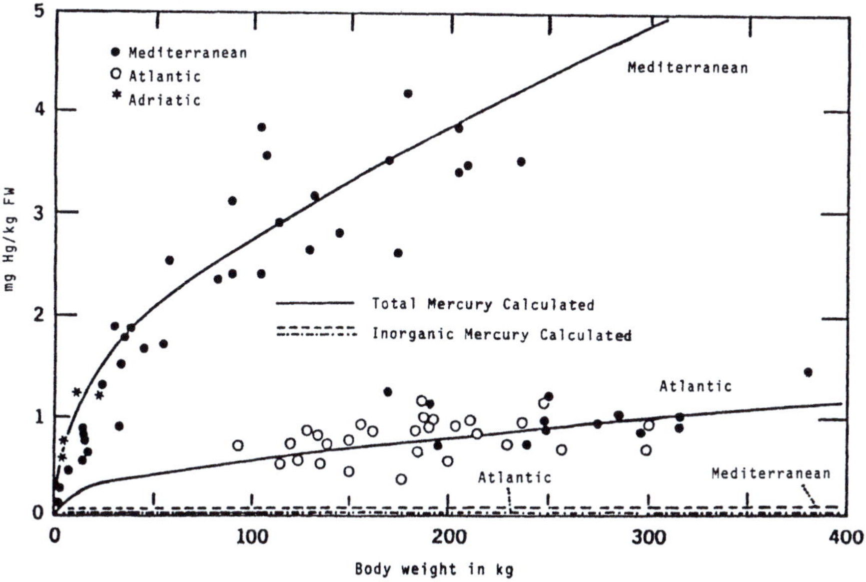

Fig. 4. Model calculations of total mercury (continuous line) and inorganic mercury (dotted line) versus body weight in *Thunnus thynnus* from the Strait of Gibraltar (A) and the Tyrhennian Sea (M) (Bernhard, 1985)

The model also shows that methylation of mercury by higher organisms is not necessary to explain the high relative MeHg enrichments observed; the differences in the uptake and loss kinetics between MeHg and Hg_I are sufficient.

Similar models could be constructed for other food chains. Mussels accumulate MeHg twice as fast as Hg_I and lose $HgCl_2$ with a biological half-time ($Tb_{1/2}$) of 140 days in the lab and 82 days in nature. On the other hand, they lose MeHg with a $Tb_{1/2}$ of 380 days in the lab and 65 days in nature. The difference in food availability to the mussels which grew in nature but lost weight in the lab culture is offered as an explanation for the difference between the $Tb_{1/2}$ in the lab and that in nature (Fowler et al., 1978). For the shrimp *L. seticaudata* no differences in loss rate were observed between laboratory and field: MeHg $Tb_{1/2} = 530$ days and Hg_I $Tb_{1/2} = 110$ days. Since stable data are not supplied in the paper, a quantitative comparison is not really possible.

The uptake of Hg_I can be further influenced by the physicochemical form in which it is present in seawater. Loama (1974, cited in Windom and

Kendall, 1979) showed that complexing agents (e.g., cysteine) and seawater containing high amounts of organic matter will reduce the Hg uptake. If Hg is bound to the organic matter present in sediments, its uptake by deposit feeders is strongly reduced (Kendall, 1978, cited in Windom and Kendall, 1979). The accumulation of both Hg_I and organic mercury is reduced by glutathione (Corner and Rigler, 1958). The availability of metals and hence their absorption from the gut depends on their chemical form, as has been shown by supplying metals to marine organisms in different matrixes (starch, gelatine, the food organism *Neris diversicolor*) (Pentreath, 1973). All these parameters could result in a further reduction of Hg_I versus MeHg concentrations.

In the food chain model mentioned above it is assumed that about 2% MeHg is present in seawater. This raises the question as to its origin. So far the only known sources of MeHg are sediments. Since sediments are too far away from the habitat of pelagic fishes, a pelagic source of MeHg must exist. Therefore, it may be interesting to investigate further whether marine pelagic bacteria can methylate Hg or even whether phytoplankton may play a role in the methylation of Hg similar to the methylation of As. As the model indicates, it will be necessary to methylate only a few percent of the total mercury present in seawater.

Ruthenium

Ruthenium concentration in the Gulf of Mexico: 0.01 pM.

The distribution of ruthenium species is very complicated since it possesses numerous valencies and can exist in many different physicochemical

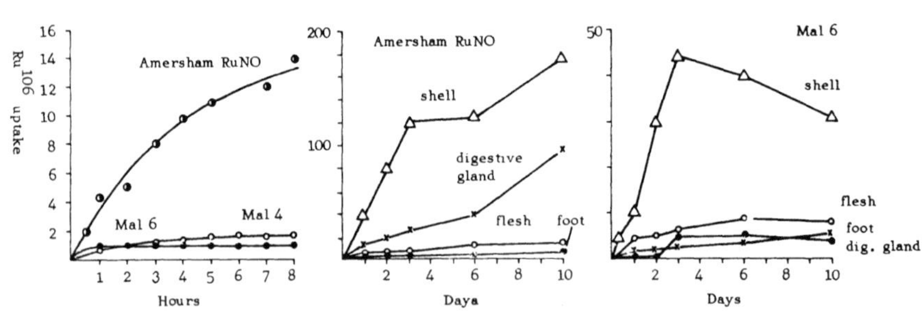

Fig. 5. Uptake of nitrosyl ruthenium (Amersham) and ruthenium from radioactive waste solution (Mal 4 and 6) by the alga *Porphyra laciniata* and the mussel *Mytilus edulis* exposed to solutions with the same radioactivity (Jones, 1960). Note the different y-scales

forms. The occurrence of numerous species has caused many difficulties in the treatment of radioactive wastes. Comparing the uptake of Ru from radioactive waste solutions and from nitrosyl ruthenium showed that ten times more Ru was taken up from the ruthenium nitrosyl than from the waste solutions (Fig. 5). There are also large differences between the uptake of Ru as chloride and as nitrate: the concentration factor of ruthenium chloride is about ten times greater than that of ruthenium nitrate (Keckes et al., 1967).

Zinc

Zinc concentration in open seawater: 10–600 ng Zn-TL^{-1}.

Zinc concentrations in the pelagic areas have recently been revised to much lower levels. Previous concentrations in open seawater are now considered to have been too high because of sample contamination (review by Florence, 1982b). This revision also casts doubt on previously reported distribution of zinc forms, although such data concern mainly coastal waters where zinc concentrations could be much higher and hence less subject to sample contamination.

The uptake experiments of Bernhard and Zattera (1969) were carried out on seawater collected with special precautions in the coastal zone of La Spezia. Even if the seawater used in these experiments could have been contaminated and therefore the results are not directly applicable to natural conditions, the seawater was extensively characterized. Thus, the experiments can still serve as a model. This characterization is briefly summarized as follows (Bernhard et al., 1975; Piro et al., 1973):

Hydrolyzing the seawater by lowering its pH showed that the electroactive zinc concentrations (not the peak height) increased (Fig. 6). The curve has a very characteristic plateau at pH 5 to 6. This plateau was observed in all samples collected in the coastal waters of the Ligurian Sea and in the Gulf of Taranto. This same plateau can be observed in model solutions of artificial seawater or NaCl to which EDTA or cysteine has been added in appropriate amounts. The amount of zinc determined at pH 1.4 was about 10% lower than the zinc extracted from seawater with Ca-Chelex-100. By centrifugation and dialysis it could be shown that the fraction corresponding to the difference in electroactive zinc concentrations determined between pH 6 and pH 8 was particulate zinc. Going up and down the pH scale revealed that returning from pH 1.4 to pH 6, the concentration of electroactive zinc was higher than when the pH was lowered, indicating a cer-

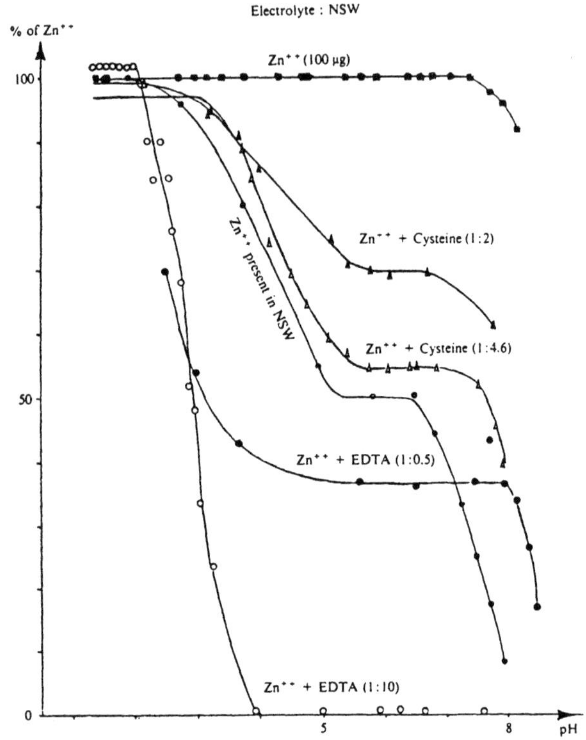

Fig. 6. Electroactive zinc concentrations versus pH for coastal seawater collected offshore near La Spezia (Mediterranean Sea) with and without the addition of EDTA and cysteine. Note that not the peak height but rather the zinc concentrations determined at various pH by standard additions are shown (Piro, 1970)

tain inertia in reforming a zinc complex which is not electroactive. If, however, the electroactive zinc was redetermined after several hours, the amounts of electroactive zinc were approximately the same as the amount determined while lowering the pH. Therefore, this fraction was tentatively called complexed zinc.

The amounts of electroactive zinc at three pHs (1.4, 6, and 8) in the port and the Gulf of La Spezia and on an off-coast transect for one year showed that the total hydrolyzable zinc decreased rapidly with distance from the port (Fig. 7). The ratios between the amounts of electroactive zinc at pH 1.4 and pH 6.5 and at pH 1.4 and pH 8 are very stable at about 50% and 8% of the total hydrolyzable zinc, respectively. There were only variations in

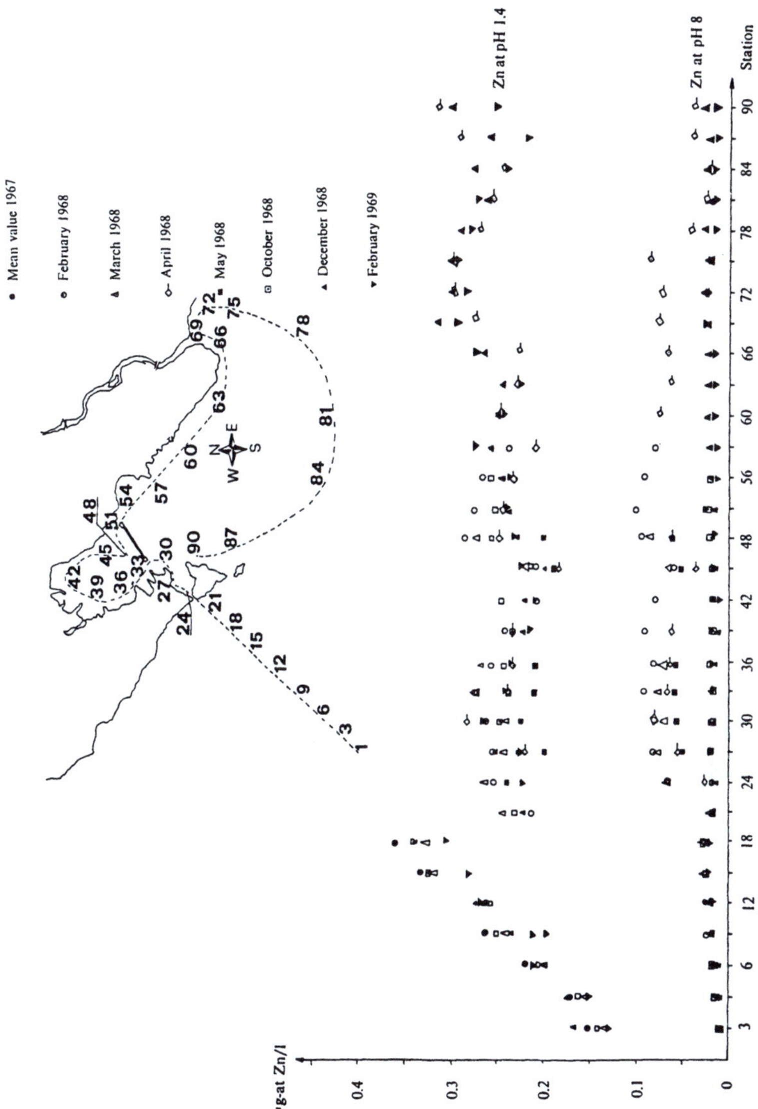

Fig. 7. Horizontal distribution of electroactive Zn at pH 1.4 and pH 8.0 in the Gulf of La Spezia. Mean values for 1967, February 1968, March 1968, April 1968, May 1968, October 1968, December 1968, February 1969 (Piro, 1970)

the ratio between cruises in the inner Gulf. It may be worthwhile mentioning that although the zinc determinations were not carried out under ultraclean conditions, all precautions were taken to avoid sample contamination (all-plastic sampling bottles and messengers; all-plastic hydrographic wire on stainless steel winch drums and pulleys; small plastic-covered wooden ship; sample bottles, storage bottles, and glassware were acid-cleaned and rinsed with sample water; etc.). The rapid decrease of the total zinc concentration with distance from the shore also seems to indicate that the total zinc concentration should decrease to background values with further distance from the zinc source, the shipyards.

Investigation of the uptake and loss of zinc by the unicellular alga *Phaeodactylum tricornutum* from natural coastal seawater with and without the addition of chelators showed that the algae preferentially accumulated ionic (i.e., electroactive) zinc (Bernhard et al., 1975; Bernhard and Zattera, 1969). When radioactive ionic zinc is added to natural seawater (Fig. 8), about 80% of the radioactive zinc (dotted line) is removed from the culture medium and accumulated in the algae (curve not shown). At the same time, about 45% of the stable zinc (dashed line) is removed from the medium and

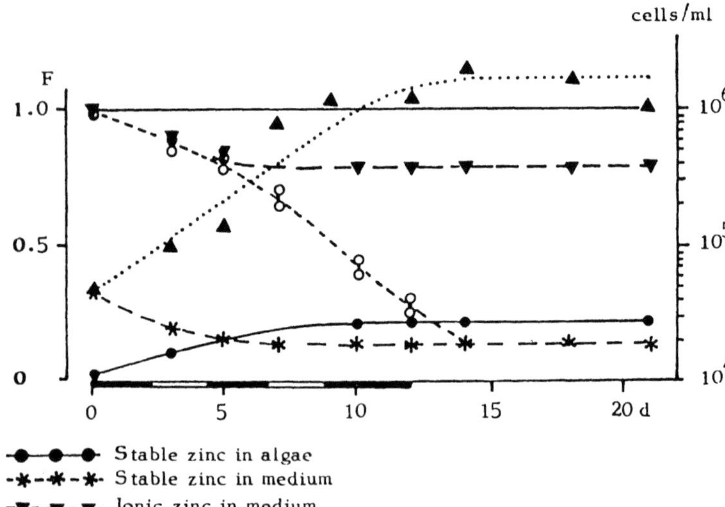

Fig. 8. Uptake of radioactive zinc and natural zinc from coastal seawater by the alga *Phaedactylum tricornutum* (Bernhard and Zattera, 1969)

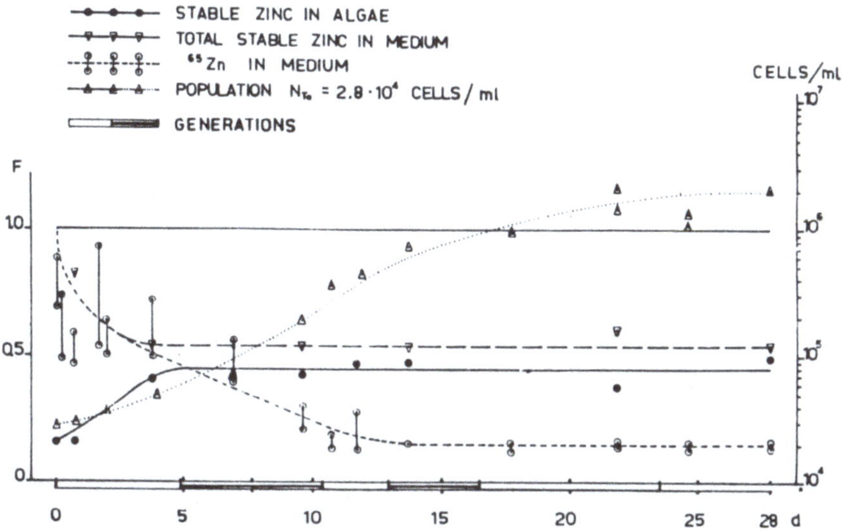

Fig. 9. Uptake of radioactive zinc and natural zinc from coastal seawater with an addition of cysteine by the alga *Phaedactylum tricornutum* (Bernhard and Zattera, 1969)

the same amount (continuous line) is accumulated in the algae. Adding enough cysteine to natural seawater (Fig. 9) so that about 30% of the stable zinc remained electroactive, and then adding radioactive ionic zinc, showed that the algae accumulated about 87.5% of the radioactive zinc while at the same time reducing the initial 30% of electroactive stable zinc to 12.5% and accumulating about the same amount (i.e., 21%) of stable zinc. Because of this sequence of additions to the seawater, the stable ionic zinc present in the seawater was complexed by cysteine. The ionic radioactive zinc later added mostly remained uncomplexed. This explains why more radioactive zinc than stable unbound zinc has been accumulated. These experiments also show that the radioactive zinc is only very slowly exchanged with the stable zinc, whether the latter is bound to natural or artificial ligands. These observations were confirmed by demonstrating that even one year after radioactive zinc had been added to natural seawater at pH 8, the radioactive zinc exchanged only with the ionic and particulate fractions but not with the stable zinc contained in an inert fraction which made up about 50% of the total zinc present (Piro et al., 1973). Suspending radioactive algae in nonactive medium revealed that practically no radioactive zinc was released. When, however, EDTA is added repeatedly to the medium, portions of the radioactive zinc are released into the medium. If sufficient EDTA is added such that no electroactive zinc can be detected in the medium, the algae take up small but measureable quantities of radioactive zinc. Stable zinc uptake by the algae or a decrease of electroactive stable zinc from the medium could not be detected. Thus, it appears that either all or a part of the electroactive zinc is accumulated by these algae and only very small amounts of zinc are taken up when all of the ionic stable zinc present is chelated.

In uptake experiments with higher organisms, differences between Zn species have also been observed. Harrison (1979) reports higher uptake of Zn^{65} from unfiltered water by oysters placed in the outfall canal of a nuclear plant than in filtered seawater during summer. The differences measured in winter were insignificant. A 24 h exposure of oysters to Zn^{65} in the presence of EDTA, glycine, "yellow stuff" (natural humic acids) and kaolin showed that the EDTA strongly reduced Zn^{65} activity in the oyster compared to the controls in all tissues examined. The Zn^{65} radioactivity in the presence of glycine was more or less the same: it increased in all tissues in the presence of kaolin and "yellow stuff," probably because of the particulate nature of these two substances.

Kuenzler (1969) and Small et al. (1974), with the help of ASV, filtration, and ion-exchange resins, showed that marine zooplankton and crustaceans labeled experimentally and naturally, through nuclear test fallout released inorganic and particulate zinc as well as zinc which was not exchangeable. The nonexchangeable zinc is presumably organically bound.

Extrusion of Metal-Complexing Substances by Marine Organisms

Marine organisms release organic matter which can complex metals. Fisher and Fabris (1982) reported that materials excreted by three species of unicellular algae can complex more Cu than Zn and more Zn than Cd. The materials released during the stationary-growth phase have much greater chelating capacity for Zn than the material released during the log-growth phase. This seems to indicate that different materials are released during the active-log phase (waste products) than during the stationary phase of senile cells (decay products). Not all algal species seem to extrude these complexing substances. Swallow et al. (1978) found that one out of eight algal species and McKnight and Morel (1979) found that nine out of 14 species produced metal-complexing substances. In a growth bioassay Hardstedt-Romeo and Gnassia-Barelli (1980) showed that the culture medium of a phytoplankton algae contained both Cd- and Cu-complexing ability. If *Chlorella* is exposed to toxic Cu concentrations it extrudes complexing agents during an initial lag phase until the concentration of the free Cu ion is reduced to tolerable levels, whereupon logarithmic growth starts (Fig. 10). This lag phase could very well be the so-called culture shock observed commonly when algae are introduced into a new medium (Fogg, 1977).

Several authors observed that exudates or extracts from macroalgae can also complex metals (e.g., Ragan et al., 1980; Sueur et al., 1982). Sueur et al. (1982) calculated the conditional stability constants for the extudates from three macrophytes. Their values lay between 9.77 and 10.15.

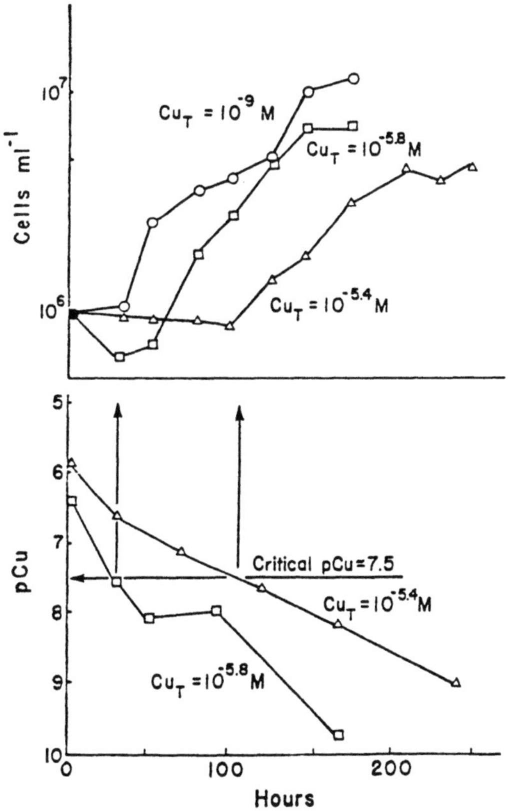

Fig. 10. Evolution of cupric ion activity during the growth of copper-sensitive *Chlorella*. A: growth curve; B: cupric ion activity (Morel and Morel-Laurens, 1983)

Iron metabolism and the excretion of iron-complexing substances have received special attention. Marine prokaryotic microorganisms (bacteria and cyanobacteria) can produce extracellular Fe(III)-specific chelating compounds (siderophores) that enable them to solubilize and therefore acquire the Fe present in the Fe(III)-hydroxide aggregates (Goyne and Carpenter, 1974). This has led to a search for similar substances in eukaryotic algae. McKnight and Morel (1979) tested 13 axenic eukaryotic phytoplankton species for the production of Fe(III)-specific chelators but could not find any evidence for their presence. Later Trick et al. (1983) showed that three out of five unicellular *Procentrum* species produce extracellular siderophores but only in iron-deficient medium and at certain very specific time intervals of the stationary-growth phase. This may be easily missed if the frequency of sampling of the medium is low. During a total of ten days growth the extracellular siderophores appear in the medium for only two to four days during the stationary phase (Fig. 11).

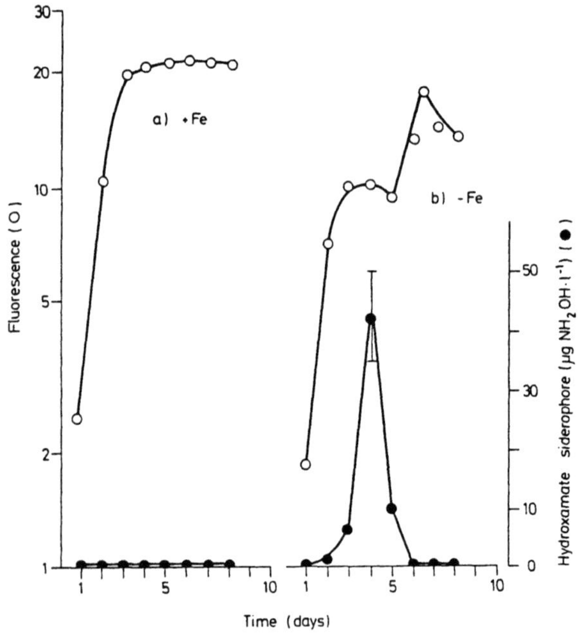

Fig. 11. Comparison of growth measured by *in vivo* fluorescence and extracellular hydroxamate siderophore concentration of *Thalassiosira pseudonana* (Trick et al., 1983)

The ecological significance of these observations is not clear. The extrudates may protect marine organisms in nature and may be responsible for the complexing capacity of seawater. Florence et al. (1983) maintain that these chelating substances are produced only when the organisms are exposed to toxic or subtoxic metal concentrations, as has also been shown by Morel and Morel-Laurens (1983). Sharp (1977) proposed that healthy cells will not excrete large amounts of organic matter. Not all species tested produced siderophores. The extracellular material produced may not always protect the algae from toxic effects (Hall et al., 1979; Sueur et al., 1982). On the other hand, many algologists believe that algae do release significant amounts of organic matter (e.g., Fogg, 1977). Huntsman and Barber (1975) observed that the addition of either organic material excreted by algae or EDTA can reduce the initial lag phase in media prepared from seawater collected at great depth.

In recent years several papers have been published on the complexing capacity of seawater. Mainly Cu-complexing capacity has been determined (see Andreae and Lund, both this volume), but lately chelation has also been observed for Fe, Cd, Ni, Mn, and Pb (Mackey, 1983a). It is also interesting that the Cu-complexing capacity is much higher than the measured Cu concentration (Mackey, 1983b; Van der Berg, 1984). Van der Berg (1984) found that 94–98% of the dissolved copper was complexed with organic compounds and that the conditional stability constants were about 10. Also, Florence (1982a) observed that only a very small fraction of the total amounts of Cu, Pb, Cd, and Zn present in seawater collected in the coastal areas of Australia could be removed with either Chelex-100 or thiol resins. On the other hand, since more than 90% of the complexed Cu was always removed by complexing Cu with several model ligands, Florence concluded that a large fraction of Cu is bound to unknown inert complexes.

Conclusion

The influence of chemical species on the uptake of many trace elements and their toxicity to marine organisms is well documented, although most observations and experiments have been performed with model compounds. Valence state, complexation with natural or artificial organic substances, and adsorption to or inclusion in particles have considerable influence on the amounts taken up or the toxicity experienced. For trace metals such as Cd, Cu, Zn, etc., the uptake and toxicity are related to the "electroactive form"; however, this electroactive form is not identical with the form or forms which are actually actively accumulated or which are toxic. The uptake of iron by algae appears to be mediated by siderophores, although there may be exceptions. In the first phase of uptake, particulate iron can be ingested directly by unicellular phytoplankton or by pinocytosis in the gills and gut of molluscs. A similar process has also been observed for colloidal Au and Th and implicated for Hg and Pb. However, pinocytosis of mucus-bound metal may be occurring. Marine organisms intervene by the formation of chemical ligands producing otherwise thermodynamically unstable metal species. They also extrude organic substances which can chelate various trace metals other than copper. It seems that marine organisms can also distinguish between different dissolved inorganic chemical species, but no direct evidence yet exists. If future research shows that 50 to 95% of a trace metal is bound to organic substances and associated with particles, these forms have to be given more attention in future experimental work.

References

Anderson MA, Morel FMM (1982) The influence of aqueous iron chemistry on the uptake of iron by the coastal diatom Thallassiosira weissflogii. Limnol Oceanog 27:789–813

Andreae MO, Klumpp DW (1979) Arsenic uptake and metabolism by marine phytoplankton. Envir Sci Technol 13:738–741

Andren AW, Harriss RC (1975) Observations on the association between mercury and organic matter dissolved in natural waters. Geochim Cosmochim Acta 39:1253–1257

Batley SE, Gardner D (1978) A study of copper, lead and cadmium speciation in some estuarine and coastal marine waters. Est Coast Mar Sci 7:59–70

Beijer K, Jerneloev A (1979) Methylation of mercury in aquatic environments. In: Nriagu JO (ed) The Biogeochemistry of Mercury in the Environment, Amsterdam: Elsevier/North-Holland, pp 203–210

Bell MV, Cally KF, Sargent GR (1983) The transport of orthovaladate and similar oxyanions in relation to salt and water transport across the isolated intestine of the common eel (*Anguilla anguilla*). J Exp Mar Biol Ecol 102:295–305

Benson AA, Summons RE (1981) Arsenic accumulation in Great Barrier Reef invertebrates. Science 211:482–483

Bernhard M (1955) Die Kultur von Seeigellarven (*Arbacia lixula*) in kuenstlichem und natuerlichem Meerwasser mit Hilfe von Ionenaustauschsubstanzen und Komplexbildner. Pubb Staz Zool Napoli 14:80–95

Bernhard M (1985) Mercury accumulation in a pelagic foodchain. In: Martell AE, Irgolic K (eds) US-Italian Workshop on Environmental Inorganic Chemistry, pp 349–358. Weinheim: Chemie Verlag, pp 349–358

Bernhard M, Goldberg ED, Piro A (1975) Zinc in seawater – an overview 1975. In: Goldberg ED (ed) The Nature of Seawater, Berlin: Dahlem Konferenzen, pp 43–68
Bernhard M, Zattera A (1969) A comparison between the uptake of radioactive and stable zinc by a marine unicellular alga. In: Nelson J, Evan D (eds) II National Symposium on Radioecology, Washington, DC: USAEC, pp 389–398
Brewer PG (1975) Minor elements in seawater. In: Riley JP, Sharrow G (eds) Chemical Oceanography, London: Academic Press, pp 415–496
Brooks PJ, Evans WH (1981) Determination of total inorganic arsenic in fish, shellfish and fish products. Analyst 106:514–520
Butler ECV, Smith JD, Fisher NS (1981) Influence of phytoplankton on iodine speciation in seawater. Limnol Oceanog 26:382–386
Carr RS, McCulloch WL, Neff JM (1982) Bioavailability of chromium from a used chrome lignosulphonate drilling mud to five species of marine invertebrates. Mar Envir Res 6:189–203
Chambers EL, Whiteley AH (1966) Phosphate transport in fertilized sea urchin eggs. I. Kinetic aspects. J Cell Physiol 68:289–308
Chapman P (1983) Changes in iodine speciation in the Benguela Current upwelling system. Deep-Sea Res 30:1247–1259
Chipman WA (1966) Uptake and accumulation of chromium-51 by the clam *Tapes decussatus* in relation to physical and chemical form. In: Disposal of Radioactive Wastes into Seas, Oceans and Surface Waters, IAEA Symposium, Vienna: IAEA, pp 571–582
Chipman WA (1967) Some aspects of the accumulation of 51-Cr by marine organisms. In: Aberg A, Hungate FP (eds) Radioecological Concentration Processes, Oxford: Pergamon Press, pp 931–941
Conney RV, Mumma RO, Benson AA (1978) Arseniumphospholipid in algae. Proc Natl Acad Sci USA 75:4262–4264
Corner EDS, Rigler FH (1958) The modes of action of toxic agents. III. Mercury chloride and N-amylmercuric chloride on crustaceans. J Mar Biol Ass UK 37:85–96
Cossa D (1976) Sorption du cadmium par une population de la diatomee *Phaeodactylum tricornutum* in culture. Mar Biol 34:163–167
Cowey CB, Know D, Andron JW, George SG, Pirie B (1977) The production of renal calcinosis by magnesium deficiency in rainbow trout (*Salmo gairdnari*). J Nutr 38:127–135
Cranston RE, Murray JW (1980) Chromium species in the Columbia River and estuary. Limnol Oceanog 25:1104–1112
De Bartoli M, Galione P, Malvicini A, Polvani C (1968) Concentration factors for strontium and caesium in fish of lakes in the region of Varese (N. Italy). J Fisica Sanitaria Prot Contr Radiazioni 12:324–331
Edmonds JS, Francesconi KA (1981) Arseno-sugars from brown kelp (*Ecklonia radiata*) as intermediates in cycling of arsenic in a marine ecosystem. Nature 289:602–604
Edmonds JS, Francesconi KA (1982) Isolation and crystal structure of an arsenic-containing sugar sulphate from the kidney of the giant clam, *Tridacna maxima,* X-ray crystal structure of (2S)-3-[5-deoxy-(dimethylarsinoyl)-&-D-ribofuranosyloxyl]-2-hydrooxypropyl hydrogen sulphate. J Chem Soc Perkin Trans 1:2989–2993

Egawa H, Kuroda T, Shiraishi N (1982) Determination of the trace amount of "methyl mercury" in sea water using chelating resins. J Chem Soc: Chem Ind Chem 4:685–691
Fisher NS, Fabris JG (1982) Complexation of Cu, Zn and Cd by metabolites excreted from marine diatoms. Mar Chem 11:245–255
Flanjak J (1982) Inorganic and organic arsenic in some commercial East Australian Crustacea. J Sci Food Agric 33:579–583
Florence TM (1982a) Development of physico-chemical speciation procedures to investigate the toxicity of copper, lead, cadmium and zinc towards aquatic biota. Anal Chim Acta 141:73–94
Florence TM (1982b) The speciation of trace elements in waters. Talanta 29:345–364
Florence TM, Lumsden BG, Fardy JJ (1983) Evaluation of some physico-chemical techniques for the determination of the fraction of dissolved copper toxic to the marine diatom *Nitzschia closterium*. Anal Chim Acta 151:281–295
Fogg GE (1977) Excretion of organic matter by phytoplankton. Limnol Oceanog 22:576–577
Foster PL, Morel FMM (1982) Reversal of cadmium toxicity in a diatom: an interaction between cadmium activity and iron. Limnol Oceanog 27:745–752
Fowler SW, Heyraud M, La Rosa J (1978) Factors affecting methyl and inorganic mercury dynamics in mussels and shrimp. Mar Biol 46:267–276
Fowler SW, Unlu MY (1978) Factors affecting bioaccumulation and elimination of arsenic in the shrimp *Lysmata seticaudata*. Chemosphere 9:711–720
Fujita M, Iwashima K (1981) Estimation of organic and total mercury in seawater around the Japanese Archipelago. Envir Sci Technol 15:929–933
George SG (1982) Subcellular accumulation and detoxication of metals in aquatic animals. In: Vernberg WB, Calabrese A, Thurberg FP, Vernberg FJ (eds) Physiological Mechanisms of Marine Pollutant Toxicity, New York: Academic Press, pp 1–60
George SG, Coombs TL (1977) The effects of chelating agents on the uptake and accumulation of cadmium by *Mytilus edulis*. Mar Biol 39:261–268
George SG, Pirie BJS, Coombs TL (1976) Absorption, accumulation and excretion of iron-protein complexes by *Mytilus edulis* (L.). In: Proceedings of the International Conference on Heavy Metals in the Environment, Toronto: National Research Council of Canada, pp 611–634
George SG, Pirie BJS, Coombs TL (1982) Characterization of metal containing granules from the kidney of the common mussel, *Mytilus edulis*. Biochem Biophys Acta 716:61–71
Goyne ER, Carpenter EJ (1974) Production of iron-binding compounds by marine algae. Limnol Oceanog 19:840–841
Hall A, Fielding AH, Butler M (1979) Mechanisms of copper tolerance in the marine fouling alga *Ectocarpus siliculosus* – Evidence for an exclusion mechanism. Mar Biol 54:195–199
Hardstedt-Romeo M, Gnassia-Barelli M (1980) Effect of complexation by natural phytoplankton exudes on the accumulation of cadmium and copper by the Haptophycee *Cricosphaera elongata*. Mar Biol 59:79–84
Harrison FL (1979) Effect of the physicochemical form of trace metals on their accumulation by bivalve molluscs. In: Jenne EA (ed) Chemical Modeling in Aqueous Systems, Washington: American Chemical Society, pp 611–634

Hirano S, Ishii T, Naramura R, Matsuba M, Koyanagi T (1983) Chemical forms of radioactive iodine in seawater and its effects upon marine organisms. Radioisotopes 32:319–322

Huckabee JW, Janzen AS, Blaylock BG, Talmi Y, Beauchamp J (1978) Methylated mercury in Brook Trout (*Salvelinus fontinalis*): Absence of an *in vivo* methylation process. Trans Am Fish Soc 107:848–852

Hung Y-W (1982) Effects of temperature and chelating agents on the cadmium uptake in the American oyster. Bull Envir Contam Toxicol 28:546–551

Huntsman SA, Barber RT (1975) Modification of phytoplankton growth by excreted compounds in low-density populations. J Phycol 11:10–13

Jones RF (1960) The accumulation of nitrosyl ruthenium by fine particles and marine organisms. Limnol Oceanog 5:312–325

Kägi JHR, Nordberg M (1979) Metallothionein. In: Proceedings of the First International Meeting on Metallothionein and Other Low Molecular Weight Metal-binding Proteins, Zurich, July 17–22, 1978. Experientia (Suppl.) 34

Keckes S, Pucar Z, Marazovic L (1967) The influence of the physico-chemical form of 106-Ru on its uptake by mussels from seawater. In: Aberg B, Hangate FP (eds) Proceedings of the International Symposium on Radioecological Concentration Processes, New York: Pergamon Press, pp 993–994

Klumpp DW (1979) Arsenic Accumulation in an Estuarine Food Chain. Ph.D. Thesis, London: University of London, p 269

Klumpp DW, Peterson PJ (1979) Arsenic and other trace elements in the waters and organisms of an estuary in SW England. Envir Poll 19:11–20

Klumpp DW, Peterson PJ (1981) Chemical characteristics of arsenic in a marine food chain. Mar Biol 62:297–305

Knauer GA, Martin JH (1972) Mercury in a marine pelagic food chain. Limnol Oceanog 17:868–876

Kuenzler EJ (1969) Elimination of iodine, cobalt, iron and zinc by marine zooplankton. In: Nelson DT, Evans EC (eds) Symposium on Radioecology. Proceedings of the 2nd National Symposium on Radioecology, Washington, DC: USAEC, pp 462–473

Lunde G (1972) The absorption and metabolism of arsenic in fish. In: Report on Technological Research Concerning the Norwegian Fish Industry, vol 5, pp 1–16

Lunde G (1973) Separation and analysis of organic-bound and inorganic arsenic in marine organisms. J Sci Food Agric 24:1021–1027

Mackey DJ (1983 a) Metal organic complexes in seawater – an investigation of naturally occurring complexes of Cu, Zn, Fe, Mg, Ni, Mn and Cd using high performance liquid chromatography with atomic fluorescence detection. Mar Chem 13:169–180

Mackey DJ (1983 b) The strong complexing capacity of seawater – an investigation of south-eastern Australian coastal waters. Mar Chem 14:73–87

Mantoura RFC (1981) Organo-metallic interactions in natural waters. In: Duursma EK, Dawson R (eds) Marine Organic Chemistry, Amsterdam: Elsevier, pp 179–223

McKnight DM, Morel FMM (1979) Release of weak and strong copper-complexing agents by algae. Limnol Oceanog 24:823–837

McLeese DW, Ray S (1984) Uptake and excretion of cadmium, CdEDTA, and zinc by *Macoma balthica*. Bull Envir Contam Toxicol 32:85–92

Millero FJ (1981) Chemical speciation of ionic components in estuarine systems. In UN River Imputs to Ocean Systems, New York: United Nations, pp 116–131
Morel FMM, Morel-Laurens NML (1983) Trace metals and plankton in the oceans: facts and speculation. In: Wong CS, Boyle E, Bruland KW, Burton JD, Goldberg ED (eds) Trace Metals in Seawater, New York: Plenum Press, pp 841–869
Morris OP, Russel l G (1973) Effect of chelation on toxicity of copper. Mar Poll Bull 4:159–160
Murphy JP, Lean DR, Nalewajko C (1976) Blue-green algae: their excretion of iron selective chelators enables them to dominate other algae. Science 192:900–902
Oshida PS, Mearns AG, Reach DG, Word CS (1976) The effects of hexavalent and trivalent chromium on *Neanthes arenaceodentata*. South California Coast Water Research Project PM-225, p 58, El Segondo, CA
Pan-Hou HSK, Imura N (1981) Biotransformation of mercurials by intestinal microorganisms isolated from yellowfin tuna. Bull Envir Contam Toxicol 26:359–363
Penrose WR (1975) Biosynthesis of organic arsenic compounds in brown trout (*Salmo trutta*). J Fish Res Bd Can 32:2385–2390
Pentreath RJ (1973) The role of food and water in the accumulation of radionuclides by marine teleost and elasmobranch fish. In: Radioactive Contamination of the Marine Environment, Vienna: IAEA, STI/PUB/313, pp 421–436
Pentreath RJ (1976a) The accumulation of mercury from food by the plaice, *Pleuronectes platessa* L. J Exp Mar Biol Ecol 25:51–65
Pentreath RJ (1976b) The accumulation of inorganic mercury from sea water by the plaice, *Pleuronectes platessa* L. J Exp Mar Biol Ecol 25:103–120
Pentreath RJ (1976c) The accumulation of organic mercury from sea water by the plaice, *Pleuronectes platessa* L. J Exp Mar Biol Ecol 25:121–132
Phillips DJH (1980) Toxicity and accumulation of cadmium in marine and estuarine biota. In: Nriagu JO (ed) Cadmium in the Environment, pp 425–569. New York: Wiley-Interscience, pp 425–569
Piro A (1970) Chemical environmental factors in marine radiocontamination. In: Bernhard M (ed) Studies on the Radioactive Contamination of the Sea. Annual Report 1968-69, Rome: CNEN Report No. RT/BIO (70)-11, pp 11–22
Piro A, Bernhard M, Branica M, Verzi M (1973) Incomplete exchange reaction between radioactive ionic zinc and stable natural zinc in seawater. In: Radioactive Contamination of the Marine Environment, Vienna: IAEA STI/PUB/313, pp 29–46
Ragan MA, Ragan CM, Jensen A (1980) Natural chelators in seawater: detoxification of Zn^{2+} by brown algal polyphenols. J Exp Biol Ecol 44:261–267
Rainbow PS, Scott AG, Wiggins EA, Jackson RW (1980) Effect of chelating agents on the accumulation of cadmium by the barnacle *Semibalanus balanoides*, and complexation of soluble Cd, Zn and Cu. Mar Ecol Prog Ser 2:143–152
Ray S, McLeese DW, Pezzack D (1979) Chelation and interelemental effects on the bioaccumulation of heavy metals by marine invertebrates. In: Proceedings of the International Conference on Management Control, Heavy Metals in the Environment, Edinburgh: CEP Consultants Ltd, pp 35–38
Reinke J, Uthe JF, Freeman HC, Johnston JR (1975) The determination of arsenite and arsenate ions in fish and shellfish by selective extraction and polarography. Envir Lett 8:371–380

Roesijadi G (1980) The significance of low molecular weight, metallothionein-like proteins in marine invertebrates. Current status. Mar Envir Res 4:167–179
Rowland IR, Davies MJ, Evans JG (1980) Transformation of mercury compounds by intestinal flora. Arch Envir Health 35:155
Sanders JG (1979a) Effects of arsenic speciation and phosphate concentration on arsenic inhibition of *Skeletonema costatum* (*Bacillariophyceae*). J Phycol 15:424–428
Sanders JG (1979b) The concentration and speciation of arsenic in marine macroalgae. Est Coast Mar Sci 9:95–99
Sanders JG, Vermersch PS (1982) Response of marine phytoplankton to low levels of arsenate. J Plankt Res 4:881–893
Sanders JG, Windom HL (1980) The uptake and reduction of arsenic species by marine algae. Est Coast Mar Sci 10:555–567
Schreiber E (1935) Ueber Kultur und Geschlechtsbestimmung von *Dictyota dichotoma*. Planta 24:266–275
Scudlark JR, Johnson DL (1982) Biological oxidation of arsenite in seawater. Est Coast Shelf Sci 14:693–706
Sharp JH (1977) Excretion of organic matter by marine phytoplankton: Do healthy cells do it? Limnol Oceanog 22:381–399
Shinagawa A, Shiomi K, Yamanaka H, Kikuchi T (1983) Selective determination of inorganic arsenic (III), (V) and organic arsenic in marine organisms. Bull Jpn Soc Fish 49:75–78
Shiomi K, Shinagawa A, Azuma M, Yamanaka H, Kikuchi T (1983) Purification and comparison of water-soluble arsenic compounds in a flatfish *Limanda herzensteini*, sea squirt *Halocynthia rorentzi* and sea cucumber *Stichpus japonicus*. Comp Biochem Physiol 74c:393–396
Simkiss K (1976) Intracellular and extracellular routes in biomineralization. In: Symposia of the Society of Experimental Biologists, ed. J.C. Duncan, vol 30, pp 423–444
Simkiss K (1981) Calcium, pyrophosphate and cellular pollution. Trends Biochem Sci 3:111–113
Small LF, Keckes S, Fowler SW (1974) Excretion of different forms of zinc by the prawn *Palaemon serratus* (Pennant). Limnol Oceanog 19:789–793
Smith JD, Butler ECV (1979) Speciation of dissolved iodine in estuarine waters. Nature 277:468–469
Steeman-Nielsen E, Wium-Anderson S (1970) Copper ions as poison in the sea and fresh-water. Mar Biol 6:93–97
Stephenson RR, Taylor D (1975) The influence of EDTA on the mortality and burrowing activity of the clam (*Venerupis decussata*) exposed to sub lethal concentrations of copper. Bull Envir Contam Toxicol 14:305–308
Stone H, Overnell J (1985) Non-metallothionein cadmium binding proteins. Minireview. Comp Biochem Physiol 80C:9–14
Sueur S, Van den Berg CMG, Riley JP (1982) Measurement of the metal complexing ability of exudes of marine macroalgae. Limnol Oceanog 27:536–543
Sugawara K, Terada K (1967) Inf Bull Planktol Jap Commemoration No. of Dr. Y. Matsue, pp 213–218 (cited by Smith and Butler, 1979)
Sunda WG, Engel DW, Thuotte RM (1978) Effect of chemical speciation on toxicity of cadmium to grass shrimp, Palaemonetes pugio: Importance of free cadmium iron. Envir Sci Technol 12:409–413

Sunda WG, Gillespie PA (1979) The response of a marine bacterium to cupric ion and its use to estimate cupric ion activity in seawater. J Mar Res 37:761–777
Sunda W, Guillard RRL (1976) The relationship between cupric ion activity and the toxicity of copper to phytoplankton. J Mar Res 34:511–529
Suzuki Y, Sugimura Y (1985) Total and organic mercury in sea water in the western north Pacific. In International Chemistry Congress of Pacific Basin Societies, Abstract Honolulu, Dec. 16–21, 1984
Swallow KC, Westall JC, McKnight DM, Morel NM, Morel FMM (1978) Potentiometric determination of copper complexation by phytoplankton exudates. Limnol Oceanog 23:538–542
Trick CG, Andersen RJ, Price NM, Gillam A, Harrison PJ (1983) Examination of hydroxamate-siderophore production by neritic eukaryotic marine phytoplankton. Mar Biol 75:9–17
Truesdale VW (1978) Iodine in inshore and off-shore marine waters. Mar Chem 6:1–13
Tsunogai S, Sase T (1969) Formation of iodine in the ocean. Deep-Sea Res 16:489–496
Tyler A (1953) Prolongation of life-span of sea urchin spermatozoa, and improvement of the fertilization reaction, by treatment of spermatozoa and eggs with metal-chelating agents (amino acids, Versene, DEDTC, oxine, cupron). Biol Bull 104:104–114
Van den Berg CMG (1984) Organic and inorganic speciation of copper in the Irish Sea. Mar Chem 14:201–212
Wallace GT (1982) The association of copper, mercury and lead with surface-active organic matter in coastal seawater. Mar Chem 11:379–394
Wells ML, Zorkin NG, Lewis AG (1983) The role of colloid chemistry in providing a source of iron to phytoplankton. J Mar Res 41:731–746
Windom HL, Kendall DR (1979) Accumulation and biotransformation of mercury in coastal and marine biota. In: Nriagu JO (ed) The Biogeochemistry of Mercury in the Environment, Amsterdam: Elsevier/North-Holland, pp 303–323
Wood JM, Wang H-K (1983) Microbial resistance to heavy metals. Envir Sci Technol 17:582A–590A
Yamamoto J, Kaneda Y, Hikasa Y (1983) Picogram determination of methylmercury in seawater by gold amalgamation and atomic absorption spectrophotometry. Intl J Envir Anal Chem 16:1–16
Zamuda CD, Sunda WG (1982) Bioavailability of dissolved copper to the American oyster *Crassostrea virginica*. I. Importance of chemical speciation. Mar Biol 66:77–82

Standing, left to right:
Greg Olson, Simon Silver, Bruce Fowler, Ira Krull, David Russell

Seated (center), left to right:
Ulrich Förstner, Wim Salomons, Fred Brinckman, John Wood

Seated (front), left to right:
Peter Craig, Ulrich Herms, Don Mackay

The Importance of Chemical "Speciation" in Environmental Processes,
eds. M. Bernhard, F. E. Brinckman, and P. J. Sadler, pp. 425–441.
Dahlem Konferenzen 1986. Springer-Verlag Berlin, Heidelberg
© Dr. S. Bernhard, Dahlem Konferenzen

Chemical Species in Systems under Stress

Group Report

J. M. Wood, Rapporteur
F. E. Brinckman
A. M. Chakrabarty
P. J. Craig
U. Förstner
B. A. Fowler
U. Herms
I. S. Krull
D. Mackay
G. J. Olson
D. H. Russell
W. Salomons
S. Silver

Introduction

It is understood that any chemical species in the biosphere may move from its source through the physical environment and into biota where it may cause either a beneficial or a toxic effect at the cellular level. The pathways taken by each element involve many chemical species, and the distribution of these species is determined by thermodynamic and kinetic considerations. Those species which are stable thermodynamically can often be isolated and quantitated. Ideally the precise chemical structures of each of these stable compounds, or complexes, should be analyzed in that matrix where they exist naturally. Unfortunately, in the stressed system entropy increases and there is a concomitant increase in instability in both the physical and biological context. Therefore, stress is defined as a perturbation of natural abiological or biological systems (ecosystems). In the abiological system it involves a change in physical or chemical composition (cause), and in the biological system it results in a reduction of the diversity of organisms (effect), with a possible compromise of cell/organism viability or health.

In that which follows, we have examined the effect of stress on the biosphere within the constraints of our ability to analyze it. We have considered both physical, chemical, and biological processes, and where possible developed some guidelines for future endeavors.

Determination of Species and Monitoring Indicators in the Stressed System

At the outset it should be recognized that there is no central reference for methods for determining the structures of inorganic or organometallic species, as has been developed by organic chemists for the characterization of organic compounds. Also, there have been no interlaboratory comparative studies and there are no available textbooks which provide a state-of-the-art approach for the precise determination of inorganic chemical structures.

Consideration must be given to thermodynamic and kinetic factors before analyzing chemical species and before assuming a stable equilibrium. The greater the stress in the environment, the less the stability and the greater the difficulty in sample handling and storage prior to analysis. In fact, many of the analytical techniques are handicapped by disruptive preparation techniques which may alter the species distribution of inorganic components or lead to loss of analyte before analysis (e.g., freezing, lyophilization, evaporation, oxidation, changes in pH, light-catalyzed reactions, reactions with the sample container, time delays before analysis with biologically active samples, sample contamination, etc.). In order to minimize these adverse effects, attention must be paid at least to statistically valid sampling, extraction in close to 100% yield, validation of analytical methodology with authentic samples in the same matrix, etc.

However, despite all of these constraints in sample acquisition and handling, the methodology for determining chemical species is very promising in a number of areas. Nevertheless, the task of the analytical chemist would be made much easier if concentrations of chemical species could be increased in the natural environment prior to analysis. In this context there is a need for the judicious selection of indicator organisms which specifically bioconcentrate chemical species prior to analysis and which can be used in a dynamic approach for studying the changing (stressed) situation. This could facilitate in understanding the flux of individual chemical species, and help to identify at least the bioavailable fraction in a given system.

The first phase of any analytical procedure is the development of appropriate separation techniques. Such techniques have been developed using gas chromatography, liquid chromatography, or ion chromatography which are specifically designed for inorganic analysis. For the structural determination of inorganic species the connection of these techniques with mass spectrometry has been slow in developing even though such techniques are well advanced in the determination of organic structures (see Wolf, Chau, Krull, Vestal, Russell, and Irgolic and Brinckman, all this volume).

Similarly, there is a need to develop sensitive nuclear magnetic resonance (NMR) methods for *in vivo* and *in situ* monitoring of inorganic complexes, which would be noninvasive, nondestructive, sensitive, and selective for particular species. At the same time there are some very practical problems inherent in any type of NMR approach, especially with regard to sensitivity, isotope abundances in natural samples, and nonapplicability to most elements including metals (see Sadler, this volume). NMR can be used *in vivo* to determine intracellular pH and to study C-metabolism by using ^{13}C-enriched substrates.

In addition to studying carbon metabolism, other elements offer promise in the application of NMR techniques. For example, Si^{39} and Al^{27} give good NMR signals and these elements have not yet been used for the determination of chemical species *in vivo*. The major problem with NMR resides in its lack of sensitivity which demands the examination of high concentrations of individual species in dense cell suspensions or tissues.

Each of the methods described above represent separation techniques and analytical techniques primarily for the homogeneous matrix, although NMR has been used with solid materials. The determination of surface structures on particulates requires a different approach and is extremely difficult due to heterogeneity (see Kane, this volume). However, techniques such as ESCA, Mössbauer Spectroscopy, NMR, and EXAFS have been used to obtain information on metals and ligands in their insoluble component.

Particulate-Surface Interactions and Mobilization of Chemical Species

Several thousand years ago the beginning of agricultural practices at the surface of the Earth imposed stress on the biosphere by increasing particulate concentrations in the atmosphere and the aqueous environment. Since the industrial revolution the distribution of particulates has increased and is very significant since particulates are the major vectors of pollutants.

Particles play a significant role in strongly stressed atmospheric and aquatic systems not only as scavengers and carriers but also as potential sources of toxic chemicals. Particularly high concentrations of pollutants were observed on combustion products (see Förstner, this volume) and wastewater particles. Accessibility to the environment (via washout, rainout, groundwater, leaching, lung fluids, etc.) is governed by both pollutant surface accessibility (extent of surface enrichment) and pollutant surface solubility (surface speciation). Methods for determining the reactivity of particle-associated pollutants are therefore mainly related to surface species. In this respect there are still problems with direct determination of trace-metal species. This is mainly valid for mixed atmospheric and aquatic particulates, whereas recent studies on coal fly-ash particles suggest that the oxidation state and chemical composition can be inferred from the chemical behavior. Sequential leaching techniques, while being problematic for the assessment of bioavailability, can be useful for the estimation of potential remobilization effects (e.g., by changes of pH, redox conditions, salinity, and increasing concentrations of organic chelators) on toxic metals from polluted solid material and were mainly used to classify dredged substances (e.g., for ocean dumping, upland disposal, or storage in confined water bodies).

Similar problems exist in understanding distribution of species in soil, especially since only a small fraction of pollutants are present in the hydrous or gaseous phases of soil with the majority being bound at the particle surface. The bioavailability of bound materials in these soil/sediment systems is known to be determined qualitatively by the biological cycles of the major elements (e.g., C, H, N, O, P, S). However, there is little quantitative information on the dynamic aspects of reactions which are at the interface between these biotic and abiotic compartments. This is probably due to our inability to connect the trace-element cycles with the major-element cycles. Some biological perspectives on this will be presented later.

Thermodynamic and Kinetic Considerations in the Biological Transport and Magnification of Chemical Species

Microorganisms have a profound influence on the concentrations of elements in the atmosphere, in the oceans, and in the terrestrial environment. Both organic and inorganic substrates are mobilized by microbial activities in soils, sediments, and water. The uptake of nutrients (and their utilization by microorganisms) depends not only on their physical and chemical properties but also on the precise chemical species of each element. Of the elements in the Periodic Table, thirty have been shown to be essential for life, although not all of these thirty elements are necessary for the growth and cell division of every biologic species. There is a correlation between the abundance of elements in the Earth's crust and the nutritional requirements for living cells. Furthermore, there is a correlation between the solubilities of these essential elements (in the absence of oxygen) and life. The rates at which elements are cycled in the aqueous environment depend on the availability of the bulk nutritional elements C, H, N, O, P, and S. The cycling of trace elements follows the cycling of the bulk elements, and the uptake of trace elements requires the input of considerable energy by living cells and, therefore, the biochemical processes for transport are carefully regulated at the gene level. Consequently, living cells are not at thermodynamic equilibrium with their external environment. For those elements which are nonessential for microbial life a variety of resistance mechanisms has evolved to prevent toxicity. These resistance mechanisms require energy also, and they represent nonequilibrium processes. Resistance mechanisms have evolved for the removal of toxic elements by the processes of precipitation, complexation, volatilization, alkylation, hydrolysis, oxidation, reduction, and by specific efflux mechanisms. These strategies for microbial resistance to toxic elements differ greatly between microbial cells (prokaryotes) and the more advanced forms of life.

The more complex eukaryotes have many membranous intracellular organelles which can be used to partition unwanted substances not only in the domain of time but also in the domain of space. In most cases, but not all, microorganisms have few intracellular membranes to act as barriers to toxic elements. The availability of elements for transport into unicellular organisms is first restricted by abundance and by the solubility of each chemical species in water. Solubility in water is profoundly influenced by pH, temperature, and E^0, as well as by the presence of competing anions, cations, and surface-active polyelectrolytes such as humic acids, fulvic acids, clays, etc.

An experiment on the exchange of twelve different ions including H^+ with humic acid has been used to model surface-to-inorganic ion complexation. This model was shown to be predictable mathematically and also chemically, since metal ion-substitution reactions followed the well-established Irving-Williams series.

In the microenvironment the pH and E^0 can vary widely from outside the cell to inside the cell. For example, many essential transition-metal ions such as iron, manganese, copper, cobalt, etc., occur at higher oxidation states outside the cell but at lower oxidation states inside the cell.

Clearly such changes in pH and E^0 affect the nature, solubility, and reactivities of chemical species as well as the net charge on many substances. These parameters determine the rates of uptake of both essential and nonessential elements. The failure of cells to transport and utilize required elements could arise as follows:

1) Low availability;
2) Excessive competition from other elements, both essential and nonessential, with similar chemical properties, e.g., Co(II) and Ni(II), Ca(II) and Cd(II), PO_4^{3-} and AsO_4^{3-}, Na^+ and Li^+, etc.;
3) Inadequate synthesis of carrier molecules by the cell;
4) Excessive excretions of elements by the cell;
5) Failure of the energy-driven uptake systems which pump essential elements up a concentration gradient.

Excessive element uptake can occur through the reversal of the above five factors. Factors 3 through 5 are dependent on cellular energy, and as such are subject to regulation by the nutritional status of the cell which in turn is responsible for a nonequilibrium with the external environment. Therefore we can only understand the impact of living organisms in a specific environmental situation by studying the kinetics for the transport, binding, metabolism, or resistance to toxicity of each individual chemical species. However, there are some exceptions to this situation. A thermodynamic approach to bioaccumulation is satisfactory for those organic compounds which are not metabolized rapidly and which partition into membranes (i.e., provide accurate partition coefficients) as well as for those components which pass through membranes by passive diffusion (e.g., the diffusion of CH_3HgCl through membranes).

For the organometallic compounds it should be recognized that most of these intermediates in metabolic pathways are thermodynamically unstable, but they do have sufficient kinetic stability and lack of polarity to diffuse rapidly through membranes and interact inside cells.

In living cells the important transition-metal ions bind best to sulfur-rich bases, followed by nitrogen/oxygen bases, followed by oxygen bases alone, and finally by coordination to water molecules alone. Such bases are readily available to metal ions in living cells due to the fast-exchange kinetics of protons in the pH range 7.0 to 7.2.

Therefore, on the basis of both thermodynamic stability and kinetic-exchange rates, metal ions in biology can be divided into three classes:

1) Ions in fast exchange,
2) Ions in intermediary exchange, and
3) Ions in slow exchange with biological ligands.

Examples of those ions in fast exchange include the alkali metals Na^+ and K^+, the alkaline-earth metals Ca^{2+} and Mg^{2+} and, of course, H^+. Those which can sometimes be in intermediary exchange are Fe^{2+} and Mn^{2+}, while examples of those in slow exchange are generally in the active sites of metalloenzymes or storage proteins (e.g., Fe^{3+}, Zn^{2+}, Cu^{2+}, Ni^{2+}, etc.).

These generalizations are enhanced by a more detailed examination of metal transport in microbial cells. For prokaryotic cells the Periodic Table of elements breaks into three classes: a) there are elements such as K, Mg, Fe, P, and S which living organisms need in all cells for growth and metabolism; b) there are nonessential elements to some microorganisms such as Na, Ca, and Cl, which are commonly abundant. Although many bacteria can grow happily in the absence of detectable levels of these materials, some bacteria require them (and use them for specialized functions). All bacteria have developed means of coping with these elements; c) the third class of elements include, for example, Hg, Cd, Pb, and As, for which there is no conceivable useful function in microorganisms.

For each of these three classes of elements, highly specific membrane-transport systems have evolved. For each required element one or more inwardly oriented membrane-transport system exists, consisting of one or several membrane proteins whose functions are to a) discriminate carefully for each specific substrate and against chemically related ions. (For example, the Mn^{2+} transport system is unaffected by a 10^6-fold molar excess of Mg^{2+}). b) The second function of these specific transport systems is to build and maintain a high intracellular concentration of a given elemental species even if the external concentration approaches vanishingly low levels. (For example, the K^+ transport system can maintain [0.3 M] internal potassium while the external [K^+] is as low as 10^{-7} M.) For each needed cation and anion there appears to be one or more transport system. Frequently there are two: a) a constitutively synthesized system of relatively low affinity (i.e., high K_m) and one of relatively low specificity when discriminating against analogues, and b) an inducibly synthesized system of high affinity (low K_m) and high specificity, that is produced during starvation for the needed ion or when toxic amounts of an analogue are present. An example is an inducibly synthesized system for phosphate transport, which is developed under conditions of phosphate starvation and also during arsenate toxicity. Fe^{3+} is a special case for membrane-transport systems, first, because Fe^{3+} is extremely insoluble in aerobic environments, and secondly, because much cellular metabolism is dependent upon iron-containing proteins. There are currently known to be five separate Fe^{3+}

transport systems in *E. coli,* all but one of which are made under Fe-stress and some of which can only be detected when the others are removed by physiological or genetic manipulations.

For the nonessential elements there are also transport proteins in the cell membrane, but these are oriented outwardly and assure that the intracellular concentration is lower than the extracellular concentration. For example, there are Na^+/H^+ exchange systems that establish an Na^+ gradient across cell membranes, and this gradient can be utilized to drive amino acids and sugar transport "uphill" along a concentration gradient. Ca^{2+} is excreted through either a $Ca^{2+}/2H^+$ exchange system or by a Ca^{2+}-ATPase. Ca^{2+} is utilized extracellularly to activate enzymes and is brought intracellularly in specialized situations such as bacterial sporulation.

For toxic elements bacteria have evolved specific resistance mechanisms. For example, the mechanisms of resistance to As(III) and As(V) are separate. There are a few exceptions and it seems that Cd^{2+}-resistance mechanisms can sometimes but not always bring about resistance to high Zn^{2+} concentrations as well, and of course Zn is a required element at low levels but a toxic element when too much is present. As(III)-resistance mechanisms seem concomitantly to confer gratuitous resistance to Sb(III). Not all of the prokaryotic resistance mechanisms involve transport systems, but some do. Chromosomally determined resistance to arsenate results from a switchover to a phosphate-transport system that has a 100-fold more favorable discrimination against arsenate than does the low-specificity phosphate-transport system. A *Bacillus* mutant has resistance to Cd^{2+} due to changes in the affinity constants of the membrane Mn^{2+}-transport system so that it will no longer accept Cd^{2+}. In both of these cases the resistance mechanism is to block cellular uptake before it occurs. For plasmid-mediated resistance to arsenate or cadmium, however, the chemical species is allowed in through the appropriate "good" element system and then is excluded by a highly specific efflux pump. Of the two best-studied examples, the Cd^{2+} system functions as a $Cd^{2+}/2H^+$ electroneutral exchange system, driven by the membrane-pH gradient, while the arsenate-efflux pump appears to be a direct ATPase system which specifically removes arsenate, leaving phosphate intracellularly.

For those toxic elements whose mechanisms of resistance are *not* membrane-transport systems, two alternative mechanisms exist. Enzymatic conversion of the chemical species metabolizes highly toxic methylmercury and phenylmercury into less toxic inorganic Hg^{2+}. (However, the enzyme responsible is a poorly understood organomercurial lyase; there seem to be two kinds, one for alkylmercurials and the other for aromatic mercurials.)

The mercuric reductase is a 563 amino-acid complex, FAD-containing, NADPH-dependent oxido-reductase whose complete sequence is known from DNA sequencing and whose reaction mechanism is quite similar to that of glutathione reductase and lipoamide dehydrogenase – in fact, the three enzymes share 26–28% of identical amino acids in the sequence. Arsenite is oxidized to arsenate (less toxic) by a soluble enzyme, which is only beginning to be studied. The last resistance mechanisms known are those of intracellular or extracellular sequestration (see Olson, this volume). Both mechanisms occur, but in no case has one been studied in biochemical and genetic detail; therefore, the overall pictures of the sequestration mechanisms are incomplete.

At this point it is important to provide some historical perspective to the problem of microbial metabolism and resistance. The metabolic pathways for the metabolism of essential elements and the evolution of resistance mechanisms depend on the ancestral history of microbial evolution. Microorganisms can only use that which is in the historical records. For example, some microorganisms have inherited the ability to tolerate high concentrations of heavy metals through their evolution under extreme environmental conditions. It is not unusual to find bacteria and algae which tolerate high concentrations of metal ions at low pH and at high temperatures. Hot springs, volcanic lakes, and deep-sea vents provide lithotrophic bacteria with built-in resistance mechanisms. Also, microorganisms can adapt to stress from man-made organic compounds by utilizing metabolic segments which evolved for the degradation of natural products with structural similarity. Quite often such organisms have only the ability partially to metabolize man-made substances such as halogenated aromatic compounds (e.g., PCBs, 2-4-5T, pentachlorophenol, etc.). However, advantage can be taken of the total microbial gene pool to engineer bacteria in the laboratory by using plasmids and transposons to produce strains capable of the complete degradation of molecules which were found to be only slowly degraded by mixed populations of bacteria in the natural environment. Recently this was accomplished for the herbicide 2-4-5T (agent orange) and for the wood preservative pentachlorophenol (PCP). These laboratory-engineered strains have been found quite effective at reducing concentrations of these pollutants which have been shown to persist in the natural microbial communities found in soils and sediments.

Nonpolar organic compounds such as PCBs can undergo cycling between air and water and between air and soil. Their binding to small particles in the air and water phases is crucial in determining the fluxes of these compounds. It is essential to appreciate the relative concentrations and residence times in each phase. Usually the soil and sediment compartment contains the bulk of such chemicals. Binding to either elemental carbon or organic carbon of natural origin predominates for such compounds.

Therefore, since these nonpolar organic substances are complexed with particulates as well as being water-insoluble in their own right, it is of great interest to determine how such substances penetrate microbial membranes so that biodegradation can occur. The enzymes which degrade insoluble compounds such as 2-4-5T are intracellular, and therefore the rate-limiting step for biodegradation resides in the solubilization. Microorganisms overcome this difficulty by producing specific emulsifiers which facilitate uptake into cells. Such emulsifiers are only produced by bacteria which are exposed to insoluble substrates which they can use as an energy source.

This discussion of individual processes and individual species reactivity needs to be expanded to a view of the interactions between element cycles. Interfaces between the element cycles occur where oxidized or reduced species of major elements (often biologically catalyzed) chemically react with metals to cause precipitation (e.g., metal sulfides), solubilization, redistribution (i.e., alkyl metals), alkylation, complexation, oxidation/reduction reactions, etc. Other interfaces are the situations where minerals representing components of different element cycles are metabolized by organisms (e.g., *Thiobacilli* and *Sulfolobus*).

The synthesis of H_2S in sediment provides a kinetically complex situation where S^{2-} competes in oxidation reactions with O_2, metals, humic acids, etc. The oxidation of S^{2-} to SO_4^{2-} is also an important process for both the direct and indirect mobilization of metals from sediment, soils, and even from ores. Microorganisms derive energy from the preceding processes, though the exact solubilization mechanisms remain to be determined. The pH can vary widely outside the cell because the external hydrogen-ion concentration can be changed rapidly through microbial metabolism of the bulk elements. For example, the fixation of atmospheric nitrogen requires the concomitant reduction of two protons to hydrogen gas, and therefore the nitrogen-fixing blooms of blue-green algae may contribute to a rapid increase of pH in lakes. This increase in pH certainly transforms species of metals in the water column, causing hydrolysis and precipitation of metal hydroxides. Another component of the nitrogen cycle, the oxidation of ammonia to nitrate, tends to reverse this process by producing soluble metal nitrates. For the carbon cycle the fixation of carbon dioxide by algal blooms can increase the pH and cause metal ions to precipitate, but this is counterbalanced by the release of organic compounds by the algae (secondary metabolites), including sugar acids and polysaccharides, which tend to keep complexed metals in solution.

While these direct and indirect connections between the cycles of the major elements and the trace elements appear to be obvious, very little has been done to integrate them. This is an important need if we are to derive holistic models of ecosystems.

Whereas most of the research effort is rightly devoted to detailed studies of the behavior of elemental species in defined biological systems (such as uptake by cells or organisms from water) and in physical systems (such as migration from water to solid surfaces), there is a complementary need to examine, understand, and quantify the entire system of interacting processes. This may involve treatment of processes within and between compartments of air, water, soils, sediments, and biota of various types.

Effective environmental management must be based on understanding these pathways and how they are facilitated, blocked, or altered by species transformations. Furthermore, a qualitative understanding of these phenomena, while an essential starting point, is inadequate because there is a need for quantitative description. This task is essentially that of mathematical modelling, a task of synthesis of disparate information into an overall description of the system. When this can be accomplished the necessity for, and efficacy of, various remedial strategies for the stressed environment can be assessed.

Once the impact of toxic elements on the environment has been measured, estimated, or predicted by direct or indirect methods (e.g., bioassays), a management plan can be formulated which usually includes engineering activities. Examples have been given for the treatment of solid-waste materials, where valuable elements are sometimes enriched to the extent of economically feasible recycling (e.g., by bacterial leaching and/or biological concentration), for acidic mine effluents (thermodynamic and kinetic methods; see Olson, this volume) and for mercury-polluted sediments (i.e., the use of physical and chemical barriers between the sediment surface and water). A common aspect of most remedial measures in such highly polluted environments is either to extract toxic species from distinct compartments or to reduce their mobilities and transfer rates into biological systems. Even stronger emphasis, however, should be given to technical measures at the source of potentially toxic chemical species.

The problem facing engineers and modellers is not lack of mathematical ability or of computing capacity, rather it is lack of understanding of the basic physical, chemical, and biological process rates as well as equilibria and, correspondingly, an inability to write down suitable descriptive equations. Even when suitable equations can be written there is often a lack of parameter data such as rate or stability constants. Ultimately the validation of the model requires detailed analytical data. Often such data are obtained on the basis of what is easiest or cheapest to obtain and not what is most useful to the modeller. It must be recognized, however, that it is the duty of the modeller to define which data are needed and with what precision.

Transport in Eukaryotic Organisms: Regulatory Mechanisms

In contrast to prokaryotes, eukaryotic organisms possess external membranes which have few agent-specific transport systems that regulate intracellular concentrations of a given toxicant or its metabolites. Membrane transport of these agents seems to occur primarily either via passive diffusion of lipophilic compounds or secondary to endocytosis of the membrane following binding of cationic species to anionic surface groups (Fig. 1). Some anionic, toxic chemical species have been shown to be transported via membrane-transport systems, mistaken for required anions due to structural similarities.

Homeostatic control over the intracellular bioavailability of transported toxic chemical species seems to occur primarily either through se-

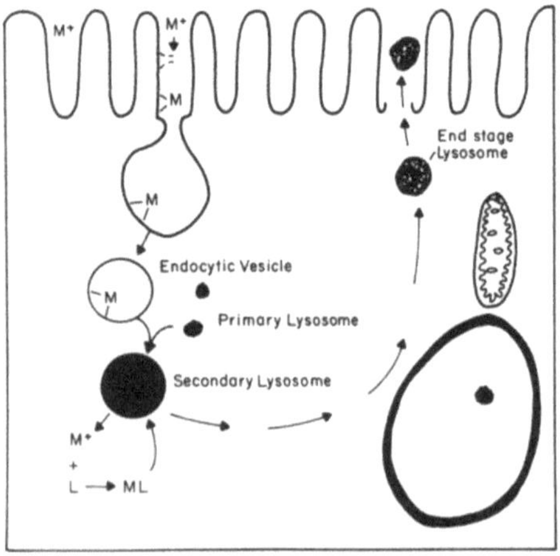

Fig. 1. Example of membrane transport in eukaryotic organisms

questration on a bulk scale within structures such as lysosomes, concretions, inclusion bodies, or via binding with high-affinity proteins (metallothioneins, lead-binding proteins, etc.). For many organic compounds or their derivatives, conjugation with low-molecular-weight compounds such as glucuronides, sulfate, or glutathione is a common mechanism for controlling reactivity. The nature and extent of control exerted by a given complexation mechanism is dependent upon a host of factors such as the given chemical species, organism, organ or cell type, and conditions of toxicant exposure. It should also be noted that none of the above control mechanisms are necessarily exclusive of one another and hence binding of a metal such as Cd^{2+} to a metallothionein, or to metallothionein-like proteins, may ultimately result in cell injury due to turnover of this protein and recomplexation of the metal to other ligands. While the above mechanisms are also highly efficient, they are not perfect and in general it is the non-complexed fraction of a given chemical species which is of greatest concern from the view of biological activity.

Intracellular Species and Biological Effects

As noted above, it is generally the uncomplexed fraction of a given chemical species which is of greatest concern with respect to biological activity (toxic potential) since this component is not under regulatory control and is presumably available to react with sensitive organelle systems or metabolic processes. Among the most sensitive cellular organelle systems or processes which have been identified as playing a primary role in chemical-induced injury are cell/organelle membranes, mitochondrial energy production, DNA/RNA, protein synthetic capacity, heme biosynthesis, and lysosome biogenesis. Available data indicate that unregulated chemical species which perturb any one or more of these systems may seriously compromise cell function and viability.

Chromate (Cr^{+6}) or mercuric ion (Hg^{2+}) damage to brush-border membranes of kidney proximal tubule cells or carbon tetrachloride and polychlorinated biphenyl-induced alteration of the endoplasmic reticulum membranes are examples of cellular injury primarily mediated via perturbation of membrane structure and function. Membrane injury from these agents occupy at least three mechanisms: 1) peroxidation of membrane lipids, 2) reaction of chemical species with membrane proteins, and 3) physical disruption of membrane structure by partitioning of lipophilic chemical species into the lipid-membrane matrix. Examples of toxicants which act primarily on cellular energy production are trivalent or pentavalent arsenicals which disrupt mitochondrial production of ATP via at least three possible mechanisms (Squibb and Fowler, 1983). The exact relationship between the demonstrated effect of these arsenicals on cellular energy production and *in vivo* oxidation/reduction or methylation is not known. Lead and methylmercury are known to produce increases and decreases in cellular DNA/RNA and protein synthesis *in vivo,* respectively. The exact chemical species of these agents which are responsible for the observed effects are presently unknown. In the case of methylmercury, the situation is further complicated by the *in vivo* demethylation of methylmercury by some organisms to yield inorganic mercury (Hg^{2+}) which is itself a highly toxic chemical species (although much less toxic than methylmercury).

The heme biosynthetic pathway is highly susceptible to perturbation by both organic (hexachlorobenzene) and inorganic (lead) agents. Again, the exact chemical species of these agents which are responsible for the observed inhibitory effects on the system are not well defined. Finally, as noted in the eukaryote transport section above, lysosomes are known to play a major role in the intracellular uptake, storage, and excretion of many essential and several major groups of toxicants. In the course of this activity, the lysosome system may sustain a compromise in functionality as reported for Cd^{2+}.

In summary, there is a clear need from the perspective of a chemical-induced cell injury for the intracellular identification of nonsequestered chemical species which play a central role in a variety of toxic processes. At present, we have little information on the exact chemical nature of these species or how they react at a molecular level to produce the observed intracellular effects.

Recommendations for Future Research

Section on Determination of Speciation and Monitoring Indicators in the Stressed System

1) There is a need for the development of low-cost, reliable, sensitive, specific, accurate, and precise element-selective detectors for high-performance liquid chromatography (HPLC) and gas chromatography. At the present time there are no commercially available, element-selective, HPLC-detector, complete systems doing element-specific or species analyses. Such detection systems would greatly aid future chemical-speciation studies, especially for inorganic and organometallic complexes.

2) Sample-handling and preparation procedures prior to analysis are recognized as being a major problem with respect to accurate determination of species in any matrix. Future studies involving chemical species analysis should pay particular attention to validating sample-handling or preparation procedures and standard reference materials for validating the effects could be developed by agencies such as the National Bureau of Standards.

Recommendations for Future Research

Section on Particulate-Surface Interactions and Mobilization of Chemical Species

1) Further research is needed for the characterization of the organic surface layers of particles with respect to the binding of organic and organometallic compounds.
2) Evaluation of particle concentration and particle size is needed to understand their interactions with pollutants.
3) Further research is needed to assess the role of the particulate as mediator of microbial-pollutant interaction.
4) The time-dependent accessibility of pollutants onto or into particles needs to be evaluated.
5) The development of improved analysis of chemical species of particulates is recommended.

Recommendations for Future Research, Microbial

Section on Thermodynamic and Kinetic Considerations in the Biological Transport and Magnification of Chemical Species

1) There is a need for more basic research on the processes and mechanisms for the transport, metabolism, and resistance pathways in the uptake of both essential and nonessential elements by microbial populations.
2) The flow of trace elements in cycles follows the flow of the bulk elements (C, H, N, O, P, S). Therefore it is important to understand the connections between these cycles in order to predict the flux of important chemical intermediates in the individual pathways of each element in the biosphere. A basic understanding of these pathways is important so that they can be connected to provide a total picture.

3) There is a need for more sensitive, noninvasive methods for measuring intermediates in these cycles (e.g., the formation and degradation of organometallic species).
4) There is a need to evaluate cycles for other trace elements (e.g., B, V, Mo, Ge, Sb, etc.) other than those which are presently receiving attention.
5) There is a need to encourage more modelling with respect to the movement of chemical species using different approaches in order that the most suitable techniques for model development will emerge.
6) There is a need to encourage post-audit of modelling efforts when a model has been used in a predictive mode and so that it is later possible to validate (or otherwise) predictions derived from the model.
7) There is a need to encourage publication of case studies in which models (even very simple models) have been successfully used to suggest environmental management strategies.

Recommendations for Future Research

Section on Transport in Eukaryotic Organisms: Regulatory Mechanisms

There are relatively few data concerning the exact nature of nonsequestered toxic chemical species *in vivo* during initiation of cellular injury. There is hence a clear need for the intracellular identification of such chemical species, preferably under noninvasive and real-time conditions.

There is a need for increasing the sensitivity and specificity of analytical techniques (e.g., NMR) capable of measuring molecular interactions between exogenous chemical species and biological molecules.

Final Recommendations

The very nature of many of the current problems related to chemical species demands the use of highly sophisticated and expensive analytical methods/instrumentation which are currently in development or testing stages. It is inevitable that strong collaborative relations between the applications chemist and the instrument specialist will facilitate progress in both areas. However, these collaborative efforts (projects) are dependent upon effective interdisciplinary communication. There is a need for future interdisciplinary workshops to increase communication between those concerned with instrumentation/analytical technique development and those involved in application of these tools to specific problems of all aspects of chemical species analysis.

References

Friello DA, Chakrabarty AM (1985) In: Suttard C, Rozel KR (eds) Transposable Mercury Resistance in *Pseudomonas putida* Plasmids and Transposons

Silver S (1984) Bacterial transformations of and resistance to heavy metals. In: Nriagu JO (ed) Changing Metal Cycles and Human Health. Dahlem Konferenzen. Berlin, Heidelberg, New York, Tokyo: Springer-Verlag pp 199–223

Squib SK, Fowler BA (1983) Biochemical mechanisms of arsenical toxicity. In: Fowler BA (ed) Environmental Effects of Arsenic. Amsterdam: Elsevier/North Holland pp 233–270

Williams RJP (1981) Physicochemical aspects of inorganic element transfer through membranes. Phil Trans Roy Soc Lond B 294:57

Wood JM (1984) Circulation of metals in the environment. In: Sigel H (ed) Metal Ions in Biology Series. New York, Basel: Marcel Dekker, Inc pp 223–237, 333–351

The Importance of Chemical "Speciation" in Environmental Processes, eds. M. Bernhard, F. E. Brinckman, and P. J. Sadler, pp. 443–464. Dahlem Konferenzen 1986. Springer-Verlag Berlin, Heidelberg
© Dr. S. Bernhard, Dahlem Konferenzen

Chemical Species in Industrial Discharges and Effluents

P. J. Craig

School of Chemistry
Leicester Polytechnic
Leicester LE1 9BH, England

Abstract. The main chemical species present in industrial effluents and discharges are noted, and their pathways and fate in the environment are discussed. Identification of actual chemical forms is made wherever possible. In many cases for metals the ultimate environmental fate is unknown, e.g., the species distribution of metals discharged (inorganic, cation, or anion forms) may be known, and water and sediment concentrations after discharge may also be measured, but in most cases species distribution and species transformation processes are not known. For only a few examples have detailed species distribution studies been carried out, and these will be reviewed as examples of possible chemistries that may occur for other metallic elements discharged into the aqueous environment.

Introduction

It is now well understood that the specific chemical identity of a polluting material is of critical importance in determining its environmental impact. Identification under a generic elemental term is insufficient, e.g., the term arsenic may cover the toxic inorganic forms and the relatively nontoxic organometallic forms; the term mercury may include the inorganic species mercuric(II) chloride with a four-day biological half time (Tb) in man and it may also include organometallic monomethylmercury(II) where the Tb is over seventy days. For such reasons current research is much involved with the determination of the chemical identity of materials released to or found in the environment and the species transformations involved during the chemical processes that occur in the environment.

This paper will consider the wide area of the identity, behavior, chemistry, and final fate of the more common industrial discharges. In this context the main area of consideration is the aqueous environment. This paper will be concerned primarily with the chemical identification of important liquid effluents. Microbiological intervention will be mentioned as necessary but detailed coverage is given elsewhere (Olson, this volume). The problems of solid waste are also discussed in another paper (Förstner, this volume).

Petroleum, surfactants, synthetic polymers, and sewage processes are also relevant to the environment, as a consequence of use or disposal of these materials, not of their production. These will be covered elsewhere (Förstner, Olson, and Chakrabarty, all this volume).

Elements whose species distribution and species transformation (and their pathways) in the environment have been most closely studied are those which are known to enter the environment in their organometallic form, or which form organometallics in the natural aqueous environment. There are a number of important organometallic products (e.g., those of tin and mercury), and inorganic forms of these and other elements are also known to form methyl-metal derivatives following introduction of their inorganic forms to the environment. Formation of organometallic compounds in the environment is important in many cases because these organometallic species are often more toxic than the inorganic substrates from which they derive. It can be seen that the extent of environmental stress can only be ascertained if the individual chemical species of the elements involved are known.

Various discussion points related to each of the topics will be highlighted in the appropriate section. The main thrust will be the identification of the chemical species involved at critical locations in the environment.

Some Discharge and Effluent Problems

In the following the terms inorganic, metal-organic, and organometallic are used. An inorganic species has no bonding to any organic ligand or molecule; a metal-organic complex involves bonding between the metal and an organic group but does not involve a direct metal carbon bond; an organometallic species has a metal-to-carbon bond.

Use and Species Identification of Some Bulk Inorganic Metal Species

The main uses of heavy metals in major industries are shown in Table 1 in summary form. Table 2 gives the appropriate effluents and some average concentrations of metals from various industries.

Other important discharges for metals include steel industries where levels of thousands of kilograms per day of iron, manganese, and zinc may be discharged. In most cases the precise chemical species have not been investigated and the metals may be present in soluble forms (cation, oxyanion, metal-anion or metal-organic complexes) or bound in charged or neutral forms to surfaces of the suspended solid fraction. The main process metals having important environmental implications are discussed below.

Table 1. The main uses of the bulk industrial metals

	Cd	Cr	Cu	Fe	Hg	Pb	Ni	Sn	Zn	Ti	Al
Pulp and Paper		×	×		×	×	×		×		
Organic, Petrochemicals	×	×		×	×	×		×	×	×	×
Inorganic, Petrochemicals	×	×	×	×	×	×	×	×	×	×	×
Fertilizers	×	×	×	×	×	×	×		×		
Petroleum Refining	×	×	×	×		×	×		×		
Foundries, Metal Works	×	×	×	×	×	×	×	×	×		
Vehicles, Aircraft, Finishing	×	×	×		×		×			×	×
Leather		×									

Table 2. Average concentration ($\mu g\ L^{-1}$) of some industrial effluents

Industry	Cu	Cr	Ni	Zn	Cd
Meat, Fish, Fat Processing	200	200	150	500–4000	10
Baking	240	230	140	1600	15
General Foods	350	150	110	1100	6
Brewing	400	60	40	470	5
Drinks, Ice Cream	2500	150	150	800–3000	30
Dyeing	40	800	250	500	30
Fur Finishing	7000	20000	740	1700	110
General Chemicals	160	280	100	800	27
Laundry	1700	1200	100	1750	134
Car Wash	180	140	190	920	18
Pickling, Dipping of Metals, Tanning	60–90000	9–140000		20–30000	

(From Andren and Nriagu, 1979; Kirk-Othner Encyclopaedia of Chemical Technology, 1978; Lead and Health, 1980)

Lead

The average crustal level is 13 μg g^{-1} DW, mainly as lead(II) carbonate, sulfide, or oxide. Uncontaminated soils have around 10 μg g^{-1} DW of lead, although contaminated industrial or urban soils may have more than ten times this level. Soils near roads or smelters may have greater than 200 μg^{-1} g^{-1} DW, and levels between 2 and 8000 μg g^{-1} DW are found in sewage sludges. Uncontaminated open ocean and fresh surface waters have about 0.1 μg L^{-1} and 0.5 μg L^{-1}, respectively,but industrialized rivers may reach levels of 100 μg L^{-1}. Overall annual emission to the world atmosphere of all forms of lead has been estimated as about 300,000 (Kirk-Othner Encyclopaedia of Chemical Technology, 1978).

The species of lead reaching the environment (Table 3) generally occur as lead(0), inorganic lead(II) and -(IV) salts, and volatile hydrophobic alkyllead species. Lead is normally toxic to microorganisms so there is less scope for biologically mediated changes, e.g., oxidation of lead(0) or -(II). Lead has no established role as a micronutrient, although there is some recent evidence to the contrary (Reichlmayr and Kirchgessner, 1981). Despite this, many plants will tolerate high lead levels and will transport divalent lead ions through the root system. Some grasses contain lead levels on the order of thousands of μg g^{-1}. Retention from food by animals is about 10% and from air about 50%. The direct toxicological activities of inorganic and organic lead have been intensively studied in recent years (Grandjean and Neilson, 1979).

Table 3. Lead in the environment

Situation	Chemical Species	World Annual Tonnage
Ores, Scrap	PbS, $PbCO_3$, $PbSO_4$, Pb	4.3 m
Batteries, Cables, Sheets	Pb, PbO	2.0 m
Antiknock Compounds	$Pb(CH_3)_4$, $Pb(C_2H_5)_4$, and mixtures converted to PbX_2, Pb(OH)X, PbO, $PbSO_4$ in use [vented to atmosphere]	340 000
Paint Pigments	$PbCO_3$, $PbCrO_4$	400 000
Rust Inhibitors	Pb_3O_4, Ca_2PbO_4	} 480 000
Glazing	Lead bisilicate	} 480 000
PVC Stabilizer	Tribasic $PbSO_4$	} 480 000
Alloys		300 000
Fossil Fuels to Atmosphere – Burning	Various unknown forms	300 tonnes (USA)

The chief environmental possibilities for widespread diffusion and mobilization of lead are through physical movement of lead-containing aerosols, oxidation and solubilization of lead(0), acid solubilization of inorganic lead(II), and hydrophobic volatilization of saturated alkyllead compounds. A biogeochemical cycle for the mobilization of lead has been described (Nriagu, 1978).

It has been estimated that annual, overall lead input into the oceans is as follows (pristine levels in tonnes are given in parentheses for comparison): aerosols from gasoline = 37,000 tonnes (zero); aerosols from forest fires and smelters = 3000; soluble lead in rivers and sewers = 60,000 (13,000); lead solids in rivers and sewers 200,000 (100,000); overall lead input into the atmosphere (USA) has been estimated at 142,000 tonnes annually. Lead content of coals varies, e.g., typical USA coal has 0.6 to 33 $\mu g\, g^{-1}$ lead; UK coal has about 17 $\mu g\, g^{-1}$. Emission to the atmosphere from fossil-fuel burning may be around 1000 tonnes worldwide per year (Craig, 1980).

A number of significant questions arise from the use and occurrence of lead components. It is still not established whether lead(0) or inorganic lead(II) or -(IV) are significantly methylated in the aqueous or sediment environments, and hence the proportion of organic lead(IV) arising from methylation compared to gasoline evaporation and/or decay is not known. The organometallic chemistry of lead in the environment is discussed in the section on Organometallic Species. Equally unknown, and of some controversy, is the possible role of the excess of process-derived lead over natural lead species in abnormal medical and behavioral patterns in humans. Although the latter question has been investigated in some detail, the extent of causation by lead over other factors (including social ones) is not fully understood (Lead and Health, 1980).

Mercury

The average crustal level for mercury is 0.5 $\mu g\, g^{-1}$ DW, mainly as mercury(II) sulfide, with a variation in topsoils from 0.01 to 2.0 $\mu g\, g^{-1}$ DW under natural conditions. Levels may be much higher around areas of industrial and agricultural usage. Open-seawater and uncontaminated rainwater levels are around 3×10^{-4} ngL^{-1} (see Bernhard and George, this volume). Coastal waters subject to contamination may have levels ten times higher. Because mercury is a volatile material of low solubility (b.p. 39 °C), atmospheric diffusion is an important transport route for this element. Mercury(0) enters the atmosphere from vulcanism and evaporation from waters and soil (Andren and Nriagu, 1979). Background atmospheric levels are less than 1 ng m^{-3}, but mineral deposits may lead to local levels twenty times greater than this, and factory levels may be around 0.1 mg m^{-3} or above. For alkylmercury compounds the USA limit is 0.01 mg m^{-3} in air.

Mercury is subject to microbial oxidation and reduction of both its inorganic and organic forms. It is well understood that inorganic mercury may be methylated in the environment to monomethyl- and dimethylmercury(II) species under certain conditions, and the latter may be demethylated and reduced. Biologically generated sulfide species also intervene in the mercury cycle. The environmental organometallic chemistry of mercury has been more closely investigated than that of any other element and more is known for mercury of the importance and role of speciation, but the ecological importance of the individual species transformations is still unclear (see Bernhard and George, this volume) (Table 4). These factors are discussed in the section on Organometallic Species. The presumed chief environmental importance of inorganic mercury products or effluents is in the ability of the various inorganic species to become transformed to the more toxic methylmercury(II) cation. The role of species in inorganic mercury(II) methylation is discussed in the section on Organometallic Species. In the remainder of this section aspects of mercury chemistry and species transformation not involving biomethylation are discussed.

Table 4. Mercury in the environment

Situation	Chemical Species	World Annual Tonnage
Ores	HgS, some Hg, HgO, $HgCl_2$	6000
Chlor-alkali Plants	Hg	385[a]
Electrical Equipment	Hg	619[a]
Paints	$PhHg^+$	309[a]
Dental, Instruments		138[a]
Agriculture, Paper and Pulp, Preservatives	$PhHg^+$, Hg_2^{2+}, $CH_3OCH_2CH_2Hg^+$, RHg^+	14[a,b]
General Laboratory	Various	
Catalysts for PVC Urethane and Acetaldehyde Mfg.	$HgCl_2$, $HgSO_4$	29[a]
Pharmaceuticals	HgO, $HgCl_2$, $Hg(CN)_2$, Hg_2Cl_2 plus organic derivatives	15[a]
Detonator	$Hg(CNO)_2$	

[a] Values for USA, 1978.
[b] See also section on Organometallic Species.
(From Andren and Nriagu, 1979; Kirk-Othner Encyclopaedia of Chemical Technology, 1978)

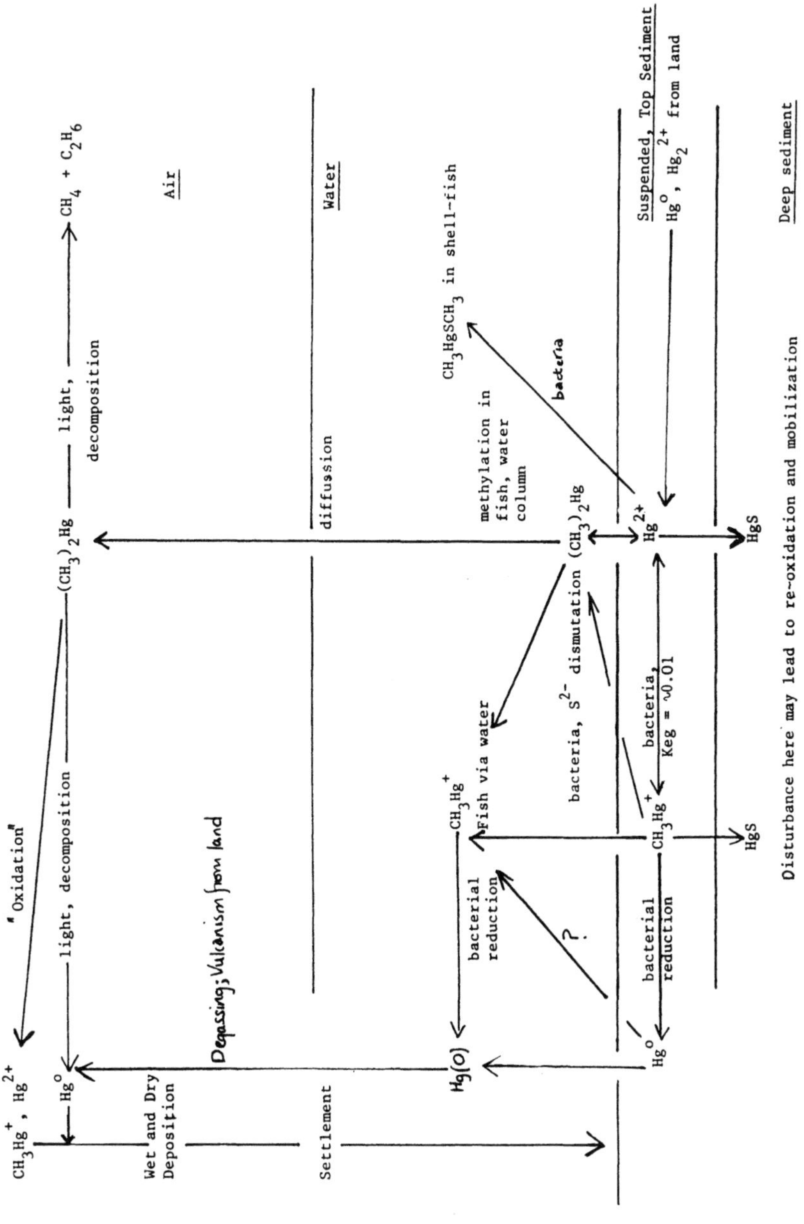

Fig. 1. A biogeochemical cycle for mercury

There appears to be little systemic uptake of agricultural mercury by the plants involved; added soil levels in agricultural use are on the order of 1–200 g Ha^{-1} (5×10^{-4}–1 μg g^{-1}) and plant concentrations after treatment are generally less than 0.1 μg g^{-1}. The ability of animals to bioconcentrate mercury is well known, up to levels more than three orders of magnitude greater than in the treated soil source. Some predatory birds have ultimately accumulated more than 100 μg g^{-1}, for example, in liver and kidneys from alkylmercury-treated seed dressings. Inorganic mercury is much less bioaccumulated owing to its lesser affinity for sulfhydryl linkage formation in biochemical environments. Methylmercury(II) has a Tb in man of more than 70 days owing to strong binding to sulfur residues in human tissue (e.g., brain), whereas inorganic mercury(II) has a Tb of four days (Craig, 1982; International Atomic Energy Agency, 1972) owing to reduced binding affinity. In addition, the ability of methylmercury(II) to solubilize in lipid media reduces its rate of elimination from organisms compared to the inorganic forms. These observations may be satisfactorily rationalized on the basis of the theory of hard and soft acids and bases (Pearson, 1973).

The main biogeochemical pathways in the mercury cycle, including a knowledge of the main mercury species involved, have by now been determined. A number of questions are posed by the biogeochemical cycle shown in Fig. 1. It is necessary for a full understanding to know the proportion of the inorganic mercury present that becomes methylated, and the roles of inorganic mercury substrate species and their oxidation states are also important. Because of its volatility mercury(0) is an easily transportable species, but the ability of other mercury(II) and methylmercury(II) species to undergo environmental transport is of equal significance. The proximity of sulfur to mercury in many aqueous and sediment environments and its affinity for mercury(II) and methylmercury(II) also suggest that sulfide species may mediate mercury chemistry and transport in the aqueous and atmospheric environments. These points are discussed in detail later. Figure 1 shows the directions of the natural mercury transformations occurring in the aqueous, sediment, and atmospheric environments (Craig, 1985b).

Tin

Soils usually contain less than 200 mg kg^{-1} DW of tin which occurs mainly as cassiterite SnO_2, although in mineralized areas the levels may be around 1000 mg kg^{-1} DW. Seawater averages 3 μg L^{-1} mainly as chloro-complexes, and drinking water may have up to 30 μg L^{-1}. The average value in food is 1 mg kg^{-1} FW. The Earth's crust averages 0.3 μg m^{-3} and contaminated air can contain around 5 μg m^{-3}. Inhalation of inorganic tin(0) particles does not produce adverse effects in man, but extended exposure to tin(IV)-oxide dust and fumes can produce a pneumoconiosis (termed stannosis). There is evidence that inorganic tin is methylated in the aquatic environment to various methyl forms (see section on Organometallic Species). From the available information it seems that bioaccumulation of tin and organotin compounds does not occur to a practically significant level in the environment (see Table 5) (Blunden and Chapman, 1982). An estimation of the mean daily intake of all forms of tin in the diet has been made; it ranges from 200 μg to 17 mg $person^{-1}$. The daily intake from water per person is probably less than 30 μg day^{-1}, and from air, less than 1 μg (WHO, 1980). Absorbed tin is not retained in the vascular system; the main site of deposition is in bone, lung, kidney, and liver. There is little penetration of the blood, brain, and placental barriers by inorganic tin (Blunden and Chapman, 1982). In general tin seems to have a low rate of migration in soil and water. Like the other trace elements, tin occurs in coal and oil (average concentrations of 2000 μg kg^{-1} and 0.01 μg kg^{-1}, respectively). The species distribution of inorganic tin in food, biota, and other sources

Table 5. Tin in the environment

Situation	Chemical Species	Annual World Tonnage
Ore	SnO_2	200000
Scrap	Sn, alloys	40000
Tin-plated Steel	Sn	100000
Alloys (Bronze, Pewter, Solder)	Sn	60000
Organic Forms; Stabilizers, Biocides	R_3Sn^+, R_2Sn^{2+}	30000[a]
Glazes, Pigments	SnO_2	4000
Catalysts	$Sn(OAc)_2$, $SnCl_2$, $Sn(COO)_2$	2000
Gas Plating of Tin	SnH_4	
Tin Coating, Plating	$SnCI_2$, $SnSO_4$, $Sn(BF_4)_2$	1000
Toothpaste	SnF_2	
Glass Manufacture	SnO	
Dyeing, Printing	Sn (tartrate), $SnCl_4$	3000
Soft Drink Stabilizer	$SnCl_2$	[b]
Alkaline Electroplating	Na_2SnO_2	3000

[a] Full details of organic use are given in Table 8.
[b] Unknown, may be added at the 0.05% level.
(From Blunden et al., 1983; Craig, 1980; Kirk-Othner Encyclopaedia of Chemical Technology, 1978)

has not been investigated in detail, but currently much attention is being given to the organometallic forms of tin in the above matrices. Identification of the organometallic species present will also give concomitant information concerning inorganic tin in the same matrix. The main question in connection with inorganic tin species concerns the extent to which they are methylated in the aqueous environment; methyltin forms have definitely been detected but the proportion formed there is unknown (Blunden et al., 1983) (see section on Organometallic Species).

Cadmium

Cadmium occurs in the Earth's crust (mainly as sulfide) (Table 6) at between 0.1 and 0.5 μg g^{-1}. Most cases of toxicity due to this element result from the inhalation of dust or fumes, and the current USA atmospheric limit is 100–200 μg m^{-3}, dependent on the species. Cadmium is not known to be an essential trace element. An increase in airborne cadmium is more dangerous than a similar rise in foodstuff content. In the body the critical organ is the kidney where one third of the accumulation takes place. Most release from industrial processes to the environment are in solid wastes such as coal ash, sewage sludge (75 μg g^{-1}), flue dust, and fertilizers (2–20 μg g^{-1}). The USA limit in wastewaters is 1.0 μg g^{-1}. Cadmium in potable waters may cause hypertension – average levels in natural waters are 0.4 μg L^{-1}.

The protein metallothionein contains cadmium and so, despite its toxicity in higher concentrations, trace amounts of cadmium might possibly be

Table 6. Cadmium in the environment

Situation	Chemical Species	World Annual Tonnage
Ores	CdS, also in ZnS	20000
Electroplating	CdO/NaCN, $CdCl_2$, $Cd(SO_4)_2$	1000 (US consumption of CdO)
Batteries, Alloys	$Cd(OH)_2$, Cd, CdS, CdO	
Pigments, Colors	CdS, $Cd(NO_3)_2$, CdS-ZnS-$BaSO_4$, $Cd(OAc)_2$	2600 (US consumption of CdS for colorants)
Stabilizers for PVC, Catalysts	R_2Cd, $CdCO_3$, $Cd(OCOR)_2$	20000 (USA)
Lithography, Engraving	CdI_2, $CdBr_2$, $CdCl_2$	
Semiconductors	CdS, CdSe, CdTe	
TV Tube Phosphors	CdS	

(From Cole and Volpe, 1983; Craig, 1980; Craig, 1985d; Kirk-Othner Encyclopaedia of Chemical Technology, 1978)

physiologically necessary. There is no mechanism for controlling cadmium levels in humans, which may therefore accumulate following intake. This, together with its long Tb of 10–30 years, makes cadmium one of the most toxic metals. This retention may be caused by strong bonding to cysteine residues in a manner similar to mercury, but cadmium(II) has a lesser affinity for sulfur than does mercury(II). The species to be considered for cadmium are those of divalent cadmium only; cadmium(0) does not occur physiologically and there is little likelihood of the formation and stability of organocadmium(II) species in the environment or in organisms. There is one report suggesting methylcadmium(II) formation, but transport phenomena rather than methylation were probably responsible (Robinson and Kiesel, 1981).

Recent work has suggested that while cadmium(II) does cause kidney damage, the levels required are higher than previously thought (approximately 400–600 $\mu g\ g^{-1}$ FW in the renal cortex compared to earlier estimates of 200 $\mu g\ g^{-1}$ FW). It is generally accepted that the kidney is the most critical organ in terms of toxic effects of cadmium(II) absorption; however, there is probably diminishing evidence that species of this element are carcinogenic. In addition, it seems now that cadmium exposure was probably not responsible for itai-itai disease in Japan (Cole and Volpe, 1983).

The environmental properties of cadmium also appear to be less serious than has been assumed. While cadmium(II) is toxic to quatic organisms at quite low concentrations, laboratory studies suggest that the bioaccumulation of cadmium(II) by aquatic biota is not very significant. Median concentration factors for both freshwater and marine organisms are <100 and for vertebrates <20. There seems to be no evidence that cadmium(II) is biomagnified along food chains. For marine molluscs and some crustacea bioaccumulation (bioconcentration through the lifetime of the organisms) does occur, leading to storage of cadmium(II) in unknown chemical forms (Taylor, 1983). The availability of these forms to the mammalian system through human consumption may be less than for some simpler cadmium(II) species. The U.S. Environmental Protection Agency recommended maximum allowable concentration of cadmium(II) is 6.3 $\mu g\ L^{-1}$ in fresh water and 59 $\mu g\ L^{-1}$ in marine waters. In fresh water 90% of the cadmium appears as cadmium(II) cationic species, but in marine zones anionic and neutral chlorocomplexes predominate.

Recent work suggests that cadmium species are less hazardous than had previously been thought. Further work is required to clarify these points in connection with health aspects (hypertension, renal toxicity) and environmental properties (concentration and accumulation, establishment of non-methylation for cadmium). Further work is also needed in connection with the fate and species distribution resulting from the use of cadmium as stabilizers in plastics and in establishing which species of cadmium are more toxic to biota. Further research here regarding the importance of pH and pCO_2 in cadmium-containing waters for toxic effects is needed. (For marine biota see also Bernhard and George, this volume.)

Chromium and Nickel

The crustal abundance of chromium (Table 7) is 125 μg g^{-1} DW, and of nickel, 160 μg g^{-1} DW. Chromium is very insoluble and leaches little into natural waters, average levels being 0.7–84 ng g^{-1} in rivers, and in potable water 0 to 36 ng g^{-1} (average = 0.43 ng g^{-1}). Foodstuffs contain chromium in the range 20 to 590 ng g^{-1} FW, resulting in daily intakes for humans of 10 to 400 μg with an average of 80 μg per person. The normal city range

Table 7. Chromium and nickel in the environment

Saturation	Chemical Species	World Annual Tonnage
Ores	$NiS \cdot 2FeS$, $NiMgSiO_3 \cdot nH_2O$, $FeO \cdot Cr_2O_3$	1.5 m (Cr) 750000 (Ni)
Refining	$Ni(CO)_4$	
Fossil-fuel Burning	Various	70000 (Ni)
Alloys	*with* Fe, C, Cu, Zn, Mn, Au, Mo	
Leather, Tanning	$Cr(OH)SO_4 \cdot 5H_2O$, $CrCl_3 \cdot nH_2O$	
Plating, Coatings	"CrO_3", $CrCl_3$, $Cr_2O_7^{2-}$, CrO_4^{2-}, $NiSO_4$, $NiCl_2 \cdot 6H_2O$, $Ni(BF_4)_2 \cdot 6H_2O$	
Pigments	$PbCrO_4$, Cr_2O_3, $Cr_2O_3 \cdot nH_2O$, $K_2O \cdot 4ZnO \cdot 4CrO_3 \cdot 3H_2O$, $SrCrO_4$ $NiCO_3$, NiO	40000
Textiles	$Na_2Cr_2O_7$	
Fungicide	NiX_2, $NiSO_4$	
Drilling Muds	Na_2CrO_4	1 ton yr^{-1} (for each well)
Batteries		
Wood Preservatives	Cu, $Cu_2O_7 \cdot 2H_2O$	10000
Furnace Refractory	$FeO \cdot Cr_2O_3$	
Corrosion Inhibition Antifouling	CrO_4^{2-}, $Cr_2O_7^{2-}$	
Catalysts	NiAl alloy, NiO, $NiSO_4$, $Ni(NO_3)_2$	
Ceramics	NiO, Ni_3S, Ni_2Si	

(From Craig, 1980; Kirk-Othner Encyclopaedia of Chemical Technology, 1978; Sax, 1974)

for chromium in air is from 0.01 to 0.02 $\mu g\,m^{-3}$. The multiple valence states for chromium should make it suitable for microbial oxidation but there is no evidence for this. Chromium is in fact an essential element in plant and animal nutrition. Chromium(III) compounds have no established toxicity, but in the hexavalent state toxicity through skin ulceration and nasal perforation as well as lung cancer are well known (calcium, zinc, lead, and chromic chromates are carcinogenic). Nickel and nickel compounds may cause allergic dermatitis. The maximum permissible exposure level (USA) has been set at 1 $mg\,m^{-3}$; for $Ni(Co)_4$ the level is 1 $ng\,g^{-1}$.

There are no possibilities for the formation or stable existence of organometallic derivatives of these metals in the natural environment. The organic derivatives are generally very unstable to air or moisture; the quasi-stability of $CH_3Cr(H_2O)_5^{2+}$ ($t_{1/2}=0.05$ min) has no environmental implication.

Special Case of Species Transformation of Metals – Acidic Mine Drainage Water

One particular effluent situation that has caused widespread environmental concern is the low pH of drainage waters from both working and abandoned coal and metal mines. The chemistry of this topic is dealt with in detail elsewhere (Olson, this volume) but is covered here from the point of view of the species involved. This phenomenon demonstrates the role that microorganisms can play in the species transformation and mobilization of metals; microorganisms are responsible here for the oxidation of iron(III) in acid solutions – not a favored chemical process.

Coal seams often have iron(II)-sulfide minerals associated with them and these are left behind after extraction work. Parallel oxidation of the iron(II) and sulfur(II) occurs chemically and microbially and produces drainage water from mine workings with a pH of 1.5 to 4.0. Microbial oxidation of iron and sulfur is mainly achieved by the acidophilic chemolithotrophs *Thiobacilli, Thiooxidans,* and *Ferrooxidans.*

The transformations giving rise to acidic water from mines have been summarized by Higgins and Burns (1975).

These acidic effluents then render the water courses into which they flow unsuitable for growth of plants, fish, and microorganisms and for use as potable or irrigation water. The problem then shades into that of acid rainfall in general. Recuperative measures which have been attempted include stimulation of sulfate reducers (e.g., *Desulphovibrio*) to reverse sulfur oxi-

Table 8. Chemistry of mine water pollution

$FeS_2 + 3O_2 \rightarrow FeSO_4 + SO_2$	[absence of water]
$2FeS_2 + O_2 + 2H_2O \rightarrow 2FeSO_4 + 2H_2SO_4$	[presence of water – slow]
$4FeSO_4 + 2H_2SO_4 + O_2 \rightarrow 2Fe_2(SO)_3 + 2H_2O$	[oxidation by microorganisms, low pH rate – determining phase]
$2Fe_2(SO_4)_3 + 12H_2O \rightarrow 2Fe_2(OH)_6 + 6H_2SO_4$	[hydrolysis]
$FeS_2 + Fe_2(SO_4)_3 \rightarrow 3FeSO_4 + 2S$	[oxidation of pyrites – Fe(II) produced is reoxidized by microorganisms]
$2S + 3O_2 + 2H_2O \rightarrow 2H_2SO_4$	[acid production]
$4FeS_2 + 150_2 + 2H_2 \rightarrow 2Fe_2(SO_4)_3 + 2H_2SO_4$	[overall process]

(From Olson, this volume; Higgins and Burns, 1975)

dation, addition of *Thiobacilli* inhibitors, or addition of limestone to precipitate sulfur at higher pH (Higgins and Burns, 1975). It is not known whether the use of anaerobic reducing conditions to counter acid production leads to other significant environmental consequences.

Organometallic Species

Probably the most significant species transformation possibility for metal effluents or products is the natural formation in the environment of organometallic derivatives (i.e., species having metal-to-carbon bonds). Most are more toxic and persistent in organisms than the inorganic species from which they are formed (arsenic is an exception in this respect). In addition, a large number of organometallic products are introduced to the environment either directly (e.g., as biocides) or indirectly through use in another mode (e.g., as a plastics additive). The consequences to biota and the environment of the introduction or formation of organometallic compounds is covered in two recent books (Craig, 1985c; Thayer, 1984).

Table 9 shows the main organometallic products, their uses and fate, and the approximate annual tonnages. Table 10 shows the species of those organometallic compounds which have been identified as being formed in the environment from inorganic precursors. It seems that the natural formation of organometallics in the environment is restricted to methylation, as other groups have not been reported as being added to inorganic metallic species in natural matrices, but this may be due to analytical procedures used which can detect only stable species.

Table 9. The main organometallic products introduced to the environment

Compound, Species	Use	Environmental Fate	Approx. Annual Tonnage
RHgX (R = CH_3, C_2H_5; X = Cl^-, PO_4^{3-} etc)	Seed dressing	Slow degradation to Hg(0), Hg(II)	21 (USA, 1978)
C_6H_5HgX (X = CH_3COO^-)	Seed dressing, fungicide, slimicide	Slow degradation to Hg(0) Hg(II)	28 (UK, 1976)
$ROCH_2CH_2HgX$ (R = CH_3, C_2H_5; X = CH_3COO, Cl)	Seed dressings Fungicides	Degradation to Hg(0) and Hg(II)	
$nC_4H_9)_2SnX_2$ (X = anionic grouping)	PVC stabilization, polyurethane catalyst, feed additive	Degrades to SnO_2[a]	30–35000; 65% of above used in PVC stabilization and about 10% for biocidal use (total usage for organotins)
$(ROCOCH_2)_2SnX_2$, $ROCOCH_2CH_2SnX_3$	PVC stabilization	Degrade to SnO_2[a]	
$(nC_4H_9)_3SnX$ (X = F, $\frac{1}{2}$O)	Slimicide, antifouling additive	Degrades to SnO_2[a]	
$(nC_8H_{17})_2SnX$ (X = maleate)	PVC stabilization	Degrades to SnO_2[a]	
CH_3SnX_3	PVA stabilization	Degrades to SnO_2[a]	
$(C_6H_{11})_3SnX$ (X = OH)	Biocides	Degrades to SnO_2[a]	
$(C_6H_5)_3SnX$ (X = OH, CH_3COO)	Biocides	Degrades to SnO_2[a]	
$(CH_3)_4Pb$, $(C_2H_5)_4Pb$ and mixtures	Gasoline antiknock additives	In water decay via R_3Pb^+ and R_2Pb^+, also by radical attack in atmosphere; final product is inorganic lead, most is burnt in air.	350000
$(CH_3)_2AsOOH$ $CH_3AsO(ONa)_2$	Herbicides	Bacterial demethylation	15000
$pNH_2C_6H_4AsO(OH)_2$ (and other substituted phenyls)	Animal feed additives		

[a] Estimates vary on the rate and extent of degradation
(Adapted from Craig, 1985c)

Table 10. Some organometallic species found in the environment

Species	Matrix	Typical Concentration Levels	General Reference
CH_3Hg^+ (bound to S-linkages in protein amino acids, etc.)	Sediments, water, some organisms	ng g^{-1} (sediments) ng L^{-1} (water)	Craig, 1985b
$(CH_3)_2Hg$	Sediments; diffuses to atmosphere		
$(CH_3)_nSn^{(4-n)+}$	Sediments, micro-organisms, natural waters	mg L^{-1}	National Research Council of Canada, 1985
$(CH_3)_3SnH_{4-n}$	Anaerobic sediments, waters	mg L^{-1}	National Research Council of Canada, 1985
$(CH_3)_4Pb$, $(CH_3)_nPb^{(4-n)+}$	Waters, sediments	ng g^{-1} (sediments) ng L^{-1}	National Research Council of Canada, 1985
$(nC_4H_9)_nSn(CH_3)_{4-n}$	Harbor waters, sediments	mg L^{-1}	Maguire, 1984
$(CH_3)_2As^+$, CH_3As^{2+} (as methylarsonate and dimethylarsinate and derivatives)	Natural waters, soils	ng L^{-1}	Andreae and Klumpp, 1979
$(CH_3)_3As$, $(CH_3)_2AsH$, $(CH_3)_3AsO$	Atmosphere after biomethylation	trace	
$(CH_3)_3As^+$ (as arsenobetain)	Marine invertebrates, fish		Craig, 1985c
$CH_3SbO(OH)_2$, $(CH_3)_2SbO(OH)$	Natural water	ng L^{-1}	Andreae et al., 1981
$(CH_3)_3Ge^{(4-n)+}$	Natural waters	ng L^{-1}	Lewis et al., 1985
CH_3CoB_{12} and derivatives (methylcobalamin and B_{12} coenzymes)	Natural waters, sediments, foods, biota	ng g^{-1} µg g^{-1}	Craig, 1985d

Concentration levels shown indicate generally observed ranges detected.
(Adapted from Craig, 1985c)

Organometallic compounds are not persistent in the natural environment. The compounds reported in Table 9 eventually decay to inorganic elements and this is an important criterion for their use. Organotin compounds, for example, are reported to decay by sequential loss of the organic group to inorganic tin – as tin(IV)-oxide.

The unknown factor in this type of work is the realization that such decomposition may not go to 100% completion. The biological route to decomposition of alkyl organometallic compounds is usually initiated by hydroxylation of side-chain carbon followed by tin-carbon cleavage. Organolead compounds also decay through sequential loss of alkyl group, initiated *in vivo* and in the atmosphere by hydroxyl attack as for tin (Harrison and Laxen, 1978). Reaction with ozone and other radicals also occurs in the atmosphere. Daylight atmospheric decay rates have been estimated: 88% per hour for $PbEt_4$ and 21% per hour for $PbMe_4$ (Craig, 1985a). Decay in water is promoted by coordination to silica and the presence of Cu(II) or Fe(II), sunlight, and sulfides. Adsorption onto sediments promotes decomposition of tetraalkyllead species in the environment. For arsenic compounds the arsenic-carbon bond is stable in plants, but microbial oxidation takes place in soils producing 10% yields of CO_2 in 30 days, together with arsenate. Losses from soil as volatile organoarsenic compounds may also occur (methylation processes). Estimates of half-lives for organoarsenicals in soil range from 6.6 to 34.3 months. Most organoarsenic compounds used in cattle feeds as growth promoters are excreted unchanged. Following loss of organoarsenic compounds to the environment, fate and species transformation and distribution then become part of the normal biogeochemical arsenic cycle (Andreae, 1985).

There is little doubt that organometallic compounds decay in the environment, but rates and the extent are often unknown or uncertain. If the decay rate in the laboratory is compared to that which occurs under natural conditions, it may be assumed that atmospheric decomposition is more rapid than simple oxidation under laboratory conditions because of the large potential for chemical attack in the atmosphere. However, complexation of organometallic species to sediments will lead to greater stability of the metal-carbon moiety than that which exists for that simple species in aqueous solution alone. Laboratory studies of rates of decay and final products may often not replicate in rates and extent the natural processes in the environment. Specifically, breakdown may not always be to the inorganic form and it may not be to 100% completion (Craig, 1985a).

Species distribution and transformation under natural conditions is of great complexity. Methyl derivatives of a number of metallic or metalloidal elements have been found in the natural environment, apparently having been formed there (Table 10). Methyl metals of the following elements may be formed under environmental or model environmental conditions: mercury, lead, tin, arsenic, germanium, antimony, sulfur, selenium, tellurium, and thallium. Cobalt, as methylcobalamin, is a special case. These methyl derivatives are probably present at their observed concentration levels as a result of equilibria between natural methylation and demethylation processes. The case of mercury has been most closely studied, and bacteria capable of methylation and demethylation have been identified. Demethylation occurs via detachment of the organic group to produce a mercury(II) species followed by further reduction to mercury(0) (Silver, 1984). Similar processes occur for other elements capable of methylation in the natural environment. Apart from a single example, methylation has always been reported to occur on the inorganic metal, but recently butylmethyltin(IV) derivatives have been found (Maguire, 1984) suggesting methylation of partially decomposed butyltin(IV) species. A list of species capable of methylation and demethylation of mercury is given in Table 11.

Table 11. Methylation and demethylation of mercury by bacteria

Species Capable of Methylation	Species Capable of Demethylation of CH_3Hg^+
P. fluorescens	Serratia marcescens
M. phlei	Providencia sp.
B. megaterium	Pseudomonas fluorescens
E. coli	Citrobacter freundii
E. coli w/B_{12}	Proteus mirabilis
A. aerogen	Enterobacter aerogenes
A. aerogenes w/B_{12}	Enterobacter cloacae
E. aerogenes	Paracoloacterium coliforme
	Achrombacter pestifer
C. cochlearium[a]	Serratia plymuthica
	Staphylococcus sp.
	Pseudomonas aeginosa
	Bacillus subtilis
	Flavobacterium marinotypicum
	Citrobacter intermedius
	Pseudomonas fragi
	Desulfovibrio desulfuricans[a]

[a] Anaerobic.
(From Craig, 1985b, 1985c; Silver, 1984; Summers and Silver, 1978)

The influence of different substrates on methylation and demethylation is shown in Fig. 2. Different rates and extents for methylation are found for different inorganic mercury(II) substrates. In the example shown, methylation appears to be governed by the ability of microorganisms to convert covalently bonded mercury species to an essentially ionic, free mercury(II) form, which is the species actually methylated. The greater facility of S-bound compared to chloromercury(II) complexes reflects their faster conversion to mercury(II) ion (Craig and Moreton, 1985). The complete methylation for the mercury(II)-acetate species is almost certainly due to methylation occurring from the acetate-methyl group (Craig and Moreton, 1985).

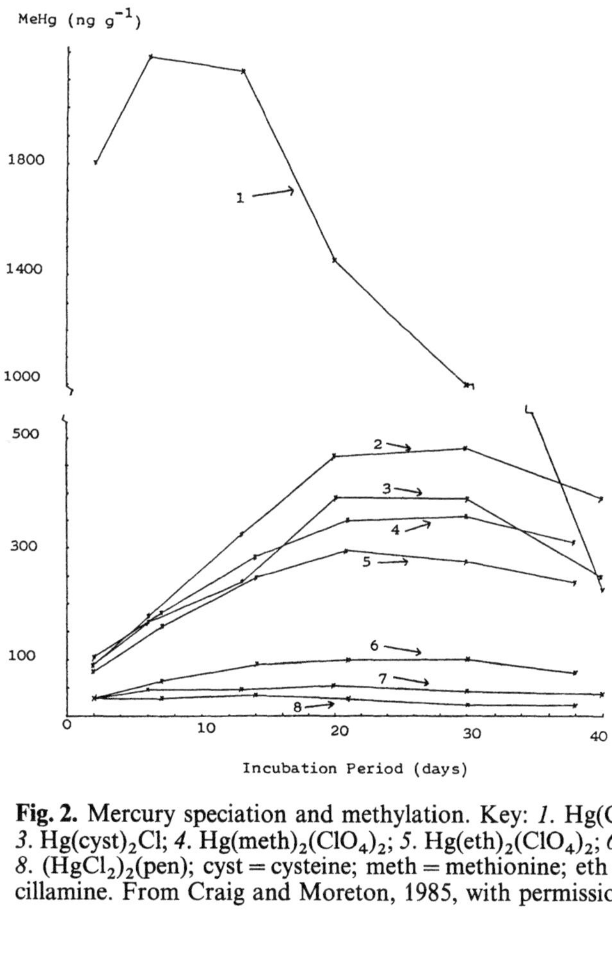

Fig. 2. Mercury speciation and methylation. Key: *1.* $Hg(CH_3CO_2)_2$; *2.* $Hg(cyst)_2$; *3.* $Hg(cyst)_2Cl$; *4.* $Hg(meth)_2(ClO_4)_2$; *5.* $Hg(eth)_2(ClO_4)_2$; *6.* $HgCl_2(cyst)$; *7.* $HgCl_2$; *8.* $(HgCl_2)_2(pen)$; cyst = cysteine; meth = methionine; eth = ethionine; pen = penicillamine. From Craig and Moreton, 1985, with permission

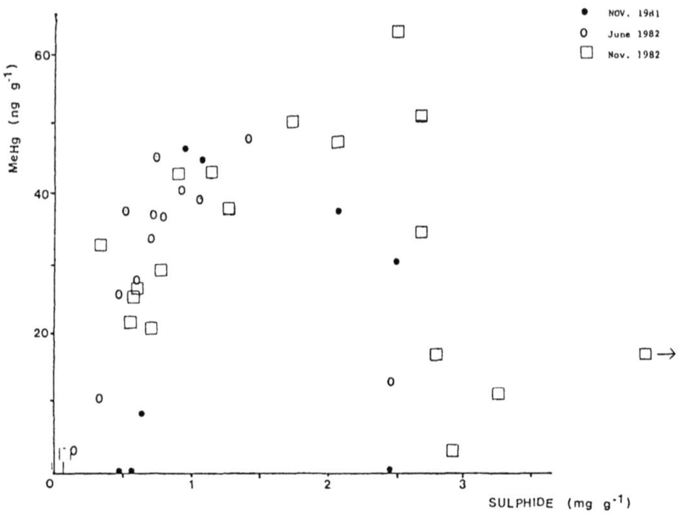

Fig. 3. Relationship between sulfide and methylmercury (from Craig and Moreton, 1984, with permission)

The importance of species for the fate and behavior of organometallic derivatives from effluents or discharges is shown by the complex interaction between mercury(II) and sulfide species in a medium capable of methylation (e.g., a sediment). Up to a level on the order of 1.8–2.2 mg g^{-1} of sulfide species, the amount of methyl-mercury produced from inorganic mercury is directly proportional to the sulfide content. This probably reflects a convergence of conditions appropriate for methylating bacteria as sulfide levels increase. However, above the levels of sulfide indicated previously the amount of methylmercury declines absolutely. This is due to increased formation of mercury(II) sulfide, which is barely capable of being methylated, and also to removal of monomethylmercury to the atmosphere as follows:

$$2CH_3Hg^+ + S^{2-} \rightarrow (CH_3Hg)_2S \rightarrow (CH_3)_2Hg + HgS. \quad (1)$$

The dimethylmercury species is volatile and hydrophobic and slowly diffuses to the atmosphere through the sediments and water layer. By this mechanism it has been calculated that up to 12% annually of the monomethyl-mercury present may be removed from sediment to the atmosphere following increased sulfide levels in the system. This reaction is illustrated in Fig. 3.

Conclusions

It can be seen from the above that the chemical form of an element is vital in the determination of the actual properties of stability, toxicity, and transport of that element in the natural environment. Future research should be directed not merely to the analysis of the bulk concentration of the element in various environmental matrices but also to the identification of chemical forms and the determination of their properties.

References

Andreae MO (1985) Organoarsenic compounds in the environment. In: Craig PJ (ed) Organometallic Compounds in the Environment. London: Longman LTD

Andreae MO, Asmode J-F, Foster P, Van't dack L (1981) Determination of antimony(III), antimony(V), and methylantimony species in natural waters. Anal Chem 53:1766–1771

Andreae MO, Klumpp DW (1979) Envir Sci Technol 13:738–741

Andren AW, Nriagu JO (1979) The global cycle of mercury. In: Nriagu JO (ed) The Biogeochemistry of Mercury in the Environment. Amsterdam: Elsevier/North Holland, pp 1–23.

Blunden SJ, Chapman AH (1982) The environmental degradation of organotin compounds – a review. Envir Technol Lett 3:267–273

Blunden SJ, Hobbs LA, Smith PJ (1983) The environmental chemistry of organotin compounds. In: Bowen HJM (ed) Environmental Chemistry, vol 3. Special Publication. London: Royal Society of Chemistry

Cole JF, Volpe R (1983) The effect of cadmium on the environment. Ecotox Envir Safety 7:151–159

Craig PJ (1980) Metal cycles and biological methylation. In: Hutzinger O (ed) The Handbook of Environmental Chemistry, Part A, vol 1. Berlin: Springer-Verlag, pp 169–227.

Craig PJ (1982) Environmental aspects of organometallic chemistry. In: Wilkinson G, Stone FGA, Abel EW (eds) Comprehensive Organometallic Chemistry, vol 2(18). Oxford: Pergamon Press, pp 979–1020.

Craig PJ (1985a) Occurrence and pathways of organometallic compounds in the environment – general considerations. In: Craig PJ (ed) Organometallic Compounds in the Environment. London: Longman Ltd

Craig PJ (1985b) Organomercury compounds in the environment. In: Craig PJ (ed) Organometallic Compounds in the Environment. London: Longman Ltd

Craig PJ (ed) (1985c) Organometallic compounds in the environment. London: Longman Ltd

Craig PJ (1985d) Other organometallic compounds in the environment. In: Craig PJ (ed) Organometallic Compounds in the Environment. London: Longman Ltd

Craig PJ, Bartlett PD (1978) The role of hydrogen sulphide in environmental transport of mercury. Nature 275:635–637

Craig PJ, Moreton PA (1984) The role of sulphide in formation of dimethyl mercury in river and estuary sediments. Mar Poll Bull 15:406–408
Craig PJ, Moreton PA (1985) The role of speciation in mercury methylation in sediments and water. Envir Poll B 10:141–158
Grandjean P, Neilson T (1979) Toxicology of organolead compounds. Residue Rev:97–148
Harrison RM, Laxen DPH (1978) Sink processes for tetraalkyllead compounds in the atmosphere. Envir Sci Technol 12:1384–1392
Higgins IJ, Burns RG (1975) The chemistry and microbiology of pollution. London: Academic Press
International Atomic Energy Agency (1972) Mercury contamination in man and his environment. Vienna: IAEA
Jarvie AWP, Markall RN, Potter HR (1981) Decomposition of organolead compounds in aqueous systems. Envir Res 25:241–249
Kirk-Othner Encyclopaedia of Chemical Technology, third ed. 1978. New York: John Wiley
Lead and Health (1980) The Lawther Report, DHSS. London: Her Majesty's Stationery Office
Lewis FL, Froelich PN, Andreae MO (1985) Methylgermanium in natural waters. Nature 313:303–304
Maguire (1984) Butyltin compounds and organic tin in sediments in Ontario. Envir Sci Technol 18:291–294
National Research Council of Canada (1985) Organotin compounds in the aquatic environment. Ottawa: National Research Council of Canada
Nriagu JO (1978) In: Nriagu (ed) The biogeochemistry of lead in the environment. Amsterdam: Elsevier/North Holland, p 11.
Pearson RG (1973) Hard and soft acids and bases. In: Benchmark Papers in Inorganic Chemistry. New York: Academic Press
Reichlmayr A, Kirchgessner M (1981) Z Tierphys Tiernährg Futterm 46:1–9
Robinson JW, Kiesel EL (1981) Methylation of cadmium with B_{12} – possible method of detoxification. J Envir Sci Health A16(3):391–452
Sax NI (1974) Industrial Pollution. New York: John Wiley
Sderquist CJ, Crosby DG (1980) Degradation of triphenyltin hydroxide in water. J Agric Food Chem 28:111–117
Silver S (1984) Bacterial transformations of and resistances to heavy metals. In: Nriagu JO (ed) Changing Metal Cycles and Human Health. Dahlem Konferenzen. Berlin, Heidelberg, New York, Tokyo: Springer-Verlag, pp 199–223.
Summers AO, Silver S (1978) Ann Rev Microb 32:637–672
Taylor D (1983) The significance of the accumulation of cadmium by aquatic organisms. Ecotox Envir Safety 7:33–42
Thayer JS (1984) Organometallic compounds and Living Organisms. New York: Academic Press
WHO (1980) Environmental Health Criteria. 15, Tin. Geneva: World Health Organization

The Importance of Chemical "Speciation" in Environmental Processes,
eds. M. Bernhard, F. E. Brinckman, and P. J. Sadler, pp. 465–491.
Dahlem Konferenzen 1986. Springer-Verlag Berlin, Heidelberg
© Dr. S. Bernhard, Dahlem Konferenzen

Chemical Forms and Environmental Effects of Critical Elements in Solid-Waste Materials – Combustion Residues

U. Förstner

Technische Universität Hamburg-Harburg
2100 Hamburg 90, F.R. Germany

Abstract. Combustion residues, especially coal fly-ash particles, represent characteristic matrices for the study of potentially toxic metal species in solid-waste products. Appropriate preconcentration procedures and advanced analytical techniques involving single mineral determinations, surface-sensitive methods, and solvent-leaching experiments provide information on specific sources, reactions, and the possible environmental impact of particle-associated pollutants. Typical changes of toxic metal species from combustion-derived particles into more bioavailable forms occurring during coal fly-ash disposal (cropland application, ocean dumping, ash-pond disposal, landfill leachates) and atmospheric deposition (sea-surface microlayer, vegetation, human respiration) are described. Toxic effects result both from metal-surface accessibility (enrichment) and surface solubility (speciation). There is a tendency that elements introduced with solid waste materials are less stably bound than those in natural systems. Even at relatively small proportions of these materials, therefore, mobilization (and subsequent transfer to biota) of potentially toxic elements by acidity, complexing agents, or redox changes may be significantly increased.

Introduction

Over the 200 years following the beginning of industrialization huge changes in the distribution and solubilization of metal ions at the Earth's surface have occurred, challenging those regulatory systems for the transport of ions through cell membranes that took billions of years to evolve (Wood and Wang, 1983). Changes of the global budget of a specific element by man's activities, i.e., the degree of its utilization, its "technogenic migration" (Nikiforova and Smirnova, 1975), or its "relative pollution potential" (Förstner and Müller, 1973) is indicated, for example, in the ratio of the annual output of the element to its natural concentration in unpolluted sediments, soils, or waters or to its "Clarke," i.e., the mean concentration in the Earth's crust. In this way the Technophility Index (Nikiforova and Smirnova, 1975) has been calculated for selected metal examples (values in 5×10^7):

$$\overset{\mathrm{Mn}}{1} = \overset{\mathrm{Fe}}{1} < \overset{\mathrm{Ni}}{2} < \overset{\mathrm{Cr}}{4} < \overset{\mathrm{Zn}}{10} < \overset{\mathrm{Cu}}{20} = \overset{\mathrm{Ag}}{20} < \overset{\mathrm{Hg}}{30} = \overset{\mathrm{Pb}}{30} < \overset{\mathrm{Au}}{60} < \overset{\mathrm{Cd}}{140}$$

Even if the dispersion on the terrestrial or aquatic environment takes place only after a certain residence time within specific objects, metals such as Cd, Pb, Hg, and Zn will be accumulated to a larger extent on the Earth's surface than are others such as Ni, Fe, and Mn. Direct comparison between the natural and anthropogenic impact on a certain environment has been made for atmospheric emissions (Galloway et al., 1982; Lantzy and MacKenzie, 1979); here, the highest "mobilization factors," i.e., the ratio of anthropogenic/natural output (both in 10^8 g/yr), were found for Pb, Ag, Mo, Sb, and Zn, whereas, for example, Co, Mn, and Hg are more strongly emitted from natural than from anthropogenic sources to the atmosphere (Salomons and Förstner, 1984):

$$\overset{\mathrm{Hg}}{0.44} < \overset{\mathrm{Mn}}{0.53} < \overset{\mathrm{Co}}{0.71} < \overset{\mathrm{Cr}}{1.6} < \overset{\mathrm{As}}{3.3} < \overset{\mathrm{Cd}}{19} < \overset{\mathrm{Zn}}{23} < \overset{\mathrm{Sb}}{39} < \overset{\mathrm{Mo}}{45} < \overset{\mathrm{Ag}}{83} < \overset{\mathrm{Pb}}{100}$$

In the discussion of environmental problems with metals it is increasingly recognized that the ecotoxicological significance of anthropogenic inputs is determined by the specific form and reactivity of a metal compound rather than by the accumulation rate of the total metal. Among the criteria for assessing which element may be of major concern one question deserves primary attention (Andreae et al., 1984): Is the element mobile in geochemical processes because of either its volatility or its solubility in natural water, so that the effect of geochemical perturbations can propagate through the environment? Here, too, characteristic examples on the relationship between solid speciation and environmental effects have been designed from the study of combustion residues. It has been suggested that aerosol toxicity

– e.g., from entrapping of microscopic particles in alveolar regions of the lung – probably is due in part to trace elements present on the surface of the particles (Natusch et al., 1974) as well as sorbed organics such as polynuclear aromatic hydrocarbons and their chemical derivates (Cabaniss and Linton, 1984). Also, accessibility to the aquatic environment (via washout, rainout, groundwater leaching) is governed by both metal-surface accessibility (extent of surface enrichment) and metal-surface solubility (surface speciation) (Farmer and Linton, 1984).

Methods

Since adsorption of pollutants onto airborne and waterborne particles is a primary factor in determining the transport, deposition, reactivity, and potential toxicity of these materials (Keyser et al., 1978), analytical methods should be related to the chemistry of the particle's surface and/or to the trace-metal species highly enriched on the surface. Basically there are three methodological concepts for determining the distribution of an element within or among small particles (Hansen and Fisher, 1980):

1) Analysis of single particles by X-ray fluorescence using either a scanning electron microscope (SEM) or an electron microprobe can identify differences in the matrix composition between individual particles. The total concentration of the element can be determined as a function of particle size. If the concentration increases as particle size decreases, it is often assumed that the element is located on the surface of the particle (Hansen and Fisher, 1980). Other physical fractionation and preconcentration methods include density (Dossis and Warren, 1980) and magnetic separation (Linton et al., 1980).

2) Direct species determinations – mainly on preconcentrated particles – can be performed, e.g., by X-ray diffractometry (XRD), Fourier Transform Infrared Spectroscopy (FT-IR), and Proton-Induced X-ray Emission (PIXE) analysis. The surface of the particles can be studied directly by the use of electron microprobe X-ray emission spectrometry (EMP), X-ray photoelectron spectroscopy (XPS), electron spectroscopy for chemical analysis (ESCA), Auger electron spectrometry (AES), and secondary ion-mass spectroscopy (SIMS). Depth-profile analysis is used to determine the variation of chemical composition with depth below the original surface (Keyser et al., 1978).

3) Solvent leaching – apart from the characterization of the reactivity of specific metals – can provide information on the behavior of metal pol-

lutants under typical environmental conditions. Common, single-reagent leachate tests, e.g., U.S. EPA, ASTM, and IAEA, use either distilled water or acetic acid (Theis and Padgett, 1983); a large number of sequential test procedures have been designed particularly for soil and sediment studies (Förstner and Wittmann, 1981; Salomons and Förstner, 1980). There are many problems associated with the interpretation of such data (Salomons and Förstner, 1984), particularly for tests on widely different waste materials (Ham et al., 1979). Proper attention must be paid to the pH, volume of the leachate, and equilibrium vs. nonequilibrium conditions (Calmano and Förstner, 1983; Hansen and Fisher, 1980). Combinations of sequential-filtration and chemical-leaching techniques (e.g., with physiological solutions, HCl, $NaBH_4$, chelating agents, and serum) were used by Hansen and co-workers (1984) to study the association of cations and anions on the surface of ash particles and the oxidation states of nonmetal and transition metals, mainly to predict biological effects on animals and man.

Trace-Element Species on Fossil-Fuel Fly-Ash Particles

Residues from coal-fired power plants may soon constitute one of the largest tonnages of solid waste material. World consumption of coal is estimated at approximately 3.4 billion of tons/yr, two thirds of which is utilized by the electric power industry (Adriano et al., 1980). Production of collected residues (bottom/boiler slag and fly ash) is estimated at 300 million tons/yr, of which approximately 70% is discharged to landfills. Recycling rates (for soil amendment and construction materials) are less than 10% in the U.S. (Page et al., 1979) and 40–50% in some major European countries (Adriano et al., 1980; Bamber, 1980). In this respect the cement and concrete industry has been a high-volume market for fly ash (Bakker, 1982). Increasing tonnages of flue-gas desulfurization sludge (FGD sludge) can be expected from the equipment of power plants with lime/limestone desulfurization systems (estimation for 1985 in U.S. plants: 35 million tons (Bern, 1976)). Residual fuel oils used in electrical generation-combustion sources have low ash contents and do not present the large-tonnage disposal problems of coal by-products; however, since most oil-fired power plants operate without particulate control systems the quantities of fly ash emitted from their stacks can be significant (Henry and Knapp, 1980).

Particle Morphology and Mineralogy

Morphology of coal fly-ash particles is affected, apart from combustion-zone temperatures and the residence time in the combustion zone, by the differences in "maceral" (the organic equivalent to inorganic "mineral" components) composition of the coal (Kautz, 1984; Shibaoka and Ramsden, 1978). For most fly ashes the most predominant forms include small, solid, spherical particles, hollow particles, or "cenospheres". Hollow spheres encapsulate small, solid particles ("plerospheres"). The occurrence of large numbers of small particles inside the plerospheres is not thought to be of much significance from the standpoint of environmental impact, but if released from their encapsulating hosts such particles are difficult to control, have a long atmospheric lifetime, provide substantial surface area for heterogenous reactions, produce significant light-scattering, and may penetrate the innermost regions of the human lung when inhaled (see Natusch and Taylor, 1980, for references). Fly ashes emitted from oil-fired power plants differ physically from those emitted from coal-fired power plants, the former having a generally more open and porous structure (Henry and Knapp, 1980).

Mineralogical analyses of coal fly ash show that between 70 and 90% of the particles are glassy spheres, the remainder consisting of quartz (SiO_2), mullite ($3Al_2O_3 \cdot 2SiO_2$), hematite (Fe_2O_3), magnetite (Fe_3O_4), and a small proportion of unburned C (1–2%). Since carbon appears to be the predominant element in the nonmagnetic particles, that fraction apparently is composed mostly of carbonaceous soot (Cabaniss and Linton, 1984). Titanium is present as rutile, zirconium as zircon, and barium as barite; tourmaline and other boron minerals have been identified or suggested (Adriano et al., 1980). Also, surface-sorbed species such as carbonates sulfates, nitrates, water, and carbon dioxide are likely to contribute to X-ray photoelectron spectroscopy-detected oxygen (Fisher et al., 1979). Crystallites in oil-fired fly ash consist mainly of magnesium and vanadium sulfates (Henry and Knapp, 1980).

Elemental Fractionation during Coal Combustion Processes

Trace elements in coal are present in aluminosilicates ("lithophilic" elements) as inorganic sulfides ("chalcophilic" elements), or as organic complexes (Klein et al., 1975): on combustion the aluminosilicates are not decomposed; rather, they melt and coalesce to form the slag and fly ash. During the initial stages of combustion the conditions in a coal particle are probably reducing and the chemical bonding between metallic elements and sulfur in sulfide minerals or the organic matrix is broken; these elements initially volatilized or dispersed in the flue-gas stream may then be oxidized to form less-volatile species which may condense or be adsorbed on the fly ash as the temperature of the flue gas drops. Metals such as Al, Ca, Ce, Fe, K, Mg, Mn, Sc, Sr, and Ti show little preferential partitioning between the slag and the collected or discharged fly ash (Group I: mainly "lithophiles"); on the other hand, As, Cd, Cu, Mo, Pb, Se, and Tl were quite concentrated in fly ash compared to the slag, and were more concentrated in the ash discharged through the stack than in that collected by the precipitator (Group II: mainly "chalcophiles"); most Hg, S, Cl, and Br was discharged to the atmosphere as gases (Lindberg et al., 1975). The relative distribution of elements between bottom ash, fly ash, and particulates in the flue gas, obtained from a mass-balance study in a 350 MW power plant, is shown in Table 1 (Page et al., 1979): an element is enriched when the percentage of this element exceeds that of the residue itself. There is no example of element enrichment in the bottom ash; iron shows practically no partitioning between the phases; Cu and – to an extreme extent – As are enriched in the precipitator ash, whereas the other examples – Pb, Se, Hg – are progressively enriched in the flue gas.

Elevated factors of enrichment between coal and slag have been noted by Klein et al. (1975) in the range of ten for Al, Ca, Ce, Cr, Fe, K, Mg, Mn, Sc, and Si (i.e., for the "lithophilic" elements), whereas the enrichment between slag and fly ash was significant for Se (> 300), follwed by Sb (× 19), Ga (× 16), Pb(× 13), Cd, Cu, Zn (all × 7), and As (× 6). The tendency of elements As, Cd, Ga, Ge, Pb, Sn, Tl, and Zn to be highly partitioned in the various ash depositories has been confirmed by analyses using spark source

Table 1. Fraction of elements entering with coal discharged in various coal residues [after Page et al. (1979) from U.S. Environmental Protection Agency (1975)]

Element	Sluice ash (22.2%) %	Precipitator ash (77.1%) %	Flue gas (0.7%) %
Iron	27.9	71.3	0.8
Copper	12.7	86.5	0.8
Arsenic	0.8	99.1	0.05
Lead	10.3	82.2	7.5
Selenium	1.4	60.9	27.7
Mercury	2.1	0.1	97.9

Table 2. Boiling points of possible inorganic species envolved during coal combustion (from Davison et al., 1974)

Species boiling or subliming <1 550° C	Species boiling or subliming >1 550° C
As, As_2O_5, As_2O_3, As_2S_3	Al, Al_2O_3
Ba	Be, BeO
Bi	Bi_2O_3
Ca	C
Cd, CdO, CdS	CaO
$Cr(CO)_6$, $CrCl_3$, CrS (1 550° C)	Co, CoO, CoS
K	Cr, Cr_2O_3
Mg	Cu, CuO
$Ni(CO)_4$	Fe, Fe_2O_3, Fe_3O_4, FeO
Pb	MgO, MgS
Rb	Mn, MnO, MnO_2
S	Ni, NiO
Se, SeO_2, SeO_3	Si, SiO_2
Sb, Sb_2S_3, Sb_2O_3 (1 550° C)	Sn, SnO_2
SnS	Ti, TiO_2, TiO
Sr	U, UO_2
Tl, Tl_2O, Tl_2O_3	
Zn, ZnS	

mass spectroscopy (SSMS) (Conzemius et al., 1984). In sediment cores, e.g., from southern Lake Michigan, the inputs of these "enriched elements" characteristically covary with increased numbers of magnetic spherules of fly ash and with the fluxes of charcoals whose morphologies and surface characteristics are indicative of different combustion processes – oil, coal, and wood burning (Goldberg et al., 1981).

The extent of fractionation between the various "phases" and the elemental composition of fly ash depends on the composition of the parent coal, conditions during coal combustion, and thermochemical properties of the ash (Quann et al., 1982). The temperature of the coal-combustion zone is of prime significance (Table 2 from Natusch and Taylor (1980), after Davison et al. (1974)). Brumsack et al. (1984) have stressed that the "uncoupling of mass- and element mass-flow" at higher temperature during "wet-bottom" firing (up to 1650 °C in cyclone furnaces) with fly-ash recycling is responsible for the high accumulation of volatile elements in fly ash and concomitant stack emissions.

Particle-size Differentiation of Elements

Preferential concentration in smaller particles is particularly obvious for the elements of Class II (mainly comprising the "chalcophilic" elements), whereas for the "matrix" elements (Class I) and elements such as Ag, Au, Be, Mn, Ni, U, and Zn the inverse concentration/particle-size dependence is less well established. A group of elements which includes Ca, Sr, Y, and the rare earths show small changes in their concentration profiles with a maximum concentration at approximately 5 μm (Campbell et al., 1978); probably the latter elements are related to accessory minerals such as apatite (Campbell et al., 1978; Hansen and Fisher, 1980). X-ray fluorescence analyses on 29 elements in a fly-ash sample which was separated into 17 well-defined size fractions (Smith et al., 1979) indicate that the concentrations of volatile trace elements increase as particle size decreases in the 1–10 μm size range. For submicron particles the concentration becomes independent of particle size (Fig. 1). Emission-control devices such as mechanical dust collectors (e.g., cyclone filters, bag filters), electrostatic precipitators, scrubbers, etc., remove larger fly-ash particles but are less efficient for vapors and finer particles, which enter the terrestrial and aquatic environ-

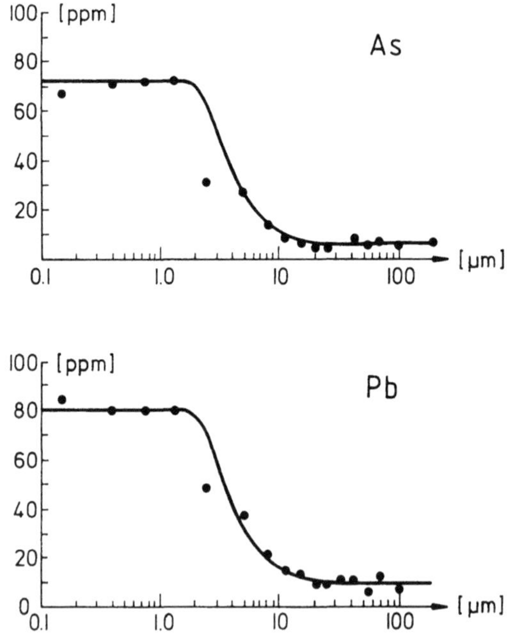

Fig. 1. Average concentrations of arsenic and lead vs. particle size of coal fly ash (Smith et al., 1979)

ment by wet or dry deposition. A comparison of a Venturi wet-scrubber system with a cold-side electrostatic precipitator – both with an overall comparable removal efficiency – showed that the concentrations of elements such as As, Br, Cr, Se, Sb, U, and V were up to 170 times greater in aerosols emitted from the scrubber system than in aerosols from the electrostatic precipitator system (Ondov et al., 1979).

Recent investigations on the fly-ash properties from an experimental atmospheric pressure fluidized-bed combustor (FBC) (Weissman et al., 1983) showed that in these emissions spherical particles will rarely be found, due to the lower combustion temperature of FBC (800–1000 °C) compared to that of conventional, pulverized-coal combustors (PCC). It is suggested that the difference in morphology may affect their relative toxicity. While ashes of the two combustion technologies do not differ greatly in bulk elemental concentration, the FBC ash showed less size-dependence of elements than has been reported for PCC ash. An extrapolation of these data to calculations of the emission rates for full-scale FBCs is not yet possible (Weissman et al., 1983).

Surface Concentration of Element Species

Direct information concerning trace-element associations, e.g., from electron spectroscopy, is as yet not available. However, elemental surface predominance was indicated from microprobe depth profiles (SIMS) on fly ashes for Fe, Na, Li, K, S, Pb, Tl, Mn, Cr, and V (Linton et al., 1976); the findings were confirmed by surface analytical techniques (e.g., AES) and solvent leaching of the fly-ash particles (see below; Natusch and Taylor, 1980). Both FT-IR analysis and time-resolved solvent leaching (TRSL) provided evidence for the presence of Cd, Cr, Cu, Mo, and Ni sulfates in coal fly ash (Natusch and Taylor, 1980). Combination of thermal treatment with acid evolution procedures suggests the presence of carbonate species for Li, Ni, K, Rb, Mg, Sr, Ba, and Cu in fly-ash samples; because flue gas is highly concentrated with CO_2 and H_2O a surface reaction is quite likely (Natusch and Taylor, 1980). Conditions for direct determinations of metal compounds are particularly favorable in oil fly ashes, which contain approximately 10% vanadium and 2–3% nickel. By means of X-ray diffraction, FT-IR, and chemical-phase or valence-state analysis the presence of V_2O_5 and $VSO_4 \cdot 2H_2O$ was identified in total and water-soluble fractions of oil-fired fly-ash samples (Henry and Knapp, 1980). Eatough et al. (1984) identified VO_2^+ as the principal vanadium species in fly-ash samples by means of combined calorimetric, ion chromatography (IC) and PIXE analysis; these results are particularly remarkable since the VO_2^+-cation – presumably as $(VO_2)MSO_4$ or $(VO_2)_2SO_4$ species in acidic oil fly ashes – is highly toxic.

In using solvent-leaching techniques one must be aware that the aqueous leachates significantly depend on four factors (Natusch and Taylor, 1980): i) the composition of the fly ash, ii) the ratio of fly ash to volume of solution, iii) the type of contact between the aqueous and solid phases, and iv) the chemical characteristics of the leaching solutions.

Filtration studies using nuclepore membranes in eight pore sizes from 5 μm to 0.03 μm were performed by Fisher et al. (1979) on neutron-activated fly-ash samples which were dispersed in an aqueous buffer solution at physiological pH. The elements Zr, Cs, Rb, Hf, and Ta were not solubilized; the elements Mo, Ca, Se, Ba, Na, and Co were partially solubilized being dependent on membrane-pore size. A combination of partial solubility and pore-size dependency was observed for Sb, As, Zn, W, and Cr. For the latter elements a relatively large proportion of the mass is associated with particles of <0.4 μm. A combined approach to quantify the relative concentration of elements in the aluminosilicate matrix and in nonmatrix or surface material of coal fly ash was proposed by Hansen and Fisher (1980). Particle size (via surface-area functions) and acid solubility (with 0.029 to 0.57 M hydrofluoric acid) of the elements were normalized to the percentage of nonmatrix fraction and the two sets of data were compared. Major differences between the nonmatrix concentrations were found for Ca, Ni, Sr, Sc, La, and the rare earths (too low size-dependence factor) and for Pb, Ba, Ga, and W (too low solubility dependence). The former effect can possibly be explained by the presence of an accessory apatite phase (Davison et al., 1974), whereas the latter anomalies are probably caused by insoluble phases either present in the ash (WO_3) or formed during the extraction (Hansen and Fisher, 1980). The results indicate that more than 70% of the Ti, Na, K, Mg, Hf, Th, and Fe is associated with the aluminosilicate matrix; more than 70% of the As, Se, Mo, Zn, Cd, W, V, U, and Sb are associated with surface material on the ash particles; the elements Mn, Be, Cr, Cu, Co, Ga, Ba, and Pb are intermediate in behavior, i.e., are distributed about equally between matrix and nonmatrix material (Hansen and Fisher, 1980).

Leaching of fly-ash samples with various reagents after differentiation into size classes indicates typical relations regarding anion/cation composition and oxidation states of surface-enriched elements (Hansen et al., 1984): surface concentrations of major anions other than silicate, i.e., fluoride, phosphate, and sulfate, as well as of cationic metals, i.e., Zn, Cd, Co, Cu, and Cr suggest the existence of typical combinations between these constituents. Following the basic findings of a mixture of arsenic(V) and arsenic(III) on the surface of the ash particles, the probable states of the surface-associated trace elements are predicted to be (Hansen et al., 1984): Mo(VI) as molybdates and MoO_3, W(VI) as tungstates; WO_3, V as vanadate(V) and vanadyl(IV); U(V) and U(VI) as U_3O_8 and uranates; Se and Se^0 and selenites; and Cu(II), Cr(III), Co(II), Mn(II), and Mn(III) as fluorides, sulfates, and phosphates.

A summary of the combined data of solvent-leaching experiments and surface microanalysis has been given by Natusch and Taylor (1980). Typical elements can be categorized into four general groups (Table 3). The group A elements (Si and Al) are major constituents of the glassy fly-ash particle matrix exhibiting low leachabilities; due to its close association with complex silicates Ti is incorporated as a minor component in the fused fly ash. The surface predominance observed for the Group B and C elements apparently is related to their typical behavior during volatilization and subsequent deposition (Davison et al., 1974). However, the higher leachabilities in the Group B elements coupled with the relatively minor changes in observed surface concentration could well represent an artifact due to surface migration of Na, Li, K, and S under ion bombardment (Natusch and Taylor, 1980). The Group D elements (Ca, Mg) exhibit similar behavior to that of the Group A elements. The alkaline earths are often present as carbonates in coal and subsequently decompose to form oxides during combustion which may be transferred into quite soluble sulfates.

The effects of different combustion conditions, coal types, and fly-ash collection devices on the chemical nature of the ash were investigated by Wu et al. (1980), using a geochemical fractionation scheme which differentiated water-soluble, acetic acid leachable, easily reducible, H_2O_2-extractable, moderately reducible and residual fractions (see below). The "leach-

Table 3. Summary of results – surface characterization (Natusch and Taylor, 1980)

Elements	Surface concentration	Effect of leaching in the immediate surface region	Leachability
A) Al, T, Si	No	Insignificant	Low
B) Fe, Na, Li, K, S	Yes	Insignificant	Moderate-high
C) Pb, Tl, Mn, Cr, V	Yes	Depletion after solvent leaching	Moderate-high
D) Mg, Ca	No	Insignificant	Moderate

able fraction," comprising metals soluble in water or being ion-exchangeable, together with those associated with carbonates and amorphous oxides, was generally highest for boron, cadmium, vanadium, selenium, and arsenic, followed by chromium, copper, nickel, lead, and zinc, whereas lead, iron, and aluminum were leached to a much lesser extent by water, 1 M acetic acid, and a mixture of 0.1 M hydroxyl-amine hydrochloride and 0.01 M nitric acid ("easily reducible fraction"). Up to 30–50% of the non-residual concentrations of Cu, Zn, V, and Ni were found in the H_2O_2-extractable (organic/sulfidic) fraction. Incomplete combustion, e.g., at larger coal-particle size (stoker furnace) and lower combustion temperature (lower heat content of coal, fluidized-bed combustion) may result in higher leachable trace metals from ash, mainly in the acetic acid-extractable and H_2O_2-extractable phases – probably associated with unburned carbonate and organic matter in the ash (Wu et al., 1980).

Complex Mixtures of Atmospheric Particulate Species

To facilitate the discussion, the contributions of typical emissions of mercury and lead are compared (e.g., Greenberg et al., 1981) (see Table 4).

It is clear that incineration of sewage sludge contributes only minor emissions of both lead and mercury, whereas the major inputs of lead can be expected from combustion of urban refuse. Municipal incinerators are responsible for major fractions of masses of Zn, Cd, Sb, Sn, In, and Ag found on aerosols in many cities (Greenberg et al., 1978). Admixtures of municipal refuse to coal may increase lead emissions, but elevated levels of Zn, Cd, Sb, and Ag have also been found in refuse-derived fuel (RDF) plants (Greenberg et al., 1981; Taylor et al., 1982); it has been suggested that printing inks are sources of lead and zinc, paints sources of titanium, chromium, and lead, and plastic stabilizers are contributing tin and cadmium to considerably higher emissions of these metal contaminants from

Table 4. Emissions (per person and day) of mercury and lead from different combustion sources (Greenberg et al., 1981)

	Coal-fired power plant	Municipal incinerator	Seawage-sludge incinerator	Gasoline exhaust
Mercury	2300 μg	600 μg	150 μg	–
Lead	3000 μg	240000 μg	6 μg	1000000 μg

refuse-derived fuel plants. Discrimination of the inputs from the two major sources of fly-ash particles – coal-fired power plants and municipal waste incinerators – can be tried on the basis of typical elements, which are highly concentrated on either one of these particulate emissions in mixed dry depositions, i.e., from the ratio of As or Se on the one hand, and Sn, Ag, or Cd on the other (Table 5):

Table 5. Average element concentration factors in stack fly-ash particles emitted from coal-fired power plants [dry-bottom and wet-bottom furnace (Brumsack et al., 1984)] and municipal incinerators [three examples (Greenberg et al., 1978)] compared to crustal abundance

	Arsenic	Selenium	Tin	Silver	Cadmium
Power plants	500	5000	10	500	750
Incinerators	130	750	5700	7000	7500

Solid-phase Analysis

With respect to the analytical methods applied in source discrimination of airborne emissions, major progress has been achieved from X-ray diffraction analysis of Hi-Vol filter samples from smelting operations (Foster and Lott, 1980; Harrison et al., 1981 b). In some cases identification of a chemical phase specific to a single operation was possible, e.g., for elemental Pb aerosol at the bullion floor in a primary smelter at Avonmouth, U.K. (Foster and Lott, 1980). It has been shown that ore handling is a major particulate contribution, whereas sintering operations – e.g., indicated by $PbSO_4$ – are of minor importance (Harrison et al., 1981 b). To identify lead species in street dust, samples were analyzed by X-ray diffractometry after magnetic and density-gradient separation as well as leaching experiments (Biggins and Harrison, 1980). The most frequently observed crystallized compound – XRD reflects only a minor proportion of lead in street dust – was $PbSO_4$: its presence is explained by weathering of $PbSO_4 \cdot (NH_4)_2SO_4$; reactions of airborne PbBrCl with neutral and acid sulfates may also be involved. Since lead sulfate has a significant water solubility, the fixation of ionic lead is suspected to be caused by reactions involving essentially insoluble organic material, with precipitation by carbonate and sorption by hydrous oxides apparently being of secondary importance (Biggins and Harrison, 1980).

A differentiated separation scheme was applied to characterize urban dusts, initially for microanalytical lead tracing (Linton et al., 1980), then

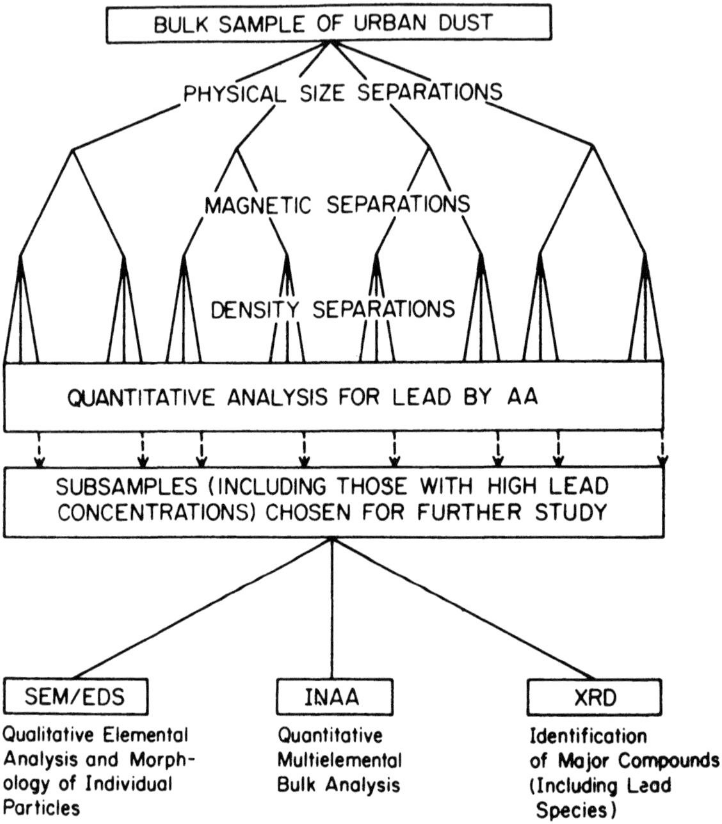

Fig. 2. Flow chart for urban-dust fractionation and lead solid speciation (Linton et al., 1980)

also to the multi-element analysis (Hopke et al., 1980) (Fig. 2). In the "lead" study (Linton et al., 1980) the identities of individual particles (automobile exhaust, paint chips) were established on the basis of characteristic morphology and chemical composition. It is shown that automobile exhaust – apart from its dominance in street-dust samples – is contributing substantially to dusts collected in the vicinity of buildings which have lead-painted trim. Surface-sensitive analytical techniques indicated that automobile-exhaust particles contain the elements Br, Cr, Cu, Mn, Ni, P, Pb, and Tl on their outer surface (Linton et al., 1980).

Element concentrations from 30 physically fractionated subsamples (each having unique particle size, density, and ferromagnetic characteristics) of a sample from the intersection of two moderately travelled (~7000 cars/day) street in Urbana, Illinois, were subjected to target transformation-factor analysis (Hopke et al., 1980). Distribution profiles, for instance, show three distinct sources of zinc; comparison with the normalized elemental concentrations for several specific sources (input test vectors) suggests that these sources are tire wear, soil, and cement. The results indicate that the composition of this roadway-dust sample is 76% soil, 5% cement, 0.3% salt, 1.5% automobile-exhaust particles, 7.7% iron (mainly rust particles), and 7.2% automobile tire-wear particles, with 2.3% of the mass unaccounted for by this analysis.

Solvent-leaching Experiments

Sequential leaching techniques, while being problematic for the assessment of "bioavailability," can be useful for the estimation of potential remobilization effects (e.g., by changes of pH, redox conditions, and salinity and increasing concentrations of organic chelators) on toxic metals from contaminated solid material, and were mainly used to classify dredged substances (e.g., for ocean dumping, upland disposal, or storage in confined water bodies). One of these procedures which has been more widely accepted than many others in the study of aquatic particulate matter (for discussion see Salomons and Förstner, 1984) was originally designed by Tessier et al. (1979):

Fraction 1: Exchangeable metals
The sediment samples was extracted for 10 min with 1 M $MgCl_2$ at pH 7.

Fraction 2: Metals bound to carbonate
The residue from fraction 1 was leached for 5 h with 1 M sodium acetate adjusted to pH 5.0 with acetic acid.

Fraction 3: Metals bound to Fe-Mn oxides
The residue from fraction 2 was extracted for 6 h at 96 °C with 0.04 M $NH_2OH \cdot HCl$ in 25% (v/v) HOAc.

Fraction 4: Metals bound to organic matter
The residue from fraction 3 was extracted at 85 °C for 5 h with 30% H_2O_2 adjusted to pH 2 with HNO_3 and then at room temperature with 3.2 M NH_4OAc in 20% HNO_3.

Fraction 5: Residual metals
The residue from fraction 4 was digested with a 5:1 mixture of hydrofluoric and perchloric acids.

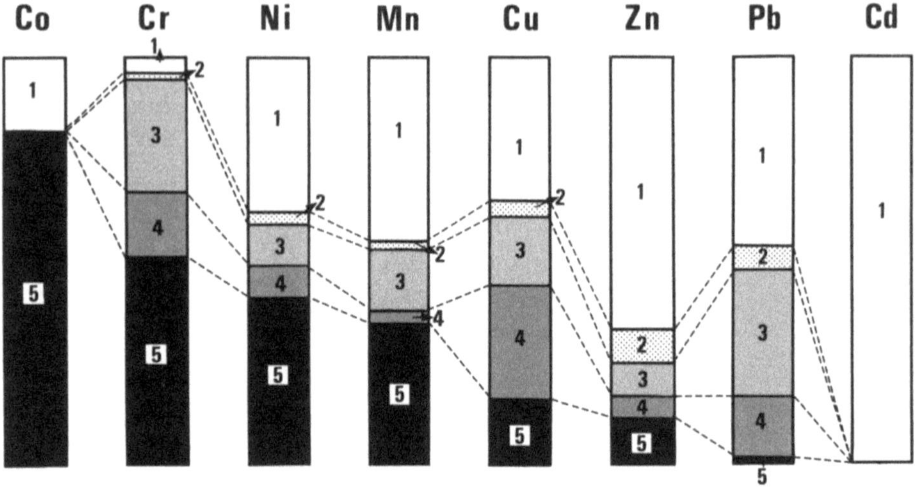

1 Exchangeable fraction, 2 Surface oxide and carbonate-bound fraction, 3 Associated with Fe-Mn oxides, 4 Organically bound fraction, 5 Residual fraction

Fig. 3. Mode of occurrence of trace metals in Standard Reference Material NBS 1648 ("urban particulate matter") (Salomons and Förstner, 1984, after data from Lum et al., 1982). For fractionation scheme see text

A practical example applying this leaching sequence to an urban-dust sample is given in Fig. 3 (Lum et al., 1982). The high proportions of Cd, Zn, Pb, and Ni in soluble and/or exchangeable forms suggest that these elements may have immediate impact on the quality of receiving waters or may be easily available for organisms in soils (Lum et al., 1982). Respective data on the fractionation of solid Cd and Pb are given in Table 6 for matrices such as street dust (A) and road-side soil (B), dredged sediment (C) and sewage sludge (D, E). A study made by Harrison et al. (1981 a) with six street-dust (A in Table 5) and four roadside-soil (B in Table 5) samples indicate that cadmium is there to an appreciable extent in the exchangeable fraction while lead is predominantly associated with carbonate and Fe-Mn oxides (copper was largely in organic associations). The high cation and chloride concentrations (1 M $MgCl_2$) in the exchange solution (Step 1) may reflect conditions in soils contaminated with deicing salt as well as transformations that might occur on mixing with seawater.

In comparison to the street-dust leaching data, the experiments on roadside soils indicate significant lower extraction rates in the more labile phases and an increase of the residual metal percentages. This is connected with the lower total-metal contents in the roadside samples compared to the dust

Table 6. Chemical extraction of cadmium and lead from different fractions of solid-waste materials. A = street dust; B = roadside soil [Harrison et al., 1981 a]; C = dredged sediment; D = fresh sewage sludge [Förstner et al., 1981]; E = incinerated sewage sludge ash [Fraser and Lum, 1982]

Fraction (Salomons and Förstner, 1984; Tessier et al., 1979)	Cadmium					Lead				
	A %	B %	C %	D %	E %	A %	B %	C %	D %	E %
Step 1	20	26	27	17	–	2	1	< 1	–	–
Step 2	38	24	63	61	–	43	26	10	11	< 1
Step 3	28	25	8	6	1	38	44	48	29	1
Step 4	8	8	2	11	10	7	12	3	16	1
Step 5	6	17	< 1	5	89	10	17	39	44	98
μg/g d.m.	4.0	2.2	8.0	1.8	10.9	1500	600	330	110	2275

specimen. On the other hand, the higher contents of cadmium and lower concentrations of lead in the dredged sediment from Hamburg harbor (sample C in Table 5) compared to the roadside-soil material seem to result in an increase of the more labile fraction of cadmium (carbonate association) and the more stable, mainly residual forms of lead. Despite the relatively low concentration of cadmium in the sewage-sludge sample (D; sample from the treatment plant at Landau/Pfalz (Förstner et al., 1981)) there is only a slight shift to more stable forms compared to the dredged sediment. Lead is typically enriched in the organic fraction of the sewage-sludge sample D, whereas the high percentage of residually bound lead is quite unexpected. The common experience that the more enriched elements are also the more reactive ones does not fit to all waste materials. This is exemplified by the data from incinerated sludge (sample E; Hamilton Municipal Incinerator), where the elements are highly enriched but rather immobile with potentially toxic metals Cd, Cu, Pb, and Zn comprising approximately 90% or more in "unavailable" forms.

Environmental Hazards from Combustion Residues

Environmental impact results from either disposal of massive quantities of fly-ash residues captured in the filter devices or from fine particulates emitted to the atmosphere and eventually reaching aquatic systems, plants, animals, and man (Bern, 1976). For the various fields typical examples were given with respect to the pathways and changes of toxic metal species from combustion-derived particles into more bioavailable forms, e.g., in lung fluids or groundwater.

Coal Fly-ash Disposal

Environmental impact considerations on coal fly-ash disposal in dicate several aspects:

i) When applied to cropland, fly ash can exhibit positive effects in neutralizing soil acidity (also with respect to aeration, water-holding capacity, hydraulic conductivity) and as an additional supply of Ca, S, B, Mo, and Se (Adriano et al., 1980). Nevertheless, because of potential problems of salinity and of elevated concentrations of boron, molybdenum, and selenium arising from application of coal ash to soils, these elements should be critically evaluated before disposal of the by-product on agricultural soils is recommended (Page et al., 1979).

ii) Ocean disposal of coal-combustion waste has been suggested as one of several disposal strategies (Lunt et al., 1977). Physical problems occur with regard to the zoobenthos because of the limiting particle-size range reflected in the low permeability and porosity of the mud substrate. Also, there is no organic content for faunistic exploitation (Bamber, 1980). Effects from metal contaminants are still considered somewhat controversial (Crecelius, 1981). Stabilization of scrubber sludges and fly ash (SSFA) with 4% quicklime reduces the rapid mobilization and dissolution of minor components such as Fe, Cu, and Ni. Blocks of SSFA could remain stable in the ocean as an artificial reef for extended periods of time provided that organism colonization does not substantially reduce their structural integrity (Seligman and Duedall, 1979).

iii) Fly ash – particularly acidic types – is toxic to aquatic ecosystems when initially slurried to disposal ponds (Suloway et al., 1983). In actual and laboratory-simulated disposal situations (Turner et al., 1982) it has been shown that elements with anionic aqueous speciation such as S, As, Se, Cr, and Mo are appreciably solubilized in ash sluicing/ponding systems that are neutral to alkaline in pH. Acidic systems appear to favor solubilization of most elements from fly ash, but especially of Fe, Mn, Cu, Zn, and Cd. Obviously there is a correspondence between those elements exhibiting elevated concentrations in the effluent waters and those most extractable in water (Table 7).

Molybdate, borate, fluoride, selenate, and possibly arsenate, chromate, and vanadate occur in soluble forms in the alkaline environments, whereas

Table 7. Trace-element concentrations in ash-pond effluent water relative to cooling lake intake and outlet [Dreesen et al., 1977]

Concentration ratio	Ash-pond effluent/ cooling lake intake	Ash-pond effluent/ cooling lake outlet
>50	B, F, Mo, Se, V	Se
10–50	As	B
2–10	–	As, Cr, F, Mo, V
< 2	(Be), Cd, Cu, Zn	(Be), Cd, Cu, Zn

the more metallic cations Be, Cd, Cu, and Zn are either precipitated after dissolution or not initially extracted from the ash (Dreesen et al., 1977).

For predictions on the formation of secondary minerals after deposition of fly ash, Mattigod (1983) determined the concentrations of Na, K, Ca, Mg, SO_4, and Si in eight successive extracts. Computation of saturation indices using the chemical equilibrium computer program GEOCHEM suggests that the precipitation-dissolution reactions of brucite, ettringite (Ca-Al-sulfate), sepiolite and zeolitic minerals (laumontite, phillipsite) may control various ion activities in the extracts (Mattigod, 1983).

If potential contaminants achieve steady-state conditions in the disposal pond they may have long residence times in the ash effluent, thus increasing the probability of bioaccumulation by aquatic organisms. The six chemical constituents most commonly accumulated from fly-ash extracts in fish were Al, B, Cd, Mn, Mo, and Ni; of these six chemical constituents Cd appeared to be of greatest importance because of its highly toxic nature (Suloway et al., 1983).

iv) Among the major effects which coal fly ash has on the groundwater environment and particularly on the release of metals are changes in pH. The change may be either basic or acidic. According to Theis and Wirth (1977) those properties which appeared to be most responsible were oxalate-extractable (amorphous) Fe-oxide and water-soluble calcium (at pH 3). Figure 4 shows the equilibrium pH as a function of the oxalate iron-to-calcium ratio; it is suggested that oxalate iron is a measure of the acid content while soluble calcium, which is associated with the lime fraction, represents the basic component of fly ash.

Extraction studies involving washing various fly ashes in distilled water at pH's of 3, 6, 9, and 12 (Theis and Wirth, 1977) indicate that desorption of metals in aqueous solutions follows a predictable pattern of decreasing

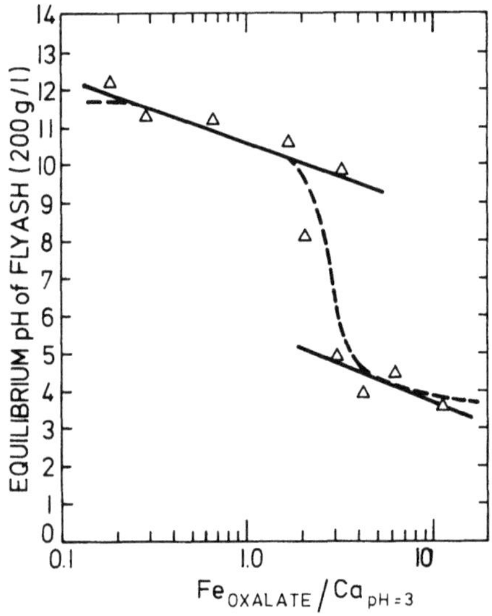

Fig. 4. pH relationship in fly-ash samples – composite pH vs. oxalate iron/soluble calcium (pH 3) ratio (after Theis and Wirth, 1977)

release with increasing pH. The U-shaped curves generated (Fig. 5) suggest a solubility-controlled reaction; the large amounts of both chloride and sulfate typically released by fly ash enhance the solubility of certain elements such as cadmium and lead by formation of inorganic complexes. The sudden increase of arsenic (possessing an anionic chemistry) release at pH 12 is probably due to the unavailability of free-metal ions to cause its precipitation, e.g., in the form of $FeAsO_4$ (Theis and Wirth, 1977).

Trace-metal migration in groundwater from ash-pond seepage has been studied by Theis et al. (1978, 1979) on a ponding system of the Northern Indiana Public Service Company. Metals were released at low concentrations into the groundwater, depending uppon the ash-loading rates and procedures. Rapid attenuation of metal contents occurred close to the pond; most metals were effectively scavenged by iron and manganese oxides or were precipitated. These effects have been investigated from thermodynamic equilibrium data and field measurements; Fig. 6 gives examples of the metal speciation according to distance from the ash pond as calculated with the computer program MINEQL (Theis et al., 1978). For Cu the formation of the basic copper carbonate (malachite) is favored for all distances from the pond. While lead hydroxide may precipitate in the vicinity of the pond due to slightly elevated pH levels, lead carbonate does not precipitate until the groundwater inorganic-carbon levels have increased substantially. Nickel is relatively nonreactive and does not adsorb readily onto oxide

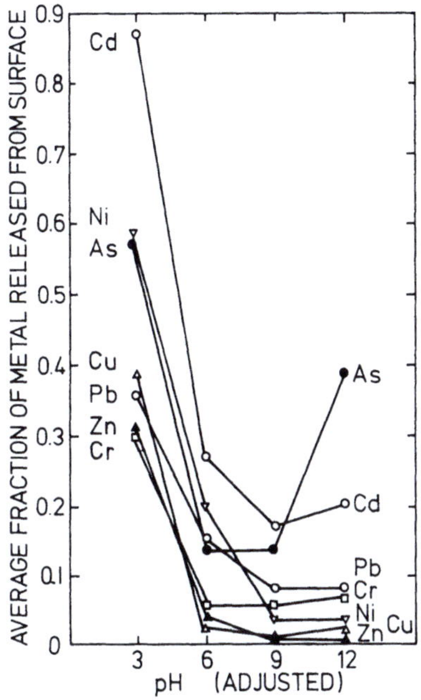

Fig. 5. Average release of heavy metals from fly ash in water as a function of surface-available metals vs. pH (after Theis and Wirth, 1977)

minerals until pH values of 8–9 are reached. For zinc conditions favor precipitation as silicate throughout the seepage zone. It has been shown that sudden increases in ash-metal concentrations result in unfavorable conditions for the oxides to exert control. This could possibly be regulated by adding lime or limestone to raise the pH (Theis et al., 1978). Mobility of arsenic in landfills is affected not only by initial leaching but also by adsorption, reactions mediated by redox and pH conditions, and biological activity (Blakey, 1984). Since trivalent arsenic is likely to be the predominant As species in ash pore water and groundwater and is also the more toxic form in water, it has been stressed by Turner (1981) that the interaction of As(III) with soils and landfill-liner material should be more closely examined to ensure the protection of drinking-water aquifers in critical areas.

Atmospheric Depositions in Typical Environments

In the marine environment surface microlayers are characteristically enriched in trace metals and organic matter compared with bulk seawater, due to inputs from the water column by rising bubbles (flotation) and from the atmosphere (Theis et al., 1978). From model calculations Lion and Leckie

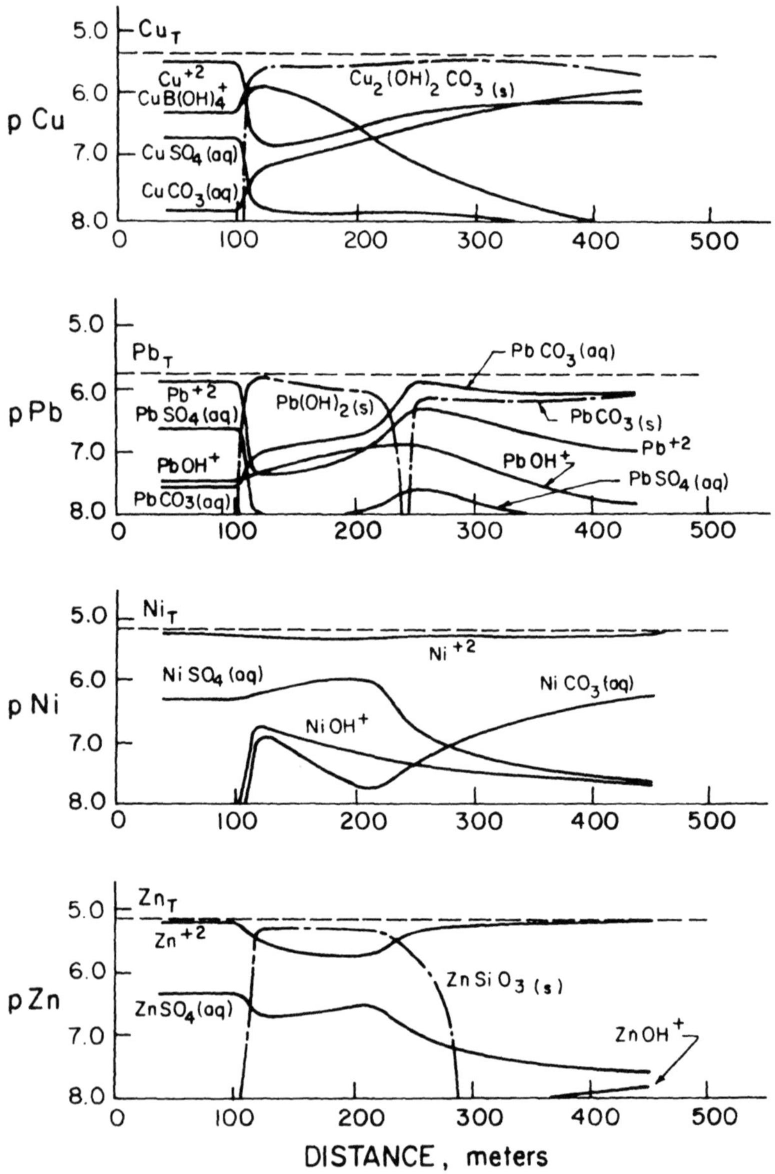

Fig. 6. Dominant inorganic complexes for copper, lead, nickel, and zinc as determined by equilibrium calculations for a coal fly-ash pond/groundwater system (Theis et al., 1978)

(1981) have suggested that trace-metal species in aerosols will undergo significant changes at the point of contact with sea-surface microlayer: while Cd and Hg are strongly complexed by chloride, copper and lead more effectively compete for dissolved complexing agents.

In the terrestrial environment trace metals introduced by both dry and wet deposition will interact with the vegetation. Some metals appear to be taken up through the leaves, whereas others are leached from leaves (foliar leaching); these processes are promoted by simultaneous decreases in pH. It was shown by Lindberg and Harris (1980) that relative solubility increases with decreasing particle size of combustion products. However, the physiological effects of surface-deposited materials on vegetation either in particulate or dissolved forms are still poorly understood (Lindberg and MacLaughlin, 1983).

In the human environment the smallest particles (<1 μm), which deposit in the pulmonary region of the respiratory tract, are of greatest concern. Toxic species which predominate in submicrometer-sized particles come into intimate contact with body fluids and tissue and thereby enter the bloodstream (Cabaniss and Linton, 1984; Natusch et al., 1974). Comparative leaching experiments on coal fly ash, including (physiological) pH 7.4 solutions of chelating agents and canine blood serum (used instead of alveolar serum, whose composition is not well established), indicate that the serum is more effective as a leaching agent than one would predict on the basis of its concentrations of citrate and histidine (Harris and Silberman, 1983). It has been suggested that other biological chelators, possibly cysteine in albumin and other serum proteins, are important leaching agents, particularly for removal of Zn, V, Cu, and Fe from fly-ash particles *in vivo*. The time-dependent drop in leaching rates probably reflect the depletion of the surface concentrations of soluble metal compounds such as sulfates, halides, and phosphates in fly ash (Harris and Silberman, 1983). Since it has been shown that the respirable-size fraction of a particular set of coal fly-ash samples contained at least two components which are mutagenic in the Ames test – one comprising of organic compounds while the other appears to be inorganic (Hansen et al., 1984) – further research should focus mainly on the recognition of these toxic chemical species.

Acknowledgement. W. Salomons from the Delft Hydraulics Laboratory kindly provided material for this review.

References

Adriano DC, Page AL, Elseewi AA, Chang AC, Straughan I (1980) Utilization and disposal of fly ash and other coal residues in terrestrial ecosystems: A review. J Envir Qual 9:333–344

Andreae MO et al. (1984) Changing biogeochemical cycles. In: Nriagu JO (ed) Changing Metal Cycles and Human Health. Dahlem Konferenzen. Berlin, Heidelberg, New York, Tokyo: Springer-Verlag, pp 359–373.

Bakker WT (ed) (1982) Workshop Proceedings: Research and development needs for use of fly ash in cement and concrete. Electric Power Research Institute, Spec. Report CS-2616-SR, Sept 1982

Bamber RN (1980) Properties of fly ash as a marine sediment. Mar Poll Bull 11:323–326

Bern J (1976) Residues from power generation: Processing, recycling, and disposal. In: Land Application of Waste Materials, pp 226–248. Ankeny, IA: Soil Conservation Society of America

Biggins PDE, Harrison RM (1980) Chemical speciation of lead compounds in street dusts. Envir Sci Technol 14:336–339

Blakey NC (1984) Behavior of arsenical wastes co-disposed with domestic solid wastes. J Wat Poll Control Fed 56:69–75

Brumsack H, Heinrichs H, Lange H (1984) West German coal power plants as sources of potentially toxic emissions. Envir Technol Lett 5:7–22

Cabaniss GE, Linton RW (1984) Characterization of surface species on coal combustion particles by X-ray photoelectron spectroscopy in concert with ion sputtering and thermal desorption. Envir Sci Technol 18:271–275

Calmano W, Förstner U (1983) Chemical extraction of heavy metals in polluted river sediments in Central Europe. Sci Total Envir 28:77–90

Campbell JA, Laud JC, Nielson KK, Smith R (1978) Separation and chemical characterization of finely-sized fly-ash particles. Anal Chem 59:1032–1040

Conzemius RJ, Welcomer TD, Svec HJ (1984) Elemental partitioning in ash depositories and material balances for a coal burning facility by spark source mass spectrometry. Envir Sci Technol 18:12–18

Crecelius EA (1981) Coal fly ash disposal in the Ocean: An Alternative Worth Considering. Report prepared for the U.S. Department of Energy, Contr. DE-AC06-76RLO 1830. Richland, WA: Pacific Northwest Laboratory

Davison RL, Natusch DFS, Wallace JR, Evans CA Jr (1974) Trace elements in fly ash – dependence of concentration on particle size. Envir Sci Technol 8:1107–1113

Dossis P, Warren LJ (1980) Distribution of heavy metals between the minerals and organic debris in a contaminated marine sediment. In: Baker RA (ed) Contaminants and Sediments, vol 1. Ann Arbor: Science Publ, pp 119–138.

Dreesen DR, Gladney ES, Owens JW, Perkins BL, Wienke CL, Wangen LE (1977) Comparison of levels of trace elements extracted from fly ash and levels found in effluent waters from a coal-fired power plant. Envir Sci Technol 11:1017–1019

Eatough DJ, Eatough NL, Hill MW, Mangelson NF, Hansen LD (1984) Identification of VO_2^+ in particles from the flue lines of oil-fired power plants. Env Sci Technol 18:124–126

Farmer ME, Linton RW (1984) Correlative surface analysis studies of environmental particles. Envir Sci Technol 18:319–326
Fisher GL, Silberman D, Prentice BA, Heft RE, Ondov JM (1979) Filtration studies with neutron-activated coal fly ash. Envir Sci Technol 13:689–693
Förstner U, Calmano W, Conradt K, Jaksch H, Schimkus C, Schoer J (1981) Chemical speciation of heavy metals in solid waste materials (sewage sludge, mining wastes, dredged materials, polluted sediments) by sequential extraction. Proceedings International Conference on Heavy Metals in the Environment, Amsterdam, pp 698–704
Förstner U, Müller G (1973) Heavy metal accumulation in river sediments: A response to environmental pollution. Geoforum 14:53–61
Förstner U, Wittmann G (1981) Metal Pollution in the Aquatic Environment, 2nd ed. Berlin: Springer
Foster RL, Lott PF (1980) X-ray diffractometry examination of air filters for compounds emitted by lead smelting operations. Envir Sci Technol 14:1240–1244
Fraser JL, Lum KR (1982) Availability of elements of environmental importance in incinerated sludge ash. Envir Sci Technol 17:52–54
Galloway JN, Thornton JD, Norton SA, Volchok HL, McLean RAN (1982) Trace metals in atmospheric deposition: a review and assessment. Atmos Envir 16:1677–1700
Goldberg ED, Hodge VF, Griffin JJ, Koide M, Edgington DN (1981) Impact of fossil fuel combustion on the sediments of Lake Michigan. Envir Sci Technol 15:466–471
Greenberg RR, Gordon GE, Zoller WH, Jacko RB, Neuendorf DW, Yost KJ (1978) Composition of particles emitted from the Nicosia municipal incinerator. Envir Sci Technol 12:1329–1332
Greenberg RR, Zoller WH, Gordon GE (1981) Atmospheric emissions of elements on particles from the Parkway sewage-sludge incinerator. Envir Sci Technol 15:64–70
Ham RK, Anderson MA, Stanforth R, Stegmann R (1979) The development of a leaching test for industrial wastes. U.S. EPA. EPA-600/2-79-109, May 1979
Hansen LD, Fisher GL (1980) Elemental distribution in coal fly ash particles. Envir Sci Technol 14:1111–1117
Hansen LD, Silberman D, Fisher GL (1981) Crystalline components of stack-collected, size fractionated coal fly ash. Envir Sci Technol 15:1057–1062
Hansen LD, Silberman D, Fisher GL, Eatough DJ (1984) Chemical speciation of elements in stack-collected, respirable-size, coal fly ash. Envir Sci Technol 18:181–186
Harris WR, Silberman D (1983) Time-dependent leaching of coal fly ash by chelating agents. Envir Sci Technol 17:139–145
Harrison RM, Laxen DPH, Wilson SJ (1981 a) Chemical associations of lead, cadmium, copper, and zinc in street dust and roadside soils. Envir Sci Technol 15:1378–1383
Harrison RM, Williams CR, O'Neill IK (1981 b) Characterization of airborne heavy metals within a primary zinc-lead smelting works. Envir Sci Technol 15:1197–1204
Henry WM, Knapp KT (1980) Compound forms of fossil fuel fly ash emissions. Envir Sci Technol 14:450–456

Hopke PK, Lamb RE, Natusch DFS (1980) Multi-elemental characterization of urban roadway dust. Envir Sci Technol 14:164–172
Kautz K (1984) Mineralogical aspects of the combustion of hard coal in power plants – from coal to fly ash. Fortschr Miner (Stuttgart) 62:51–72
Keyser TR, Natusch DFS, Evans CA Jr, Linton RW (1978) Characterizing the surface of environmental particles. Envir Sci Technol 12:768–773
Klein DH, Andren AW, Carter JA, Emery JF, Feldman C, Fulkerson W, Lyon WS, Ogle JC, Talmi Y, Van Hook RI, Bolton N (1975) Pathways of thirty-seven trace elements through coal-fired power plant. Envir Sci Technol 10:973–979
Lantzy RJ, MacKenzie FT (1979) Atmospheric trace metals: global cycles and assessment of man's impact. Geochim Cosmochim Acta 43:511–525
Lindberg SE, Andren AW, Raridon RJ, Fulkerson W (1975) Mass balance of trace elements in Walker Branch Watershed: Relation to coal-fired steam plants. Envir Health Persp 12:9–18
Lindberg SE, Harriss RC (1980) Emissions from coal combustion: use of aerosol solubility in hazard assessment. In: Singh JJ, Deepak A (eds) Environmental and Climatic Impact of Coal Utilization. New York: Academic Press
Lindberg SE, MacLaughlin SB (1983) Air pollutant interactions with vegetation: research needs in data aquisition and interpretation. In: Krupa SV, Legge AH (eds) Air Pollutants and Their Effect on Terrestrial Ecosystems. New York: Wiley
Linton RW, Loh A, Natusch DFS, Evans CA Jr, Williams P (1976) Surface predominance of trace elements in airborne particles. Science 191:852–854
Linton RW, Natusch DFS, Solomon RL, Evans CA Jr (1980) Physicochemical approach to lead tracing. Envir Sci Technol 14:159–164
Lion LW, Leckie JO (1981) The biochemistry of the air-sea interface. Ann Rev Earth Planet Sci 9:449–486
Lum KR, Betteridge JS, Macdonald RR (1982) The potential availability of P, Al, Cd, Co, Cr, Cu, Fe, Mn, Ni, Pb and Zn in urban particulate matter. Envir Technol Lett 3:57–62
Lunt RR, Cooper CB, Johnson SL, Oberholtzer JE, Schimke GR, Watson WI (1977) An Evaluation of the Disposal of Flue Gas Desulfurization Wastes in Mines and in the Ocean, Initial Assessment. EPA Rept. 600/7-77-051. Research Triangle Park, NC
Mattigod SV (1983) Chemical composition of aqueous extracts of fly ash: Ionic speciation as a controlling factor. Envir Technol Lett 4:485–490
Natusch DFS, Taylor DR (1980) Environmental effects of western coal combustion: Part IV – Chemical and Physical Characterization of Coal Fly Ash. Final Report. Contract No. R803950, U.S. EPA 600/3-80-093
Natusch DFS, Wallace JR, Evans CA Jr (1974) Toxic trace elements: Preferential concentration in respirable particles. Science 183:202–204
Nikiforova EM, Smirnova RS (1975) Metal technophility and lead technogenic migration. Abstract. International Conference on Heavy Metals in the Environment, Toronto, C-94/96
Ondov JM, Ragini RC, Biermann AH (1979) Elemental emissions from a coal-fired power plant. Comparison of a Venturi wet scrubber system with a cold-side electrostatic precipitator. Envir Sci Technol 13:598–607

Page AL, Elseewi AA, Straughan IR (1979) Physical and chemical properties of fly ash from coal-fired power plants with reference to environmental impact. Residues Rev 71:83–120
Quann RJ, Neville M, Janghorbani M, Mims CA, Sarofim AF (1982) Mineral matter and trace-element vaporization in a laboratory-pulverized coal combustion system. Envir Sci Technol 16:776–781
Salomons W, Förstner U (1980) Trace metal analysis on polluted sediments. II. Evaluation of environmental impact. Env Technol Lett 1:506–517
Salomons W, Förstner U (1984) Metals in the Hydrocycle. Berlin: Springer
Seligman JD, Duedall IW (1979) Chemical and physical behavior of stabilized scrubber sludge and fly ash in seawater. Envir Sci Technol 13:1082–1087
Shibaoka M, Ramsden AR (1978) Microscopic investigations of the behaviour of inorganic material in coal during combustion. In: Bryers R (ed) Ash Deposits and Corrosion due to Impurities in Combustion Gases. London: Hemisphere Publ
Smith RD, Campbell JA, Nielson KK (1979) Concentration dependence upon particle size of volatilized elements in fly ash. Envir Sci Technol 13:553–558
Suloway JJ, Roy WR, Skelly TM, Dickerson DR, Schuller RM, Griffin RA (1983) Chemical and toxicological properties of coal fly ash. Illinois Dept of Energy and Natural Resources, Environmental Geology Notes 105
Taylor DR, Tompkins MA, Kirton SE, Mauney T, Natusch DFS (1982) Analysis of fly ash produced from combustion of refuse-derived fuel and coal mixtures. Envir Sci Technol 16:148–154
Tessier A, Campbell PGC, Bisson M (1979) Sequential extraction procedure for the speciation of particulate trace metals. Anal Chem 51:844–851
Theis TL, Padgett LE (1983) Factors affecting the release of trace metals from municipal sludge ashes. J Wat Poll Control Fed 55:1271–1279
Theis TL, Richter RO (1979) Chemical Speciation of Heavy Metals in Power Plant Ash Pond Leachate. Proceedings of the 33rd Industrial Waste Conference, Purdue University, pp 487–496. Ann Arbor, MI: Ann Arbor Science Publ
Theis TL, Westrick JD, Hsu CL, Marley JJ (1978) Field investigation of trace metals in groundwater from fly ash disposal. J Wat Poll Control Fed 50:2457–2469
Theis TL, Wirth JL (1977) Sorptive behavior of trace metals on fly ash in aqueous systems. Envir Sci Technol 11:1096–1100
Turner RR (1981) Oxidation state of arsenic in coal ash leachate. Envir Sci Technol 15:1062–1066
Turner RR, Lowry P, Levin M, Lindberg SE, Tamura T (1982) Leachability and aqueous speciation of selected trace constituents of coal fly ash. Final Report, Research Project 1061-1/EA-2588, Electric Power Research Institute, EPRI-EA-2588
U.S. Environmental Protection Agency (1975) Coal fired power plant, trace element study, vol 1. A Three Station Comparison. Rocky Mountain-Prairie Region. Region VIII, Denver, CO
Weissman SH, Carpenter RL, Newton GJ (1983) Respirable aerosols from fluidized bed coal combustion. 3. Elemental composition of fly ash. Envir Sci Technol 17:65–71
Wood JM, Wang H-K (1983) Microbial resistance to heavy metals. Envir Sci Technol 17:582A–590A
Wu EJ, Choi WW, Chen KY (1980) Chemical affiliation of trace metals in coal ash. AIChE Symp Ser 210(76):177–187

The Importance of Chemical "Speciation" in Environmental Processes,
eds. M. Bernhard, F. E. Brinckman, and P. J. Sadler, pp. 493–512.
Dahlem Konferenzen 1986. Springer-Verlag Berlin, Heidelberg
© Dr. S. Bernhard, Dahlem Konferenzen

Microbial Intervention in Trace Element-containing Industrial Process Streams and Waste Products

G. J. Olson

Chemical and Biodegradation Processes Group
National Bureau of Standards
Washington, DC 20234, USA

Abstract. Microorganisms are important agents in solubilization, precipitation, accumulation, and alkylation-dealkylation reactions involving heavy elements in environments associated with industrial process streams and wastes. Such microbial processes may be harmful or beneficial. Microbial resistance to toxic heavy elements often involves metabolic mechanisms causing chemical species transformation. With certain bacteria heavy elements may serve as metabolic energy sources. The presence of chemical species of trace elements in these environments is critical for understanding the mechanisms of microbial heavy-element transformations and optimizing or inhibiting these processes for industrial application and environmental assessment.

Introduction

This paper considers reactivity and distribution of chemical species in systems under stress, and some heavy-element-containing industrial process streams and waste products create unusual gradients of pH and heavy metals. Although these environments are stressed when compared to clean systems, the stress selects for microbial populations which grow under contaminated conditions. Often these microorganisms have heavy-element resistance mechanisms which are transmitted through the populations by extrachromosomal genetic elements, which may cause species transformations of toxic elements. In other cases heavy elements serve as cellular energy sources or electron acceptors and are oxidized or reduced. Extreme environments have selected for organisms with unusual physiology. The very existence of such microorganisms depends on extremes of pH, temperature, or metal content, and the organism's metabolism may help to maintain or propagate such conditions. An example is *Thiobacillus ferrooxidans,* an iron-oxidizing acidophilic bacterium which accelerates metal sulfide oxidation, generating acidity and soluble metal species. This organism flourishes in acidic, metal-rich mine-drainage waters, certainly a stressed environment from our point of view.

Natural environments other than those impacted by man may present heavy-element or pH stresses, such as in areas of mineral weathering, or hot springs. These environments also select for resistant microbial populations. In some cases organisms from such environments possess properties making them potentially useful for processing industrial waste or raw materials. For instance, *Sulfolobus,* discovered in hot springs in Yellowstone National Park, is a promising organism for thermophilic industrial metal-ore bioleaching. Organisms from other "stressed" environments, such as deep oceanic hydrothermal vents, may find future applications in industrial heavy-element transformations.

Microbe – heavy-element interactions are both beneficial and detrimental to man, playing important commercial roles in metal extraction and recovery, yet generating costly acidic, metallic wastes and toxic organometallic compounds. In each of these cases species measurements play a key role in understanding the trace-element transformations that are catalyzed by microorganisms, factors influencing the rate and identity of intermediate and final products, and the underlying biochemical mechanisms. Understanding of these processes permits prediction and process optimization or prevention.

Microbial Solubilization of Heavy Elements

Coal Mining

Acidic drainage from coal and metal mines is a serious environmental pollution problem occurring worldwide. In the United States acidic drainage from active and abandoned mines pollutes thousands of miles of rivers and streams, especially those surrounding Eastern bituminous coalfields that contain high levels of iron-disulfide minerals (pyrite and marcasite).

Microorganisms are prominent factors in the production of acidic mine drainage by accelerating the oxidation of metal sulfides exposed during mining. Colmer and Hinkle (1947) provided the first conclusive evidence that microorganisms participate in the oxidation of iron and sulfur compounds to generate acidity in waters associated with bituminous coal mines. The oxidation-reduction transformation of iron species in this aquatic environment was a clue to the involvement of microorganisms, because ferrous iron is relatively chemically stable in acidic solutions. Therefore, a simple chemical mechanism could not explain the rapid oxidation of iron in samples of acidic mine water. These observations led to the discovery of bacterial involvement in pyrite oxidation and the description of a new organism, *Thiobacillus ferrooxidans* (Temple and Colmer, 1951). Singer and Stumm (1970) later showed that the rate-limiting step in the weathering of pyrite to produce acid was indeed the oxidation of ferrous ions, a reaction that bacteria accelerated 10^6 times over the abiotic rate.

Iron- and sulfur-oxidizing members of the genus *Thiobacillus* are the most commonly implicated microorganisms in the production of acidic mine drainage. The basic abiotic and biological reactions in pyrite weathering are shown below.

Pyrite oxidizes on exposure to air and water to produce acid and ferrous ions:

$$2FeS_2 + 7O_2 + 2H_2O \rightarrow 2FeSO_4 + 2H_2SO_4. \quad (1)$$

Microorganisms, especially *T. ferrooxidans*, readily oxidize ferrous iron to ferric iron at low pH:

$$4FeSO_4 + O_2 + 2H_2SO_4 \rightarrow 2Fe_2(SO_4)_3 + 2H_2O. \quad (2)$$

Reaction 2 is the rate-determining step in the overall pyrite oxidation and acidity-formation process (Singer and Stumm, 1970).

Ferric ions react with additional pyrite:

$$FeS_2 + Fe_2(SO_4)_3 \rightarrow 3FeSO_4 + 2S^0. \quad (3)$$

Ferric iron is the major oxidant of pyrite in the natural oxidation sequence, and bacteria are the major factors in regenerating the ferric iron in mine water (Singer and Stumm, 1970). The ferrous iron in Reaction 3 is oxidized by *T. ferrooxidans* and the elemental sulfur is oxidized to sulfuric acid by various thiobacilli. Without bacterial oxidation a layer of sulfur would build up on the pyrite surfaces, inhibiting further reactions. Reduced iron- and sulfur-compound oxidation serve as the metabolic energy sources for these organisms.

The overall biological pyrite-oxidation reaction is summarized:

$$4FeS_2 + 15O_2 + 2H_2O \rightarrow 2Fe_2(SO_4)_3 + 2H_2SO_4. \quad (4)$$

Abiotic reactions between ferric sulfate and oxygen and water produce additional acidity and various ferric hydroxy-compounds which coat and color the banks and bottoms of streams receiving acidic drainage. Some of these products may also coat pyrite crystals, interfering with biological oxidation. *T. ferrooxidans* also is capable of reducing ferric ions when growing on elemental sulfur. This reaction may be of significance where oxygen is limited.

In addition to accelerating chemical attack on pyrite by generating ferric iron at low pH, *T. ferrooxidans* attaches to specific sites on the surface of pyrite crystals where it enzymatically attacks the solid substrate, forming holes or pits on the surface (Bennett and Tributsch, 1978). The distribution of bacteria on the surface seems to be dependent on the pyrite crystal structure. *T. ferrooxidans* may selectively attach at defects in the crystal structure where energy (S, Fe) extraction is more favorable (Bennett and Tributsch, 1978). Andrews and Maczuga (1982) found that bacterial degradation of pyrite does not follow the expected stoichiometry (Reaction 4), rather that sulfur is released preferentially. They suggested that since cells preferentially attach to definite sites on the pyrite surfaces, and since diffusion of atoms in crystalline solids is much faster along grain boundaries and dislocations, thiobacilli attached at such locations would have a better supply of sulfur substrate as they dissolve their way into the solid pyrite. However, it is not yet certain whether pyrite attack is site-selective (involving chemotaxis) or if bacteria attach randomly but only grow at favorable crystal-surface sites (Brierley, 1978). Thus, the chemical species on the surface of metal-sulfide particles are important to degradability, affecting weathering and acidity formation in the mining environment.

Iron sulfides also occur in anoxic soils and sediments, and upon exposure (i.e., following reclamation of swamps or coastal areas) microbial pyrite oxidation occurs. This process lowers soil pH and mobilizes metals, especially Al, causing agricultural problems (Bloomfield and Coulter, 1973).

Bacterial attack on pyrite has generally been quantitated by the determination of total solubilized iron. Studies on the complicated distribution pattern of iron species leached from pyrite by *T. ferrooxidans* have just begun (Vuorinen et al., 1983). Varying levels of dissolved and complexed species of ferrous and ferric iron are released from pyrite by microbial action, with the species distribution pattern varying depending on cultural conditions, including pH and presence of organic compounds. Since ferric iron is so important in metal-sulfide oxidation in mining environments, more studies evaluating iron distribution patterns of iron species during biological pyrite degradation are warranted (Vuorinen et al., 1983).

Uncontrolled pyrite oxidation in mining environments is, from an environmental and materials viewpoint, a harmful process. However, the controlled biological oxidation of pyritic and organic sulfur species in coal is receiving serious consideration as a potential cost-effective means of desulfurization and metals removal prior to combustion.

The removal of pyritic sulfur from coal by solubilization has received the most attention because many of the coals mined in the eastern United States contain high levels of this form of sulfur and the basic biological reactions involved in pyrite oxidation have been studied. Removal of >90% of pyritic sulfur from coal within days has been achieved in laboratory cultures of *T. ferrooxidans* or mixed natural populations (Dugan and Apel, 1978; Hoffman et al., 1981). Thermophilic bacteria have also been shown to oxidize pyrite and may offer the advantages of faster reaction rates and less chance of biological contamination (Kargi, 1982).

Much less is known about the potential for microbial degradation of organic sulfur in coal-processing streams or reactors. In some cases organic sulfur species comprise a large portion of total sulfur in coal and are present as R-S-R, R-S-S-R, and R-O-SO_3 linkages (Kargi, 1982). A significant portion of organic sulfur in bituminous coal occurs as thiophenes. Some studies on bacterial degradation of thiophenes have been done, and recent work suggests that *Thiobacillus*-like moderate thermophiles degrade organic sulfur in coal (Gokcay and Yurteri, 1983). Unfortunately, the analysis of sulfur species in many of these studies was neglected. Much more work is needed to identify pathways for microbial degradation of organic sulfur in coals. Studies of the coal organic-sulfur metabolism of thiobacilli and other bacteria present in the mining environment may reveal that both organic and pyritic sulfur species could be removed by microorganisms. The major chemical species of sulfur in feedstock coal will be important in choosing candidate organisms for potential biodesulfurization processes. Coal bioprocessing will also require measurements of key chemical species in process streams as diagnostics of metabolic processes for optimized conditions.

Metals Bioextraction

As with the exposure of pyrite during coal mining, the mining of metal ores creates special environments in which metal-solubilizing bacteria proliferate. Man has recovered metals from ores for centuries using bacterially assisted leaching reactions, but only in the past three decades has the microbial participation in the process been recognized. Bacterially assisted hydrometallurgical leaching of copper ore was responsible for 11.5%–15% of U.S. copper production in the past decade (Brierley, 1978; Lundgren and Malouf, 1983). During the current metals-market slump, such recovery processes account for a greater share of total production. Bacterially assisted leaching processes have also been used to recover uranium from mines in Canada (Brierley, 1978; Kelly et al., 1979; Lundgren and Malouf, 1983).

There is significant interest worldwide in exploring the microbial extraction of Cu, U, Au, and other heavy elements from ores as evidenced by the growing literature and number of international conferences in this area. It is apparent that significant advancements in developing microbial metals extraction processes will require interdisciplinary research and communication between microbiologists, chemists and engineers so that the most effective bioprocessing systems can be designed, operated, and monitored. The determination of chemical species will be important in investigating metal biotransformations and in process monitoring.

Most attention has been directed toward acid leaching of sulfide ores as catalyzed by iron- and sulfur-oxidizing bacteria, especially *T. ferrooxidans*. Pyrite is often found in association with metal-sulfide-ore deposits and its biologically catalyzed oxidation produces acidic ferric-sulfate solutions which are powerful oxidants. These attack metal-sulfide minerals, releasing the soluble metal (M) as shown by the generalized question:

$$MS + 2Fe^{3+} \rightarrow M^{2+} + 2Fe^{2+} + S^0. \tag{5}$$

Additionally, bacterial attachment and enzymatic attack occur with some metal sulfides producing soluble metal sulfates:

$$MS + 2O_2 \rightarrow MSO_4. \tag{6}$$

Uranium deposits occurring as oxide ores are also solubilized by ferric iron produced by microbial oxidation of pyrite associated with the deposit. Ferric iron oxidizes insoluble tetravalent uranium to soluble hexavalent species:

$$UO_2 + 2Fe^{3+} \rightarrow UO_2^{2+} + 2Fe^{2+}. \quad (7)$$

Laboratory cultures of *T. ferrooxidans* oxidize U(IV) to U(VI) (DiSpirito and Tuovinen, 1982) and Cu(I) to Cu(II) (summarized by Ehrlich, 1978), obtaining metabolic energy from the reactions. Thus, enzymatic microbial attack on reduced uranium and copper species may also occur in the mining environment. Bacterial leaching is especially valuable for leaching of U from low-grade ores (0.03%) which are uneconomical to work by conventional chemical extraction processes.

Microbial leaching of copper ores in the western United States has been the most important application of metal-extraction biotechnology. As with uranium, bioleaching of low-grade copper deposits is attractive because it permits the recovery of copper from ores and wastes which are uneconomical to mine by conventional processes. Enormous piles of waste and low-grade ($<0.5\%$ Cu) copper ore are processed by percolation of leaching solutions down through the piles. Biological degradation of ore occurs, probably by both enzymatic and ferric iron mechanisms, generating acid and soluble metals (Reactions 5, 6).

The leaching environment is complex, often having significant temperature, oxygen, and heavy-element concentration gradients, and contains a variety of acidophilic and metal-tolerant bacteria (Brierley, 1978; Kelly et al., 1979; Lundgren and Malouf, 1983). It is becoming apparent that *T. ferrooxidans* is not the only organism of importance in metals bioextraction; rather, interactions occur between different organisms in the leaching environment which are not yet completely understood. Mixed-culture reactions are often more effective than monocultures in environmental and industrial microbial transformations, and this is also true in the extraction and bioaccumulation of metals by microorganisms (Norris and Kelly, 1982).

Undoubtedly, new organisms and metal biotransformations await discovery. Additionally, the role of extreme thermophiles such as *Sulfolobus* (capable of metal-sulfide degradation) in bioleaching processes in self-heating ore dumps remains to be determined.

Toxic trace elements occur in ores; therefore, potential resistance mechanisms in ore-leaching bacteria are of importance. *T. ferrooxidans* is very sensitive to low levels of soluble silver, mercury, and molybdenum (Brierley, 1978). Mercuric ions at $\mu g\ L^{-1}$ levels inhibit growth of *T. ferrooxidans*, as measured by ferrous-iron oxidation at $g\ L^{-1}$ levels (Fig. 1).

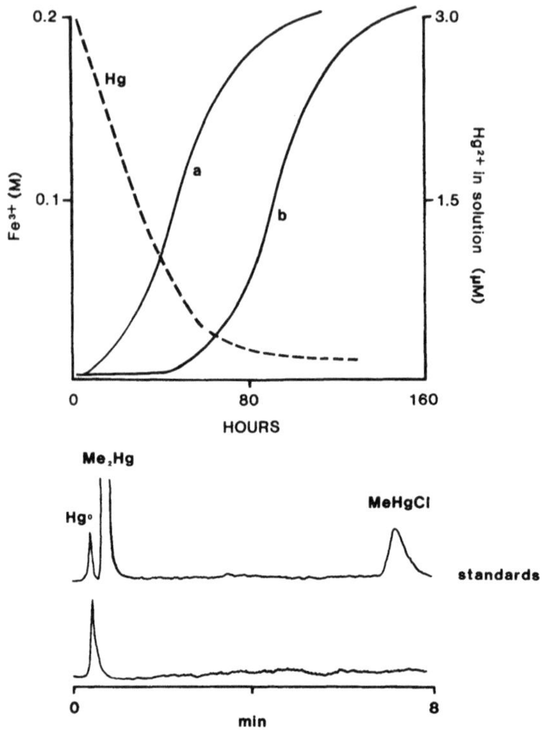

Fig. 1. Decrease in soluble-mercury concentration (*Hg*) compared to iron oxidation (*b*) by a culture of *T. ferrooxidans*. Soluble mercuric ions (3.0 µM) cause a lag in the onset of measurable iron oxidation (initially 0.2 M). Curve (*a*) shows iron oxidation in a mercury-free control. Culture headspace gas analysis showed that mercuric ions are reduced to volatile elemental mercury by the organism as shown in the bottom chromatogram (coupled gas chromatography-atomic absorption spectrophotometry)

However, some strains of this organism show enhanced mercury resistance and convert mercuric ions to volatile elemental mercury. Biochemical studies show that the mechanism of resistance is the synthesis of mercuric reductase enzyme (Olson et al., 1982; Booth and Williams, 1984). *T. thiooxidans* is quite sensitive to vanadium(V), molybdenum(VI), and chromium(VI) (Jack et al., 1980). When the reducing capacity of the culture is not exceeded, added vanadium(V) is reduced to less-toxic vanadium(IV) by *T. thiooxidans*. A process to recover V from petroleum cokes by microbially assisted acid leaching is based on this reaction which generates soluble vanadyl(IV) ions (Jack et al., 1980).

Toxic metals may inhibit the activity of these or other ore-leaching organisms, inhibiting or preventing metal extraction or coal desulfurization. As shown above, resistance to toxic metals by metabolic conversion of toxic-metal species can occur in strains of some ore-leaching bacteria. However, our knowledge of such mechanisms is very limited. Microbial populations in other metal-enriched environments can develop resistance to some toxic metals by exchange of extrachromosomal genetic material (plasmids) which confers the ability to synthesize metal-detoxifying enzymes (Silver, 1983). Virtually nothing is known about the mechanisms of toxic-metal resistance and the genetic factors involved in microorganisms from the ore-leaching environment. Plasmids have been found in *T. ferrooxidans* (Mao et al., 1980); however, their function has not yet been determined, largely because of the inherent difficulty in working with acidophilic bacteria which are difficult to grow on solid media, and the lack of suitable plasmid-curing techniques under these conditions. However, work has begun on cloning *T. ferrooxidans* plasmids into *E. coli* to facilitate genetic investigations (Holmes et al., 1984). Development of microbial resistance to toxic metals is an important issue in microbial ore-leaching development. Defining species transformations arising from toxic metal-microbe interactions coupled with genetic and biochemical studies will be needed for more progress in this area.

In addition to leaching of low-grade ores in large heaps or dumps, the development of other bioleaching techniques is likely. *In situ* bioleaching processes are especially desirable, allowing economic recovery of deep deposits and preventing some mining-related pollution problems. With *in situ* leaching, solutions are introduced by injection wells, and after reaction with the ore body the metal-containing solution is recovered by pumping from surrounding wells. Microorganisms could be employed *in situ* or could generate leaching solutions in surface ponds or vats. *In situ* bioprocessing of oil shale has been proposed (Yen and Chilingar, 1976) since certain oil shales occur in carbonate-containing rock. Microbial degradation of pyrite or sulfur produces acid which dissolves the carbonate rock matrix and releases the oil.

Another potential area of metals biotechnology is the bioleaching of ore concentrates in tanks or vats (Brierley, 1978; Kelly et al., 1979; Lundgren and Malouf, 1983). In this case more control could be exercised over processing conditions and organisms than in large, outdoor leaching systems. For example, more precise temperature control could be maintained for use of thermophilic organisms.

Biogenic metabolites other than mineral acids or ferric ions also have potential for the solubilization of metals. Table 1 lists some examples of these agents and shows that in many cases product-metal species were not identified and are listed as "soluble metal."

Since many of these metabolites may be widely synthesized in nature, their identity and reaction products are of interest from an environmental as well as industrial-process standpoint.

Table 1. Examples of metal solubilization by extracellular metabolites

Metabolite	Source organism	Insoluble substrate	Soluble-product molecules	Ref.
Methyl iodide	Marine algae	Metal sulfides	Soluble metal	Thayer et al. (1984)
		SnS	CH_3SnI_3	Manders et al. (1984)
		Pb^0	$(CH_3)_nPb^{(4-n)}$	Craig and Rapsomanikis (1982), Jarvie and Whitmore (1981)
		Sn^0	$(CH_3)_4Sn$	Craig and Rapsomanikis (1982)
Hydroxymate, catechol siderophores	Bacteria	$Fe(OH)_3$	Iron chelates	Raymond and Carrano (1979)
Unidentified	Bacteria	Au	Soluble gold	cited in LeRoux (1970)
Ammonia	Bacteria	Brass, Cu^0	$Cu(NH_3)_4$	cited in Iverson (1972)
Unidentified	Penicillium	Fe^0, steel	Soluble Fe	Siegel et al. (1983)
Unidentified	Soil heterotrophic bacteria	ZnS, CdS, CuS	Soluble Zn, Cd, Cu	Cole (1979)
Unidentified	Penicillium	TiO_2	Soluble Ti	Silverman and Munoz (1971)
Unidentified	Penicillium	Chalcocite ore	Soluble Cu	Wenberg et al. (1971)
Methylcobalamin	Bacteria	PbO_2	Soluble Pb	Thayer and Brinckman (1982)
		SnO_2	Methyltin	

Microbial Precipitation and Concentration of Trace Elements

Microorganisms have potential application in the recovery of metals in process and waste streams for recycling and for environmental protection. Many laboratory studies have shown that pure and mixed cultures of microorganisms can precipitate a wide range of trace elements from solution by a variety of mechanisms.

One mechanism of metal precipitation is the generation of extracellular metabolic products which form insoluble compounds with metals. Metals are precipitated in natural anaerobic environments by hydrogen sulfide which is widely produced in nature by bacterial respiratory-sulfate reduction (Postgate, 1979). Metal sulfides are also produced by bacteria and algae, apparently as resistance mechanisms toward heavy metals. Pan-Hou and Imura (1981) found that a mercury-resistant strain of *Clostridium* produced hydrogen sulfide which bound mercury as a sulfide. A strain of the acidophilic, thermophilic alga *Cyanidium* precipitates metal sulfides at the cell surface and may be useful in precipitation of toxic elements in acidic drainage (Wood and Wang, 1983).

Hydrogen sulfide does not always immobilize metals. Metal-organic species formed even under anoxic, sulfidic conditions are important in metal mobilization. For example, in the presence of certain kinds of sedimentary organic material, metals such as copper are chelated, preventing precipitation of the copper by hydrogen sulfide (Hallberg et al., 1980). A similar process occurs with mercury (Craig, 1980). Hydrogen sulfide also reacts with relatively nonvolatile methylmercury to generate highly volatile dimethylmercury (Craig, 1980).

Microorganisms also adsorb and accumulate heavy elements. A variety of anionic ligands are available on the surface of microbial cells which adsorb metal cations (Beveridge, 1984; Kelly et al., 1979). Cell-wall components such as chitin in fungi and peptidoglycan in bacteria contribute to cell-surface binding of metals. Such interactions may be important in metal transport in aquatic environments and result in heavy-element partitioning into and immobilization in sediments as an early step in mineral formation (Beveridge, 1984). Extracellular polymers produced by microorganisms, such as the polysaccharide matrix of *Zoogloea* in sewage also bind metals (Brown and Lester, 1979). Such surface-binding phenomena are typically rapid, reversible, and nonmetabolic.

Micronutrient (i.e., Ni, Zn, Mn) transport systems in microorganisms function to bring essential metals into the cell. These generally function at trace levels of metals and so would be unlikely for metal recovery in metal-process streams (Kelly et al., 1979). In some cases toxic trace elements which structurally resemble nutrients (i.e., arsenate-phosphate) are accumulated by energy-driven membrane-transport systems (Silver, 1983). As yet undefined uptake processes occur on a more significant scale with some organisms. A strain of *Pseudomonas,* for example, rapidly accumulated uranium from solution to 10%–15% of cell dry weight, immobilizing U by intracellular precipitation (Strandberg et al., 1981).

Precipitation of metals may occur following various microbial transformations of trace elements (Kelly et al., 1979). For instance, organic phosphorus degradation by certain bacteria releases phosphate at cell surfaces, resulting in precipitation of lead phosphates. Magnetotactic bacteria produce intracellular magnetite crystals after uptake of soluble iron (Blakemore, 1982). Enzymatic and nonenzymatic manganese oxidation by microorganisms results in precipitation of manganese(IV) oxide and may contribute to formation of ferromanganese concretions found in nature. An interesting and unusual precipitation of uranium occurs with the yeast *Saccharomyces* which rapidly accumulates uranium to nearly 50% of cell dry weight – far in excess of expected stoichiometry (Strandberg et al., 1981). A better understanding of the chemical species at cell surfaces would help to explain the mechanisms of uranium deposition.

Trace elements are also reduced to their elemental states by microorganisms. This usually results in precipitation, as with selenium, but with mercury results in volatilization.

Industrial processes use these microbial precipitation and accumulation reactions for element recovery. Some mining operations use and even encourage (by design or nutrient addition) microbial processes for precipitation of metals in effluent sedimentation ponds (Sterritt and Lester, 1979). Often combinations of algal, higher-plant, and bacterial growth (especially anaerobic degradation of organic compounds to generate hydrogen sulfide) are most effective. Such ponds have been used in association with uranium mines to reduce levels of U, Ra, Se, and Mo in effluents (Brierley, 1983). The growth of photosynthetic organisms in effluents from lead mining and milling operations in Missouri removes lead, zinc, copper, and other heavy metals from wastewater and has permitted this largest lead-producing area in the world to operate within a national forest without creating serious environmental disturbances (Gale and Wixson, 1978).

Wastewater-treatment plants can be effective for removal of toxic metals from sewage, though their primary function is to digest organic wastes. In general, the secondary, or activated sludge, treatment process is more efficient than primary settling for removal of most trace metals (Brown and Lester, 1979). Bacteria-mediated flocculation and settling of activated sludge is responsible for metal removal. The metals may be bound to the extracellular polymers produced by the sludge bacteria or to microbial cell walls, or be accumulated intracellularly. The efficacy of metal removal is variable and depends on many physical, chemical, and biological factors including metal species as influenced by pH, organic content and chelators, and the activity of the sludge bacteria.

Some attempts have been made in the laboratory to understand metal precipitation by sludge bacteria. Nelson et al. (1981) found that pH was the single most important factor in metal uptake in a bacterial activated-sludge batch reactor. The pH affects distribution of metal species and the degree of protonation of heavy-metal-binding sites, affecting adsorption to organic and inorganic surfaces. Activated-sludge removal of metals was rapid (essentially complete in an hour) and no evidence was obtained that bacterial active transport was important in the binding. Apparently, chemical and physical factors were most responsible for the distribution of metals between bacterial solids and solution (Nelson et al., 1981).

The importance of mixed bacterial populations in sludge in the precipitation of metals was illustrated by a study of a stable, ten-member community of sludge bacteria which effectively accumulated copper from solution (Dunn and Bull, 1983). The community produced extracellular slime and bound more metal as a community than as individual pure cultures, and suspended biomass bound much less metal than when attached to container walls. The interaction of organisms and surface-mediated reactions to accelerate metal precipitation in such attached communities deserves more study. Such a community attached to substrates and placed in an effluent stream could effectively function in metal recovery.

Doses of heavy metals in wastewater can be toxic to certain populations of bacteria in the activated-sludge environment and in severe cases may largely inhibit biodegradation and ruin plant operations. With time, acclimation to elevated metals levels occurs (Brown and Lester, 1979). Despite the general concentration of metals in sludge at wastewater-treatment plants, some plant effluents still discharge significant levels of metals to receiving waters. Also, the disposal of metal-enriched sludge is becoming a serious problem since toxic metals can leach from sludges. Prior to disposal it may be possible to treat sludges as low-grade ores for metal extraction, depending on the concentration of elements of interest and their species. For example, acidophilic thiobacilli can solubilize metals bound as sulfides in certain sewage sludges (Schonborn and Hartmann, 1978). Quantities of toxic metals in different sludges vary widely, reflecting local industry and geology (Furr et al., 1976).

Many laboratory studies have evaluated metal binding my microorganisms potentially applicable to metal recovery in specific industrial effluents. For example, photographic industry effluents represent a major source of recyclable silver, and continuous flow, multistage bioreactors could be used for efficient recovery of silver. Silver-resistant microorganisms inhabit photographic wastes, sludges, and soils (Belly and Kydd, 1982). A silver-resistant *Pseudomonas* accumulated silver up to 38% of the dry weight of the cell. Charley and Bull (1979) found that a defined community of silver-resistant bacteria was an effective silver-accumulating system, with up to 300 mg silver per g dry weight of biomass.

Laboratory studies have also been directed toward microbial accumulation of heavy elements in nuclear-fuel-industry waste effluents including mining, process, and tailings wastewaters. A variety of microorganisms including algae, fungi, and bacteria (Kelly et al., 1979) precipitate uranium and other radionuclides and may be applied to recovery of such elements in process streams and wastewaters. Other industries also have potential uses for microbial recovery of heavy elements. For example, a design for algal removal of arsenic from smelting wastewater has been outlined (Spisak, 1978).

Trace-Element Alkylation and Dealkylation by Microorganisms

Perhaps nowhere is the need for understanding the transformation of chemical species in the environment more dramatically illustrated than with organometallic compounds, because of their much higher toxicity compared with their component elements.

Industrial process streams and wastes discharge various species of trace elements into the environment. Awareness of the importance of environmental methylation of these elements has increased in recent years. A variety of microorganisms and microbial products methylate inorganic species of mercury, tin, lead, arsenic, and selenium in aquatic environments (Craig, 1980; Thayer and Brinckman, 1982; Wood et al., 1978).

The biosynthesis of methyl metals by microorganisms is also important in the specialized environments that are intimately associated with higher forms of life. For example, intestinal contents of some fish and humans harbor mercury-methylating bacteria (Craig, 1980). An aquatic environment devoid of methyl metals may not reflect methylation occurring on or in resident organisms.

The high toxicity of some organometals has made them attractive biocides, and they are used in exterior house paints (phenylmercury), antifoulant coatings (butyltin), and as pesticides (methylarsenicals). Other industrial uses discharge organometals into the environment including organotins in plastics and organoleads in gasoline.

The levels of organometals in the environment will be influenced by degradation as well as industrial discharge and biosynthesis. Microorganisms can cleave carbon-mercury bonds enzymatically, for example, producing methane and inorganic mercury from methylmercury, and benzene and inorganic mercury from phenylmercury. The inorganic mercury is generally reduced to elemental mercury by cell mercuric reductase. However, very little work has been done with other organometals regarding microbial processes for degradation and the chemical species produced. This knowledge is extremely important for modeling and predicting the impact of use or disposal of organometals in the environment.

The identification of organometal species in the environment may also be a clue to their origin. For instance, it is likely that butyltin species in the environment are anthropogenic, since bioalkylations other than biomethylation have not been conclusively demonstrated. Relatively little work has

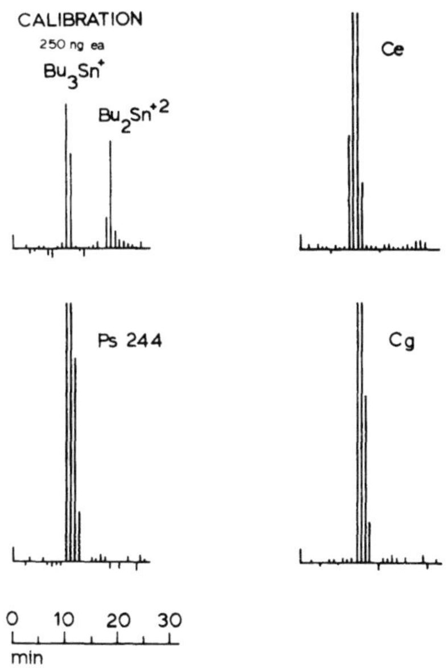

Fig. 2. HPLC-GFAA chromatograms of methanol extracts from three tributyltin-resistant estuarine bacteria exposed for 2.5 hr to 10 mg/L tributyltin chloride. Similar results were obtained after 8 days exposure

been done to measure the uptake of organometals on microorganisms and the way in which this may relate to food-chain metal bioaccumulation. We have determined the species of organometals bound to microorganisms using a coupled high-performance liquid chromatograph-atomic absorption system (HPLC-GFAA, Fig. 2). This system was used to determine that tin-resistant microorganisms isolated from the Chesapeake Bay accumulated but did not metabolize tributyltin, a toxic compound that is used in antifouling paints (Blair et al., 1982).

These measurements were supported by measurements of culture medium and headspace gases using tin-selective flame photometric gas chromatography. Increasing use of organometals requires studies to assess their possible biotransformations and experimental food-chain bioaccumulations. Without such supporting analytical schemes for species analysis the use of radioisotopes has very limited application because by themselves they give no information on chemical species involved.

Recent progress has been made in predicting organotin toxicity to crab larvae and algae using total molecular surface-area calculations (Brinckman, 1984). It may be possible to extend these measurements to predict microbial uptake and food-chain bioaccumulation (Brinckman, 1984).

Conclusion

It is evident that the analysis of chemical species will continue to play an important role in our understanding of microbial processes in heavy-element-stressed environments. The chemical species may be important at ultratrace or molar concentrations in terms of oxidation state, alkylation, complexation, or distribution on a surface.

Microorganisms can transform trace-element species. They may obtain useful energy from oxidation of trace elements. They may alkylate, dealkylate, complex, oxidize, or reduce for detoxification. In a more passive role microorganisms may cause changes in chemical species by the metabolic products they secrete.

By measuring the transformation of chemical species of trace elements we can begin to understand the biological and chemical transformation mechanisms involved. These processes then can be predicted and controlled more precisely so that more rational judgements on potential hazards or choices for biotechnological applications can be made. Such information is also prerequisite for planning genetic improvements in organisms of industrial significance. This is especially true in potentially trace-element and fuels biotechnology where a more complete understanding of basic microbial processes is needed.

Acknowledgements. I thank F. E. Brinckman for his encouragement and helpful discussions. The research at NBS was supported in part by the Office of Naval Research.

References

Andrews GF, Maczuga J (1982) Bacterial coal desulfurization. In: Scott CD (ed) Biotechnology and Bioengineering Symposium No. 12, New York: Wiley Interscience, pp 337–348.

Belly RT, Kydd GC (1982) Silver resistance in microorganisms. Dev Ind Microbiol 23:567–577

Bennett JC, Tributsch H (1978) Bacterial leaching patterns on pyrite crystal surfaces. J Bacteriol 134:310–317

Beveridge TJ (1984) Mechanisms of the binding of metallic ions to bacterial walls and the possible impact on microbial ecology. In: Klug MJ, Reddy CA (eds) Current Perspectives in Microbial Ecology, Washington, DC: American Society for Microbiology, pp 601–607.

Blair WR, Olson GJ, Brinckman FE, Iverson WP (1982) Accumulation and fate of tri-n-butyltin chloride in estuarine bacteria. Microb Ecol 8:241–251

Blakemore RP (1982) Magnetotactic bacteria. Ann Rev Microbiol 36:217–238

Bloomfield C, Coulter JK (1973) Genesis and management of acid sulfate soils. In: Brady NC (ed) Advances in Agronomy, vol 25, New York: Academic Press, pp 265–326.

Booth JE, Williams JW (1984) The isolation of a mercuric ion-reducing flavoprotein from *Thiobacillus ferrooxidans*. J Gen Microbiol 130:725–730

Brierley CL (1978) Bacterial leaching. CRC Crit Rev Microbiol 6:207–262

Brierley JA (1983) Biological accumulation of some heavy metals – biotechnological applications. In: Westbroek P, de Jong EW (eds) Biomineralization and Biological Metal Accumulation, D Reidel Pub Co, pp 499–509.

Brinckman FE (1984) Environmental effects of organotins. Paper presented at the Fourth International Conference on Germanium, Tin, and Lead, Montreal, Aug. 8–12, 1983

Brown MJ, Lester JN (1979) Metal removal in activated sludge: The role of bacterial extracellular polymers. Wat Res 13:817–838

Charley RC, Bull AT (1979) Bioaccumulation of silver by a multispecies community of bacteria. Arch Microbiol 123:239–244

Cole MA (1979) Solubilization of heavy metal sulfides by heterotrophic soil bacteria. Soil Sci 127:313–317

Craig PJ (1980) Metal cycles and biological methylation. In: Hutzinger O (ed) The Handbook of Environmental Chemistry. New York: Springer-Verlag, pp 169–227

Craig PJ, Rapsomanikis S (1982) A new route to tris(dimethylsulfide) with tetramethyltin as co-product; the wider implications of this and some other reactions leading to tetramethyltin and -lead from iodomethane. J Chem Soc, Chem Commun 114

DiSpirito AA, Tuovinen OH (1982) Uranous ion oxidation and carbon dioxide fixation by *Thiobacillus ferrooxidans*. Arch Microbiol 133:28–32

Dugan PR, Apel WA (1978) Microbiological desulfurization of coal. In: Murr LE, Torma AE, Brierley JA (eds) Applications of Bacterial Leaching and Related Microbiological Phenomena. New York: Academic Press, pp 223–250.

Dunn GM, Bull AT (1983) Bioaccumulation of copper by a defined community of activated sludge bacteria. Eur J Appl Microbiol Biotechnol 17:30–34

Ehrlich HL (1978) Inorganic energy sources for chemolithotrophic and autotrophic bacteria. Geomicrobiol J 1:65–83

Furr AK, Lawrence AW, Tong SSC, Grandolfo MC, Hofstader RA, Bache CA, Gutenmann WH, Lisk DJ (1976) Multielement and chlorinated hydrocarbon analysis of municipal sewage sludges of American cities. Envir Sci Technol 10:683–687

Gale NL, Wixson BG (1978) Removal of heavy metals from industrial effluents by algae. Dev Ind Microbiol 20:259–273

Gokcay CF, Yurteri RN (1983) Microbial desulfurization of lignites by a thermophilic bacterium. Fuel 62:1223–1224

Hallberg RO, Bubela B, Ferguson J (1980) Metal chelation in sedimentary systems. Geomicrobiol J 2:99–113

Hoffman MR, Faust BC, Panda FA, Koo HH, Tsuchiya HM (1981) Kinetics of the removal of iron pyrite from coal by microbial catalysis. Appl Envir Microbiol 42:259–271

Holmes DS, Lobos JH, Bopp LH, Welch GC (1984) Cloning of a Thiobacillus ferrooxidans plasmid in *Escherichia coli*. J Bacteriol 157:324–326

Iverson WP (1972) Biological corrosion. In: Fontana MG (ed) Advances in Corrosion Science and Technology, vol 2. New York: Plenum Press

Jack TR, Sullivan EA, Zajic JE (1980) Growth inhibition of *Thiobacillus thiooxidans* by metals and reductive detoxification of vanadium(V). Eur J Appl Microbiol 9:21–30

Jarvie AW, Whitmore AP (1981) Methylation of elemental lead and lead(II) salts in aqueous solution. Envir Technol Lett 2:197–204

Kargi F (1982) Microbiological coal desulphurization. Enzyme Microbiol Technol 4:13–19

Kelly DP, Norris PR, Brierley CL (1979) Microbiological methods for the extraction and recovery of metals. In: Bull AT, Ellwood DC, Ratledge C (eds) Microbial Technology: Current State, Future Prospects. Cambridge: Cambridge University Press, pp 263–308.

LeRoux NW (1970) Mineral attack by microbiological processes. In: Miller JDA (ed) Microbial Aspects of Metallurgy. New York: American Elsevier, pp 173–182.

Lundgren DG, Malouf EE (1983) Microbial extraction and concentration of metals. Adv Biotechnol Proc 1:223–249

Manders WR, Olson GJ, Brinckman FE, Bellama JM (1984) A novel synthesis of methyltin triiodide with environmental implications. J Chem Soc, Chem Commun 1984:538–540

Mao MWH, Dugan PR, Martin PAW, Tuovinen OH (1980) Plasmid DNA in chemoorganotrophic *Thiobacillus ferrooxidans* and *T. acidophilus*. FEMS Microbiol Lett 8:121–125

Nelson PO, Cheng AK, Hudson MC (1981) Factors affecting the fate of heavy metals in the activated sludge process. J Wat Poll Control Fed 53:1323–1333

Norris PR, Kelly DP (1982) The use of mixed microbial cultures in metal recovery. In: Bull AT, Slater JH (eds) Microbial Interactions and Communities. London: Academic Press, pp 443–474.

Olson GJ, Porter FD, Rubenstein J, Silver S (1982) Mercuric reductase enzyme from a mercury-volatilizing strain of *Thiobacillus ferrooxidans*. J Bacteriol 151:1230–1236

Pan-Hou HKS, Imura N (1981) Role of hydrogen sulfide in mercury resistance determined by plasmid of *Clostridium cochlearium* T-2. Arch Microbiol 129:49–52

Postgate JR (1979) The Sulphate Reducing Bacteria. Cambridge: Cambridge University Press

Raymond KN, Carrano CJ (1979) Coordination chemistry and microbial iron transport. Acc Chem Res 12:183–190

Schonborn W, Hartmann H (1978) Bacterial leaching of metals from sewage sludge. Eur J Appl Microbiol 5:305–313

Siegel SM, Siegel BZ, Clark KE (1983) Bio-corrosion: solubilization and accumulation of metals by fungi. Water Air Soil Poll 19:229–236

Silver S (1983) Bacterial transformations of and resistances to heavy metals. In: Changing Metal Cycles and Human Health. Dahlem Konferenzen. Berlin, Heidelberg, New York, Tokyo: Springer-Verlag

Silverman MP, Munoz EF (1971) Fungal leaching of titanium from rock. Appl Microbiol 22:923–924

Singer PC, Stumm W (1970) Acidic mine drainage: The rate-determining step. Science 167:1121–1123

Spisak JF (1978) Metallurgical effluents – growing challenges for second generation treatment. Dev Ind Microbiol 20:249–257

Sterritt RM, Lester JN (1979) The microbiological control of mine waste pollution. Min Envir 1:45–47

Strandberg GW, Shumate SE, Parrott JR (1981) Microbial cells as biosorbents for heavy metals: Accumulation of uranium by *Saccharomyces cerevisiae* and *Pseudomonas aeruginosa*. Appl Envir Microbiol 41:237–245

Temple KL, Colmer AR (1951) The autotrophic oxidation of iron by a new bacterium, *Thiobacillus ferrooxidans*. J Bacteriol 62:605–611

Thayer JS, Brinckman FE (1982) The biological methylation of metals and metalloids. In: Stone FGA, West R (eds) Advances in Organometallic Chemistry, vol 20. New York: Academic Press, pp 313–356.

Thayer JS, Olson GJ, Brinckman FE (1984) Iodomethane as a potential metal mobilizing agent in nature. Envir Sci Technol 18:726–729

Vuorinen A, Hiltunen P, Hsu JC, Tuovinen OH (1983) Solubilization and speciation of iron during pyrite oxidation by *Thiobacillus ferrooxidans*. Geomicrobiol J 3:95–120

Wenberg GM, Erbisch FH, Volin ME (1971) Leaching of copper by fungi. Soc Mining Eng AIME 250:207–212

Wood JM, Cheh A, Dizikes LJ, Ridley WP, Rakow S, Lakowicz JR (1978) Mechanisms for the biomethylation of metals and metalloids. Fed Proc 37:16–21
Wood JM, Wang HK (1983) Microbial resistance to heavy metals. Envir Sci Technol 17:582A–590A
Yen TF, Chilingar GV (1976) Introduction to oil shales. In: Yen TF, Chilingarian GV (eds) Oil Shale. Amsterdam: Elsevier, pp 1–12.

The Importance of Chemical "Speciation" in Environmental Processes,
eds. M. Bernhard, F. E. Brinckman, and P. J. Sadler, pp. 513–531.
Dahlem Konferenzen 1986. Springer-Verlag Berlin, Heidelberg
© Dr. S. Bernhard, Dahlem Konferenzen

Microbial Interactions with Toxic Elements in the Environment

A. M. Chakrabarty

Dept. of Microbiology and Immunology
University of Illinois Medical Center
Chicago, IL 60612, USA

Abstract. A major purpose of this Dahlem Workshop was to analyze and assess the chemical species of various elements that influence the biogeochemical cycle of the world. We have therefore chosen to discuss the nature of toxic effects of individual biotic or abiotic species or elements on living systems and the environment as a whole, stressing the importance of chemical, biochemical, or molecular mechanisms that affect such interaction. The primary focus of this paper is to evaluate the mode of genetic interaction of microorganisms with two chemical species, viz., metals or metalloids and a group of synthetic chemicals, chlorinated aromatics, that have been widely used as herbicides and pesticides, solvents, refrigerants, fire retardants, and in general as industrially useful compounds. Since both of these groups of chemicals exert useful as well as toxic effects on natural microflora, microorganisms either use them as constituents of their basic metabolic processes and food supply or try to develop resistance to them to circumvent the toxicity associated with their presence. Either event occasionally necessitates the evolution of new genetic characteristics that endow the host cells with such interactive abilities or entail appreciable genetic rearrangements for such events to occur at a high frequency.

Interaction of Microorganisms with Metals and Metalloids

Different metals interact with microorganisms in different ways. Some metals such as aluminum have little interaction, whereas others such as Fe,

Mg, Mn, and Zn are an integral part of many enzymes as cofactors or activators, therefore constituting an essential ingredient in the life cycle of the microorganisms. Still others, heavy metals such as Hg, Cd, Sb, and Ag in particular, are quite toxic when present in appreciable amounts and somehow must be detoxified or prevented from interacting with essential microbial cellular constituents. Because the biochemical and genetic bases of resistance against heavy metals have been discussed in a number of reviews (Iverson and Brinckman, 1978; Silver and Misra, 1984; Wood and Wang, 1983) and the role of microbial envelope proteins in binding iron has been reviewed (Neilands, 1982), this paper merely touches on the recent developments that throw considerable light on the nature or regulation of genes governing metal-microbe interaction.

Interaction with Fe

Although it is quantitatively abundant in the environment, the insolubility of iron at biological pH in the aerobic environment has necessitated the evolution of specific uptake and transport systems for this element in microorganisms. Since iron, like most metals, is also toxic in higher amounts, the specific uptake systems are regulated at the cell-envelope level by high or low concentrations of iron. The microbial uptake system consists of low-molecular-weight, Fe(III)-specified ligands, termed siderophores, and their cognate membrane receptors. Both components of the high-affinity system appear to be regulated by the available iron concentration (Neilands, 1982). To date, the genetic organization of two siderophore-associated iron-transport systems has been described – the high-affinity enterobactin (enterochelin) and the aerobactin systems of enteric bacteria. Enterobactin is a cyclic trimer of 2,3-dihydroxyl-N-benzoyl-L-serine and nine genes are thought to be involved in the overall enterobactin-dependent iron-transport system in *E. coli* (Laird et al., 1980). Seven out of the nine genes are involved in the biosynthesis of enterobactin. One gene (*fep*) encodes an 81,000-dalton outer-membrane receptor protein for ferric enterobactin while another gene (*fes*) product is involved in the release of iron from ferric enterobactin.

Aerobacter aerogenes is known to produce a siderophore, aerobactin, which is a conjugate of 6-(N-acetyl-N-hydroxamino)-2-amino-hexanoic acid and citric acid. In contrast to the synthesis of enterobactin, the high-affinity iron-uptake system common to most enteric bacteria and which is chromosomally coded, aerobactin is known to be coded by ColV-type plasmids (Warner et al., 1981). Recently the complete gene complex including the biosynthesis genes for aerobactin as well as the synthesis of the ferric aerobactin outer-membrane receptor was cloned (Bindereif and Neilands, 1983). The production of this siderophore, as specified by the ColV plasmid, is believed to be an important component in the virulence of invasive strains of *E. coli* (Warner et al., 1981).

Another interesting siderophore called pseudobactin is produced by the *Pseudomonas fluorescens-P. putida* group of bacteria, often known as plant growth-promoting rhizobacteria. These bacteria rapidly colonize plant roots of a variety of crops, significantly increasing their growth yield (Schroth and Hancock, 1982). This growth-promoting activity of rhizobacteria is primarily achieved by depriving certain native microflora of iron(III), thereby reducing microbial root colonization (Kloepper et al., 1980 a). Thus *Pseudomonas sp.* strain B10, which produces the siderophore pseudobactin, has been effectively used as a biological control agent of Fusarium wilt and take-all diseases caused by the soil-borne fungi *Fusarium oxysporum* f. sp. lini and *Gaeumannomyces graminis* var. tritici, respectively (Kloepper et al., 1980 b). The pseudobactin produced by this strain is believed to sequester the iron, thus denying iron(III) to these pathogens and inhibiting their growth.

Pseudobactin is a linear hexapeptide, L-lysine-D-thero-β-hydroxyaspartic acid-L-alanine-D-allo-threonine-L-alanine-D-N^8-hydroxyornithine, in which the N^8-OH nitrogen of the ornithine is cyclized with the C-terminal carboxyl group and the $N\varepsilon$-amino group of the lysine is linked via an amine bond to a quinoline derivative which is responsible for the yellow-green fluorescence of pseudobactin. The iron-chelating groups consist of a hydroxamate group derived from N^8-hydroxy-ornithine, an α-hydroxyacid from β-hydroxyaspartic acid, and an *o*-dihydroxy aromatic group derived from the quinoline moiety. The combination of metal-chelating ligands and the alternating L- and D-amino acids as found in pseudobactin is normally rare.

Recently cloning of the genes involved in the biosynthesis of pseudobactin was achieved by utilizing a gene bank of *Pseudomonas sp.* strain B10 DNA and a large number of mutants defective in pseudobactin production. Eight recombinant plasmids were constructed in 154 nonfluorescent pseudobactin-nonproducing mutants. The pattern of complementation suggests that a minimum of 12 genes arranged in 4 clusters is required for the biosynthesis of pseudobactin (Moores et al., 1984).

Interaction with Heavy Metals

Unlike iron, heavy metals are toxic to microorganisms even at relatively low concentrations. Microorganisms have therefore evolved very efficient systems that allow them to detoxify a number of heavy metals, particularly several that comprise major constituents of industrial or municipal effluents. Depending upon the nature of the metals, or even of a single metal ion, microorganisms may employ a variety of biochemical reactions such as oxidation, reduction, and methylation that will reduce the toxicity associated with the entry of the metal ions inside the cell. Similar to the siderophores that may be chromosomally or plasmid-encoded, heavy-metal detoxification mechanisms are either chromosomal or plasmid-specified (Silver and Misra, 1984). The clearest example of a plasmid-specified detoxification mechanism is that of mercury, where inorganic Hg^{2+} is reduced to a less-toxic metallic Hg by the plasmid-specified enzyme mercuric reductase (Rinderle et al., 1983; Silver and Misra, 1984). This enzyme has quite similar physicochemical properties when isolated from various gram-negative bacteria harboring different plasmids that specify mercury resistance. Inorganic mercuric ion or organomercurial resistance is determined by an inducible operon of closely linked genes. A typical mer operon consists of the genes merR, a regulatory gene encoding a positively acting activator protein, the operator-promoter sequence; the gene merT, which specifies a membrane protein involved in the uptake of the mercury ions; merC, a gene whose function is not understood at present; merA, the gene for mercuric reductase; and finally merD, a gene of unknown function (NiBhriain et al., 1983). This operon sequence is present also as a transposable element Tn501, originally isolated from *P. aeruginosa,* whose nucleotide sequence demonstrates appreciable homology (approximately 85%) with that of the operon on *Shigella* plasmid R100 (Silver and Misra, 1984). The product of the merA gene, mercuric reductase, belongs to a class of NAD(P)H-dependent, FAD-containing flavoproteins including glutathione reductase and lipoamide dehydrogenase. Bacterial mercuric reductase contains a stretch of 15 amino acids which is essentially identical to sequences at the active sites of human-erythrocyte glutathione reductase or the pig-heart lipoamide dehydrogenase (Brown et al., 1983), thereby implying a common ancestry. Organomercurials are also enzymatically detoxified to volatile compounds via organomercurial lyases. For example, benzene is produced from phenylmercury and methane from methylmercury.

Besides the reduction of inorganic or organomercurials, methylation of mercury also occurs in nature. As a matter of fact, methylation of various metals other than Hg is well known and has been extensively studied, particularly for tin (Brinckman et al., 1982; Hallas et al., 1982; Wood et al., 1978). Bacteria from Chesapeake Bay have been reported to synthesize dimethyl- and trimethyltin (Hallas et al., 1982); similarly a *Pseudomonas* species has been reported to form unstable and volatile methyltin derivatives (Hansen et al., 1985). Formation of vitamin B_{12}-dependent volatile alkylmetal derivatives may therefore be a detoxification mechanism for bacteria. A strain of *Clostridium cochlearium* T-2C has been reported to develop resistance against high concentrations of mercury by employing a vitamin B_{12}-dependent methylation process (Pan-Hou and Imura, 1982). Although a plasmid is known to encode organomercury decomposition and hydrogen-sulfide formation in *C. cochlearium* T-2P, the plasmid-cured cells of strain T-2C will carry out the mercury-methylation reactions, suggesting that the methylation genes are presumably chromosomal (Pan-Hou and Imura, 1982). Very little is otherwise known on the organization or regulation of biomethylation genes for heavy metals.

Another mechanism of bacterial resistance to heavy metals is a potent efflux mechanism by which the resistant cells drive out the highly toxic concentrations of the metals. For example, resistance to Cd^{2+} by cells of *Staphylococcus aureus* is achieved by a plasmid-encoded efflux system which is energy-dependent and is abolished by dinitrophenol or at low temperatures (Tynecka et al., 1981). In staphylococci, the membrane electrical potential is utilized to achieve Cd^{2+} accumulation through a membrane-transport system that normally functions to accumulate low levels of non-toxic Mn^{2+}. Two separate plasmid genes are responsible for Cd^{2+} resistance, cadA and cadB; the cadA gene confers high levels of resistance. Both genes also confer increased resistance to Zn^{2+} indicating the involvement of the same type of system for Zn^{2+} resistance.

A similar type of energy-dependent plasmid-mediated efflux mechanism is also responsible for resistance to arsenate (Silver and Keach, 1982). Arsenate is normally accumulated in cells via their phosphate-transport system. The presence of the resistance plasmid does not alter the kinetic properties of the phosphate-transport system but simply facilitates the efflux of the excess arsenate via an ATPase-linked transport system.

Oxidation is also involved as a mechanism of resistance, particularly against arsenite. *Alcaligenes* species are known to oxidize toxic arsenite to less-toxic arsenate (Osborne and Ehrlich, 1976). The plasmid arsenite-resistance gene is closely linked to the arsenate-resistance determinant and forms an inducible operon-like system in both *S. aureus* and *E. coli* (Silver et al., 1981). However, the plasmid-mediated resistance to arsenite does not involve its oxidation to arsenate, and its basic mechanism is unknown. Whether arsenite oxidation by the *Alcaligenes sp.* is specified by plasmid-borne genes or by chromosomal genes is also unknown at present.

Yet another mechanism of resistance to heavy metals is interaction at the surface. For example, the green alga *Cyanidium caldarium* has been reported to be tolerant to high concentrations of Cu^{2+} or Ni^{2+} present in acidic mine water from a copper mine by precipitation of the metals at the surface, thus preventing their entry inside the cell (Wood, 1983). Nickel resistance in the blue-green alga *Synecococcus* is mediated through production of polymeric cyanophycin granules that bind both Ni^{2+} and Cu^{2+} (Wood, 1983). Cells of *Aeromonas* species have been reported to accumulate organic lead on their surface (Silverberg et al., 1976). Many bacteria, including *E. coli,* can produce H_2S in an anaerobic environment; H_2S production is sometimes encoded by plasmid-borne genes. Metals can be precipitated by such microbially produced H_2S, and such metal sulfides, particularly that of Hg, may be slowly methylated (Iverson and Brinckman, 1978). Silver resistance in enteric bacteria, which has been reported to be plasmid-mediated, is expressed only in the presence of halide ions. Both sensitive and resistant cells bind Ag^+ tightly. However, the sensitive cells are killed due to the binding of Ag^+; the resistant cells allow the bound Ag^+ to be coupled with the external halide ions, thereby effecting their removal as silver halide at the surface (Silver and Misra, 1984).

Apart from the above examples, microorganisms are known to elaborate plasmids that encode resistance to other heavy metals, but their mechanisms of action are poorly understood. For example, the mechanisms of plasmid-determined bismuth, antimony, and arsenite resistances are not known. A plasmid has been characterized from a *Pseudomonas* (isolated from heavily chromate-contaminated areas of the Hudson River) that affords resistance to toxic concentrations of chromate (Bopp et al., 1983). Little is known about the enzyme system involved in the reduction of Cr(VI) to Cr(III) present in such strains.

A summary of the various modes by which microorganisms interact with metals is given in Table 1.

Another area where metal-microbe interaction has attracted considerable attention is the application of microorganisms in the leaching of important metals such as copper or uranium (Brierley, 1978). The leaching is a result of the growth of acidophilic, chemolithotrophic bacteria that derive their energy from the oxidation of reduced or partially reduced sulfur compounds and iron or copper salts. The iron-oxidizing *Thiobacilli,* as well as others, occasionally demonstrate resistance to various heavy metals, and

Table 1. Representative examples of metal-microbe interactions

Metal	Microbe	Nature of interactions	Genetic basis	References
Fe	*E. coli*, *Salmonella*, *Pseudomonas*	Binding of the metal by siderophores	Chromosomal, plasmid	Bindereif and Neilands (1983), Hansen et al. (1985), Neilands (1982)
Hg, Cr	*E. coli*, *Pseudomonas*	Reduction $Hg^{2+} \rightarrow Hg^0$ $Cr^{6+} \rightarrow Cr^{3+}$	Plasmid Plasmid	Bopp et al. (1983), Rinderle et al. (1983), Silver and Misra (1984)
As	*Alcaligenes*	Oxidation, $AsO_3^{3-} \rightarrow AsO_4^{3-}$	Unknown	Osborne and Ehrlich (1976)
As, Cd, Zn	*S. aureus*, *E. coli*	Efflux of the metal from the cell	Plasmid	Silver and Keach (1982), Tynecka et al. (1981)
Sn, Hg, Pb	*E. coli*, *Pseudomonas*	Methylation	Chromosomal	Brinckman et al. (1982), Hallas et al. (1982), Silver and Misra (1984)
Cu, Ni, Ag	Blue-green algae, enteric bacteria	Precipitation or binding at the surface	Chromosomal, plasmid	Silver and Misra (1984), Wood (1983)

such resistance has been postulated to be due to the presence of resistance plasmids (Chakrabarty, 1978). Although plasmids have been demonstrated in *Thiobacillus* (Martin et al., 1981), and thiobacilli plasmid genes have been cloned in *E. coli* (Holmes et al., 1984), no information is currently available regarding the role of the plasmid or chromosomal genes in heavy-metal resistance or metal oxidation, or regarding the regulation of such processes. A detailed description of chemolithotrophy has therefore been omitted and is considered beyond the scope of this review.

Practical Considerations in Metal-microbe Interactions

It is clear from the above discussion that different microorganisms interact differently with various metals, either producing specific binding agents for scavenging metals (as in siderophore-bound Fe) or preventing the metal from entering the cells as toxic agents (as in cyanophycin-granule production by blue-green algae). Because in most cases such bindings are extremely specific, it is possible to use such binding agents for specific removal of the metal from a mixture as is often the case with ores or industrial effluents. It may not be an economically viable process for low-cost metals such as Fe but could be quite important commercially for metals such as nickel, gold, platinum, or uranium, if such specific binding agents can be characterized and produced in bulk quantities at a competitive price. The volatilization of a metal or its alkyl derivative from municipal sewage or industrial effluents by intact bacteria or the bacterial-enzyme system could similarly be an effective way of removing toxic metals from their sources of contamination. For metals such as Hg it might be possible to condense the volatile mercury for its ultimate recovery. Indeed, it has been reported that greater than 98% of the mercury present at 70 mg per liter can be removed from municipal sewage by treatment with plasmid-containing bacteria (Hansen et al., 1985).

An interesting role of specific metal-binding agents produced by bacteria is their contribution towards enhanced virulence of such bacteria. For example, siderophores such as enterobactin are chromosomally coded, whereas the siderophore aerobactin is specified by ColV-type plasmids. The ColV-specified siderophore is critical to the invasive cells since it allows sequestering of the iron from tissue fluids for growth and colonization of the cells. An understanding of the mechanisms by which the available iron is sequestered by ColV-specified ligands may allow development of anti-binding agents that may be useful in the treatment of infections caused by such ColV-harboring, invasive bacteria.

The need of an iron-binding ligand in bacterial growth has interesting ramifications for agriculture. For example, the *Pseudomonas sp.* strain B10 is known to promote growth of a variety of plants by inhibiting growth of other plant-invasive bacteria and fungi through specific binding of the available iron with pseudobactin, thus denying iron to these microorganisms. The recent success in cloning the pseudobactin biosynthetic genes may allow future transfer of this capability to selective root-nodule bacteria specific for cereals or other important crop plants, thereby greatly promoting the growth of such plants. Genetic engineering with enhanced nodulation, nitrogen-fixation and metal-binding ability may therefore go hand in hand to improve future agricultural productivity.

Microbial Interaction with Toxic Chemicals

Similar to many toxic metals that are used widely in industry and are highly prevalent in the environment, many toxic chemicals have become ubiquitous in nature because of their release in massive amounts as herbicides, pesticides, refrigerants, industrial solvents, etc. One group of such chemicals, the chlorinated aromatics, is of particular concern because of their persistence in the environment (Alexander, 1981). This persistence appears to be a reflection of the inability of most microorganisms to degrade them effectively. Unlike metals, a few of which are only used by a small group of bacteria as their source of energy (Brierley, 1978), toxic chlorinated compounds may act as potential sources of carbon and energy if their interaction with microorganisms can be enhanced. This section deals with the mode of degradation of toxic chlorinated compounds in nature, the role of plasmids in the biodegradation of simple chlorinated compounds, and the scope of genetic manipulations in broadening the substrate specificity of microorganisms to include toxic, persistent chemicals through evolution of new degradative functions.

Microbial Biodegradation of Chlorinated Compounds

Although a few simple chlorinated compounds are known to be used by microorganisms as their sole source of carbon and energy, the bulk of these compounds are usually degraded by a process called co-metabolism (Alexander, 1981). This means that the compounds usually undergo slow transformation while the microorganisms grow at the expense of other carbon sources. Thus the co-metabolizing microorganisms do not derive any energy out of such transformation and the chemical undergoing metabolism does not contribute towards buildup of the bacterial cell mass. The metabolism of DDT (1,1-bis[p-chlorophenyl]2,2,2-trichloroethane) and many other insecticides represents this sort of metabolism (Slater and Bull, 1982). In addition to co-metabolism, the microbial community as a whole may participate in the total mineralization of a synthetic compound. For example, Senior et al. (1976) have isolated a complex community growing on the herbicide Dalapon (2,2′-dichloro-propionic acid) using chemostatic selection. A stable microbial community of six microbes was originally characterized, three of which were primary organisms able to grow on Dalapon in pure culture, whereas the three secondary organisms could not. The community was very stable for a long period of time, and later one of the secondary organisms was found to have acquired the capability to utilize Dalapon. One of the primary organisms, a strain of *P. putida,* was shown to harbor two dehalogenases for effective dechlorination of Dalapon. Co-metabolic transformations and biodegradation by individual microorganisms and microbial communities represent the major routes of dissimilation of most chlorinated compounds in nature.

The Role of Plasmids in the Biodegradation of Chlorinated Compounds

In contrast to highly chlorinated compounds that are only slowly degraded by co-metabolism, many mono- or dichloro-compounds are known to be degraded by pure cultures. In some cases the degradation of such simple chlorinated compounds is known to be mediated by plasmids (Chakrabarty et al., 1984). A list of representative plasmids specifying degradation of chlorinated compounds is given in Table 2.

Degradative plasmids for chlorinated compounds may or may not specify a complete degradative pathway. For example, plasmids pAC21 and pKF1 only specify a partial pathway for the conversion of chlorobiphenyls to the corresponding chlorobenzoic acids, but they cannot effect dechlorination of the chlorobenzoates. Plasmid pAC25, on the other hand, encodes a complete degradative pathway for 3-chlorobenzoates with release of the chloride ions. A complete pathway for 2,4-D(2,4-dichloro-phenoxyacetic acid) degradation is also believed to be encoded by the pJP2 or pJP4 plasmid.

The enzymes of the 3-chlorobenzoate degradative pathway (Fig. 1), particularly pyrocatechase II, muconate cycloisomerase II, and 4-carboxymethylene-but-2-en-4-olide hydrolase, appear to have high affinity towards the chlorinated substrates and are only weakly active against the nonchlorinated parents. The hydrolase in particular is highly specific (Knackmuss,

Table 2. Plasmids specifying biodegradation of halogenated compounds

Plasmid	Degradative pathway	Molecular size (Mdal)	References
pUO1	Fluoroacetate	43	Kawasaki et al. (1981)[a]
pAC21	p-chlorobiphenyl	65	Kamp and Chakrabarty (1979)[b]
pKF1	p-chlorobiphenyl	53	Furukawa and Chakrabarty (1982)
pJP2	2,4-D		Don and Pemberton (1981)[c]
pJP4	2,4-D 3-chlorobenzoate	56	Don and Pemberton (1981)[c]
pWR1 (pB13)	3-chlorobenzoate	72	Chatterjee and Chakrabarty (1983), Knackmuss (1981)
pAC25	3-chlorobenzoate	76	Chatterjee and Chakrabarty (1982)[d]
pAC31	3,5-Dichloro-benzoate	72	Chatterjee and Chakrabarty (1982)[d]
No designation	2,6-Dichlorotoluene	63	Vandenbergh et al. (1981)[e]

[a] Agric Biol Chem 45:1477–1481
[b] Plasmids of Medical, Environmental and Commercial Importance, eds. K.N. Timmis and A. Puhler. Amsterdam: Elsevier/North-Holland Biomedical Press
[c] J Bacteriol 145:681–686
[d] Molec Gen Genet 188:279–285
[e] Appl Envir Microbiol 42:737–739

1981). Thus, if these genes evolved from the ancestral β-ketoadipate pathway genes that encode degradation of nonchlorinated catechol (Fig. 1), then they must have undergone substantial changes, at least in the coding region for substrate recognition and binding. Indeed, the chlorobenzoate plasmid pAC25 is known to complement only the catB mutation (the gene for muconate-lactonizing enzyme) but not the catA, catC, or catD mutations in the β-ketoadipate pathway (Chakrabarty, 1982). Thus this plasmid presumably evolved after substantial genetic alterations in the structural genes derived from ancestral β-ketoadipate pathway genes. It is interesting to note that a cloned fragment of the pAC25 plasmid containing the catB-complementing genes and a cloned fragment containing the β-ketoadipate pathway catB gene from the *P. putida*-strain RB1 chromosome demonstrates some DNA homology as indicated by DNA-DNA hybridization, perhaps implying extensive changes in the pAC25-borne chloromuconate-lactonizing enzyme gene (Kilbane and Chakrabarty, 1984). It is also interesting that two different 3-chlorobenzoate-degrading *Pseudomonas* species, isolated in two different parts of the world, harbored essentially identical

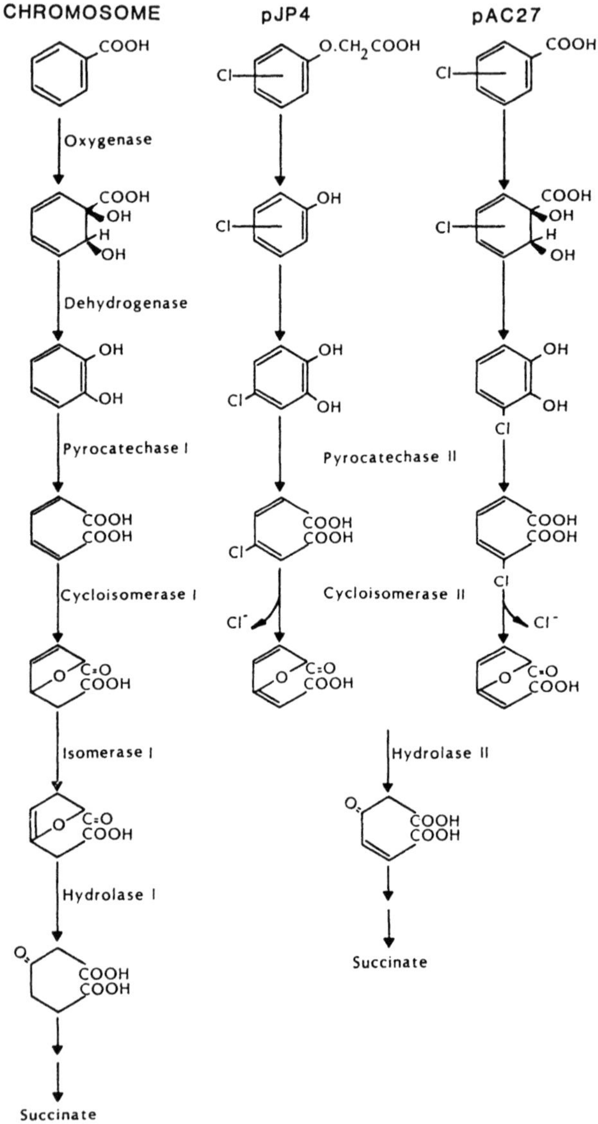

Fig. 1. Pathways for the degradation of benzoate, chlorophenoxyacetates, and chlorobenzoates. The benzoate-degradative pathway is known to be chromosomally coded (indicated by chromosome on top) whereas the 2,4-dichlorophenoxyacetic acid-degradative and the 3-chlorobenzoate-degradative genes are known to be borne on plasmids pJP4 and pAC27, respectively

plasmids that encode 3-chlorobenzoate degradation (Chatterjee and Chakrabarty, 1983). Such genetic similarity presumably indicates a rapid spread of the plasmid-borne characters throughout microorganisms all over the world.

Another interesting property of the degradative plasmids for chlorinated compounds is the clustering of the degradative genes. For the 2,4-D-degradative plasmid pJP4, transposon mutagenesis of the plasmid genes has located most of the structural genes in a single EcoRl fragment (Weightman et al., 1984). Cloning of the chlorocatechol-degradative genes from the plasmid pJP4 has demonstrated the clustering of these genes on the 15 kb EcoRl-B fragment (Ghosal et al., 1985). Fragments EcoRl-E and presumably EcoRl-F are also necessary for efficient expression of the chlorocatechol-structural genes present on the EcoRl-B fragment. The chlorocatechol-degradative genes are present on a single 4.2 kb BglII-E fragment in the plasmid pAC27, which demonstrates strong genetic homology with the pJP4 EcoRl-B and EcoRl-E fragments (Ghosal et al., 1985). Both the pAC27 BglII-E and the pJP4 EcoRl-B fragments allow only slow growth with chlorocatechols in the absence of neighboring regulatory genes which are believed to control the expression of the structural genes in a positive manner. In the absence of such activator genes, the pAC27 BglII-E and the pJP4 EcoRl-B fragments undergo tandem amplification to about eight to ten copies when selected on 3-chlorobenzoate plates, suggesting that the activator genes also somehow repress the amplification process.

Evolution of New Degradative Functions in Bacteria

Most chlorinated aromatic compounds are degraded in nature via formation of chlorocatechols. Simple or highly chlorinated aromatic compounds must therefore be converted to their corresponding chlorocatechols before their final degradation by the "modified ortho" pathway (Knackmuss, 1981). Total degradation of mono- or dichlorocatechols is known to be specified by such transmissible plasmids as pWRl (Knackmuss, 1981), pAC25, pAC27 or pAC31 (Chakrabarty et al., 1984). It is therefore possible to introduce the degradative genes for chlorocatechols into bacterial strains that may otherwise lack these genes but can undergo simple mutations in the existing genes to allow conversion of a chlorinated substrate to chlorocatechol. This sort of simple genetic manipulation has allowed the construction of new strains that can utilize chloroaromatics such as chlorophenols, chlorosalicylates (Reineke et al., 1982), and chloroaniline (Kilbane and Chakrabarty, 1984).

Apart from endowing the host cells with the ability to degrade chlorocatechols, plasmids and their gene products may also interact in a variety of ways either for total mineralization of a chlorinated compound or to allow evolution of a new degradative function. For example, following the demonstration by Knackmuss and his colleagues (Knackmuss, 1981; Reineke et al., 1982) that inclusion of TOL in a continuous culture along with the 3-chlorobenzoate-degrading *Pseudomonas sp.* B13 is essential for the development of cultures capable of utilizing 4-chloro- or 3,5-dichlorobenzoate, we have demonstrated that a portion of the TOL plasmid undergoes recombination with the chromosome to provide a broad substrate-specific benzoate-oxygenase function, while another part of the TOL plasmid is used to supply the replication functions for the evolution of a new plasmid that harbors mutationally altered gene segments from the chlorobenzoate-degradative plasmid so as to allow efficient degradation of a dichlorobenzoate such as 3,5-dichlorobenzoate (Chakrabarty et al., 1984). Interaction among plasmids is therefore an important component in the evolution of new degradative functions against other chlorinated compounds. Different plasmids, even without interaction, are also sometimes necessary for the total mineralization of a recalcitrant compound. For example, the plasmid pKF1 specifies conversion of mono- or polychlorinated biphenyls to the corresponding chlorobenzoic acids, but it cannot effect any further conversion of these compounds. The plasmids pAC27 or pAC31, on the other hand, encode total degradation of the mono- or dichlorobenzoic acids. A combination of strains harboring these two types of plasmids allows complete mineralization of mono- or dichlorobiphenyls (Furukawa and Chakrabarty, 1982). A different kind of role for plasmids has been implicated in the evolution of a degradative pathway for the herbicide 2,4,5-trichlorophenoxyacetic acid (2,4,5-T). During prolonged selection with soil microorganisms in the chemostat in the presence of a large number of plasmids and with 2,4,5-T as a major source of carbon and energy, a strain of *P. cepacia* ultimately emerged that could utilize 2,4,5-T as its sole source of carbon and energy (Chakrabarty et al., 1984). Selection in the absence of the plasmid genes did not produce any 2,4,5-T-degrading strain, indicating that the 2,4,5-T-degradative capability perhaps evolved by recruitment and genetic rearrangement of at least some of the plasmid genes that were supplied to the chmostat (Karns et al., 1984). The role of mutations, gene duplications, and genetic rearrangements in the evolution of new degradative functions has previously been emphasized (Clarke, 1978). Since no natural microorganism, pure or mixed, was previously known to degrade 2,4,5-T as its sole source of carbon and energy, the development of such a strain raises the hope that other strains capable of totally degrading other recalcitrant compounds can also be developed. The 2,4,5-T-degrading strain shows some interesting properties. It is not only capable of completely dechlorinating 2,4,5-T, but it can also dechlorinate a variety of chlorophenols such as various di-, tri-, tetra-, and pentachlorophenols. The strain converts 2,4,5-T to 2,4,5-trichlorophenol which it can use as its sole source of carbon and energy. This step is constitutive. The degradation of 2,4,5-trichlorophenol is mediated by an inducible enzyme system which is induced by 2,4,5-trichlorophenol but not 2,4,5-T. Since 2,4,5-trichlorophenol is extremely toxic and does not exert any feedback inhibition on its formation from 2,4,5-T, growth of the 2,4,5-T-degrading *P. cepacia* cells with succinate and 2,4,5-T leads to accumulation of toxic concentrations of the chlorophenol, thereby inhibiting the growth of the cells (Karns et al., 1984).

Practical Considerations in Microbial Interaction with Toxic Chemicals

Directed evolution and natural gene-transfer mechanisms have not only allowed development of microorganisms capable of degrading chlorophenols, chlorosalicylates, and 2,4,5-T, but this coupled with our ability to clone entire degradative-gene clusters with cosmid vectors harboring broad host-range plasmid replication genes (Ghosal et al., 1985) will in the future greatly accelerate the rate of directed evolution of new degradative functions against other toxic chemicals. The absence of any alternate technology for the cleanup of toxic chemicals from waste-dump sites as well as places such as Love Canal in the U.S. and Seveso in Italy dictates that appropriate microbial technology should be developed for this purpose. It should be emphasized that, although limited in scope, studies with the 2,4,5-T-degrading strain have demonstrated the effectiveness of microorganisms in removing more than 90% of the chemical present at very high concentrations (20,000 ppm) from contaminated soil in a few weeks (Karns et al., 1984). Removal of the 2,4,5-T from contaminated soil allows the soil to support the growth of plants sensitive to low (10 ppm) concentrations of 2,4,5-T. From a practical point of view, laboratory-developed strains capable of degrading a toxic chemical may therefore find use in combating environmental pollution caused by the deliberate or accidental release of such chemicals. However, several questions remain to be answered. Would microbial technology allow development of strains for the complete mineralization of such recalcitrant compounds as chlorinated dioxins, polychlorinated biphenyls, ethylene dibromide, and a host of other potent carcinogens? Even if it were possible to develop such strains, would the strains be effective in removing very low concentrations (less than 5 ppm) of the chemicals that are normally found in contaminated sites? This is a pertinent question since it has been demonstrated (Karns et al., 1984) that although the 2,4,5-T-degrading strain can effectively remove 2,4,5-T when it is present in high concentrations, it does not appear to have any effect when the 2,4,5-T concentration is less than 10 ppm. On the other hand, oligotrophic bacteria are known that can utilize the substrates only at low concentrations. Would introduction of 2,4,5-T- or dioxin-degradative genes in such oligotrophic bacteria enable them to utilize the toxic chemical and remove it from the environment?

Another important question in this context is the role of the government regulatory apparatus in formulating defined guidelines that will ensure the safety of the application of genetically improved microorganisms in an open environment. Although the removal of a toxic chemical from the contaminated sites is most desirable from the public-health point of view, one has to ask if such application may lead to new problems affecting the health of nearby people and the ecology in general. It should be emphasized again that the microorganisms being developed are derived from the natural surrounding, and the only discernible changes that occur concern the evolution of a new capability for the degradation of a hazardous compound. It has been demonstrated that such microorganisms have little competitive ability to survive in nature, in the absence of the toxic chemical which only they can utilize, and die off quickly (Karns et al., 1984). Since fear of the unknown is an emotional issue, no amount of assurance in the absence of hard scientific evidence will satisfy the opponents of this technology. Such scientific evidence has to be obtained under the careful watch of a government-sponsored supervisory committee, in order to ensure that such a technology, if proven effective, will be harmless, inexpensive, and ecologically acceptable.

Acknowledgements. The work in our laboratory was supported by grants from the National Science Foundation (DMB85-14671) and Public Health Service grant ES04050 from the National Inst. of Environmental Health Sciences.

References

Alexander M (1981) Biodegradation of chemicals of environmental concern. Science 211:132–138

Bindereif A, Neilands JB (1983) Cloning of the aerobactin-mediated iron assimilation system of plasmid ColV. J Bacteriol 153:1111–1113

Bopp LH, Chakrabarty AM, Ehrlich HL (1983) Plasmid-determined resistance to Cr(VI) and reduction of Cr(VI) to Cr(III). J Bacteriol 155:1105–1109

Brierley CL (1978) Bacterial leaching. Crit Rev Microbiol 6:207–267

Brinckman FE, Olson GJ, Iverson WP (1982) The production and fate of volatile molecular species in the environment: metals and metalloids. In Atmospheric Chemistry, ed. E.D. Goldberg, pp 231–249. Dahlem Konferenzen. Berlin, Heidelberg, New York: Springer-Verlag

Brown NL, Ford SJ, Pridmore RD, Fritzinger DC (1983) Nucleotide sequence of a gene from the Pseudomonas transposon Tn501 encoding mercuric reductase. Biochemistry 22:4089–4095

Chakrabarty AM (1978) Genetic mechanisms in metal-microbe interactions. In Metallurgical Applications of Bacterial Leaching and Related Microbiological Phenomena, eds. L.E. Murr, A.E. Toema, and J.A. Brierley, pp 137–149. New York: Academic Press

Chakrabarty AM (1982) Genetic mechanisms in the dissimilation of chlorinated compounds. In Biodegradation and Detoxification of Environmental Pollutants, ed. A.M. Chakrabarty, pp 127–139. Boca Raton, FL: CRC Press

Chakrabarty AM, Karns JS, Kilbane JJ, Chatterjee DK (1984) Selective evolution of genes for enhanced degradation of persistent, toxic chemicals. In Genetic Manipulation: Impact on Man and Society, eds. W. Arber, W.J. Peacock, K. Illmensee, and P. Starlinger, pp 43–54. Miami: ISCU Press

Chatterjee DK, Chakrabarty AM (1983) Genetic homology between independently isolated chlorobenzoate-degradative plasmids. J Bacteriol 153:532–534

Clarke PH (1978) Experiments in microbial evolution. In The Bacteria, eds. J.R. Sokatch and L.N. Ornston, vol 6, pp 137–218. New York: Academic Press

Furukawa K, Chakrabarty AM (1982) Involvement of plasmids in total degradation of chlorinated biphenyls. Appl Envir Microbiol 44:619–626

Ghosal D, You I-S, Chatterjee DK, Chakrabarty AM (1985) Genes specifying degradation of 3-chlorobenzoic acid in plasmids pAC27 and pJP4. Proc Natl Acad Sci USA 82:1638–1642

Hallas LE, Means JC, Cooney JJ (1982) Methylation of tin by estuarine microorganism. Science 215:1505–1507

Hansen CL, Zwolinski G, Martin D, Williams JW (1986) Bacterial removal of mercury from sewage. Biotechnol Bioeng (in press)

Holmes DS, Lobos JH, Bopp LH, Welch GC (1984) Cloning of a Thiobacillus ferrooxidans plasmid in Escherichia coli. J Bacteriol 157:324–326

Iverson WP, Brinckman FE (1978) Microbial metabolism of heavy metals. In Water Pollution Microbiology, ed. R. Mitchell, vol 2, pp 201–232. New York: John Wiley and Sons

Karns JS, Kilbane JJ, Chatterjee DK, Chakrabarty AM (1984) Microbial biodegradation of 2,4,5-trichlorophenoxyacetic acid and chlorophenols. In Genetic Control of Environmental Pollutants, eds. G.S. Omenn and A. Hollaender, pp 3–21. New York: Plenum Press

Kilbane JJ, Chakrabarty AM (1984) Evolution of new degradative functions for halogenated aromatics. In Genetic Control of Environmental Pollutants, eds. G.S. Omenn and A. Hollaender, p 373. New York: Plenum Press
Kloepper JW, Leong J, Teintze M, Schroth MN (1980a) Enhanced plant growth by siderophores produced by plant growth-promoting rhizobacteria. Nature 386:885–886
Kloepper JW, Leong J, Teintze M, Schroth MN (1980b) Pseudomonas siderophores: a mechanism explaining disease-suppressive soils. Curr Microbiol 4:317–320
Knackmuss H-J (1981) Degradation of halogenated and sulfonated hydrocarbons. In Microbial Degradation of Xenobiotics and Recalcitrant Compounds, eds. T. Leisinger, A.M. Cook, P. Hutter, and J. Nuesch, pp 189–212. London: Academic Press
Laird AJ, Ribbons DW, Woodrow GC, Young IG (1980) Bacteriophage Mu-mediated gene transposition and in vitro *cloning of the enterochelin gene cluster of Escherichia coli*. Gene 11:347–357
Martin PAW, Dugan PR, Tuovinen OH (1981) Plasmid DNA in acidophilic, chemolithotrophic thiobacilli. Can J Microbiol 27:850–853
Moores JC, Magazin M, Ditta GS, Leong J (1984) Cloning of genes involved in the biosynthesis of pseudobactin, a high-affinity iron transport agent of a plant growth-promoting Pseudomonas strain. J Bacteriol 157:53–58
Neilands JB (1982) Microbial envelope proteins related to iron. Ann Rev Microbiol 36:285–309
NiBhriain N, Silver S, Foster TJ (1983) Tn*5* insertion mutations in the mercuric ion resistance genes derived from plasmid R100. J Bacteriol 155:690–703
Osborne FH, Ehrlich HL (1976) Oxidation of arsenite by a soil isolate of *Alcaligenes*. J Appl Bacteriol 41:295–305
Pan-Hou HS, Imura N (1982) Involvement of mercury methylation in microbial mercury detoxification. Arch Microbiol 131:176–177
Reineke W, Wessels SW, Rubio MA, Latorre J, Schwien U, Schmidt E, Schlomann M, Knackmuss H-J (1982) Degradation of monochlorinated aromatics following transfer of genes encoding chlorocatechol catabolism. FEMS Microbiol Lett 14:291–294
Rinderle SJ, Booth JE, Williams JW (1983) Mercuric reductase from R-plasmid NR1: Characterization and mechanistic study. Biochemistry 22:869–876
Schroth MN, Hancock JG (1982) Disease-suppressive soil and root-colonizing bacteria. Science 216:1376–1381
Senior E, Bull AT, Slater JH (1976) Enzyme evolution in a microbial community growing on the herbicide Dalapon. Nature 263:476–479
Silver S, Budd K, Leahy KM, Shaw WV, Hammond D, Novick RP, Willsky GR, Malamy MH, Rosenberg H (1981) Inducible plasmid-determined resistance to arsenate, arsenite and antimony(III) in Escherichia coli and *Staphylococcus aureus*. J Bacteriol 146:983–996
Silver S, Keach D (1982) Energy-dependent arsenate efflux: the mechanism of plasmid-mediated resistance. Proc Natl Acad Sci USA 79:6114–6118
Silver S, Misra RK (1984) Bacterial transformations of and resistances to heavy metals. In Genetic Control of Environmental Pollutants, eds. G. Omenn and A. Hollaender, pp 23–46. New York: Plenum Press

Silverberg BA, Wong PTS, Chau YK (1976) Ultrastructural examination of Aeromonas cultures in the presence of organic lead. Appl Envir Microbiol 32:723–725
Slater JH, Bull AT (1982) Environmental microbiology: biodegradation. Phil Trans Roy Soc Lond B 297:575–597
Tynecka Z, Gos Z, Zajac J (1981) Energy-dependent efflux of cadmium coded by a plasmid resistance determinant in Staphylococcus aureus. J Bacteriol 147:313–319
Warner PJ, Williams PH, Bindereif A, Neilands JB (1981) ColV plasmid-specified aerobactin synthesis by invasive strains of Escherichia coli. Infect Immun 33:540–545
Weightman AJ, Don RH, Lehrbach PR, Timmis KN (1984) The identification and cloning of genes encoding haloaromatic catabolic enzymes and the construction of hybrid pathways for substrate mineralizations. In Genetic Control of Environmental Pollutants, eds. G.S. Omenn and A. Hollaender, pp 47–80. New York: Plenum Press
Wood JM (1983) Selected biochemical reactions of environmental significance. Chem Scr 21:150–166
Wood JM, Cheh A, Dizikes LJ, Ridley WP, Rakow S, Lakowicz JR (1978) Mechanisms for the biomethylation of metals and metalloids. Fed Proc 37:16–21
Wood JM, Wang HK (1983) Microbial resistance to heavy metals. Envir Sci Technol 17:582A–590A

The Importance of Chemical "Speciation" in Environmental Processes,
eds. M. Bernhard, F. E. Brinckman, and P. J. Sadler, pp. 533–561.
Dahlem Konferenzen 1986. Springer-Verlag Berlin, Heidelberg
© Dr. S. Bernhard, Dahlem Konferenzen

Electrochemical Methods and Their Limitations for the Determination of Metal Species in Natural Waters

W. Lund

Dept. of Chemistry
University of Oslo, Blindern
0315 Oslo 3, Norway

Abstract. The use of ion-selective electrodes, polarography, anodic stripping voltammetry, and related techniques for studies of metal species is reviewed. Ion-selective electrodes measure the concentration of free metal and are particularly useful for toxicity studies and for the determination of the complexation capacity of fresh water. The polarographic and voltammetric methods measure the concentration of free metal as well as easily reducible (labile) metal complexes, whereas inert complexes are not detected. Anodic stripping voltammetry is particularly useful for the study of trace metal species in seawater at natural concentration levels. Several speciation schemes are discussed in which anodic stripping is used.

Introduction

In the past, electrochemical methods have frequently been used for the determination of metals in natural waters. However, among the many different electroanalytical techniques (Meites et al., 1976; Turner and Whitfield, 1981) only a few have gained wider acceptance for the study of chemical species. Among these one can distinguish between two principally different approaches, namely, those based on potentiometry and those based on voltammetry.

In the potentiometric method the equilibrium potential of an ion-selective electrode is measured at zero current and the potential is proportional to the logarithm of the activity or concentration of the ion in question.

In the voltammetric approach redox reactions take place as a result of the potential which is applied to the working electrode; the measured current is proportional to the concentration of the free metal ions and easily reducible metal complexes (labile metal) in solution. Strictly speaking, the term voltammetry is reserved for the measurement of current-voltage curves at a stationary working electrode, while the term polarography indicates the use of a dropping mercury electrode.

Related to voltammetry are also some other techniques in which the metal to be determined is first electrodeposited at a suitable working electrode and then determined by some other means than the recording of current-voltage curves, such as by potentiometric stripping or atomic absorption spectrometry.

Direct potentiometry has mostly been confined to the determination of the major components of natural waters, whereas potentiometric titration with copper has been widely used for the determination of the complexing capacity of humic material. In contrast, the voltammetric methods have been much used for the determination of trace metals, owing to a high degree of selectivity and favorable detection limits.

The electrochemical techniques mentioned above rank among the most frequently used methods for the study of metal species in natural waters. Even so, the interpretation of the results of electrochemical measurements may be rather complicated, for a number of reasons which will be discussed below.

Potentiometry

In direct potentiometry the potential between an ion-selective electrode (ISE) and a reference electrode is measured with a high-impedance voltmeter. The ISE responds to the activity of the ion in question, according to the equation

$$E = K - \frac{RT}{nF} \ln a_M, \tag{1}$$

where E is the measured potential of the ISE and a_M is the ionic activity. The potential of the reference electrode is assumed to remain constant during the measurements. However, the liquid junction potential of the reference electrode may pose a problem because it can change with time. Generally it depends on the construction and/or history of the liquid junction, the solution composition, and the rate of stirring (Culberson, 1981).

Although the ISE responds to the ionic activity, the concentration of the ion can be determined with an ISE, provided that the activity coefficient is kept constant. This is normally achieved by ionic strength buffering, which has the additional advantage of stabilizing the liquid junction potential of the reference electrode. Equation 1 can then be rewritten

$$E = K' - \frac{RT}{nF} \ln C_M, \tag{2}$$

where C_M is the ionic concentration.

Owing to problems related to the solubility of the ion-selective membrane, the detection limit of ISEs is usually in the 10^{-6}–10^{-7} M range, in terms of the total-metal concentration. This prevents the use of ISEs for the direct determination of most trace metals in natural waters. However, in the presence of complexing agents the electrode may show a Nernstian response for free metal far below the detection limit, provided that the total metal concentration is above this limit. Thus, a Nernstian response of a copper ISE down to pCu 19 has been observed for solutions containing 1 mM of copper and 0.01 M ethylenediamine (Avdeef et al., 1983).

ISEs have been used to determine alkali and alkaline-earth metals and halide ions plus a few other ions in natural waters. Furthermore, copper, lead, and cadmium electrodes have been used in titrations to determine the complexation capacity of such samples.

Unfortunately, the ISEs are subject to interference from the major ions present in natural waters. Thus, each of the alkali or alkaline-earth metal electrodes suffers from interference from the other alkali and alkaline-earth metals. For instance, the potential of the calcium electrode has been described by the equation (Culberson, 1981)

$$E_{Ca} = E^{O}_{Ca} + \frac{RT}{2F} \ln(a_{Ca} + K_{Mg} a_{Mg} + K_{Na} a^{2}_{Na}), \tag{3}$$

where K_{Mg} and K_{Na} are the respective selectivity constants. Similarly, the copper and lead electrodes are subject to interference from chloride, particularly in seawater (Culberson, 1981; Zirino and Seligman, 1981).

However, by matching standards and unknown solutions the species of the major cations in seawater have been studied by means of ISEs. Thus,

Thompson (1966) and Thompson and Ross (1966) found that 80%–90% of the total magnesium and calcium in seawater was present in ionized form, in fair agreement with the chemical model for seawater proposed by Garrels and Thompson (1962). Still, the association constants obtained from such studies are somewhat ambiguous owing to limitations in the experimental setup (Culberson, 1981).

As already mentioned, the detection limits of ISEs are too high to measure the concentration of trace metals such as copper, lead, and cadmium in most natural water samples. Moreover, ISE experiments cannot be carried out at very low ionic strength. However, copper, lead, and cadmium ISEs have been much used to measure the complexation of metal ions by humic matter at elevated metal-ion concentrations (Saar and Weber, 1982).

An ISE has also been used to measure the free metal-ion concentration in studies of the toxicity of copper towards biological test organisms. Thus, Sunda and Lewis (1978) found that the toxicity of copper towards algae is a function of the free-metal concentration and not of the total concentration of metal. However, an observed correlation between the ISE response and toxicity does not prove that only free metal is responsible for the toxic effect. If experiments are performed with metal-ion buffers and other variables are kept constant (pH, alkalinity, salinity, and competing organic ligands), there will be a fixed ratio of free-metal ions to all complexes which do not contain the metal-buffer ligands, and all of these species could equally well contribute to the toxicity. For instance, to evaluate which of the species Cu^{2+}, $CuOH^{+}$, or $CuCO_3$ is toxic both pH and the total carbonate concentration must be varied (Guy and Ross Kean, 1980). In any case, there appears to be no simple relationship between the free-metal concentration and toxicity (Guy and Ross Kean, 1980; Florence et al., 1983). For instance, in a moderately humic-influenced eutrophic lake, an increased cadmium toxicity was observed during the growth season which far exceeded what would be expected according to the free-ion activity (Lagreid et al., 1983).

The majority of studies with ISEs have been concerned with the characterization of humic material by titrations with metals. The conditional formation constant and complexation capacity of the humic material, which are obtained from these titrations, will depend on the metal used in the titration. Normally, copper has been used because it is strongly bound to natural organic matter.

In their most simple form the complexation capacity calculations are based on 1:1 stoichiometry between the metal M and ligand L (Perdue and

Lytle, 1983):

$$M+L=ML. \tag{4}$$

For multiligand molecules such as humic material, the metal can be attached to n sites per molecule, which gives

$$n \cdot C_L=[L]+[ML], \tag{5}$$

where C_L is the concentration of humic material. By introducing the formation constant K, Eq. 5 can be rearranged in the following way:

$$\frac{C_M-[M]}{C_L}=\frac{nK[M]}{1+K[M]}, \tag{6}$$

where C_M is the total concentration of metal and

$$C_M=[ML]+[M]. \tag{7}$$

The free-metal concentration [M] is measured with an ISE.

Several linear transformations of Eq. 6 have been used, the one suggested by Scatchard (1949),

$$\frac{C_M-[M]}{C_L[M]}=nK-\frac{C_M-[M]}{C_L}\cdot K, \tag{8}$$

leads to a graphical solution, a "Scatchard plot," in which the left side of Eq. 8 is plotted against $(C_M\text{-}[M])/C_L$: K is estimated from the slope, and n from the intercept (Bresnahan et al., 1978).

An alternative linearization is given by

$$\frac{C_L}{C_M-[M]}=\frac{1}{n}+\frac{1}{nK[M]}. \tag{9}$$

By plotting the left side of Eq. 9 vs. 1/[M], n and K can be obtained (Buffle et al., 1977).

If C_L is known in terms of g L^{-1} of total organic matter, the average molecular weight, instead of n, is determined from the plots described above.

If C_L is unknown, the following linearization can be used to calculate C_L, assuming $n=1$:

$$\frac{[M]}{C_M-[M]}=\frac{[M]}{C_L}+\frac{1}{K\cdot C_L}. \tag{10}$$

Here, a plot of the left side of this equation vs. [M] should produce a straight line, with slope $1/C_L$ and intercept $1/KC_L$ (Hart, 1981). C_L is normally called the complexation capacity of the water sample.

The complexation capacity can also be obtained more directly from the endpoint of the titration curve, particularly if the titration data are transformed to give a Gran's plot (Gran, 1952; Shuman et al., 1983), or the data are plotted as bound metal vs. total metal (Tuschall and Brezonik, 1983).

It should be stressed that the complexation capacity as determined above is really a measure of the ability of the chosen metal ion (usually copper) to replace any cation (e.g., calcium) that might already be bound to the ligand L. Some metals (e.g., Fe(III)), which can occupy most of the complexing sites of the ligand (Reuter and Perdue, 1977), may not be easily replaced by copper. Furthermore, any measurable amount of copper present in the sample before the titration starts should be recorded.

Little attention has been focussed on the relationship between the amounts of free metal, bound metal, and the complexation capacity for additional metal.

The significance of the experimentally determined K value has been disputed (Perdue and Lytle, 1983). Strictly speaking, it is not a constant at all but varies with the solution composition. Particularly the C_M/C_L ratio and pH influence the degree of complexation, whereas regarding the origin of organic matter (whether derived from soil or formed in water, etc.) and the ionic strength and temperature, these parameters seem to have less influence (Buffle et al., 1984).

Furthermore, Scatchard plots are often nonlinear, owing to the fact that the humic material contains a mixture of ligands with a large number of nonequivalent binding sites. For nonlinear Scatchard plots a modified equation can be used (Perdue and Lytle, 1983):

$$\frac{C_M-[M]}{C_L}=\sum_{i=1}^{m}\frac{n_iK_i[M]}{1+K_i[M]}. \tag{11}$$

Using $m=2$ will result in an improved fit of the experimental data (Sposito et al., 1981). However, the four parameters K_1, K_2, n_1, and n_2

which are obtained would only be chemically meaningful if the humic material contained only two types of binding sites, which is considered by most workers to be an oversimplification for metal binding with humic materials. Even so, these parameters may still be used to predict the degree of complexation of a given metal by humic substances within the experimental range of C_M and C_L values. These questions are more fully discussed in a recent symposium volume about humic materials (Langford et al., 1983; Perdue and Lytle, 1983; Shuman et al., 1983).

Many authors have discouraged the use of the term complexation capacity. Thus Langford et al. (1983) recommend that this parameter as determined by blind titration be abandoned because it is such a sensitive function of many factors.

As in the case of the complexation constant, the complexation capacity will be a function of experimental parameters such as pH and concentration of reactants. Furthermore, it will depend on the metal which is used for titrating the organic material.

Adding to the confusion is the fact that some authors have used the term complexation capacity for a parameter which is quite different from the one discussed above. Thus, McCrady and Chapman (1979) defined the copper-complexation capacity as the percentage of complexed copper, including solid forms within the water column, at any given total-copper concentration in the water column. The authors used a copper ISE and a computer program to measure and estimate the free copper-ion activity at different total concentrations of copper.

Giesy et al. (1978) defined the term binding capacity, which is often used synonymously with the complexation capacity, as the reduction of cadmium, copper, and lead in the metal standards resulting from titration with the water sample. Again, the free metal was measured with ISEs.

Polarography

The polarographic and voltammetric techniques are particularly well suited to the determination of trace metals in natural waters. Even high concentrations of alkali and alkaline-earth salts do not interfere; on the contrary, such salts are needed as a supporting electrolyte. Normally, little or no addition of reagents is needed for the analysis.

In polarography the current arising from the reduction or oxidation of a chemical species under diffusion-controlled conditions is recorded as a function of the applied potential. In classical polarography the current is recorded while a linearly increasing negative potential is applied to a dropping mercury electrode. In the resulting polarogram the diffusion current is proportional to the concentration of the species in solution, while the half-wave potential ($E_{1/2}$) is related to the standard potential of the redox reaction and can be used to identify the species. Complex formation with a ligand L according to the equation

$$M + xL \rightleftharpoons ML_x \tag{12}$$

shifts $E_{\frac{1}{2}}$ to more negative values,

$$\Delta E_{\frac{1}{2}} = -\frac{RT}{nF}\ln K - x\frac{RT}{nF}\ln[L], \tag{13}$$

thus the formation constant K and ligand number x can be determined from such measurements (Crow, 1969). It is advisable that the ligand be present in a marked excess; if not, the release of uncomplexed ligand at the electrode surface may give rise to erroneous results (Stolzberg, 1977). If the complexation reaction involves protons or hydroxyl ions, the number of these ions can be obtained from the variation of $E_{1/2}$ with pH.

The detection limit of classical polarography is around 10^{-5} M, whereas that of the modern polarographic techniques such as differential pulse polarography and alternating current polarography is in the range 10^{-7}–10^{-8} M (Bond, 1980). The metals which are most suitable for polarographic analysis are copper, lead, cadmium, and zinc. However, these metals are normally present in natural waters at concentrations even below the detection limit of the modern techniques mentioned above. Even so, these techniques may provide valuable information about complexation reactions in synthetic solutions at elevated metal concentrations (Nürnberg, 1983).

A limitation in the use of polarographic methods for the study of metal-complex equilibria is that the reduction of the metal complex should be electrochemically reversible. That is, the rate constant for the complex formation/dissociation should be high, relative to the time scale of the experiment (drop time of the electrode), so that the polarographic curve is governed by Nernstian behavior. Such complexes are called labile.

Many metals form labile complexes with inorganic ligands such as Cl^-, OH^-, CO_3^{2-}, and SO_4^{2-}. Also, many organic molecules form electrochemically labile complexes. However, larger organic molecules of natural origin, such as humic material, normally form complexes with heavy metals that are not reducible and therefore not directly accessible to polarographic analysis. Such complexes are termed inert. In this case polarography can only be used to measure the amount of free metal remaining in solution, as in the case of measurements with ion-selective electrodes.

When polarographic methods are used to study the complexation of metals by humic material, this material may adsorb on the electrode surface and cause changes in both the $E_{1/2}$ and diffusion current values (Buffle et al., 1976). These effects may be particularly marked when modern polarographic techniques are used because these are high-frequency methods which are particularly sensitive to the presence of surface-active substances. Such substances can alter the rate of the electrode reactions (Florence, 1982) or even give rise to tensammetric waves (Jacobsen and Lindseth, 1976).

When differential pulse polarography is used, one should also take into account that instrumental artifacts can be observed (Christie et al., 1973).

Anodic Stripping Voltammetry

Anodic stripping voltammetry (ASV) is the electroanalytical method that offers the best sensitivity and detection limits. It has become the predominant electroanalytical method for studies of metal species because the trace metals are frequently present in sub-ppb concentrations in natural waters. The ASV determination consists of two steps: the metal is first deposited for a given time by reduction at a constant potential on the electrode, and then stripped from the electrode during a positive scanning of the potential by reoxidation. In the latter step the current is recorded as the reoxidation of each metal gives rise to a current vs. potential peak. Typical voltammograms are shown in Fig. 1.

Normally either a hanging mercury drop electrode (HMDE) or a mercury film electrode (MFE) is used. The HMDE gives better results in complicated matrices, while the MFE has better detection limits. A graphite or glassy carbon substrate is the most common support for the mercury film.

The HMDE is usually preferred because it is more easily reproducible and renewable than the preformed MFE. A third type of mercury electrode, prepared by adding mercuric ions to the sample and forming the mercury film *in situ,* is less attractive for determination of species because the addition of mercuric ions may disturb the chemical equilibria in the sample (Plavšić et al., 1982; Skogerboe et al., 1980).

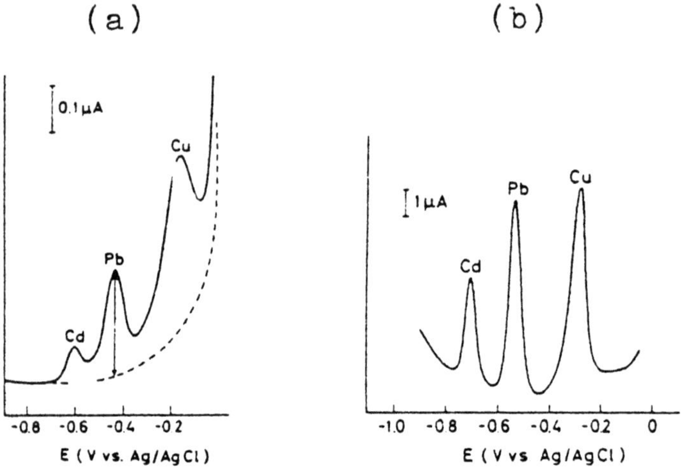

Fig. 1 a, b. Differential pulse anodic stripping voltammograms of polluted seawater at pH 8.0, obtained with a hanging mercury drop (**a**) and a mercury-film electrode (**b**) (Lund and Onshus, 1976)

The electrochemical stripping is usually carried out by the differential pulse technique (DPASV) because it gives better detection limits than linear sweep stripping. Alternatively, alternating current stripping can be used which gives detection limits comparable with differential pulse stripping (Batley and Florence, 1974).

For environmental samples the studies by DPASV have been focussed on the metals which form soluble amalgams, particularly copper, lead, cadmium, and zinc, but some other metals such as bismuth, antimony, and thallium can also be determined. The detection limit for these metals is usually below 0.01 μg L^{-1}. It depends on the deposition time of the ASV measurement, but for practical reasons a deposition time on the order of minutes is rarely exceeded.

The interpretation of an ASV curve is more complicated than that of a polarogram. The peak current is a function of the concentration of a reducible species in solution, but the peak potential does not relate directly to this species but to the oxidation of the corresponding metal amalgam. To overcome this problem, so-called stripping-polarographic curves (also called pseudopolarographic) can be recorded. Here the peak current, which is normally obtained by DPASV, is recorded as a function of the deposition potential; the resulting current-voltage curve looks like an ordinary d.c. polarogram.

Labile-metal complexes can be studied by varying the ligand concentration. From the shift in the half-wave potential of the stripping polarogram the formation constant and ligand number can be obtained with Eq. 13 (Kounaves and Zirino, 1979; Zirino, 1981). In this way, the complex equilibria of lead carbonates in seawater have been studied and the predominance of these lead complexes has been demonstrated (Valenta, 1983).

For inert complexes it is not the shift in the half-wave potential but the reduction in the peak current which indicates the complexation reaction.

Titration with a Ligand

In a series of papers Nürnberg and co-workers (1983, 1981, 1983) have studied the complexation of lead, cadmium, and zinc by titration with nitrilotriacetic acid (NTA), which forms irreversible complexes with these metals. The competing equilibria of inorganic cations (Na^+, Ca^{2+}, Mg^{2+}) and anions (Cl^-, OH^-, CO_3^{2-}) for the NTA were examined, and the conditional formation constants and concentration of NTA required to reach a given degree of chelation of the metals in seawater were established (Nürnberg, 1983; Valenta, 1983).

The studies with synthetic ligands have also revealed that owing to a ligand exchange mechanism, the reaction between the heavy metals and a ligand such as EDTA is kinetically slow in seawater. The alkaline-earth ions which occupy the binding site of the organic ligand must be displaced by the more strongly complexing heavy metals, when these are added in trace concentrations. The formation-rate constants are in the range $2 \cdot 10^2$–$5 \cdot 10^3$ $\mathrm{Lmol^{-1}\ g^{-1}}$ for cadmium, lead, and zinc, which implies that the peak height corresponding to labile metal may change even after 1 hr (Nürnberg, 1983). These results agree with those of other workers who have found that equilibrium between ionic metal spikes and organic ligands naturally present in seawater may be reached after 20 min to 3 hr (Duinker and Kramer, 1977; Plavšić et al., 1982).

In addition to kinetic information of this type, there are other kinetic effects which should be taken into consideration. In voltammetric measurements there is always the possibility of a partial dissociation of the inert metal complex during the experiment (Davison, 1978; Nürnberg and Raspor, 1981; Whitfield and Turner, 1979). Such effects could obscure the interpretation of the results. Fortunately, when a large excess of ligand is used and/or the time scale of the measurement is short, the kinetic effects are generally negligible (Nürnberg and Raspor, 1981).

Even so, some authors discourage the use of synthetic complexers for studying trace-metal chemistry in seawater (Piotrowicz et al., 1983). They point out that natural metal-organic interactions are easily disturbed, and hence our perception of these interactions depends upon how quickly samples can be analyzed because true oceanic conditions cannot be duplicated under experimental conditions.

Titration with Metals

Instead of titrating the metals with a ligand, titration with a metal is often preferred in studies on metal species, for example, when investigating the complexation properties of naturally occurring humic material. Again it is assumed that inert complexes are formed, so that an eventual voltammetric response must be due to the free-metal or labile complexes in solution. This approach then becomes identical to that described above for ISE, where Scatchard plots or a similar treatment may be used for determining the complexing capacity and formation constant.

Under favorable conditions the complexing capacity can be obtained directly from the ASV titration curve (Fig. 2). This may occur for high concentrations of humic material when the formation constant is also high, so that no measurable metal peak is observed even after a few additions of metal.

However, the interpretation of the ASV measurements are frequently complicated by a number of factors. The observed decrease in current as

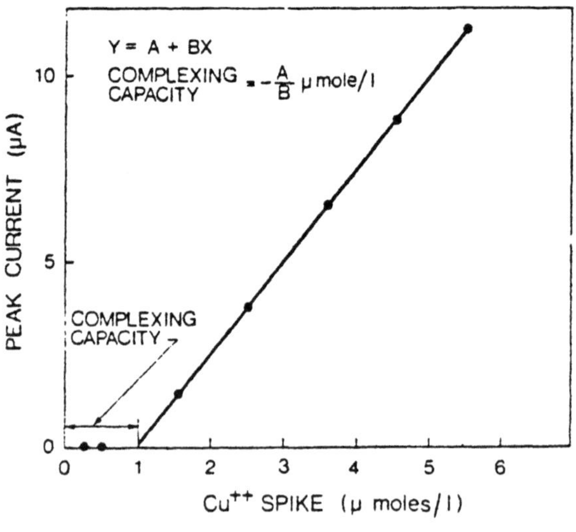

Fig. 2. Determination of the complexation capacity of a water sample under favorable conditions. The peak current of the anodic stripping voltammogram is recorded as a function of added copper (Chau et al., 1974)

compared to the calibration curve may be due to any of the following processes (Buffle et al., 1984).

a) ML is inert, i.e., dissociates slowly, and hence the metal is only partly or not at all reduced during the deposition step.
b) ML is labile but diffuses slowly, therefore it contributes less than M to the reduction current during the deposition step.
c) ML is labile, but the total concentration of L in the bulk solution is not much larger than the total concentration of M, so that L is saturated with M at the electrode surface during the stripping step and not in the bulk of the solution.

The possible role of the surface saturation of L can be examined by varying the deposition time; the effect of surface saturation varies with this parameter, whereas the bulk complexation should not.

No doubt the possible reactions in the diffusion layer should be taken into account when interpreting results based on ASV measurements. As shown by Hanck and Dillard (1977), low values for the complexing capacity may be obtained, even for simple ligands such as EDTA, from complexometric titrations with a voltammetric endpoint. In principle such titrations violate the rule that polarographic studies of metal complexes should be carried out in the presence of a marked excess of the ligand (Crow, 1969). This condition is evidently not fulfilled in the region of the equivalence point. However, under favorable conditions (i.e., high formation constants) the error introduced is relatively small (Chau et al., 1974).

It is also possible to correct for the kinetic-current contribution arising from the dissociation of the complex (Bhat and Weber, 1982; Saar and Weber, 1982; Shuman and Cromer, 1979; Shuman and Woodward, 1973). On the other hand, instead of being considered a source of interference, the fact that the voltammetric methods are sensitive to the rate of dissociation of the metal complexes can be used to advantage in studies on reaction kinetics of metal species. Thus, the dissociation rate of the complexes can be determined from rotating-disk electrode experiments, as discussed below.

Adsorption of Organic Compounds

Another possible source of interference in voltammetric measurements is the adsorption of dissolved organic compounds at the electrode surface. As pointed out by many authors (Brezonik et al., 1976; Florence, 1982; Saar and Weber, 1982), both the peak current and the potential of the ASV peak can be influenced by adsorption processes caused by the presence of surface-active substances. No doubt such compounds may seriously affect the results obtained in natural waters, particularly for organic-rich waters at natural pH levels. In this respect ASV is more vulnerable than polarography because of the long deposition time relative to the drop time of the dropping mercury electrode. However, when free metal-ion concentrations are calculated from *in situ* calibration curves, the effect of adsorption processes on the complexometric titration curves is minimized (Saar and Weber, 1982). Also, Plavšić et al. (1982) found that relatively high concentrations of surface-active substances, such as 10 mg L^{-1} of Triton X-100, are needed before any noticeable effect is observed. The authors concluded that adsorption of organic material did not interfere with the determination of the complexation capacity.

Comparison of Methods for Complexation Capacity

Several authors have compared different methods for determining complexation capacity. Thus, Tuschall and Brezonik (1983) compared the five methods ASV, ISE, fluorescence spectroscopy, continuous ultrafiltration, and competing-ligand differential spectroscopy for the determination of copper-binding capacities and conditional stability constants for swamp waters. The ASV results were found to yield consistently low values compared with the other four methods, which gave results that agreed reasonably well. The authors suggest that a possible reason for the low ASV results is that the metal-organic complexes were reducible under the experimental conditions used.

Jones and Hart (1984), when studying Island Billabong water, also found that the ASV method gave smaller complexing capacity and larger stability constants than the ISE method.

Buffle et al. (1984) have compared literature values for the complexation capacity, and they demonstrated that most of the values determined by ASV are lower than those determined by other methods. The authors suggest that the release of metal at the electrode surface during the stripping step may be responsible for the low results.

However, Jardim and Allen (1984) state that reasonable agreement among stability-constant and complexation capacity measurements were achieved for both algal-exudate and sewage-effluent samples when the following experimental techniques were compared: ASV, ISE, MnO_2 ion exchange, dialysis, and fluorescence. Also, Truitt and Weber (1981) found agreement between the results obtained by dialysis and Cu-ISE.

The determination of complexation capacity by different experimental techniques has recently been reviewed (Hart, 1981; Neubecker and Allen, 1983).

Effect of pH

The variation in the voltammetric peak potential and peak height with pH has some relevance to the determination of trace metal species (Bernhard et al., 1975; Duinker and Kramer, 1977; Ernst et al., 1975). The shift in the peak potential or, preferably, the half-wave potential of d.c. – or stripping-polarographic curves may be used to study the formation of hydroxo-complexes in synthetic solutions (Bilinski et al., 1976; Ernst et al., 1975). For natural waters this approach can hardly be used (Florence and Batley, 1980) because of both the complexity of such samples and the possible interference from organic compounds which may adsorb on the electrode surface.

The usefulness of following the variation in peak height with pH has also been questioned (Florence and Batley, 1980) for the same reason. For synthetic seawater the peak heights of copper, lead, cadmium, and zinc are independent of pH below pH 6 (Ernst et al., 1975; Sagberg and Lund, 1982). In alkaline solution the peak height of copper and lead decreases with increasing pH, owing to the formation of carbonato-complexes, whereas the peak heights of cadmium and zinc are less affected.

For natural waters increased peak currents have been reported for copper, lead, and zinc, when the sample was acidified below pH 6 (Zirino, 1981; Zirino and Yamamoto, 1972), whereas the cadmium peak remained constant.

Of course, an increase in peak current with decreasing pH does not necessarily mean an increase in concentration. However, for copper and lead the presence of more labile metal at pH 2 than at pH 6 has been clearly demonstrated (Duinker and Kramer, 1977; Hasle and Abdullah, 1981), whereas no such effect was observed for cadmium. The results obtained for zinc by Bernhard et al. (1975) also suggest that ionic zinc at lower pH is released from particulate and organically complexed forms present in seawater.

There has been some controversy about the optimal pH for determining the concentration of labile metal in fresh water. It has been argued that the addition of an acetate buffer, which is normally employed in voltammetry of fresh water, may release more metal than would be obtained by a direct

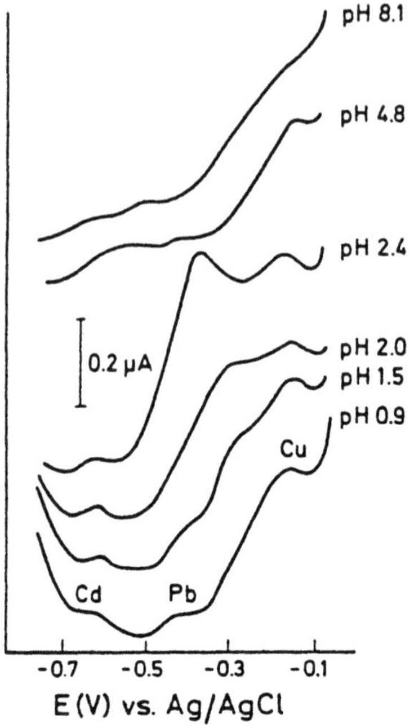

Fig. 3. Effect of pH on a differential pulse anodic stripping voltammogram for a seawater sample where interferences are observed (Nilsen and Lund, 1982)

analysis. Although this may be true, voltammetric measurements in unbuffered solutions can hardly be recommended.

For some samples it may be difficult to carry out a direct ASV analysis both at natural pH and in the presence of an acetate buffer (Hoffmann et al., 1981; Nilsen and Lund, 1982). In this case it is still possible to determine the metals in a more strongly acidified solution (Nilsen and Lund, 1982), as shown in Fig. 3.

Schemes for the Determination of Metal Species

A number of schemes have been developed which use various combinations of separation and measurement techniques. By necessity, the classification of metal species according to such schemes will be operationally defined, but the schemes may still be useful for the characterization of natural water. ASV is used in a majority of the schemes developed. There are various reasons for this choice. The detection limits of ASV allow a direct determination of heavy metals at the ppt level, and its selectivity makes direct analysis in seawater possible. Also, it is possible to discriminate between ASV-labile and total metal in solution, the latter being determined after a digestion of the sample.

The Batley and Florence scheme (1976) involves four sample treatments: filtration, passage through a Chelex-100 column, irradiation with UV light, and UV irradiation followed by a Chelex-100 separation. For each treatment both labile and total metal are measured. From these eight measurements seven groups of species are quantified (Fig. 4).

The operational character of the scheme is evident from the fact that labile metal is determined after addition of an acetate buffer and 10^{-5} M mercury(II) ions, which are required for the *in situ* formation of the mercury-film electrode.

The effect of these additions is discussed by Skogerboe et al. (1980). The *in situ* mercury-film electrode can be substituted with a preformed mercury-film electrode or a hanging mercury-drop electrode, which avoids the use of mercury(II) ions. However, ASV measurements in unbuffered freshwater samples at pH around 7 may not be advisable. Some buffer system should be used, even though this may shift the chemical equilibria in the solution.

Figura and McDuffie (1979) observed that the labile fractions determined by the Chelex method were generally larger than the fractions obtained by ASV. Their scheme (1980) uses both column and batch procedures to vary the contact time of the water sample with the Chelex-100 resin (Fig. 5). In this way they were able to distinguish between "very labile," "moderately labile," "slowly labile," and "inert" species, depending on the time scale of the measurements. The ASV-labile fractions were measured at pH 6.3 with a hanging mercury-drop electrode after addition of an acetate buffer.

Laxen and Harrison (1981) used ultrafiltration, Chelex-100, and ASV at natural pH and after acidification (Fig. 6), while Hoffman et al. (1981) developed an ultrafiltration procedure which involved the measurement of total metal, ASV-labile metal, and dissolved organic carbon in each fraction.

Florence (1982b) recently compared the effect of Chelex-100 resin, an oxine porous glass, and various thiol ion-exchange materials. Bio-Rad SM2 resin and hexane-butanol extraction were used to estimate lipid-soluble metal. The thiol resin, which was supposed to be more relevant for determining biologically reactive metal than Chelex-100, yielded similar results to Chelex-100 for zinc, cadmium, and lead but gave a higher percentage of

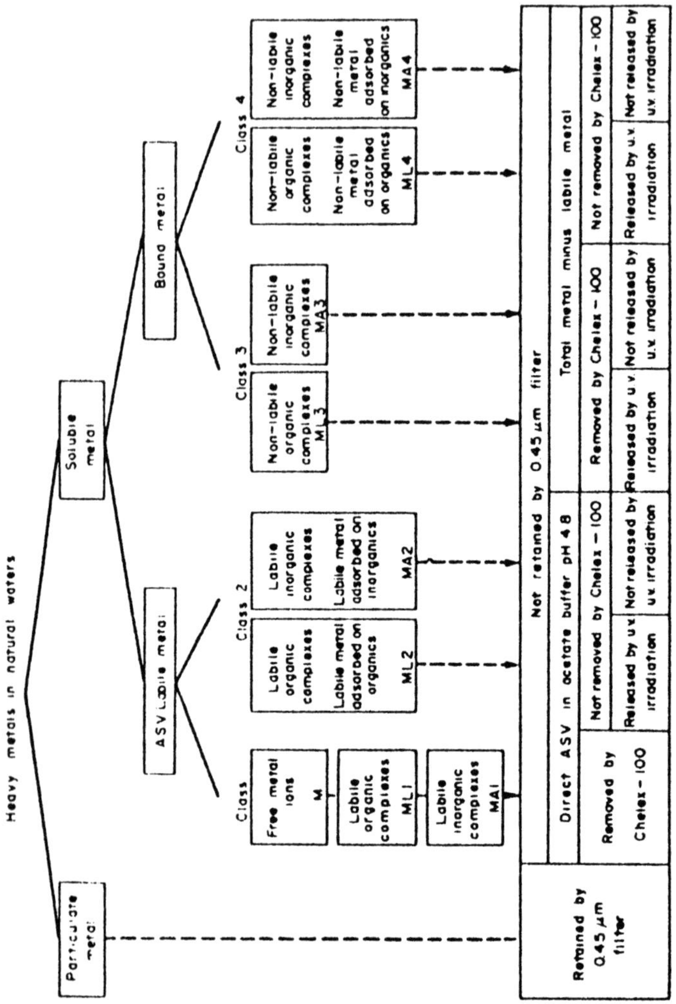

Fig. 4. The scheme for analysis of chemical species from Batley and Florence (1976)

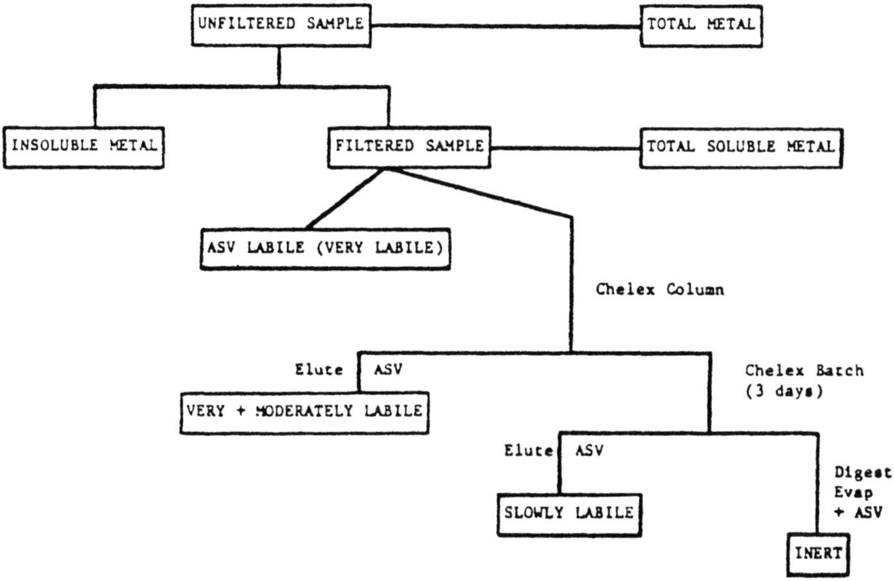

Fig. 5. The scheme for analysis of species from Figura and McDuffie (1979)

resin-labile copper. A hanging mercury-drop electrode was used in these experiments.

The measurement of total metal in these schemes must be carried out after digestion of the sample. Various digestion procedures have been used, such as boiling with nitric acid, fuming with a nitric-perchloric acid mixture, digestion with persulfate, or exposure to ultraviolet irradiation (Florence, 1982 a). The treatment necessary for complete digestion varies with the type of water sample (Florence, 1977; Nilsen and Lund, 1982). UV irradiation gives better blank values than acid digestion. However, for the determination of total metal it must be used in combination with acids. A quantitative release of bound metal can then be obtained (Batley and Farrar, 1978).

For a selective destruction of organic matter without decomposing inorganic material UV irradiation at the natural pH of the sample has been used. However, Florence (1977, 1982 a) found that it could not be applied to anoxic fresh waters, because oxidation of iron(II) leads to the precipitation of hydrous iron(III)-oxide, which co-precipitated other trace metals. Laxen and Harrison (1981) and Blutstein and Smith (1978) reported that

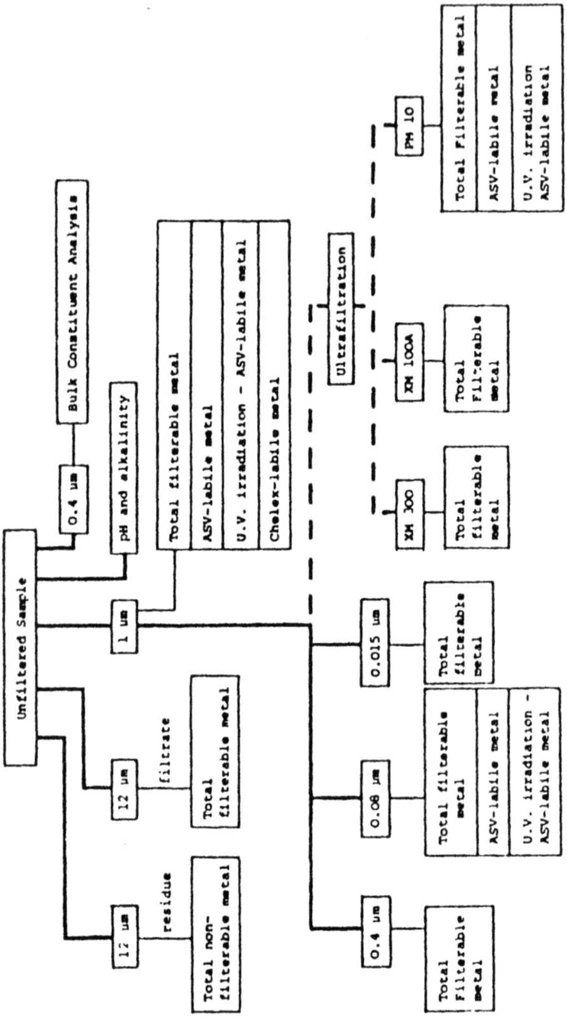

Fig. 6. The scheme for analysis of species from Laxen and Harrison (1981)

brown deposits formed on the bottom of the tubes when river water samples were irradiated. Therefore, uncritical use of UV irradiation in species analysis schemes cannot be recommended.

Rotating-Disk Electrode Measurements

A rotating-disk electrode has been used to estimate the dissociation-rate constant of copper-organic complexes (Shuman et al., 1983; Shuman and Michael, 1978). The topic is of particular interest because the kinetic dissociation of metal complexes may be related to the bioavailability and toxicity of the metals. ASV is combined with a mercury-plated rotating-disk electrode which allows kinetic measurements at very low metal concentrations. The effect of electrode-rotation rate during the deposition step is measured by integrating the stripping current. It is assumed that at very slow rotation rates all complexes have time to dissociate, whereas at high rotation rates there is no time for dissociation and only uncomplexed metal is reduced at the disk. Kinetic information is obtained from data between these two limits. The range of rate constants that can be measured by this technique is typically 10^{-3} to $10^{2} s^{-1}$.

The dissociation-rate constants were found to decrease with the molecular size of the ligand and also with the salinity of the samples from which the organics were obtained. Less than half of the total copper dissociated slowly enough to be distinguished kinetically from uncomplexed copper.

Potentiometric Stripping Analysis

Potentiometric stripping analysis (PSA) was introduced by Jagner in 1976. It is similar to ASV in that the metals to be determined are preconcentrated as amalgams at a mercury electrode. However, the metals are stripped from the electrode by a chemical oxidation agent which is present in the solution, e.g., oxygen. No external potential is applied to the cell in the stripping step. During the oxidation process the variation in the potential of the working electrode is recorded as a function of time; a stripping curve superficially resembling a d.c. polarogram is obtained (Fig. 7). The time needed for the stripping of each metal is proportional to the concentration of the metals in solution because the rate of oxidation is constant and controlled by the transport of oxidant to the electrode surface (Fig. 7).

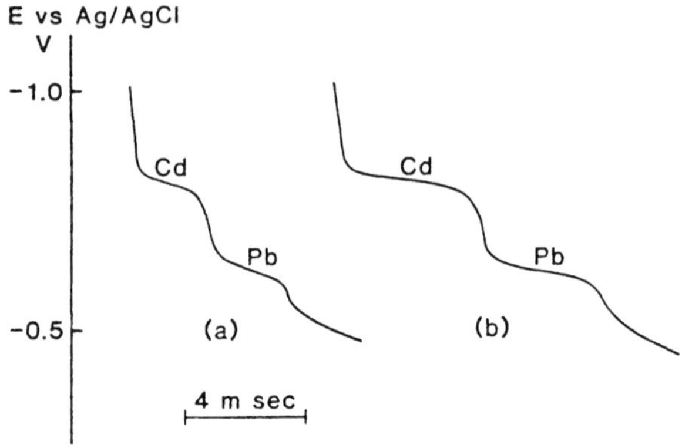

Fig. 7 a, b. Potentiometric stripping curve for a seawater sample, before (**a**) and after (**b**) the standard addition of cadmium and lead (Jagner et al., 1981)

In PSA the use of a mercury-film electrode is normally essential. When the concentration of metals is low or the concentration of oxidant is high, the stripping will take place so fast (in the course of milliseconds) that a microcomputer has to be used to record the stripping curve. The signal can then be displayed at a suitable rate when desired (Granéli et al., 1980). In addition, background subtraction and other manipulations of the signal are easily performed with computerized PSA.

In comparison with ASV, the PSA technique has the advantage that the oxygen in the sample must not be removed prior to analysis. PSA is also less sensitive to reducible or surface-active organics which may be present in the sample because, unlike ASV, no current passes through the working electrode during the stripping step. Therefore, it has been suggested that PSA may be more suitable than ASV for measurement of electrochemically labile metals in solution (Florence, 1982 a).

However, interferences from other species that can be reduced/oxidized may also be observed in PSA, because such species will affect the potential vs. time behavior of the working electrode when present in sufficiently high concentrations (Christensen et al., 1982).

Like ASV, PSA has also mainly been applied to the determination of metals which form soluble amalgams. So far PSA has not been used for analysis of species.

Electrodeposition – Atomic Absorption Spectrometry

The use of electrochemical preconcentration in combination with atomic absorption spectrometry offers an interesting possibility for those metals which do not form soluble amalgams and thus are unsuitable for determination by ASV. Batley and Matousek (1977, 1980) electrodeposited lead, cobalt, nickel, and chromium from seawater onto a tubular graphite electrode. Mercury(II) ions were added to the sample to facilitate the deposition. The electrochemical preconcentration step permitted the discrimination between labile and total metal in the water sample. Labile metal was determined after the addition of an acetate buffer (pH 4.6), while total metal was measured after boiling the sample with nitric acid or after UV irradiation in the presence of hydrogen peroxide for 6 hr.

For chromium, use was made of the fact that at pH 4.7 both Cr(VI) and Cr(III) are reduced and accumulated at a potential of -1.8 V vs. SCE, whereas at -0.3 V only Cr(VI) is selectively reduced to Cr(III), which accumulates by adsorption. In this way labile Cr(VI) and Cr(III) could be determined, as well as the total amount of chromium in the samples. Cr(VI) was found to predominate in the seawater samples, and most Cr(VI) was present in labile form.

Electrochemistry and Biological Availability

An important goal of species measurements is to obtain information about the bioavailability and toxicity of metals. As pointed out by some authors (Batley, 1983; Florence, 1982b; Whitfield and Turner, 1979), there are similarities between the processes taking place across a cell membrane and those occurring at an electrode surface. In voltammetric techniques the metal must be dissociated from its carrier during the reduction step. In biological systems such a dissociation may facilitate the transport through membranes by means of metal-binding proteins. However, to assume that there is a relationship between ASV lability and aquatic toxicity may be an oversimplification.

The assumption that labile metal is always more toxic to aquatic organisms than electrochemically inert forms of the metal is incorrect. The complexation of a heavy metal by an organic ligand may either hinder or favor its uptake by an organism, depending on the specific uptake mechanism (Guy and Ross Kean, 1980; Florence et al., 1983; and see Bernhard and George, this volume).

A large fraction of the trace metals in natural waters is probably bound to pseudocolloids such as fulvic and humic substances and humate-coated inorganic particles. A variety of these species may coexist in solution. It is possible that some metal which may be inert in some measurements is trapped inside the structure of these colloids without really being strongly bound in a chemical sense. A change in the composition of the sample solution may cause a change in the structure of these pseudocolloids, which makes the trapped metal more ASV- or bioavailable.

Conclusion

Under certain conditions electrochemical methods can be used directly to determine operationally defined species.

The ISEs are selective to free (hydrated) metal ions but do not have sufficiently low detection limits for the determination of most trace metals in natural waters. However, ISEs can be used in toxicity studies at elevated metal concentrations. They can also be used to determine the complexation capacity of natural waters.

ASV has the required sensitivity and selectivity for the determination of the trace metals copper, lead, cadmium, and zinc. The technique usually determines the sum of free-metal ions and easily reducible metal complexes. Metal ions which are bound to complexing agents such as dissolved humic material are normally not detected by the ASV technique. Therefore, a discrimination between ASV-labile and total metal is possible. The concentration of total metal can be determined after digesting the sample. ASV is used in various analytical schemes which determine operationally defined species. It can also be used to determine the complexation capacity of natural waters. Although in general, toxicity of a trace metal is related to its electroactivity, there seems to be no simple relationship between the concentration of ASV-labile metal and the toxicity of metal to aquatic organisms.

Various other electroanalytical methods based on voltammetry, rotating-disk electrode measurements, potentiometric stripping, and electrochemical preconcentration can be used for more specific purposes in studies on chemical species.

References

Avdeef A, Zabronsky J, Stuting HH (1983) Calibration of copper ion selective electrode response to pCu 19. Anal Chem 55:298–304

Batley GE (1983) The current status of trace element speciation studies in natural waters. In Trace Element Speciation in Surface Waters and Its Ecological Implications, ed. G.G. Leppard, pp 17–36. New York: Plenum Press

Batley GE, Farrar YJ (1978) Irradiation techniques for the release of bound heavy metals in natural waters and blood. Anal Chim Acta 99:283–292

Batley GE, Florence TM (1974) An evaluation and comparison of some techniques of anodic stripping voltammetry. J Electroanal Chem Interfac Electrochem 55:23–43

Batley GE, Florence TM (1976) A novel scheme for the classification of heavy metal species in natural waters. Anal Lett 9:379–388

Batley GE, Florence TM (1976) The effect of dissolved organics on the stripping voltammetry of seawater. J Electroanal Chem Interfac Electrochem 72:121–126

Batley GE, Matousek JP (1977) Determination of heavy metals in seawater by atomic absorption spectrometry after electrodeposition on pyrolytic graphite-coated tubes. Anal Chem 49:2031–2035

Batley GE, Matousek JP (1980) Determination of chromium speciation in natural waters by electrodeposition on graphite tubes for electrothermal atomization. Anal Chem 52:1570–1574

Bernhard M, Goldberg ED, Piro A (1975) Zinc in seawater – an overview 1975. In The Nature of Seawater, ed. E.D. Goldberg, pp 43–68. Berlin: Dahlem Konferenzen

Bhat GA, Weber JH (1982) Cadmium(II) binding by soil-derived fulvic acid measured by anodic stripping voltammetry. Anal Chim Acta 141:95–103

Bilinski H, Huston R, Stumm W (1976) Determination of the stability constants of some hydroxo and carbonato complexes of Pb(II), Cu(II), Cd(II) and Zu(II) in dilute solutions by anodic stripping voltammetry and differential pulse polarography. Anal Chim Acta 84:157–164

Blutstein H, Smith JD (1978) Distribution of species of Cu, Pb, Zu and Cd in a water profile of the Yarra River estuary. Water Res 12:119–125

Bond AM (1980) Modern Polarographic Methods in Analytical Chemistry. New York: Marcel Dekker

Bresnahan WT, Grant CL, Weber JH (1978) Stability constants for the complexation of copper(II) ions with water and soil fulvic acids measured by an ion selective electrode. Anal Chem 50:1675–1679

Brezonik PL, Branner PA, Stumm W (1976) Trace metal analysis by anodic stripping voltammetry: Effect of sorption by natural and model organic compounds. Water Res 10:605–612

Buffle J, Greter FL, Haerdi W (1977) Measurement of complexation properties of humic and fulvic acids in natural waters with lead and copper ion-selective electrodes. Anal Chem 49:216–222

Buffle J, Greter FL, Nembrini G, Paul J, Haerdi W (1976) Capabilities of voltammetric techniques for water quality control problems. Fresenius Z Anal Chem 282:339–350

Buffle J, Tessier A, Haerdi W (1984) Interpretation of trace metal complexation by aquatic organic matter. In Complexation of Trace Metals in Natural Waters, eds. C.J.M. Kramer, M. Branica, J.C. Duinker, J. Hart-Stam, R.F.C. Mantoura, H.W. Nürnberg, and W. Stumm. The Hague: M. Nijhoff/W. Junk
Chau YK, Gächter R, Lum-Shue-Chan K (1974) Determination of the apparent complexing capacity of lake waters. J Fish Res Board Can 31:1515–1519
Christensen JK, Kryger L, Pind N (1982) Effect of electroactive substances forming water-soluble reduced species on the potentiometric stripping determination of mercury soluble metals. Anal Chim Acta 136:39–50
Christie JH, Osteryoung J, Osteryoung RA (1973) Instrumental artifacts in differential pulse polarography. Anal Chem 45:210–215
Crow DR (1969) Polarography of Metal Complexes. London: Academic Press
Culberson CH (1981) Direct potentiometry. In Marine Electrochemistry, eds. M. Whitfield and D. Jagner, pp 203–261. New York: John Wiley
Davison W (1978) Defining the electroanalytically measured species in a natural water sample. J Electroanal Chem Interfac Electrochem 87:395–404
Duinker JC, Kramer CJM (1977) An experimental study on the speciation of dissolved zinc, cadmium, lead and copper in river Rhine and North Sea water, by differential pulsed anodic stripping voltammetry. Mar Chem 5:207–228
Ernst R, Allen HE, Mancy KH (1975) Characterization of trace metal species and measurement of trace metal stability constants by electrochemical techniques. Water Res 9:969–979
Figura P, McDuffie B (1979) Use of Chelex resin for determination of labile trace metal fractions in aqueous ligand media and comparison of the method with anodic stripping voltammetry. Anal Chem 51:120–125
Figura P, McDuffie B (1980) Determination of labilities of soluble trace metal species in aqueous environmental samples by anodic stripping voltammetry and Chelex column and batch methods. Anal Chem 52:1433–1439
Florence TM (1977) Trace metal species in fresh waters. Water Res 11:681–687
Florence TM (1982a) The speciation of trace elements in waters (review). Talanta 29:345–364
Florence TM (1982b) Development of physico-chemical speciation procedures to investigate the toxity of copper, lead, cadmium and zinc towards aquatic biota. Anal Chim Acta 141:73–94
Florence TM, Batley GE (1980) Chemical speciation in natural waters. CRC Crit Rev Anal Chem 9:219–296
Florence TM, Lumsden BG, Fardy JJ (1983) Evaluation of some physico-chemical techniques for the determination of the fraction of dissolved copper toxic to the marine diatom Nitzschia closterium. Anal Chim Acta 151:281–295
Garrels RM, Thompson ME (1962) A chemical model for sea water at 25 °C and one atmosphere total pressure. Am J Sci 260:57–66
Giesy JP, Briese LA, Leversee GJ (1978) Metal binding capacity of selected Maine surface waters. Envir Geol 2:257–268
Gran G (1952) Determination of the equivalence point in potentiometric titrations. Part II. Analyst 77:661–671
Granéli A, Jagner D, Josefson M (1980) Microcomputer system for potentiometric stripping analysis. Anal Chem 52:2220–2223
Guy RD, Ross Kean A (1980) Algae as a chemical speciation monitor – a comparison of algal growth and computer calculated speciation. Wat Res 14:891–899

Hanck KW, Dillard JW (1977) Evaluation of micromolar compleximetric titrations for the determination of the complexing capacity of natural waters. Anal Chim Acta 89:329–338

Hart BT (1981) Trace metal complexating capacity of natural waters: a review. Envir Tech Lett 2:95–110

Hasle JR, Abdullah MI (1981) Analytical fractionation of dissolved copper, lead and cadmium in coastal seawater. Mar Chem 10:487–503

Hoffmann MR, Yost EC, Eisenreich SJ, Maier WJ (1981) Characterization of soluble and colloidal-phase metal complexes in river water by ultrafiltration. A mass-balance approach. Envir Sci Tech 15:655–661

Jacobsen E, Lindseth H (1976) Effects of surfactants in differential pulse polarography. Anal Chim Acta 86:123–127

Jagner D, Granéli A (1976) Potentiometric stripping analysis. Anal Chim Acta 83:19–26

Jagner D, Josefson M, Westerlund S (1981) Determination of zinc, cadmium, lead and copper in sea water by means of computerized potentiometric stripping analysis. Anal Chim Acta 129:153–161

Jardim WF, Allen HW (1984) Measurement of copper complexation by naturally occurring ligands – a comparison of experimental techniques. In Complexation of Trace Metals in Natural Waters, eds. C.J.M. Kramer, M. Branica, V.C. Duinker, J. Hart-Stam, R.F.C. Mantoura, H.W. Nürnberg, and W. Stumm. The Hague: M. Nijhoff/W. Junk

Jones MJ, Hart BT (1984) Measurement of the trace metal complexing capacity of Magela Creek waters. In Complexation of Trace Metals in Natural Waters, eds. C.J.M. Kramer, M. Branica, V.C. Duinker, J. Hart-Stam, R.F.C. Mantoura, H.W. Nürnberg, and W. Stumm. The Hague: M. Nijhoff/W. Junk

Kounaves SP, Zirino A (1979) Studies of cadmium-ethylenediamine complex formation in seawater by computer-assisted stripping polarography. Anal Chim Acta 109:327–339

Laegreid M, Alstad J, Klaveness D, Seip HM (1983) Seasonal variation of cadmium toxicity towards the alga Selenastrum capricornutum Printz in two lakes with different humus content. Envir Sci Technol 17:357–361

Langford CH, Gamble DS, Underdown AW, Lee S (1983) Interaction of metal ions with a well characterized fulvic acid. In Aquatic and Terrestrial Humic Materials, eds. R.F. Christman and E.T. Gjessing, pp 219–238. Ann Arbor: Ann Arbor Science

Laxen DP, Harrison RM (1981) A scheme for the physico-chemical speciation of trace metals in freshwater samples. Sci Total Envir 19:59–82

Lund W, Onshus D (1976) The determination of copper, lead and cadmium in sea water by differential pulse anodic stripping voltammetry. Anal Chim Acta 86:109–122

McCrady JK, Chapman GA (1979) Determination of copper complexing capacity of natural river water, well water and artificially reconstituted water. Water Res 13:143–150

Meites L, Nürnberg HW, Zuman P (1976) Classification and nomenclature of electroanalytical techniques. Pure Appl Chem 45:83–97

Neubecker TA, Allen HE (1983) The measurement of complexation capacity and conditional stability constants for ligands in natural waters. Water Res 17:1–14

Nilsen SK, Lund W (1982) The determination of weakly and strongly bound copper, lead and cadmium in Oslofjord samples. Mar Chem 11:223–233
Nürnberg HW (1983) Investigations on heavy metal speciation in natural waters by voltammetric procedures. Fresenium Z Anal Chem 316:557–565
Nürnberg HW, Raspor B (1981) Applications of voltammetry in studies of the speciation of heavy metals by organic chelators in sea water. Envir Tech Lett 2:457–483
Nürnberg HW, Valenta P (1983) Potentialities and applications of voltammetry in chemical speciation of trace metals in the sea. In Trace Metals in Sea Water, eds. C.S. Wong, E. Boyle, K.W. Bruland, J.D. Burton, and E.D. Goldberg, pp 671–697. New York: Plenum Press
Perdue EM, Lytle CR (1983) A critical examination of metal ligand complexation models: Application to defined multiligand mixtures. In Aquatic and Terrestrial Humic Materials, eds. R.F. Christman and E.T. Gjessing, pp 295–314. Ann Arbor: Ann Arbor Science
Piotrowicz SR, Harvey GR, Springer-Young M, Courant RA, Boran DA (1983) Studies of cadmium, copper and zinc interactions with marine fulvic and humic materials in seawater using anodic stripping voltammetry. In Trace Metals in Sea Water, eds. C.S. Wong, E. Boyle, K.W. Bruland, J.D. Burton, and E.D. Goldberg, pp 699–717. New York: Plenum Press
Plavšić M, Krznatić D, Branica M (1982) Determination of the apparent copper complexing capacity of seawater by anodic stripping voltammetry. Mar Chem 11:17–31
Reuter JH, Perdue EM (1977) Importance of heavy metal-organic matter interactions in natural waters. Geochim Cosmochim Acta 41:325–334
Saar RA, Weber JH (1982) Fulvic acid: modifier of metal-ion chemistry. Envir Sci Technol 16:510A–517A
Sagberg P, Lund W (1982) Trace metal analysis by anodic-stripping voltammetry. Effect of surface-active substances. Talanta 29:457–460
Scatchard G (1949) The attraction of proteins for small molecules and ions. Ann NY Acad Sci 51:600–672
Shuman MS, Collins BJ, Fitzgerald PJ, Olson DL (1983) Distribution of stability constants and dissociation rate constants among binding sites on estuarine copper-organic complexes: Rotated disk electrode studies and an affinity spectrum analysis of ion-selective electrode and photometric data. In Aquatic and Terrestrial Humic Materials, eds. R.F. Christman and E.T. Gjessing, pp 349–370. Ann Arbor: Ann Arbor Science
Shuman MS, Cromer JL (1979) Copper association with aquatic fulvic and humic acids. Estimation of conditional formation constants with a titrimetric anodic stripping voltammetry procedure. Envir Sci Technol 13:543–545
Shuman MS, Michael LC (1978) Application of rotating disk electrode techniques to the measurement of copper-organic complex dissociation rate constants in marine coastal samples. Envir Sci Technol 12:1069–1072
Shuman MS, Woodward GP (1973) Chemical constants of metal complexes from a complexometric titration followed with anodic stripping voltammetry. Anal Chem 45:2032–2035
Skogerboe RK, Wilson SA, Osteryoung JG, Florence TM, Batley GE (1980) Exchange of comments on scheme for classification of heavy metal species in natural waters. Anal Chem 52:1960–1963

Sposito G, Holtzclaw KM, LeVesque-Madore CS (1981) Trace metal complexation by fulvic acid extracted from sewage sludge: I. Determination of stability constants and linear correlation analysis. Soil Sci Soc Am J 45:465–468
Stolzberg RJ (1977) Potential inaccuracy in trace metal speciation measurements by differential pulse polarography. Anal Chim Acta 92:193–196
Sunda WG, Lewis JAM (1978) Effect of complexation by natural organic ligands on the toxicity of copper to a unicellular alga, Monochrysis Lutheri. Limnol Oceanogr 23:870–876
Thompson ME (1966) Magnesium in sea water: an electrode measurement. Science 153:866–867
Thompson ME, Ross JW (1966) Calcium in sea water by electrode measurement. Science 154:1643–1644
Truitt RE, Weber JH (1981) Determination of complexing capacity of fulvic acid for copper(II) and cadmium(II) by dialysis titration. Anal Chem 53:337–342
Turner DR, Whitfield M (1981) The classification of electroanalytical techniques. In Marine Electrochemistry, eds. M. Whitfield and D. Jagner, pp 67–97. New York: John Wiley
Tuschall JR, Brezonik PL (1983) Complexation of heavy metals by aquatic humus: A comparative study of five analytical methods. In Aquatic and Terrestrial Humic Materials, eds. R.F. Christman and E.T. Gjessing, pp 275–294. Ann Arbor: Ann Arbor Science
Valenta P (1983) Voltammetric studies on trace metal speciation in natural waters. Part I: Methods. In Trace Element Speciation in Surface Waters and Its Ecological Implications, ed. G.G. Leppard, pp 49–70. New York: Plenum Press
Whitfield M, Turner DR (1979) Critical assessment of the relationship between biological, thermodynamic and electrochemical availability. In Chemical Modeling in Aqueous Systems, ed. E.A. Jenne, pp 657–682. Washington, DC: American Chemical Society
Zirino A (1981) Voltammetry of natural seawater. In Marine Electrochemistry, eds. M. Whitfield and D. Jagner, pp 421–503. New York: John Wiley
Zirino A, Seligman PF (1981) A note on the polarographic behaviour of the Cu(II) ion selective electrode in seawater. Mar Chem 10:249–255
Zirino A, Yamamoto S (1972) A pH-dependent model of the chemical speciation of copper, zinc, cadmium, and lead in seawater. Limnol Oceanogr 17:661–671

The Importance of Chemical "Speciation" in Environmental Processes, eds. M. Bernhard, F. E. Brinckman, and P. J. Sadler, pp. 563–578. Dahlem Konferenzen 1986. Springer-Verlag Berlin, Heidelberg
© Dr. S. Bernhard, Dahlem Konferenzen

Multinuclear NMR Methods for the in situ Characterization of Chemical Species

P. J. Sadler

Dept. of Chemistry
Birkbeck College, University of London
London WC1E 7HX, England

Abstract. The application of NMR methods to the study of chemical species in solutions, biofluids, cells, tissues, organs, polymers, and minerals is described. The technique is noninvasive and nondestructive. Sensitivity is highest for spin-½ nuclei such as ^{1}H, ^{19}F, and ^{31}P, giving signals for millimolar solutions in a few minutes. Less-sensitive nuclei can also be studied. The problems of quadrupolar nuclei are discussed. The rates of metal-ligand exchange reactions can often be calculated. Well-resolved spectra can be obtained from solids using magic-angle spinning, especially from ^{27}Al and ^{29}Si in aluminosilicates.

Introduction

Nuclear magnetic resonance (NMR) has continued to develop ever since it was discovered over 40 years ago. Today much NMR developmental work is concerned with noninvasive, in situ detection and analysis of chemical and biochemical species. Once the technique was applied by chemists almost exclusively to clear, homogeneous liquids. Now, highly resolved spectra can be obtained from inhomogeneous biofluids, cells, tissues, organs, coal, minerals, and even whole animals.

NMR can therefore be applied to a very wide variety of species analyses. There are several types of instruments doing different jobs: high-field (up to 11.7 T), narrow-bore spectrometers for high-resolution work on liquids;

low-field, wide-bore machines using field gradients to obtain spatial images of sections through whole bodies; topical NMR using surface coils for localized detection of signals; and cross-polarization, magic-angle-spinning instruments for solids. Gases will not be considered here. The sensitivity for these is poor. Vapor pressures of about one atmosphere are required.

With such a range of specimens and equipment, and the knowledge that all elements in the periodic table except argon and cerium have at least one isotope suitable for NMR detection, a very large number of experiments are possible. The limitations of the method require careful thought: the large differences in sensitivity between nuclei, the broadening of signals from macromolecules or immobilized small molecules or nuclei near paramagnetic centers. With careful design of the NMR experiment even disadvantages can be turned to advantage, as we shall see.

The NMR Experiment

The NMR phenomenon depends on the possession by many nuclei of spin-angular momentum. They behave like tiny magnets and in a magnetic field precess around this field at radio frequencies. They absorb energy from a secondary radio-frequency field applied perpendicular to the main field. The amount of energy absorbed is strictly proportional to the concentration of nuclei in the detection coil (Akitt, 1983; Harris, 1983). The areas under the absorption signals can therefore be used quantitatively. The experiment is noninvasive and nondestructive.

The magnetic fields currently in use range from 0.15 T for imaging work to 11.7 T for high-resolution studies on liquids. The upper limit is governed by superconducting magnet technology. The higher the field, the higher the sensitivity of detection (signal intensity α field$^{3/2}$) and greater the dispersion of signals.

The highest fields are not currently used for imaging animals for fear of damage caused by field gradients. The signals of heavy nuclei are sometimes optimum at intermediate fields because of line-broadening mechanisms at higher fields (Harris, 1983; Lambert and Riddell, 1983). Most chemistry departments have at least one machine operating at ca. 4.7 T (200 MHz for ^{1}H): typical cost U.S. $ 140,000, and there are many regional service instruments at 9.4 and 11.7 T.

The energies absorbed by the nuclei are small. NMR is therefore a relatively insensitive technique. At best, for protons (^{1}H), the limit is 0.3 ml of a 10 μM solution. This would require several hours of accumulation time. At worst several mL of molar solutions are required (e.g., ^{33}S).

Which Nuclei Can Be Detected?

Isotopes with even mass and even atomic number do not possess nuclear spin and cannot be detected by NMR. These include (unfortunately!) ^{12}C, ^{16}O, and ^{32}S.

Nuclei with spin-quantum number $I = \frac{1}{2}$ have the greatest receptivity for NMR. Receptivity is a function of size of the magnetic moment and natural abundance (see Table 1). Thus ^{1}H, ^{19}F, ^{31}P, and ^{205}Tl are the easiest nuclei to study. ^{187}Os and ^{57}Fe are more than six orders of magnitude less sensitive to detection than ^{1}H. The position of ^{13}C and ^{15}N, for example,

Table 1. Classification of some spin-½ nuclei according to natural abundance and magnetic strength (Harris and Mann, 1978)

Magnetic strength	Natural abundance		
	High (>90%)	Medium	Low (<10%)
Strong	^{1}H, ^{19}F, ^{31}P	^{205}Tl	^{3}H, ^{3}He
Medium		^{113}Cd, ^{129}Xe, ^{171}Yb, ^{195}Pt, ^{199}Hg, ^{207}Pb	^{13}C, ^{15}N, ^{29}Si, ^{77}Se, ^{119}Sn, ^{125}Te
Weak	^{89}Y, ^{103}Rh, ^{169}Tm	^{109}Ag, ^{183}W	^{57}Fe, ^{187}Os

Table 2. Classification of some quadrupolar nuclei according to natural abundance and magnetic strength (Harris and Mann, 1978)

Magnetic strength	Natural abundance		
	High (>90%)	Medium	Low (<10%)
Strong	^{7}Li		
Medium	^{9}Be, ^{23}Na, ^{27}Al, ^{45}Sc, ^{51}V, ^{55}Mn, ^{59}Co, ^{75}As, ^{93}Nb, ^{115}In, ^{127}I, ^{133}Cs, ^{181}Ta, ^{209}Bi	^{11}B, ^{35}Cl, ^{63}Cu, ^{65}Cu, ^{71}Ga, ^{81}Br, ^{87}Rb, ^{121}Sb, ^{137}Ba, ^{139}La, ^{187}Re	^{2}H, ^{6}Li, ^{17}O, ^{21}Ne, ^{99}Tc
Weak	^{14}N, ^{39}K, ^{197}Au	^{25}Mg, ^{37}Cl, ^{83}Kr, ^{95}Mo, ^{131}Xe, ^{189}Os, ^{201}Hg	^{33}S, ^{43}Ca, ^{47}Ti, ^{49}Ti, ^{53}Cr, ^{67}Zn, ^{73}Ge, ^{87}Sr

with natural abundances of 1.1 and 0.4%, respectively, can be considerably improved by enrichment.

Table 2 shows a classification of nuclei with I > ½, quadrupolar nuclei. The overall receptivity of several of them (including ^{7}Li, ^{11}B, ^{23}Na, ^{25}Mg, ^{51}V, ^{55}Mn, and ^{59}Co) is greater than that of ^{31}P. When present in symmetrical coordination environments they give rise to sharp, readily detectable signals. However, all quadrupolar nuclei in asymmetrical environments suffer from line-broadening problems, sometimes broadened beyond detection. For example, no useful chemical work has ever been done on ^{197}Au.

Detection frequencies are proportional to the applied field. Thus for ^{1}H this is 60 MHz at 1.4 T, 200 MHz at 4.7 T, and 500 MHz at 11.7 T. At 4.7 T the frequencies for some other nuclei are: ^{13}C, 50.2; ^{29}Si, 39.8; ^{31}P, 80.8; ^{59}Co, 47.2; ^{113}Cd, 44.4; and ^{195}Pt, 42.8 MHz. For a more complete list see Harris (1983) and Harris and Mann (1978).

NMR Parameters

The main parameters used to characterize NMR signals are (Akitt, 1983; Harris, 1983):

1) the chemical shift (δ in parts per million, ppm relative to a standard, e.g., tetramethylsilane for ^{1}H);
2) spin-spin coupling constant (J in Hz);
3) spin-lattice or longitudinal relaxation time (T_1 in seconds); and
4) spin-spin or transverse relaxation time (T_2 in seconds – inversely proportional to linewidth).

All of these provide information about the local environment of the nucleus and can therefore be used to characterize chemical species.

Solution Studies of Spin-½ Nuclei

NMR studies can be made either directly on the element of interest or indirectly via bound ligands. The latter will usually involve ^{1}H, ^{13}C, ^{19}F, or ^{31}P. The choice will depend on the sensitivity of the element and the strengths and dynamics of the ligand binding.

We illustrate direct-element detection with examples from platinum chemistry and cadmium biochemistry. Figure 1 shows ^{195}Pt NMR spectra for aqueous solutions containing mixtures of Pt(II) and Pt(IV) bromide and chloride complexes (Kerrison and Sadler, 1978). The oxidation state of platinum can be determined by NMR, signals for Pt(II) occurring at lower fre-

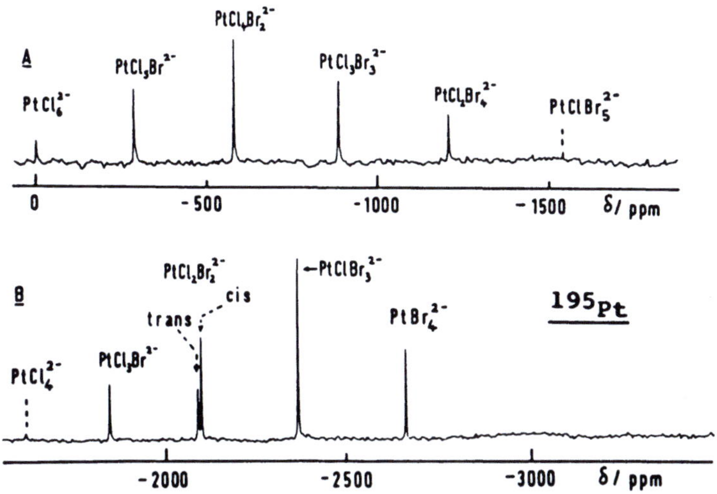

Fig. 1 A, B. Mixtures of platinum(II) and platinum(IV) halides. 12.8 MHz ^{195}Pt NMR spectra of **A** $Na_2PtCl_6 + 2NaBr$ and **B** $K_2PtCl_4 + 4NaBr$ in D_2O (Kerrison and Sadler, 1978)

quency (higher field) than those of Pt(IV). The different halide species are also distinguishable. The shift change for substitution of Cl^- by Br^- of more than 200 ppm follows a predictable pattern, enabling assignments to be made. Even isomers are distinguishable. With greater expansion, the *cis-trans* and *mer-fac* isomers of Pt(IV) halides are also seen. Platinum-ligand bonds usually show a strong degree of covalency and ligands exchange relatively slowly on the NMR time scale, defined as (approximately) the inverse in frequency difference between signals. Separate signals are therefore seen for each Pt species.

For cadmium, a labile metal ion, the opposite is usually the case. If a mixture of cadmium complexes is present in solution, only a single averaged signal is seen. However, an exception is the protein metallothionein where exchange reactions are slowed down by strong binding (Nicholson et al., 1983 b). As seen in Fig. 2, the mixed Cd-Zn protein contains many slightly different coordination environments for Cd(II). The shift range for ^{113}Cd covers a total of ca. 900 ppm and that from 600–700 ppm (shown in Fig. 2) is typical of tetrahedral CdS_4 sites. This can be contrasted with CdO_6 sites wich chemical shifts of 100 to -100 ppm and CdN_3O sites from 100 to 300 ppm.

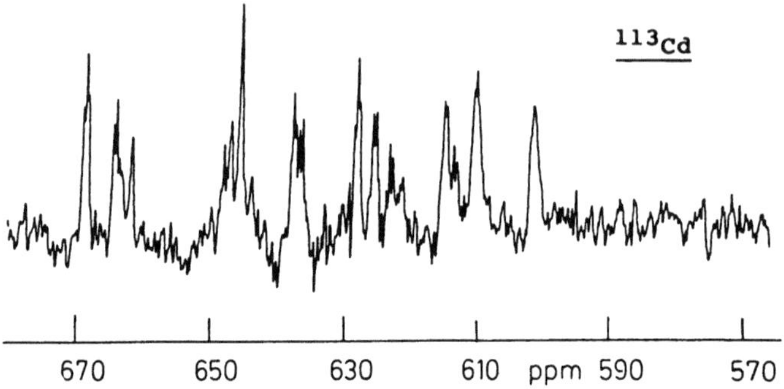

Fig. 2. Characterization of cadmium-binding sites in a protein. 88 MHz ^{1}H-decoupled ^{113}Cd NMR spectrum of rat liver ^{113}Cd,Zn-metallothionein (ca. 4 mM) (Nicholson et al., 1983 b)

Ligand-exchange Reactions

We can use an example of thiolates binding to Au(I) in aqueous solution (Isab and Sadler, 1982) to illustrate the detection of dynamic processes by NMR. As can be seen in Fig. 3, at low temperature separate ^{13}C NMR signals are seen for bound and free thiomalate and N-acetyl-L-cysteine. As the solution is warmed up the signals for each type of carbon move together, coalesce, and sharpen. Exchange processes of intermediate rates (on the order of the shift difference between species in Hz) cause severe broadening of resonances. An example of this is the spectrum taken at 50 °C shown in Fig. 3. At high temperature a rapid exchange between the bound and free states is occurring with a rate of $> 120\ s^{-1}$. Activation energies for these exchanges can be calculated. In general, rate processes occurring on time scales of 10^{-1} to $10^5\ s^{-1}$ can be detected, depending on the chemical shift differences involved.

When exchange rates are fast on the NMR time scale, the analysis of species must rely on chemical-shift changes alone. Mixtures of PO_4^{3-}, HPO_4^{2-}, $H_2PO_4^-$, and H_3PO_4, for example, give rise to a single ^{31}P resonance at all pH values. The chemical shift follows a titration curve, from which the percentage of each species can be deduced.

Spin-spin Coupling

Heteronuclear coupling can be a valuable assignment aid. This is likely to be present where strong bonds exist and ligand exchange is slow. Examples of one-bond couplings are given in Table 3 (see Harris and Mann, 1978).

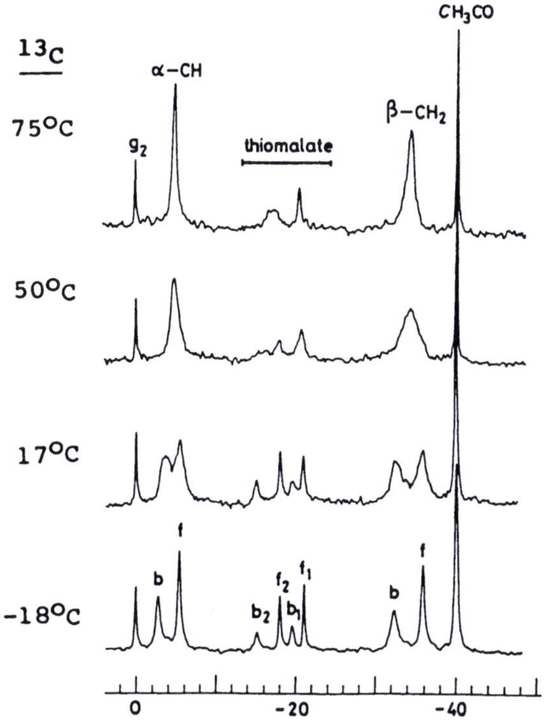

Fig. 3. Effect of metal-ligand exchange on NMR spectra. At low temperature separate ^{13}C peaks are seen for free (f) and bound (b) thiomalate and N-acetyl-L-cysteine (α-CH, β-CH_2 and CH_3CO) in an aqueous solution containing gold(I). At high temperature these thiolates are exchanging rapidly ($>120\ s^{-1}$) between free and gold-bound states (Isab and Sadler, 1982) and averaged resonances are seen. The peak labelled g_2, used as a reference, is from glycerol

Table 3. One-bond coupling constants for ^{13}C-X bonds. Approximate ranges are given. Values are often strongly dependent on bond-order and other ligands present (Harris and Mann, 1978)

X	^1J/Hz	*X*	^1J/Hz
^1H	110–150	^{77}Se	(−90)–100
^{29}Si	(−113)–(−37)	^{125}Te	170
^{119}Sn	(−900)–200	^{195}Pt	600–2000
^{207}Pb	30–950	^{199}Hg	600–1800

They usually decrease in the order $^1J > ^2J > ^3J$, where the superscript is the number of bonds between the coupled nuclei. Small splittings of the ^{113}Cd signals in Fig. 2 are due to two-bond ^{113}Cd-S-^{113}Cd couplings of ca. 33 Hz.

It is often worth designing the NMR experiment so as to introduce couplings between $I = 1/2$ nuclei, e.g., by ^{15}N enrichment to detect ^{195}Pt-^{15}N bonds. Couplings to quadrupolar nuclei are detected in symmetrical species, e.g., NH_4^+, BF_4^-.

Paramagnetic Complexes

Unpaired electrons introduce large, local magnetic fields which can cause shifts and broadenings of NMR resonances (Dwek, 1973). These may arise from free radicals or paramagnetic metal ions.

The metal ions giving rise to the greatest broadening effects are those which give well-resolved electron-spin resonance spectra at room temperature: V(IV), Cr(III), Mn(II), Cu(II), and Gd(III). Others give both shifts and broadenings, e.g., Fe(II), Fe(III), Co(II), Ni(II), Nd(III), and Dy(III), whilst the ability of Eu(III) and Pr(III) to shift resonances has made them popular as "shift reagents" amongst organic chemists.

Paramagnetic centers can cause severe problems in "unknown" samples. With more well-defined systems the shifts and broadenings can be advantageous. Shifts show an approximate $1/r^3$ and broadening a $1/r^6$ dependence on distance between the paramagnetic ion and the nucleus of interest. These can be used for mapping the position of nuclei, e.g., in proteins, or of substrates and inhibitors bound to enzymes. Paramagnetic probes are often introduced specifically for this purpose: Mn(II) as a replacement for Mg(II), Gd(III) for Ca(II) (Dwek, 1973).

Mobility

The linewidths of NMR resonances increase as the mobility of molecules decreases (Harris, 1983). Thus globular proteins of molecular weight > ca. 30,000 give rise to broad, unresolvable ^{1}H NMR signals. For solutions they cannot be sharpened. This will also be the case for small molecules bound to macromolecules, e.g., fatty acids bound to albumin. Sharp resonances can be obtained from high-molecular-weight polymers when there is significant local molecular mobility. Such is often the case for polymers bearing pendant tri-n-butyltin moieties or constructed from poly(dialkylsiloxo) units. These can give sharp ^{1}H, ^{13}C, ^{119}Sn, or ^{29}Si spectra even at low temperature (F. E. Brinckman, personal communication).

Biological Fluids and Cells

The ^1H NMR spectra shown in Fig. 4 are from intact blood plasma and serum (Nicholson et al., 1983 a). There are two major problems in obtaining such spectra: first, the broad envelope from high-molecular-weight species such as albumin and immunoglobulins, and second, the intense resonance from H_2O (110 M protons). The protein signals can be filtered out of the spectrum by the use of spin-echo pulse sequences. As described above, they have short T_2 values and disappear during the delays of the spin-echo pulse sequence. Only resonances from the mobile CH_3 and CH_2 end groups of fatty acids (P_2, P_3) and small molecules such as valine and alanine remain in the spectrum. If EDTA is used as anticoagulant, the concentrations of EDTA-chelatable Ca(II) and Mg(II) can be determined directly and simultaneously from the spectrum (ca. 2 and 1 mM, respectively). We have used

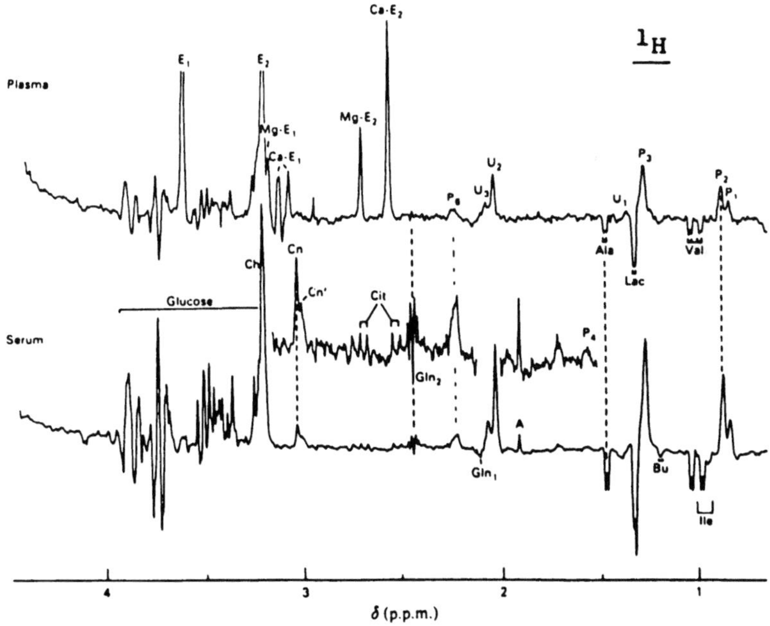

Fig. 4. Calcium and magnesium in biofluids. 400 MHz ^1H spin-echo NMR spectrum of human serum (bottom) and plasma (top) (Nicholson et al., 1983a). The peaks Mg-E_2 and Ca-E_2 are due to magnesium and calcium complexes of EDTA, added to the blood as anticoagulant. The spin-echo sequence leads to a phase modulation of some resonances

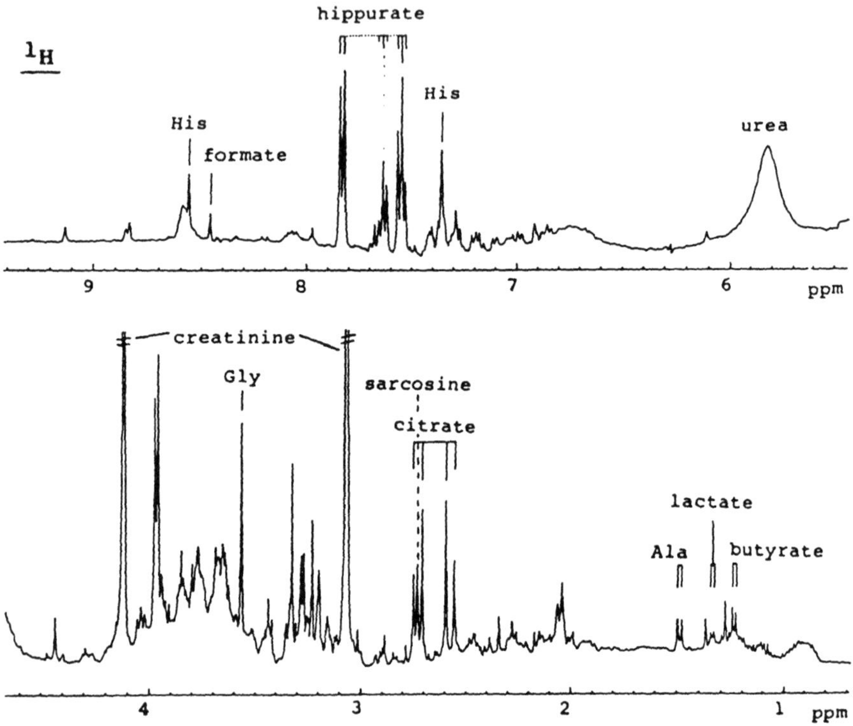

Fig. 5. 400 MHz ^{1}H NMR spectrum of intact human urine (Bales et al., 1984) obtained in ca. 4 min

such EDTA resonances for measuring the rates of removal of Zn(II) and Cd(II) from proteins and bacteria by NMR (Higham et al., 1984).

Spin-echo techniques have also been applied to the study of cells. Rabenstein and co-workers (1982) have studied the entry of a variety of metals into red cells, including methylmercury, $HgCl_2$, and $PbMe_3^+$. It is possible to identify the sulfhydryl group of glutathione as one of the specific intracellular binding sites. We have carried out similar studies with gold and cadmium. The reversal of binding using chelating agents can then be investigated.

Only H_2O suppression is needed to obtain highly resolved ^{1}H NMR signals from urine (Fig. 5). One of our current tasks is to identify metal complexation by citrate in urine based on its shifts and couplings. The biochemical consequences of drug ingestion, kidney damage caused by Cd exposure,

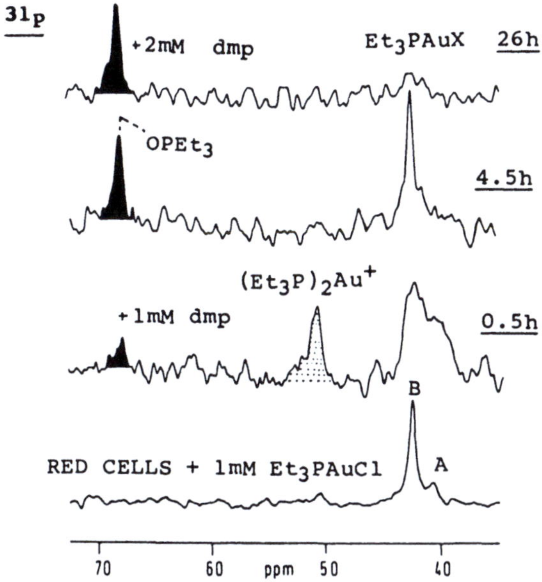

Fig. 6. Distribution of phosphorus species in cells. 80 MHz ^{1}H-decoupled ^{31}P NMR spectra of human red blood cells containing 1 mM Et_3PAuCl *in vitro* and the subsequent release of phosphine as Et_3PO on challenge with dimercaptopropanol (dmp) (Razi et al., 1983). The chemical shift reference was H_3PO_4

and metabolic diseases are readily monitored by ^{1}H NMR studies of urine (Bales et al., 1984).

With nuclei such as ^{19}F and ^{31}P there is little or no background interference in comparison to ^{1}H. Two examples illustrate this. The first concerns studies of fluorine-labelled chelating agents by Metcalf and coworkers (1983) to probe intracellular Ca(II) concentrations. Difluoro-derivatives of 1,2-bis(o-aminophenoxy)ethane-N,N,N′,N′-tetraacetic acid exhibit large ^{19}F shifts on chelation of metal ions. They show rapid (<10 ms) responses to changes in free Ca(II) concentrations of >0.1 μM inside cells. In thymocytes, for example, this was measured as 250 nM and increased when Ca(II) ionophores such as A23187 were added. Such studies show good future promise. In general, NMR studies with cells require $>10^6$ cells per mL.

In the second example the metabolism of gold-phosphine drugs can be probed directly inside red cells by ^{31}P NMR, as shown in Fig. 6. The phosphine signals are well shifted from those of natural phosphate metabolites such as ATP (ca. 0–20 ppm). Dimercaptopropanol causes the release of gold-bound Et_3P to give Et_3PO via an intermediate $(Et_3P)_2Au^+$ complex (Razi et al., 1983).

Whole Animals and Tissues

NMR imaging involves mapping the concentration (also T_1 and T_2) of H_2O in sections through animals or other objects (Mansfield and Morris, 1982). No direct information on the species present is obtained in the images, although a high field (>1.4 T) fat can be distinguished from H_2O. Factors such as the presence of paramagnetic metal ions affect the intensities observed. Attention is currently being devoted to other nuclei such as ^{19}F and ^{31}P which may be more useful for combining species identification with imaging.

In surface-coil experiments, so-called topical NMR, signals are detected from localized regions (Gadian, 1982). Thus ATP, phosphocreatine, inorganic phosphate, and phosphate esters can be distinguished in muscle or brain. ^{13}C studies using enriched substrates (glucose) and ^{19}F studies with labelled substrates are also feasible.

Quadrupolar Nuclei

Most of the discussion so far has been concerned with spin-½ nuclei. In principle many similar experiments are possible with quadrupolar nuclei. However, there is one big drawback. Resonances from quadrupolar nuclei in asymmetrical environments are usually very broad, often unobservable (Harris and Mann, 1978; Lambert and Riddell, 1983). The size of the effect depends on the quadrupole moment and the degree of asymmetry.

The alkali metals are all reasonably sensitive to detection, except ^{39}K. ^{23}Na images of the body can be obtained and the quadrupole moments of ^{6}Li and ^{133}Cs are especially low. The alkaline earths are not so encouraging. Molar solutions of natural-abundance isotopes are required as standards. The quadrupole moment of ^{25}Mg is so large that there is little hope of observing signals from Mg(II) bound to a macromolecule; even that for $MgEDTA^{2-}$ is 500 Hz wide. Forsén and co-workers have found enriched ^{43}Ca more useful (see Lambert and Riddell, 1983).

^{27}Al and ^{71}Ga give reasonable signals in a variety of environments, but for ^{75}As, ^{121}Sb, and ^{209}Bi in the $+3$ oxidation state lone pairs of electrons cause asymmetry and are likely to restrict studies to the $+5$ state, even then with difficulty.

Resonances from ^{33}S are often too broad to detect: compare linewidths of 4 Hz for $CsSO_4$ with 1600 Hz for Na_2S in H_2O. Isotopic enrichment can often make studies of ^{17}O worthwhile.

The halide ions are mobile and binding to macromolecules can be studied by ^{79}Br, ^{81}Br, ^{35}Cl, or ^{37}Cl NMR. Free halide ions give sharp resonances and bound halides broad ones. Since averaging occurs, the degree of broadening can be used to assess binding.

The species composition of several quadrupolar transition metals in diamagnetic complexes can be studied. ^{59}Co signals have been observed from Co(III) (shift range 18,000 ppm); for ^{51}V from V(V), V(I), and V(−I); for ^{53}Cr from Cr(VI) (CrO_4^{2-}) and Cr(O) ($Cr(CO)_6$); for ^{55}Mn from Mn(VII), Mn(I), and Mn(−I); for ^{95}Mo from Mo(VI), Mo(IV), and Mo(O); and for ^{183}W from W(VI) and W(O).

Solids

NMR resonances from solids are often broad and featureless: 2 kHz wide for ^1H signals of solid EtOH compared to 0.3 Hz in solution. They are dominated by dipole-dipole interactions which are averaged out to zero in solution. Line-broadening also results from the different chemical shieldings of nuclei caused by their different orientations with respect to the applied magnetic field, so-called chemical-shift anisotropy (CSA).

However, an experiment can be performed for solids which simulates the molecular motion of liquids: spinning the sample at 54.7° – the magic angle. The rates of rotation required to eliminate CSA (a few kHz) can usually be achieved, but not those to average out dipolar effects.

Some examples of ^{11}B, ^{15}N, ^{27}Al, and ^{207}Pb NMR spectra of solids are shown in Fig. 7. Note the advantage of an increased observation frequency for ^{27}Al. The importance of magic-angle spinning (MAS) is demonstrated for ^{207}Pb. The spectrum of the nonspinning sample is a characteristic powder pattern. At least 100 mg of material is usually required.

Dipolar interactions with protons can be removed by decoupling. An enhancement of signal intensity arises from cross-polarization (CP) techniques commonly used for ^{13}C NMR of solids. For examples of such spectra of humin and whole sapropel see Hatcher et al. (1983).

The resulting signals occur at average, isotropic chemical shifts similar to those of solutions. At best, the resolution obtainable from solids is an order of magnitude worse than for solutions. The signals from solids reflect the uniqueness of the solid-state structure as well as the isolated molecule. The easiest solids to study are those with "dilute spins" either through low natural abundance (^{13}C, ^{29}Si) or by physical dilution (^1H in ^2H, or dilute ^{31}P).

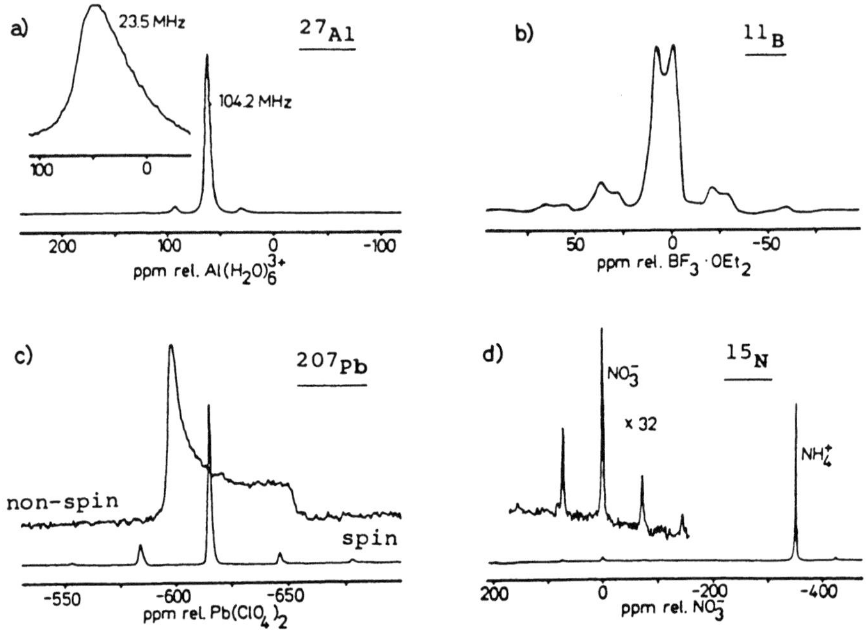

Fig. 7 a–d. Solid-state NMR spectra obtained with magic-angle spinning (Fyfe et al., 1983 b). **a** Zeolite Y: note sharper spectrum at high frequency; **b** Corning 7070 glass: splittings due to quadrupolar interaction; **c** $Pb(NO_3)_2$; **d** $^{15}NH_4{}^{15}NO_3$

Silicates

An example of the use of ^{29}Si to probe the local environments in the synthetic mineral cordierite, $Mg_4Al_4Si_5O_{18}$, is shown in Fig. 8. Signals for both chain and ring structures are seen with varying numbers of Al neighbors (Fyfe et al., 1983 a). By ^{27}Al-MAS-NMR, tetrahedral and octahedral Al sites can be distinguished.

Zeolites are particularly amenable to MAS-NMR study since the frameworks can be synthesized with NMR-active nuclei including ^{10}B, ^{17}O, ^{27}Al, ^{29}Si, and ^{31}P. Also the majority of exchangeable cations in zeolites can be studied by NMR. It is possible to probe reactions of guest molecules with aluminosilicate hosts of catalytic importance. Also using CP/MAS techniques, surface species can be probed. Si(OH), $Si(OH)_2$, and $Si(OH)_3$ have been identified at silica or zeolite internal surfaces (Fyfe et al., 1983 b). Complementary ^{29}Si NMR studies of enriched potassium-silicate solutions have been reported (Harris and Knight, 1983) and the structures of up to

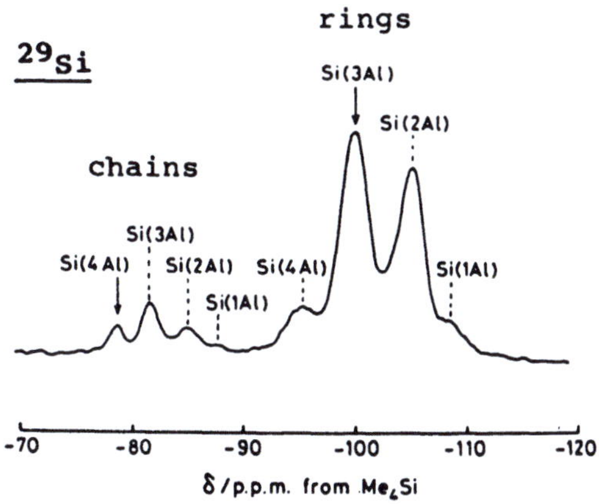

Fig. 8. Silicon species in solid silicates. 104 MHz ^{29}Si solid-state, magic-angle spinning spectrum of freshly crystallized (hexagonal) $Mg_2Al_4Si_5O_{18}$ showing $Si(OAl)_4$, $Si(OAl)_3(OSi)$, $Si(OAl)_2(OSi)_2$, etc., in chain and ring parts of the structure (Fyfe et al., 1983 a)

18 species proposed. The dramatic effect which paramagnetic impurities can have on ^{29}Si spectra should be noted. Spectra from samples using NaOH pellets and commercial silica powder showed a single broad resonance 50 Hz wide, whereas those from AristaR NaOH and SiO_2, obtained from the hydrolysis of $SiCl_4$, yielded spectra with much fine structure and sharp lines 1 Hz wide.

The Future

Further developments in NMR technology can be confidently predicted. These will include increased sensitivity of detection, routine use of multipulse sequences (e.g., for two-dimensional spectra), chemical shift imaging (spectra from spatially localized areas), and higher fields. The total scope of NMR for the determination of chemical species has yet to be defined.

Acknowledgements. The NMR work of my own group in recent years has been supported by the SERC, MRC, ARC, and University of London with industrial assistance from Rio Tinto Zinc Services, Johnson Matthey Ltd., May and Baker Ltd., and Smith Kline and French Laboratories. I thank F. E. Brinckman and J. R. Bales for helpful comments and criticism of the manuscript.

References

This list is illustrative and not exhaustive. Those sources marked with an asterisk (*) are general texts and provide good starting points for further reading on specific topics

*Akitt JW (1983) NMR and Chemistry, second ed. London: Chapman and Hall

Bales JR, Higham DP, Howe I, Nicholson JK, Sadler PJ (1984) Use of high-resolution proton nuclear magnetic resonance spectroscopy for rapid multicomponent analysis of urine. Clin Chem 30:426–432
*Dwek RA (1973) NMR in Biochemistry. Oxford: Oxford University Press
Fyfe CA, Gobbi GC, Klinowski J, Putnis A, Thomas JM (1983 a) Characterization of local atomic environments and quantitative determination of changes in site occupancies during the formation of ordered synthetic cordierite by ^{29}Si and ^{27}Al magic angle spinning NMR spectroscopy. Chem Comm 556–558
Fyfe CA, Thomas JM, Klinowski J, Gobbi GC (1983 b) Magic-angle-spinning NMR (MAS-NMR) spectroscopy and the structure of zeolites. Angew Chem 22:259–275
*Gadian D (1982) NMR of Living Biological Systems. Oxford: Clarendon Press
*Harris RK (1983) Nuclear Magnetic Resonance Spectroscopy. London: Pitman
Harris RK, Knight CTG (1983) Silicon-29 nuclear magnetic resonance studies of aqueous silicate solutions. J Chem Soc, Farad Trans 79:1525–1538
*Harris RK, Mann BE (eds) (1978) NMR and the Periodic Table. London: Academic Press
Hatcher PG, Spiker EC, Szeverenyi NM, Maciel GE (1983) Selective preservation and origin of petroleum-forming aquatic kerogen. Nature 305:498–501
Higham DP, Nicholson JK, Sadler PJ (1985) The environmental chemistry of metals with examples from studies on cadmium speciation. In Environmental Inorganic Chemistry, Proceedings, U.S.-Italy International Workshop on Environmental Inorganic Chemistry, eds. K.J. Irgolic and A.E. Martell, pp 249–272. Deerfield Beach, FL: VCH Publishers, Inc
Isab AA, Sadler PJ (1982) A carbon-13 nuclear magnetic resonance study of thiol-exchange reactions of gold(I) thiomalate ('Myocrisin') including applications to cysteine derivatives. J Chem Soc Dalton 135–141
Kerrison SJS, Sadler PJ (1978) The convention for referencing platinum-195 NMR shifts and the cooperative shift effect for chloro-bromo complexes of Pt(IV) and Pt(II). J Magnet Res 31:321–325
*Lambert JB, Riddell FG (eds) (1983) The Multinuclear Approach to NMR Spectroscopy. Dordrecht: D. Reidel Publishing Co
*Mansfield P, Morris PG (1982) NMR Imaging in Biomedicine. London: Academic Press
Nicholson JK, Buckingham MJ, Sadler PJ (1983 a) High resolution ^{1}H n.m.r. studies of vertebrate blood and plasma. Biochem J 211:605–615
Nicholson JK, Sadler PJ, Cain K, Holt DE, Webb M, Hawkes GE (1983 b) 88 MHz ^{113}Cd-n.m.r. studies of native rat liver metallothioneins. Biochem J 211:251–255
Rabenstein DL, Isab AA, Reid RS (1982) A proton nuclear magnetic resonance study of the binding of methylmercury in human erythrocytes. Biochim Biophys Acta 720:53–64
Razi MT, Otiko G, Sadler PJ (1983) Ligand exchange reactions of gold drugs in model systems and in red cells. ACS Symp Ser 209:371–384
Smith GA, Hesketh RT, Metcalf JC, Feeney J, Morris PG (1983) Intracellular calcium measurements by ^{19}F NMR of fluorine labelled chelators. Proc Natl Acad Sci USA 80:7178–7182

The Importance of Chemical "Speciation" in Environmental Processes,
eds. M. Bernhard, F.E. Brinckman, and P.J. Sadler, pp. 579–611.
Dahlem Konferenzen 1986. Springer-Verlag Berlin, Heidelberg
© Dr. S. Bernhard, Dahlem Konferenzen

Analysis of Inorganic Species by Ion Chromatography and Liquid Chromatography

I. S. Krull

The Barnett Institute
and Dept. of Chemistry, Northeastern University
Boston, MA 02115, USA

Abstract. This paper is a review of the currently available column liquid chromatographic approaches with specific and nonspecific detection for the determination of inorganic species. Most of the possible liquid chromatography-detection techniques already used for the analysis of inorganic species are described, and the advantages or disadvantages of each of these are mentioned and compared. The intent here is not to summarize everything that has been reported with column liquid chromatography for the analysis of species, but rather to give the reader an overview of what is possible today and perhaps tomorrow. Sufficient recent references are provided to allow the interested reader to determine which hyphenated techniques have already been used for the study and determination of specific inorganic species.

Introduction: An Overview of Currently Available Methods Using Column Liquid Chromatography-Detection Interfacing Basic Requirements

The environmentalist, toxicologist, biologist, physician, or similarly interested researcher must have an analytical approach for the determination of inorganic species that fulfills at least the following minimum criteria: 1) a high degree of accuracy, precision, and reproducibility of the final qualitative and quantitative results; 2) a high degree of analyte selectivity and specificity for an individual element or species; 3) excellent (low) minimum detection limits (MDLs); 4) ease of interfacing between separation mode and detection method; 5) continuous, on-line, real-time operation; 6) ease of operation, maintenance, and repair for daily use; 7) inexpensive instrumentation for initial purchase and day-to-day operation; 8) an approach which does not require too high a degree of expertise in setup of instrumentation and its operation, repair, or maintenance; 9) easily understandable operational principles; 10) compatibility with aqueous or organic solution samples; 11) ease of recovery of analytical data at the end of each analysis or each day's set of analyses. Each of the approaches described below meets some or all of the above requirements or desired conditions, and some will meet more than others. In general, those hyphenated or direct analytical methods that meet more of these conditions have already or will soon become the more widely practiced for the analysis of inorganic species, and those that meet too few of these requirements will eventually be discarded.

Background and Introduction to Column Liquid Chromatography (CLC), Comparison with Thin-Layer Chromatography (TLC) and Gas Chromatography (GC)

Although modern column liquid chromatography [also known as liquid chromatography (LC) or high-performance liquid chromatography (HPLC), in some instances comprising ion chromatography (IC)] has been available for about twenty years now, it is perhaps only within the past ten years that it has received prominence in metal and nonmetal (inorganic) analysis of species. HPLC books of all types have already appeared in print, but there does not yet appear to be one specifically devoted to the topic of this paper (Grob, 1983; Grob and Kaiser, 1982). We are considering ion chromatography [single or dual (suppressed) column] to be just another part of CLC, similar to paired-ion reversed-phase (RP) HPLC rather than something totally separate. Within the past five years a large number of review papers have appeared summarized most, if not all, of what has been reported in the field of analysis of inorganic species, especially for metals, by CLC interfaced (off-line or on-line, continuous or discontinuous) with various specific or nonspecific detection methods (Cassidy, 1981; Florence and Batley, 1980; Krull, 1984a, 1984b; Krull and Jordan, 1980; Schwedt, 1981a, 1981b; Veening and Willeford, 1983). This paper addresses the questions as to what specific types of CLC and what types of CLC-compatible detectors are available commercially or otherwise for either off-line or on-line interfacing, and which is designed for and compatible with both metal and nonmetal analysis and detection of individual species. Although the paper by Irgolic and Brinckman (this volume) discusses interfacing of CLC with element-specific detection systems, mention is also made here of such valuable approaches. Vestal (this volume) has addressed the current state of interfacing CLC with mass spectrometers, but since very little has been done with this approach, it can only be suggested here as another viable method for future studies. There must be some overlap in any discussion of CLC interfacing with various detection systems, be these ultraviolet-visible (UV-VIS), fluorescence (FL), electrochemical (EC), mass spectrometry (MS), inductively coupled plasma emission spectroscopy (ICP), or others. Detectors for CLC were recently reviewed by Vickrey (1983), and the reader is referred to this book for a complete, up-to-date discussion of modern detectors for CLC.

Advantages and Disadvantages of GC and TLC

Before beginning a thorough discussion of the various forms of CLC available for interfacing and analysis of species, it is worthwhile to discuss why we are even considering CLC as opposed to TLC, GC, electrophoresis, or related separation techniques. If one reviews the literature before 1975, most references dealing with inorganic analysis utilized such separation methods, especially GC and TLC, but there were always serious disadvantages in such approaches. In general, most inorganic metal or nonmetal species, other than those organically bound and nonionic, are not sufficiently volatile for GC determinations. Although some work has been reported in the area of high-temperature GC, except for a single research group in Germany (Schwedt, 1981 a), such methods have not been widely accepted or utilized. Thus, GC has been mostly used for volatile or semivolatile organometals, metal halides, metal carbonyls, and related inorganics, but almost never for species such as arsenate, arsenite, chromate, dichromate, chromic, cuprous, cupric, ferrous, ferric, and related nonchelated/noncomplexed cations or anions of metals or nonmetals. Though GC continues to be used, often with electron capture detection (ECD) or flame photometric detection (FDP) and, at times, with plasma emission detection (inductively coupled, direct current, or microwave-induced plasmas), for organometals such as organolead and organomercury species, such GC-based approaches have limited applications for total-metal analyses in real-world samples.

However, there is much literature dealing with GC-MS or GC-plasma or GC-atomic absorption spectroscopic techniques, and such work continues in a number of laboratories around the world (Krull and Jordan, 1980; Krull et al., 1982 b; Jordan et al., 1982 a, 1982 b). Indeed, we ourselves have reported on the use of GC-microwave-induced plasma (MIP) emission spectroscopy for the analysis of organoboron species, and such approaches do have some real value, especially when a single, volatile derivative species is the only analyte of interest in a real-world sample. However, it is our belief that GC has serious limitations and that it may never serve as a truly universal, practical, and comprehensive analytic tool for inorganic species, no matter what particular detector might be employed. If all of the particular inorganic species of interest could be converted to volatile derivatives, such as by hydride-generation techniques (before or after the GC step) together with some form of element-selective detection, then GC-element-selective detectors may yet realize a high degree of importance and widespread application for species analysis. Others may argue that they have already reached such a level of significance, but this reviewer has serious reservations.

TLC has been used even less than GC for the determination of inorganic species, perhaps because all other conventional detector schemes except for UV or FL require nonselective spray reagents or off-line methods that are cumbersome, unwieldy, impractical, insensitive, and time/engery/money-demanding. Very few people have reported using TLC with atomic absorption spectroscopy (AAS), flame (FAA) or flameless (graphite furnace AA = GFAA), inductively coupled plasma (ICP), direct current plasma (DCP), or related truly element-selective detection methods. Such approaches would require removing every spot or band from the developed TLC plate, concentrating this material where possible, and then introducing each recovered spot or streak in an aqueous solution into the detector.

Column Liquid Chromatography

It would appear that most of these separation methods would or could meet all or most of the initial requirements listed above for the practical and valuable determination of inorganic species. This, of course, also assumes that the detector half of the hyphenated technique (CLC-detector) would meet such requirements at the same time. Thus, both volatile and nonvolatile inorganic species can be separated by CLC, and both aqueous and organic as well as aqueous:organic solutions are compatible with various forms of

CLC. Automation for sample injection, separation, column switching, preconcentration (on-line), detection, data acquisition, data manipulation, and final data interpretation are all techniques and methods that are available today. As soon as a particular sample is made compatible with the particular CLC separation approach chosen for that desired separation, it is compatible with full CLC-detector automation steps. This means that analytical chemistry today is prepared for routine analysis of inorganic species, with virtually complete system optimization and automation from injection point to data interpretation and printout. Indeed, even sample preparation and preconcentration steps can now be fully automated, often with robotic assistance. Such determinations can now be done almost free of any operator attendance and interference, overnight if desired, with a high degree of reproducibility, accuracy, precision, and freedom from matrix interferences or sample complexity.

CLC is completely different from open-column, atmospheric pressure LC, but these latter approaches are almost no longer used today. Other than for the Dionex approach to IC, modern CLC utilizes high-pressure solvent delivery systems, high-pressure valve-type injectors, thick-walled stainless-steel chromatographic (analytical) columns, small-diameter packing materials (all types), narrow-bore (internal diameter, i.d.) stainless-steel connecting tubing of minimum dead volume, and small-volume cell-detection systems with high sensitivity (UV, FL, EC, etc.). Solvents are pumped through the entire system at pressures often in the 1000–5000 psi range, and such solvents can be aqueous, organic, or any mixture of two, three, or even four such solvents as long as these are compatible, nonreactive, and homogeneous. The sample is injected under equilibrium conditions in the CLC system, usually in volumes of 5–200 μL, in mobile-phase compatible solvents or the mobile phase itself, and is swept along onto the separation column wherein the desired separations occur. All of this is usually done in less than 10 min. With fast LC using 3 μm, short packed columns, separations of organics or inorganics can now be routinely done in less than 3–5 min, often in less than 30–60 seconds, with full baseline resolution. We have seen a separation of five organic standards done with fast LC in less than 1.5 seconds. It is thus entirely conceivable that the analysis of metal or nonmetal species could be done some day by fast LC-detector methods in less than one minute for all species of interest, and with full instrumentation-data automation (DiCesare et al., 1981 a, 1981 b; Dong and Gant, 1984). Relatively little has been reported with regard to detection of inorganic species that involve fast LC, but there is every reason to hope that this area will develop more fully within the coming years. The ability to vary both the packing materials and the mobile-phase solvents from one run to another as well as the chance to use gradient-elution and flow-programming methods, together with box-car chromatography or column switching, provides an incredible degree of possible separations. And, again, both volatile and nonvolatile inorganic species can be routinely resolved by current and future CLC approaches.

Microbore or Capillary CLC Detection

Variations on conventional, 2.0 mm or 4.6 mm i.d. packed columns in CLC have evolved over perhaps the past decade or less, and they clearly have certain inherent advantages with regard to solvent consumption, sample consumption, time of analysis (at times), flow rates, etc. With conventional concentration-sensitive detectors, especially UV and FL, microbore LC or capillary LC appear to provide improved sensitivity and lowered MDLs, especially for organics. Indeed, Brinckman's group at the U.S. Bureau of Standards has recently developed and introduced a novel microbore CLC-fluorescence approach using microscopy, that provides incredibly good MDLs for the analysis of certain inorganic species. On the other hand, there has also been a good deal of work already done on the interfacing of micro-LC with MS, such as the outstanding work of Henion at Cornell, but all of this work has involved only organics. No studies on the determination of inorganic species have ever been reported for these methods. For mass-flow-sensitive detectors including virtually all element-selective detectors for CLC, micro-LC methods have a very serious shortcoming: mass loadability on the column itself. With very narrow i. d. columns and very slow flow rates, maximum sample loadings can only reach a few microliters (μL) at best, which severely limits the amount (mass) of sample that can be placed onto the column for a single analysis. With mass-flow-sensitive type detectors, including the element-selective type detectors (AAS, AES, AFS, etc.), that means one is at the very start limiting the amount (mass) of inorganic species which can reach the detector. With conventional LC (4.1–4.6 mm i.d.) sample loadings can often reach several hundreds of microliters without serious performance or resolution loss of the column; this is especially true for simple sample mixtures. This simply means that one at times has hundreds of times more mass of each species reaching the element-selective detector. Apparently Fassel's recent work with microbore LC and micro-torch ICP only highlights this inherent problem with regard to the analysis of inorganic species (Lawrence et al., 1984).

We feel that the future of inorganic species analysis with a high degree of element selectivity may never rely on micro-CLC approaches unless a way is found to load 100–300 μL of sample solution onto the column at the very start of the analysis. It may also be the case that micro-LC-MS is also not a very viable approach, except in those cases where one is severely sample-limited (1–20 μL maximum). Recent discussions with Henion seem to support these ideas, at least for micro-LC-MS interfacing (Covey and Henion, 1983; Eckers et al., 1982; Henion, 1981; Henion and Maylin, 1980; Henion and Wachs, 1981; Henion, personal communication, 1984). We disagree with Chau's suggestion (this volume) that future research and development work should involve micro-LC, perhaps with element-selective detection interfacings. In view of the basic theory and principles behind micro-LC and mass-flow-sensitive detection devices, except for very sample-limited cases, such hyphenated approaches will suffer from poor detection limits without further sample preconcentration steps. Such preconcentration steps should ideally be done on-line, in a continuous manner, with automated sample introduction. These techniques are already available with practically all other forms of CLC, as described below.

Currently Available Column Liquid Chromatography Techniques for Inorganic Analysis of Species

Unfortunately, there are no instruments for species analyses now commercially available, although there are several commercial ion chromatographs on the market today. We believe that true procedures for the detection of inorganic species, especially for metals, must rely on more than simple chromatographic retention times, capacity factors, or elution volumes for such determinations. Thus, UV-VIS, RI, FL, and even EC detection methods do not really provide anything more than chromatographic retention properties, at least in the ways they are commonly employed. Conductivity detection (CD), which is perhaps the most popular detector for ion chromatography, is a totally general approach and does not provide element-specific information by the detection process. It is impossible to base element-species specificity on chromatographic properties alone, though some colleagues have suggested this, and one must have a good deal more analyte information via the detection method employed. We believe that only element-specific detectors such as AAS or AES or AFS (atomic fluorescence spectroscopy) can provide such information, and this is not possible by UV,

FL, EC, or CD, or even by post-column derivatization reactions. True inorganic analysis and detection of species must therefore combine a suitable chromatographic separation of each species containing the same element, followed by some type of element-selective detection step for one or more elements (AES/AFS). Commercial systems today do not provide such interfaced capabilities, and it therefore falls to the interested environmentalist, toxicologist, biologist, or analytical chemist to develop such an interfaced or hyphenated technique from instrumentation already available on the market from various manufacturers. Perhaps in the not-too-distant future a single instrument will be available within a reasonable price range that is capable of doing automated, continuous inorganic analysis of species with a high degree of analyte identification. We are not quite certain that the commercial LC-MS systems available today can be routinely used for inorganic analysis and the detection of species, and surely not enough work in this area has been described in the available literature. Perhaps with more recent sample introduction and ionization approaches in LC-MS, this situation will soon change (Blakley and Vestal, 1983; Hardin et al., 1984; Hirata et al., 1979; Kenyon et al., 1983; Kratos Instruments Bulletin, 1983; McFadden, 1980; Takeuchi et al., 1981; Vestal, 1983). It is also possible that ICP-MS will become a viable and perhaps practical approach to inorganic species analysis, although there are no reports on interfacing this with HPLC (HPLC-ICP-MS) as yet (Heppner, 1983; Houk et al., 1980; Houk and Thompson, 1982, 1983). This would be an extremely expensive system, and although it could provide extremely selective and sensitive identification of inorganic species, its overall cost, together with computerization for data acquisition and handling, might make it an approach that would never be widely utilized.

Commercially Available CLC Systems

With regard to specific CLC approaches and systems that are commercially available today for the routine, continuous, real-time separation of various inorganic species including cations, anions, organometals (charged or not), metal chelates/complexes, etc., it is possible to separate virtually all such species other than for insoluble, zerovalent metals. All CLC approaches require the total dissolution of the sample components in either the mobile phase or another solvent that is fully compatible with this mobile phase (Krull, 1982). Since most relevant inorganic species are charged in some way as naked or organically derived ions, CLC approaches make use of such ionic properties in their separation mechanisms and modes. We consider here ion chromatography in its two basic forms (suppressed vs. nonsuppressed) as just another type of CLC, which is to be compared with conventional HPLC-derived ion-exchange or paired-ion approaches (Cassidy, 1981; Fritz et al., 1982; Krull, 1984a, 1984b; Krull and Jordan, 1980; Mulik and Sawicki, 1979; Proceedings of the Eighth International Symposium on Column Liquid Chromatography, 1984; Schwedt, 1981a, 1981b; Smith and Chang, 1983). Table 1 summarizes these four basic approaches to inorganic separations by modern CLC methods, along with commercial availability, mechanisms of operation, applications, and specific references. Commercial instrumentation for single-column (suppressed) ion chromatography has been available since 1975 from the Dionex Corporation, a firm in California that manufactures ion chromatographs and related/ancillary equipment for both inorganic and organic ionic-type separations.

This has clearly become a widely accepted instrumental approach for mainly inorganic-type separations, especially for nonmetal species, but more lately it has acquired capabilities for doing neutral organic analyses with paired-ion approaches and reversed-phase packings. Today the basic Dionex system, with special reversed-phase polymeric columns and paired-ion reagents in the mobile phase, is capable of doing inorganic/organic ionic-type analyses as well as neutral organic/inorganic-type separations. In essence, then, this current, latest model Dionex system can do just about all the separations possible with conventional HPLC equipment (though gel-permeation packings may not yet be available from Dionex) as well as all of the original ionic-type separations demonstrated between 1975 and 1982. Dual-column IC requires, for the initial separations in the first (separation) anion- or cation-exchange column, highly conductive mobile-phase eluents that perform the basic resolutions of ions. This background ionic mobile-phase constituent must therefore be almost entirely removed from the eluent before the usual conductivity detector, in order to allow for reasonable MDLs of the ionic analytes. This was initially accomplished by Dionex with the second stripper or suppressor column, which altered the initial mobile-phase constituents so that they became much less conductive, but left the desired analyte ions in a form that was still quite conductive. More modern variations by Dionex have led to hollow fiber suppression, which does not use the second stripper column at all but rather depends on membrane separation by an osmotic process. This has overcome almost all of the effective dead volume contributed by the stripper column in the original designs of the Dionex system and has vastly improved overall peak

Table 1. Summary of column liquid chromatography approaches for the determination of inorganic species

Name of method	Operating principles	Advantages or disadvantages
Dual-column, suppressed ion chromatography (DCIC)	Analytical column (first) preforms the desired ionic separations, second stripper (suppressor) column removes background ions from eluent before final detection	Requires an all-glass/plastic system, low-pressure pumping system, special packing materials for separation and suppression, available from but one supplier at high prices, not always compatible with other commercial HPLC equipment, suppression step introduces extra column variances (dead volume)
Single-column, non-suppressed ion chromatography (SCIC)	One analytical separation column, low-capacity packing material, low-conductivity eluent mobile phase, conductivity detection	Can be used with almost all commercial HPLC equipment, available at reasonable prices from a number of manufacturers, uses modular system for interchanging parts, can be used with variety of detectors from many suppliers, latest chromatography has improved greatly over initial peak shapes/asymmetry factors, resolutions, etc.
Ion-exchange HPLC	Single analytical ion-exchange column, uses silica or polymeric ion-exchange packings, separations by ion-ion interactions, used with most commercial detectors	Compatible with all commercial HPLC equipment, inexpensive instrumentation/columns, aqueous or aqueous/organic mobile phases, can be used with almost all detectors, chromatographic performance with latest, small-particle packings is excellent, reasonably priced from numerous suppliers
Paired-ion reversed-phase HPLC (PIRP)	Single analytical reversed- or bonded-phase column, silica or polymeric packings, separates both ionic and neutral species at same time, compatible with all commercial detectors, separations based on hydrophilic-hydrophobic interactions between aqueous-based mobile phase and hydrocarbonaceous stationary phase, equilibrium distribution of organic/inorganic, ionic/neutral species	Compatible with all commercial HPLC equipment, inexpensive instrumentation/columns, aqueous or aqueous/organic mobile phases, can be used with just about all detectors, chromatographic performance and peak shapes generally excellent, large variety of commercially available bonded phases (bulk or packed) for varying separations, can do both neutral and charged separations in one run for organics and inorganics, compatible with post-column reactions for derivatization (on-line)

shapes, asymmetry factors, resolutions, column efficiencies, and total system performance. With the now available small-diameter-particle packing materials in the analytical separation column, IC chromatograms on a modern Dionex system are hardly distinguishable from any HPLC-generated chromatogram using ion-exchange, paired-ion RP, gel permeation, or even normal-phase modes.

Although modern-day suppressed (Dionex) IC appears to have overcome almost all of the original problems inherent in the dual-column approach introduced in the late 1970s, it has not overcome the basic problem of price, which I feel was always a bit exorbitant.

Recently Wescan commercialized the single-column IC (SCIC) approach, also known as nonsuppressed IC, because it uses but a single analytical ion-exchange column, usually silica-based, followed by an electronically suppressed conductivity detector to reduce background noise levels arising from salts in the mobile phase and eluent. It is the combination of low-conductivity mobile phases, low-capacity, weak ion-exchange packings, and finally an electronically suppressed conductivity detector that provides all that is needed for the single-column approach. As originally developed by Fritz and colleagues at Iowa State, the method has shown great promise for both organic and inorganic IC, although it has mainly been used in the past five years for inorganic ions (Technical Publications on SCIC). Wescan has commercialized SCIC as a completely viable alternative to the Dionex system, at prices that are considerably lower than Dionex's for systems that can do just about the same things. If one compares the most advanced instrumentation and resultant chromatograms from Dionex and Wescan, it is very hard to distinguish these or to see significant chromatographic differences (peak shape, asymmetry, resolutions, efficiencies, etc.). Although such a direct comparison of the two commercial systems is hard to find in the literature, it is our impression that chromatographically and detection-wise they are very similar for inorganic ionic-type analyses. Another serious advantage of the Wescan approach is that it is designed in a modular fashion so that an individual researcher can purchase one or another part, or the entire coordinated system. The Wescan background suppression conductivity detector has become widely accepted as the leader in this field, although the newer Bio-Rad CD appears to offer promise as well (Technical Publications on Conductivity Detectors). Many people purchase just the conductivity detector and then set up their own commercially available (Vydac packings) ion-exchange analytical columns, low-conductivity mobile phases, conventional HPLC pumping system and injector/tubing, and are practically set to do SCIC. A number of other commercial suppliers have recently entered the IC field, such as Waters Associates, Perkin-Elmer Corporation, and Environmental Science Associates. In most instances these approaches use variations of SCIC as initially commercialized by Wescan, or conventional ion-exchange HPLC with modern conductivity detection. It is possible to estimate at least a half-dozen firms which are now firmly committed to the commercialization of various IC approaches, all of these using single-column methods and equipment. This type of SCIC system can be used for just about all other types of conventional HPLC: gel permeation, paired-ion RP, normal phase, etc. Other detection methods can also be used with SCIC as these are now in use for dual-column IC, such as electrochemical, UV-VIS following post-column derivatizations (if necessary), or refractive index (RI). Also, there is no reason why SCIC cannot be successfully interfaced in most labs with element-selective detection, hopefully in an on-line manner. Indeed, some labs have already taken this approach, but we feel that other conventional HPLC separation modes are even better than SCIC. In SCIC the use of low-capacity ion-exchange packing material, be this silica-based (Vydac) or polymer-based/derived, means that it is difficult both to separate nonionic and ionic analytes at the same time within a single run and to retain all species. Paired-ion approaches are almost never done with an ion-exchange packing, since that would defeat the basic mechanism inherent to ion-exchange HPLC. On the other hand, paired-ion methods can be used with all types of reversed-phase packings (bonded phases) for the simultaneous separation of both ionic and neutral species. We feel that this is a significant advantage of the PIRP-HPLC approach, regardless of the particular detection mode.

Paired-Ion, Reversed-Phase HPLC

Within the past four to five years paired-ion, RP-HPLC methods have gained more popularity and acceptance (Chiu, 1980; de Kleijn, 1982; Girard and Glatz, 1981; Iskandarani and Pietrzyk, 1982a, 1982b; Molnar et al., 1980; Newburger and Day, 1983; Reeve, 1979; Skelly, 1982; Stevenson and Harrison, 1981). This is an area with which we have been directly involved and one which we believe deserves more serious consideration on the part of the analytical community for inorganic ionic analyses. Paired-ion RP-HPLC separation methods are more recent in development than conventional ion-exchange HPLC, and such methods were initially utilized entirely for ionically charged organics, especially drugs such as amine derivatives. Only within the past four to five years have researchers and commercial firms recognized the inherent possibilities for conventional paired-ion RP-HPLC for the detection of inorganic species (Bushee et al., 1982, 1984; Ding and Krull, 1983; Krull and Panaro, 1984; Krull et al., 1982a, 1983b, 1983c). We and others have realized that conventional, inexpensive HPLC instrumentation, bonded-phase columns (CN, C-8, C-18, C-4, etc.), and selective or nonselective detectors can provide the interested researcher with a simple and straightforward approach to the determination of inorganic species for a wide variety of anions and cations. Most inorganic anions have rather poor detection properties for UV, FL, or EC, although this latter approach has not as yet been fully evaluated. Conductivity detection with conventional paired-ion reagents is difficult, in view of the rather high background noise level due to the reagent present in the mobile phase at all times. However, Cassidy and Elchuk have apparently overcome this problem by the judicious choice of a novel paired-ion reagent and counterion (1982, 1983a, 1983b). Paired-ion RP-HPLC approaches offer some advantages in contrast to the previously described single- or dual-column IC ones, some of which have already been discussed. In addition, the method allows one to analyze for both charged and uncharged species within the same single chromatographic run. This is generally not possible with other methods in IC or ion-exchange HPLC, as described above. In using nonselective detectors, improved detection limits are now possible by the use of UV-absorbing or FL ion-pairing reagents, with or without post-column, liquid-liquid extraction techniques on-line (Denkert et al., 1981; Hackzell and Schill, 1982; Helboe, 1983). Indirect photometric detection with paired-ion RP-HPLC for inorganics is yet another approach to nonselective determination of inorganic species (Dreux et al., 1982; Larson and Pfeiffer, 1983; Small and Miller, 1982). We must emphasize that a great deal remains to be done in the area of paired-ion RP-HPLC species analysis for inorganics, and it is our hope that others will pick up on such methods for eventual optimization and application. These methods are almost trivial to use, a large variety of chromophoric or fluorophoric ion-pairing reagents are now commercially available, and the commercial instrumentation is relatively inexpensive to assemble and operate (Frei, 1981). In general, separations for inorganic anions appear to offer better chromatographic performance than those for cations, but the precise reasons for these observations have not yet been delineated. Some further work is obviously needed specifically for inorganic cations, especially for metals and organometals, in order to improve peak shapes, asymmetry factors, resolutions, efficiencies, elution times, etc. In addition, for most of the published work with bare cations, it has only been possible with a single RP-18 column to separate all +1 cations from +2 species, but it is not yet that simple to separate all +1 cations from one another (Bushee et al., 1982; Molnar et al., 1980).

Conventional, Ion-Exchange HPLC

Ion-exchange HPLC has long been used for organic ionic separations, but for some reason less work has developed, until very lately, for the detection of inorganic species. With the commercialization of such approaches together with conductivity detection, this should soon change for the better. A large variety of silica- and polymer-based ion-exchange packings are available for both cations and anions, in bulk and prepacked form, and these serve admirably for a large variety of organic or inorganic ionic separations (Cochrane and Hillman, 1982; Dadone et al., 1981; Davenpor and Johnson, 1974; Kamiura and Tanaka, 1979; Mackey, 1982; Roden and Tallman, Snyder and Kirkland, 1979; Woolson and Aharonson, 1980). A large number of procedures for the detection of inorganic species, especially using element-selective detection, have already employed ion-exchange HPLC methods. Conventional, open-column, ion-exchange methods have been with us for many decades, the mechanisms of organic and inorganic ionic separations by such approaches are well studied and understood, and these are directly transferable to ion exchange in HPLC. Many such applications already exist with regard to both organics and inorganics, and when such approaches can be used with AAS, AES, or AFS, they would appear virtually ideal for further development. Ion exchange relies on an ion-exchange packing material that interacts with inorganic ionic solutes to varying extents, and together with ionic mobile phases such species can be readily separated from one another. Both cation- and anion-exchange packings have long been available for HPLC, both silica-based and polymeric; these are generally stable for long periods of time. Currently designed background-suppressed conductivity detectors are ideal partners for ion-exchange HPLC, except for the fact that they are totally general as opposed to specific detectors. What is clearly needed is an inexpensive, highly sensitive and specific, low-cost, accurate, precise, reliable, and reproducible detector for ion-exchange or paired-ion RP-HPLC that would provide element-selective detection of species. In lieu of such a detector, which does not quite appear within our grasp, one is left with element-selective detection derived from AAS, AES, and/or AFS methods, as above. We believe that this is an area for immediate development, especially with regard to electrochemical, thin-layer, flow-through, single or dual transducers for HPLC (see below).

We believe that the above adequately summarizes most of the more widely available and employed CLC approaches for performing half of the desired inorganic species determination. With currently available 3 μm to 10 μm packing materials for single- and dual-column IC, ion-exchange HPLC, and reversed-phase HPLC, striking improvements in the chromatographic step may not take place as rapidly in the next decade as they did in the past decade. CLC separations for just about all inorganic ionic materials have probably already been published in one form or another, though some improvements in overall resolutions, elution time, efficiencies, peak shape, etc., may be possible in the years ahead. Such improvements may appear minor in view of what has been accomplished in the past five to ten years for inorganic separations. If this idea is correct, as we believe that it is, then significant advances in CLC species analysis approaches must come from two areas: 1) inexpensive, reliable, sensitive, selective, and easy-to-use detectors for CLC; and 2) interfacings and post-column reactions for inorganics that are selective for an element or certain species. We believe that this is where the emphasis should be placed within the analytical community in the future, as well as further optimization of the chromatographic separations already developed for most inorganic species.

Comparison and Discussion of Currently Available Detection Techniques

Detection in CLC for inorganic species is at present a rapidly evolving and developing area which promises greatly simplified and improved element-selective detection for the determination of inorganic species in the future. We do not believe that it is possible to analyze true inorganic species with a nonspecific (general) type detector such as UV-VIS, FL, CD, and perhaps EC. In general most inorganic species, except for some anions (nonmetal), do not show any significant chromophoric or fluorophoric properties. Those that do, such as mercury species, often do not have high epsilons, and there is still no true element species selectivity possible other than chromatographic retention/elution parameters. Conductivity detection is also an approach that responds to virtually all conducting species. Thus all inorganic anions have differing sensitivities on this detector, but again there is no real species selectivity present. Refractive index (RI) detection is a totally general method that responds to changes in the refractive index of the HPLC eluent as an inorganic analyte species elutes through the detector. No determination of inorganic species has ever been possible via RI, and it is very unlikely that this will occur in the future. The latest widely embraced detector for HPLC, electrochemistry (EC), has been employed for organic species, especially biologically active substrates, but very little has yet been described for inorganics. The work of Johnson and colleagues (Maitoza and Johnson, 1980; MacKoul et al., 1984; Polta and Johnson, 1983; Sherwood and Johnson, 1981) with regard to LCEC for inorganics is of importance here (see also Bond and Wallace, 1981, 1982; Bond et al., 1982; Cohen and Bartak, 1983; Fogg et al., 1983; Girard, 1979; Hanekamp et al., 1982; Kauffmann et al., 1982; Krull et al., 1983a; Lewis et al., 1983; Shoup, 1982; Tjaden et al., 1983). We hasten to add that virtually all of the LCEC (liquid chromatography-electrochemical detection) approaches for inorganic determinations have not involved true element species, and it is generally true that EC does not have the ability to determine selectively one elemental species from another. Anodic stripping voltammetry can do total-metal analyses under certain conditions, but it too will never be able to perform on-line LCEC analysis for inorganic species. MacCrehan's recent work in reductive LCEC with differential pulse methods, specifically for metal cations, also cannot distinguish between element species, though improved chromatographic fractionations are evident (MacCrehan, 1981; MacCrehan and Durst, 1981; MacCrehan et al., 1979). Though differential pulse LCEC provides simplified chromatograms and the elimination of otherwise interfering peaks, it provides virtually no analyte identification other than chromatographic parameters (capacity factor, retention time, elution volume, etc.). True element species identification is only possible in LC with an element-specific detector, such as AAS, AES, or AFS, and Irgolic and Brinckman (this volume) adequately describe such approaches in great detail. Single-electrode, fixed-potential, or differential-pulse LCEC provides little more analyte information than UV or FL or RI methods, though it could be correctly argued that some specificity is provided in the differential-pulse mode, but not in the identification of element species. Chromatograms are simplified, true separation of species is still absent. Some scanning voltammograms have been reported by Wightman and Osteryoung in LCEC of HPLC with dropping mercury electrode (DME), and here some real improvements in the detection of inorganic species or specificity may be possible. However, very little has yet been reported for inorganic species analysis with such methods, and thus this is yet another area for further development and improvement (Barnes and Nieman, 1983;

Caudill et al., 1983; O'Dea and Osteryoung, 1980; Trojanek and De Jong, 1982). Elemental ratios and true elemental content is also quite difficult (actually impossible) by all current EC methods, and this is a significant disadvantage in comparison with atomic absorption or atomic emission/fluorescence methods.

Electrochemical Detection Approaches in CLC

Perhaps because of the above reasons, relatively few studies have reported on the utilization of LCEC (amperometric or coulometric) with thin-layer, flow-through, or dropping mercury for the analysis of inorganic species of metals or nonmetals. Perhaps one of the most serious hindrances to more widespread studies on inorganic species is the general inaccessibility and unavailability of inexpensive, reliable, sensitive, selective, accurate, precise, and reproducible, perhaps automatable, methods of analysis. If LCEC could be turned into such a viable approach, in view of its much lower cost in comparison with HPLC-GFAA or HPLC-ICP/DCP methods, it might be much more widely accepted and utilized. We have already made some suggestions above as to how this might be accomplished, but surely the use of parallel/series dual-electrode approaches in LCEC should be much more widely introduced and applied for inorganic species analysis. Though a large number of papers and technical publications have described the use of single-electrode electrochemical detection for inorganics, virtually none of these has described the use of dual-electrode steps (see Bratin et al., 1984). Substantial species identification of organics has already been demonstrated by the dual-electrode methods, especially for various catecholamines in biological samples (Andrews et al., 1982; Krull et al., 1984a; Roston and Kissinger, 1981, 1982; Roston et al., 1982). We feel that similar approaches would be just as useful and practical for inorganic species detection, and this is currently an area of interest and activity within our research group. Dual-electrode response ratios as a function of working electrode potentials provide for greatly improved analyte identification, perhaps not elemental species identification, but surely more qualitative data than is ever possible by single-electrode approaches for the same injection. We have recently demonstrated this type of analyte identification for certain organics, organic nitro-compounds, in dual-electrode LCEC studies (Krull et al., 1984a, 1984b).

Dionex Corporation has recently introduced its own variety of flow-through EC detector for IC, which is placed just after the first separator column and just before the second stripper column in the dual-column apparatus (see Rocklin and Johnson, 1983). Other commercial HPLC firms have also introduced EC detectors for inorganic/organic ionic determinations, following ion-exchange HPLC or SCIC. We would expect that in the immediate future there will be a much greater emphasis on EC detection for inorganics and inorganic species, and surely this trend should be encouraged, for it could lead to some significantly different and very useful analytical approaches.

Element-selective Detection in CLC: AAS

We have thus far been concerned with the most important, already commercially available detectors for HPLC: UV-VIS, FL, RI, EC, CD, and the operating parameters or principles of these and other detectors are summarized in Table 2. None of these five separate approaches really is capable of providing information on true inorganic species. Although there are a number of post-column derivatization approaches that have already been described for various inorganic (especially metal) species, none of these, followed by UV-VIS or FL, really provide for the determination of element species (Cassidy, 1981; Cassidy and Elchuk, 1980, 1981; Elchuk and Cassidy, 1979; Florence and Batley, 1980; Schwedt, 1981 a, 1981 b; Veening and Willeford, 1983). This is because the final detection step is still a nonselective, general approach. There are no really inexpensive, easy-to-operate and maintain, element-selective detectors for HPLC commercially available today, and this is just what is demanded in the future for all applications to inorganic species analysis. We and others have recently reviewed the area of element-selective detectors for HPLC, and the recent book edited by Vickrey summarizes this area very well (Cassidy, 1981; Florence and Batley, 1980; Krull, 1984a, 1984b; Krull and Jordan, 1980; Schwedt, 1981 a, 1981 b; Veening and Willeford, 1983; Vickrey, 1983). The paper by Irgolic and Brinckman (this volume) is also designed in part to review this important field, and we hesitate therefore to duplicate their efforts and tax the reader. However, in view of our own interest and experience in this field, we shall briefly discuss this subject.

A large number of reports on metal and nonmetal species analysis have involved the direct or indirect interfacing of various forms of CLC with element-selective detection, AAS, AES, or AFS. In the case of flame AA, HPLC-FAA or IC-FAA systems are continuous, on-line, real-time, and compatible with all reversed-phase mobile phases, but the final MDLs are

Table 2. List of detectors for column liquid chromatography: principles of operation, selectivity, and advantages versus disadvantages

Type of detector	Operating principles	Selective or nonselective	Advantages vs. disadvantages
Ultraviolet-visible (UV-VIS)	Absorbance of UV-VIS radiation by chromophore present in analyte	Nonselective (general)	Inexpensive, many suppliers, easy to interface, much information/data, little structural information possible, nut useful for true species detection, at times a very sensitive approach
Fluorescence (FL)	Depends on initial absorbance of light by analyte, emission of light at longer wavelength in UV-VIS region	Nonselective (general)	Inexpensive, many suppliers, easy to interface, very sensitive, much information and data, little real structural information, not useful for true species detection
Refractive index (RI)	Depends on differences in refractive index between analyte solution and reference mobile phase	Nonselective (general)	Inexpensive, many suppliers, easy to interface, not sensitive, much information and data, little real structural information, not useful for true species detection, little sensitivity, poor detection limits
Conductivity (CD)	Depends on differences in conductance of analyte peaks compared to mobile phase	Nonselective (general)	Inexpensive, many suppliers, easy to interface, can be very sensitive for strongly conducting ions, much literature and technical info, not useful for true species detection, useful detection limits
Mass spectrometry (MS)	Depends on mass-to-charge ratios of ions formed from analyte, ability of analyte molecules to undergo ionization process within MS unit	Selective	Very expensive to purchase and to operate, excellent specificity and sensitivity, rarely used for detection of inorganic species, requires constant operator attendance maintenance, etc.
Electrochemical (EC)	Electrochemical oxidation or reduction of original analyte atom or molecule, as a function of working potential, measures current generated	Selective, but not element/metal-selective	Inexpensive, easy to operate, easy to interface with CLC, single and dual methods can lead to improved specificity, excellent detection limits/sensitivity, minimum of maintenance or operator attendance, not enough work yet done for determination of inorganic species

Table 2 (continued)

Type of detector	Operating principles	Selective or nonselective	Advantages vs. disadvantages
Flame atomic absorption (FAA)	Atomization and excitation of atoms in flame, measurement of absorbance of specific wavelength light	Element selective	Expensive, easy to operate after learning period, easy to interface with CLC, highly specific, poor detection limits and overall sensitivity, not useful for real-world samples in determination of inorganic species
Graphite furnace atomic absorption (GFAA)	Atomization and excitation of atoms in furnace, measurement of absorbance of specific wavelength light	Element selective	Expensive, easy to operate after learning period, difficult to interface in on-line fashion with CLC, very specific, very good detection limits, usable in off-line manner for real-worl samples in determination of inorganic species
Inductively coupled plasma emission (ICP)	Atomization, ionization, excitation, and emission of light from analyte at specific wavelengths	Element selective	Expensive, not so easy to operate on HPLC basis, difficult to interface with CLC for usable detection limits, very specific, with post-column hydride formation can give practical method for real-world samples
Direct current plasma emission (DCP)	Atomization, ionization, excitation, and emission of light from analyte at specific wavelengths	Element selective	Expensive, not to easy to operate on an interfaced basis for usable detection limits, very specific, with post-column hydride formation can give practical method for real-world samples, at times HPLC-DCP can be practical without derivatization post-CLC
Atomic fluorescence spectroscopy (AFS)	Atomization, ionization, excitation, and fluorescence emission of light from analyte at specific wavelengths, after initial absorbance of light	Element selective	Expensive, limited number of suppliers, not a great deal of literature in CLC-AFS mode, detection limits not usable for real-world samples, not as popular as other methods for determination of inorganic species as above

generally too high for trace inorganic species detection. Perhaps for this reason alone, more recent studies do not utilize HPLC-FAA interfacing. Graphite furnace AA (GFAA) has also been interfaced numerous times with CLC, and except for the fact that this is an off-line approach it serves admirably in inorganic species analysis. The use of a post-column hydride derivatization step for arsenic species with a heated quartz furnace AA, following IC, has provided an on-line method. Off-line, in HPLC-GFAA, means that the time scale of a single GFAA determination step is not compatible with continuous HPLC separations. Individual fractions from the HPLC eluent must therefore be collected manually or automatically and separately introduced into the graphite furnace/cup of the GFAA system. Irgolic and Brinckman have discussed this particular area at much greater length elsewhere (this volume). Apart from the noncontinuous nature of HPLC-GFAA and the histogram nature of the final data, which limits resolutions in HPLC, this is an ideal method. Detection limits are generally excellent, despite the fact that most of the HPLC eluent never even gets into the GFAA unless one wishes to use preconcentration methods within the GFAA sampling step. Specificity, element species detection, sensitivity, reliability, accuracy, precision, and reproducibility are all present in this method. The price for such a system is basically the cost of a good HPLC system or IC unit, a good GFAA system, and an automatic, carousel-type sampler, which could come to about $50,000 or less. The price for a good HPLC-FAA would come to a bit less money, but perhaps not that much less. If a firm were able to produce a high-quality, benchtop, HPLC-compatible GFAA system that could also work in the on-line, continuous fashion, we believe that there would be a waiting market for such a system, and many more applications forthcoming in the immediate future. This is another area for future development.

Atomic Emission Spectroscopy in CLC

Entering the AES field, we come to the plasma emission spectroscopic detectors, inductively coupled (ICP), direct current (DCP), or microwave-induced (MIP). The last of these may not be at all compatible or suitable for any form of HPLC, even with microbore or microcapillary methods, and until some vastly improved (more stable) MIP is developed, HPLC-MIP will not be a viable or practical approach for the analysis of inorganic species. Both ICP and DCP instruments have been successfully interfaced with various forms of HPLC and IC for the determination of inorganic species, and these appear to be ideal methods but for the singularly important matter of MDLs. Seemingly dependent on the particular metal species and operator, at times both HPLC-DCP and HPLC-ICP can provide practical trace MDLs (10–50 ppb or below). However, in our own experience direct HPLC-ICP does not routinely provide MDLs below about 10–20 ppm for most metal species. With post-column, on-line hydride derivatization in both HPLC-ICP and HPLC-DCP, one obtains MDLs always less than about 50–75 ppb for all species thus far studied. However, only a limited number of metals are capable of forming hydride derivatives and thus this approach has limited capabilities and application. Various groups are now trying to improve MDLs in HPLC-DCP and HPLC-ICP systems, some by going to a heated furnace or tantalum-ribbon introduction system instead of the more conventional cross-flow or concentric tube nebulizer for ICP/DCP or their interfaced counterparts. Again, a heated tube or ribbon interface between HPLC and AES will have the same problems as in HPLC-GFAA: noncontinuous, delayed operation, histogram data, limited overall chromatographic resolution, off-line operation, non-real-time data production, etc. It would be an engineering nightmare to construct such a system that would allow this type of sample introduction approach to be used in an on-line manner for HPLC-GFAA or HPLC-ICP/DCP. We have considered just this type of approach for HPLC-ICP for quite some time and have eventually recognized full well the inherent engineering problems. A graphite furnace or tantalum-ribbon HPLC-ICP system would/could provide the best detection limits yet possible, at least equal to HPLC-GFAA, along with a continuous, on-line, real-time method, thus being somewhat superior to HPLC-GFAA.

There may very well be other viable approaches for interfacing HPLC-ICP/DCP in a continuous, on-line manner, without a heated furnace/cup device, wherein species eluted from the HPLC are derivatized or tagged with a reagent that converts them into more volatile species. These could then be more effectively delivered to the plasma region of the ICP/DCP unit, perhaps as efficiently as the volatile hydrides (100%), and thus final MDLs could be again in the low ppb region. Such work is now in progress in our laboratories, but it is too early to know if such chemical approaches will be completely successful. Other approaches that might work here could utilize some sort of moving-belt interface between the HPLC and ICP or DCP detector, which would operate in a continuous, almost realtime manner, similar to the moving-belt (Finnigan) HPLC-MS system available commercially for many years. There are virtually no reports of attempts to interface HPLC with either ICP or DCP using a moving belt, but there are some very good reasons why this might be quite successful and of practical utility. We are about to initiate this type of study program within our university using a novel type of ICP instrument commercially available in the U.S., and we are quite hopeful that this particular HPLC-ICP system may prove to overcome all of the major problems indicated above for the more conventional HPLC-ICP interface. Of course, it would probably raise some problems not already present in our already described HPLC-ICP system for analysis of metal species (Krull, 1984a).

Atomic fluorescence spectroscopy has also been interfaced with HPLC, but far less often than ICP or DCP or GFAA (Krull, 1984b). In general it too suffers from the same limitations as HPLC-ICP, namely, the nature of the interface leading to less than ideal MDLs. In all of this work, as in that using other element-selective detectors, problems of interfacing have arisen because analytical chemists have attempted to force a system originally designed and engineered for batch-type operation to function with a continuously moving HPLC eluent stream of mobile phase. At times, as in FAA, these two instruments are fairly compatible, especially when the organic or organic/aqueous mobile phase can be fully burned in the FAA. At other times, as in HPLC-GFAA, the two instruments have very different operating requirements, and they cannot reach a true compromise in interfaced operation. In general, with HPLC-ICP or HPLC-DCP there is a constant problem, in the direct interface without hydride generation, of a nebulizer system that takes too long to equilibrate on the HPLC time scale. We believe that this is a major factor in the abnormally poor detection limits that we have obtained in HPLC-ICP analysis for most metals. Everyone has tried to force a round peg into a square hole, or *vice versa,* and to make a conventional, static or batch AES instrument function as a continuous detector for HPLC. This will apparently not be feasible until some improvements in the final interface are available and implemented. Once this ideal arrangement comes about, and it is possible in the near future, we could very well have HPLC-ICP and HPLC-DCP and HPLC-AFS instruments that would meet all of the ideal operating requirements for continuous and automated analysis of inorganic species. At this point in time, we believe that we know how to do this; it is only a matter of time (and of course money) until we are able to realize our goals and ideas.

Summary and Conclusions: IC and HPLC for Analysis of Inorganic Species

Basically it is our feeling that inorganic separations can now be adequately and inexpensively accomplished by a wide variety of commercially available instrumental approaches, almost all of which employ some type of column liquid chromatography. Though there may be some improvements in the near future, specifically with regard to inorganic separations and resolutions, these may not be as significant and obvious as those which have occurred in the past five or ten years. More significant advances in analytical methods for determination of inorganic species should, indeed must, come in the detector and separation-detection interface areas. What is clearly needed is an inexpensive, on-line, easy-to-use, reliable, accurate, precise, automatable, real-time, continuous, sensitive, and element-selective detector for all inorganic elements and species. If, as has already been suggested, some firm could manufacture an inexpensive, stable, and reproducible plasma emission detector readily interfaced with CLC, which also provided trace MDLs, the entire problem would be solved. An inexpensive, benchtop GFAA that could operate in a continuous mode with CLC interfacing would be another totally usable and practical approach/instrument. Improved interfacings for current ICP or DCP units with currently available CLC separations is another approach that could lead to overall success. All of these areas need further research and development before we have an "ideal" or semi-ideal CLC-detector system that meets all of our requirements for true determination of inorganic species. It is hoped that these efforts will reach fruition in the very near future and that everyone involved in monitoring for specific inorganic species will then be able, once and for all, to perform true analysis of inorganic species from a variety of samples. We also need a final species analysis system that is easy to set up, operate, maintain, service, and automate, so that researchers who are not analytical chemists by training or desire can successfully operate their systems with a high degree of confidence in the final analytical data. Too many so-called studies for determination of species have been reported that were apparently not 100% rigorous or correct/accurate or precise. This situation must change in the future.

Acknowledgement. This is contribution number 213 from The Barnett Institute at Northeastern University.

References

Andrews RW, Schubert C, Morrison J, Zink EW, Matson WR (1982) Dual electrode cells for LCEC: recent developments. Am Lab 140

Barnes AC, Nieman TA (1983) Coulostatic pulse amperometry for liquid chromatography/electrochemistry detection. Anal Chem 55:2309

Blakley CR, Vestal ML (1983) Thermospray interface for liquid chromatography/mass spectrometry. Anal Chem 55:750

Bond AM, Heritage ID, Wallace GG, McCormick MJ (1982) Simultaneous determination of free sulfide and cyanide by ion chromatography with electrochemical detection. Anal Chem 54:582

Bond AM, Wallace GG (1981) Determination of copper as a dithiocarbamate complex by reversed-phase liquid chromatography with electrochemical detection. Anal Chem 53:1209

Bond AM, Wallace GG (1982) Simultaneous determination of copper, nickel, cobalt, chromium(VI) and chromium(III) by liquid chromatography with electrochemical detection. Anal Chem 54:1706

Bratin K, Blank CL, Krull IS, Lunte CE, Shoup RE (1984) Fourth international symposium on LCEC and voltammetry: Recent advances in instrumentation, methodology, and applications. Am Lab (May)

Buchholz AE, Verplough CI, Smith JL (1982) A method for the simultaneous measurement of less than a part-per-million of chloride, nitrate, and sulfate in aqueous samples by nonsuppressed ion chromatography. J Chrom Sci 20:499

Bushee DS, Krull IS, Demko PR, Smith SB Jr (1984) Trace analysis and speciation for arsenic anions by HPLC-hydride generation-inductively coupled plasma emission spectroscopy. J Liquid Chrom 7

Bushee DS, Krull IS, Savage RN, Smith SB Jr (1982) Metal cation/anion speciation via paired-ion, reversed phase HPLC with refractive index and/or inductively coupled plasma emission spectroscopic detection methods. J Liquid Chrom 5:563

Cassidy RM (1981) The separation and determination of metal species by modern liquid chromatography. In Trace Analysis, ed. J.F. Lawrence, vol 1, p 122. New York: Academic Press

Cassidy RM, Elchuk S (1980) Trace enrichment methods for the determination fo metal ions by high performance liquid chromatography. J Chrom Sci 18:217

Cassidy RM, Elchuk S (1981) Application of high performance liquid chromatography to the analysis of nuclear materials. J Liquid Chrom 4(3):379

Cassidy RM, Elchuk S (1982) Dynamically coated columns for the separation of metal ions and anions by ion chromatography. Anal Chem 54:1558

Cassidy RM, Elchuk S (1983a) Dynamic and fixed-site ion-exchange columns with conductimetric detection for the separation of inorganic anions. J Chrom Sci 21:454

Cassidy RM, Elchuk S (1983b) Dynamically coated columns and conductivity detection for trace determination of inorganic anions by ion chromatography. J Chrom 262:311

Caudill WL, Ewing AG, Jones S, Wightman RM (1983) Liquid chromatography with rapid scanning electrochemical detection at carbon electrodes. Anal Chem 56:1877
Chiu G (1980) Determination of ethyl sulfate by reversed-phase ion pair chromatography. Anal Chem 52:1157
Cochrane RA, Hillman DE (1982) Analysis of anions by ion chromatography using ultraviolet detection. J Chrom 241:392
Cohen SA, Bartak DE (1983) An electrochemical flow cell for the detection of metal complexes. Anal Lett 16(A6):429
Cotter R (1984) Practical applications of microbore column technologies. Am Lab 80
Covey T, Henion J (1983) Direct liquid introduction/thermospray interface for liquid chromatography/mass spectrometry. Anal Chem 55:2275
Dadone A, Baffi F, Frache R, Cosma B, Mazzucotelli A (1981) Organic acid solutions in the chromatography of inorganic ions VIII. Chromatographia 14(1):32
Dasgupta PK (1984 a) Ion chromatographic separation of anions with ion interaction reagents and an annular helical suppressor. Anal Chem 56:769
Dasgupta PK (1984 b) Linear and helical flow in a perfluorosulfonate membrane of annular geometry as a continuous cation exchanger. Anal Chem 56:96
Davenpor RJ, Johnson DC (1974) Determination of nitrate and nitrite by forced-flow liquid chromatography with electrochemical detection. Anal Chem 46:1971
de Kleijn JP (1982) Simple, sensitive, and simultaneous determination of some selected inorganic anions by high-performance liquid chromatography. Analyst 107:223
Denkert M, Hackzell L, Schill G, Sjogren E (1981) Reversed-phase ion-pair chromatography with UV-absorbing ions in the mobile phase. J Chrom 218:31
DiCesare JL, Dong MW, Atwood JG (1981 a) Very-high-speed liquid chromatography. J Chrom 217:369
DiCeasre JL, Dong MW, Vandemark FL (1981 b) Instrumentation and columns for very high-speed LC. Am Lab 13:52
Ding X-D, Krull IS (1983) Dual electrode LCEC analysis of platinum derived anticancer drugs. J Liquid Chrom (Special Issue) 6(12):2173
Dong MW, Gant JR (1984) Short three-micron columns. Applications in high-speed liquid chromatography. LC Mag 2(4):294
Downey SH, Hieftje GM (1983) Replacement ion chromatography with flame photometric detection. Anal Chim Acta 153:1
Dreux M, Lafosse M, Pequignot M (1982) Separation of inorganic anions by ion-pair, reversed phase liquid chromatography monitored by indirect photometry. Chromatographia 15(10):653
DuVal DL, Fritz JS, Gjerde DT (1982) Indirect determination of cyanide by single-column ion chromatography. Anal Chem 54:831
Eckers C, Skrabalak DS, Henion J (1982) On-line direct liquid introduction for micro-liquid chromatography/mass spectrometry: application to drug analysis. Clin Chem 28:1882
Edwards P, Haak KK (1983) A pulsed amperometric detector for ion chromatography. Am Lab 78
Elchuk S, Cassidy RM (1979) Separation of the lanthanides on high-efficiency bonded phases and conventional ion-exchange resins. Anal Chem 51:1434

Engelhardt H (1979) High Performance Liquid Chromatography. Chemical Laboratory Practice. Berlin: Springer-Verlag
Fitchett AW, Woodruff A (1983) Determination of polyvalent anions by ion chromatography. LC Mag 1(1):48
Florence TM, Batley GE (1980) Chemical speciation in natural waters. CRC Crit Rev Anal Chem 219
Fogg AG, Chamsi AY, Abdalla MA (1983) Flow injection voltammetric determination of nitrate after reduction to nitrite. Analyst 108:464
Frei RW (1981) Reaction detectors in liquid chromatography. In Chemical Derivatization in Analytical Chemistry: Chromatography, eds RW Frei and JF Lawrence, vol 1. New York: Plenum Press
Fritz JS, Gjerde DT, Pohlandt C (1982) Ion Chromatography. Heidelberg: AH Verlag
Gilbert R, Rioux R, Saheb SE (1984) Ion chromatographic determination of morpholine and cyclohexylamine in aqueous solutions containing ammonia and hydrazine. Anal Chem 56:106
Girard JE (1979) Ion chromatography with coulometric detection for the determination of inorganic ions. Anal Chem 51:836
Girard JE, Glatz JA (1981) Ion chromatography with conventional HPLC instrumentation. Am Lab 26
Gjerde DT, Fritz JS, Schmuckler G (1979) Anion chromatography with low-conductivity eluents. J Chrom 186:509
Gjerde DT, Schmuckler G, Fritz JS (1980) Anion chromatography with low-conductivity eluents II. J Chrom 187:35
Grob RL (ed) (1983) Chromatographic Analysis of the Environment, Second ec. New York: Marcel Dekker
Grob RL, Kaiser MA (1982) Environmental Problem Solving Using Gas and Liquid Chromatography. Amsterdam, New York: Elsevier Scientific Publishing Co
Guiochon G (1981) Conventional packed columns vs. packed or open tubular microcolumns in liquid chromatography. Anal Chem 53:1318
Hackzell L, Schill G (1982) Detection by ion-pairing probes in reversed-phase liquid chromatography. Chromatographia 15(7):437
Hancock WS, Sparrow JT (1984) HPLC Analysis of Biological Compounds. A Laboratory Guide. New York: Marcel Dekker
Hanekamp HB, Bos P, Frei RW (1982) Electrochemical detectors in flowing liquid systems: an assessment of their use. Trends Anal Chem 1(6):135
Hardin ED, Fan TP, Blakley CR, Vestal ML (1984) Laser desorption mass spectrometry with thermospray sample deposition for determination of nonvolatile biomolecules. Anal Chem 56:2
Hearn MTW, Regnier FE, Wehr CT (eds) (1983) High-Performance Liquid Chromatography of Proteins and Peptides. New York: Academic Press
Helboe P (1983) Separation and quantitative determination of long-chain alkyltrimethylammonium ions by reversed-phase ion-pair liquid chromatography using ultraviolet-absorbing counter ions. J Chrom 261:117
Henion JD (1981) Continuous monitoring of total micro LC eluent by direct liquid introduction LC/MS. J Chrom Sci 19:57
Henion JD, Maylin GA (1980) Drug analysis by direct liquid introduction micro liquid chromatography mass spectrometry. Biomed Mass Spec 7(3):115

Henion JD, Wachs T (1981) Micro liquid chromatography/mass spectrometry diaphragm probe interface. Anal Chem 53:1963
Heppner RA (1983) Elemental detection with a microwave-induced plasma/gas chromatograph-mass spectrometer system. Anal chem 55:2170
Hewlett-Packard LS/MS Interface for HP 5985B GC/MS Systems. Technical Bulletin. Palo Alto, CA: Hewlett-Packard Corporation
Hirata Y, Takeuchi T, Tsuge S, Yoshida Y (1979) The application of a new sampling technique using an atomizer for chemical ionization mass spectrometry for free amino acids, drug components, higher phthalates and oligomers of styrene and ethyleneglycol. Org Mass Spec 14(3):126
Hoover TB, Yager GD (1984) Determination of trace anions in water by multidimensional ion chromatography. Anal Chem 56:221
Houk RS, Fassel VA, Flesch GD, Svec HJ, Gray AL, Taylor CE (1980) Inductively coupled argon plasma as an ion source for mass spectrometric determination of trace elements. Anal Chem 52:2283
Houk RS, Thompson JJ (1982) Elemental and isotopic analysis of solutions by mass spectrometry using a plasma ion source. Am Mineral 67:238
Houk RS, Thompson JJ (1983) Trace metal isotopic analysis of microliter solution volumes by inductively coupled plasma mass spectrometry. Biomed Mass Spec 10(2):107
Ion Chromatography Technical Bulletins. Sunnyvale, CA: Dionex Corporation
Ishii D, Hirose A, Hibi K, Iwasaki Y (1978) Studies on micro high-performance liquid chromatography. J Chrom 157:43
Ishii D, Takeuchi T (1980) Open tubular capillary LC. J Chrom Sci 18:462
Iskandarini Z, Pietrzyk DJ (1982a) Determination of nitrite and nitrate in water and food samples by ion interaction chromatography. Anal Chem 54:2601
Iskandarini Z, Pietrzyk DJ (1982b) Ion interaction chromatography of inorganic anions on a poly(styrene-divinylbenzene) absorbent in the presence of tetraalkylammonium salts. Anal Chem 54:2427
Jenke D (1981) Anion peak migration in ion chromatography. Anal Chem 53:1535
Jenke DR, Pagenkopf GK (1984) Optimization of anion separation by nonsuppressed ion chromatography. Anal Chem 56:85
Johnson EL (1982) Modern ion chromatography. Am Lab 98
Jordan SW, Krull IS, Karger BL, Smith SB Jr (1982a) An all glass-lined, open-split, solvent-venting interface for gas chromatography-microwave induced plasma emission spectroscopy (GC-MIP). Chem Biomed Envir Instrum 12(4):263
Jordan SW, Krull IS, Smith SB Jr (1982b) The trace analysis for catechol derivatives via boronate ester formation and GC-microwave induced plasma emission spectroscopy (GC-MIP). Anal Lett 15(A14):1131
Jupille T (1983) Single-column ion chromatography: a mini-review. LC Mag 1(1):24
Jupille T, Burge D, Togami D (1984) Ion chromatography uses only one column to get all the ions. Res Devel 135
Kabra PM, Marton LJ (eds) (1981) Liquid Chromatography in Clinical Analysis. Clifton, NJ: The Humana Press
Kaiser RE, Oelrich E (1981) Optimization in HPLC. Heidelberg: AH Verlag
Kamiura T, Tanaka M (1979) Determination of nitrate in suspended particulate matter with high-performance liquid chromatography with UV detection. Anal Chim Acta 110:117

Kauffmann JM, Laudet A, Patriarche GJ, Christian GD (1982) The graphite spray electrode and its application in the anodic stripping voltammetry of bismuth. Anal Chim Acta 135:153

Kenyon CN, Goodley PC, Dixon DJ, Whitney JO, Faull KF, Barchas JD (1983) Direct liquid inlet LC/MS and negative chemical ionization MS. Am Lab 38

King JN, Fritz JS (1978) Separation of metal ions using an aromatic o-hydroxyoxime resin. J Chrom 153:507

Koch WF (1979) Complication in the determination of nitrite by ion chromatography. Anal Chem 51:1571

Kratos Instruments Bulletin No 1 (1983) Liquid Chromatography-Mass Spectrometry Using a Direct Liquid Introduction Interface on a Sector Mass Spectrometer. Ramsey, NJ: Kratos Instruments

Krstulovic AM, Brown PR (1982) Reversed-Phase High-Performance Liquid Chromatography. Theory, Practice, and Biomedical Applications. New York: Wiley Interscience, Inc

Krull IS (1982) Sample introduction requirements in analytical chemistry. In Sample Preparation Technology Manual. Hopkinton, MA: Zymark Corporation

Krull IS (1984a) Trace metal analysis and speciation via HPLC-plasma emission spectroscopic detection. Trends Anal Chem 3(3):76

Krull IS (1984b) Trace metal analysis by high-performance liquid chromatography. In Liquid Chromatography in Environmental Analysis, ed. J.F. Lawrence, ch 5. Clifton, NJ: The Humana Press

Krull IS, Bratin K, Shoup RE, Kissinger PT, Blank CL (1983a) HPLC-electrochemical detection for trace organic/inorganic analysis: Recent advances in instrumentation, methods, and applications. Am Lab 57

Krull IS, Bushee D, Savage RN, Schleicher RG, Smith SB Jr (1982a) Speciation of Cr(III) and Cr(VI) via reversed phase HPLC with inductively coupled plasma emission detection. Anal Lett 15(A3):267

Krull IS, Ding X-D, Braverman S, Selavka C, Hochberg F, Sternson IA (1983b) Trace analysis for *cis*-platinum anti-cancer agents via liquid chromatography-electrochemical detection (LCEC). J Chrom Sci 21:166

Krull IS, Ding X-D, Selavka C, Nelson R (1984a) Post-column, on-line, continuous photolytic derivatizations for improved detection in HPLC-electrochemical detection (HPLC-hv-EC). LC Mag 2(3):214

Krull IS, Jordan S (1980) GC and HPLC interfaced to plasma emission spectroscopy. Am Lab 21

Krull IS, Jordan SW, Kahl S, Smith SB Jr (1982b) The trace analysis for steroidal carboranes via gas chromatography-microwave induced plasma emission spectroscopic detection (GC-MIP). J Chrom Sci 20:489

Krull IS, Panaro KW (1984) Trace analysis and speciation for methylated organotins via HPLC-hydride generation-direct current plasma emission spectroscopy (HPLC-HY-DCP). Anal Chem

Krull IS, Panaro K, Gershman LL (1983c) Trace analysis and speciation for Cr(VI) and CR(III) via HPLC-direct current plasma emission spectroscopy. J Chrom Sci 21:460

Krull IS, Selavka C, Ding X-D, Bratin K, Forcier G (1984b) The trace analysis for organic nitro compounds and explosives via HPLC-photolysis-electrochemical detection (HPLC-hv-EC). J Forensic Sci 29(2):449

Kucera P (ed) (1984) Microcolumn High-Performance Liquid Chromatography. Amsterdam: Elsevier Scientific Publishing Co
Larson JR, Pfeiffer CD (1983) Determination of alkyl quaternary ammonium compounds by liquid chromatography with indirect photometric detection. Anal Chem 55:393
Lawrence JF (1982) High-Performance Liquid Chromatography of Pesticides. New York: Academic Press
Lawrence KE, Rice GW, Fassel VA (1984) Direct liquid sample introduction for flow injection analysis and LC with ICP spectrometric detection. Anal Chem 56:289
Lewis JY, Zodda JP, Deutsch E, Heineman WR (1983) Determination of pertechnate by liquid chromatography with reductive electrochemical detection. Anal Chem 55:708
Lurie IS, Wittwer JD Jr (eds) (1983) High Performance Liquid Chromatography in Forensic Chemistry. New York: Marcel Dekker
MacCrehan WA (1981) Differential pulse detection in liquid chromatography and its application to the measurement of organometal cations. Anal Chem 53:74
MacCrehan WA, Durst RA (1981) Dual-electrode, liquid chromatographic detector for the determination of analytes with high redox potentials. Anal Chem 53:1700
MacCrehan WA, Durst RA, Bellama JM (1979) Study of organometal speciation in water samples using liquid chromatography with electrochemical detection. In National Bureau of Standards Special Publication 519, Trace Organic Analysis: A New Frontier in Analytical Chemistry. Gaithersburg, MD: U.S. Government Printing Office
MacDonald JC (1979) Ion chromatography. Am Lab 45
Mackey DJ (1982) Cation-exchange behaviour of a range of adsorbents and chromatographic supports with regard to their suitability for investigating trace metal speciation in natural waters. J Chrom 242:275
MacKoul D, Johnson DC, Schick KG (1984) Effect of variation in flow rate on amperometric detection in flow injection analysis. Anal Chem 56:436
Macrae R (ed) (1982) HPLC in Food Analysis. New York: Academic Press
Maitoza P, Johnson DC (1980) Detection of metal ions without interference from dissolved oxygen by reverse pulse amperometry in flow-injection systems and liquid chromatography. Anal Chim Acta 118:233
McFadden WH (1980) Liquid chromatography/mass spectrometry systems and applications. J Chrom Sci 18:97
Molnar I, Knauer H, Wilk D (1980) High-performance liquid chromatography of ions. J Chrom 201:225
Mulik JD, Sawicki E (eds) (1979a) Ion Chromatographic Analysis of Environmental Pollutants, vol 2. Ann Arbor, MI: Ann Arbor Science Publishers, Inc
Mulik JD, Sawicki E (1979b) Ion chromatography. Env Sci Tech 13:804
Newburger JD, Day WR (1983) Chromatography of Ions. Paper presented at the Pittsburgh Conference on Analytical Chemistry and Applied Spectroscopy, Atlantic City, NJ, March 1983
Novotny M (1980) Microcolumn liquid chromatography: A tool of potential significance in biomedical research. Clin Chem 26:1474

Novotny M (1981) Microcolumns in liquid chromatography. Anal Chem 53:1294A
O'Dea J, Osteryoung J (1980) Rapid scan square wave voltammetric detector for high-performance liquid chromatography. Anal Chem 52:2215
Okada T, Kuwamoto T (1983) Nonsuppressor ion chromatography of inorganic and organic anions with potassium hydroxide as eluent. Anal Chem 55:1001
Pohl CA, Johnson EL (1980) Ion chromatography – state of the art. J Chrom Sci 18:442
Pohlandt C, Fritz JS (1979) Separation of metal ions on a new amide resin. J Chrom 176:189
Polta JA, Johnson DC (1983) The direct electrochemical detection of amino acids at a platinum electrode in an alkaline chromatographic effluent. J Liquid Chrom 6(10):1727
Proceedings of the Eighth International Symposium on Column Liquid Chromatography, New York, May 1984 (Abstracts)
Reeve RN (1979) Determination of inorganic main group anions by high-performance liquid chromatography. J Chrom 177:393
Rich WE, Wetzel RA (1979) Ion chromatographic analysis of trace ions in environmental samples. In Monitoring Toxic Substances, ACS Symposium Series No 94, ed. D. Scheutzle, ch 14. Washington, DC: American Chemical Society
Rocklin RD, Johnson EL (1983) Determination of cyanide, sulfide, iodide, and bromide by ion chromatography with electrochemical detection. Anal Chem 55:4
Roden DR, Tallman DE Determination of inorganic selenium species in groundwaters containing organic interferences by ion chromatography and hydride generation/atomic absorption spectrometry
Roston DA, Kissinger PT (1981) Identification of phenolic constituents in commercial beverages by liquid chromatographic with electrochemical detection. Anal Chem 53:1695
Roston DA, Kissinger PT (1982) Series dual-electrode detector for liquid chromatography/electrochemistry. Anal Chem 54:429
Roston DA, Shoup RE, Kissinger PT (1982) Liquid chromatography/electrochemistry: thin-layer multiple electrode detection. Anal Chem 54:1417A
Schwedt G (1981 a) Chromatographic Methods in Inorganic Analysis. Heidelberg: AH Verlag
Schwedt G (1981 b) In Topics in Current Chemistry, ed. F.L. Boschke, vol 85, p 160. Berlin: Springer-Verlag
Sevenich GJ, Fritz JS (1983) Addition of complexing agents in ion chromatography for separation of polyvalent metal ions. Anal Chem 55:12
Sherwood GA, Johnson DC (1981) A chromatographic determination of nitrate with amperometric detection at a copperized cadmium electrode. Anal Chim Acta 129:101
Shoup RE (ed) (1982) Recent Reports on Liquid Chromatography/Electrochemistry. W. Lafayette, In: Bioanalytical Systems, Inc/BAS Press
Skelly NE (1982) Separation of inorganic and organic anions on reversed-phase liquid chromatography columns. Anal Chem 54:712
Small H (1983) Modern inorganic chromatography. Anal Chem 55:235A
Small H, Miller TE Jr (1982) Indirect photometric chromatography. Anal Chem 54:462

Small H, Stevens TS, Bauman WC (1975) Novel ion exchange chromatographic method using conductimetric detection. Anal Chem 47:1801
Smith FC Jr, Chang RC (1983) The Practice of Ion Chromatography. New York: J Wiley & Sons
Snyder LR, Kirkland JJ (1979) Introduction to Modern Liquid Chromatography, Second ec. New York: J Wiley & Sons, Inc
Stevenson RL, Harrison K (1981) Design and performance of a modular chromatograph for chromatography of anions. Am Lab 76
Stinson SC (1983) HPLC makers strive for speed, selectivity. Chem Eng News 29 (May 23)
Sunden T, Lindgren M, Cedergren A, Siemer DD (1983) Separation of sulfite, sulfate, and thiosulfate by ion chromatography with gradient elution. Anal Chem 55:2
Suzuki K, Aruga H, Shirai T (1983) Determination of monovalent cations by ion chromatography with ion-selective electrode detection. Anal Chem 55:2011
Takeuchi T, Tsuge S, Hirata Y, Mochizuki K (1981) Method and Apparatus for Introducing Samples to a Mass Spectrometer. U.S. Patent No 4,298,795 (Nov 3, 1981)
Technical Publications on Conductivity Detectors. Richmond, CA: Bio-Rad Corporation
Technical Publications on SCIC. Santa Clara, CA: Wescan Corporation
The Separations Group. Hesperia, CA: Vydac Packing Materials Technical Publications
Thermospray LC-MS Kits and Interfaces. Houston, TX: Vestec Corporation
Tjaden UR, de Jong J, van Valkenburg CFM (1983) Gradient elution of biogenic amines and derivatives in reversed phase ion-pair partition chromatography with electrochemical and fluorometric detection. J Liquid Chrom 6(12):2255
Trojanek A, De Jong HG (1982) Fast-scan A.C. voltammetry for better resolution of chromatographically overlapping peaks. Anal Chim Acta 141:115
Tsuda T, Novotny M (1978) Packed microcapillary columns in high performance liquid chromatography. Anal Chem 50:271
Veening H, Willeford BR (1983) High-performance liquid chromatography and metal complexes. In Advances in Chromatography, eds. J.C. Giddings, E. Grushka, J. Cazes, and P.R. Brown, vol 22, ch 4. New York: Marcel Dekker
Vestal ML (1983) Studies of ionization mechanisms involved in thermospray LS-MS. Intl J Mass Spec Ion Phys 46:193
Vickrey TM (ed) (1983) Liquid Chromatography Detectors. New York: Marcel Dekker
Wee VT, Kennedy JM (1982) Determination of trace levels of quaternary ammonium compounds in river water by LC with conductometric detection. Anal Chem 54:1631
Wetzel RA, Anderson CL, Schleicher H, Crook GD (1979) Determination of trace level ions by ion chromatography with concentrator columns. Anal Chem 51:1532
Willard HH, Merrit LL Jr, Dean JA, Settle FA Jr (1981) Instrumental Methods of Analysis, Sixth ed, chs 17 and 18. New York: D Van Nostrand Co
Williams RJ (1983) Determination of inorganic anions by ion chromatography with ultraviolet absorbance detection. Anal Chem 55:851

Woolson EA, Aharonson N (1980) Separation and detection of arsenical pesticide residues and some of their metabolites by high pressure liquid chromatography-graphite furnace atomic absorption spectrometry. J Ass Anal Chem 63:523
Yamamoto M, Yamamoto H, Yamamoto Y, Matsushita S, Baba N, Ikushige T (1984) Simultaneous determination of inorganic anions and cations by ion chromatography with ethylenediaminetetraacetic acid as eluent. Anal Chem 56:832

The Importance of Chemical "Speciation" in Environmental Processes,
eds. M. Bernhard, F. E. Brinckman, and P. J. Sadler, pp. 613–629.
Dahlem Konferenzen 1986. Springer-Verlag Berlin, Heidelberg
© Dr. S. Bernhard, Dahlem Konferenzen

Mass Spectrometers as Detectors for Liquid Chromatography

M. L. Vestal

Dept. of Chemistry
University of Houston
Houston, TX 77004, USA

Abstract. Techniques for on-line coupling of liquid chromatography with mass spectrometry are reviewed with particular emphasis on those suitable for application to analysis of chemical species in environmental systems. The present status of the various techniques is summarized and the strengths and weaknesses of the various approaches are assessed. The enormous potential for future application of recently developed LC-MS techniques for species analysis is discussed.

Introduction

The potential benefits to be derived from on-line coupling of mass spectrometry to liquid chromatography have been recognized for at least fifteen years. Unfortunately, the problems associated with accomplishing this union in a practical way are substantial, and only recently have techniques emerged which appear to offer feasible solutions. This recent progress has been extensively reviewed (Arpino and Guichon, 1979; McFadden, 1979), including an exhaustive review of the literature through 1982 (Edmonds et al., 1983). The present review is aimed at summarizing the present status of the various techniques which have been developed, assessing their strengths and weaknesses, and evaluating their potential for application to problems of species analysis in environmental systems. None of these techniques have as yet been widely applied to analyses of real samples of the kinds which were of primary interest at this workshop, but it appears that some have now reached the stage of development where many important applications are currently being developed.

Mass spectrometry is widely recognized as a powerful analytical tool which can provide both qualitative and quantitative data which may not be readily available by other techniques. In particular it can provide molecular weight, empirical formula (through precise mass measurement), isotope ratios, detection of functional groups and other substituents, and elucidation of structure, including in some cases stereochemical features. For most of these determinations by classical electron impact techniques it is important to vaporize a relatively pure sample into the ion source of the mass spectrometer without causing decomposition. Production of a vapor of intact neutral molecules from nonvolatile samples is (by definition) impossible. Chemical manipulations which produce a more volatile derivative sometimes offer a way around this difficulty, but application of these techniques to trace components in complex mixtures is often unsuccessful. Recent development of ionization techniques which involve direct production of ions from the condensed phase, thus bypassing the neutral intermediate, have largely overcome this volatility barrier. These new techniques very often provide ions characteristic of the intact molecule but frequently do not provide the stable, reproducible fragmentation pattern required for structural elucidation.

The basic incompatibility between mass spectrometry and liquid chromatography stems primarily from two factors. First, the mass spectrometer requires ions in the gas phase at rather low background gas pressures (ca. 10^{-6} Torr) while the LC employs liquid solutions often containing nonvolatile solutes and buffers at atmospheric pressure and above and flow rates of ca. 1 mL min^{-1}. This mass flow corresponds typically to about one hundred times the amount that can be accommodated by the mass spectrometer vacuum system. Thus it is necessary to vaporize the LC effluent and to modify either the LC or the MS in such a way as to overcome this basic mass flow incompatibility. The second major problem is that in LC, unlike GC, sample volatility is not an issue; therefore, to be generally useful the mass spectrometer must utilize an ionization technique which produces the required gas-phase ions without necessarily requiring thermal vaporization of the sample.

Most of the early work on LC-MS interfacing focussed on the problem of the mass flow incompatibility with conventional electron or chemical ionization techniques. These approaches are necessarily limited to relatively volatile samples. More recently attention has been given to the problem of ionizing nonvolatiles. In some cases new ionization techniques have been developed in conjunction with efforts to improve LC-MS interfaces. These coincidental developments have led to some confusion in the literature between the interfacing technique and the ionization method. We will attempt to clarify this matter by separately discussing LC/MS interfacing techniques and ionization methods. Since a review of mass spectral techniques for high-molecular-mass compounds is included elsewhere (Russell, this volume), the present discussion will be limited to those techniques which are compatible with on-line LC/MS interfaces.

Techniques for LC/MS Interfacing

It is easy to state the features of the ideal interface between the liquid chromatograph and the mass spectrometer. One would like to transmit all of the sample without chemical modification and no more of the solvent vapor than the vacuum system can handle. Furthermore, neither the chromatographic nor mass-spectrometric performance should be impaired, all of the standard mass-spectrometric techniques should be available for use as needed, and no elaborate or expensive modifications of either instrument should be required. Some compromises with this ideal have been made in all practical interfaces developed to date. The nature and consequences of the compromises which have been made in the various approaches to LC/MS interfacing are described in the following discussion.

Direct Liquid Introduction

The earliest reported efforts to interface a liquid chromatograph with a mass spectrometer employed a small-bore capillary tube to couple a sample in liquid solution directly into the ion source of the mass spectrometer (McLafferty and Baldwin, 1973; Tal'roze et al., 1969). This approach has been further developed and refined (Arpino and Guichon, 1982; Eckers et al., 1982; Kenyon et al., 1981) and is now commercially available from several manufacturers. A schematic diagram of a direct liquid introduction (DLI) interface of the type in current use in a number of laboratories is shown in Fig. 1. This interface uses a diaphragm with a small aperture (ca. 5–10 μm) rather than a capillary. The LC effluent flows in through a central tube and the excess flows out through a concentric outer tube. The fraction passing through the pinhole in the diaphragm and into the mass spectrometer is controlled via a needle valve in the liquid exit channel, which allows the pressure in the flowing liquid stream to be adjusted independent of flow

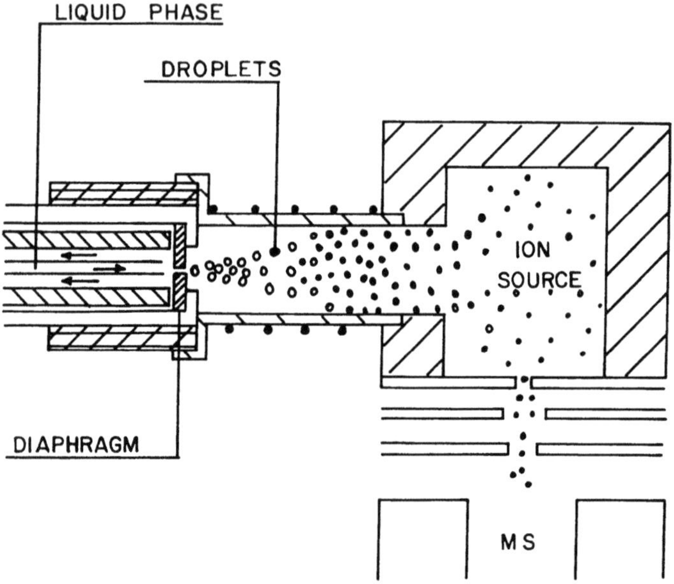

Fig. 1. Schematic diagram of a direct liquid introduction (DLI) interface

rate. The DLI probe is normally water-cooled to prevent premature vaporization. With proper adjustment of operating conditions, a high-kinetic-energy liquid jet issues from the orifice and breaks up into a spray of small droplets. In recent versions of this approach (Thomson et al., 1982), a heated desolvation chamber was employed in addition to the heated ion-source chamber to aid in desolvation and vaporization of difficult samples.

The early work of Talroze et al. (1969) employed electron-impact ionization, but the useable flow rates (<1 μL min^{-1}) were too low to be of much analytical utility. All of the later work has employed chemical ionization sources and vacuum systems. Depending on details of construction, these interfaces can generally accommodate some tens of μL min^{-1} of solvent with the upper limits being somewhat lower for water than for organic solvents.

An important variation on DLI involves vaporization of the solvent and ionization of the sample at atmospheric pressure. This approach was pioneered by Horning and co-workers (1974) and has recently been substantially improved by Thomson et al. (1982). In the latter work a dry-gas "curtain" is used to prevent most of the solvent vapor from entering the mass spectrometer while allowing reasonably efficient transmission of ions. A somewhat different but related approach has been described by Tsuge and co-workers (1979).

Thermospray

The thermospray technique has emerged from efforts to develop an LC-MS interface suitable for efficiently analyzing samples dissolved in aqueous mobile phases at typical analytical flow rates on the order of 1 mL min^{-1} (Blakley et al., 1978, 1980a, 1980b, 1980c; Blakley and Vestal, 1983). In contrast to the DLI technique, this is accomplished by application of heat to the capillary tube connecting the LC to the MS. Thermospray may be defined as the production of a supersonic jet of vapor with entrained droplets or particles by application of sufficient heat to a capillary to effect controlled partial vaporization of the liquid as it passes through. This approach has evolved from the original use of a focussed 50 watt CO_2 laser (Blakley et al., 1978), through oxy-hydrogen torches (Blakley et al., 1980c), to the use of simple electrical cartridge heaters imbedded in a copper block brazed to the end of the capillary (Blakley and Vestal, 1983). In the most recent work the "indirect" electrical heating has been replaced by "direct" electrical heating in which the power required to vaporize the liquid is supplied by passing an electrical current through the capillary tube itself (Fergusson and Vestal, 1983).

A schematic diagram of a recent version of the thermospray LC-MS interface is shown in Fig. 2. This interface and the ion source are much simpler than the earlier versions and require only minor modifications of a commercial mass spectrometer. Until recently the thermospray interface had been restricted to quadrupole mass spectrometers, but similar interfaces are now available for magnetic instruments as well (Vestal, 1984). As shown in Fig. 2, the thermospray vaporizer is mounted in a probe which may be introduced into the mass spectrometer using the direct insertion probe lock normally used for introducing solid samples. The vaporizer produces a supersonic jet of vapor containing a mist of fine droplets or particles. As these droplets travel at high velocity through the heated ion source they continue to vaporize due to rapid heat input from the surrounding hot vapor. Ions may be produced either by direct evaporation from charged droplets or by gas-phase ion molecule reactions. After the jet traverses the ion source it is pumped away by a 1 cm diameter pumping line which is connected to a dry ice-cooled trap backed by a mechanical vacuum pump. Ions

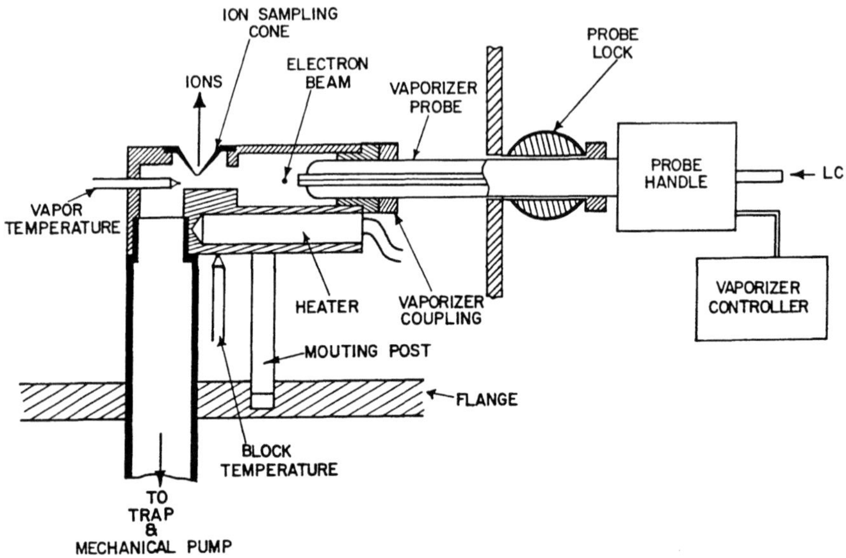

Fig. 2. Schematic diagram of a thermospray LC-MS interface

are sampled into the mass spectrometer by a small aperture (ca. 0.5 mm diameter) in a cone which intersects orthogonally with the axis of the jet. A 100 watt cartridge heater is imbedded in the source block to supply the heat necessary to complete the vaporization of the droplets as they fly through the ion source. With the addition of the vaporizer and pump-out line and these minor modifications to the ion source, any commercial quadrupole or magnetic mass spectrometer equipped for chemical ionization (CI) can be readily converted to a thermospray LC-MS system.

At first it may be somewhat surprising that this system can accommodate vaporizing more than 2 mL min^{-1} of water or other liquid directly into the ion source without overloading the pumping system since this corresponds to a gas flow about 100 times larger than that normally used in the conventional CI mode. The addition of the mechanical vacuum pump connected directly to the source would normally only account for about one order of magnitude of increased capacity, but the high velocity of the gas flow transverse to the sampling orifice introduces a Venturi effect which drastically reduces the flow out of this aperture. Initially it was thought that the pumping tube should be located directly opposite the vaporizer, but more recent work has shown that an orthogonal location such as shown in Fig. 2 is satisfactory so long as the flow is not "choked" in the vicinity of the ion-sampling aperture.

Electrospray

Efforts to employ the electrospray process for producing ions for mass spectrometry from nonvolatile samples in liquid solution was pioneered by Dole and co-workers (1968). In these early experiments ions were produced from benzene-acetone solutions of polystyrene with molecular weights ranging from 600 to 411,000 amu. Retarding potential energy spectra were obtained which were interpreted as indicating the presence of desolvated polystyrene macroions in the beam, but attempts at time-of-flight mass analysis of these ions were unsuccessful (Dole et al., 1978). Recently Whitehouse and Fenn (1984) deloped an improved version of the electrospray apparatus and obtained mass spectra from a number of nonvolatile samples. In this apparatus a high electrical field is applied to the surface of the liquid as it emerges from the capillary. The resulting charge separation causes the emission and acceleration of charged droplets, either positively or negatively charged, depending on the direction of the electrical field. These droplets vaporize as they pass through the heated-bath gas, and ions are produced by direct evaporation from the charged droplets. If vaporization of solvent is completed before the droplets reach the exit nozzle, then only ions and bath-gas molecules pass through. This effectively prevents problems of clustering of solvent molecules with the ions during the adiabatic expansion which otherwise may be quite serious with polar solvents such as water. A weak electrical field is applied between the nozzle and skimmer to improve the ion transmission, and most of the bath-gas molecules are pumped away upstream of the skimmer.

A key feature of this apparatus is the use of a contraflowing, heated-bath gas to aid in vaporizing the droplets. This allows the solvent vapor to be effectively removed before it enters the mass spectrometer, and only the ions and a relatively inert bath gas such as nitrogen or helium is transmitted along with the ions. The present version accommodates liquid flow in the 1 to 50 $\mu L\ min^{-1}$ range, and while it has not yet been extensively tested with on-line LC columns, it appears to hold significant promise, particularly for use with micro-LC.

Transport Devices

All of the above techniques involve vaporization or nebulization of the LC effluent and direct coupling of the resulting vapor or spray to the ion source of the mass spectrometer. In some cases partial or complete solvent-vapor removal may be accomplished by gas-flow techniques. The alternative approach to LC-MS interfacing is the use of a mechanical transport device in which the sample is deposited on a moving solid surface with the solvent being removed before the sample reaches the ion source of the mass spectrometer. This approach was originally developed by Scott et al. (1974) by adapting a wire transport mechanism originally developed for use in a flame ionization detector for LC. The technique was further developed and refined by McFadden et al. (1976) with the introduction of a continuous ribbon of relatively inert polyimide and improved methods for solvent removal. Interfaces based on this approach are now commercially available from several manufacturers.

This early work employed direct deposition of the liquid effluent onto the belt and thermal desorption of samples into either EI or CI ion sources. This approach was successful for a variety of samples in relatively volatile

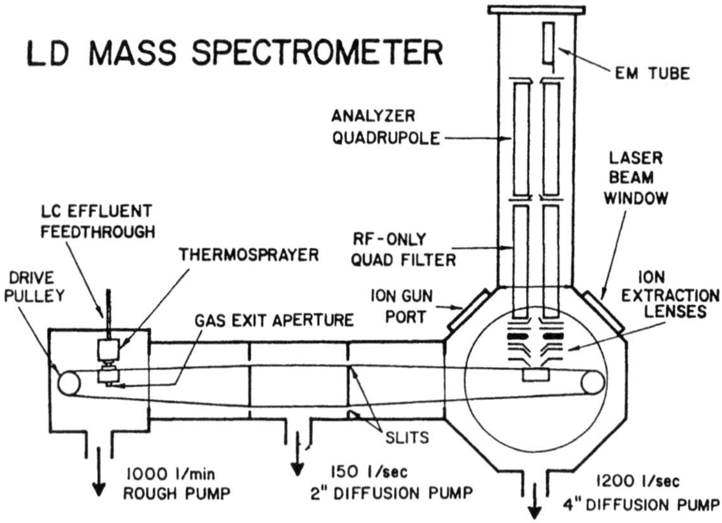

Fig. 3. Schematic diagram of a transport interface employing thermospray deposition and both laser and fast-ion desorption ionization

solvents but has not given good results for nonvolatile samples, particularly with water and other solvents used in reversed-phase chromatography. A major reason for this difficulty is associated with the problems encountered in rapidly vaporizing large quantities of polar solvents such as water from a surface. If sufficient heat is supplied to cause the water to vaporize before it reaches the ion source, then a nonequilibrium boiling mode is likely to occur rather than smooth-equilibrium vaporization. As a result a large fraction of the liquid may depart as small droplets carrying all or part of the less volatile sample along in solution. Several approaches have been developed to overcome this difficulty, including the use of a segmented-flow reactor system (Karger et al., 1979), spray deposition (Hayes et al., 1983; Smith and Johnson, 1981), and microbore LC (Games et al., 1982). Of these, the spray deposition techniques appear to be the most generally applicable since it allows, in principle, the sample to be deposited on the moving surface with more or less complete vaporization of the solvent.

One approach to spray deposition of sample onto a moving surface is shown schematically in Fig. 3. In this device (Hardin et al., 1984) samples are sprayed onto the belt under partial vacuum using a thermospray vaporizer operated so that nearly complete vaporization of the solvent occurs before the spray strikes the surface. This approach has been used routinely for vaporizing up to 2 mL min^{-1} of water or other reversed-phase solvents. Sample transfer efficiencies are typically about 50% and little or no band broadening occurs. Comparable performance has been obtained by other workers using a coaxial gas-flow nebulizer (Hayes et al., 1983).

Ionization Techniques

In the foregoing discussion we have attempted to present the various techniques for coupling liquid chromatography with mass spectrometry while ignoring, as much as possible, the ionization techniques used in the mass spectrometry. In this section the ionization techniques compatible with online LC-MS are briefly reviewed with particular emphasis on their utility for species analysis.

Electron Ionization

This classical method for producing ions in the mass spectrometer by bombarding the sample vapor with electrons of some nominal energy (ca. 70 eV) is the standard against which all newer ionization techniques must be compared. The mechanism of electron ionization and fragmentation can be represented by the reactions

$$M(g) + e^- \rightarrow M^+(g) + 2e^-, \quad (1)$$

and

$$M^+(g) \quad \rightarrow I_i^+ + N_i. \quad (2)$$

The power of this technique derives from the fact that all substances which can be vaporized can be ionized, and more importantly, the relative intensities of the parent and fragment ions are only weakly dependent on operating parameters such as the electron energy and temperature of the ion source. As a result it is relatively easy to obtain reproducible spectra on different instruments in different laboratories. This has led to development of a relatively large data base of electron-impact mass spectra with which spectra of unknowns can be readily compared using modern data systems for rapid and reliable identification of compounds already contained in the data base. Comparison with spectra of known compounds can often provide significant structural information even when the spectrum of the unknown compound has not been previously recorded.

The major difficulty with electron ionization is that it requires that the sample be present as a neutral vapor; furthermore, the total pressure in the ion source must be low enough (ca. $<10^{-4}$ Torr) that the relative intensities of the ions present are not significantly modified by ionmolecule reactions. Thus, by definition it is impossible to obtain the electron ionization spectrum of a nonvolatile molecule. The source pressure limitation requires that only very limited amounts of solvent vapor or other gases be admitted to the ion source along with the sample. Some of the early attempts at direct coupling of LC to MS attempted to provide EI spectra, but these efforts were not very successful because it is difficult to transfer the sample efficiently without also transferring a much larger quantity of solvent vapor or carrier gas. On the other hand, the transport devices can efficiently transfer sample to the EI source with little additional vapor. Thus it appears that the transport LC-MS with spray deposition is the preferred technique for EI analysis of molecules with sufficient volatility.

Chemical Ionization

An additional difficulty with electron ionization is that the odd-electron parent ions are often not stable, and even though an intact neutral molecule may be readily produced in the gas phase, the molecular ion may not be detectable in the EI mass spectrum. The production of ions by ion-molecule reactions with a reagent ion was aptly named chemical ionization by its originators, Field and Munson (Field et al., 1966). In this technique the sample to be analyzed is introduced at low partial pressure in an excess of some reagent gas which is present in the ion source at a sufficiently high pressure (ca. 1 Torr) so that a large number of ion-neutral collisions occur before an ion exits from the source. If a gas such as ammonia is used as the reagent, then the mechanism involved in ionizing a sample is as follows:

$$NH_3 + e^- \rightarrow NH_3^+ + 2e^-$$

$$NH_3 + NH_3^+ \rightarrow NH_4^+ + NH_2$$

$$NH_4^+ + M \rightarrow MH^+ + NH_3, \qquad (3)$$

where M represents the sample molecule. This ionization mechanism is, of course, effective only if the proton affinity of the sample is greater than that of ammonia. This technique very often yields ions representative of the intact molecule because inherently more stable even-electron ions are produced and because the total energy available is essentially limited to the difference in proton affinity between sample and reagent.

An advantage of CI for use with LC-MS interfacing is that it operates at relatively high ion-source pressure and can be quite sensitive even though the sample is present at very low concentration. It is particularly useful with DLI and thermospray where the solvent vapor may be used as the reagent gas. In some cases applicability of these techniques may be limited by this fact. If the proton affinity of the sample is less than that of major components in the vapor, then the protonated molecular ion will not be formed. However, the sample may be detected in the positive ion mode by the formation of a cluster ion (e.g., $M + NH_4^+$) or it may be detected in the negative ion mode either as the negatively charged molecular ion, M^-, formed by electron capture, or as the deprotonated molecular ion, $(M\text{-}H)^-$, formed by proton transfer to a stronger base. In any event, a necessary condition for ionizing a sample by CI is that the ion formed must not react with the reagent vapor. In such cases the transport devices may be preferred because they allow the CI reagent to be chosen independently of the choice of LC solvent and operating condition.

A major disadvantage of CI is that both the sensitivity and fragmentation depend strongly on operating parameters such as reagent gas composition and source temperature. These effects may be used to great advantage to ionize particular components selectively, but they also make it much more difficult to reproduce spectra reliably. As a result, compilation of a data base of CI spectra has not proven very fruitful. Like EI, chemical ionization requires the production of a neutral vapor as an intermediate and thus is not applicable to nonvolatile samples. Some very interesting results have been obtained on relatively nonvolatile molecules by a related technique known as direct chemical ionization in which samples are dispersed on the surface of a wire and rapidly vaporized directly into the ion "plasma" present in the CI ion source. The precise mechanism has not yet been determined, and it is possible that some processes such as those occurring in other desorption ionization techniques may be involved rather than gas-phase chemical ionization.

Ion Evaporation

Direct evaporation of molecular ions or small clusters from macroscopic, charged droplets was first discussed by Iribarne and Thomson (1976). It has been fairly well established that this mechanism is primarily responsible for the production of molecular ions in thermospray and electrospray and may be involved in other soft ionization techniques involving production of ions from liquids. This mechanism requires that the net charge on a droplet be sufficiently large and/or the radius of the droplet be sufficiently small such that the self-generated electrical field at the surface of the droplet can be high enough to allow evaporation of ions to compete thermodynamically with evaporation of neutrals. The techniques used in achieving this condition differ among the various approaches, but is appears that field-assisted ion evaporation is the dominant mechanism involved in the final step of ion production (Vestal, 1983). This ionization mechanism is very important for extending LC-MS to nonvolatile samples since it allows direct transfer of ions from solution to the gas phase without requiring an intact neutral molecule in the vapor phase as an intermediate. Thus, it provides an elegant bypass around the volatility barrier which until recently appeared to preclude the application of molecular mass spectrometry to nonvolatile samples. Field desorption (FD) has been used successfully for applications to a wide variety of nonvolatile organic and inorganic samples (Burlingame et al., 1984), and the similarity in mechanism between FD and ion evaporation suggests that the latter should find many important applications in these areas.

Desorption Ionization

Beginning with the pioneering work of Macfarlane and Torgerson (1976) using ^{252}Cf fission fragments, the development of techniques for producing ions from surfaces has developed with extraordinary rapidity. These techniques involve bombarding surfaces with everything from the Mev fission fragments used in the initial work, to keV ions and neutrals, to focussed, pulsed lasers providing local energy densities in the range of 10^7 to 10^8 watts cm^{-2}. While details of the resulting spectra depend strongly on the operating conditions and properties of the primary excitation techniques as well as on the sample matrix, all of these "desorption ionization" techniques (DI) have produced ions characteristic of the intact molecules from a variety of nonvolatile materials. These techniques are particularly well suited for use with transport devices and may provide a generally useful LC-MS interface applicable to nonvolatile samples. Jungclas has described an on-line LC-MS interface for use with ^{252}Cf desorption ionization (Jungclas et al., 1982), and preliminary work on laser and keV ion/neutral desorption have been reported by others (Fan et al., 1984; Smith et al., 1981). A schematic diagram of one approach to an on-line LC-MS system using either laser or particle desorption (Fan et al., 1984) is shown in Fig. 3. In this apparatus the sample is deposited on a moving stainless-steel belt using a thermospray vaporizer, and sample ions may be desorbed into a quadrupole mass spectrometer using either a Q-switched Nd:YAG laser or a beam of 3 keV xenon ions. Mass spectra containing intense molecular ions have been obtained on a number of nonvolatile organic and inorganic samples. Detection limits are typically in the low nanogram range. Preliminary work with this apparatus has demonstrated great potential for this combination of spray deposition with desorption ionization, but this approach has not yet been extensively applied to real samples.

Present Status of LC-MS Interfacing

The field of LC-MS interfacing has survived infancy and may be entering adolescence, but it is clearly not yet mature. Despite rapid progress in developing apparatus and techniques over the past ten years, the number of published accounts of applications to real samples is still quite small. In the opinion of this author the present status of LC-MS may be stated as follows:

1) The transport detector with spray deposition of sample still offers great (but as yet largely unrealized) potential, in that it can provide EI, CI,

and DI spectra and excellent decoupling between the LC and the MS. Further development of this technique is likely to occur and it holds great promise for future applications to species analysis.

2) DLI has made an important contribution to the development of LC-MS, but it is limited to CI on relatively volatile compounds and is restricted to rather low LC flow rates. These limitations along with the practical difficulties associated with the use of very small apertures make it unlikely that this technique will survive, at least in its present form.

3) Thermospray is presently the method of choice for obtaining molecular-weight information for a wide variety of volatile and nonvolatile samples. It is applicable to a broad range of flow rates and LC conditions (including micro-LC with post-column makeup) and is reliable and simple to operate.

4) Electrospray is not yet commercially available, but it appears very promising, particularly for use with micro-LC.

It is not possible at present to estimate accurately the performance to be expected from any of these techniques in specific applications. With thermospray it is usually possible to obtain reliable molecular-weight information on 10 ng or less of components present in relatively complex mixtures even if the LC separation is less than complete. Some structural information may be obtained, but its reliability and utility remains to be established. In the near future transport systems, particularly with spray deposition and desorption ionization, may provide comparable performance.

Recent Applications

Most of the applications of thermospray LC-MS to date have focussed on biological samples such as amino acid and peptides (Kim et al., 1984; Pilosoff et al., 1984), L-carnitine and its short-chain acyl derivatives (Yergey et al., 1983), and glucuronides (Liberato et al., 1983), but some recent work on environmental samples has been reported. The latter includes primarily pesticides (Voyksner et al., 1984) and dyes (Betowski and Ballard, 1984). Applications to various drugs and drug metabolites have also been recently reported (Covey and Henion, 1983; Garteiz and Vestal, 1985).

Future Directions

LC-MS using either thermospray or transport interfaces should provide very powerful techniques for species analysis in environmental systems. While these applications have not yet been developed, the present state of the apparatus and techniques has evolved to the point that such applications are clearly practical. With the mass spectrometric techniques presently available molecular weight can be determined for almost any volatile or nonvolatile, organic or inorganic sample. When combined with the information available from the LC separation, this may be sufficient in many cases. The qualitative identification and structural information available from EI spectra is presently limited to volatile molecules. Future development of the combination of LC with tandem MS/MS may provide similar information for nonvolatile samples. Measurement of isotope ratios and determination of empirical formula (through precise mass measurement) are now practical for nonvolatile samples, and it is now possible, in principle, to perform such measurements with on-line LC separation. The recent development of ICP-MS makes mass spectrometry competitive for determination of elemental composition of inorganic samples. It appears that an on-line LC-ICP-MS is very likely in the near future, and it is possible that such an instrument might operate in both elemental and molecular modes to provide both elemental analysis and molecular weight on individual peaks separated by LC.

Acknowledgement. Supported in part by the National Institute of General Medical Sciences, NIH, under grants GM 24031 and GM 29151.

References

Arpino PJ, Guichon G (1979) LC/MS coupling: report. Anal Chem 51:682A–701A

Arpino PJ, Guichon G (1982) Optimization of the instrumental parameters of a combined liquid chromatograph-mass spectrometer coupled by an interface for direct liquid introduction. III. Why solvent should not be removed in liquid chromatograph-mass spectrometric interfacing methods. J Chrom 251:153–164

Betowski LD, Ballard JM (1984) Identification of dyes by thermospray ionization and mass spectrometry/mass spectrometry. Anal Chem 56:2604–2607

Blakley CR, McAdams MJ, Vestal ML (1978) Crossed beam liquid chromatograph-mass spectrometer combination. J Chrom 158:261–275

Blakley CR, Carmody JJ, Vestal ML (1980a) A new soft ionization technique for mass spectrometry of complex molecules. J Am Chem Soc 102:5931–5933

Blakley CR, Carmody JJ, Vestal ML (1980 b) Combined LC-MS for involatile biological samples. Clin Chem 26:1467–1473pBlakley CR, Carmody JJ, Vestal ML (1980 c) Liquid chromatograph-mass spectrometer for analysis of nonvolatile samples. Anal Chem 52:1636–1644
Blakley CR, Vestal ML (1983) Thermospray interface for liquid chromatography/mass spectrometry. Anal Chem 55:750–754
Burlingame AL, Whitney JO, Russell DH (1984) Fundamental reviews: mass spectrometry. Anal Chem 56:417R–467R
Covey T, Henion J (1983) Direct liquid introduction/thermospray interface for liquid chromatography/mass spectrometry. Anal Chem 55:2275–2284
Dole M, Cox HL Jr, Geniec J (1978) Electrospray mass spectrometry. Adv Chem Ser 125:73–84
Dole M, Mack LL, Mobley RC, Ferguson LD, Alice MB (1968) Molecular beams of macroions. J Chem Phys 49:2240–2244
Eckers C, Skrabalak D, Henion J (1982) On-line direct liquid introduction interface for micro-liquid chromatography/mass spectrometry: application to drug analysis. Clin Chem 28:1882–1886
Edmonds CG, McCloskey JA, Edmonds VA (1983) A bibliography of combined liquid chromatography mass spectrometry. Biomed Mass Spec 10:237–252
Fan TP, Hardin ED, Vestal ML (1984) Direct comparison of SIMS and LDMS in a moving belt LC-MS system. Presented at the 32nd Annual Conference on Mass Spectrometry and Allied Topics, San Antonio. Abstract FPA 18
Fergusson GJ, Vestal ML (1983) Mechanism for producing pseudomolecular ions from nonvolatile molecules. Presented at the 31st Annual Conference on Mass Spectrometry and Allied Topics, Boston. Abstract, pp 609–610
Field FH, Munson MSB, Becker DA (1966) Chemical ionization mass spectrometry: paraffin hydrocarbons. In Advances in Chemistry Series No 58, pp 167–192
Games DE, Lant MS, Westwood SA, Cocksedge MJ, Evans N, Williamson J, Woodhall BJ (1982) Biomed Mass Spec 9:215–234
Garteiz DA, Vestal ML (1985) Thermospray LC/MS interface: Principles and applications. LC 3:334–346
Hardin ED, Fan T-P, Vestal ML (1984) Laser desorption mass spectrometry with thermospray sample deposition for determination of nonvolatile biomolecules. Anal Chem 56:2–7
Hayes MJ, Lankmayer EP, Vouros P, Karger BL (1983) Moving belt interface with spray deposition for liquid chromatography/mass spectrometry
Horning EC, Carroll DI, Dzidic I, Haegele KD, Horning MG, Stillwell RN (1974) J Chrom Sci 12:725–729
Iribarne JV, Thomson BA (1976) J Chem Phys 64:2287–2294
Jungclas H, Danigel H, Schmidt L (1982) Quantitative Californium-252 plasma desorption mass spectrometry for pharmaceuticals. A new approach to coupling liquid chromatography with mass spectrometry. Org Mass Spec 17:86–90
Karger BL, Kirby DP, Vouros P, Foltz RL, Hidy B (1979) On-line reversed phase liquid chromatography-mass spectrometry. Anal Chem 51:2324–2328
Kenyon CN, Melera A, Erni F (1981) Utilization of direct inlet LC/MS in studies of pharmacological and toxilogical importance. J Anal Toxicol 5:216–230

Kim HY, Pilosoff D, Dyckes DF, Vestal ML (1984) On-line peptide sequencing by enzymatic hydrolysis, high performance liquid chromatography, and thermospray mass spectrometry. J Am Chem Soc 106:7304–7309

Liberato DJ, Fenselau CC, Vestal ML, Yergey AL (1983) Characterization of glucuronides with a thermospray liquid chromatography/mass spectrometry interface. Anal Chem 55:1741–1746

Macfarlane RD, Torgerson DF (1976) Californium-252 plasma desorption mass spectroscopy. Science 191:920–925

McFadden WH (1979) Interfacing chromatography and mass spectrometry. J Chrom Sci 17:2–16

McFadden WH, Schwartz HL, Bradford DC (1976) Direct analysis of liquid chromatographic effluents. J Chrom 122:389–396

McLafferty FW, Baldwin MA (1973) Liquid chromatography-mass spectrometry interface. I. the direct introduction of liquid solutions into a chemical ionization mass spectrometer. Org Mass Spec 1111–1112

Pilosoff D, Kim HY, Dyckes DF, Vestal ML (1984) Determination of nonderivatized peptides by thermospray liquid chromatography/mass spectrometry. Anal Chem 56:1236–1240

Scott RPW, Scott CG, Munroe M, Hess J (1974) Interface for on-line liquid chromatography-mass spectroscopy analysis. J Chrom 99:395–405

Smith RD, Burger JF, Johnson AL (1981) Liquid chromatography-mass spectrometry with electron impact and fast ion bombardment with a ribbon storage interface. Anal Chem 53:1603–1611

Smith RD, Johnson AL (1981) Deposition method for moving ribbon liquid chromatograph-mass spectrometer interface for on-line chromatography-mass spectroscopy analysis. Anal Chem 53:739–740

Tal'roze VL, Grishin VD, Skurat VE, Tantsyrev GD (1969) In Recent Devęlopments in Mass Spectrometry. eds. K. Ogata and T. Hagakawa, pp 1218–1225. Baltimore: University Park Publications

Thomson BA, Iribarne JV, Dziedzic PJ (1982) Liquid ion evaporation/mass spectrometry for the detection of polar and labile molecules. Anal Chem 54:2219–2224

Tsuge S, Hirata Y, Takeuchi T (1979) Vaccum nebulizing interface for direct coupling of micro-liquid chromatography and mass spectrometry. Anal Chem 51:166–169

Vestal ML (1983) Ionization techniques for nonvolatile molecules. Mass Spec Rev 2:447–480

Vestal ML (1984) Thermospray liquid chromatograph interface for magnetic mass spectrometers. Anal Chem 56:2590–2592

Voyksner RD, Bursey JT, Pellizzari ED (1984) Postcolumn addition of buffer for thermospray liquid chromatography/mass spectrometry identification of pesticides. Anal Chem 56:1507–1514

Whitehouse CM, Fenn JB (1984) The electrospray ion source (ESPIC as an LC-MS interface). Presented at the 32nd Annual Conference on Mass Spectrometry and Allied Topics, San Antonio. Abstract MPA 6

Yergey AL, Liberato DJ, Millington DF (1983) Thermospray liquid chromatography/mass spectrometry for the analysis of L-carnitine and its short-chain acyl derivatives. Anal Biochem 139:278–283

The Importance of Chemical "Speciation" in Environmental Processes,
eds. M. Bernhard, F. E. Brinckman, and P. J. Sadler, pp. 631–641.
Dahlem Konferenzen 1986. Springer-Verlag Berlin, Heidelberg
© Dr. S. Bernhard, Dahlem Konferenzen

Mass Spectrometry of Biomolecules

D. H. Russell

Dept. of Chemistry
Texas A & M University
College Station, TX 77843, USA

Abstract. This paper reviews the present status of mass spectrometry of large molecules. Emphasis is given to the use of ionization methods for nonvolatile, thermally labile molecules and to the utilization of tandem mass spectrometry for structural characterization. The combination of desorption ionization and tandem mass spectrometry is relatively new and must undergo additional development before it is a routine analytical method. This review discusses areas where further developments are required and the potential effects that such developments could have on the problems of molecular characterization by mass spectrometry-based methods.

Introduction

Like so many areas of analytical chemistry, mass spectrometry has undergone many changes since the introduction of modern laboratory computers and high-grade solid-state electronics. In spite of the rapid developments made during the 1970s, the major growth in biomolecule mass spectrometry was stimulated by the development of new ionization methods for nonvolatile, thermally labile molecules and instruments with extended mass ranges. Concurrent with these developments was the introduction of nonchromatographic methods for direct-mixture analysis, e.g., mass spectrometry-mass spectrometry or tandem mass spectrometry (TMS). Prior to these developments it was generally recognized that mass spectrometry alone was not a useful analytical method for analyzing large, polar biomolecules or relatively complex mixtures. Although desorption ionization (DI) methods and tandem mass spectrometry (TMS) were developed independently, the potential analytical utility of desorption ionization coupled to tandem mass spectrometry (DI-TMS) was recognized very early. The major advantage of this instrumental method is that (in principle) molecular weight and structural information can be obtained without the requirement for extensive sample purification (Cooks and Glish, 1981; McLafferty, 1981). In addition, the sensitivity of DI-TMS is comparable to other mass spectrometry methods, e.g., sample sizes from a few micrograms to nanograms are required.

In the following sections some general aspects of ionizing large molecules and the basic instrumentation requirements for DI-TMS will be examined. There are numerous commercially available instruments which can be used for DI and TMS. The emphasis of this paper will be on sector magnetic field instruments. Although DI-TMS can be performed on other types of instruments (e.g., multistage quadrupoles, time-of-flight instruments, and Fourier transform mass spectrometers), these instruments do not have the general capabilities of the sector instruments. The basic limitations of these other instruments include mass range, mass resolution, and limitations on the types of experiments that can be performed. A possible exception is Fourier transform mass spectrometry (FTMS); however, at this point it is too early to evaluate the general capabilities of this method for large-molecule DI-TMS. There are many reasons, however, to be optimistic about the potential of this new method, and much progress has been made in the last two years in adapting new ionization methods and TMS experiments to the Fourier transform systems.

Ionization of Polar Biomolecules

Until recently the major obstacle to analyzing polar molecules by mass spectrometry has been the difficulty in volatilizing and ionizing the sample. Although polar molecules have been successfully ionized by secondary ion mass spectrometry (Benninghoven and Sichtermann, 1978; Westmore et al., 1982), a major breakthrough occurred when Barber and co-workers (1981) found that the bombardment of liquid matrices with a high-energy (keV energy) neutral beam-afforded abundant molecular ions and highly informative fragment ions. This new method, commonly referred to as fast-atom bombardment (FAB) ionization, is easily amenable to conventional high-resolution mass spectrometers of the type used for tandem MS work. The FAB ionization method is very sensitive. Samples ranging from micrograms to nanograms can be analyzed routinely, and this method has roughly the same technical difficulty as electron-impact ionization (EI).

The FAB experiment is performed by dissolving the sample in a polar, viscous liquid (typically glycerol or thioglycerol) and placing the solution on the tip of a direct insertion probe which is inserted into the FAB source. It is important that the sample is soluble in the liquid matrix; suspensions do not give good, stable, secondary ion yields. After the sample has been placed in the ion source, it is irradiated with a high-energy (3–10 keV) beam of neutral (Ar or Xe) or charged (Cs^+) particle beam. Excitation of the liquid matrix by the energetic particles results in sputtering of the sample and solvent molecules. Although the major fraction of the sputtered molecules are uncharged, there are sufficient ionic species to produce stable ion beams which can be analyzed by the mass spectrometer. The mechanism of the ionization process is quite complex and is still a very active research area. Suffice it to say that regardless of the actual mechanism for ion production, the dominant molecular ions produced are of the form $[M+H]^+$ or $[M-H]^-$, and that the most structurally significant fragment ions are formed by fragmentation of the molecular ions. Ions of the type $[M+Na]^+$ and $[M+xNa\text{-}(x-1)H]^+$ are also often observed in the FAB mass spectrum. However, these ions can be suppressed by taking precautions to minimize the salt content of the sample, glassware, and solvents. If the salt content of the sample is very high it is possible to suppress totally the ionization of the sample. We have found that most samples which cannot be ionized by FAB have very high salt contents; a frequent source of the salts is buffers used in liquid chromatography.

The results of studies on FAB ionization of large molecules are very impressive. Molecules which are highly polar and/or thermally labile are readily ionized by FAB, giving both intense fragment ions and molecular ions. A large variety of biomolecules have been successfully ionized by FAB, e.g., peptides, sugars, glycopeptides, and naturally occurring and synthetic antibiotics (for a review see Burlingame et al., 1984). The FAB mass spectra are generally useful for both molecular weight and structural information. However, in general the abundance of fragment ions in the FAB spectra decreases as the size of the molecule increases. A similar phenomenon is observed in the mass spectra of large molecules ionized by ^{252}Cf-plasma desorption mass spectrometry (Macfarlane, 1982).

Metal Compounds

This general class of compounds is often difficult to analyze by FAB-MS, especially organometallic compounds containing metals in high oxidation states (Cerny et al., 1983; Russell and Holmes, 1974). The general problem with organometallic compounds is ligand-exchange reactions with the glycerol solvent, e.g., reactions of the type shown in Eq. 1:

$$ML_x + G \rightarrow ML_{x-y}G_y, \tag{1}$$

where G = glycerol. Such reactions are generally not encountered with macrocyclic systems (e.g., porphyrins, corrins) which have strong metal-ligand bonds. However, ligands with binding energies less than the metal-glycerol binding energy are readily displaced (presumably) in solution prior to ionization. Also, in some cases redox reactions occur, especially with metals in high oxidation states (Russell and Holmes, 1984). We frequently encounter ligand-exchange reactions with metal complexes involving ethylenediamine, trietyhlenediamine, and ethylenediaminetetraacetic acid as well as ionic ligands or counterions. However, the problem is most often found with complexes having metals in the higher oxidation states.

We have found that liquid matrices which are less polar and less reactive than glycerol can be used for the analysis of labile organometallic compounds. In particular we have found o-nitrophenyl octyl ether and sulfolane to be a very useful matrix for many organometallic compounds. For example, chlorophyll a in glycerol is very difficult to analyze by the FAB method, but a reasonably good FAB spectrum can be obtained by using thioglycerol, and excellent results are obtained by using the o-nitrophenyl octyl ether matrix. We have successfully used the latter for analyzing transition metal-carbonyl clusters of the type $Os_x(CO)_y$ ($x = 2$–6; $y = 9$–36) and similar mixed-metal clusters containing Os and Co, Co and Bi, and W and Co. For all systems studied to date the results have been quite impressive. The FAB mass spectra contain relatively abundant (>10%) molecular ions, typically $[M+H]^+$, and structurally informative fragment ions. Similar results have been obtained from protected peptides and nonpolar derivatives of polysaccharides. Due to their low solubility, these relatively nonpolar molecules do not ionize readily from glycerol nor in some cases from thioglycerol. On the other hand, they are readily ionized from the o-nitrophenyl octyl ether matrix and give abundant (>10%) molecule ions $[M+H]^+$ and informative fragment ions.

A major point to be noted from the above is that while FAB is a routine ionization method for mass spectrometric analysis, the quality of the mass spectrum, and even the successful acquisition of a mass spectrum, depends on the sample purity and the solubility and stability of the sample in the liquid matrix. Thus, it is not surprising that factors such as pH strongly affect the results. Also, some molecules which are strong electrolytes will be ionized only in a single ion mode, e.g., positive or negative ion spectrum. For example, most peptides give good positive and negative ion spectra. However, molecules which contain phosphate groups or sulfate groups give very poor positive ion spectra, but the sensitivity in the negative ion mode is spectacular. Many polysaccharides and glycoconjugates are easily ionized and analyzed in the negative ion mode; however, the positive ion spectrum is very complex. The positive ion FAB spectra of these molecules contain intense signals corresponding to organo-alkali metal-cluster ions. For unknown samples we find it very useful to have information on the polarity of the molecule obtained by electrophoresis. Such information greatly reduces the number of trial-and-error attempts to obtain the FAB mass spectrum. Also, it is generally preferable to dissolve the sample in a solvent such as methanol, chloroform, or trifluoracetic acid prior to introducing it to the liquid FAB matrix. Dissolving the sample prior to introduction to the liquid matrix simplifies the sample-handling process, and in many cases this simple procedure greatly increases the FAB-MS sensitivity.

Tandem Mass Spectrometry (TMS)

The tandem MS experiment is performed as the name implies, i.e., two mass spectrometers (MS-1 and MS-2) are connected together in such a manner that MS-1 separates a particular ion, denoted M_a^+, formed by direct ionization of the sample, and the fragment ions formed by dissociation of M_a^+ are mass-analyzed by scanning MS-2 over the appropriate mass range. A number of instrument configurations can be used to accomplish the experiment, and the various instrument types have been discussed in several review articles (Cooks and Busch, 1982; McLafferty, 1980; Yost and Enke, 1979). McLafferty (1983) has also edited a volume which gives a review of both instrumentation and applications of tandem MS methods.

The advantage of the TMS experiment is the ability to ionize a complex sample and use the mass spectrometer to perform both the separation and analysis steps. The instrument shown in Fig. 1 is typical of sector instruments. This instrument is capable of performing high-resolution mass sep-

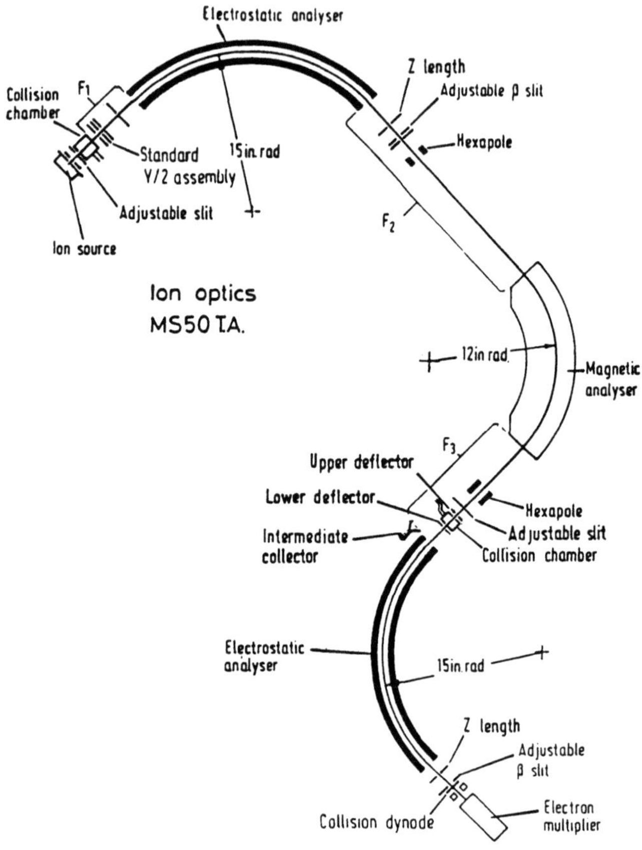

Fig. 1. Typical sector instrument for tandem mass spectrometry

aration (MS-1) (mass resolution exceeds 100,000 in the TMS mode) with low-resolution (MS-2) (ca. 100–1000) analysis of the fragment ions. In addition, it has a full-sensitivity mass range of ca. 3500 amu and this mass range can be extended by reducing the sensitivity, e.g., 5000–7500 amu can be measured with an approximately 50% reduction in sensitivity.

The most important step in the TMS experiment occurs in the region between MS-1 and MS-2. Following mass selection of M_a^+ by MS-1, the ion must be activated to energies above the dissociation threshold. Typically the activation process involves a high-energy (keV) collision between the mass-selected ion and a target neutral atom. Thus the term collisional activation is frequently used to describe this process. Following the activation step, ions with excess internal energy undergo a spontaneous dissociation reaction; thus the overall process is referred to as collision-induced dissociation (CID) or collisionally activated dissociation (CAD). The process of collisional activation involves converting some fraction of the translational energy of the ion into internal energy which drives the dissociation reaction. Although the dissociation efficiency (defined as the percent conversion of incident ions to fragment ions) is relatively poor (only a few percent at best), the sensitivity of the method for small molecules is adequate. For example, several workers have demonstrated picogram and subpicogram detection levels (Biemann, 1982; Morris, 1982; Morris et al., 1982). Also, these low detection levels have been demonstrated for a wide range of compound classes, ranging from relatively simple molecules to rather complex pharmaceuticals and organic dyes (for a review see Burlingame et al., 1984).

To date only a few examples of collision-induced dissociation of large molecules have appeared in the literature. McLafferty and co-workers (Amster et al., 1983) have reported desorption ionization-tandem MS studies (DI-TMS) of relatively large biomolecules, e.g., vitamin B_{12}, 15-Met-gastrin, Lys-bradykinin, and Met-Lys-bradykinin. The DI-TMS spectra of these molecules contain the desired information, e.g., fragment ions indicative of amino acid sequence and other structurally significant fragment ions. Although the absolute abundance of the fragment ions are low, signal-averaging of multiple mass scans gives adequate signal-to-noise ratios. We have used DI-TMS to study the fragmentation reactions of vitamin B_{12}, chlorophyll a, and other biologically important metal compounds. As an example the DI-TMS spectrum of chlorophyll a is shown in Fig. 2. This spectrum contains the same structural information obtained using desorption and mass analysis by time-of-flight mass spectrometry. The dominant fragment ions in the DI-TMS spectrum correspond to fragmentations of the phytol side chain and other substituents attached to the porphyrin ring. Gross and co-workers (Tomer et al., 1984) have also reported interesting results on the DI-TMS of cyclic peptides. The problem with this work, as well as with the work reported by McLafferty and ourselves, is that many of the most important fragmentation reactions, i.e., those fragment ions indicative of amino acid sequence and other important structural features, are either not observed at all or are very weak signals in the DI-TMS spectrum. This problem is related to the dynamics of unimolecular dissociation reactions. For example, simple cleavage reactions, which are the most structurally informative (e.g., cleavage of the peptide-amide bond), have large activation energies. On the other hand, rearrangement reactions have low

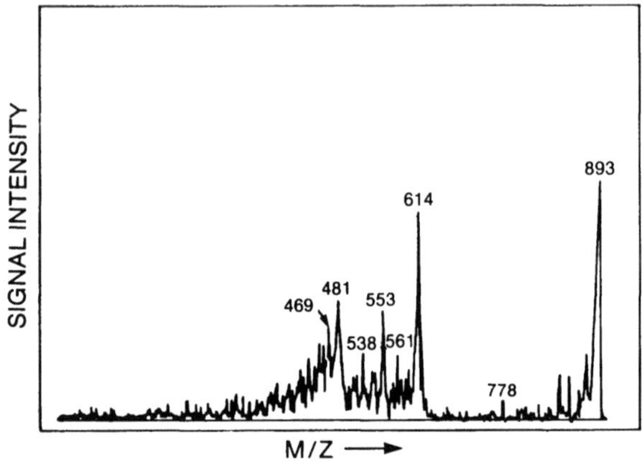

Fig. 2. Spectrum of chlorophyll a obtained by desorption-ionization tandem mass spectrometry (DI-TMS)

activation energies and are the dominant fragment ions formed by dissociation of near-threshold-energy ions. The energy transfer function for collisional activation maximizes at low energy (1–2 eV), and only a small fraction of the ions have large excess energies. Therefore, it is not surprising that the resultant spectrum is dominated by low-energy dissociation reactions.

Although the feasibility of collision-induced dissociation studies of biomolecules has been demonstrated, the sensitivity of the method for large molecules is not as good as that for small molecules. Derrick and coworkers (Neumann and Derrick, 1984) examined in detail the dynamics of collisional activation of large molecules and measured the amount of energy deposited upon collision (the fraction of translational energy converted into internal energy of the ion). In our laboratory we studied a number of relatively large molecules and concluded that the method is limited for structural characterization of large molecules (Bricker and Russell, 1985). It is apparent from our work as well as Derrick's that the incident ions suffer appreciable energy loss (as much as 20–50 eV depending on the mass of the incident ion) due to the high-energy collision, but the efficiency of collision-induced dissociation is quite low (less than 1%). Additional work in this area is needed in order to optimize this technique for large molecules. In our laboratory we are using laser-ion beam photodissociation to probe the dynamics of the dissociation of large molecules. The principle objective of this work is to answer questions concerning the dynamics of both activation and dissociation. Is the relatively poor sensitivity for induced dissociation of large molecules due to the large number of vibrational degrees of freedom of such systems (i.e., are these systems good energy "sinks"?), or is it related to the dynamics of activation? It is also highly probable that this method of performing TMS will add a new dimension to the characterization of large molecules.

An additional area that has not received much attention is the low mass resolution of MS-2. It would be advantageous if the resolution of this sector were higher. This limited resolution can be overcome by converting MS-2 to a double-focussing, high-resolution instrument. Such instruments have been constructed (Neumann and Derrick, 1984) and are also available commercially.* To date the highest mass resolution obtained by a double-focussing MS-2 is ca. 10,000, which is much better than the resolution obtained by instruments such as that shown in Fig. 1. The full utility of high resolution (both MS-1 and MS-2) will not be realized until a more efficient method of inducing the dissociation reaction is developed; however, the advantages of increased resolution for MS-2 are obvious.

New Developments in Large-Molecule MS

It is evident that additional changes will be forthcoming in the area of high-mass mass spectrometry. This field is less than a decade old and the introduction of FAB in 1980 has had a profound effect. Already new designs for instruments are in progress and are being tested. It was once felt that high mass resolution was not of any great significance to this area. Now it is now becoming increasingly apparent that high resolution, at least nominal mass resolution, is essential. Ideally one would like to be able to perform precise mass measurements. Clearly this is not feasible in the near future using present instruments.

One method that could have a major impact on high resolution, high-mass mass spectrometry in the foreseeable future is Fourier transform mass spectrometry (Wilkins and Gross, 1981). This method has capabilities for ultra-high resolution and mass range comparable to state-of-the-art magnetic sector instruments. Like sector instruments, the Fourier transform instrument is highly versatile and can be used in several experimental modes, e.g., tandem MS, desorption ionization, and in combination with laser experiments. In our laboratory we have demonstrated desorption ionization

* Instruments of this nature are available from two commercial manufacturers: V.G. Instruments and Kratos Analytical Instruments.

by Cs^+ ion secondary ion MS, and several groups have performed laser desorption ionization Fourier transform experiments with biomolecules (McCrery et al., 1982).

There is much work to be done before Fourier transform MS will become routine for biomolecules. However, the progress made in this area has already had an impact. The second generation of instruments are more generally applicable to work on biomolecules, and it is only a matter of time before the full range of mass spectrometric experiments will be adapted to this new technology.

References

Amster IJ, Baldwin MA, Cheng MT, Proctor CJ, McLafferty FW (1983) Tandem mass spectrometry of higher molecular weight compounds. J Am Chem Soc 105:1654
Barber M, Bordoli RS, Sedgwick RD, Tyler AN (1981) Fast-atom bombardment of solids (FAB): A new ion source for mass spectrometry. J Chem Soc Chem Comm 325
Benninghoven A, Sichtermann WK (1978) Detection, identification, and structural investigation of biologically important compounds by secondary ion mass spectrometry. Anal Chem 50:1181
Biemann K (1982) Sequencing of proteins. Intl J Mass Spec Ion Proc 45:183
Bricker DL, Russell DH (1985) Energy transfer processes in keV energy collisional activation. J Phys Chem
Burlingame AL, Whitney JO, Russell DH (1984) Mass spectrometry: Fundamental reviews. Anal Chem 56:417R
Cerny RL, Sullivan BP, Bursey MM, Meyer TJ (1983) Comparison of fast-atom bombardment and field desorption mass spectrometry of coordination complexes. Anal Chem 55:1954
Cooks RG, Busch KL (1982) Counting molecules by desorption ionization and mass spectrometry/mass spectrometry. J Chem Ed 59:926
Cooks RG, Glish GL (1981) Mass spectrometry-mass spectrometry. Chem Eng News 59:40
Macfarlane RD (1982) Surface chemistry in ion-induced desorption. Nucl Inst Meth 198:75
McCrery DA, Ledford EB, Gross ML (1982) Laser desorption Fourier transform mass spectrometry. Anal Chem 54:1435
McLafferty FW (1980) Tandem mass spectrometry (MS/MS): A promising new analytical technique for specific component determination in complex mixtures. Acc Chem Res 13:33
McLafferty FW (1981) Tandem mass spectrometry. Science 214:280
McLafferty FW (ed) (1983) Tandem Mass Spectrometry. New York: J Wiley & Co, Inc
Morris HR (1982) Biomolecule mass spectrometry. Intl J Mass Spec Ion Proc 45:331

Morris HR, Panico M, Karplus A, Lloyd PE, Riniker B (1982) Elucidation by FAB-MS of the structure of a new cardioactive peptide from *Aplysia*. Nature 300:643

Neumann GM, Derrick PJ (1984) Energy transfer in the collision-induced decomposition of peptide ions formed by field desorption. Org Mass Spec 19:165

Russell DH, Holmes G (1984) Preliminary Results Presented at the 32nd Annual Conference on Mass Spectrometry and Allied Topics, May 27–June 1, 1984, San Antonio, TX, p 797

Tomer KB, Crow FW, Gross ML, Kopple KD (1984) Fast-atom bombardment combined with tandem mass spectrometry for the determination of cyclic peptides. Anal Chem 56:880

Westmore JB, Ens W, Standing KG (1982) Secondary ion mass spectra of the neuropeptides leucine-enkephalin and methionine-enkephalin. Biomed Mass Spec 9:119

Wilkins CL, Gross ML (1981) Fourier transform mass spectrometry for analysis. Anal Chem 53:1661A

Yost RA, Enke CG (1979) Triple quadrupole mass spectrometry for direct mixture analysis and structure elucidation. Anal Chem 51:1251A

The Importance of Chemical "Speciation" in Environmental Processes,
eds. M. Bernhard, F. E. Brinckman, and P. J. Sadler, pp. 643–665.
Dahlem Konferenzen 1986. Springer-Verlag Berlin, Heidelberg
© Dr. S. Bernhard, Dahlem Konferenzen

Surface Characterization Techniques Applied to Environmental Problems

P. F. Kane

Texas A & M University
College Station, TX 77843, USA

Abstract. A number of techniques for the study of gas-solid interfaces are reviewed, particularly those involving the interaction of photons and particles with surface atoms. Although these techniques require ultra-high vacuum environments and therefore tend to modify the specimen, they yield valuable information at submicron geometries not available by optical methods. Some applications to environmental samples are given, with emphasis on the identification of chemical species.

Introduction

In the broadest sense of the word, surfaces – or better, interfaces – are profoundly involved in environmental processes ranging from wave formation on the surface of the ocean to photosynthesis in plants. Obviously characterization of such an enormous variety of interfaces involves techniques of equally bewildering diversity. My intention in this paper is to restrict the topic to gas-solid and liquid-solid interfaces and to concentrate on microsamples such as particles or porous substrates, i.e., adsorbents.

Gaseous adsorbates have been of considerable interest in pollution studies. Oxides of sulfur, nitrogen, and carbon plus hydrocarbons from man and nature undergo constant reactions in the atmosphere which is basically in a nonequilibrium condition. While many of these reactions are in the gas phase, airborne particles may catalyze such reactions by the formation of radicals on the surface. Dust deposited from the atmosphere, e.g., fly ash, may concentrate such pollutants in the soil; conversely, the use of charcoal, for example, can remove them from effluents. Some metallic compounds (e.g., tetraethyllead) can form solid particles whose composition must be determined.

Adsorption from solution can involve both organic and inorganic compounds. The metal ions are of importance because they are readily exchanged in many naturally occurring compounds, e.g., zeolites, and concentrate on colloidal particles found in all natural waters. Many of these metal ions are essential to life, some are toxic, and a few can occur either at different levels or in different species.

Characterization of these particles requires a knowledge of the surface structure, the substrate composition and crystal form, and the identification of the adsorbate, its oxidation state and, for some compounds, its orientation to the surface. A meaningful characterization often demands submicron resolution.

Surface Characterization

Characterization methods can roughly be categorized into structural and compositional. Among the first group, microscopy and diffraction figure prominently; optical microscopy is still one of the first techniques to be employed although often only to prepare samples for further examination. It can, however, in the hands of skilled microscopists, deliver a great deal more information on the structure, shape, and crystalline form of the sample than commonly realized. The work of McCrone (1973) is an outstanding example of the wealth of information that can be obtained in just a few minutes by an expert. However, there is a limitation on the resolution of about 1.5 µm. This resolution is obtained only when multi-beam interferometry is used. The electron microscope improved this resolution to the subnanometer range and paved the way for the many electron and ion spectroscopies which have proliferated in the past decade. Unfortunately, these techniques require high vacuum environments which may alter the surface characteristics of the sample.

The optical techniques have also developed, chiefly by extending the spectrum into the ultraviolet and infrared regions and by the introduction of lasers as sources. Some of these techniques can be applied to particles *in situ* and thus have an important advantage over the vacuum techniques. They also contribute important additional information on the valence of atoms and on the structure of organic molecules. Infrared and Raman spectroscopy in particular have been of value. A micro-Raman method, commercialized by Jobin Yvon and termed by them the MOLE (Molecular Optical Laser Examiner), has been applied to the topographical determination of organic contaminants on surfaces. Although these optical techniques are not generally as sensitive as the UHV techniques, a more recent development, surface-enhanced Raman spectroscopy (SERS), has reported order-of-magnitude improvement in the case of some organics adsorbed on metal. A review of these techniques has been given by Allara (1982).

Because the UHV methods, despite the fundamental problem of the "real" surface, are more sensitive and more generally applicable, this review will concentrate on those.

Electron Microscopy

Conventional transmission electron microscopy (TEM) is the electron analog of the light microscope and uses similar optics, i.e., objective, intermediate, and projector lens. The image is formed by transmission and the sample thickness is limited to about 60 μm. Any technique involving particles requires that the mean free path be such that enough of them reach and are measured by the detector, in this case a photographic plate. For this instrument a vacuum of about 10^{-6} Torr is sufficient but essential. Because the electrons travel through the sample TEM is, strictly speaking, not a surface characterization technique. However, it can give a great deal of information about the size and shape of particles and, since the contrast varies with the element, also about the distribution of elements. By changing the focal length of the intermediate lens, an electron diffraction pattern can be obtained which will identify the crystal structure of a particle or, in some cases, of part of a particle.

The scanning electron microscope (SEM) operates on a somewhat different principle. The electron beam scans the surface of the sample on a raster and the image is formed from secondary electrons originating in the sample. This is truly a near surface imaging technique; the signal is limited to the escape depth of electrons (about 5 nm). Since the voltages used are less than those for TEM (25 kV as against 100 kV or higher), damage to the sample is reduced and this can be important in some cases. The beam diameter is about 50 Å (an order of magnitude more than that of the TEM) which is adequate for most particle work. When electrons strike a solid they not only cause secondary electrons to be emitted but also generate X-rays. These emitted X-rays (XRE) are characteristic of the atoms in the target and are analyzed either by energy-dispersive (ED) or wavelength-dispersive (WD) spectrometers. The ED detector is usually preferred since its efficiency is much higher than the WD and allows rapid (less than 1 min) scans of the spectrum for identification of elements. At selected wavelengths elemental distributions can be obtained. The energy resolution of 150 eV is seldom a limitation except for lighter elements. The WD detector has a resolution of about 10 eV and is used for quantitative work, especially for elements with lower atomic numbers. This SEM-XRE combination is frequently referred to as the electron microprobe.

A more recent development is the Scanning Transmission Electron Microscope (STEM). This very high-resolution instrument combines the best features of both electron microscopes. To achieve maximal resolution it uses a field-emission source which requires a higher vacuum; in fact, about 10^{-11} Torr are necessary for these instruments. STEMs are much more powerful than SEMs and TEMs. Not only can transmission micrographs be obtained with STEMs at a resolution of 2–3 Å, but microdiffraction data, electron energy loss spectra, and XRE data as well. To date STEM represents the ultimate in analytical microscopy.

Electron Spectroscopy

Reference to electron energy loss spectroscopy leads us naturally into a number of techniques arising from the interaction of electrons with solids. Figure 1 summarizes these interactions as well as those for X-rays and ions (Sparrow and Drummond, 1981). All of these techniques require ultrahigh vacuum (UHV) for good signal to noise. The analysis chamber should be at about 10^{-11} Torr. The sample treatment chamber can operate a higher pressure but is typically around 10^{-9} Torr. Figure 2 shows a typical multipurpose instrument consisting of two UHV chambers connected through a Gate Valve (GV). The samples are set up in the Sample Treatment Chamber (STC) and are pumped down and transferred to the Sample Analysis Chamber (SAC) via the GV port. The particular instrument shown is set up for several different functions.

X-ray Photoelectron Spectroscopy (XPS)

This technique was pioneered by Siegbahn (1967) who christened it Electron Spectroscopy for Chemical Analysis (ESCA). This acronym is still used occasionally. It depends for its effect on the ejection of core electrons from inner shells by X-ray excitation (Fig. 1 a). The ejected electron is limited in its escape from the solid by its kinetic energy which is dependent

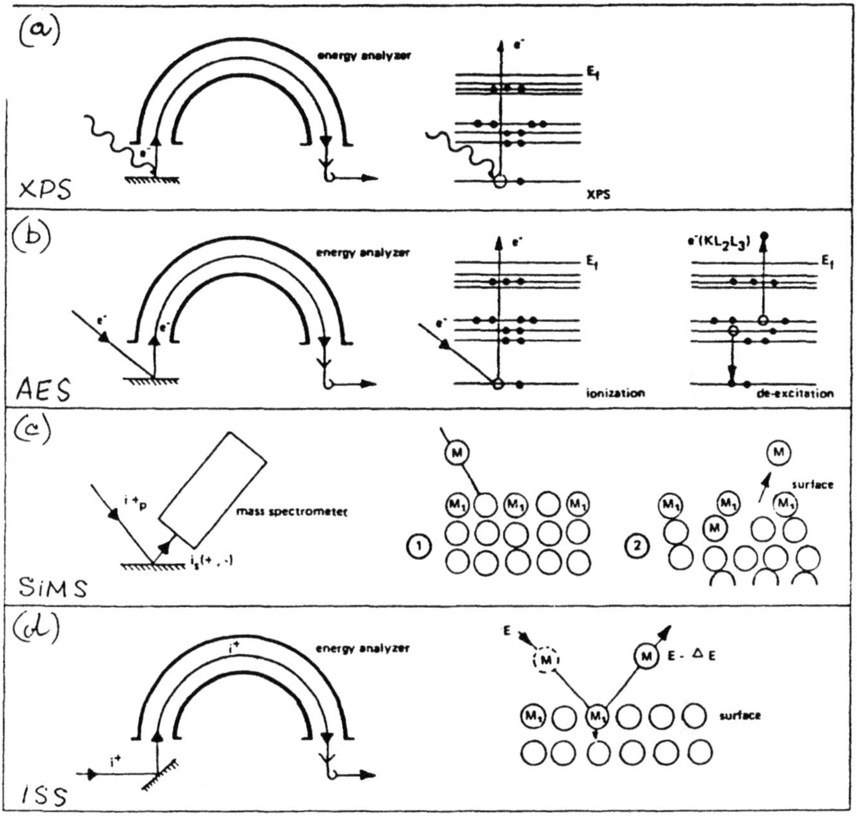

Fig. 1. Surface analysis techniques

on the particular core level involved. The escape depth is generally about 20 to 50 Å. ESCA can be regarded as a surface characterization technique.

The instrument in Fig. 2 has the ability to function as an XPS instrument. The most important component is the electron spectrometer. Two types are commonly used: the hemispherical analyzer (HSA) and the cylindrical mirror analyzer (CMA). Both give good resolution, better than 1 eV over most of the range, although the HSA, shown in Fig. 2, has become more general in multitechnique instruments because of its more flexible geometry. The CMA is still favored for dedicated Auger systems by virtue of its increased sensitivity. The rest of the instrument is an UHV chamber with

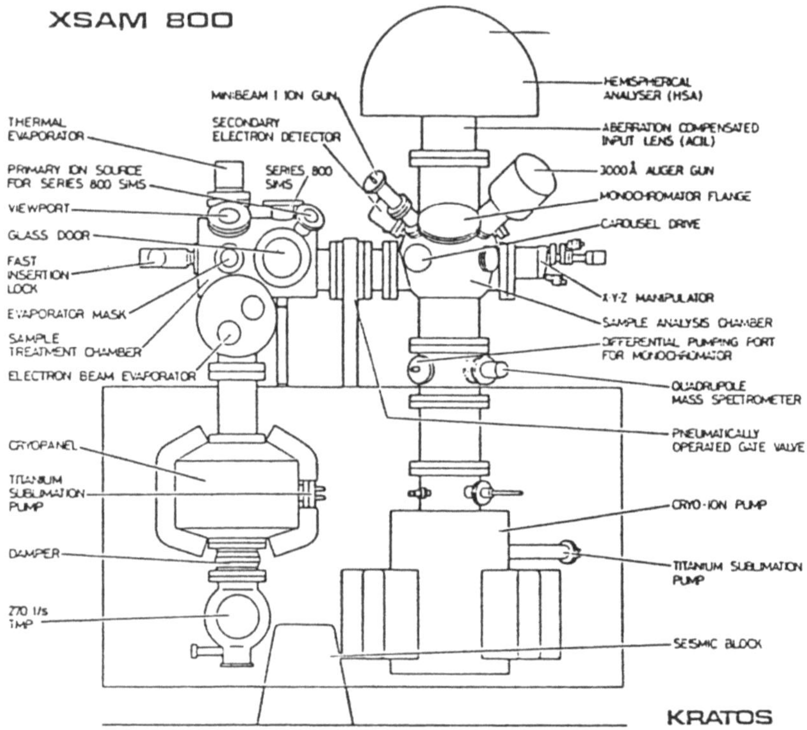

Fig. 2. Surface analysis system

ports for various sources and detectors and, in the system shown, a connecting chamber for preliminary sample mounting or preparation.

For XPS the source of X-rays is either a magnesium or aluminum target; for most purposes a dual anode system is used. The kinetic energy of the emitted electrons will depend on the excitation energy, 1254 eV for Mg K_α, 1487 for Al K_α. Spectra obtained with these two sources will be different when plotted in kinetic energy (KE). However, there is a simple relationship between the X-ray energy and the energies characteristic of the electron:

$$h\nu = KE + BE + \varnothing$$

$h\nu$: excitation energy, KE: kinetic energy, BE: binding energy, and $\varnothing$: work function. Spectra plotted against BE will not depend on the excitation en-

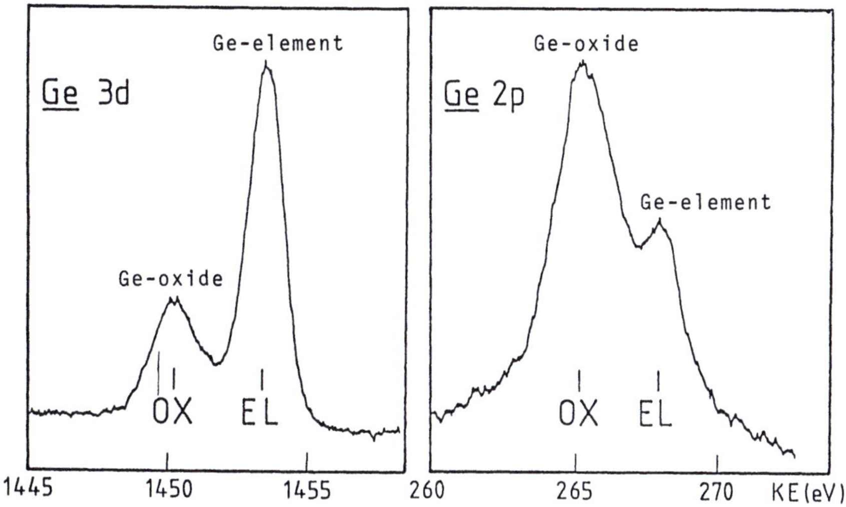

Fig. 3. XPS spectra for oxide film on germanium

ergy. Ø is a small correction, less than 10 eV typically, dependent largely on the instrument parameters. The peaks in the resulting BE spectra are related, of course, to the particular orbital from which the electrons originate and are identified as such, e.g., N_{1s}, Si_{2p}, etc. Some complications can arise. For example, X-rays can excite Auger electrons (see later) but the kinetic energy of these is not dependent on the excitation energy. Auger electrons can be identified by the fact that their signals do not shift in either Mg or Al KE spectra. Because the X-rays are not monochromatic, satellites may appear on major peaks; these can be removed by using monochromatic X-rays.

The energy levels of the core electrons will vary with their chemical environment, that is, with the number of valence electrons in the outer shell. In other words, the BE changes slightly with the compound in which the element is present. This change is the so-called "chemical shift." Over the past decade very many shifts have been measured. Figure 3 shows spectra of 3 d and 2 p germanium electrons. The chemical shift for germanium in Ge-oxide is about 5 eV relative to elemental germanium. These spectra illustrate another useful feature. The germanium peaks are from two different orbitals with very different energies and considerably different escape depths. Consequently, the surface layer emission is much higher for the low a 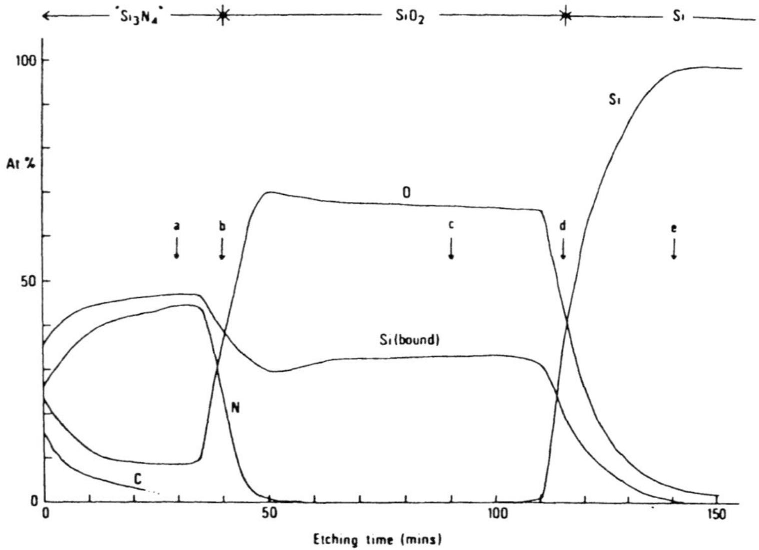

b 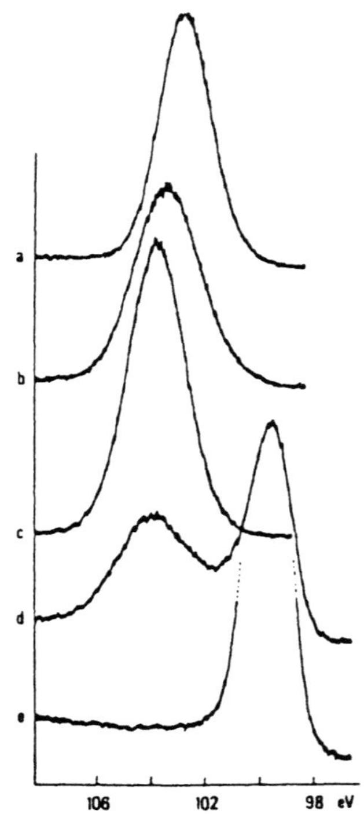

Fig. 4. a Depth profile of a MNOS device surface produced by XPS and a mechanically scanned Argon-ion beam (sample size 20 × 6 (mm)). Note the oxygen profile through the "Si N_x" outer layer. **b** Si2p spectra recorded at points a–e in Fig. 4(a). Note the chemical shift which distinguishes Si in nitride, oxide, and elemental states energy electrons. By suitable calculation the film thickness can be estimated.

The position of the source relative to the detector is, of course, set by the geometry of the instrument. For a homogeneous sample the angle made by the analyzer with the surface has no effect on the relative peak heights of the constituents. However, for a sample in which the surface is enriched (or depleted) by an element, the angle can make a difference. At very low angles only surface or near-surface atoms are seen because electrons coming from depper in the solid are inelastically scattered and lost to background. This angular distribution has been studied extensively by Fadley (1978) and others. The instrument shown (Fig. 2) can rotate the sample about an axis in the plane of the sample holder for the measurement of angular distribution.

An alternative approach to surface analysis involves alternate measurement and etching. The instrument shown is fitted with an ion gun which bombards the surface and removes successive layers of sample. XPS spectra are collected at various times corresponding to various depths. A typical elemental distribution plot is given in Fig. 4 (Briggs, 1979).

Quantitative treatments have been based mostly on empirical calibration. One of the more widely used series of relative atomic sensitivies reported by Wagner (1972) can be applied, at least semiquantitatively, to a variety of compounds.

It is necessary to treat some of the results from ESCA analyses with caution. ESCA is a surface technique, and the surface can be modified by the measurement. The UHV alone will tend to remove all but the most firmly bound volatile compounds. Photons, e.g., X-rays, are among the least damaging forms of radiation, but even photons may induce some chemical changes. If ions are used for etching, chemical reduction can be induced even by inert gas ions such as argon. Despite these problems, important information can be obtained by this technique as long as the potential limitations are borne in mind.

Ultraviolet Photoelectron Spectroscopy (UPS)

If the X-ray source is replaced by a UV source the excitation is, of course, reduced in energy. Instead of energies in excess of 1000 eV, a helium discharge lamp delivers between 20 and 40 eV. The emitted photoelectrons come from the outer valence shells and from depths of no more than 5–10 Å. One very important application of UPS is the identification of adsorbate resonances, the ability to distinguish between alternative bonding mechanisms. A comparison of the UPS and XPS techniques is given by Feuerbacher and Fitton (1977). The same instrument shown in Fig. 2 can be used.

Auger Electron Spectroscopy (AES)

XPS and UPS are basically broad-area techniques. X-rays are difficult to collimate and most sources give beams about 1 mm in diameter. Although UV light could be focussed to about 1 μm, the usual helium discharge lamps collimate a beam down a capillary tube. This beam usually has a diameter of 1 mm. By substituting electrons it is possible to focus a high-energy beam to a diameter of 30 nm. X-rays primarily generate photoelectrons from core orbitals but also produce some Auger electrons. Auger electrons arise (Fig. 1 b) from a secondary effect, de-excitation. Ejection of the photoelectron leaves a vacancy in an inner orbital with a high binding energy. An ion in this highly excited, unstable state becomes a more stable ion by loss of energy. One path for de-excitation drops an electron from an orbital with lower binding energy, i.e., further from the nucleus, into the inner-orbital vacancy. The excess energy is transferred to an electron in an orbital with even lower binding energy. If the energy transferred exceeds this binding energy, the electron is ejected, leaving a relatively stable ion. The ejected electron is termed an Auger electron and its energy is dependent not on the primary energy but on the differences in energy levels involved in the secondary de-excitation. For example, a KL_2L_3 Auger electron is ejected when a K-level vacancy is filled by an L_2 electron and energy is transferred to an L_3 electron. Although some chemical information can be obtained from AES, this technique is not as powerful as XPS. Reviews of AES have been written by Chang (1974) and McGuire and Holloway (1981).

The term AES is usually applied to the static mode in which a broad area of the surface is bathed with electrons. However, one of the most attractive features of the technique is the high spatial resolution, as small as 30 nm. The instrument shown in Fig. 2 has an electron beam of about 300 nm and is fitted with a Secondary Electron Detector (SED). The beam is scanned on a raster and can generate an SED image for topographical examination or, through the energy analyzer, a scanning Auger image at one particular energy. This technique is called scanning Auger microscopy (SAM). By suitable electronics the beam can be held stationary at a selected point and the energy scanned. This allows very small areas (or particles) to be analyzed.

Again, caution must be exercised in interpretation of the results since electrons are considerably more damaging than photons. Crystal defects are generated along with heat and this may cause diffusion to occur either across or into the surface. Salts and organic materials may decompose. Charging effects may distort the image. Nevertheless, unique information can result from the technique, particularly as to the distribution of elements across the surface and, by ion etching, in three dimensions.

Other Electron Techniques

A number of other techniques involving electrons have been applied to surfaces, in particular diffraction. High-energy electron diffraction (HEED) has been applied for many years by electron microscopists and, with the development of the new UHV scanning transmission electron microscope, has become a very important tool. Microdiffraction has identified areas as small as 30 Å in diameter although strictly speaking this is not a surface technique. Low-energy electron diffraction (LEED), however, using backscattered electrons of less than 500 eV, has been used, mostly on single crystal surfaces, to analyze the arrangement of atoms in the surface layer. In fact, a whole subdiscipline of surface crystallography has grown up around this technique. An excellent treatment of the subject has been made by Somorjai (1981). Electron Energy Loss Spectroscopy (EELS) is another technique used in the STEM. Electrons of 50 KeV lose energy to the atoms in the sample. This energy loss is characteristic of the particular elements. A more specific technique which can see surface vibrational modes uses monochromatic, low-energy (<1 KeV) electrons and a spectrometer with a resolution better than 10 meV. The technique is described by Froitzheim (1977). The necessary additional equipment can be fitted to the instrument of Fig. 2.

Ion Spectroscopy

Secondary Ion Mass Spectrometry (SIMS)

Ion etching is used to remove surface layers for depth profiling by XPS or SAM. Ions are generated by electron bombardment and accelerated to about 5 KeV in the UHV chamber. As a result of collisions with atoms or molecules in the surface, secondary ions are formed which can be analyzed in a mass spectrometer (MS). The process is shown schematically in Fig. 1c.

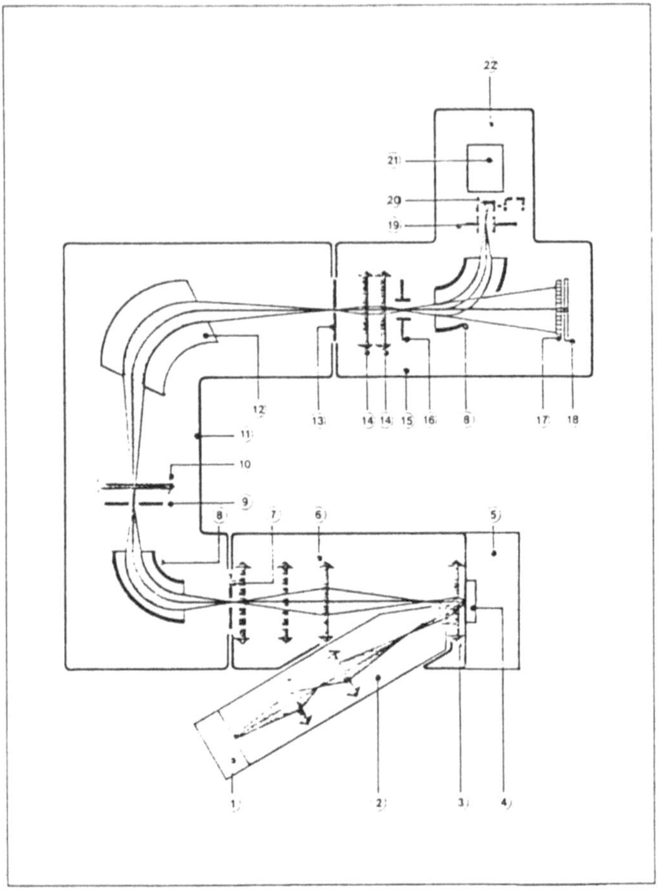

Fig. 5. Ion microscope. 1 Ion gun; 2 Primary column; 3 Immersion lens; 4 Specimen; 5 Specimen chamber; 6 Transfer optics; 7 Entrance slit; 8 Electrostatic analyzer; 9 Energy slit; 10 Image transfer lens; 11 Spectrometer; 12 Electromagnet; 13 Final slit; 14 Projection lenses; 15 Projection display and detection system; 16 Deflector; 17 Channel-plate, 18 Fluorescent screen; 19 Deflector; 20 Remote controlled Faraday cup; 21 Electron multiplier; 22 Detection system

The incident ions are usually from an inert gas such as argon although for some applications more reactive ions such as oxygen or cesium are used.

The SIMS analysis can be carried out in the system shown in Fig. 2. For reasons of geometry it is common to fit the SIMS system to the treatment chamber rather than the analysis chamber. For most purposes, quadrupole mass spectrometers have adequate mass resolution and will separate masses differing by one up to about 1000 amu. Because ions can be focussed just like electrons, this method has good spatial resolution, down to about 1 μm, and can be rastered by the same electronics that operate the SAM.

An alternative approach to SIMS is shown in Fig. 5. The instrument consists of an ion microscope in which the sample surface is bathed with ions. The image is formed just as in an electron microscope. This instrument has a superior MS using a double focussing system and can discriminate molecular ions with very close masses, e.g., $^{40}Ca^{+}$ can be separated from $^{40}MgO^{+}$. By using appropriate apertures, selected areas of the image can be surveyed for a complete spectrum.

In either system the ion beam generates secondary ions from the first one or two atomic layers so that this is a very sensitive technique and in some ways less damaging than AES. However, it does remove material from the surface, so it can hardly be described as nondestructive. The mass spectra obtained carry not only elemental information but chemical information too. By a simple reversal of the field, either positive or negative ions can be detected. The spectra will contain signals not only from atomic ions but also from fragmentary ions from both organic and inorganic compounds. Figure 6 shows the spectra obtained from steel-furnace dust particles (Van Craen et al., 1982) using the ion microscope of Fig. 5. The positive ion spectrum shows several metals and their oxides. The negative ion spectrum is even more interesting. Peaks corresponding to CN^{-}, Cl^{-}, and SO^{-} and phosphates indicate the presence of metal compounds. The SO^{-} is particularly revealing since the relative ratios of the oxides of sulfur suggest that sulfate is not the primary sulfur-containing anion.

Ion Scattering Spectrometry (ISS)

A deceptively simple technique is illustrated in Fig. 1 d. Incident ions, about 1.5 KeV, in essence bounce from the surface by eleastic scattering. In doing so they lose some energy to the atom which is impacted. The energy loss is a function of the mass of that atom. In fact, if the angle between the in-

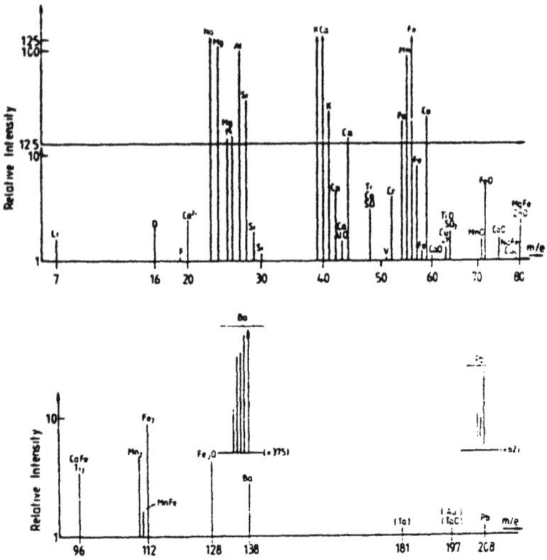

(Top) Positive ion mass spectrum obtained from the surface of electric furnace dust [m/e 5–80] (↑ denotes off-scale intensities). (Bottom) Extension of the mass spectrum presented [m/e 90–210].

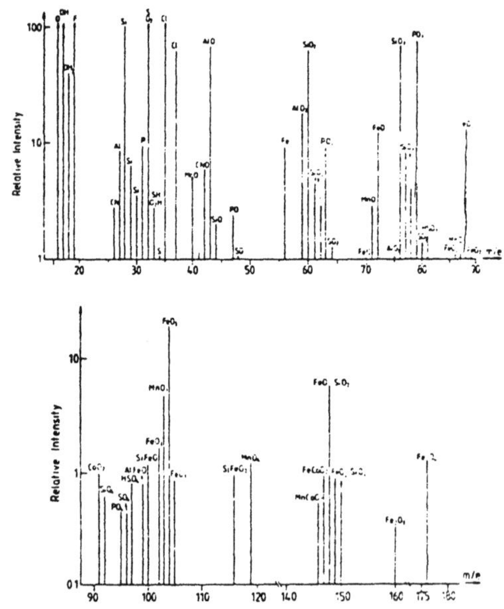

(Top) Negative ion mass spectrum obtained from the surface of electric furnace dust [m/e 15–90] (↑ denotes off-scale intensities). (Bottom) Extension of the mass spectrum [m/e = 90–180].

Fig. 6. Mass spectra of furnace dust

cident and scattered beam is 90°, the following relationship holds (Smith, 1967):

$$\frac{E_1}{E_0} = \frac{M_s - M_o}{M_s + M_o},$$

where E_1 and E_0 are the incident and scattered energies and M_s and M_o are the sample and incident masses, respectively, provided $M_s > M_o$. An instrument with this geometry has been built (Goff and Smith, 1970). In practice any fixed scattering angle will do. The instrument shown in Fig. 2 can be

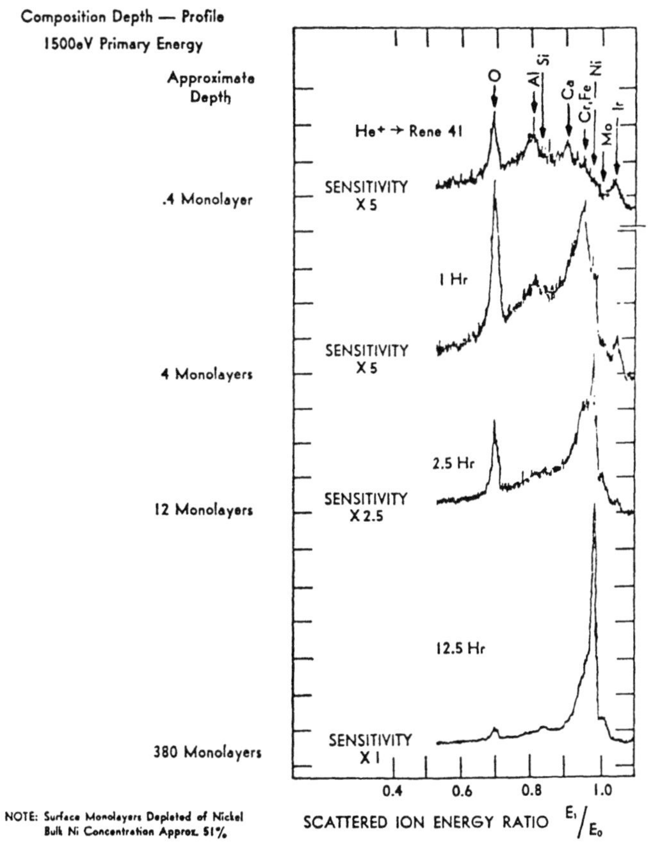

Fig. 7. ISS spectra at successive depths for rene 41 nickel catalyst

easily modified by using the ion source in the SAC and reversing the polarity of the analyzer.

For samples consisting of low-mass atoms He^+ ions are normally used. For atoms with higher masses heavier ions such as Ar^+ may be preferable to keep E_1/E_0 as small as possible for better resolution; however, beams of heavy ions will cause more surface damage.

Because scattering in ISS occurs only from the first atomic layer, ISS is a highly surface-sensitive technique. Figure 7 shows spectra of a nickel catalyst, in which the impurities decrease with depth. Note the dramatic change after removal of only 0.4 monolayers.

Other Ion Techniques

ISS studies the scattering phenomenon with ions of about 1.5 KeV energy. If the He^+ ions are energized in an accelerator to over 1 MeV, the scattering mechanism is termed Rutherford Back-Scattering (RBS). Since the ions are so much more energetic, they penetrate deeper into the surface. This technique has been applied mostly to the analysis of thin films and to studies of defects in single crystals. Consequently, it has been applied particularly

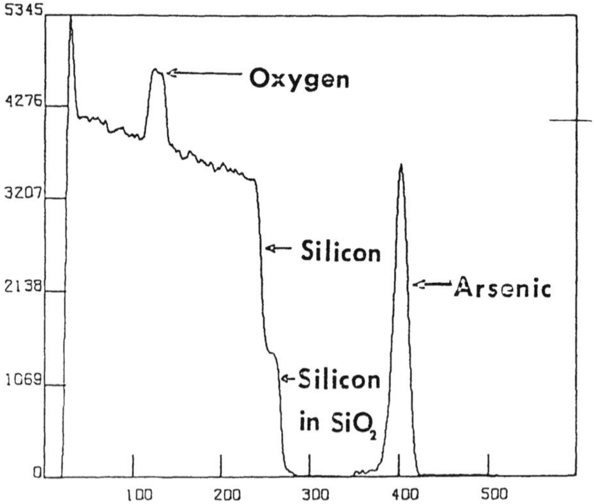

Fig. 8. Backscatter spectrum for 150 KeV arsenic ion implant into 1250 Å of SiO_2 on silicon. The oxide thickness is measured as 1216 Å based upon the width of the oxygen peak. The implant dose is measured as 1.12×10^{16} atoms/cm^2 as compared to 1.04×10^{16} atoms/cm^2 by neutron activation analysis

to electronic materials. The energy losses in this technique are due to the mass of the atom which is struck and to its depth in the sample. The higher the mass of the sample atom, the higher the energy of the scattered particle. Figure 8 (Keenan, private communication) shows an RBS spectrum from an arsenic implant into a thin oxide film on silicon. The peak for arsenic is sharp because all the ions are scattered from a relatively thin region below the oxide. The silicon forms a plateau with the high-energy edge representing the mass function. The plateau is caused by ions scattered from atoms located deep in the sample. The little step is due to the ions scattered by silicon in the oxide. Note the oxygen peak superimposed on the plateau. Since the scattering cross-sections of atoms are well established, the thickness of the oxide film can be calculated. This method is particularly valuable for single crystals. If the crystal is correctly oriented, the ions can travel longer distances without collision as compared with a random orientation and less return by scattering to the surface. Substitutional impurities (or dopants) can be detected as buried layers because they prevent this channeling.

If low-energy protons (low for physicists – about 1–3 MeV) are used as the source, prompt gamma rays and X-rays are emitted characteristic of the target. The more commonly used information is from the X-ray spectrum. This technique is called proton-induced X-ray emission (PIXE). It has been applied, for example, to the determination of sulfur in ferromanganese nodules (Goff, 1972).

A number of sources have been used to generate ions from a surface, chiefly to form molecular ions from labile materials. These techniques are often called soft ionization methods and include thermal and field desorption (TD and FD), fast atom bombardment (FAB), and laser desorption (LD). The latter in particular lends itself to the analysis of small areas. A commercial instrument, the LAMMA-500, uses a laser with a 1 μm resolution in conjunction with a time-of-flight (TOF) mass spectrometer. The intensity of the laser can be controlled to vary the fragmentation. A series of mass spectra (Denoyer et al., 1982) for an oil fly ash is shown in Fig. 9. Not only are a number of elements identified, but information is also obtained about oxides of sulfur and vanadium as well as about potential organic compounds. By reducing the intensity of the laser, less fragmented ions of greater mass can be detected and their structures deduced.

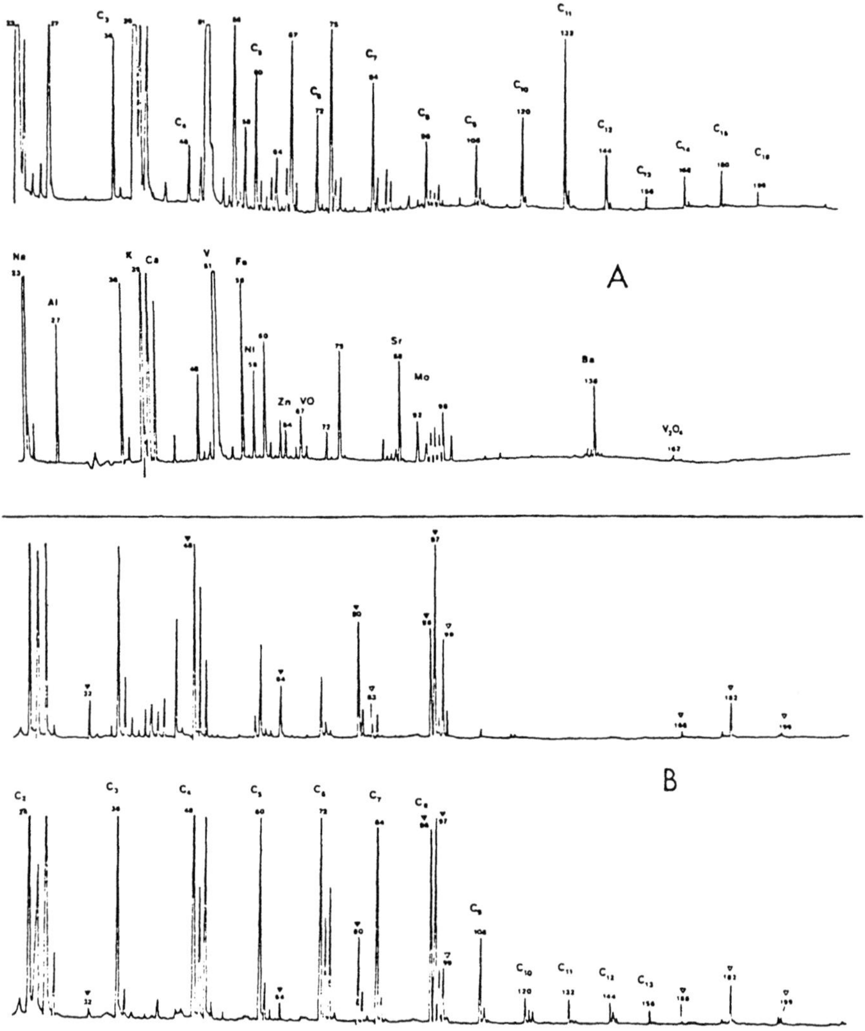

Fig. 9. Typical positive and negative LAMMA spectra of individual oil fly ash particles (A) positive ion spectra; (B) negative ion spectra. (▽= peaks associated with oxides of vanadium; ▼= peaks associated with oxides of sulfur)

Applications to Environmental Samples

This review concentrates on particulate matter in air or water. The microbeam techniques discussed are particularly suited to these types of sample. Problems involving such particles usually require a knowledge of their structure and composition and of the transport of various species across the interface between the particle and its environment, air or water.

Air-Solid Interface

One of the earliest, and still one of the most important, problems in environmental science has been the identification of dust particles collected on the filters of air-sampling units. A considerable amount of effort has been devoted to their microscopy, diffraction, and X-ray spectroscopy. The identification of most of these particles can now be accomplished by these and similar microbeam techniques. Many examples are given by McCrone and his associates (1973) and in other publications (see Gooley, 1983; Heinrich, 1980, 1982). The many studies of asbestos particles, of lead-bearing automobile exhausts, and of fly ash are typical of these analyses. However, they are generally not surface studies. The techniques such as TEM, SEM, XRE, XRD, etc. identify the particle *per se*. Relatively few studies have been concerned with adsorbents.

One of the earliest workers to investigate the surface of airborne particles was Novakov. Both he (1982) and Hulett (1972) reported as early as 1972 the application of XPS to the identification of adsorbed gases on aerosols. Novakov et al. (1983) have recently reviewed such studies concerned with compounds of nitrogen and sulfur. They found that nitrogen exists on the surface not only as nitrate but also as ammonium ion and other reduced species, probably amines, amides, and/or nitriles (Fig. 10). Oxides of sulfur can be similarly identified.

One of the most significant sources of atmospheric pollution is the fly ash from coal- and oil-fired power plants. The surface composition of these particles is important because particle surfaces serve as interfaces with the human lung. XPS has been widely applied to the determination of elemental species. AES is used in conjunction with ion etching to determine the profile of composition into the surface. Application of these two techniques has been reviewed by Powell (1980) and by Linton et al. (1983). Identification of both organic and inorganic compounds on particle surfaces by SIMS has already been mentioned for furnace dust (Van Craen et al., 1982). The same technique has been applied by a number of workers to fly ash and automo-

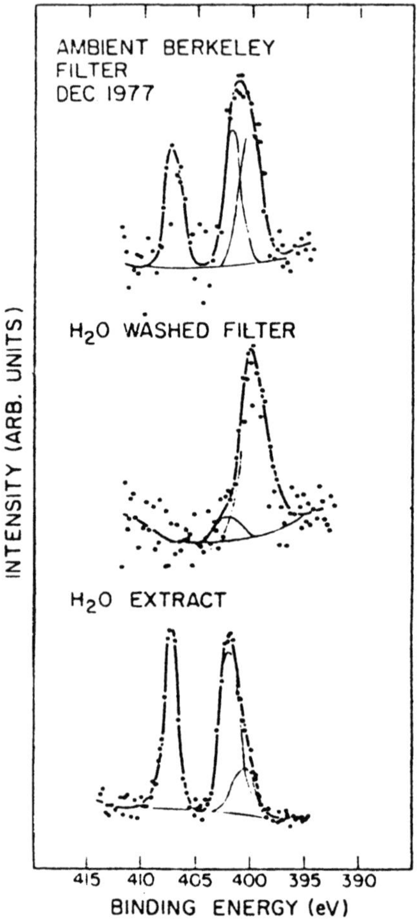

Fig. 10. XPS spectra of the nitrogen (ls) region of an ambient Berkeley, California, sample, H_2O washed and dried H_2O extract. The binding energies for NO_3^-, NH_4^+, and N_x are 407, 402, and 400 eV, respectively

bile exhausts. The use of laser-induced MS for this purpose has already been described (Denoyer et al., 1982).

Water-Solid Interface

This interface is of importance when suspended or sedimentary material is in contact with water in lakes, streams, and the ocean. Less obviously, it is significant in percolation, in water treatment, and in the fate of pollutants in groundwater.

Clays are one of the most widely distributed sedimentary materials and are known to concentrate trace metals from aquatic systems. Recently it was found that the presence of metal oxides, particularly iron, in these minerals increases the adsorbance for some toxic metals. Several investigators (e.g., Harvey and Linton, 1981) have applied XPS to the study of iron-oxide surfaces using chemical shifts to identify the type of oxide or hydroxide involved. Other XPS studies have been aimed at determining the specific ions adsorbed on the surface. Koppelman and Dillard (1980), for example, have carried out extensive studies on a number of minerals, including the adsorption of chromium ions on clay.

Conclusion

The electron and ion spectroscopies described in this review have the proven ability to identify and analyze particulate matter. Electron microscopy, particularly STEM, can examine these particles with a spatial distribution approaching atomic dimensions, determining not only topography but elemental and crystallographic distributions with the same resolution.

Surface characterization at the micron level is possible using SAM, ISS, and SIMS. The latter has the potential for providing information on the compounds present on the surface. Chemical shifts in AES have been used to identify compounds. However, Auger shifts are often difficult to interpret. XPS and UPS, despite having poor spatial resolution, are the most useful techniques for providing chemical information.

All the ion and electron spectroscopies can be carried out in the same system of Fig. 2. Complementary information is readily available without the necessity of exposing the sample to contamination by transferring it to different instruments. Some treatments, such as ion etching, can be carried out on the same sample.

The prime disadvantage of these techniques is the necessity to work in an ultrahigh vacuum. In such an artifical environment surfaces, particularly those from an aquatic environment, are subject to change. It is always necessary to carry out some initial studies with other techniques, e.g., thermal desorption, optical absorptiometry, XRD, to determine whether the changes are significant. Reactions can also be initiated either by heat or by the particles or photons used for excitation. However, important information is available from these techniques even if the results must be interpreted with care.

References

Allara DL (1982) Analysis of surfaces and thin films by IR, Raman, and optical spectroscopy. ACS Symp Ser 199:33–41

Briggs D (1979) Analytical applications of XPS. In Electron Spectroscopy: Theory, Techniques and Applications, eds. C.R. Brundle and A.D. Baker, vol 3, pp 305–358. London: Academic Press

Chang CC (1974) Analytical Auger electron spectroscopy. In Characterization of Solid Surfaces, eds. P.F. Kane and G.B. Larrabee, pp 509–575. New York: Plenum Press

Denoyer E, Manney T, Natusch DFS, Lee JH (1982) Laser microprobe mass analysis of coal and oil fly ash particles. In Microbeam Analysis 1982, ed. K.F.J. Heinrich, pp 191–196. San Francisco, CA: San Francisco Press

Fadley CS (1978) Basic concepts of photoelectron spectroscopy. In Electron Spectroscopy: Theory, Techniques and Applications, eds. C.R. Brundle and A.D. Baker, vol 2, pp 1–156. London: Academic Press

Feuerbacher B, Fitton B (1977) Photoemission spectroscopy. In Electron Spectroscopy for Surface Analysis, ed. H. Ibach, pp 151–203. Berlin: Springer-Verlag

Froitzheim H (1977) Electron energy loss spectroscopy. In Electron Spectroscopy for Surface Analysis, ed. H. Ibach, pp 205–250. Berlin: Springer-Verlag

Goff RF (1972) Reported by P.F. Kane and G.B. Larrabee. Trace analysis techniques for solids. In Annual Reviews of Materials Science, ed. R.A. Higgins, vol 2, p 54. Palo Alto, CA: Annual Reviews, Inc

Goff RF, Smith DP (1970) Surface composition analysis by binary scattering of noble gas ions. J Vac Sci Technol 7:72–75

Gooley R (ed) (1983) Microbeam Analysis 1983. San Francisco, CA: San Francisco Press

Harvey DT, Linton RW (1981) Chemical characterization of hydrous ferric oxides by X-ray photoelectron spectroscopy. Anal Chem 53:1684–1688

Heinrich KFJ (ed) (1980) Characterization of Particles. NBS Special Publication 533. Washington, DC: National Bureau of Standards

Heinrich KFJ (ed) (1982) Microbeam Analysis 1982. San Francisco, CA: San Francisco Press

Hulett LD, Charlson TA, Fish BR, Durham JL (1972) Studies of sulfur compounds adsorbed on smoke particles and other solids by photoelectron spectroscopy. In Determination of Air Quality, eds. G. Mamantov and W.D. Shults, pp 179–187. New York: Plenum Press

Koppelman MH, Dillard JG (1980) Adsorption of $Cr(NH_3)_6{}^{3+}$ and $CR(en)^{3+}$ on clay minerals and the characterization of chromium by X-ray photoelectron spectroscopy. Clays Clay Min 28:211–216

Linton RW, Harvey DT, Cabaniss GE (1983) Environmental applications of surface analysis techniques: PAS, XPS, Auger electron spectroscopy, SIMS. In Analytical Aspects of Environmental Chemistry, eds. D.F.S. Natusch and P.K. Hopke, pp 137–189. New York: John Wiley & Sons

McCrone WC (ed) (1973) The Particle Atlas, 2nd ed. Ann Arbor, MI: Ann Arbor Science Publishers

McGuire GE, Holloway PH (1981) Applications of Auger spectroscopy in material analysis. In Electron Spectroscopy: Theory, Techniques and Applications, eds. C.R. Brundle and A.D. Baker, vol 4, pp 1–84. London: Academic Press

Novakov T, Chang S-G, Dod RL, Gundel L (1983) ESCA in environmental chemistry. In Analytical Aspects of Environmental Chemistry, eds. D.F.S. Natusch and P.K. Hopke, pp 191–217. New York: John Wiley & Sons

Novakov T, Mueller PK, Alcocer AE, Otvos JW (1982) Chemical composition of Pasadena aerosol particles, size and time of day. III. Chemical states of nitrogen and sulfur by photoelectron spectroscopy. J Colloid Interf Sci 39:225–234

Powell CJ (1980) Application of Auger-electron spectroscopy and X-ray photoelectron spectroscopy to the characterization of pollutant particles. In Characterization of Particles, ed. K.F.J. Heinrich, pp 131–137. NBS Special Publication 533. Washington, DC: National Bureau of Standards

Siegbahn K et al. (1967) ESCA: Atomic, Molecular and Solid State Structures Studied by Means of Electron Spectroscopy. Uppsala: Almquist and Wiksells

Smith DP (1967) Scattering of low-energy noble gas ions from metal surfaces. J Appl Phys 38:340–347

Somorjai GA (1981) Chemistry in Two Dimensions: Surfaces. Ithaca, NY: Cornell University Press

Sparrow GR, Drummond IW (1981) Surface analysis: selecting a technique and an instrument. Ind Res Devel 23:112–115

Van Craen M, Natusch DFS, Adams F (1982) Quantitative surface analysis of steel furnace dust particles by secondary ion mass spectrometry. Anal Chem 54:1786–1792

Wagner CD (1972) Sensitivity of detection of the elements by photoelectron spectroscopy. Anal Chem 44:1050–1053

The Importance of Chemical "Speciation" in Environmental Processes,
eds. M. Bernhard, F. E. Brinckman, and P. J. Sadler, pp. 667–684.
Dahlem Konferenzen 1986. Springer-Verlag Berlin, Heidelberg
© Dr. S. Bernhard, Dahlem Konferenzen

Liquid Chromatography Element-specific Detection Systems for Analysis of Molecular Species

K. J. Irgolic* and F. E. Brinckman**

* Dept. of Chemistry
Center for Energy and Mineral Resources
Texas A & M University
College Station, TX 77843
** Chemical and Biodegradation Processes Group
Center for Materials Science
National Bureau of Standards
Gaithersburg, MD 20899, USA

Abstract. Systems for the separation of inorganic compounds of environmental interest are described. They consist of a high-pressure liquid chromatograph linked to a graphite furnace atomic absorption or inductively coupled plasma spectrometer as the element-specific detector. The possibility of correlating the elution rates of compounds from columns with their biological activity is discussed.

Introduction

The need for methods applicable to the identification and quantitation of trace-element compounds arises from the realization that environmental samples cannot be adequately characterized with respect to effects of trace elements solely by the determination of total trace-element concentrations. The beneficial and detrimental properties now usually assigned to a trace element are really the properties of trace-element compounds. Biologically important molecules such as enzymes interact with trace-element compounds, which may be labile or stable coordination compounds with a metal ion surrounded by ligands, and include organometallic compounds in which at least one covalent metal- or metalloid-carbon bond is present. The extent to which these interactions take place and their rates are governed by the chemical nature of the trace-element compounds.

Nature cycles trace elements. During passage through these cycles trace-element compounds are transformed by abiological and biologically mediated reactions. These cycles can be understood only when the transformation rates are known. The transformation rates cannot be determined unless the participating compounds are identified and quantified. Trace-element compounds that are volatile, or can be changed into volatile derivatives without major modification of their structures, can be separated by gas chromatographic techniques. However, many trace-element compounds of environmental interest are not volatile and cannot be converted easily to volatile compounds without loss of vital chemical information (species transformation). These nonvolatile, often macromolecular, substances can best be separated by liquid chromatography. Commercial column substrates that permit efficient molecular separations based on diagnostic polar, ionic, or molecular size properties of metal- and metalloid-containing molecules occurring at trace levels are now widely available for high-performance liquid chromatographs (HPLC). Proper combinations of mobile phase, column type, and detector will provide analytical results that promise to be invaluable for predicting useful physicochemical and biological properties of analytes.

Environmental samples contain in most cases many other organic and inorganic compounds in addition to the trace-element compounds of interest. On-line detectors commonly employed with chromatographs, such as refractive index and conductivity detectors, are generally not specific for particular trace elements but respond to many other substances co-eluting from the sample matrix. Also the detection limits achievable with "general" detectors are often not adequate for the low concentrations of trace elements frequently encountered in environmental work. Element-specific detectors, based on light absorbed by or emitted from atoms, afford high sensitivity and drastically simplify the chromatogram by producing signals only for compounds that contain a particular element. In addition, their use makes chromatography much easier. The chromatographic procedure needs to separate only compounds that contain the element to be detected. Overlap of the analytical bands with bands from compounds not containing this element is of no consequence to such a system but is a certainly undesirable complication intrinsic with a "general" detector.

A detector is element-specific when it responds reproducibly to a unique property of an element or a particular compound of that element. Detectors that measure atomic transitions are element-specific and have the additional advantage of relative insensitivity to matrix effects. The most useful detectors for liquid chromatography include flame and graphite furnace absorption spectrometers, atomic fluorescence spectrometers, and plasma emission spectrometers. Although other detectors possess element-specificity (Jewett and Brinckman, 1983), the discussion in this paper will be restricted to automated systems consisting of liquid chromatographs coupled with graphite furnace atomic absorption spectrometers or with inductively coupled argon plasma emission spectrometers. The use of direct current plasma emission spectrometers as element-specific detectors is described in a recent review (Uden, 1981) and by Krull (this volume).

The Liquid Chromatograph

Almost any liquid chromatograph can be coupled to a graphite furnace atomic absorption spectrometer (GFAA) or an inductively coupled argon plasma emission spectrometer (ICP). The major requirements for a successfully operating system include a flow rate through the column matching the demands of the detector and compatibility of the mobile phase with the detector.

Recent emergence of "microbore HPLC" merits consideration, whenever very low eluent flow (1–10 $\mu L\ min^{-1}$), limited sample size, and high separation efficiency with rapid analysis are important (Scott, 1983). At such very low flow rates, direct introduction of mobile phases into detectors such as mass spectrometers, flame ionization or flame photometric detectors, and microwave plasma detectors that can operate with very small volumes, will become more popular. The most commonly used chromatographs are high-pressure liquid chromatographs and ion chromatographs. All types of columns (normal-phase, reversed-phase, ion-exchange, size-exclusion) can be used (Table 1; see also Krull, this volume).

Table 1. Applications of HPLC-GFAA and HPLC-ICP systems

Compounds	System[a]	Column	Ref.[b]
$(C_6H_5)_3As$	HPLC-PE/GFAA	C-18 Reversed-Phase	1
Arsenite, Arsenate, $CH_3AsO_3H_2$, $(CH_3)_2AsOOH$, Arsenobetaine	HPLC-ICP	Anion-Cation-Exchange	2
Arsenite, Arsenate, $CH_3AsO_3H_2$, $(CH_3)_2AsOOH$, $C_6H_5AsO_3H_2$	HPLC-ICP	Reversed-Phase	3
Arsenite, Arsenate, $CH_3AsO_3H_2$, $(CH_3)_2AsOOH$	HPLC-PE/GFAA	Lichrosorb SAX	4
Arsenite, Arsenate, $CH_3AsO_3H_2$, $(CH_3)_2AsO_2H$	HPLC-HZ/GFAA	C-18 Reversed-Phase	4
Arsenite, Arsenate, $CH_3AsO_3H_2$, $(CH_3)_2AsOOH$	HPLC-PE/GFAA	Dionex	5
Arsenite, Arsenate, $CH_3AsO_3H_2$, $(CH_3)_2AsOOH$, 4-$H_2NC_6H_4AsO_3H_2$	HPLC-HG/QFAA	Dionex	6
Arsenate, $CH_3AsO_3Na_2$, $(CH_3)_2AsOONa$, 4-$H_2NC_6H_4AsO_3H_2$, $C_6H_5AsO_3H_2$	HPLC-ICP	Hypersil, Reversed-Phase	7
4-$H_2NC_6H_4AsO_3H_2$	HPLC-PE/GFAA	Lichrosorb SAX	4
Arsenite, Arsenate, Arsenobetaine, Arsenocholine	HPLC-HZ/GFAA	C-18 Reversed-Phase	8
Arsenic Compounds (oil, shale process waters)	HPLC-PE/GFAA	Dionex	9
Arsenic Compounds (Green River Oil Shale)	HPLC-PE/GFAA	Dionex	10
Arsenic-Containing Lipids	HPLC-HZ/GFAA	Microporosil	11
Arsenobetaine (Crabs)	HPLC-ICP	C-18 Reversed-Phase	12
$(C_6H_5CH_2)_2Se_2$, 1,1-Dimethylthiourea	HPLC-HZ/GFAA	Reversed-Phase	13
Selenite, Selenate	IC-HZ/GFAA	Dionex 30589	14
Selenite	HPLC-ICP	Reversed-Phase	3
Hg^{2+}, RHg^+ ($R=CH_3$, C_2H_5, C_3H_7, C_6H_5)	HPLC-PE/GFAA	C-8 Reversed-Phase	1
$HgCl_2$, CH_3Hg^+, $C_2H_5Hg^+$, $C_3H_7Hg^+$	HPLC-ICP	Hypersil, Reversed-Phase	7
R_3Sn^+, R_2Sn^{2+}	HPLC-PE/GFAA	SCX	15
R_3Sn^+ ($R=C_6H_5$, C_4H_9, C_3H_7)	HPLC-PE/GFAA	C-2 Reversed-Phase	1
Organotin Polymers	HPLC-PE/GFAA	Size-Exclusion	16
Si, Sn in Tricyclohexylstannoxy Derivative of Ortho-pentasilicate	HPLC-PE/GFAA	C-18 Reversed-Phase	16
$(C_4H_9)_3Sn$ in Estuarine Bacteria	HPLC-PE/GFAA	Cation-Exchange	17

$(C_6H_5)_4Pb$	HPLC-HZ/GFAA	C-18 Reversed-Phase	18
$(C_6H_5)_6Pb_2$	HPLC-PE/GFAA	Lichrosorb SI-100	1
Pb in Lubricating Oils	HPLC-PE/GFAA	Silica	19
Pb Compounds in Petrol	HPLC-ICP	Zorbax C-8 Reversed-Phase	7
V Compounds in Heavy Crudes	HPLC-GFAA	Spherogel, Aminocyano	20
Cr(III), Cr(VI)	HPLC-HZ/GFAA	Cation-Exchange	21
Ferrocene Compounds	HPLC-ICP	Silica Gel-60	7
Iron Carbonyls	HPLC-ICP	Zorbax C-8 Reversed-Phase	7
Molybdenum Carbonyls	HPLC-ICP	Zorbax C-8 Reversed-Phase	7
Co (Vitamin B_{12})	HPLC-ICP	Gel Permeation	22
Co, Cu, Fe, Mn, P, Zn, C in Protein Mixture	HPLC-ICP	Gel Permeation	22
Copper(II) Complexes of EDTA and NTA	HPLC-ICP	Aminex-A-14 Anion-Exchange	23
Phosphate	HPLC-ICP	Reversed-Phase	3
Organic Phosphorus Compounds (oil additives)	HPLC-PE/GFAA	C-18 Reversed-Phase	24
Macromolecular Complexes of Mg, Ca, Ti, Cu, Zn (coal)	HPLC-ICP	Off-line silica gel, then on-line size-exclusion with poly-styrene-divinylbenzene copolymer	25

[a] GFAA: Graphite Furnace Atomic Absorption Spectrometry; HG: Hydride Generation; HPLC: High-Pressure Liquid Chromatography; HZ: Hitachi Zeeman; IC: Ion Chromatography; ICP: Inductively Coupled Argon Plasma Emission Spectrometry; PE: Perkin Elmer; QFAA: Quartz Furnace Atomic Absorption Spectrometry.

[b] 1: Brinckman et al., 1977. 2: Morita, Uehiro, and Fuwa, 1981. 3: Irgolic et al., 1983. Brinckman et al., 1980. 5: Woolson and Aharonson, 1980. 6: Ricci et al., 1981. 7. Gast et al., 1979. 8: Stockton and Irgolic, 1979. 9: Fish, Brinckman, and Jewett, 1982. 10: Fish et al., 1983. 11: Bottino et al., 1978. 12: Francesconi et al., 1985. 13: Vickrey, Buren, and Howell, 1978. 14: Chakraborti et al., 1982. 15: Jewett and Brinckman, 1981. 16: Parks, Brinckman, and Blair, 1979. 17: Blair et al., 1982. 18: Vickrey, Howell, and Paradise, 1979. 19: Brinckman and Blair, 1979. 20: Fish and Komlenic, 1984. 21: Vickrey and Eue, 1979. 22: Morita, Uehiro, and Fuwa, 1980. 23: Fraley, Yates, and Manahan, 1979. 24: Tittarelli and Muscherpa, 1981. 25: Taylor, Hausler, and Squires, 1981.

High-Pressure Liquid Chromatography-Graphite Furnace Atomic Absorption Systems

An HPLC-GFAA system consists of the chromatograph as the separator of species, the spectrometer as the element-specific detector and the interface between the two components. The interface must reproducibly and automatically transfer an adequate aliquot of the column effluent, which flows at a constant rate, to the discontinually operating detector. The graphite furnace AA then proceeds through the predetermined analysis cycle: deposition of the analyte in the graphite cup or tube, evaporation of the solvent (drying), decomposition of organic material (ashing), atomization of the remaining substances, and cooling of the graphite furnace prior to the injection of the next sample. The interphase must accurately time the delivery of the analyte to the furnace. Three HPLC-GFAA systems are available.

The HPLC-Perkin Elmer GFAA System

(Brinckman et al., 1977). In the HPLC-Perkin Elmer GFAA system an autosampler and a Teflon "well sampler" provide the connection between the column and the graphite furnace. The column exit is connected to the bottom of the well sampler via a zero dead-volume screw (A) (Fig. 1). The sampler is placed into an appropriate position of the fixed autosampler carousel. The effluent continuously flows through the 50 μL sample well (B). Excess liquid is withdrawn by suction at C. The sampling pipette of the autosampler dips into the sample well, collects 10 to 50 μL of effluent, and transfers this aliquot into the graphite tube for analysis. Detectors other than GFAA may be placed between the column and the autosampler.

The HPLC-Hitachi Zeeman GFAA System

(Irgolic, 1982; Stockton and Irgolic, 1979; Vickrey et al., 1978; Vickrey and Eue, 1979). The Hitachi Zeeman spectrometer offers the advantage of inherent background correction. The column effluent is routed to an eight-port, Teflon slider injection valve with two attached actuators. The effluent passes through a 40 μL sample loop to a fraction collector or to waste. When an aliquot of the effluent is to be injected into the graphite cup, the injection valve is switched, and the 40 μL sample in the sample loop is pushed by nitrogen into the cup through a Teflon tube and a stainless steel injection tube, which has been lowered into the furnace by a linear actuator.

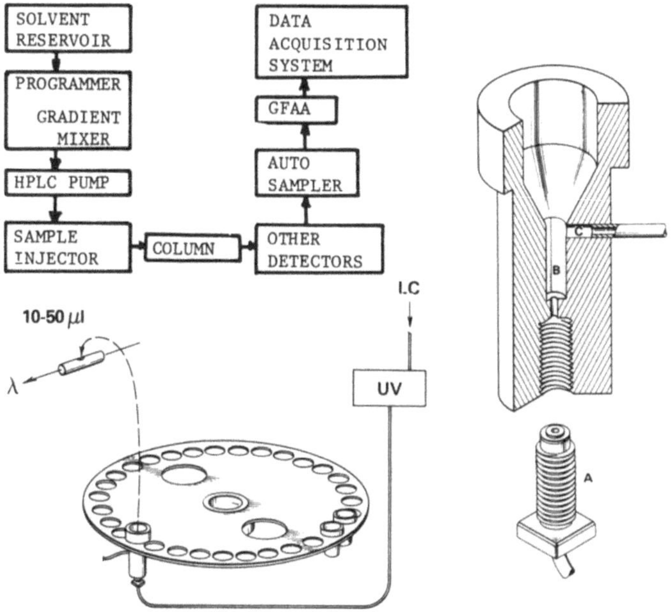

Fig. 1. The high-pressure liquid chromatography – Perkin Elmer graphite furnace atomic absorption spectrometer system (adapted from Brinckman et al., 1977)

After the injection valve has been switched to the flow-through position and the injection tube retracted from the furnace, the normal analysis cycle of the GFAA is started. To prevent pressure buildup during the switching of the valve, several feet of 0.3 mm ID Teflon tubing connect the line coming from the column with the line leading from the valve to the injector. The sequence of events is determined by the timing and control circuits. The results of the analyses are collected on a strip-chart recorder. The interval between sequential analyses is adjustable and can be made as short as thirty seconds. With excessively short cycle times the cooling system for the magnets becomes overloaded and the protective heat sensor shuts down the instrument. A detailed description of the HPLC-Hitachi Zeeman GFAA system schematically shown in Fig. 2 is available in the literature (Irgolic, 1982). Alternatively, the column effluent can be stored in a capillary in suitable aliquots separated by nitrogen plugs for later analysis by GFAA (Vickrey and Eue, 1979). This HPLC-GFAA system can be controlled by a microprocessor (Vickrey et al., 1978).

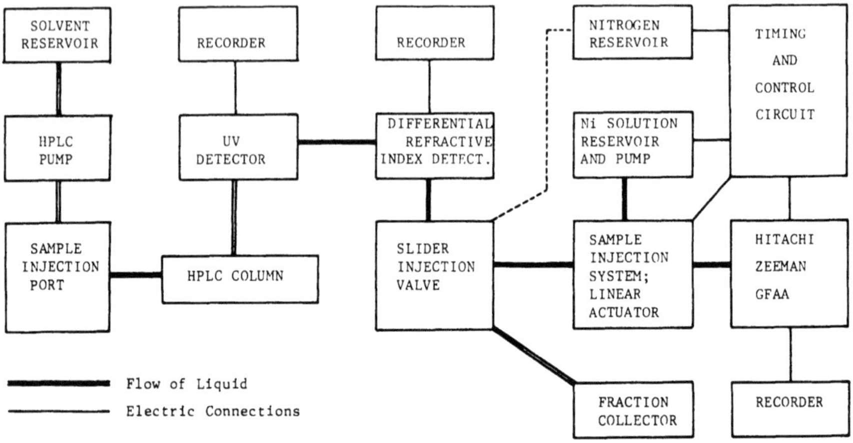

Fig. 2. Block diagram for the high-pressure liquid chromatography – Hitachi Zeeman graphite furnace atomic absorption spectrometry system. Reproduced from the Journal of Chromatographic Science by permission of Preston Publishers, Inc.

The HPLC-IL 555 GFAA System

(Haswell et al., 1986). An HPLC-GFAA system was designed for an instrumentation laboratory Model IL 555 graphite furnace atomic absorption spectrometer. The injector for this system is a specially mounted hypodermic needle that is moved into the graphite tube by a solenoid and retracted by a spring. The system is controlled by an Ohio Scientific Superboard II 6502 microprocessor-based, single-board computer that also collects data, integrates the peaks generated by the GFAA, and monitors the temperature of the graphite tube.

An HPLC-GFAA system operable either manually or under computer control can incorporate any GFAA. The principal parts of the interface will be the same. However, the injector system and the electrical connections between the interface and the spectrometer must be adapted to the requirements of each type of spectrometer. Construction of such an interface is not complicated and the necessary parts are inexpensive. Spectrometers fitted with autosamplers are probably the easiest to modify and place into service as element-specific detectors for liquid chromatographs.

The HPLC Simultaneous Inductively Coupled Argon Plasma Emission Spectrometer (ICP) System

Multi-element-specific detection can be achieved with GFAA as detectors by splitting the column effluent and routing the resulting streams to two (or more) AA spectrometers in tandem (Weiss et al., 1983). However, simultaneous plasma emission spectrometers – designed for multi-element operation – are easier to use as multi-element-specific detectors for HPLC and are capable of providing much more information. The choice of elements to monitor is only limited by the analytical lines on the spectrometer array. The interface consists of a piece of tubing connecting the exit of the column to the entrance of the nebulizer. The flow rate through the column must not be too different from the uptake rate of the nebulizer. The software operating the ICP and collecting the signals from the various channels must probably be modified to provide a choice of elements among those on the array. A program was written for an ARL Model 34000 ICP and the standard software modified to allow the chromatograms of several elements to be displayed graphically on-line. In addition, the retention times and intensities of the signals are stored on a magnetic disk for later recall. This HPLC-ICP was used successfully to separate and determine organic and inorganic arsenic compounds, phosphate, and selenite (Irgolic et al., 1983).

Detection Limits of HPLC-GFAA and HPLC-ICP Systems

Graphite furnace atomic absorption and plasma emission spectrometers are probably the most sensitive instruments for the routine determination of environmentally interesting trace elements (Table 2). When these spectrometers serve as element-specific detectors for HPLC they do not lose their excellent sensitivities which approach part-per-trillion concentrations. However, the detection limits of the HPLC-GFAA and HPLC-ICP systems are ten-to-one-hundredfold higher than the GFAA or ICP detection limits. When a small volume of a sample is chromatographed, each compound is spread into a band. The detector (GFAA, ICP) takes aliquots from these bands. For an element to be detectable, the aliquot from the center of the band, where the concentration of the analyte is highest, must have a concentration above the detection limit of the spectrometer. Although the sample injected into the column might have a concentration sufficient for the determination of an element, spreading the substance into a band during chromatography might reduce the concentrations in the aliquots taken for analysis below the detection limit. Table 2 contrasts some of the typically ob-

Table 2. Detection limits of graphite furnace atomic absorption spectrometers, inductively coupled plasma emission spectrometers, HPLC-GFAA systems, and HPLC-ICP systems

Element	Detection limit, µg/L			
	GFAA[a]	HPLC-GFAA[a]	ICP	HPLC-ICP[b]
As	1	5[c]	20	130
Se	10	20[d]	80	5,000
P	–	–	200	10,000
Sn	>20	>50[e]	80	–

[a] The GFAA detection limits depend on the quality and age of the graphite tube or cup, presence of "enhancers," and individual organoelements (Brinckman et al., 1977, 1983; Weiss et al., 1983)
[b] Irgolic et al., 1983
[c] Brinckman et al., 1980
[d] Chakraborti et al., 1982
[e] Brinckman et al., 1983; Scott, 1983

served "system detection limits" with the "spectrometer detection limits." The ICP may consume almost all of the column effluent within a wide range of flow rates and may therefore be more sensitive as a detector than a GFAA that can analyze only a rather small volume (<100 µL) approximately every minute. At flow rates higher than 100 µL min, on-line analysis with total consumption of the effluent by the GFAA is not possible. The GFAA detector may therefore become less sensitive than ICP and may completely or partially miss a narrow chromatographic band.

Applications of HPLC-GFAA and HPLC-ICP Systems

During the past few years the HPLC-GFAA and HPLC-ICP systems have been used to separate and determine arsenic, selenium, mercury, tin, lead, iron, and molybdenum compounds among others (see Table 1 for a summary of these applications). With the exception of arsenic and tin compounds, most investigations have been sporadic and were carried out with clean standards. The analysis of "real" samples such as oil shale extracts, shale oil process waters, and extracts from organisms is much more complicated than work with solutions of pure compounds in distilled water. Although element-specific detectors tremendously simplify the chromatogram of a complex mixture, they do not solve all the problems. Retention times, generally used to identify substances, are not always reliable. Arsenobetaine, for instance, produces a sharp peak when a distilled water solution is chromatographed on a reversed-phase column with 2.5% aqueous acetic acid containing 0.05 M heptanesulfonic acid as mobile phase. When arsenobetaine is spiked into an extract from crab meat, the retention time is drastically shorter and several broad, arsenic-containing peaks appear. Apparently the extracts contain materials that compete with the ion-pairing reagent (heptanesulfonic acid). When the concentration of the ion-pairing reagent was increased or the quantity of extract placed on the column decreased, the arsenobetaine peak could again be used for the identification of this compound (Francesconi et al., 1985). Much more work needs to be carried out with "real" samples.

In spite of the great utility of HPLC-GFAA and HPLC-ICP systems in environmental chemistry, relatively little use has been made of them. A major factor contributing to this underutilization is the unavailability of commercial HPLC-GFAA systems. Investigators searching for trace-element compounds in complex environmental systems have to use the best available techniques. The HPLC-GFAA and HPLC-plasma emission systems are the best, state-of-the-art techniques for the identification and determination of the broadest range of trace-element compounds.

Examples of the Application of HPLC-GFAA to the Study of the Biotransformations of Arsenic

The pervasiveness of the biogeochemistry of arsenic in both primordial and modern times has recently become evident. HPLC-GFAA has proven to be a powerful tool for demonstrating the presence of organoarsenicals in oil shale, shale oil, and process waters. Methylarsonic acid, phenylarsonic acid, and inorganic arsenic species were found and quantitated in process waters of an *in-situ* oil-shale retort (Fish et al., 1982). The fossil precursors of these arsenic compounds were recently discovered in the biogenic kerogen (Fish et al., 1983). Using dual-tandem GFAA detectors coupled with size-exclusion HPLC, organic arsenic moieties were found also in shale oils. These arsenic moieties are bound to macromolecules ($MW > 3000$) via Fe-O-As linkages (Weiss et al., 1983). Iron/arsenic ratios and molecular weight distributions of arsenic were highly dependent on the location of the oil shale. This dependence implies local differences in the biogeochemical history of the shale.

Arsenic-containing herbicides are anthropogenic sources of arsenic in the environment. Disodium methylarsonate and sodium dimethylarsinate

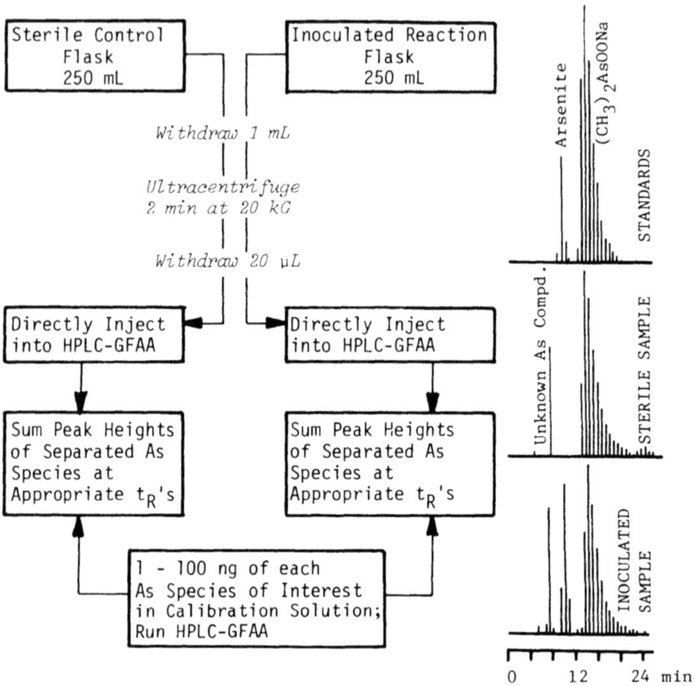

Fig. 3. Protocol adopted by NBS for direct determination of arsenic species by HPLC-GFAA of cultures of soil microorganisms treated with methylarsenic herbicides. Reproduced from the Journal of Chromatographic Science by permission of Preston Publishers, Inc.

are commercially used on cotton crops. We used the HPLC-GFAA procedure, previously shown to be effective for the determination of traces of methylarsenic acids (Brinckman et al., 1980), to develop a protocol (Fig. 3) for the routine determination of methylarsenic herbicides and for the exploration of the metabolic pathways of these arsenic compounds in soil microorganisms (Brinckman et al., 1984). Typical results obtained with this method are shown in Fig. 3. Sodium dimethylarsinate was demethylated to arsenite in inoculated soil cultures. Disodium methylarsonate, which has a retention volume of 26 mL, was not detected, nor was it demethylated to arsenite under these conditions. These results suggest that dimethylarsinate is not demethylated stepwise.

New approaches for the determination of species use post-column derivatization with borohydride to convert methylarsenic acids to volatile methylarsines, which are then detected by GFAA or ICP. Other metalloids and post-transition metals that form relatively stable volatile hydrides or alkyl derivatives can be similarly derivatized. Among the molecules that might be determined by these methods in biota and other environmental samples are those containing mercury, antimony, bismuth, silicon, tin, lead, selenium, and tellurium.

Estimation of Physicochemical and Biological Properties by HPLC-GFAA

Basic theory of the chromatographic separation process predicts that under "poised" (near equilibrium) or equilibrium conditions, attainable with a well-controlled HPLC system, the column capacity factor, or retention index k′, is a function of the rentention time:

$$k' = t_R/(t_R - t_o) \quad (1)$$

t_R: retention time (or volume) for a separated molecule,
t_o: void retention time (or volume) of the column.

In practice, ln k′ is a temperature-dependent free energy term for HPLC columns that separate molecules efficiently and mainly by a single process such as ion exchange, solvolysis, or polar affinity. This term is directly proportional to the equilibrium constant for the retention-determining reaction between the analyte and the column substrate in the moving front of the mobile phase. Substituent constants useful for the prediction of molecular shapes and effects have been available for some time for organic molecules (Hansch and Leo, 1979; Tomlinson, 1975) and were recently reported for organometals and organometalloids (Weiss et al., 1981). The chromatographic capacity factor, k′, is correlated with these substituent constants as expressed in Eq. 2:

$$\ln k' = mS + \text{constant} \quad (2)$$

S: linear sum of some molecular property such as aqueous solubilities, Hansch's pi or log P parameters derived from water/n-octanol partition coefficients (Nahum and Horvath, 1980), and σ^{Ph} as Taft-Hammett substituent parameters derived from a series of pK_a values for several hundred organophosphorus acids in water (Mastryukova and Kabachnik, 1969).

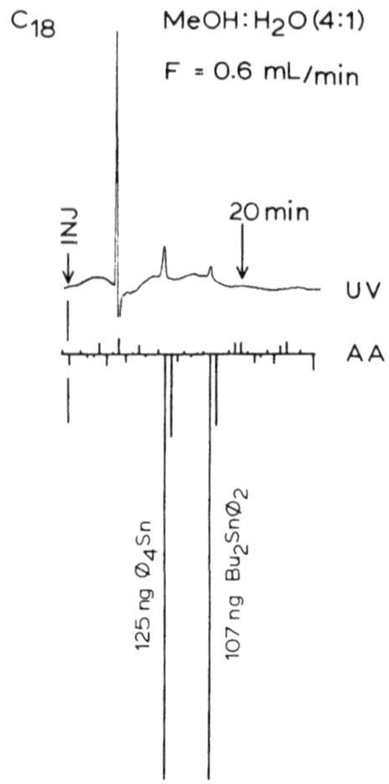

Fig. 4. Dual HPLC-UV-GFAA chromatograms of tetraphenyltin and dibutyldiphenyltin separated on a C-18 reverse-bonded phase column

The selection of parameters (pi, log P, σ^{Ph}) for the calculation of S, frequently called Quantitative Structure-Activity Relationship (QSAR), depends upon the principal mode of the HPLC separation. For example, when solvophobic and lipophilic columns with reverse-bonded phases are used, the QSAR for neutral analytes are obtained from log P or Hansch's pi values, whereas σ^{Ph} is employed for charged analytes separated on ion-exchange columns.

The HPLC-GFAA and HPLC-ICP systems can be used to measure k′ values at low concentrations in complex media with which conventional detectors do not function. Experimentally determined k′ values, especially those for homologous series of xenobiotics, may then be correlated with the toxicities of these compounds in organisms. Independently derived QSARs can be used to predict k′. Figure 4 shows an HPLC-UV-GFAA chromatogram of tetraphenyltin and dibutyldiphenyltin separated on a reverse-bonded phase column. The k′ values for these compounds and other neutral butylethyltin compounds that are not detectable by UV are plotted versus Hansch's pi (log P) parameter in Fig. 5. Good, straight-line relationships exist for each one of these two series of tin compounds. Based on work recently reported (Laughlin et al., 1984), topological models using only independently determined bond angles, bond lengths, and Van der Waals radii

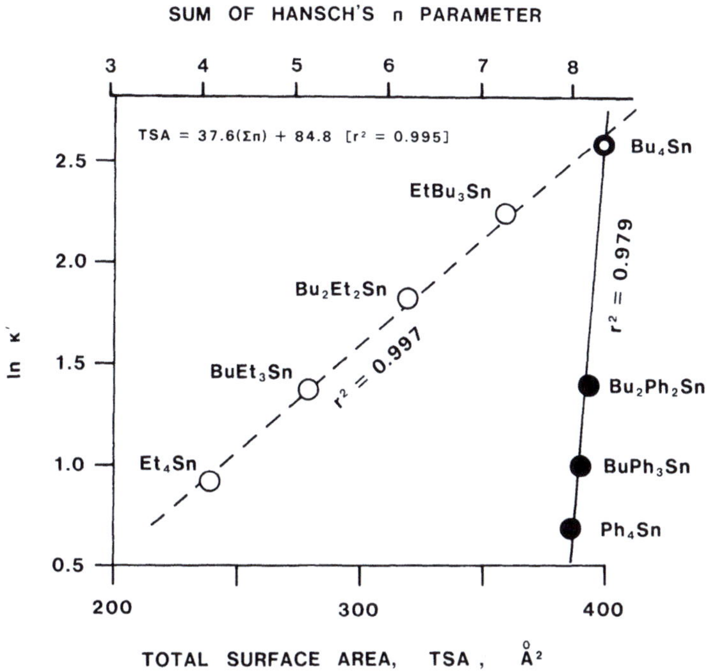

Fig. 5. The correlation of column capacity factors (ln k′) for mixed alkyltins and alkylphenyltins with log P (pi) (Hansch's n parameter) or TSA parameters derived from independent sources (from McLaughlin et al., 1984)

for the organotin molecules produced "total surface areas" (TSA) for these molecules. These TSAs are sensitive predictors of the conformations that these molecules are envisioned to adopt in the HPLC solvent system. TSA and pi can distinguish alkyltins from phenyl-substituted organotins on the basis of their k′ values. Presumably the increased steric constraints imposed by rigid, planar phenyl moieties allow considerably less retention (solvolysis) on the C-18 reverse-bonded phase than the conformationally flexible alkyl groups in alkyltin compounds. This aspect of detailed QSAR relating to specific, sterically active column sites suggests future utility in modeling regiospecificity of xenobiotic organoelements by high-pressure liquid chromatography and element-specific detection. Recently it was demonstrated that "total surface areas" accurately predict organotin toxicity toward marine crab larvae (Laughlin et al., 1984) and microorganisms (Wong et al., 1982). It is apparent that k′ values measured for other organometal and

-metalloid compounds by these methods will merit evaluation as new and rapid laboratory predictors of their biological activity.

Acknowledgements. The experimental work from the laboratory of K.J.I. presented in this paper was supported by the U.S. Environmental Protection Agency, Grant No. R 8047 740 10, and the Robert A. Welch Foundation of Houston, Texas. F.E.B. gratefully acknowledges partial support of the Office of Naval Research in some of the cited experimental work done at NBS and in preparation of this paper. Contributions of the U.S. National Bureau of Standards are not subject to copyright.

References

Blair WR, Olson GJ, Brinckman FE, Iverson WP (1982) Accumulation and fate of tri-n-butyltin cation in estuarine bacteria. Microb Ecol 8:241–251

Bottino NR, Cox ER, Irgolic KJ, Maeda S, McShane WJ, Stockton RA, Zingaro RA (1978) Arsenic uptake and metabolism by the alga *Tetraselmis chui*. ACS Symp Ser 82:116–129

Brinckman FE, Blair WR (1979) Speciation of metals in used oils: recent progress and environmental implications of molecular lead compounds in used crankcase oils. NBS Spec Publ 556:25–38

Brinckman FE, Blair WR, Jewett KL, Iverson WP (1977) Application of a liquid chromatograph coupled with a flameless atomic absorption detector for speciation of trace organometallic compounds. J Chrom Sci 15:493–503

Brinckman FE, Jackson JA, Blair WR, Olson GJ, Iverson WP (1983) Ultratrace speciation and biogenesis of methyltin transport species in estuarine waters. In Trace Metals in Sea Water, eds. C.S. Wong, E. Boyle, K.W. Bruland, J.D. Burton, and E.D. Goldberg, pp 39–72. New York: Plenum Press

Brinckman FE, Jewett KL, Iverson WP (1984) U.S. National Bureau of Standards, unpublished results

Brinckman FE, Jewett KL, Iverson WP, Irgolic KJ, Ehrhardt KC, Stockton RA (1980) Graphite furnace atomic absorption spectrometers as automated element-specific detectors for high-pressure liquid chromatography: the determination of arsenite, arsenate, methylarsonic acid and dimethylarsinic acid. J Chrom 191:31–46

Cassidy RM (1981) The separation and determination of metal species by modern liquid chromatography. In Trace Analysis, ed. J.F. Lawrence, vol 1, pp 122–192. New York: Academic Press

Chakraborti D, Hillman DCJ, Irgolic KJ, Zingaro RA (1982) Hitachi Zeeman graphite furnace atomic absorption spectrometer as a selenium-specific detector for ion chromatography: separation and determination of selenite and selenate. J Chrom 249:81–92

Fish RE, Brinckman FE, Jewett KL (1982) Fingerprinting inorganic arsenic and organoarsenic in *in-situ* oil shale retort and process waters using a liquid chromatograph coupled with an atomic absorption spectrometer as a detector. Envir Sci Technol 16:174–179

Fish RH, Komlenic JJ (1984) Molecular characterization and profile identifications of vanadyl compounds in heavy crude petroleums by liquid chromatography/graphite furnace atomic absorption spectrometry. Anal Chem 56:510–517

Fish RH, Tannous RS, Walker W, Weiss CS, Brinckman FE (1983) Organometallic geochemistry. Isolation and identification of organoarsenic compounds from Queen River formation oil shale. J Chem Soc Chem Commun 490–492

Fraley DM, Yates D, Manahan SE (1979) Inductively coupled plasma emission spectrometric detection of simulated high performance liquid chromatographic peaks. Anal Chem 51:2225–2229

Francesconi KA, Micks P, Stockton RA, Irgolic KJ (1985) Quantitative determination of arsenobetaine, the major water soluble arsenical in Alaskan King, Alaskan Snow, and Dungeness Crab, by high pressure liquid chromatography with an inductively coupled argon plasma emission spectrometer as the arsenic-specific detector. Chemosphere

Gast CH, Kraak JC, Poppe H, Maessen FJMJ (1979) Capabilities of on-line element-specific detection in high performance liquid chromatography using an inductively coupled argon plasma emission source detector. J Chrom 185:549–561

Hansch C, Leo A (1979) Substituent Constants for Correlation Analysis in Chemistry and Biology. New York: J Wiley & Sons

Haswell SJ, Stockton RA, Bancroft KCC, O'Neill P, Rahman A, Irgolic KJ (1986) A computerized, element-specific detection system for high pressure liquid chromatography. J Automat Chem, in press

Irgolic KJ (1982) Speciation of arsenic compounds in water supplies. Report 1982, EPA/600/1-82/010; Order No. PB82-257817, NTIS, pp 12–32

Irgolic KJ, Stockton RA, Chakraborti D, Beyer W (1983) Simultaneous inductively coupled argon plasma emission spectrometer as a multi-element-specific detector for high pressure liquid chromatography: the determination of arsenic, selenium, and phosphorus compounds. Spectrochim Acta 38B:437–445

Jewett KL, Brinckman FE (1981) Speciation of trace di- and tri-organotins in water by ion exchange HPLC-GFAA. J Chrom Sci 19:583–593

Jewett KL, Brinckman FE (1983) The use of element-specific detectors coupled with high-performance liquid chromatographs. In Detectors in Liquid Chromatography, ed. T.M. Vickrey, pp 205–241. New York: Marcel Dekker Publishers

Laughlin RB, French W, Johannesen RB, Guard HE, Brinckman FE (1984) Predicting toxicity using computed molecular topologies: the example of triorganotin compounds. Chemosphere 13:575–584

Mastryukova TA, Kabachnik MI (1969) The application of the Hammett equation with the constant σ in the chemistry of organophosphorus compounds. Russ Rev Chem 38:795–811

Morita M, Uehiro T, Fuwa K (1980) Speciation and elemental analysis of mixtures by high performance liquid chromatography with inductively coupled argon plasma emission spectrometric detection. Anal Chem 52:349–351

Morita M, Uehiro T, Fuwa K (1981) Determination of arsenic compounds in biological samples by liquid chromatography with inductively coupled argon plasma-atomic emission spectrometric detection. Anal Chem 53:1806–1808

Nahum A, Horvath C (1980) Evaluation of octanol-water partition coefficients using HPLC. J Chrom 192:315–322

Parks EJ, Brinckman FE, Blair WR (1979) Application of a graphite furnace atomic absorption detector automatically coupled to a high pressure liquid chromatograph for separation of metal-containing macromolecules. J Chrom 185:563–572

Ricci GR, Shepard LS, Colovos G, Hester NE (1981) Ion chromatography with atomic absorption spectrometric detection for determination of organic and inorganic arsenic species. Anal Chem 53:610–613

Scott RPW (1983) Small-bore columns in liquid chromatography. In Advances in Chromatography, eds. J.C. Giddings, E. Grushka, J. Cazes, and P.R. Brown, vol 22, pp 247–294. New York: Marcel Dekker, Inc

Stockton RA, Irgolic KJ (1979) The Hitachi graphite furnace Zeeman atomic absorption spectrometer as an automated element-specific detector for high pressure liquid chromatography: the separation of arsenobetaine, arsenocholine and arsenite/arsenate. Intern J Envir Anal Chem 6:313–319

Taylor LT, Hausler DW, Squires AM (1981) Organically bound metals in a solvent-refined coal: metallograms for a Wyoming subbituminous coal. Science 213:644–646

Tittarelli P, Mascherpa A (1981) Liquid chromatography with graphite furnace atomic absorption detector for speciation of organophosphorus compounds. Anal Chem 53:1466–1469

Tomlinson E (1975) Chromatographic hydrophobic parameters in correlation analysis of structure-activity relationships. J Chrom 113:1–45

Uden PC (1981) Plasma emission spectroscopic detectors in chromatography: a review of a developing field. In Developments in Atomic Plasma Spectrochemical Analysis, ed. R.M. Barnes, pp 302–320. London: Heyden & Son Ltd

Vickrey TM, Buren MS, Howell HE (1978) A simple and inexpensive interface between a liquid chromatograph and a flameless atomic absorption spectrometer detector. Anal Lett A11:1075–1095

Vickrey TM, Eue W (1979) A microprocessor-controlled liquid chromatograph/atomic absorption sampling system. J Automat Chem 1:198–202

Vickrey TM, Howell HE, Paradise MT (1979) Liquid chromatogram peak storage and analysis by atomic absorption spectrometry. Anal Chem 51:1880–1883

Weiss CS, Jewett KL, Brinckman FE, Fish RH (1981) Application of molecular substituent parameters for the speciation of trace organometals in energy-related process fluids by element-selective HPLC. NBS Spec Publ 618:197–210

Weiss CS, Parks EJ, Brinckman FE (1983) Speciation of arsenic in fossil fuels and their conversion process fluids. In Arsenic: Industrial, Biomedical, Environmental Perspectives, eds. W.H. Lederer and R.J. Fensterheim, pp 309–326. New York: Van Nostrand Reinhold Co

Wong PTS, Chau YK, Kramar O, Bengert GA (1982) Structure-toxicity relationship of tin compounds on algae. Can J Fish Aquat Sci 39:483–488

Woolson EA, Aharonson N (1980) Separation and detection of arsenical pesticide residues and some of their metabolites by high pressure liquid chromatography-graphite furnace atomic absorption spectrometry. J Ass Off Anal Chem 63:523–528

The Importance of Chemical "Speciation" in Environmental Processes,
eds. M. Bernhard, F. E. Brinckman, and P. J. Sadler, pp. 685–710.
Dahlem Konferenzen 1986. Springer-Verlag Berlin, Heidelberg
© Dr. S. Bernhard, Dahlem Konferenzen

The Basis for Quality Assurance of Chemical Measurements and Standards

R. A. Velapoldi and H. S. Hertz

Center for Analytical Chemistry
National Bureau of Standards
Gaithersburg, MD 20899, USA

Abstract. The attainment of reliable accuracy and precision in chemical analyses at trace and ultratrace levels is difficult but can be obtained by implementation of a good quality assurance program and by designing and controlling the total analytical measurement process. Quality assurance, the steps in the total analytical measurement process, the impact of these steps on the determination of chemical species, and two case studies, which illustrate many of the requisite procedures to obtain reliable data, are discussed.

Introduction

Analytical measurement technology is used in virtually all industrial sectors, private testing laboratories, and local, state, and Federal laboratories. The data produced are used at all interfaces of the economy: for example, buyer-seller specifications, regulatory systems, monitoring programs, and health care. It is critical that these measurements be accurate to establish a firm basis for the decisions that have to be made on an everyday basis.

To go one step further, to establish a causal relationship between a chemical entity and its effect, e.g., a human health effect, it is frequently essential to know not only the concentration but also the molecular form of that particular species. However, the mere process of sampling, converting the material to a measurable state, or any step in the analytical process may convert the species of interest to a different one.

The sample-analysis interface creates additional havoc with the reliability of measurements for specific chemical species. Although analysts have developed measurement procedures for determining "stable" chemicals in complex matrices, little, if any, data exist for measuring potentially labile chemical species in those matrices. Only through exhaustive research and testing can the analyst have confidence in the analytical results and be certain that the species was not transformed or lost in any of the steps of the measurement process. Thus we will review generic measurement concepts that can be applied to the determination of chemical species. References to specific determinations will be made at the appropriate places.

Major problems arise when attempts are made to determine trace and ultratrace concentrations of elements and compounds (Table 1). Today analysts are required to determine species at lower and lower concentrations; the lower limit for trace analysis has been decreased by a factor of 10^{11} (from $10^{-6}\,g\,g^{-1}$ to $10^{-17}\,g\,g^{-1}$) during the last decade or so (Currie, 1982). Additionally, although trace and ultratrace concentrations of a chemical species may not cause a measurable effect directly, various organisms can concentrate specific chemical species by factors of 10^4–10^6 and concentrations that are toxic to the organism are reached.

Table 1. Typical concentration levels analyzed

	$g_{analyte}/g_{sample}$	%
Major	$>10^{-2}$	>1
Minor	10^{-3}–10^{-2}	0.1–1
Trace	10^{-7}–10^{-3}	0.00001–0.1
Ultratrace	$<10^{-7}$	<0.00001

Compositional measurements are seldom made on an absolute basis. They are usually made on a comparative basis, relying heavily on the calibration/standardization of instruments and the use of matrix-matched reference materials. It is difficult, time consuming, and expensive to produce Standard Reference Materials (SRMs) [called Certified Reference Materials (CRMs) internationally when produced by national laboratories other than NBS] with certified concentrations of trace and ultratrace species. Relative measurements are frequently made using CRMs that are poorly matched to the matrix being examined. The impact of this poor match on the analytical results frequently is not understood and, often, not even considered. Many of the chemicals in the sample may cause interferences in the analytical measurement process for any one specific species. The extent of these interferences frequently is not determined and can differ between sample and reference material even with matrix matching. This is a current and future matter of great concern to environmental studies of species.

Other requirements have to be considered. Qualitative and quantitative data on chemical species must be comparable from laboratory to laboratory. Only then can data acquired in different studies on different populations be meaningfully evaluated.

Many excellent symposia and workshops have been held to define accuracy in trace analysis and to explore measurement methods and reference materials for environmental measurements (see for example Branica et al., 1980; Hertz and Chesler, 1979; Kirchhoff, 1977; LaFleur, 1976). Particular problems arise from the rapid development and use of modern instrumentation and the resulting production of myriad data – often produced without reliable reference materials or validated methods. An excellent review on the principles of environmental measurements has recently been published (Keith et al., 1983) and should be required reading for all analysts before addressing any analytical problem.

The scientist is faced with the difficult task of having to determine accurately species-specific information over wide concentration ranges combined with any number of possible interferents – not an enviable position.

How can the analyst accomplish these tasks? It is critical for the analyst to look at the total analytical measurement process, plan each step, have it under adequate control, and then with an integrated quality assurance program produce data that are reliable with a stated uncertainty for an intended end use. In this paper, we shall discuss briefly the concept of the total analytical measurement process and quality assurance and then discuss each step in the total analytical measurement process. We will also give specific examples of the current state of the art and mention problems yet to be solved in measuring stable chemicals in complex matrices as well as the more difficult area of measuring labile chemical species in complex matrices. Finally, we will give two examples of studies carried out at the National Bureau of Standards (NBS) that illustrate some of the points discussed.

The Total Analytical Measurement Process

The total analytical measurement process (TAMP) is pictorially represented in Fig. 1. It can be seen from the individual steps that it is interdisciplinary in nature, requiring knowledgeable analysts as well as statisticians.

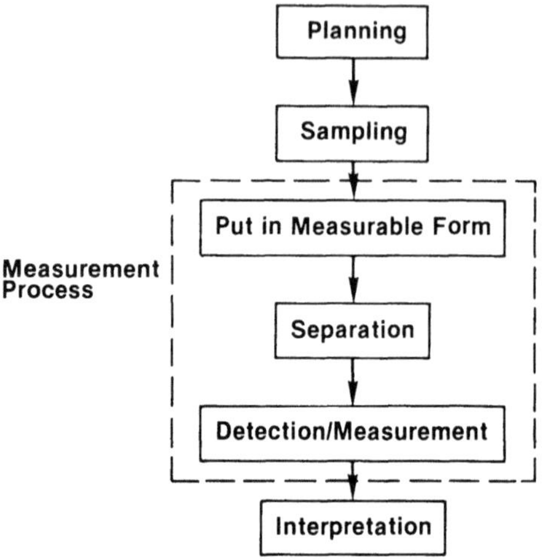

Fig. 1. The total analytical measurement process (TAMP). It should also include sociological and policy decisions but these will not be discussed here (Currie, 1982)

The first step, simplified as planning, includes identifying the measurement problem, designing a model, and establishing precision and accuracy goals. Sampling includes obtaining, storing, and transporting the sample and can be separated from the measurement process.

The measurement process includes getting the sample into a measurable state, separation, and detection/measurement. Getting the sample into a measurable state may be the most difficult step to accomplish. Usually it is more difficult for solids than for liquids or gases. The major question is whether the act of dissolution or extraction of the solid changes the concentrations and/or recovers all of the species in the sample. Separation may be made by physical, chemical, or even spectroscopic means, and detection/measurement involves obtaining an analytical signal and relating that signal to concentration. Interpretation in this section means data reduction and final error/uncertainty assessment. Each of these steps has to have feedback loops so that changes in each step can be made as information on the process increases. Integrated into the process is the quality assurance program required to produce quality data from a total analytical measurement system that is in statistical control.

Quality Assurance

Quality assurance is as critical for scientists who do research and put values on the concentrations of chemical species, thermodynamic parameters, and physical constants as it is for the industrial manufacturing process. The basic objective of quality assurance is to estimate the uncertainty of measured or calculated values. Through the use of a quality assurance program, we, as producers or users of data, have some indication of the quality of the data and can use the data as a solid basis to make informed decisions. The quality of analytical results (Currie, 1982), the principles of quality assurance of chemical measurements (Taylor, 1984 b), and general guidelines for establishing a quality assurance program (Taylor, 1981) have been discussed extensively. These are reviewed below.

Quality assurance has two components (Taylor, 1984 b): quality control and quality assessment. Quality control is the process used to control (reduce) errors. The elements of a quality control program include: good laboratory practices; good measurement practices; standard operating procedures; developed protocols for all steps in the TAMP; qualified personnel with a "measurement" philosophy and attitude; reliable, well-maintained, calibrated/standardized equipment; and reference materials to calibrate, standardize, and test the equipment and process.

Quality assessment is the process used to see how effectively quality control is working. Control charts developed from the measurement of reference materials are an ideal way to assess whether a measurement system is in statistical control. An example of a calibration control chart for a spectrophotometer showing the transmittance of a standard sample versus time is given in Fig. 2 (Mavrodineanu and Baldwin, 1975). Control limits (dashed lines) are set at several σ (usually 2 or 3) from the mean (Taylor, 1981). Data obtained from February 5–18 would certainly be suspect, or even worthless. Although this control chart checks only the instrument calibration in the measurement process, a chemical compositional reference material, homogeneous, complex or spiked samples, or surrogate materials could be measured periodically and a control chart produced by plotting concentration or residuals versus time. This control chart would test the analytical measurement process and the calculated data could be used to determine measurement uncertainty.

Control charts are easy to maintain in laboratories where repetitive analytical measurements are made for a specific analyte in the same matrix. They constitute considerably more work when the analytes and measure-

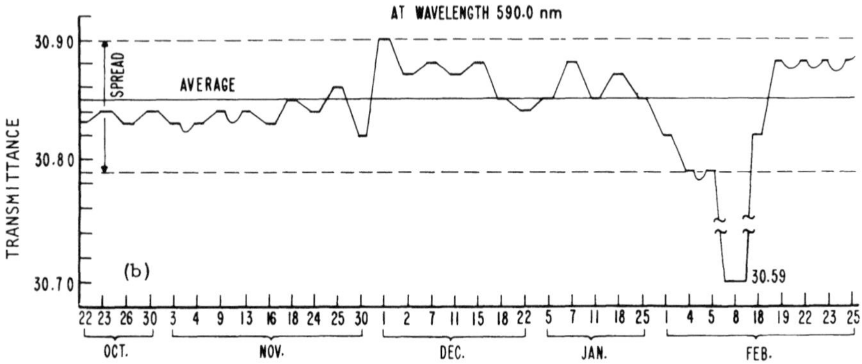

Fig. 2. Control chart for a spectrophotometer showing the variation of transmittance as a function of time for a neutral glass filter at 590.0 nm and 24.0 °C (from Mavrodineanu and Baldwin, 1975)

ment process are continually changing. Nevertheless, well-kept control charts lend confidence to the analyst that the analytical measurement process is in statistical control and reliable data are likely to be obtained.

Interlaboratory test programs assess the capabilities and quality of data from various laboratories. From the data produced in a well-designed interlaboratory test, measures of repeatability (repetitive measurement process), reproducibility (repetitive TAMP), intralaboratory and interlaboratory variabilities, and measurement accuracy are determined for that particular population of laboratories. Examples of interlaboratory tests will be presented under the measurement process section. (Note: Other definitions exist for quality assurance and quality control; see, for example, American National Standard, ANSI/ASQC Document A3-1978, American Society for Quality Control, 161 West Wisconsin Ave., Milwaukee, Wisconsin, 53203).

Planning

Planning is a critical step in the total analytical measurement process, because to plan adequately one must consider each step and make decisions that will affect the work performed. Planning does not guarantee reliable data but does provide the best estimates of the route to it. In addition, planning will give a realistic estimate of the analyst's inventory and capabilities. Taylor (1984b) has described this inventory, presented in Fig. 3. A wide

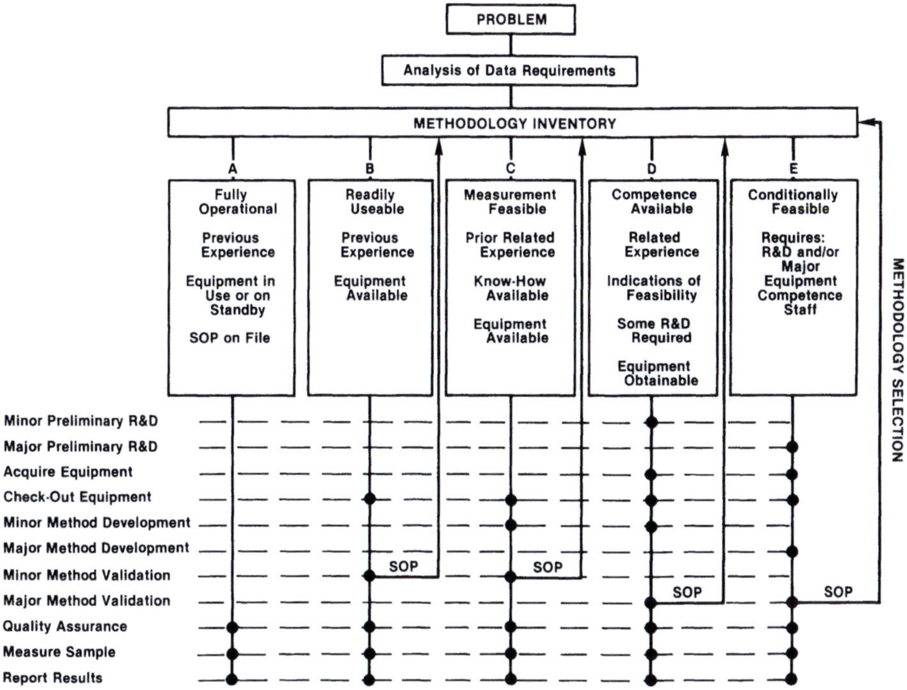

Fig. 3. Steps necessary to achieve reliable measurements considering expertise of personnel and use of specific instruments for an analytical procedure (from Taylor, 1984b); SOP = Standard Operations Procedures)

spectrum of inventories is defined, and the amount of research, method development, and method validation – in essence, the amount of work that has to be done to obtain reliable data – is defined by the particular methodology inventory.

Thus, planning defines the problem and considers the sample, the chemical species to be measured (including its stability), the required precision and accuracy of the data, the analytical method, and the data reduction and interpretation. The quality assurance plan is an integral part of planning. To achieve quality results one has to determine how each step will be accomplished, and what materials/procedures will be used to attain quality control and quality assessment.

In any plan, it is especially important to define what species will be measured. The method you select has to be able to measure the species you define!

Sampling

Too many samples are given to chemists for chemical analysis without thought to the sampling process – which frequently provides a larger source of uncertainty than the measuring process. The sample used for analysis should be sufficiently homogeneous to give requisite precision and at the same time be representative of the material sampled. No analytical result can be better than the sample taken, and no measurement process can improve the outcome for a poor sample. For random errors, the total standard deviation for the total analytical measurement process (σ_t) has contributions from the standard deviations of sampling (σ_s) and the measurement process (σ_m), Eq. 1:

$$\sigma_t = (\sigma_s^2 + \sigma_m^2)^{1/2}. \quad (1)$$

Other models including multiplicative can also be used. A prerequisite for obtaining quality data is that the measurement process be in a state of statistical control (Eisenhart, 1963) (i.e., statistical concepts can be used to define the uncertainty of the measurements). Thus, σ_m can be estimated using reference or homogeneous materials, σ_t can be estimated by carring several samples from the same "population" through the TAMP, and σ_s can be calculated from Eq. 1. If either σ_m or σ_s is larger than the other by a factor of three or more, that standard deviation is the major contributor to σ_t and to the uncertainty of the process. If σ_t is not within the uncertainty goals set under planning, the standard deviation of the larger step should be reduced. If reduction cannot be accomplished, then the original goals have to be revised. Kratochvil and Taylor (1981) have extensively reviewed sampling for chemical analysis and have also provided a bibliography for sampling (1982).

An ideal situation would be to have a sensitive, selective analytical technique that could be used in the field to obtain reliable and accurate data. In some cases, such as electrochemical detection, this has been accomplished (Frant and Oliver, 1980; Koryta and Stulik, 1983). Attempts to miniaturize instruments (putting a Fourier transform-infrared spectrophotometer in a "shoe box" or an ultraviolet-visible spectrophotometer and gas chromatograph on a chip) are making great strides (Callis, 1984). The development of chemical sensors to accomplish analysis in the field is progressing (Seitz, 1984). However, field samples are generally brought to the laboratory for more sophisticated analysis. Questions to be addressed before sampling in the field or laboratory include: where, when, and how to take samples; how many, how much, and what type of samples (gas/liquid/solid/combination or random/composite/stratified) should be taken; what size particles are to be included/excluded; what constitutes foreign bodies and what to do with them; how to label, stabilize, store, and transport the samples; and finally, what happens to the chemical species in the sampling and transporting steps. Many of these questions have been discussed (Currie, 1982; Kratochvil and Taylor, 1981; Moody, 1982b, 1984). Generic guidelines may be developed and are useful in developing specific protocols for each sample type. Questions concerning the contamination (positive or negative) of the sample during the sampling, storage, or handling steps must also be addressed (Currie, 1982; Koch et al., 1984; Kratochvil and Taylor, 1981; Moody, 1982b, 1984; Nurnberg, 1982). Four general areas have to be considered to reduce contamination: containers and apparatus, the reagents, the chemist, and the environment. Apparatus and containers for sampling and storage must be scrupulously cleaned, transported, and stored without being contaminated, and must consist of the right material so that contamination from the walls or loss by adsorption onto the walls does not occur. The sampling containers should be equilibrated in the medium to be sampled so that no adsorption losses to the "clean" surface occur (Moody, 1982a). Purified reagents should be used during any chemical handling step. Scientists at NBS have produced nonebullient distilled reagents (Moody, 1982a) with concentrations of potential contaminant factors of 10 below those found in typical Reagent Grade chemicals (Moody, 1984). This is extremely important when there is a preconcentration step in the procedure. Reagent Grade chemicals must be checked to keep blank levels low. For trace and ultratrace analyses the analysts should wear protective clothing to minimize contamination from, e.g., hair, skin flakes, clothes, fibers, and perspiration – and must refrain from coughing, sneezing, and smoking when handling samples.

Airborne particles in the sampling environment and in the laboratory are a major source of contamination. Moody has reviewed the design and use of clean laboratories (1983) and has stated that typical laboratory air contains about 3×10^7 particles in the 0.5 to 100 μm size range per cubic meter of air (Moody, 1982b).

It is generally accepted that the use of clean laboratories, purified reagents, clean containers and apparatus, and personal care are all necessary to control the analytical blank and reduce contamination for trace and ultratrace analyses (Moody, 1982a). While these parameters can be controlled in the laboratory, they are frequently difficult to control in the field. Sampling should be accomplished as cleanly as possible, and the level of the sampling blank should always be determined, generally employing statistical blank correction procedures (Taylor, 1984b). Appropriate corrective procedures should be used if the blank is too high.

Measurement Process

In the general case, as presented in Fig. 4, quality assurance plays a major role in determining whether reliable, precise, and accurate data will be obtained. The first step is to determine whether the measurement process is in statistical control (at the interpretation step using data evaluation, con-

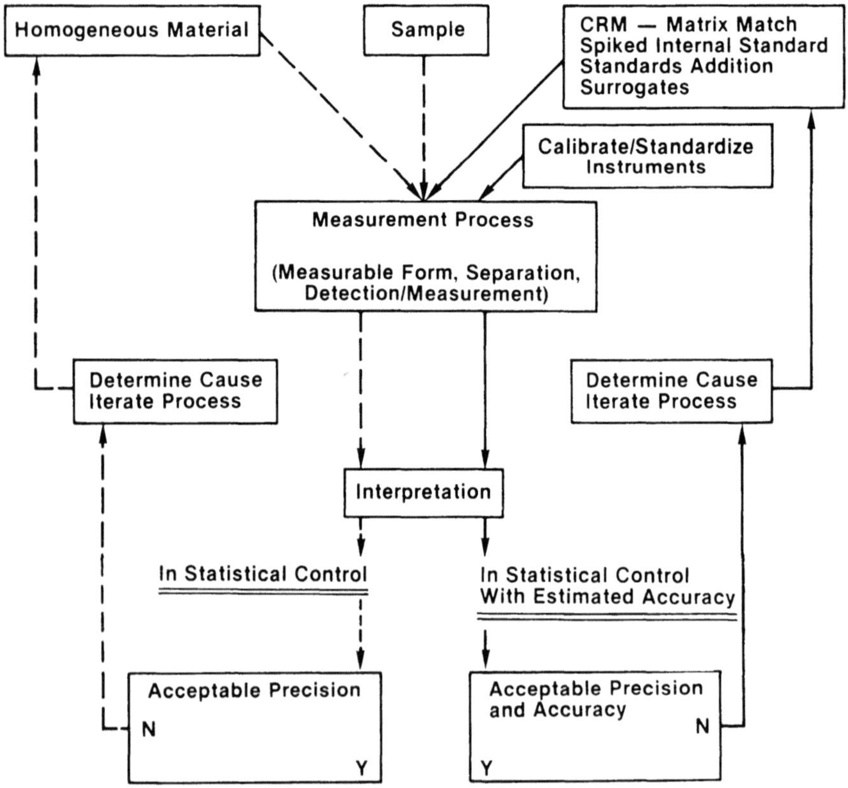

Fig. 4. Methods for determining whether the measurement process is in statistical control and for estimating measurement accuracy

trol chart averages; see Eisenhart, 1963) and whether the precision is sufficient for the goals of the project. The precision of the measurement process is determined by following the pathways delineated by either the dashed or solid lines and using any of the materials (assuming the sample is homogeneous). The accuracy of the measurement system is inferred by following the solid line pathways. Certified Reference Materials (CRMs) are well characterized, homogeneous materials with certified chemical compositions (Alvarez et al., 1982) that are well suited, albeit sometimes expensive, for determining the precision of a measurement process. Other materials (e.g., samples, surrogates) can also be used provided that they are homogeneous and stable. If they are not, a precision that is a combination of the sample inhomogeneity and measurement process variability will be determined.

If the precision is not acceptable, then an iterative process has to be undertaken to identify the cause(s) of imprecision. If the requisite precision is still not attainable, a different technique or method has to be chosen and the same tests run. Although precision and accuracy are both goals, precision is a prerequisite, since it demonstrates that the measurement process is in statistical control and efforts at achieving accuracy are meaningful.

The second step is assessing the accuracy of the measurement process. First, the physical scales of the instruments and related apparatus (e.g., absorbance, temperature, flow) should be calibrated – usually periodically, but definitely before and after critical measurements – to check system stability. Reference Materials should be used to standardize instrument response as a function of compound concentration (see Keith et al., 1983, for discussion of standardization and calibration). While the measurement process is in a state of statistical control, a Certified Reference Material of the same or similar matrix as the sample is analyzed as an unknown to the participating laboratories and an estimate of the accuracy of the measurement process is made. A decision is then made as to whether the objectives of precision and accuracy, established in the planning portion of the TAMP, can be met.

If the accuracy objective is not met, the steps in the measurement process have to be studied by an iterative process such that reasons for the inaccuracy are defined and corrections made until the accuracy objective is met (or changed).

A planned analytical measurement scheme consisting of, e.g., CRM, samples, CRM, samples, CRM (or control materials for the CRM) may then be carried out with the confidence that the measurement process will yield measured values with definable precision and accuracy. Even then, however, different levels of accuracy may be attained. If the CRM has a similar or even the same matrix as the sample, a statement can be made as to the accuracy of the measured value for the sample. However, one must always realize that use of a similar matrix does not guarantee that the same interferences/problems/precisions will occur to the same extent. If the matrix is similar, statements of accuracy can be made, but the uncertainty in the measured value will be larger because of potential interferents from the matrix. If the matrices are quite different, few (if any) reliable inferences of accuracy of the measured value can be made.

The recognition of the need for and the production of CRMs for biological and other complex matrices have increased greatly in the last two decades. Expensive, exhaustive, and time-consuming research has to be performed to certify the trace elements or compounds in these matrices. There has been a dramatic increase in the production of Standard Reference Materials (SRMs) for use in the areas of clinical, biological, botanical, and environmental measurements (Alvarez et al., 1982). Although methods exist for determining As(III) and As(V) in aqueous solution, no SRM has been produced with certified concentrations of inorganic chemical species, undoubtedly due to possible changes in element oxidation states upon dissolution with acids (Diamondstone and Burke, 1977). [A more definitive listing can be found in the new SRM catalog (NBS Standard Reference Materials Catalog, 1984–1985).]

Assuming that the analyst can attain statistical control for the measurement process, what can the analyst do if no CRMs are available to assess accuracy? An idealized accuracy-based system is given in Fig. 5. In general, the analytical technique used by the analyst is not a definitive method. Definitive methods are tied to basic SI units, are highly precise, and have been exhaustively studied to identify, eliminate, reduce, or correct identifiable sources of bias (Uriano and Gravatt, 1977). By elimination of bias, the accuracy can truly be tied to the measurement precision. Because of this, definitive methods are considered to be of the highest accuracy. Unfortunately, they are usually extremely expensive with respect to money, time, instrumentation, and manpower. Several definitive methods [isotope dilution/mass spectrometry (ID/MS), gravimetry, coulometry] have been developed at the National Bureau of Standards (NBS) to determine the concentrations of trace organic and inorganic compounds in complex matrices (Table 2).

Fig. 5. An idealized accuracy-based measurement system used to develop reference methods, (From Alvarez et al., 1982; for more detailed description see Uriano and Gravatt, 1977).

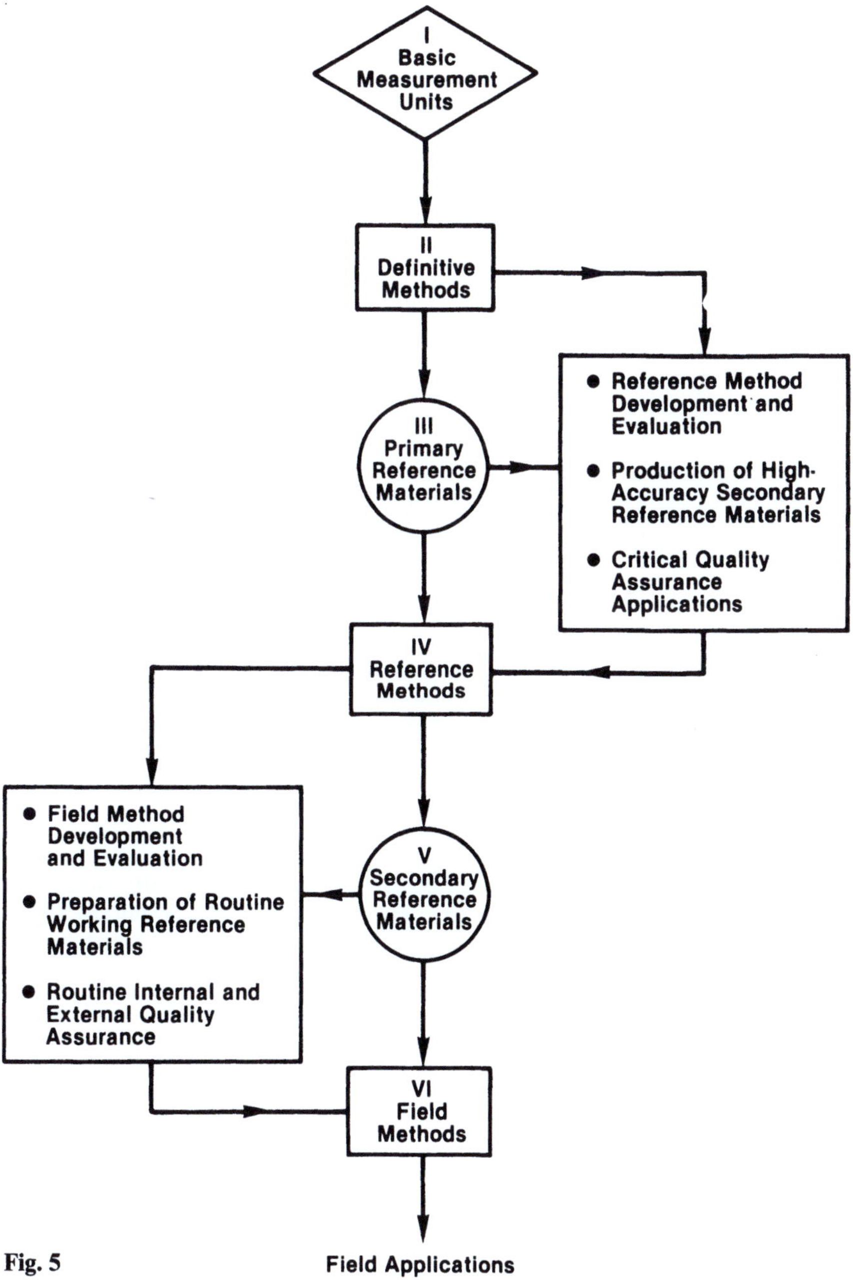
I
Basic
Measurement
Units
II
Definitive
Methods
III
Primary
Reference
Materials
• Reference Method Development and Evaluation
• Production of High-Accuracy Secondary Reference Materials
• Critical Quality Assurance Applications
IV
Reference
Methods
• Field Method Development and Evaluation
• Preparation of Routine Working Reference Materials
• Routine Internal and External Quality Assurance
V
Secondary
Reference
Materials
VI
Field
Methods
Field Applications

Fig. 5

Table 2. Definitive analytical methods developed/applied at NBS

Procedure	Accuracy estimate[a]	Element/ compound/area	Matrix	Reference (representative)
ID/MS[b] (organic)	<1.0%	Glucose[c], cholesterol, urea	Serum	Velapoldi et al. (1978)
ID/MS[b] (inorganic)	<0.5%	Li, B, Mg, S, Cl, K, Ca, V, Cr, Fe, Ni, Cu, Zn, Ga, Br, Rb, Sr, Mo, Rh, Ag, Cd, I, Cs, Ba, La, Nd, Sm, Hf, W, Re, Tl, Pb, Th, U, Pu	Metals, salts, biologicals, botanicals, liquids	Barnes et al. (1982)
Coulometry	<0.01%	Redox (U, Cr, As) Acidimetry, Alkalimetry, Compleximetry	Oxides, acids, bases	Marinenko et al. (1983)
Gravimetry	<0.2%	Na, F, Cu	Serum, minerals, pure salts, gases	Velapoldi et al. (1980)

[a] The values given actually represent estimated inaccuracy at the 95% confidence limits: <0.1% for well-defined inorganic systems
[b] ID/MS = isotope dilation/mass spectrometry
[c] Methods developed for creatinine and uric acid – not published. Method under development for triglycerides

Definitive methods are developed to be precise and accurate but usually are not designed to have high sample throughput. We have used them to certify SRMs, determine target value concentrations in transfer standards for proficiency testing, and develop/test reference methods. Thus, because of the resource requirements, the analyst will not develop a definitive method to ensure accurate measurements in a particular measurement program.

If a reference method (step IV in Fig. 5) has been developed for a particular chemical species in a particular matrix, then the analyst, after "practice," can use the reference method and expect comparable precision and accuracy assuming that statistical control and an appropriate quality assurance program are incorporated. Many examples of reference method development have been published and one developed at NBS (Velapoldi et al., 1980) is discussed below. The reference method was developed with the help of a panel of experts in accord with the accuracy-based measurement system concept and tested by interlaboratory exercises.

The lithium concentrations in several homogeneous serum pools were determined precisely and accurately by a definitive method. The analyzed pools then were used as unknown samples to test the "candidate" reference method. Solutions of NBS SRM 924 (lithium carbonate) were prepared and used to standardize the instrument response. Several laboratories analyzed the serum pools using the candidate reference method and the measured lithium concentration values were used to determine the precision and accuracy components. The results for two protocols (manual and semiautomated pipetting) are summarized in Table 3. The components of variability were calculated; only σ_{total} is given in which total refers to the uncertainty expected by a single laboratory running the complete tested protocol.

Table 3. Summary of imprecision and bias results in mM for lithium in serum, final interlaboratory exercise

Li level	Estimated imprecision		
	Manual pipetting protocol σ_{total}	Goal	Semiautomated pipetting protocol σ_{total}
0.5	0.008	0.1	0.005
1.0	0.014	0.1	0.010
1.8	0.028	0.1	0.017
2.9	0.047	0.1	0.029

Li level	Bias		
	Manual pipetting protocol Interlaboratory exercise composite bias $(X_{obs}-X_{DM})$[a]	Goal	Semiautomated pipetting protocol Interlaboratory exercise composite bias $(X_{obs}-X_{DM})$[a]
0.5	−0.012	±0.2	−0.012
1.0	−0.008	±0.2	−0.012
1.8	−0.030	±0.2	−0.036
2.9	−0.016	±0.2	−0.033

[a] Li concentrations for method (X_{obs}) and for Definitive Method (X_{DM}).

Although there is a small negative bias for the reference method, both precision and accuracy values were well within the goals set up in the planning step by a panel of experts.

If reference methods or materials are not available, the analyst could use two or more independent methods with independent analytical techniques (with appropriate quality control) such as high-performance liquid chromatography (HPLC), gas chromatography (GC), and gas chromatography/mass spectrometry (GC/MS) to infer accuracy. NBS frequently uses this approach to certify the chemical composition of SRMs.

The use of several independent methods (or even a definitive method, for that matter) does not guarantee accurate results. While consistency among independent methods and different laboratories is an indication of accuracy, it may not be sufficient. However, with careful investigations of the steps in the procedure, with a quality assurance program, and with agreement of the results from the independent methods and laboratories, the analyst can infer with some confidence that the measurement systems are under control and that there are no major biases in the measured chemical composition value.

Multilaboratory analysis with the same or different techniques leading to consensus values frequently has been used as a means of determining accuracy. However, such a study must have good matrix-matched reference materials and an effective quality assurance program to be able to make some accuracy statement about the consensus value. In general, however, this process will usually lead to significant uncertainty in the measured value, and thus to less accuracy compared to the other procedures because of the between-lab variability component. There is no doubt, however, that accuracy is attainable, requires a good quality assurance program, and takes effort and commitment.

The first step of the measurement process, getting the sample into a measurable state, often provides a barrier and challenge to obtaining accurate determinations of the concentrations of specific chemical species in complex samples. If the sample is a liquid or a gas, several procedures exist for performing direct analysis of the components. If necessary, the sample may be put through a preconcentration step; the species then may be directly determined, e.g., by electrochemical techniques, or can be separated by some form of chromatography with the effluent being fed directly into some measuring instrument (see, e.g., Olsen et al., 1983; Zeisler et al., 1983; Appendix III). Procedures for determining V(IV) and V(V), As(III) and As(V), Se(IV) and Se(VI), and the "neutral" complexed forms of these metals have been developed for a water matrix while vanadium species have been determined in blood and arsenic species have been determined in biological matrices (Iyengar et al., 1978). For the determination of trace organics in viscous matrices such as crude or shale oils, simple dilution or

HPLC fractionation can be used (May et al., 1981; Rebbert et al., 1984). New techniques such as supercritical fluid chromatography may provide additional quantitative, simultaneous dilution/dissolution and separation of organic components from complex samples.

A major problem arises when the sample is a solid. Several possibilities exist for getting organic species into a measurable form. Usually one must extract the compounds from the matrix without changing chemical forms of the analytes. There is no guarantee that all of the analyte has been removed from the matrix. Furthermore, quantitative recovery of spiked internal standards does not guarantee equivalent recovery of natively present compounds, since the internal standards are not necessarily bound to the matrix in the same manner as the naturally occurring analytes. Frequently analysts will conduct extractions with two different extraction solvents and infer accuracy (with a degree of confidence) if equivalent results are obtained. The "specter" of incomplete recovery still exists, possibly due to species distribution.

Organometallic compounds have been extracted successfully from several matrices. This is undoubtedly due to their stability and the mild conditions that were used. The NBS has produced some pure organometallic standards (1984–1985). Although some measurement expertise exists, few efforts have been directed towards the certification of these materials in complex matrices.

The determination of inorganic species in solid, complex matrices poses more difficult questions. Essentially nondestructive procedures, such as neutron activation analysis, provide excellent methods for determining elemental concentrations but do not provide information on the actual species present. The classical use of oxidizing acids and heat dissolution gives accurate values for total elemental concentrations but no information about the type of chemical species which were in the sample. How do we ensure the stability or efficacy of the elemental oxidation states and/or coordination geometries upon dissolution? One way may be cryogenic grinding (as discussed below in the Pilot Environmental Specimen Bank Program) of the sample to exceedingly fine particles, which increases the surface area enormously. Using chelating agents or supercritical fluid chromatography (e.g., using NH_3 as fluid), can we then quantitatively extract or separate the species and not disturb the ratios of elemental oxidation states? Research needs to be done to assess these possibilities.

The use of pulse distribution analysis rather than sparking analytical conditions may offer some advantage for studying particles in solid samples in spark source emission spectrometry (Lukas, 1983).

Care must be taken in using chemicals in complex matrices to calibrate instruments. Interferences of other constituents in the matrix can cause a signal-concentration response relationship that is different from that produced by a pure chemical calibrant. It is generally accepted that a close matching of the concentration of species in the sample with their concentrations in the reference materials gives analytical signals that yield the most accurate results. Currie (1982) reviews extensively the use of sound statistical methods to combine, compare, and interpret data from the point of view of the analyst.

There are pitfalls that must be considered, however, to obtain accurate results in trace and ultratrace analyses from an analytical signal. These considerations are: the limit of detection (LoD), the limit of quantification (LoQ), the measurement of the blank, and interferences. The importance of the analytical blank has been discussed extensively (Currie, 1982; Taylor, 1984a), and variations due to contamination have to be eliminated before the definitions of LoD and LoQ are meaningful. Care in sampling and the use of clean rooms and pure reagents play major roles in decreasing the blank. The LoD has been defined as the "lowest concentration level that can be determined to be statistically different from the blank" (Keith et al., 1983) and is usually defined as $\geqq 3\sigma$ higher than the blank mean. Discussion of the LoD has been done in terms of detection theory and hypothesis testing (Currie, 1982), and the assumptions underline the necessity for interference and bias-free measurements to obtain the lowest detection limits. The LoQ is defined to be $\geqq 10\sigma$ higher than the blank mean and defines the lower limit of the useful range of the measurement process.

Interferences consist of two types – those inherent in complex samples (e.g., arising from co-elution of two compounds in a chromatogram) and contaminants added during sample handling and measuring. Reduction of contamination as previously discussed addresses the latter. Modifying the methodology usually addresses the former (e.g., change elution solvents, chromatographic conditions, or even element specificity of detectors; see Irgolic and Brinkman, this volume). Chromatographic peak elution monitoring as a function of time with techniques such as Fourier transform infrared spectroscopy or mass spectrometry also can identify two co-eluting components.

Interpretation of DATA

The interdisciplinary nature of most analytical problems requires at a minimum consultation with a statistician and, ideally, direct involvement of the statistician in planning, quality assurance, and data manipulation/interpretation. Discussion of analytical problems and the total analytical measurement process with statisticians generally leads to a better design of the study, data manipulation, and final data interpretation.

Statistical treatment of the data is important for deducing the a) analyte concentration (measured value), b) the random components of uncertainty in measurement, c) bias, and d) an overall uncertainty in the measured value. Depending on the statistical model used and the nature of the data, the estimates of uncertainty may be asymmetric about the measured value.

In summary, the progression from a general analytical model in which the gross analytical signal y(E) is a function of various experimental parameters plus an error term

$$y(E) = f(B; x_1 ---- x_n; E) + e(E) \quad (2)$$

(in which E represents a controllable experimental variable, B is the blank, x_1, ---- x_n are concentrations of analytes, and e is the random error component) to a single, observable reading

$$y_i = B_i + x_1 A_{1i} + e_i \quad (3)$$

(in which x_1 is the net signal for the analyte of interest and A_{1i} is the calibration factor for x_1) is fraught with chemical and statistical pitfalls (Currie, 1982). Care must be taken to avoid them.

Two Illustrations of the Total Analytical Measurement Process

Two studies give excellent examples of many of the points discussed in this paper.

Trace Elements in Chesapeake Bay

This was a joint interdisciplinary study with the States of Maryland and Virginia and the U.S. Environmental Protection Agency (EPA) concerned with measuring toxic trace elements in the Chesapeake Bay (Kingston et al., 1983), and was part of a larger study to assess the condition of the Bay. The study consisted of the four major phases presented in TAMP, Fig. 1. One hundred and two samples were obtained: one each at the top and bottom of the water column at 51 locations in the Bay. The samples and blanks were filtered to separate the samples into particulate material and aqueous samples. Fifteen trace elements at the 10^{-6} to 10^{-12} g g^{-1} concentration levels were determined in the 204 analytical samples (102 aqueous and 102 particle). More than 5000 elemental concentrations were determined for samples, blanks, and standards. The filtered, aqueous samples were preconcentrated on chelating resin (Kingston et al., 1978) to reduce the blank and interferences of the alkali, alkaline-earth, and halogen elements in the analytical techniques used (neutron activation analysis – NAA, isotope dilution/mass spectrometry – ID/MS, voltammetry – VOL, and graphite furnace atomic absorption spectrometry – GFAAS). A procedure was developed to determine the trace elements directly on the chelating resin by NAA (Greenberg and Kingston, 1983). Exhaustive method development and research were performed and quality assurance was demonstrated by carrying

Table 4. Trace elements in water – SMR 1643a analyzed using chelation and NAA, GFAAS techniques. (From Kingston et al., 1983)

Element	Total element concentration [ng g^{-1}]		
	This work[a]		Certified[b]
	NAA	GFAAS	
Cd	10.1 ± 0.5	11 ± 2	10 ± 1
Co	19 ± 1	–	19 ± 2
Cr	16 ± 2	–	17 ± 2
Cu	19.1 ± 0.6	16 ± 4	18 ± 2
Pb	–	28 ± 4	–
Fe	88 ± 16	–	88 ± 4
Mn	30.9 ± 0.6	–	31 ± 4
Mo	97 ± 6	–	95 ± 6
Ni	56 ± 8	52 ± 7	55 ± 3
V	52 ± 1	–	53 ± 3
Zn	68 ± 5	–	72 ± 4

[a] Uncertainties are 2s.
[b] Uncertainties are at the 95% confidence limit.

samples of a Standard Reference Material (SRM 1643 a, Trace Elements in Water) through the measurement process to establish statistical control (determine standard deviation values) and assess accuracy (agreement with the SRM certified concentration values). The results of these tests are summarized in Table 4. Blank analyses and rigorous statistical analysis of the blank data were carried out. The effects of the blanks on the reported concentrations were also discussed extensively.

The absolute concentrations of elements in the particulate samples showed extensive variability (more than two to three orders of magnitude) because of differences in total particle loading from the various sampling locations. Interpretation of the data was difficult; also, similar to previous studies, the meaning of the highly variable results was unclear. However, use of elemental ratios (used extensively in geochemistry), incorporating one nonanthropogenic element such as Ce or Sc, gave results from which reliable baseline data and trends could be identified.

Results of Ni analyses for the 102 particle samples at the 51 sampling locations using the crustal enhancement normalization technique shows increased surface Ni concentrations relative to the bottom sample. These results suggest a pollution source from a specific tributary. Cobalt, Mn, Zn, and Cu concentrations also show the same pattern as Ni.

Pilot National Environmental Specimen Bank

The second study consisted of establishing a Pilot Environmental Specimen Bank at NBS (Zeisler et al., 1983) with the support of the EPA. Using stored biological specimens, the study has the objectives of 1) detecting changes in the environment, 2) separating natural pollution sources from anthropogenic sources, and 3) providing a bank of well-preserved and -characterized environmental samples for retrospective analysis. (An additional result from this effort is that the analyst is challenged and develops new analytical methods that can be used in a generic sense.) Extensive planning and protocol development for all aspects of this integrated systems study included sample selection, storage, and analysis (see appropriate sections, Zeisler et al., 1983). To obtain noncontaminated samples chemists visited the hospitals, gave instructions and specific sampling and storage protocols to the staff who would take the samples, and provided equipment which included: non-talced gloves; special titanium scalpels (rather than stainless steel which would contaminate the sample with Cr, Ni, etc.); purified water; and Teflon sheets/bags and containers for cryogenic storage and shipping.

In sample storage and preparation for analysis, extensive studies were carried out to demonstrate protocol efficacy. For example, reduction of the bulk sample to a size suitable for the chosen analytical procedure was an important step in the TAMP. Cryogenic grinding of the approximately 150 g bulk sample was performed using Teflon mills to produce a homogeneous material so that representative 1 g subsamples could be used for analysis. The requirement was that the variabilities due to the grinding (inhomogeneity) and the analytical technique should not exceed 1 percent at 1σ. Using ^{24}Na as a tracer, it was found that cryogenic grinding by a Teflon ball mill did not achieve this goal. A Teflon disc mill was designed, built, and tested, and met the homogeneity goal.

Elemental analyses were carried out by NAA where feasible because of the sensitivity, small sample-size requirements, and elemental coverage. Analyses were also carried out by other techniques including inductively coupled plasma spectrometry, atomic absorption spectrometry, ID/MS, voltammetry, and X-ray fluorescence. Biological or botanical SRMs were used as control materials in the quality assurance program. The data from this study demonstrated good precision. Average values and ranges for the essential trace elements were generally lower than those reported by Iyengar (1978), probably indicative of better control of blank and contamination. A point-by-point comparison of the values produced for the elements using the various analytical techniques showed no apparent biases of the results, giving confidence that the values reported are reliable and can be used to accomplish the objectives of the study.

Analyses for organochlorine residues (pesticides) were carried out by GC and GC/MS using methods based on internal standard spiking. Striking differences were seen in the pesticide residues in samples from the USA and the Federal Republic of Germany, probably a direct result of different pesticide-use patterns. The drawing of comparative conclusions was made possible by use of a good quality assurance program, which included interlaboratory comparisons between the American and German laboratories.

It was mentioned previously that programs such as the Specimen Bank can involve analytical challenges. For example, Zeisler and Greenberg (1983, Appendix II) developed an NAA scheme based on radiochemical separation for the determination of Pt and Au in liver samples at the 10^{-11} g g^{-1} level, the lowest concentration determined to date on these types of samples. Previous analyses of Pt were limited by the analytical blank and incomplete separation of gold and platinum. However, use of the NAA technique with post-irradiation chemistry provided a blank-free analytical step with requisite separation and concentration determinations that were not possible before. The development of this method provides an accurate procedure and permits certification of biological SRMs for assessing the concentration of platinum in biological matrices, giving support to the research now being performed on the measurement of the species derived from *cis*-$Pt(NH_3)_2Cl_2$ and related drugs being used in cancer chemotherapy.

Conclusions

By integrating a credible quality assurance program with the total analytical measurement process, the analytical chemist can effectively handle the expanding requirements for the analysis of trace compounds in complex samples. Following these concepts, work at the National Bureau of Standards has effectively increased the accuracy in the analysis of trace inorganic elements, organic species, and selected organometallic species in complex samples. This same concept must now be applied more widely to the determination of inorganic species. Accurate analytical procedures and standards for species must be developed to include the analysis of the oxidation states of elements and species in complex materials.

Only through a concerted effort, in which needs and analytical information are supplied to National Laboratories by associations, scientists, and organized conferences such as this one, can the scientific community hope to have available the relevant Certified Reference Materials for species. Reference materials will provide precision and accuracy in critical analytical measurements and provide a sound basis for safe and cost-effective regulatory, health, and industrial decisions.

Acknowledgements. The authors gratefully acknowledge L. A. Currie and J. K. Taylor for excellent discussions and suggestions.

References

(Useful reviews for new readers in this field are marked with an asterisk.)

Alvarez R, Rasberry SD, Uriano GA (1982) NBS Standard Reference Materials: Update 1982. Anal Chem 54:1226A–1244A

Barnes IL, Murphy TJ, Michiels EAI (1982) Certification of lead concentration in Standard Reference Materials by isotope dilution mass spectrometry. J Ass Off Anal Chem 65:953–956

*Branica M et al. (1980) Organizing Committee of the Chemistry of the Mediterranean. Sixth International Symposium, Rovinj, Yugoslavia

Callis JB (1984) A miniature spectrophotometer for process analytical chemistry. Anal Chem (and personal communication)

*Currie LA (1982) Quality of analytical results, with special reference to trace analysis and sociochemical problems. Pure Appl Chem 54:715–754

Diamondstone BI, Burke RW (1977) Some difficulties encountered in speciation studies of arsenic. Analyst 102:613–614

*Eisenhart C (1963) J Res NBS (USA) 67:161–187

Frant MS, Oliver RT (1980) Process analytical measurements. Anal Chem 52:1252A–1262A

*Gallorini M, Orvini E, Sabbioni E, Pietra R (1984) Contribution of neutron activation analysis and radioanalytical techniques to trace element speciation in environmental analysis. In Proceedings of the Fifth International Conference on Nuclear Methods in Environmental and Energy Research, Mayaguez, Puerto Rico, April 2–6, 1984, ed. J.R. Vogt, pp 141–153. U.S. Dept. of Energy Technical Information Center, Conf-840408

Greenberg RA, Kingston HM (1983) Trace element analysis of natural water samples by neutron activation analysis with chelating resin. Anal Chem 55:1160–1165

Hertz HS, Chesler SN (eds) (1979) Trace Organic Analysis: A New Frontier in Analytical Chemistry. Proceedings of the 9th Materials Research Symposium, Gaithersburg, MD, 1978. NBS Spec Publication 519

Iyengar GV, Kollmer WE, Bowen HJM (1978) The Elemental Composition of Human Tissue and Body Fluids. Weinheim, New York: Verlag Chemie

Keith LH, Crummett W, Deegan J Jr, Libbey RA, Taylor JK, Wentler G (1983) Principles of environmental analysis. Anal Chem 55:2210–2218

Kingston HM, Barnes IL, Brady TJ, Rains TC, Champ MA (1978) Separation of eight transition metals from alkali and alkaline earth metals in estuarine and seawater with chelating resin and their determination by graphite furnace atomic absorption spectrometry. Anal Chem 50:2064–2070

Kingston HM, Greenberg RR, Beary ES, Hardas BR, Moody JR, Rains TC, Liggett WS (1983) The Characterization of the Chesapeake Bay: A Systematic Analysis of Toxic Trace Elements. NBSIR 83-2698. Gaithersburg, MD: National Bureau of Standards

*Kirchhoff WH (ed) (1977) Methods and Standards for Environmental Measurements. Proceedings of the 8th Materials Research Symposium, Gaithersburg, MD, 1976. NBS Spec Publication 464

Koch WF, Marinenko G, Wu YC (1984) The development of reference materials for acid rain research. Envir Int, in press

Koryta J, Stulik K (1983) Ion Selective Electrodes, 2nd ed. Cambridge: Cambridge University Press

Kratochvil BG, Taylor JK (1981) Sampling for chemical analysis. Anal Chem 53:924A–938A

Kratochvil BG, Taylor JK (1982) A Survey of the Recent Literature on Sampling for Chemical Analysis. NBS Tech Note 1153. Gaithersburg, MD: National Bureau of Standards

*La Fleur PD (ed) (1976) Accuracy in Trace Analysis: Sampling, Sample Handling, Analysis. Proceedings of the 7th Materials Research Symposium, Gaithersburg, MD, 1974. NBS Spec Publication 422

Lukas M (1983) Development of a Particle Size Independent Method for Wear Metal Analysis. International Oil Analysis Symposium, Pensacola, FL, May 1983

Marinenko G, Koch WF, Etz ES (1983) High precision coulometric titration of uranium. J Res NBS (USA) 88:117–124

Mavrodineanu R, Baldwin JR (1975) Glass Filters as a Standard Reference Material for Spectrophotometry – Selection, Preparation, Certification, Use SRM 930. NBS Spec Publication 260-51. Gaithersburg, MD: National Bureau of Standards

May WE, Brown-Thomas JM, Chesler SN, Guenther FR, Hilpert LR, Parris RM, Richie KL, Wise SA, Hertz HS (1983) Interlaboratory comparisons of quantitative analyses of individual compounds in simple and complex mixtures. In Advanced Techniques in Synthetic Fuels Analysis, eds. W.C. Weiner and W.D. Felix, pp 381–404. U.S. Dept of Energy Technical Information Center

May WE, Brown-Thomas JM, Hilpert LR, Wise SA (1981) The certification of selected polynuclear aromatic hydrocarbons in Standard Reference Material 1580 "Organics in Shale Oil". In Polynuclear Aromatic Hydrocarbons: Fifth International Symposium on Chemical Analysis and Biological Fate, pp 1–16. Columbus, OH: Battelle Press

May WE, Chesler SN, Hertz HS, Hilpert LR, Rebbert RE, Vogt CR, Wise SA (1984) Characterization of polycyclic aromatic hydrocarbons in air particulate extracts by liquid/gas chromatographic methods. In Identification and Analysis of Organic Pollutants in Air, ed. L.H. Keith, pp 197–230. Boston: Butterworth Publishers

Moody JR (1982a) NBS clean laboratories for trace element analysis. Anal Chem 54:1358A–1376A

Moody JR (1982b) The sampling, handling, and storage of materials for trace analysis. Phil Trans Roy Soc Lond A 305:669–680

Moody JR (1983) Sampling and storage of materials for trace elemental analysis. Trends Anal Chem 2:116–118

*Moody JR (1984) Sampling, storage, and handling of materials for trace metal analysis. In Sample Preparation Technology, vol 3, in press

NBS Standard Reference Materials Catalog (1984–1985) Office of Standard Reference Materials, Chem. B311, National Bureau of Standards, Gaithersburg, MD

Nurnberg HW (1982) Voltammetric trace analysis in ecological chemistry of toxic metals. Pure Appl Chem 54:853–878

Olson GJ, Brinckman FE, Jackson JA (1983) Purge and trap flame photometric gas chromatography technique for the speciation of trace organotin and organosulfur compounds in a Human Serum Standard Reference Material (SRM). Intl J Envir Anal Chem 15:249–261

Rebbert RE, Chesler SN, Guenther FR, Parris RM (1984) Liquid chromatography-gas chromatography procedure to determine the concentration of dibenzothiophene in a crude oil matrix. J Chrom 284:211–217

Seitz WR (1984) Chemical sensors based on fiber optics. Anal Chem 56:16A–33A
*Taylor JK (1981) Quality assurance of chemical measurements. Anal Chem 53:1588A–1596A
Taylor JK (1984a) Guidelines for evaluating the blank correction. J Test Eval 12:54–55
*Taylor JK (1984b) Principles of Quality Assurance of Chemical Measurements. NBSIR 84-xxxx. Gaithersburg, MD: National Bureau of Standards
*Uriano GA, Gravatt CC (1977) The role of reference materials and methods in chemical analysis. CR Anal Chem 6:361–410
Velapoldi RA, Paule RC, Schaffer R, Mandel J, Machlan LA, Garner EL, Rains TC (1980) A Reference Method for the Determination of Lithium in Serum. NBS Spec Publication 260-69. Gaithersburg, MD: National Bureau of Standards
Velapoldi RA, Paule RC, Schaffer R, Mandel J, Moody J (1978) A Reference Method for Determination of Sodium in Serum. Appendix C. NBS Spec Publication 260-60. Gaithersburg, MD: National Bureau of Standards
Welch MJ, Cohen A, Hertz HS, Ruegg FC, Schaffer R, Sneigoski LT, White EV (1984) Determination of serum urea by isotope dilution mass spectrometry as a candidate definitive method. Anal Chem 56:713–719
Youden WJ (1960) The sample, the procedure, and the laboratory. Anal Chem 32:23A–37A
Zeisler R, Harrison SH, Wise SA (eds) (1983) The National Pilot Environmental Specimen Bank – Analysis of Human Liver Specimens. NBS Spec Publication 656. Gaithersburg, MD: National Bureau of Standards

The Importance of Chemical "Speciation" in Environmental Processes, eds. M. Bernhard, F.E. Brinckman, and P.J. Sadler, pp. 711–715. Dahlem Konferenzen 1986. Springer-Verlag Berlin, Heidelberg
© Dr. S. Bernhard, Dahlem Konferenzen

Ad hoc Analytical Report

Group Report

I.S. Krull, Moderator
R.A. Velapoldi, Rapporteur
Y.K. Chau
F.H. Frimmel
K.J. Irgolic
P.F. Kane
W. Lund
F.P. Prosi
D.H. Russell
P.J. Sadler
M.L. Vestal
W.R. Wolf

Introduction

An *ad hoc* group of analytical chemists met independently to discuss what we could specifically contribute to solving the major measurement problems of determining the concentration of chemical species in the environment, upon which identification of causal agents, conclusions, solutions, and theoretical models are based. To solve these complex measurement problems, it is imperative that investigations be carried out with a rigorous analytical strategy and with the total measurement capability available.

No comprehensive texts exist that deal with available, optimal, ideal, or even possible analytical methods for determining chemical species. There are texts that discuss specific techniques such as gas, thin-layer, and high performance liquid chromatographies, x-ray fluorescence, anodic stripping voltammetry, and other approaches that could be used for various speciation problems. Thus, there is a need for a comprehensive textbook of methods to determine chemical species.

An alternative would be to hold an international conference with the specific objective of summarizing precise and accurate methods for measuring inorganic and organic species in the environment and publishing the information as Conference Proceedings. The analytical methods included would have to be validated and/or information supporting the total analytical measurement process with uncertainty estimates would have to be provided.

The Proceedings of this Dahlem Conference will provide some background analytical information on determining chemical species, but will not provide a 'how-to' manual of methods and techniques. Some of the articles that have appeared in recent years dealing with comprehensive analytical approaches for measuring inorganic and organic species are listed in Appendix A.

We have thus drafted four specific recommendations and some research areas/techniques that should be investigated to establish applicability in the determination of chemical species in the environment.

Recommendations

1) A total analytical measurement process should be used, including objective and operational planning, sampling, sample preparation, separation, measuring, and interpretation of data to investigate environmental problems and identify key chemical species in the environment.

2) The problem of determining chemical species in the environment should be addressed through interdisciplinary research that includes input from environmental scientists, statisticians, kineticists, and analytical chemists in the relevant steps of the analytical process.

3) Periodic interdisciplinary meetings should be held to exchange information on experimental/theoretical results and discuss state-of-the-art analytical techniques/methods that could be used to determine chemical species in the environment.

4) Independent analytical research should be undertaken to develop techniques and methods for determination of specific chemical species.

Techniques/Methods

The following are techniques/methods that should be investigated/developed:

1) *In situ,* real-time monitors for inorganic species in complex environmental and biological matrices. Techniques should be selective, sensitive, nondestructive, continuous, precise, and accurate.

2) Automated, continuous analyses to remove labor intensiveness of measurements and improve precision/accuracy. Systems have computer control and feedback loops to modify automated systems.

3) High-pressure liquid chromatography (HPLC): Develop HPLC substrates and columns (e.g., microbore) ideal for and compatible with organometal macromolecules.

4) Improved coupling of HPLC with selective, sensitive detection techniques:

a) Mass spectrometry (MS) has been used extensively for trace organic analyses ($<10^{-9}$ g g^{-1}) but must be extensively investigated and evaluated for determining trace inorganic species in complex matrices.

b) Spectrofluorimetry – use of continuous flow systems for on-line pre- or post-column derivatization of inorganic species with fluorescent organic/inorganic ligands to improve selectivity/sensitivity (e.g., sensitivities of up to 10^{-18} g g^{-1} available using laser excitation). This method is also applicable to UV-VIS absorbance and electrochemical, atomic absorption, and plasma techniques.

c) Sensitive/selective techniques without derivatization – the detection techniques considered here are graphite furnace atomic absorption and plasma spectrometries, as well as new electrochemical techniques.

d) Electrochemical techniques – vibrating electrode, pulse polarography, fast-scanning voltammetry, differential pulse amperometry, etc.

5) Develop selective preconcentration devices (e.g., chelating resins, solid-phase adsorbents, affinity supports, etc.) to isolate and preserve chemical species.

6) Spectroscopic and other techniques such as LAMMA, IETS, EELS, EXAFS, solid sample NMR, FTIR, and RAMAN to assess applicability.

Application Areas

Specific measurement problem areas have been identified for which a variety of analytical techniques should be tried.

1) Particulate analyses: analyses for specific chemical species in the bulk of particulate matter, but especially analyses for the adsorbed species on the surface of the particles which are major contributors to the transport across phase boundaries. This includes adsorbed organic layers as well as adsorbed inorganic species.

2) Determination of species that take part in the transport of elements across membranes.

3) Determination of metal macromolecule-bound species (e.g., metal-proteins, metal-enzymes, metal-nucleic acids, etc.).

4) Sample preparation techniques should be developed that will be representative of the bulk sample and lead to 100% recovery for stable and unstable species.

5) Methods should be developed for analysis of chemical species in aqueous, colloidal, and particulate phases without separation (this task may be impossible).

6) Certified Reference Materials – Three basic types of standards should be developed for total elemental and chemical species analysis:

a) Seawater
b) Biota
c) Sediments

Future development will require input from appropriate experts in all of the above areas to further the development and optimization of analytical methods for the determination of chemical species in real-world samples.

Appendix A

List of References that Deal with Analytical Methods for Determining Chemical Species

Inorganic

Branica M, et al. (1980) The chemistry of the mediterannean. Organizing Committee of Sixth International Symposium held in Rovinj, Yugoslavia

Cassidy RM (1981) The separation and determination of metal species by modern liquid chromatography. In: Lawrence JF (ed) Trace analysis, vol 1. Academic Press, New York, p 122

Florence TM (1982) The speciation of trace elements in waters. Talanta 29:345

Florence TM, Batley GE (1980) Chemical speciation in natural waters, CRC Critical Reviews in Analytical Chemistry, 219 (August, 1980). Chemical Rubber Company Press, Boca Raton, Florida

Grob RL (ed) (1983) Chromatographic analysis of the environment. Marcel Decker, New York

Grob RL, Kaiser MA (1982) Environmental problem solving using gas and liquid chromatography. In: Journal of Chromatography Library, vol 21. Elsevier, Amsterdam

Krull IS (1984) Trace metal analysis by high performance liquid chromatography. In: Lawrence JF (ed) Liquid chromatography in environmental analysis. Humana Press, Clifton, New Jersey, chap 5

Lafleur PD (ed) (1976) Accuracy in trace analysis: sampling, sample handling, analysis. Proceedings of the 7th Materials Research Symposium held at the National Bureau of Standards, Gaithersburg, MD 20889 in 1974. NBS Special Publication 422

Mizuike A (1983) Enrichment techniques for inorganic trace analysis. Springer, Berlin Heidelberg New York
Nurnberg HW (1982) Voltammetric trace analysis in ecological chemistry of toxic metals. Pure Appl Chem 54:853–878
O'Laughlin JW (1984) High performance liquid chromatography of metal chelates. J Liquid Chrom 7(S-1):127
Olsen GJ, Brinckman FE, Jackson JA (1983) Purge and trap flame photometic gas chromatography technique for the speciation of trace organotin and organosulfur compounds in a Human Serum Standard Reference Material (SRM). Int J Environ Anal Chem 15:249–261
Schwedt G (1979) High-performance liquid chromatography in inorganic analysis. Chromatographia 12(9):613
Schwedt G (1981) Chromatography in inorganic trace analysis. In: Boschke FL (ed) Topics in current chemistry, vol 85. Instrumental inorganic chemistry. Springer Berlin Heidelberg New York, p 160
Schwedt G (1981) Chromatographic methods in inorganic analysis. A. H. Verlag, Heidelberg
Veening H, Willeford BR (1979) High performance liquid chromatography in organometallic and coordination chemistry. Rev Inor Chem 1(3):282
Veening H, Willeford BR (1983) High performance liquid chromatography of metal complexes. In: Giddings JC, Grushka E, Cazes J, Brown PR (eds) Advances in chromatography, vol 22. Marcel Dekker, New York, chap 4

Organic

Beyermann K (1982) Organic trace analysis. Ellis Horwood Limited, Halsted Press, A Division of J. Wiley & Sons, New York, NY
Hertz HS, Chesler SN (eds) (1979) Trace organic analysis: a new frontier in analytical chemistry. Proceedings of the 9th Materials Research Symposium held at the National Bureau of Standards, Gaithersburg, MD 20899 in 1978. NBS Special Publication 519
Keith LH (ed) (1984) Identification and Analysis of Organic Pollutants in Air. Boston: Butterworth Publishers
Kirchhoff WH (ed) (1974) Methods and standards for environmental measurements. Proceedings of the 8th Materials Research Symposium held at the National Bureau of Standards, Gaithersburg, MD 20899, 1976. NBS Special Publication 464
Lawrence JF (ed) (1981, 1982) Trace analysis, vols 1,2. Academic Press, New York
Polynuclear aromatic hydrocarbons (1981) Fifth International Symposium on Chemical Analysis and Biological Fate. Columbus, Ohio, Battelle Press, pp 1–16
Weiner WC, Felix WD (eds) (1983) Advanced techniques in synthetic fuels analysis. US Department of Energy Technical Information Center

Standards

Russell DS (1984) Available standards for use in analysis of marine materials. NRCC Report No 23025, National Research Council of Canada, Ottawa, Canada

List of Participants with Fields of Research

ANDREAE, M.O.
Dept. of Oceanography
Florida State University
Tallahassee, FL 32306
USA

Biogeochemistry of trace metals and sulfur in the oceans and atmosphere

BACCINI, P.
EAWAG
8600 Dübendorf
Switzerland

Transport and transformation processes in ecosystems and man-made systems

BAULD, J.
Baas Becking Geobiological Laboratory
P.O. Box 378
Canberra City, ACT 2601
Australia

Ecology and geobiology of phototrophic microbes (primarily saline benthic habitats, especially microbial mats and stromatolites)

BERNHARD, M.
Centro Studi Ambiente Marino
ENEA
P.O. Box 316
19100 La Spezia
Italy

Marine ecology and marine pollution

BRANICA, M.
Center for Marine Research Zagreb
The Rudjer Boskovic Institute
P.O. Box 1016
41001 Zagreb
Yugoslavia

Physicochemical characterization of traces in natural and polluted waters. Development of sensitive procedures for trace metal speciation

BRINCKMAN, F.E.
National Bureau of Standards
Center for Materials Science
Ceramics Division
Gaithersburg, MD 20899
USA

Biogenesis and transformations of organometallic substances in the environment

BRÜMMER, G.W.
Institut für Bodenkunde
der Universität Bonn
Nussallee 13
5300 Bonn 1
F.R. Germany

Soil science: heavy metals in soils – solubility, bonding, extractability, availability

CALMANO, W.
Arbeitsbereich Umweltschutztechnik
Technische Universität Hamburg-Harburg
Eissendorfer Strasse 38
2100 Hamburg 90
F.R. Germany

Chemical speciation of heavy metals in sediments

CHAKRABARTY, A.M.
Dept. of Microbiology
University of Illinois Medical Center
835 South Wolcott Avenue
Chicago, IL 60612
USA

Biodegradation of xenobiotic compounds, microbial oil recovery, pseudomonas infection in cystic fibrosis

CHAU, Y.K.
National Water Research Institute
Canada Center for Inland Waters
867 Lakeshore Road
P.O. Box 5050
Burlington, Ontario L7R 4A6
Canada

Organometals in the environment: speciation, biotic and abiotic methylation; structure-toxicity relationship

CRAIG, P.J.
School of Chemistry
Leicester Polytechnic
P.O. Box 143
Leicester LE1 9BH
England

Environmental analytical and organometallic chemistry

DUINKER, J.C.
Institut für Meereskunde
der Universität Kiel
Düsternbrooker Weg 20
2300 Kiel 1
F.R. Germany

Marine chemistry

FÖRSTNER, U.
Arbeitsbereich Umweltschutztechnik
Technische Universität Hamburg-Harburg
Eissendorfer Strasse 38
2100 Hamburg 90
F.R. Germany

Geochemistry, sediments and contaminants, solid-waste disposal, environmental engineering

FOWLER, B.A.
National Institute of Environmental Health Sciences
P.O. Box 12233
Research Triangle Park, NC 27709
USA

Regulatory roles of high-affinity metal-binding proteins in mammals and marine invertebrates with respect to cell injury

FRIMMEL, F.H.
Institut für Wasserchemie
Technische Universität München
Marchioninistrasse 17
8000 München 70
F.R. Germany

Complexation reactions in aquatic systems

GAMBLE, D.S.
Chemistry and Biology Research Institute
Agriculture Canada
Ottawa, Ontario K1A 0C6
Canada

The physical chemistry of humic materials and their interactions with cations and pesticides

GEORGE, S.G.*
Unit für Aquatic Biochemistry
University of Stirling
Stirling FK9 4LA
Scotland

Mechanisms of trace metal accumulation and detoxication in marine organisms and effects of toxic metals on organic pollutant metabolism

HERMS, U.
Niedersächs. Landesamt für Bodenforschung
Bodentechnologisches Institut Bremen
Friedrich-Missler-Strasse 46–50
2800 Bremen 1
F.R. Germany

Crop uptake of heavy metals from dredged harbor mud; influence of redox variations on heavy metal solubility

IRGOLIC, K.J.
Dept. of Chemistry
Center for Energy and Mineral Resources
Texas A & M University
College Station, TX 77843
USA

Transformation of arsenic, selenium, tellurium compounds by organisms; organometallic synthesis (As, Se, Te); element-specific detection systems for liquid chromatography; trace element analysis

JENNE, E.A.
Battelle Pacific Northwest Laboratories
Box 999
Richland, WA 99352
USA

Trace element and radionuclide speciation in waters and partitioning between phases

KABATA-PENDIAS, A.
Laboratorium Mikroelementów
I.U.N.G.
24-100 Pulawy
Poland

Behavior of trace elements in soils under stress; assessment of safe threshold levels of heavy metals in soils

KANE, P.F.*
19122 Kahala Drive
RRI Box 198
Galveston, TX 77551
USA

Surface science

KESTER, D.R.
Graduate School of Oceanography
University of Rhode Island
Kingston, RI 02881
USA

Physical chemistry of metals in seawater and the relationship to biological activity

---
* Present address

KRULL, I.S.
The Barnett Institute
341 Mugar Building
Northeastern University
360 Huntington Avenue
Boston, MA 02115
USA

Trace metal analysis and speciation by high-performance liquid chromatography with plasma emission spectroscopic detection (HPLC-ICP, HPLC-DCP, and HPLC-MIP)

LECKIE, J.O.
Environmental Engineering and Science Dept.
Stanford University
Stanford, CA 94305
USA

Surface chemical phenomena in aqueous systems; low-temperature, low-pressure aqueous geochemistry

LUDWICKI, K.J.
National Institute of Hygiene
Dept. of Food Research
ul. Chocimska 24
00-971 Warsaw
Poland

Food contamination by heavy metals and their health aspects

LUND, W.
Dept. of Chemistry
University of Oslo
Box 1033 Blindern
Oslo 3
Norway

Trace metals in environmental samples; electrochemical methods of analysis

LUOMA, S.N.
U.S. Geological Survey
Water Resources Division
345 Middlefield Road
Mailstop-65
Menlo Park, CA 94025
USA

Bioavailability of trace metals; metal chemistry in sediments; metal metabolism in invertebrates

MACKAY, D.
Chemical Engineering Dept.
University of Toronto
Toronto, Ontario M5S 1A4
Canada

Environmental dynamics of toxic organic chemicals

MEHLHORN, R.J.
Membrane Bioenergetics Group
Lawrence Berkeley Laboratory
Top of Hearst
Berkeley, CA 94720
USA

Membrane biophysics, oxidative free-radical damage, mitochondrial bioenergetics

MERTZ, W.
USDA, ARS
Beltsville Human Nutrition Research Center
Rm. 223, Bldg. 308
BARC-East
Beltsville, MD 20705
USA

Trace elements, nutrition

MILLS, C.F.
Rowett Research Institute
Bucksburn
Aberdeen AB2 9SB
Scotland

Investigation of the metabolic consequences of essential trace element deficiencies and of variables influencing their utilization by animals and man

MÜLLER, G.K.
Institut für Sedimentsforschung
der Universität Heidelberg
Postfach 10 30 20
6900 Heidelberg 1
F.R. Germany

Environmental geochemistry, especially removal of heavy metals from contaminated materials; influence of artificial complexing agents on sediments and sludges

OEHMICHEN, U.
Institut für Geowissenschaften, FB 22
Postfach 3980
6500 Mainz
F.R. Germany

Speciation of organometallic compounds in biological samples

OLSON, G.J.
National Bureau of Standards
A-331 Materials Bldg.
Washington, DC 20234
USA

Microbiology, biogeochemistry, environmental chemistry

PISCATOR, M.
Dept. of Environmental Hygiene
Karolinska Institute
P.O. Box 60400
104 01 Stockholm
Sweden

Metal/cadmium toxicity

PROSI, F.P.
Zoologisches Institut I
der Universität Heidelberg
Im Neuenheimer Feld 230
6900 Heidelberg
F.R. Germany

Interactions of heavy metals on the sediment-water interface and their availability for benthic invertebrates; bioindicator systems for heavy metals on the level of tissues and cells

RUSSELL, D.H.
Dept. of Chemistry
Texas A & M University
College Station, TX 77843
USA

Development of mass spectral methods for identification/characterization of large molecules

SADLER, P.J.
Dept. of Chemistry
Birkbeck College
University of London
Malet Street
London WC1E 7HX
England

Biological chemistry

SALOMONS, W.
Delft Hydraulics Laboratory
Haren Branch
c/o Institute for Soil Fertility
P.O. Box 30003
9740 Haren (Gr)
Netherlands

Aquatic geochemistry, in particular the behavior of trace metals in lakes and estuaries; research on the fate and transport of contaminants from solid materials dumped on land and in the aquatic environment

SILVER, S.
Biology Dept.
Washington University
St. Louis, MO 63130
USA

Molecular biology and biochemistry of bacterial resistance to toxic heavy metals

THORNELEY, R.N.F.
AFRC
Unit of Nitrogen Fixation
University of Sussex
Brighton BN1 9RQ
England

Nitrogen fixation, electron transfer with metalloproteins

TOBSCHALL, H.J.
Institut für Geowissenschaften
Johannes Gutenberg-Universität
Saarstrasse 21
6500 Mainz 1
F.R. Germany

Biogeochemistry of tin and platinum group elements, organometallic compounds of mercury, tin, and lead in aquatic systems

TRAMIER, B.
Elf Aquitaine
Direction Environment
Les Miroirs
92096 Paris La Defense Cedex 27
France

Oil pollution and impact, heavy metals pollution and impact

VELAPOLDI, R.A.
National Bureau of Standards
Center for Analytical Chemistry
Chemistry Bldg., Rm. A309
Gaithersburg, MD 20899
USA

Luminescence/analytical chemistry

VESTAL, M.L.
Dept. of Chemistry
University of Houston
University Park
Houston, TX 77004
USA

Development and applications of new techniques for combined liquid chromatography/mass spectrometry; new ionization techniques in mass spectrometry suitable for nonvolatile and/or thermally labile compounds

WEBER, G.
Institut für Spektrochemie und
Angewandte Spektroskopie
Bunsen-Kirchhoff-Strasse 11
4600 Dortmund 1
F.R. Germany

Trace analysis and speciation of metals (especially tin) in matrices such as water, foods, body fluids

WOLF, W.R.
Beltsville Human Nutrition
Research Center
Bldg. 161 BARCE
U.S. Dept. of Agriculture
Beltsville, MD 20705
USA

Inorganic nutrient analysis of foods, development of biological reference materials

WOOD, J.M.
Gray Freshwater Biological Institute
P.O. Box 100
Navarre, MN 55392
USA

Metalloenzymes, transport and toxicity of metal ions

ZEPPEZAUER, M.
Fachrichtung Biochemie, 15.2
Universität des Saarlandes
6600 Saarbrücken
F.R. Germany

Catalytic function of metals in metalloenzymes; biosynthesis of metalloproteins; thymic hormones in immunology and cancer research

Subject Index

Author Index

Dahlem Workshop Reports

Life Sciences Research Reports (LS)

LS 1 The Molecular Basis of Circadian Rhythms
Editors: J. W. Hastings, H.-G. Schweiger

LS 2 Appetite and Food Intake
Editor: T. Silverstone

LS 3 Hormone and Antihormone Action at the Target Cell
Editors: J. H. Clark et al

LS 4 Organization and Expression of Chromosomes
Editors: V. G. Allfrey et al

LS 5 Recognition of Complex Acoustic Signals
Editor: T. H. Bullock

LS 6 Function and Formation of Neural Systems
Editor: G. S. Stent

LS 7 Neoplastic Transformation: Mechanisms and Consequences
Editor: H. Koprowski

LS 8 The Bases of Addiction
Editor: J. Fishman

LS 9 Morality as a Biological Phenomenon
Editor: G. S. Stent

LS 10 Abnormal Fetal Growth: Biological Bases and Consequences
Editor: F. Naftolin

LS 11 Transport of Macromolecules in Cellular Systems
Editor: S. C. Silverstein

LS 12 Light-Induced Charge Separation in Biology and Chemistry
Editors: H. Gerischer, J. J. Katz

LS 13 Strategies of Microbial Life in Extreme Environments
Editor: M. Shilo

LS 14 The Role of Intercellular Signals: Navigation, Encounter, Outcome
Editor: J. G. Nicholls

LS 15 Biomedical Pattern Recognition and Image Processing
Editors: K. S. Fu, T. Pavlidis

LS 16 The Molecular Basis of Microbial Pathogenicity
Editors: H. Smith et al

LS 17 Pain and Society
Editors: H. W. Kosterlitz, L. Y. Terenius

LS 18 Evolution of Social Behavior: Hypotheses and Empirical Tests
Editor: H. Markl

LS 19 Signed and Spoken Language: Biological Constraints on Linguistic Form
Editors: U. Bellugi, M. Studdert-Kennedy

Physical and Chemical Sciences Research Reports (PC)

PC 1 The Nature of Seawater (out of print)

PC 2 Global Chemical Cycles and Their Alteration by Man
Editor: W. Stumm

Distributor for LS 1–19 and PC 1+2:
Verlag Chemie, Pappelallee 3, 6940 Weinheim,
Federal Republic of Germany

Springer

Dahlem Workshop Reports

Life Sciences Research Reports (LS)

LS 37 Mechanisms of Cell Injury: Implications for Human Health
Editor: B. A. Flower (1986, in press)

LS 36 Patterns and Processes in the History of Life
Editors: D. Jablonski, D. M. Raup (1986)

LS 35 Biotechnology: Potentials and Limitations
Editor: S. Silver (1986)

LS 34 The Molecular Mechanism of Photoreception
Editor: H. Stieve (1986)

LS 32 Exploitation of Marine Communities
Editor: R. M. May (1984)

LS 31 Microbial Adhesion and Aggregation
Editor: K. C. Marshall (1984)

LS 30 Leukemia
Editor: I. L. Weissman (1985)

LS 29 The Biology of Learning
Editors: P. Marler, H. S. Terrace (1984)

LS 28 Changing Metal Cycles and Human Health
Editor: J. O. Nriagu (1984)

LS 27 Minorities: Community and Identity
Editor: C. Fried (1983)

LS 26 The Origins of Depression: Current Concepts and Approaches
Editor: J. Angst (1983)

LS 25 Population Biology of Infectious Diseases
Editors: R. M. Anderson, R. M. May (1982)

LS 24 Repair and Regeneration of the Nervous System
Editor: J. G. Nicholls (1982)

LS 23 Biological Mineralization and Demineralization
Editor: G. H. Nancollas (1982)

LS 22 Evolution and Development
Editor: J. T. Bonner (1982)

LS 21 Animal Mind - Human Mind
Editor: D. R. Griffin (1982)

LS 20 Neuronal-glial Cell Interrelationships.
Editor: T. A. Sears (1982)

Physical, Chemical and Earth Sciences Research Reports (PC)

PC 5 Patterns of Change in Earth Evolution
Editors: H. D. Holland, A. F. Trendall (1984)

PC 4 Atmospheric Chemistry
Editor: E. D. Goldberg (1982)

PC 3 Mineral Deposits and the Evolution of the Biosphere
Editors: H. D. Holland, M. Schidlowski (1982)

Springer-Verlag
Berlin Heidelberg New York London Paris Tokyo

Springer

MIX
Papier aus verantwortungsvollen Quellen
Paper from responsible sources
FSC® C105338

If you have any concerns about our products,
you can contact us on
ProductSafety@springernature.com

In case Publisher is established outside the EU,
the EU authorized representative is:
Springer Nature Customer Service Center GmbH
Europaplatz 3, 69115 Heidelberg, Germany

Printed by Libri Plureos GmbH
in Hamburg, Germany